CAMBRIDGE MONOGRAPHS ON APPLIED AND COMPUTATIONAL MATHEMATICS

32 Multivariate Approximation

The *Cambridge Monographs on Applied and Computational Mathematics* series reflects the crucial role of mathematical and computational techniques in contemporary science. The series publishes expositions on all aspects of applicable and numerical mathematics, with an emphasis on new developments in this fast-moving area of research.

State-of-the-art methods and algorithms as well as modern mathematical descriptions of physical and mechanical ideas are presented in a manner suited to graduate research students and professionals alike. Sound pedagogical presentation is a prerequisite. It is intended that books in the series will serve to inform a new generation of researchers.

A complete list of books in the series can be found at
www.cambridge.org/mathematics.
Recent titles include the following:

17. Scattered data approximation, *Holger Wendland*
18. Modern computer arithmetic, *Richard Brent & Paul Zimmermann*
19. Matrix preconditioning techniques and applications, *Ke Chen*
20. Greedy approximation, *Vladimir Temlyakov*
21. Spectral methods for time-dependent problems, *Jan Hesthaven, Sigal Gottlieb & David Gottlieb*
22. The mathematical foundations of mixing, *Rob Sturman, Julio M. Ottino & Stephen Wiggins*
23. Curve and surface reconstruction, *Tamal K. Dey*
24. Learning theory, *Felipe Cucker & Ding Xuan Zhou*
25. Algebraic geometry and statistical learning theory, *Sumio Watanabe*
26. A practical guide to the invariant calculus, *Elizabeth Louise Mansfield*
27. Difference equations by differential equation methods, *Peter E. Hydon*
28. Multiscale methods for Fredholm integral equations, *Zhongying Chen, Charles A. Micchelli & Yuesheng Xu*
29. Partial differential equation methods for image inpainting, *Carola-Bibiane Schönlieb*
30. Volterra integral equations, *Hermann Brunner*
31. Symmetry, phase modulation and nonlinear waves, *Thomas J. Bridges*
32. Multivariate approximation, *V. Temlyakov*

Multivariate Approximation

V. TEMLYAKOV
University of South Carolina,
Steklov Institute of Mathematics, Moscow
and
Lomonosov Moscow State University

CAMBRIDGE
UNIVERSITY PRESS

CAMBRIDGE
UNIVERSITY PRESS

University Printing House, Cambridge CB2 8BS, United Kingdom

One Liberty Plaza, 20th Floor, New York, NY 10006, USA

477 Williamstown Road, Port Melbourne, VIC 3207, Australia

314-321, 3rd Floor, Plot 3, Splendor Forum, Jasola District Centre, New Delhi - 110025, India

79 Anson Road, #06-04/06, Singapore 079906

Cambridge University Press is part of the University of Cambridge.

It furthers the University's mission by disseminating knowledge in the pursuit of education, learning and research at the highest international levels of excellence.

www.cambridge.org
Information on this title: www.cambridge.org/9781108428750
DOI: 10.1017/9781108689687

First published 2018

A catalogue record for this publication is available from the British Library

ISBN 978-1-108-42875-0 Hardback

Contents

Preface

The twentieth century was a period of transition from univariate problems (i.e., single-variable problems) to multivariate problems in a number of areas of mathematics. In many cases this step brought not only new phenomena but also required new techniques. In some cases even the formulation of a multivariate problem requires a nontrivial modification of a univariate problem. For instance, the problem of the convergence of the multivariate trigonometric series immediately encounters the question of which partial sums we should consider; there is no natural ordering in the multivariate case. In other words: what is a natural multivariate analog of the trigonometric polynomials? In answering this question mathematicians have studied different generalizations of the univariate trigonometric polynomials: those with frequencies from a ball, a cube, a hyperbolic cross.

Multivariate problems turn out to be much more difficult than their univariate counterparts. The main goal of this book is to demonstrate the evolution of theoretical techniques from the univariate case to the multivariate case. We do this using the example of the approximation of periodic functions. It is justified historically and also it allows us to present the ideas in a concise and clear way. In many cases these ideas can be successfully used in the nonperiodic case as well.

We concentrate on a discussion of some theoretical problems which are important in numerical computation. The fundamental problem of approximation theory is to resolve a possibly complicated function, called the target function, into simpler, easier to compute, functions called approximants. Generally, increasing the resolution of the target function can be achieved only by increasing the complexity of the approximants. The understanding of this tradeoff between resolution and complexity is the main goal of approximation theory. Thus the goals of approximation theory and numerical computation are similar, even though approximation theory is less concerned with computational issues. Approximation and computation are intertwined and it is impossible to understand fully the possibilities in numerical computation without a good understanding of the elements of approximation theory. In particular, good approximation methods (algorithms) from approximation theory find applications in image processing, statistical estimation, regularity for PDEs, and other areas of computational mathematics.

We now give a brief historical overview of the challenges and open problems in approximation theory, with emphasis on multivariate approximation. It was understood at the beginning of the twentieth century that the smoothness properties of a univariate function determine the rate of approximation of this function by polynomials (trigonometric in the periodic case and algebraic in the nonperiodic case). A fundamental question is: what is a natural multivariate analog of the univariate smoothness classes? Different function classes have been considered in the multivariate case: isotropic and anisotropic Sobolev and Besov classes, classes of functions with bounded mixed derivative, and others. The next fundamental question is: how do we approximate functions from these classes? Kolmogorov introduced the concept of the n-width of a function class. This concept is very useful in answering the above question. The Kolmogorov n-width is a solution to an optimization problem where we optimize over n-dimensional linear subspaces. This concept allows us to understand which n-dimensional linear subspace is the best for approximating a given class of functions. The rates of decay of the Kolmogorov n-width are known for the univariate smoothness classes; in some cases, exact values are known. The problem of the rates of decay of the Kolmogorov n-width for the classes of multivariate functions with bounded mixed derivatives is still an open problem. We note that the function classes with bounded mixed derivatives are not only an interesting and challenging object for approximation theory but are also important in numerical computations. M. Griebel and his group have used approximation methods designed for these classes in elliptic variational problems. Recent work of H. Yserentant on new regularity models for the Schrödinger equation shows that the eigenfunctions of the electronic Schrödinger operator have a certain mixed smoothness similar to that of the bounded mixed derivative. This makes approximation techniques developed for classes of functions with bounded mixed derivatives a proper choice for a numerical treatment of the Schrödinger equation.

Following the Kolmogorov idea of the n-width the optimization approach to approximation problems has been accepted as a fundamental principle. We follow this fundamental principle in this book. We will demonstrate the development of ideas and techniques when we pass from the univariate case to the multivariate case, and we will solve some optimization problems for the multivariate approximation. We find the correct order of decay for a number of asymptotic characteristics of smoothness function classes. We provide the rate of decay, in the sense of order, as a function of the complexity of the approximation method. For example, in the case of approximation by elements of a linear subspace the complexity parameter is the dimension of the subspace. In the case of numerical integration the complexity parameter is the number of points (knots) of the cubature formula. Solving the problem of the correct (in the sense of order) decay of an asymptotic characteristic, we need to solve two problems: (U) prove the upper bounds and (L) prove the lower bounds. Very often the technique used to solve the problem (U) is very different from the one for solving (L). We present both the classical, well-known, techniques and comparatively recent modern techniques. Usually, when we solve an approximation problem, we use a combination of fundamental results (theories) and special tricks. In this book the fundamental results from harmonic analysis, for instance, the Littlewood–Paley theorem, the Marcinkiewicz multiplier theorem, the Hardy–Littlewood inequality, the Riesz–Thorin theorem, and the Hausdorff–Young theorem are widely used. It turns out that in addition to these classical results we need new fundamental results in order to solve the multivariate approximation problems for mixed smoothness classes. We develop and demonstrate this new technique here in detail. Let us mention three examples of this new fundamental technique: (I) an embedding-type inequality proved in §3.3.3 (see Theorem 3.3.6); (II) the volume estimates of sets of coefficients of trigonometric polynomials (see §3.2.6 and Chapter 7); (III) the greedy approximation (see Chapters 8 and 9).

Motivated by practical applications we study the following theoretical problem: how do we replace in an optimal way an infinite-dimensional object (a function class) by a finite or finite-dimensional object? The theory of widths, in particular, the Kolmogorov widths, addresses the problem of approximation from a finite-dimensional subspace. In addition to the Kolmogorov width we employ the linear width and the orthowidth (Fourier width). In this study we use techniques (I) and (II). Technique (II) has proved to be very useful in the proof of the lower bounds. There are still open problems in finding the correct orders of decay of the above widths in case of the mixed smoothness classes.

Discretization is a very important step in making a continuous problem computationally feasible. The problem of the construction of satisfactory sets of points in a multidimensional domain is a fundamental problem of mathematics and, in particular, computational mathematics. We note that the problem of arranging points in a multidimensional domain is also a fundamental problem in coding theory. It is a problem on optimal spherical codes. This problem is equivalent to the problem from compressed sensing on building large incoherent dictionaries in $\mathbb{R}^d$. A very interesting and difficult problem is to provide an explicit (deterministic) construction of a large system with small coherence. The optimal rate in this problem is still unknown (see Temlyakov, 2011, Chapter 5, for further discussion).

A prominent example of a discretization problem, discussed in detail in this book, is the problem of numerical integration. It turns out that, contrary to numerical integration in the univariate case (see §2.4) and in the multivariate case of anisotropic smoothness classes (see §3.6), where regular grid methods are optimal (in the sense of order), in the case of the numerical integration of functions with mixed smoothness regular grid methods are very far from being optimal. Numerical integration in mixed smoothness classes requires deep number-theoretical results for constructing optimal (in the sense of order) cubature formulas (see Chapter 6). In addition to number-theoretical methods, technique III is also of use here.

Another example of a classical discretization problem is the problem of metric entropy (covering numbers and entropy numbers). Bounds for the ε-entropy of function classes are important in themselves and also have important connections to other fundamental problems. For instance, the problem of the ε-entropy of some classes of functions with bounded mixed derivatives is equivalent to the fundamental small ball problem from probability theory. This problem is still unresolved in dimensions greater than two (see Temlyakov, 2011, Chapter 3, and Dinh Dung *et al.*, 2016 for further discussion). We obtain the correct orders of decay of the entropy numbers of mixed smoothness classes in Chapter 7. Technique II plays a fundamental role in proving the lower bounds.

The above discussion demonstrates that multivariate approximation theory in a classical setting has close connections with other areas of mathematics and has many applications in numerical computation. Recently, driven by applications in engineering, biology, medicine, and other areas of science, new and challenging problems have appeared. The common feature of these problems is very high dimensions. Classical methods developed in multivariate approximation theory may work for moderate dimensions, say, up to 40 dimensions. Many contemporary numerical problems, however, have dimensions which are really large – sometimes in the millions. Classical methods do not work for such enormous dimensions. This is a rapidly developing and hot area of mathematics and numerical analysis, where researchers are trying to understand which new approaches may work. A promising contemporary approach is based on the concept of sparsity and nonlinear m-term approximation. We present the corresponding results in Chapters 8 and 9 of this book.

The fundamental question of nonlinear approximation is how to devise effective constructive methods (algorithms) of nonlinear approximation. This problem has two levels of nonlinearity. The first is m-term approximation with regard to bases. In this problem one can use the unique function expansion with respect to a given basis to build an approximant. Nonlinearity enters by looking for m-term approximants with terms (i.e. basis elements in the approximant) that are allowed to depend on a given function. Since the elements of the basis used in the m-term approximation are thus allowed to depend on the function being approximated, this type of approximation is very efficient. On the second level of nonlinearity, we replace a basis by a more general system, which is not necessarily minimal (for example, a redundant system, or a dictionary). This setting is much more complicated than the first (the bases case); however, there is a solid justification due to the importance of redundant systems in both theoretical questions and in practical applications. Technique III turns out to be very useful for approximation at both levels of nonlinearity. In this book we are primarily interested in the trigonometric approximation. A very interesting phenomenon was observed recently. It turns out that nonlinear algorithms, in particular the Chebyshev greedy algorithms, designed for approximation with respect to redundant systems, work better than algorithms, in particular the thresholding greedy algorithm, designed for bases when they are applied to a trigonometric system. We discuss this phenomenon in detail in Chapter 8.

Above we discussed a strategy based on the optimization principle, which we will apply for finding optimal (in the sense of order) finite dimensional subspaces (theory of the widths) and optimal (in the sense of order) discretization (numerical integration, entropy). In addition to the optimization principle we study another fundamental principle – universality. In Chapters 5 and 6 we illustrate the following general observation. Methods of approximation and numerical integration which are optimal in the sense of order for classes with mixed smoothness are universally applicable for the collection of anisotropic smoothness classes. This gives an a posteriori justification for the thorough study of classes of functions with mixed smoothness. The phenomenon of saturation is well known in approximation theory (DeVore and Lorentz, 1993, Chapter 11). The classical example of a saturated method is the Fejér operator for approximation of the univariate periodic functions. In the case of a sequence of Fejér operators, saturation means that the approximation order by Fejér operators of order n does not improve over the rate $1/n$ even if we increase the smoothness of the functions under approximation. Methods (algorithms) that do not have the saturation property are called unsaturated. The reader can find a detailed discussion of unsaturated algorithms in approximation theory and in numerical analysis in the survey paper Babenko (1985). We point out that the concept of smoothness becomes more complicated in the multivariate case than it is in the univariate case. In the multivariate case a function may have different smoothness properties in different coordinate directions. In other words, functions may belong to different anisotropic smoothness classes (see Chapter 3). It is known (see Chapter 3) that the approximation characteristics of anisotropic smoothness classes depend on the average smoothness and that optimal approximation methods depend on the anisotropy of classes. This motivated a study, in Temlyakov (1988c) of the existence of an approximation method that works well for all anisotropic smoothness classes. The problem is that of the existence of a universal method of approximation. We note that the universality concept in learning theory is very important and is close to the concepts of adaptation and distribution-free estimation in nonparametric statistics (Györfy *et al.*, 2002,

Binev *et al.*, 2005, Temlyakov, 2006). We discuss universality in approximation theory in §5.4 and universality in numerical integration in §6.8.

We now give a brief description of the book by chapters.

Chapter 1. Approximation of Univariate Functions This chapter contains classical results of approximation theory: the properties of trigonometric polynomials and approximation by trigonometric polynomials. The selection of the material for this chapter was dictated by further developments and applications in the multivariate approximation.

Chapter 2. Optimality and Other Properties of the Trigonometric System This chapter contains classical results on the Kolmogorov, linear, and Fourier widths of the univariate smoothness classes. Discretization techniques and the fundamental finite-dimensional results are discussed in detail in this chapter. Also, classical results on the convergence of Fourier series are presented here. The book by DeVore and Lorentz (1993) contains a comprehensive presentation of the univariate approximation. The presentation in Chapters 1 and 2 is aimed towards multivariate generalizations. It is somewhat close to the presentation in Temlyakov (1993b).

Chapter 3. Approximation of Functions from Anisotropic Sobolev and Nikol'skii Classes This chapter is the first step from the univariate approximation to the multivariate approximation. The approximation technique discussed here is mostly similar to the univariate technique. Results of this type are typically presented in books, such as Nikol'skii (1969), on function spaces. However, we include in this chapter some nontrivial embedding-type inequalities and estimates for the volumes of sets of Fourier coefficients of the multivariate trigonometric polynomials, which are frequently used in further chapters.

Chapter 4. Hyperbolic Cross Approximation This is one of the main chapters on the linear approximation theory of functions with mixed smoothness. In the sense of its settings it is parallel to Chapter 1. This parallelism in settings allows us to demonstrate a deep difference in technique between univariate polynomial approximation and hyperbolic cross polynomial approximation.

Chapter 5. The Widths of Classes of Functions with Mixed Smoothness This is the other main chapter on the linear approximation theory of functions with mixed smoothness. The relation between this chapter and Chapter 2 is the same as that between Chapters 4 and 1.

Chapter 6. Numerical Integration and Approximate Recovery This is the third main chapter on the linear approximation theory of functions with mixed smoothness. This chapter is important from the point of view of applications. Also, the technique for numerical integration developed in this chapter is very different from that for univariate numerical integration. It is based on deep number-theoretical constructions and on the general theory of greedy algorithms. The numerical integration of classes of functions with mixed smoothness has attracted a lot of attention recently. There are many books on discrepancy theory that are related to this chapter. However, the development in terms of numerical integration is more general than a discrepancy-type presentation. Roughly speaking, discrepancy theory corresponds to the case of smoothness equal to 1 and equal weights in cubature formulas, while numerical integration theory considers the whole range of smoothness and general cubature formulas.

Chapter 7. Entropy This is the first chapter devoted to nonlinear approximation theory. We include here classical results on the entropy numbers of finite-dimensional compacts. The main new ingredient of this chapter is a study of the entropy numbers of classes of functions with mixed smoothness.

Chapter 8. Greedy Approximation This chapter contains a brief introduction to greedy approximation in Banach spaces and a recent result on the Lebesgue-type inequality for the Chebyshev greedy algorithm (CGA) with respect to special dictionaries. In particular, this result implies that the CGA provides almost ideal (up to a $\log m$ factor) m-term trigonometric approximation for all functions. Our introduction to greedy approximation in Banach spaces follows the lines of Temlyakov (2011). The Lebesgue-type inequality is also a recent result (see Temlyakov, 2014).

Chapter 9. Sparse Approximation This is one of the most important chapters of the book. Our main interest in this chapter is to study sparse approximation problems for classes of functions with mixed smoothness. We discuss in detail m-term approximation with respect to the trigonometric system. We use techniques based on a combination of results from the hyperbolic cross approximation, which were obtained in the 1980s and 1990s (and are presented in Chapters 3–5 and 7), and recent results on greedy approximation (given in Chapter 8) to obtain sharp estimates for the best m-term approximation.

Appendix. The Appendix contains classical inequalities and results from harmonic analysis that are often used in this text.

The book is devoted to the linear and nonlinear approximation of functions with mixed smoothness. Both Temlyakov (1986c) and Temlyakov (1993b) contain results on the linear approximation theory of such classes. At present there are no books on the nonlinear approximation theory of these classes. In addition to the results treated in Temlyakov (1993b), we describe in this book substantial new developments in the linear approximation theory of classes with mixed smoothness. This makes the book the most complete text on the linear approximation theory of these classes. Further, it is the first book on the nonlinear approximation theory of such classes. The background material included in Chapters 1–3 makes the book self-contained and accessible for readers with graduate or even undergraduate level mathematical education. The theory of the approximation of these classes and related questions are important and actively developing areas of research. There are still many unresolved fundamental problems in the theory. Many open problems are formulated in the book.

Acknowledgement

The work was supported by the Russian Federation Government Grant No. 14.W03.31.0031.

1

Approximation of Univariate Functions

1.1 Introduction

The primary problem in approximation theory is the choice of a successful method of approximation. In this chapter and in Chapter 2 we test various approaches, based on the concept of width, to the evaluation of the quality of a method of approximation. We take as an example the approximation of periodic functions of a single variable. The two main parameters of a method of approximation are its accuracy and complexity. These concepts may be treated in various ways depending on the particular problems involved. Here we start from classical ideas about the approximation of functions by polynomials. After Fourier's 1807 article the representation of a 2π-periodic function by its Fourier series became natural. In other words, the function $f(x)$ is approximately represented by a partial sum $S_n(f,x)$ of its Fourier series:

$$S_n(f,x) := a_0/2 + \sum_{k=1}^{n} (a_k \cos kx + b_k \sin kx),$$

$$a_k := \frac{1}{\pi}\int_{-\pi}^{\pi} f(x)\cos kx dx, \qquad b_k := \frac{1}{\pi}\int_{-\pi}^{\pi} f(x)\sin kx dx.$$

We are interested in the approximation of a function f by a polynomial $S_n(f)$ in some L_p-norm, $1 \le p \le \infty$. In the case $p = \infty$ we assume that we are dealing with the uniform norm. As a measure of the accuracy of the method of approximating a periodic function by means of its Fourier partial sum we consider the quantity $\|f - S(f)\|_p$. The complexity of this method of approximation contains the following two characteristics. The order of the trigonometric polynomial $S_n(f)$ is the quantitative characteristic. The following observation gives us the qualitative characteristic. The coefficients of this polynomial are found by the Fourier formulas, which means that the operator S_n is the orthogonal projector onto the subspace of trigonometric polynomials of order n.

In 1854 Chebyshev suggested representing continuous function f by its polynomial of best approximation, namely, by the polynomial $t_n(f)$ such that

$$\|f - t_n(f)\|_\infty = E_n(f)_\infty := \inf_{\alpha_k, \beta_k} \left\| f(x) - \sum_{k=0}^{n} (\alpha_k \cos kx + \beta_k \sin kx) \right\|_\infty .$$

He proved the existence and uniqueness of such a polynomial. We consider this method of approximation not only in the uniform norm but in all L_p-norms, $1 \leq p < \infty$. The accuracy of the Chebyshev method can be easily compared with the accuracy of the Fourier method:

$$E_n(f)_p \leq \|f - S_n(f)\|_p .$$

However, it is difficult to compare the complexities of these two methods. The quantitative characteristics coincide but the qualitative characteristics are different (for example, it is not difficult to understand that for $p = \infty$ the mapping $f \to t_n(f)$ is not a linear operator). The Du Bois–Reymond 1873 example of a continuous function f such that $\|f - S_n(f)\|_\infty \to \infty$ when $n \to \infty$, and the Weierstrass theorem which says that for each continuous function f we have $E_n(f)_\infty \to 0$ as $n \to \infty$, showed the advantage of the Chebyshev method over the Fourier method from the point of view of accuracy.

The desire to construct methods of approximation which have the advantages of both the Fourier and Chebyshev methods has led to the study of various methods of summation of Fourier series. The most important among them from the point of view of approximation are the de la Vallée Poussin, Fejér, and Jackson methods, which were constructed early in the twentieth century. All these methods are linear. For example, in the de la Vallée Poussin method a function f is approximated by the polynomial

$$V_n(f) := \frac{1}{n} \sum_{l=n}^{2n-1} S_l(f)$$

of order $2n - 1$.

From the point of view of accuracy this method is close to the Chebyshev method; de la Vallée Poussin proved that

$$\|f - V_n(f)\|_p \leq 4E_n(f)_p, \qquad 1 \leq p \leq \infty.$$

From the point of view of complexity it is close to the Fourier method, and the property of linearity essentially distinguishes it from the Chebyshev method.

We see that common to all these methods is approximation by trigonometric polynomials. However, the methods of constructing these polynomials differ: some methods use orthogonal projections on to the subspace of trigonometric polynomials of fixed order, some use best-approximation operators, and some use linear operators.

Thus, the approximation of periodic functions by trigonometric polynomials is natural and this problem has been thoroughly studied. The approximation of functions by algebraic polynomials has been studied in parallel with approximation by trigonometric polynomials. We now point out some results, which determined the style of investigation of a number of problems in approximation theory. These problems are of interest even today.

It was proved by de la Vallée Poussin (1908) that, for best approximation of the function $|x|$ in the uniform norm on $[-1,1]$ by algebraic polynomials of degree n, the following upper *estimate* or *bound* holds:

$$e_n(|x|) \le C/n.$$

He raised the question of the possibility of an improvement of this estimate in the sense of order. In other words, could the function C/n be replaced by a function that decays faster to zero? Bernstein (1912) proved that this order estimate is sharp. Moreover, he then established the asymptotic behavior of the sequence $\{e_n(|x|)\}$ (see Bernstein, 1914):

$$e_n(|x|) = \mu/n + o(1/n), \qquad \mu = 0.282 \pm 0.004.$$

These results initiated a series of investigations into best approximations of individual functions having special singularities.

At this stage of investigation the natural conjecture arose that the smoother a function, the more rapidly its sequence of best approximations decreases.

In 1911 Jackson proved the inequality

$$E_n(f)_\infty \le Cn^{-r}\omega(f^{(r)}, 1/n)_\infty.$$

The relations which give upper estimates for the best approximations of a function in terms of its smoothness are now called the *Jackson inequalities*, and in a wider sense such relations are called *direct theorems* of approximation theory.

As a result of Bernstein's (1912) and de la Vallée Poussin's (1908, 1919) investigations we can formulate the following assertion, which is now called the *inverse theorem* of approximation theory. If

$$E_n(f)_\infty \le Cn^{-r-\alpha}, \qquad 0 \le r \text{ integer}, \qquad 0 < \alpha < 1,$$

then f has a continuous derivative of order r which belongs to the class Lip α; that is, $f \in W^r H^\alpha$ (in the notation of this book it is the class $H_\infty^{r+\alpha}$). Thus, the results of Jackson, Bernstein, and de la Vallée Poussin show that functions from the class $W^r H^\alpha$, $0 < \alpha < 1$, can be characterized by the order of decrease of its sequences of best approximations.

We remark that at that time, early in the twentieth century, classes similar to W^rH^α were used in other areas of mathematics for obtaining the orders of decrease of various quantities. As an example we formulate a result of Fredholm (1903). Let $f(x,y)$ be continuous on $[a,b]\times[a,b]$ and

$$\max_{x,y}\left|f(x,y+t)-f(x,y)\right|\le C|t|^\alpha, \qquad 0<\alpha\le 1.$$

Then for eigenvalues $\lambda(J_f)$ of the integral operator

$$(J_f\psi)(x)=\int_a^b f(x,y)\psi(y)dy$$

the following relation is valid for any $\rho>2/(2\alpha+1)$:

$$\sum_{n=1}^{\infty}\left|\lambda_n(J_f)\right|^\rho<\infty.$$

The investigation of the upper bounds or estimates of errors of approximation of functions from a fixed class by some method of approximation began with an article by Lebesgue (1910). In particular, Lebesgue proved that

$$S_n(\mathrm{Lip}\,\alpha)_\infty := \sup_{f\in\mathrm{Lip}\,\alpha}\left\|f-S_n(f)\right\|_\infty \asymp n^{-\alpha}\ln n.$$

Here and later we write $a_n\asymp b_n$ for two sequences a_n and b_n if there are two positive constants C_1 and C_2 such that $C_1b_n\le a_n\le C_2b_n$ for all n.

The problem of approximation of functions in the classes W^rH^α by trigonometric polynomials was so natural that a tendency to find either asymptotic or exact values of the following quantities appeared:

$$S_n(W^rH^\alpha)_\infty := \sup_{f\in W^rH^\alpha}\|f-S_n(f)\|_\infty, \qquad E_n(W^rH^\alpha)_\infty := \sup_{f\in W^rH^\alpha}E_n(f)_\infty.$$

We now formulate the first results in this direction. Kolmogorov (1936) proved the relation (in our notation $W^r=W^r_{\infty,r}$, see §1.4)

$$S_n(W^r)_\infty=\frac{4}{\pi^2}\frac{\ln n}{n^r}+O(n^{-r}), \qquad n\to\infty.$$

Independently, Favard (1937) and Akhiezer and Krein (1937) proved the equality

$$E_n(W^r)_\infty=K_r(n+1)^{-r},$$

where K_r is a number depending on the natural number r.

In 1936 Kolmogorov introduced the concept of the width d_n of a class F in a space X:

$$d_n(F,X) := \inf_{\{\phi_j\}_{j=1}^n}\sup_{f\in F}\inf_{\{c_j\}_{j=1}^n}\left\|f-\sum_{j=1}^n c_j\phi_j\right\|_X.$$

This concept allows us to find, for a fixed n and for a class F, a subspace of dimension n that is optimal with respect to the construction of a best approximating element. In other words, the concept of width allows us to choose from among various Chebyshev methods having the same quantitative characteristic of complexity (the dimension of the approximating subspace) the one which has the greatest accuracy.

The first result about widths (Kolmogorov, 1936), namely

$$d_{2n+1}(W_2^r, L_2) = (n+1)^{-r},$$

showed that the best subspace of dimension $2n+1$ for the approximation of classes of periodic functions is the subspace of trigonometric polynomials of order n. This result confirmed that the approximation of functions in the class W_2^r by trigonometric polynomials is natural. Further estimates of the widths $d_{2n+1}(W^r_{q,\alpha}, L_p)$, $1 \le q$, $p \le \infty$, some of which are discussed in §2.1 below, showed that, for some values of the parameters q, p, the subspace of trigonometric polynomials of order n is optimal (in the sense of the order of decay) but for other values of q, p this subspace is not optimal.

The Ismagilov (1974) estimate for the quantity $d_n(W_1^r, L_\infty)$ gave the first example, where the subspace of trigonometric polynomials of order n is not optimal. This phenomenon was thoroughly studied by Kashin (1977).

In analogy to the problem of the *Kolmogorov width*, that is, to the problem concerning the best Chebyshev method, problems concerning the best linear method and the best Fourier method were considered.

Tikhomirov (1960b) introduced the *linear width*:

$$\lambda_n(F, L_p) := \inf_{A:\operatorname{rank} A \le n} \sup_{f \in F} \|f - Af\|_p,$$

and Temlyakov (1982a) introduced the *orthowidth* (Fourier width):

$$\varphi_n(F, L_p) := \inf_{\text{orthonormal system } \{u_i\}_{i=1}^n} \sup_{f \in F} \left\| f - \sum_{i=1}^{n} \langle f, u_i \rangle u_i \right\|_p .$$

A discussion and comparison of results concerning $d_n(W_q^r, L_p)$, $\lambda_n(W_q^r, L_p)$ and $\varphi_n(W_q^r, L_p)$ can be found in §2.1. Here we remark that, from the point of view of the orthowidth, the Fourier operator S_n is optimal (in the sense of order) for all $1 \le q$, $p \le \infty$ with the exception of the two cases $(1,1)$ and (∞, ∞).

Keeping in mind the primary question about the selection of an optimal subspace of approximating functions, we now draw some conclusions from this brief historical survey.

(1) The trigonometric polynomials have been considered as a natural means of approximation of periodic functions during the whole period of development of approximation theory.

(2) In approximation theory (as well as in other fields of mathematics) it has turned out that it is natural to unite functions with the same smoothness into a class.
(3) The subspace of trigonometric polynomials has been obtained in many cases as the solution of problems regarding the most precise method for the classes of smooth functions: the Chebyshev method (which uses the Kolmogorov width), the linear method (which uses the linear width), or the Fourier method (which uses the orthowidth).

On the basis of these remarks we may formulate the following general strategy for investigating approximation problems; we remark that this strategy turns out to be most fruitful in those cases where we do not know *a priori* a natural method of approximation. First, we solve the width problem for a class of interest in the simplest case, that of approximation in Hilbert space, L_2. Second, we study the system of functions obtained and apply it to approximation in other spaces L_p. This strategy will be used in Chapters 3, 4, and 5.

1.2 Trigonometric Polynomials

Functions of the form

$$t(x) = \sum_{|k|\le n} c_k e^{ikx} = a_0/2 + \sum_{k=1}^{n} (a_k \cos kx + b_k \sin kx)$$

(c_k, a_k, b_k are complex numbers) will be called trigonometric polynomials of order n. We denote the set of such polynomials by $\mathscr{T}(n)$ and the subset of $\mathscr{T}(n)$ of real polynomials by $\mathscr{RT}(n)$.

We first consider a number of concrete polynomials that play an important role in approximation theory.

1.2.1 The Dirichlet Kernel of Order n

The classical univariate Dirichlet kernel of order n is defined as follows:

$$\mathscr{D}_n(x) := \sum_{|k|\le n} e^{ikx} = e^{-inx}(e^{i(2n+1)x} - 1)(e^{ix} - 1)^{-1}$$
$$= \frac{\sin(n+1/2)x}{\sin(x/2)}. \tag{1.2.1}$$

The Dirichlet kernel is an even trigonometric polynomial with the majorant

$$|\mathscr{D}_n(x)| \le \min(2n+1, \pi/|x|), \qquad |x| \le \pi. \tag{1.2.2}$$

The estimate

$$\|\mathscr{D}_n\|_1 \le C \ln n, \qquad n = 2, 3, \dots, \tag{1.2.3}$$

follows from (1.2.2).

We mention the well-known relation (see Dzyadyk, 1977, p. 112)

$$\|\mathscr{D}_n\|_1 = \frac{4}{\pi^2}\ln n + R_n, \qquad |R_n| \le 3, \qquad n = 1,2,3,\dots$$

For any trigonometric polynomial $t \in \mathscr{T}(n)$ we have

$$\mathscr{D}_n * t := (2\pi)^{-1}\int_{\mathbb{T}} \mathscr{D}_n(x-y)t(y)dy = t.$$

Denote

$$x^l := 2\pi l/(2n+1), \qquad l = 0,1,\dots,2n.$$

Clearly, the points x^l, $l = 1,\dots,2n$, are zeros of the Dirichlet kernel $\mathscr{D}_n$ on $[0,2\pi]$. For any $|k| \le n$ we have

$$\sum_{l=0}^{2n} e^{ikx^l}\mathscr{D}_n(x-x^l) = \sum_{|m|\le n} e^{imx}\sum_{l=0}^{2n} e^{i(k-m)x^l} = e^{ikx}(2n+1).$$

Consequently, for any $t \in \mathscr{T}(n)$,

$$t(x) = (2n+1)^{-1}\sum_{l=0}^{2n} t(x^l)\mathscr{D}_n(x-x^l). \tag{1.2.4}$$

Further, it is easy to see that for any $u,\ v \in \mathscr{T}(n)$ we have

$$\langle u,v\rangle := (2\pi)^{-1}\int_{-\pi}^{\pi} u(x)\overline{v(x)}dx = (2n+1)^{-1}\sum_{l=0}^{2n} u(x^l)\overline{v(x^l)} \tag{1.2.5}$$

and, for any $t \in \mathscr{T}(n)$,

$$\|t\|_2^2 = (2n+1)^{-1}\sum_{l=0}^{2n} |t(x^l)|^2. \tag{1.2.6}$$

For $1 < q \le \infty$ the estimate

$$\|\mathscr{D}_n\|_q \le C(q)n^{1-1/q} \tag{1.2.7}$$

follows from (1.2.2). Applying the Hölder inequality (see (A.1.1) in the Appendix) for estimating $\|\mathscr{D}_n\|_2^2$ we get

$$2n+1 = \|\mathscr{D}_n\|_2^2 \le \|\mathscr{D}_n\|_q \|\mathscr{D}_n\|_{q'}. \tag{1.2.8}$$

Relations (1.2.7) and (1.2.8) imply for $1 < q < \infty$ the relation

$$\|\mathscr{D}_n\|_q \asymp n^{1-1/q}. \tag{1.2.9}$$

Relation (1.2.9) for $q = \infty$ is obvious.

We denote by S_n the operator taking a partial sum of order n. Then for $f \in L_1$ we have

$$S_n(f) := \mathscr{D}_n * f = (2\pi)^{-1} \int_{-\pi}^{\pi} \mathscr{D}_n(x-y) f(y) dy.$$

Theorem 1.2.1 *The operator S_n does not change polynomials from $\mathscr{T}(n)$ and for $p = 1$ or ∞ we have*

$$\|S_n\|_{p\to p} \le C \ln n, \qquad n = 2,3,\dots,$$

and for $1 < p < \infty$ for all n we have

$$\|S_n\|_{p\to p} \le C(p).$$

This theorem follows from (1.2.3) and the Marcinkiewicz multiplier theorem (see Theorem A.3.6).

For $t \in \mathscr{T}(n)$,

$$t(x) = a_0/2 + \sum_{k=1}^{n} (a_k \cos kx + b_k \sin kx),$$

we call the polynomial $\tilde{t} \in \mathscr{T}(n)$, where

$$\tilde{t}(x) := \sum_{k=1}^{n} (a_k \sin kx - b_k \cos kx)$$

the polynomial conjugate to t.

Corollary 1.2.2 *For $1 < p < \infty$ and all n we have*

$$\|\tilde{t}\|_p \le C(p) \|t\|_p.$$

Proof Let $t \in \mathscr{T}(n)$. It is not difficult to see that $\tilde{t} = t * \widetilde{\mathscr{D}}_n$, where

$$\widetilde{\mathscr{D}}_n(x) := 2 \sum_{k=1}^{n} \sin kx.$$

Clearly, it suffices to consider the case of odd n. Let this be the case and set $m := (n+1)/2$, $l := (n-1)/2$. Representing $\tilde{\mathscr{D}}_n(x)$ in the form

$$\tilde{\mathscr{D}}_n(x) = \frac{1}{i}\left(\sum_{k=1}^{n} e^{ikx} - \sum_{k=-n}^{-1} e^{ikx} \right) = \frac{1}{i}\left(e^{imx} \mathscr{D}_l(x) - e^{-imx} \mathscr{D}_l(x) \right),$$

we obtain the corollary. □

A trigonometric conjugate operator maps a function $f(x)$ to a function

$$\sum_k (\operatorname{sign} k) \hat{f}(k) e^{ikx}.$$

The Marcinkiewicz multiplier theorem A.3.6 implies that this operator is bounded as an operator from L_p to L_p for $1 < p < \infty$. We denote by $\tilde{f}$ the conjugate function.

1.2.2 The Fejér Kernel of Order $n-1$

The classical univariate Fejér kernel of order $n-1$ is defined as follows:

$$\mathcal{K}_{n-1}(x) := n^{-1} \sum_{m=0}^{n-1} \mathcal{D}_m(x) = \sum_{|m| \le n} (1 - |m|/n) e^{imx}$$
$$= \frac{\big(\sin(nx/2)\big)^2}{n\big(\sin(x/2)\big)^2}.$$

The Fejér kernel is an even nonnegative trigonometric polynomial in $\mathcal{T}(n-1)$ with majorant

$$\big|\mathcal{K}_{n-1}(x)\big| = \mathcal{K}_{n-1}(x) \le \min\big(n, \pi^2/(nx^2)\big), \qquad |x| \le \pi. \tag{1.2.10}$$

From the obvious relations

$$\|\mathcal{K}_{n-1}\|_1 = 1, \qquad \|\mathcal{K}_{n-1}\|_\infty = n$$

and the inequality, see (A.1.6),

$$\|f\|_q \le \|f\|_1^{1/q} \|f\|_\infty^{1-1/q}$$

we get in the same way as we obtained (1.2.9),

$$Cn^{1-1/q} \le \|\mathcal{K}_{n-1}\|_q \le n^{1-1/q}, \qquad 1 \le q \le \infty. \tag{1.2.11}$$

1.2.3 The de la Vallée Poussin Kernels

The classical univariate de la Vallée Poussin kernel with parameters m, n is defined as follows:

$$\mathcal{V}_{m,n}(x) := (n-m)^{-1} \sum_{l=m}^{n-1} \mathcal{D}_l(x), \qquad n > m.$$

It is convenient to represent this kernel in terms of Fejér kernels:

$$\mathcal{V}_{m,n}(x) = (n-m)^{-1}\big(n\mathcal{K}_{n-1}(x) - m\mathcal{K}_{m-1}(x)\big)$$
$$= (\cos mx - \cos nx)\big(2(n-m)\big(\sin(x/2)\big)^2\big)^{-1}.$$

The de la Vallée Poussin kernels $\mathcal{V}_{m,n}$ are even trigonometric polynomials of order $n-1$ with majorant

$$\big|\mathcal{V}_{m,n}(x)\big| \le C\min\big(n,\ 1/|x|, 1/((n-m)x^2)\big), \qquad |x| \le \pi. \tag{1.2.12}$$

Relation (1.2.12) implies the estimate

$$\|\mathcal{V}_{m,n}\|_1 \le C \ln\big(1 + n/(n-m)\big).$$

We often use the de la Vallée Poussin kernel with $n = 2m$ and denote it by

$$\mathcal{V}_m(x) := \mathcal{V}_{m,2m}(x), \qquad m \ge 1, \qquad \mathcal{V}_0(x) := 1.$$

Then for $m \ge 1$ we have

$$\mathcal{V}_m = 2\mathcal{K}_{2m-1} - \mathcal{K}_{m-1},$$

which, with the properties of $\mathcal{K}_n$, implies

$$\|\mathcal{V}_m\|_1 \le 3. \tag{1.2.13}$$

In addition,

$$\|\mathcal{V}_m\|_\infty \le 3m.$$

Consequently, in the same way as above, see (1.2.9) and (1.2.11), we get

$$\|\mathcal{V}_m\|_q \asymp m^{1-1/q}, \qquad 1 \le q \le \infty. \tag{1.2.14}$$

Denote

$$x(l) := \pi l/2m, \qquad l = 1, \dots, 4m.$$

Then, analogously to (1.2.4), for each $t \in \mathcal{T}(m)$ we have

$$t(x) = (4m)^{-1} \sum_{l=1}^{4m} t\big(x(l)\big)\mathcal{V}_m\big(x - x(l)\big). \tag{1.2.15}$$

The operator $\mathcal{V}_m$ defined on L_1 by the formula

$$V_m(f) := f * \mathcal{V}_m$$

is called the de la Vallée Poussin operator.

The following theorem is a corollary of the definition of the kernels $\mathcal{V}_m$ and the relation (1.2.13).

Theorem 1.2.3 *The operator V_m does not change polynomials from $\mathcal{T}(m)$, and for all $1 \le p \le \infty$ we have*

$$\|V_m\|_{p \to p} \le 3, \qquad m = 1, 2, \dots$$

In addition, we formulate two properties of the de la Vallée Poussin kernels.

(1) Relation (1.2.12) with $n = 2m$ implies the inequality

$$\big|\mathcal{V}_m(x)\big| \le C \min\big(m, 1/(mx^2)\big), \qquad |x| \le \pi.$$

It is easy to derive from this inequality the following property.

(2) For h satisfying the condition $C_1 \leq mh \leq C_2$ we have

$$\sum_{0 \leq l \leq 2\pi/h} \left|\mathscr{V}_m(x - lh)\right| \leq Cm.$$

We remark that property (2) is valid for the Fejér kernel $\mathscr{K}_m$.

1.2.4 The Jackson Kernel

The classical univariate Jackson kernel with parameters n, a is defined as follows:

$$J_n^a(x) := \gamma_{a,n}^{-1} \left(\frac{\sin(nx/2)}{\sin(x/2)}\right)^{2a}, \qquad a \in \mathbb{N},$$

where $\gamma_{a,n}$ is selected in such a way that

$$\|J_n^a\|_1 = 1. \tag{1.2.16}$$

Let us estimate $\gamma_{a,n}$ from below. We have

$$\begin{aligned} \gamma_{a,n} &= (2\pi)^{-1} \int_{-\pi}^{\pi} \left(\frac{\sin(nx/2)}{\sin(x/2)}\right)^{2a} dx \\ &\geq \pi^{-1} \int_0^{\pi/n} \left(\frac{nx/\pi}{x/2}\right)^{2a} dx \geq Cn^{2a-1}. \end{aligned} \tag{1.2.17}$$

The Jackson kernel is an even nonnegative trigonometric polynomial of order $a(n-1)$. It follows from (1.2.17) that

$$J_n^a(x) \leq C\min(n, n^{1-2a}x^{-2a}), \qquad |x| \leq \pi. \tag{1.2.18}$$

Relation (1.2.18) implies that for $0 \leq r < 2a - 1$,

$$\int_0^{\pi} J_n^a(x) x^r dx \leq C(r) n^{-r}. \tag{1.2.19}$$

1.2.5 The Rudin–Shapiro Polynomials

We define recursively pairs of trigonometric polynomials $P_j(x)$ and $Q_j(x)$ of order $2^j - 1$:

$$P_0 := Q_0 := 1,$$

$$P_{j+1}(x) := P_j(x) + e^{i2^j x} Q_j(x), \qquad Q_{j+1}(x) := P_j(x) - e^{i2^j x} Q_j(x).$$

Then at each point x we have

$$|P_{j+1}|^2+|Q_{j+1}|^2=(P_j+e^{i2^jx}Q_j)(\overline{P}_j+e^{-i2^jx}\overline{Q}_j)$$
$$+(P_j-e^{i2^jx}Q_j)(\overline{P}_j-e^{-i2^jx}\overline{Q}_j)$$
$$=2\big(|P_j|^2+|Q_j|^2\big).$$

Therefore, for all x

$$\big|P_j(x)\big|^2+\big|Q_j(x)\big|^2=2^{j+1}.$$

Thus, for example,

$$\|P_n\|_\infty\leq 2^{(n+1)/2}. \tag{1.2.20}$$

It is clear from the definition of the polynomials P_n that

$$P_n(x)=\sum_{k=0}^{2^n-1}\varepsilon_k e^{ikx},\qquad \varepsilon_k=\pm1,\quad \varepsilon_0=1.$$

Let N be a natural number and

$$N=\sum_{j=1}^{m}2^{n_j},\qquad n_1>n_2>\cdots>n_m\geq 0,$$

its binary representation. We set

$$R'_N(x):=P_{n_1}(x)+\sum_{j=2}^{m}P_{n_j}(x)e^{i(2^{n_1}+\cdots+2^{n_{j-1}})x},$$
$$R_N(x):=R'_N(x)+R'_N(-x)-1.$$

Then $R_N(x)$ has the form

$$R_N(x)=\sum_{|k|<N}\varepsilon_k e^{ikx},\qquad \varepsilon_k=\pm1,$$

and for this polynomial the estimate

$$\|R_N\|_\infty\leq CN^{1/2} \tag{1.2.21}$$

holds.

1.2.6 A Modification of the Fejér Kernel

We consider the polynomials

$$G_n(x):=\sum_{|m|<n}\big(1-|m|/n\big)^{1/2}e^{imx}.$$

These are even trigonometric polynomials of order $n-1$ with the properties

$$(2\pi)^{-1}\int_0^{2\pi} G_n(x-a)G_n(x-b)dx = \mathscr{K}_{n-1}(a-b), \tag{1.2.22}$$

$$\|G_n\|_1 \le C. \tag{1.2.23}$$

Relation (1.2.22) is obvious. It implies that the system of polynomials $G_n(x-2\pi l/n)$, $l=1,\dots,n$ is an orthogonal system in $\mathscr{T}(n-1)$.

Let us prove the relation (1.2.23). Denote

$$\phi(u) := \left(1-|u|\right)^{1/2}, \qquad |u| \le 1.$$

Then we have on $[-1,1]$

$$\phi(u) = \sum_l a_l e^{i\pi l x}$$

and it is not hard to prove that

$$|a_l| \le C\left(|l|+1\right)^{-3/2}. \tag{1.2.24}$$

Further,

$$\begin{aligned} G_n(x) &= \sum_{|m|\le n} \phi(m/n)e^{imx} \\ &= \sum_l a_l \sum_{|m|\le n} e^{im(x+\pi l/n)} = \sum_l a_l \mathscr{D}_n(x+\pi l/n). \end{aligned} \tag{1.2.25}$$

Let us consider the function

$$g_{n,l}(x) := \mathscr{D}_n(x+\pi l/n) - (-1)^l \mathscr{D}_n(x).$$

Using the representation (1.2.1) one can obtain the estimate

$$\|g_{n,l}\|_1 \le C\ln\left(|l|+2\right). \tag{1.2.26}$$

Further, owing to the equality

$$\phi(1) = \sum_l a_l(-1)^l = 0,$$

relation (1.2.25) can be rewritten in the form

$$G_n(x) = \sum_l a_l g_{n,l}(x).$$

Using relations (1.2.24) and (1.2.26) we then obtain relation (1.2.23).

1.2.7 A Generalization of the Rudin–Shapiro Polynomials

The trigonometric polynomials considered above (see §§1.2.1–1.2.6) were constructively obtained: either they are given by a formula (§§1.2.1–1.2.4, 1.2.6) or a method of construction is supplied (§1.2.5). In this subsection we formulate a theorem that proves the existence of polynomials with given properties.

Theorem 1.2.4 *Let $\varepsilon > 0$, and let a subspace $\Psi \subset \mathscr{T}(n)$ be such that* $\dim \Psi \geq \varepsilon(2n+1)$. *Then there exists a $t \in \Psi$ such that*

$$\|t\|_\infty = 1,$$

and

$$\|t\|_2 \geq C(\varepsilon) > 0.$$

An analogous statement is valid for the multivariable trigonometric polynomials, will be proved in Chapter 3 (see Theorem 3.2.1).

We remark that the polynomial t from Theorem 1.2.4, by virtue of the inequality

$$\|t\|_2^2 \leq \|t\|_1 \|t\|_\infty,$$

satisfies the condition

$$\|t\|_1 \geq C(\varepsilon)^2 > 0. \tag{1.2.27}$$

1.2.8 An Application of the Gaussian Sums

In this subsection we construct polynomials that we will use in studying linear widths. This construction is based on properties of the Gaussian sums:

$$S(q,l) := \sum_{j=1}^{q} e^{i2\pi l j^2/q},$$

where q is a natural number and l,q are coprime; that is, $(l,q) = 1$. We confine ourselves to the case where q is an odd prime.

Theorem 1.2.5 *Let $q > 2$ be a prime, $l \neq 0$ an integer, and k an integer. Then, for*

$$S(q,l,k) := \sum_{j=1}^{q} e^{i2\pi(lj^2+kj)/q},$$

the following equality is true:

$$\big|S(q,l,k)\big| = q^{1/2}.$$

Proof We first consider the case $k = 0$. Note that the quantity $S(q,l)$ does not change if we sum over the complete system of remainders modulo q instead of the segment $[1,q]$. Consequently, for any integer h,

$$S(q,l) = \sum_{j=1}^{q} e^{i2\pi l(j+h)^2/q}. \tag{1.2.28}$$

Further,

$$\left|S(q,l)\right|^2 = \left(\sum_{h=1}^{q} e^{-i2\pi lh^2/q}\right)\left(\sum_{j=1}^{q} e^{i2\pi lj^2/q}\right).$$

Using (1.2.28), we see that this is equal to

$$\sum_{h=1}^{q} e^{-i2\pi lh^2/q} \sum_{j=1}^{q} e^{i2\pi l(j+h)^2/q} = \sum_{h=1}^{q}\sum_{j=1}^{q} e^{i2\pi l(j^2+2jh)/q}. \tag{1.2.29}$$

Taking into account that

$$\sum_{h=1}^{q} e^{i2\pi l2jh/q} = \begin{cases} q & \text{for } j = q, \\ 0 & \text{for } j \in [1,q), \end{cases}$$

we get from (1.2.29),

$$\left|S(q,l)\right|^2 = q. \tag{1.2.30}$$

Now let k be nonzero. Since q is a prime different from 2, the numbers $2lb$, $b = 1,\dots,q$, run through a complete system of remainders modulo q. Consequently, there is a b such that

$$2lb \equiv k \pmod{q}.$$

Then

$$lj^2 + kj \equiv l(j+b)^2 - lb^2 \pmod{q}$$

and, consequently,

$$\left|S(q,l,k)\right| = \left|S(q,l)\right| = q^{1/2}.$$

The theorem is proved. □

Theorem 1.2.6 *Let q be a prime and $q = 2a+1$. For any $n \in [1,a]$ there is a trigonometric polynomial $t_n \in \mathscr{T}(a)$ such that only n Fourier coefficients of t_n are nonzero and for all k we have $\left|\hat{t}(k)\right| \le 1$ and in addition*

$$t_n(0) \ge (n+1)/2, \qquad \left|t(2\pi l/q)\right| \le Cq^{1/2}, \qquad l = 1,\dots,2a.$$

Proof The proof of this theorem can easily be derived from a deep number theoretical result due to Hardy and Littlewood about estimating incomplete Gaussian sums: for any $n \in [1,q]$

$$\left|\sum_{j=1}^{n} e^{i2\pi l j^2/q}\right| \le Cq^{1/2}, \qquad (l,q)=1. \tag{1.2.31}$$

Indeed, let k_j denote the smallest nonnegative remainder of the number j^2 modulo q, $j=1,\dots,n$, and let

$$G := \{k_j - a,\ j=1,\dots,n\}.$$

We set

$$t_n(x) := \sum_{k\in G} e^{ikx}.$$

Then

$$\left|t_n(2\pi l/q)\right| = \left|\sum_{k\in G} e^{i2\pi lk/q}\right| = \left|\sum_{j=1}^{n} e^{i2\pi l j^2/q}\right|,$$

which by (1.2.31) implies the required estimates for $t_n(2\pi l/q)$. The bound $t_n(0) = n \ge (n+1)/2$ is obvious. □

For the sake of completeness we will prove Theorem 1.2.6 using Theorem 1.2.5. Instead of (1.2.31) we prove the inequality

$$\left|\sum_{j} |(1-|j-a|/n)_+ e^{i2\pi l j^2/q}\right| \le q^{1/2}, \qquad (l,q)=1. \tag{1.2.32}$$

Let $l \in [1,q-1]$. Consider the trigonometric polynomial

$$t(x) := \sum_{j=0}^{q-1} e^{i2\pi l j^2/q} e^{i(j-a)x}.$$

Then at the points $x^k = 2\pi k/(2a+1) = 2\pi k/q$ we have

$$\left|t(x^k)\right| = \left|S(q,l,k)\right| = q^{1/2}, \qquad k=0,\dots,2a. \tag{1.2.33}$$

We set

$$u_n(x) := t(x) * \mathscr{K}_{n-1}(x).$$

Then by (1.2.5),

$$u_n(x) = q^{-1} \sum_{k=0}^{2a} t(x^k)\mathscr{K}_{n-1}(x-x^k),$$

and, using (1.2.33) we find that

$$\left|u_n(x)\right| \le q^{-1/2} \sum_{k=0}^{2a} \mathscr{K}_{n-1}(x-x^k) = q^{1/2}. \tag{1.2.34}$$

Further,

$$u_n(0) = \sum_j \left(1 - |j-a|/n\right)_+ e^{i2\pi l j^2/q},$$

where $(a)_+ := \max(a,0)$. By (1.2.34) this implies (1.2.32).

Setting

$$t_n(x) := \sum_j \left(1 - |j-a|/n\right)_+ e^{i(k_j-a)x},$$

where the k_j are the same as in the beginning of the proof of this theorem, we get

$$\left|t_n(2\pi l/q)\right| = \left|\sum_j \left(1 - |j-a|/n\right)_+ e^{i2\pi l k_j/q}\right| = \left|\sum_j \left(1 - |j-a|/n\right)_+ e^{i2\pi l j^2/q}\right|,$$

which by (1.2.32) implies the conclusion of Theorem 1.2.6, with $2n-1$ nonzero Fourier coefficients instead of n.

1.3 The Bernstein–Nikol'skii Inequalities. The Marcienkiewicz Theorem

The Bernstein–Nikol'skii inequalities connect the L_p-norms of a derivative of some polynomial with the L_q-norm, $1 \le q \le p \le \infty$, of this polynomial. We obtain here inequalities for a derivative that is slightly more general than the Weyl fractional derivative. We first make some auxiliary considerations.

For a sequence $\{a_\nu\}_{\nu=0}^{\infty}$ we write

$$\Delta a_\nu := a_\nu - a_{\nu+1}; \qquad \Delta^2 a_\nu := \Delta(\Delta a_\nu) = a_\nu - 2a_{\nu+1} + a_{\nu+2}.$$

Theorem 1.3.1 *We have*

$$(\pi)^{-1} \int_{-\pi}^{\pi} \left| a_0/2 + \sum_{\nu=1}^{n} a_\nu \cos \nu x \right| dx \le \sum_{\nu=0}^{n} (\nu+1)|\Delta^2 a_\nu|.$$

Proof Applying twice the Abel transformation (see (A.1.18) in the Appendix) with $a_\nu = 0$ for $\nu > n$, we obtain

$$\begin{aligned} t(x) &:= a_0 + \sum_{\nu=1}^{n} a_\nu 2\cos \nu x = \sum_{\nu=0}^{n} \mathscr{D}_\nu(x)\Delta a_\nu \\ &= \sum_{\nu=0}^{n} \left(\sum_{\mu=0}^{\nu} \mathscr{D}_\mu(x) \right) \Delta^2 a_\nu = \sum_{\nu=0}^{n} (\nu+1)\mathscr{K}_\nu(x)\Delta^2 a_\nu. \end{aligned} \tag{1.3.1}$$

From (1.3.1), using $\|\mathscr{K}_\nu\|_1 = 1$ we find

$$\|t\|_1 \le \sum_{\nu=0}^{n} (\nu+1)|\Delta^2 a_\nu|,$$

as required. □

1.3.1 The Bernstein inequality

We first prove the Bernstein inequality. Let us consider the following special trigonometric polynomials. Let s be a nonnegative integer. We define

$$\mathscr{A}_0(x) := 1, \quad \mathscr{A}_1(x) := \mathscr{V}_1(x) - 1, \quad \mathscr{A}_s(x) := \mathscr{V}_{2^{s-1}}(x) - \mathscr{V}_{2^{s-2}}(x), \quad s \ge 2,$$

where the $\mathscr{V}_m$ are the de la Vallée Poussin kernels (see §1.2.3). Then $\mathscr{A}_s \in \mathscr{T}(2^s)$ and by (1.2.13),

$$\|\mathscr{A}_s\|_1 \le 6. \tag{1.3.2}$$

Let $r \ge 0$ and α be real numbers. We consider the polynomials

$$\mathscr{V}_n^r(x,\alpha) := 1 + 2\sum_{k=1}^{n} k^r \cos(kx + \alpha\pi/2)$$
$$+ 2\sum_{k=n+1}^{2n-1} k^r \big(1 - (k-n)/n\big)\cos(kx + \alpha\pi/2).$$

Let us prove that, for all $r > 0$ and α,

$$\big\|\mathscr{V}_n^r(x,\alpha)\big\|_1 \le C(r)n^r, \qquad n = 1,2,\dots \tag{1.3.3}$$

Since for an arbitrary α

$$\mathscr{V}_n^r(x,\alpha) - 1 = \big(\mathscr{V}_n^r(x,0) - 1\big)\cos(\alpha\pi/2) + \big(\mathscr{V}_n^r(x,1) - 1\big)\sin(\alpha\pi/2),$$

it suffices to prove (1.3.3) for $\alpha = 0$ and for $\alpha = 1$. We first consider the case $\alpha = 0$. Let v_k be the Fourier cosine coefficients of the function $\mathscr{V}_n^r(x,0)$. Then, by Theorem 1.3.1,

$$\big\|\mathscr{V}_n^r(x,0)\big\|_1 \le \sum_{k=0}^{2n-1} (k+1)|\Delta^2 v_k|. \tag{1.3.4}$$

It is easy to see that, for $1 \le k \le n-2$,

$$|\Delta^2 v_k| \le C(r)k^{r-2}. \tag{1.3.5}$$

By the identity

$$\Delta^2(a_k b_k) = (\Delta^2 a_k)b_k + 2(\Delta a_{k+1})(\Delta b_k) + a_{k+2}(\Delta^2 b_k)$$

with $a_k = k^r$ and $b_k = 1 - (k-n)/n$, we see that the inequality (1.3.5) will be valid for $n \le k \le 2n-3$ too. For the remaining values of $k \neq 0$ we have

$$|\Delta^2 v_k| \le |\Delta v_k| + |\Delta v_{k+1}| \le C(r)n^{r-1}. \tag{1.3.6}$$

From the inequality $|\Delta^2 v_0| \le C(r)$ and relations (1.3.4)–(1.3.6) we get the relation (1.3.3) for $r > 0$ and $\alpha = 0$.

Let $\alpha = 1$ and let $\tilde{\mathscr{A}}_s(x)$ denote the polynomial which is the trigonometric conjugate to $\mathscr{A}_s(x)$, which means that in the expression for $\mathscr{A}_s(x)$ the functions $\cos kx$ are substituted by $\sin kx$. We prove that

$$\|\tilde{\mathscr{A}}_s\|_1 \le C. \tag{1.3.7}$$

Clearly, it suffices to consider $s \ge 3$. It is not difficult to see that the equality

$$\tilde{\mathscr{A}}_s(x) = 2\,\mathrm{Im}\big(\mathscr{A}_s(x) * \big(\big(4\mathscr{K}_{2^{s-1}-1}(x) - 3\mathscr{K}_{2^{s-1}-2^{s-3}-1}(x)\big)e^{i(2^{s-1}+2^{s-3})x}\big)\big),$$

holds. From this equality, by virtue of the Young inequality with $p = q = a = 1$ (see A.1.16)) and the properties of the functions $\mathscr{K}_n$ and $\mathscr{A}_s$, we obtain (1.3.7).

Further, for $n = 2^m$, we have

$$\begin{aligned}\mathscr{V}_n^r(x,1) - 1 &= \big(\mathscr{V}_{2n}^r(x,0) - 1\big) * \mathscr{V}_n^0(x,1)\\ &= -\sum_{s=1}^{m+1} \mathscr{V}_{2n}^r(x,0) * \tilde{\mathscr{A}}_s(x) = -\sum_{s=1}^{m+1} \mathscr{V}_{2^s}^r(x,0) * \tilde{\mathscr{A}}_s(x).\end{aligned} \tag{1.3.8}$$

From (1.3.8) by means of the Young inequality and using (1.3.7) and relation (1.3.3), which has been proved for $\alpha = 0$, we get

$$\|\mathscr{V}_n^r(x,\alpha)\|_1 \le C(r) \sum_{s=0}^{m+1} 2^{rs} \le C(r)n^r. \tag{1.3.9}$$

Now let $2^{m-1} \le n < 2^m$; then

$$\mathscr{V}_n^r(x,1) = \mathscr{V}_{2^{m+1}}^r(x,1) * \mathscr{V}_n(x),$$

which by (1.3.9) and the Young inequality gives the required estimate for all n. Relation (1.3.3) is proved.

We define the operator D_α^r, $r \ge 0$, $\alpha \in \mathbb{R}$, on the set of trigonometric polynomials as follows. Let $f \in \mathscr{T}(n)$; then

$$D_\alpha^r f := f^{(r)}(x,\alpha) := f(x) * \mathscr{V}_n^r(x,\alpha), \tag{1.3.10}$$

and $f^{(r)}(x,\alpha)$ is called the (r,α) derivative. It is clear that for $f(x)$ such that $\hat{f}(0) = 0$ we have for natural numbers r,

$$D_r^r f = \frac{d^r}{dx^r} f.$$

The operator D^r_α is defined in such a way that it has an inverse for each $\mathscr{T}(n)$. This property distinguishes D^r_α from the differential operator and it will be convenient for us. On the other hand it is clear that

$$\frac{d^r f}{dx^r} = D^r_r f - \hat{f}(0).$$

Theorem 1.3.2 *For any $t \in \mathscr{T}(n)$ we have, for $r > 0$, $\alpha \in \mathbb{R}$, $1 \le p \le \infty$,*

$$\left\|t^{(r)}(x,\alpha)\right\|_p \le C(r) n^r \|t\|_p, \qquad n = 1,2,\dots$$

Proof By the definition (1.3.10),

$$t^{(r)}(x,\alpha) = t(x) * \mathscr{V}_n^r(x,\alpha).$$

Therefore, by the Young inequality (A.1.16) with $p = q$, $a = 1$ for all $1 \le p \le \infty$ and r we have

$$\left\|t^{(r)}(x,\alpha)\right\|_p \le \|t\|_p \left\|\mathscr{V}_n^r(x,\alpha)\right\|_1.$$

To conclude the proof we just use inequality (1.3.3). □

Let us discuss the case $r = 0$, which is excluded from Theorem 1.3.2. In the case where $r = 0$ and α is an even integer we have

$$\left|t^{(0)}(x,\alpha)\right| = \left|t(x)\right|$$

and, consequently,

$$\left\|t^{(0)}(x,\alpha)\right\|_p = \|t\|_p, \qquad 1 \le p \le \infty. \tag{1.3.11}$$

To investigate the general case it suffices to study the trigonometric conjugate operator. Theorem 1.2.1 and its corollary show that for all α and $1 < p < \infty$ the inequality

$$\left\|t^{(0)}(x,\alpha)\right\| \le C(p)\|t\|_p$$

holds.

It remains to consider the cases $p = 1, \infty$. It is sufficient to consider $\alpha = 1$. We have for $t \in \mathscr{T}(n)$,

$$t^{(0)}(x,1) = \hat{t}(0) - \tilde{t}(x) = \hat{t}(0) - t(x) * \tilde{\mathscr{D}}_{2n+1}(x).$$

Further,

$$\tilde{\mathscr{D}}_{2n+1}(x) = 2\sum_{k=1}^{2n+1} \sin kx = 2\,\mathrm{Im}\,\mathscr{D}_n(x) e^{i(n+1)x};$$

consequently,

$$\|\tilde{\mathscr{D}}_{2n+1}\|_1 \le C \ln(n+2).$$

Thus, for $t \in \mathscr{T}(n)$,

$$\left\|t^{(0)}(x,1)\right\|_p \leq C\ln(n+2)\|t\|_p, \qquad p=1,\infty. \tag{1.3.12}$$

The relation (1.3.11) with $\alpha = 0$ and (1.3.12) imply for all α the inequality

$$\left\|t^{(0)}(x,\alpha)\right\|_p \leq C\ln(n+2)\|t\|_p, \qquad p=1,\infty. \tag{1.3.13}$$

Remark 1.3.3 We have the relation

$$\sup_{t\in\mathscr{T}(n)} \|t^{(0)}(x,1)\|_p \,/\, \|t\|_p \asymp \ln(n+2), \qquad p=1,\infty.$$

The upper estimate follows from (1.3.12). Let us prove the lower estimate. We first consider the case $p=\infty$. Let $f(x) = (\pi - x)/2$, $0 < x < 2\pi$, be a 2π-periodic function; then

$$f(x) = \sum_{k=1}^{\infty} (\sin kx)/k.$$

Let $m = [n/2]$. Then

$$t(x) := f(x) * \mathscr{V}_m(x)$$

has the following properties: $t \in \mathscr{T}(n)$,

$$\|t\|_\infty \leq 3\pi/2, \qquad t^{(0)}(0,1) \geq \sum_{k=1}^{m} 1/k \geq C\ln(m+2), \tag{1.3.14}$$

which imply the required lower estimate in the case $p=\infty$.

Let $p = 1$ and $m = [n/2]$. Then the function $\mathscr{V}_m \in \mathscr{T}(n)$ has the following properties:

$$\|\mathscr{V}_m\|_1 \leq 3, \tag{1.3.15}$$

$$\left\|\mathscr{V}_m^{(0)}(x,1)\right\|_1 \geq C\ln(m+2). \tag{1.3.16}$$

Let us prove (1.3.16). For t we have from the above consideration for $p=\infty$,

$$\sigma = |\langle \mathscr{V}_m^{(0)}(x,1), t\rangle| \leq \left\|V_m^{(0)}(x,1)\right\|_1 \|t\|_\infty \tag{1.3.17}$$

and

$$\sigma \geq \sum_{k=1}^{m} 1/k \geq C\ln(m+2). \tag{1.3.18}$$

From relations (1.3.14), (1.3.17) and (1.3.18) we obtain (1.3.16). Then (1.3.15) and (1.3.16) give the required lower estimate for $p=1$.

1.3.2 The Nikol'skii Inequality

Let us now prove the Nikol'skii inequality.

Theorem 1.3.4 *For any $t \in \mathscr{T}(n)$, $n > 0$, we have the inequality*

$$\|t\|_p \le C n^{1/q-1/p} \|t\|_q, \qquad 1 \le q < p \le \infty.$$

Proof First let $p = \infty$; then

$$t = t * \mathscr{V}_n$$

and by the Hölder inequality (A.1.1) we have

$$\|t\|_\infty \le \|t\|_q \|\mathscr{V}_n\|_{q'},$$

which, by (1.2.14), implies that

$$\|t\|_\infty \le C\|t\|_q n^{1/q}. \tag{1.3.19}$$

Further, let $q < p < \infty$. Then by (A.1.6),

$$\|t\|_p \le \|t\|_q^{q/p} \|t\|_\infty^{1-q/p}. \tag{1.3.20}$$

The theorem follows from relations (1.3.19) and (1.3.20). □

We now formulate a corollary of Theorems 1.3.2 and 1.3.4.

Corollary 1.3.5 (The Bernstein–Nikol'skii inequality) *For $t \in \mathscr{T}(n)$ and arbitrary $r > 0$, α, $1 \le q \le p \le \infty$, we have the inequality*

$$\left\|t^{(r)}(x,\alpha)\right\|_p \le C(r) n^{r+1/q-1/p} \|t\|_q, \qquad n = 1, 2, \ldots$$

1.3.3 The Marcinkiewicz Theorem

The set $\mathscr{T}(n)$ of trigonometric polynomials is a space of dimension $2n+1$. Each polynomial $t \in \mathscr{T}(n)$ is uniquely defined by its Fourier coefficients $\left\{\hat{t}(k)\right\}_{|k|\le n}$, and by the Parseval identity we have

$$\|t\|_2^2 = \sum_{|k|\le n} \left|\hat{t}(k)\right|^2, \tag{1.3.21}$$

which means that the set $\mathscr{T}(n)$ as a subspace of L_2 is isomorphic to ℓ_2^{2n+1}. Relation (1.2.6) shows that a similar isomorphism can be set up in another way: by mapping a polynomial $t \in \mathscr{T}(n)$ to the vector $m(t) := \left\{t(x^l)\right\}_{l=0}^{2n}$ of its values at the points

$$x^l := 2\pi l/(2n+1), \qquad l = 0, \ldots, 2n.$$

Relation (1.2.6) gives

$$\|t\|_2 = (2n+1)^{-1/2} \left\|m(t)\right\|_2.$$

The following statement is the Marcinkiewicz theorem.

Theorem 1.3.6 *Let $1 < p < \infty$; then for $t \in \mathscr{T}(n)$, $n > 0$, we have the relation*

$$C_1(p)\|t\|_p \le n^{-1/p}\big\|m(t)\big\|_p \le C_2(p)\|t\|_p.$$

Proof We first prove a lemma.

Lemma 1.3.7 *Let $1 \le p \le \infty$; then, for $n > 0$,*

$$\left\|\sum_{l=0}^{2n} a_l \mathscr{V}_n(x-x^l)\right\|_p \le Cn^{1-1/p}\|\mathbf{a}\|_{\ell_p^{2n+1}}, \qquad \mathbf{a} := (a_0, \dots, a_{2n}).$$

Proof Let V be an operator on ℓ_p^{2n+1} defined as follows:

$$V(\mathbf{a}) := \sum_{l=0}^{2n} a_l \mathscr{V}_n(x-x^l).$$

It is obvious that (see (1.2.13))

$$\|V\|_{\ell_1^{2n+1}\to L_1} \le 3. \tag{1.3.22}$$

Using the estimate (see (1.2.12))

$$\big|\mathscr{V}_n(x)\big| \le C\min\big(n, (nx^2)^{-1}\big)$$

it is not hard to prove that

$$\|V\|_{\ell_\infty^{2n+1}\to L_\infty} \le Cn. \tag{1.3.23}$$

From relations (1.3.22) and (1.3.23), using the Riesz–Torin theorem (see Theorem A.3.2) we find that

$$\|V\|_{\ell_p^{2n+1}\to L_p} \le Cn^{1-1/p},$$

which implies the lemma. □

We now continue the proof of Theorem 1.3.6. Let S_n be the operator that takes the partial Fourier sum of order n. Using Theorem 1.2.1 we derive from Lemma 1.3.7 the upper estimate (the first inequality in Theorem 1.3.6):

$$\begin{aligned} t(x) &= (2n+1)^{-1}\sum_{l=0}^{2n} t(x^l)\mathscr{D}_n(x-x^l) \\ &= S_n\left((2n+1)^{-1}\sum_{l=0}^{2n} t(x^l)\mathscr{V}_n(x-x^l)\right). \end{aligned}$$

Consequently,

$$\|t\|_p \le C(p)n^{-1/p}\big\|m(t)\big\|_p.$$

We now prove the lower estimate (the second inequality in Theorem 1.3.6) for $1 \le p < \infty$. We have

$$\|m(t)\|_p^p = \sum_{l=0}^{2n} |t(x^l)|^p = \sum_{l=0}^{2n} t(x^l)\varepsilon_l |t(x^l)|^{p-1}$$
$$= (2\pi)^{-1} \int_0^{2\pi} t(x) \sum_{l=0}^{2n} \varepsilon_l |t(x^l)|^{p-1} \mathscr{V}_n(x - x^l)dx$$
$$\le \|t\|_p \left\| \sum_{l=0}^{2n} \varepsilon_l |t(x^l)|^{p-1} \mathscr{V}_n(x - x^l) \right\|_{p'},$$

using Lemma 1.3.7 we see that the last expression is

$$\le C\|t\|_p n^{1/p} \|m(t)\|_p^{p-1},$$

which implies the required lower estimate and the theorem is proved. □

Remark 1.3.8 In the proof of Theorem 1.3.6 we also proved the inequality

$$\|m(t)\|_1 \le Cn\|t\|_1.$$

We now prove a statement that is analogous to Theorem 1.3.6 but, in contrast to it, includes the cases $p = 1$ and $p = \infty$. Instead of the vector $m(t)$ we now consider the vector

$$M(t) := \big(t(x(1)), \dots, t(x(4n))\big), \qquad x(l) := \pi l/(2n), \qquad l = 1, \dots, 4n.$$

Theorem 1.3.9 *For an arbitrary $t \in \mathscr{T}(n)$, $n > 0$, $1 \le p \le \infty$, we have*

$$C_1\|t\|_p \le n^{-1/p} \|M(t)\|_p \le C_2\|t\|_p.$$

Proof In the same way as for Lemma 1.3.7 one can prove:

Lemma 1.3.10 *Let $1 \le p \le \infty$, then, for $n > 0$,*

$$\left\| \sum_{l=1}^{4n} a_l \mathscr{V}_n(x - x(l)) \right\|_p \le Cn^{1-1/p} \|\mathbf{a}\|_{\ell_p^{4n}}.$$

Lemma 1.3.10 with $\mathbf{a} = M(t)$ and relation (1.2.15) implies the upper estimate

$$\|t\|_p \le Cn^{-1/p} \|M(t)\|_p.$$

The corresponding lower estimate for $1 \le p < \infty$ can be proved in the same way as above for $m(t)$, substituting x^l by $x(l)$.

The lower estimate for $p = \infty$ is obvious. □

1.4 Approximation of Functions in the Classes $W^r_{q,\alpha}$ and H^r_q

1.4.1 Some Properties of the Bernoulli Kernels

For $r > 0$ and $\alpha \in \mathbb{R}$ the functions

$$F_r(x,\alpha) = 1 + 2\sum_{k=1}^{\infty} k^{-r}\cos(kx - \alpha\pi/2)$$

are called *Bernoulli kernels*.

We define the following operator in the space L_1,

$$(I^r_\alpha \phi)(x) := (2\pi)^{-1}\int_0^{2\pi} F_r(x-y,\alpha)\phi(y)dy. \tag{1.4.1}$$

Let us prove that the definition of this operator is reasonable. To establish this it suffices to prove that $F_r \in L_1$.

Theorem 1.4.1 *For $r > 0$, $\alpha \in \mathbb{R}$ we have*

$$F_r \in L_1, \qquad E_n(F_r)_1 \le C(r)(n+1)^{-r}, \qquad n = 0, 1, \ldots$$

Proof Let us consider the functions

$$f^r_s(x,\alpha) := \mathscr{A}_s(x) * \left(1 + 2\sum_{k=1}^{2^s} k^{-r}\cos(kx - \alpha\pi/2)\right),$$

where the $\mathscr{A}_s$ are defined in §1.3.

We first consider the case $\alpha = 0$. Using Theorem 1.3.1 in the same way as in the proof of inequality (1.3.3) we get

$$\left\|f^r_s(x,0)\right\|_1 \le C(r)2^{-rs}. \tag{1.4.2}$$

Further,

$$f^r_s(x,\alpha) = D^r_{-\alpha} f^{2r}_s(x,0),$$

and, consequently, from (1.4.2) and Theorem 1.3.2 we find that

$$\left\|f^r_s(x,\alpha)\right\|_1 \le C(r)2^{-rs}. \tag{1.4.3}$$

Thus the series

$$\sum_{s=0}^{\infty} f^r_s(x,\alpha)$$

converges in L_1 to some function $f(x)$ and

$$\left\|\sum_{s=m}^{\infty} f^r_s(x,\alpha)\right\|_1 \le C(r)2^{-rm}. \tag{1.4.4}$$

From the definition of the function $f_s^r(x,\alpha)$ we get

$$S_n(f) = 1 + 2\sum_{k=1}^{n} k^{-r}\cos(kx - \alpha\pi/2)$$

and

$$\begin{aligned}\|f - S_n(f)\|_1 &\le \sum_{s=0}^{\infty} \|f_s^r(x,\alpha) - S_n(f_s^r(x,\alpha))\|_1 \\ &\le \sum_{s:2^s>n} \|f_s^r - S_n(f_s^r)\|_1 \le C\ln(n+2)\sum_{2^s>n}\|f_s^r\|_1 \\ &\le C(r)n^{-r}\ln(n+2). \end{aligned} \tag{1.4.5}$$

Here we have used Theorem 1.2.1 and relation (1.4.3). Relation (1.4.5) shows that the series defining the function $F_r(x,\alpha)$ converges in L_1 to $f(x)$. The first part of the theorem is proved. The second part of the theorem follows from relation (1.4.4). □

We now proceed to formulate some properties of the operators D_α^r and I_α^r. From the equality ($\phi \in L_1$)

$$\int_0^{2\pi}\left(\pi^{-1}\int_0^{2\pi}\phi(u)\cos\big(k(y-u)+\alpha\pi/2\big)\cos\big(k(x-y)+\beta\pi/2\big)dy\right)du$$
$$= \int_0^{2\pi}\phi(u)\cos\big(k(x-u)+(\alpha+\beta)\pi/2\big)du,$$

which is valid for any nonzero k, the equalities

$$D_{\alpha_1}^{r_1}D_{\alpha_2}^{r_2} = D_{\alpha_1+\alpha_2}^{r_1+r_2}, \tag{1.4.6}$$
$$I_{\alpha_1}^{r_1}I_{\alpha_2}^{r_2} = I_{\alpha_1+\alpha_2}^{r_1+r_2}, \tag{1.4.7}$$
$$D_\alpha^r I_\alpha^r = I_\alpha^r D_\alpha^r = I \tag{1.4.8}$$

follow (we assume that the operators act on a set of trigonometric polynomials).

Denote by $W_{q,\alpha}^r B$, $r > 0$, $\alpha \in \mathbb{R}$, $1 \le q \le \infty$, the class of functions $f(x)$ representable in the form

$$f = I_\alpha^r \phi, \qquad \|\phi\|_q \le B. \tag{1.4.9}$$

For such functions, with some q and B. we define (see (1.4.8))

$$D_\alpha^r f = \phi.$$

Let $1 < q < p < \infty$, $\beta := 1/q - 1/p$. From Corollary A.3.8 of the Hardy–Littlewood inequality (see the Appendix) and the boundedness of the trigonometric conjugate operator as an operator from L_p to L_p for $1 < p < \infty$ (see Corollary 1.2.2), it follows that

$$\|I_\alpha^\beta\|_{q\to p} \le C(q,p). \tag{1.4.10}$$

Relations (1.4.7) and (1.4.10) imply the following embedding theorem.

Theorem 1.4.2 *Let $1 < q < p < \infty$, $\beta = 1/q - 1/p$, $r > \beta$; then*

$$W^r_{q,\alpha_1} \subset W^{r-\beta}_{p,\alpha_2} B, \qquad \alpha_1,\ \alpha_2 \in \mathbb{R}.$$

1.4.2 Approximation for Smoothness Classes

Let us define the classes $H^r_q B$, $r > 0$, $1 \le q \le \infty$ as follows:

$$H^r_q B := \left\{ f \in L_q : \|f\|_q \le B,\ \left\|\Delta^a_t f(x)\right\|_q \le B|t|^r,\ a = [r]+1 \right\},$$

$$\Delta_t f(x) := f(x) - f(x+t), \qquad \Delta^a_t := (\Delta_t)^a.$$

For the case $B = 1$ we simply write $H^r_q := H^r_q 1$, i.e., we drop the constant B.

Let us study these classes from the point of view of their approximation by trigonometric polynomials.

Theorem 1.4.3 *Let $r > 0$, $1 \le q \le \infty$, then*

$$E_n(H^r_q)_q \asymp (n+1)^{-r}, \qquad n = 0, 1, \dots$$

Proof Let us prove the upper estimate. Clearly, it suffices to consider the case $n > 0$. Let $f \in H^r_q$. We consider (see §1.2.4)

$$t(x) := (2\pi)^{-1} \int_{-\pi}^{\pi} \left(f(x) - \Delta^a_y f(x)\right) J^a_n(y) dy.$$

Then $t \in \mathscr{T}(an)$ and

$$f(x) - t(x) = (2\pi)^{-1} \int_{-\pi}^{\pi} \Delta^a_y f(x) J^a_n(y) dy.$$

By a generalization of the Minkowskii inequality, (A.1.9), we have

$$\|f - t\|_q \le (2\pi)^{-1} \int_{-\pi}^{\pi} \left\|\Delta^a_y f(x)\right\|_q J^a_n(y) dy,$$

which by the definition of the class H^r_q and relation (1.2.19) implies that

$$\|f - t\|_q \le C(r) n^{-r}.$$

The upper estimate is proved.

We now prove the lower estimate. We construct functions which will be used in the proof of the more general Theorem 1.4.9. Let $n > 0$ be given and s be such that

$$4n \le 2^s \le 8n.$$

We consider

$$f(x) := 2^{-(r+1-1/q)s}\mathscr{A}_s(x) \tag{1.4.11}$$

and remark that to prove the theorem it suffices to consider the simpler function $f(x) = (n+1)^{-r}e^{i(n+1)x}$. Then, for any $t \in \mathscr{T}(n)$, we have on the one hand

$$\langle f-t, \mathscr{A}_s\rangle = \langle f, \mathscr{A}_s\rangle = 2^{-(r+1-1/q)s}\|\mathscr{A}_s\|_2^2 \geq C2^{-(r-1/q)s}. \tag{1.4.12}$$

On the other hand using the definition of $\mathscr{A}_s$ and (1.2.14) we get

$$\langle f-t, \mathscr{A}_s\rangle \leq \|f-t\|_q\|\mathscr{A}_s\|_{q'} \leq C2^{s/q}\|f-t\|_q. \tag{1.4.13}$$

From relations (1.4.12) and (1.4.13) we obtain

$$E_n(f)_q \geq C2^{-rs} \geq Cn^{-r}.$$

To show that $f \in H_q^r B$, we prove the following auxiliary statement.

Lemma 1.4.4 *Let $g(x)$ be an a-times continuously differentiable 2π-periodic function. Then for all $1 \leq q \leq \infty$ we have*

$$\left\|\Delta_y^a g(x)\right\|_q \leq |y|^a \left\|g^{(a)}(x)\right\|_q.$$

Proof Clearly it suffices to consider the case $a = 1$. We have

$$\left\|\Delta_y g(x)\right\|_q = \left\|\int_x^{x+y} g'(u)du\right\|_q = \left\|\int_0^y g'(x+u)du\right\|_q \leq |y|\|g'\|_q,$$

as required. □

From (1.4.11), (1.2.14), and the Bernstein inequality (Theorem 1.3.2) we get

$$\|f^{(a)}\|_q \leq C(a)2^{(a-r)s}. \tag{1.4.14}$$

Using Lemma 1.4.4 and the simple inequality

$$\left\|\Delta_y^a f(x)\right\|_q \leq 2^a\|f\|_q,$$

we obtain

$$\left\|\Delta_y^a f(x)\right\|_q \leq C(a)\min\left(|y|^a n^{a-r}, n^{-r}\right), \tag{1.4.15}$$

which implies that $f \in H_q^r B$ with some B that is independent of n, and this proves the lower estimate. □

Let us now prove a representation theorem for the class $H_q^r B$. Let

$$A_s(f) := \mathscr{A}_s * f$$

and denote the value of $A_s(f)$ at a point x by $A_s(f,x)$.

Theorem 1.4.5 *Let $f \in L_q$, $1 \le q \le \infty$, $\|f\|_q \le 1$. For $\|\Delta_t^a f\|_q \le |t|^r$, $a = [r]+1$ it is necessary and sufficient that the following conditions be satisfied:*

$$\left\|A_s(f)\right\|_q \le C(r,q)2^{-rs}, \qquad s = 0,1,\dots.$$

(The constants $C(r,q)$ may be different for the cases of necessity and sufficiency.)

Proof

Necessity. Let $f \in H^r_q$; then for any $t_s \in \mathcal{T}(2^{s-2})$, $s \ge 2$ we have

$$A_s(f) = A_s(f - t_s)$$

and

$$\|A_s(f)\|_q \le \|\mathscr{A}_s\|_1 \|f - t_s\|_q.$$

Applying Theorem 1.4.3 and using relation (1.3.2) we get

$$\left\|A_s(f)\right\|_q \le C(r,q)2^{-rs}.$$

Sufficiency. Let

$$\left\|A_s(f)\right\|_q \le \gamma 2^{-rs}, \tag{1.4.16}$$

then using Corollary 2.2.7 we get

$$f = \sum_{s=0}^{\infty} A_s(f),$$

in the sense of convergence in L_q, and

$$\|\Delta_t^a f\|_q \le \sum_{s=1}^{\infty} \left\|\Delta_t^a A_s(f)\right\|_q. \tag{1.4.17}$$

From Lemma 1.4.4 we find, in the same way as in (1.4.15),

$$\left\|\Delta_t^a A_s(f)\right\|_q \le C(a)2^{-rs}\min\left(1, \left(|t|2^s\right)^a\right). \tag{1.4.18}$$

From (1.4.17) and (1.4.18) we obtain

$$\|\Delta_t^a f\|_q \le C(r)\gamma|t|^r,$$

which concludes the proof of the theorem if we take $\gamma < 1/C(r)$. □

Denote

$$\delta_0(f) := S_0(f), \qquad \delta_s(f) := S_{2^s-1}(f) - S_{2^{s-1}-1}(f), \qquad s = 1,2,\dots$$

Corollary 1.4.6 *In the case $1 < q < \infty$ the functions $A_s(f)$ in Theorem 1.4.5 can be replaced by $\delta_s(f)$.*

Proof For $1 < q < \infty$ the conditions

(1) $\|A_s(f)\|_q \le C(q)2^{-rs}$,
(2) $\|\delta_s(f)\|_q \le C(q)2^{-rs}$

are equivalent for all s. Indeed,

$$A_s(f) = \mathscr{A}_s * (\delta_{s-1}(f) + \delta_s(f)),$$
$$\delta_s(f) = \delta_s(A_s(f) + A_{s+1}(f)),$$

which by (1.3.2) and the boundedness of the operator δ_s as an operator from L_q to L_q, $1 < q < \infty$ (see Corollary A.3.4) implies the equivalence of conditions (1) and (2). □

Corollary 1.4.7 *Let* $1 \le q \le \infty$, $\|f\|_q \le 1$ *and*

$$E_n(f)_q \ll (n+1)^{-r}, \qquad n = 0, 1, \dots;$$

then $f \in H_q^r B$ *for some* B.

Indeed, in the same way as in the proof of the necessity in Theorem 1.4.5 we get

$$\|A_s(f)\|_q \ll 2^{-rs},$$

which by Theorem 1.4.5 (regarding the sufficiency) implies that $f \in H_q^r B$.

Statements of the type of Theorem 1.4.3 are called *direct* theorems of approximation theory, and statements of the type of Corollary 1.4.7 are called *inverse* theorems of approximation theory.

Theorem 1.4.1 and Corollary 1.4.7 imply that

$$F_r(x, \alpha) \in H_1^r B. \tag{1.4.19}$$

Consequently, for $f \in W_{q,\alpha}^r$ we have

$$\|\Delta_t^a f(x)\|_q \le \|\Delta_t^a F_r(x, \alpha)\|_1 \|D_\alpha^r f\|_q \le B|t|^r;$$

that is, $f \in H_q^r B$.

Thus, we have proved that

$$W_{q,\alpha}^r \subset H_q^r B. \tag{1.4.20}$$

Let us prove an embedding theorem for the H classes.

Theorem 1.4.8 *Let* $1 \le q \le p \le \infty$, $\beta := 1/q - 1/p$, $r > \beta$. *We have the inclusion*

$$H_q^r \subset H_p^{r-\beta} B$$

(in the case $p = \infty$ *this means that for any* $f \in H_q^r$ *there is an equivalent* $g \in H_\infty^{r-\beta} B$*).*

Proof Let $f \in H^r_q$. By Theorem 1.4.5

$$\left\|A_s(f)\right\|_q \le C(r,q)2^{-rs}.$$

Therefore, by the Nikol'skii inequality (Theorem 1.3.4) we have

$$\left\|A_s(f)\right\|_p \le C(r,q)2^{-(r-\beta)s}. \tag{1.4.21}$$

Let $g(x)$ denote the sum of the series $\sum_{s=0}^{\infty} A_s(f,x)$ in the sense of convergence in L_p. From Corollary 2.2.7 below it follows that f and g are equivalent. From (1.4.21) and the equality $A_s(f) = A_s(g)$, by Theorem 1.4.5 we obtain $g \in H_p^{r-\beta}B$.

The theorem is proved. □

With the aid of Theorem 1.4.8 we can prove the following statement.

Theorem 1.4.9 *Let* $1 \le q,\ p \le \infty$, $r > (1/q - 1/p)_+$. *Then*

$$E_n(W^r_{q,\alpha})_p \asymp E_n(H^r_q)_p \asymp n^{-r+(1/q-1/p)_+}.$$

Proof By relation (1.4.20) it suffices to prove the upper estimate for the H classes and the lower estimate for the W classes. We first prove the upper estimate. Let $1 \le q \le p \le \infty$. Then Theorems 1.4.8 and 1.4.3 give

$$E_n(H^r_q)_p \ll n^{-r+1/q-1/p}. \tag{1.4.22}$$

For $1 \le p < q \le \infty$ we have, by the monotonicity of the L_p-norms and Theorem 1.4.3,

$$E_n(H^r_q)_p \le E_n(H^r_q)_q \ll n^{-r}.$$

From this and relation (1.4.22) the required upper estimates follow.

Let us prove the lower estimate. Let n and s be the same as in the proof of the lower estimate in Theorem 1.4.3 and let f be defined by (1.4.11). Then by the Bernstein inequality,

$$\|D^r_\alpha f\|_q \le C(r),$$

and $f \in W^r_{q,\alpha}C(r)$.

Let $1 \le q \le p \le \infty$. From relation (1.4.12) and relation (1.4.13) with p instead of q we get

$$E_n(f)_p \ge Cn^{-r+1/q-1/p}. \tag{1.4.23}$$

For $1 \le p \le q \le \infty$ it suffices to consider as an example

$$f(x) = 2(n+1)^{-r}\cos(n+1)x.$$

Then $f \in W^r_{\infty,\alpha}$ and, for any $t \in \mathscr{T}(n)$,

$$\sigma = \langle f(x) - t(x),\ \cos(n+1)x\rangle = (n+1)^{-r}, \qquad \sigma \le \|f-t\|_1,$$

which implies the estimate

$$E_n(W^r_{\infty,\alpha})_1 \geq (n+1)^{-r}. \tag{1.4.24}$$

The required lower estimates follow from (1.4.23) and (1.4.24) and the theorem is proved. □

Remark 1.4.10 Theorem 1.2.3 implies that for any $f \in L_p$ the de la Vallée Poussin inequality holds:

$$\|f - V_n(f)\|_p \leq 4E_n(f)_p, \quad 1 \leq p \leq \infty. \tag{1.4.25}$$

This inequality and Theorem 1.4.9 show that, for all $1 \leq q, p \leq \infty$,

$$V_n(H^r_q)_p := \sup_{f \in H^r_q} \|f - V_n(f)\|_p \asymp E_{2n}(H^r_q)_p, \tag{1.4.26}$$

and an analogous relation is valid for the W classes.

Thus, for the classes $W^r_{q,\alpha}$ and H^r_q there exist linear methods giving an approximation of the same order as the best approximation.

Remark 1.4.11 From Theorem 1.2.1 it follows that for all $1 < p < \infty$ and $f \in L_p$,

$$\|f - S_n(f)\|_p \leq C(p)E_n(f)_p. \tag{1.4.27}$$

Consequently, if we are interested only in the dependence of the approximation of a function $f \in L_p$ on n then it suffices, in the case $1 < p < \infty$, to consider the simplest method of approximation, namely, the Fourier method.

This remains true for the classes $W^r_{q,\alpha}$ and H^r_q for all $1 \leq q, p \leq \infty$, excepting the cases $q = p = 1$ and $q = p = \infty$. For the function class F let us denote

$$S_n(F)_p := \sup_{f \in F} \|f - S_n(f)\|_p.$$

Theorem 1.4.12 *Let* $1 \leq q, p \leq \infty$, $(q,p) \neq (1,1)$ *or* (∞,∞), *and* $r > (1/q - 1/p)_+$. *Then*

$$S_n(W^r_{q,\alpha})_p \asymp S_n(H^r_q)_p \asymp n^{-r+(1/q-1/p)_+}.$$

Proof In the case $1 < p < \infty$ the theorem follows from Theorem 1.4.9 and relation (1.4.27). It remains to consider the cases $p = 1$, $q > 1$ and $1 \leq q < p = \infty$. In the case $p = 1$, $q > 1$ we have

$$S_n(H^r_q)_1 \leq S_n(H^r_{q^*})_{q^*} \ll n^{-r},$$

where $q^* = \min(q, 2)$.

Now let $1 \leq q < p = \infty$. In the case $1 \leq q < 2$, by Theorem 1.4.8 we have

$$H^r_q \subset H^{r-(1/q-1/2)}_2 B,$$

which indicates that it suffices to consider the case $2 \le q < \infty$. In this case by Theorem 1.2.1 and Corollary 1.4.6 we have for $s > s_n$, where s_n is such that $2^{s_n-1} \le n < 2^{s_n}$,

$$\left\|\delta_s(f)\right\|_q \le C(r,q)2^{-rs},$$
$$\left\|\delta_{s_n}(f) - S_n\left(\delta_{s_n}(f)\right)\right\|_q \le C(r,q)2^{-rs_n}.$$

From these inequalities, using the Nikol'skii inequality, we get

$$\begin{aligned}\left\|f - S_n(f)\right\|_\infty &\le \left\|\delta_{s_n}(f) - S_n\left(\delta_{s_n}(f)\right)\right\|_\infty + \sum_{s>s_n} \left\|\delta_s(f)\right\|_\infty \\ &\le C(r,q) \sum_{s\ge s_n} 2^{-(r-1/q)s} \le C(r,q)n^{-r+1/q},\end{aligned}$$

which concludes the proof of the theorem. □

We proceed to the cases $q = p = 1$ or ∞, which were excluded in Theorem 1.4.12. For these cases we obtain from Theorem 1.2.1 the following *Lebesgue inequality*: for $f \in L_p$, $p = 1$, or ∞,

$$\left\|f - S_n(f)\right\|_p \le C(\ln n)E_n(f)_p, \qquad n = 2,3,\dots \tag{1.4.28}$$

Theorem 1.4.13 *Let $p = 1$, or ∞ and $r > 0$; then*

$$S_n(W^r_{p,\alpha})_p \asymp S_n(H^r_p)_p \asymp n^{-r}\ln n, \qquad n = 2,3,\dots$$

Proof The upper estimates follow from Theorem 1.4.9 and the inequality (1.4.28). Owing to (1.4.20) it suffices to prove the lower estimates for the W classes. We first remark that

$$S_n(W^r_{1,\alpha})_1 = S_n(W^r_{\infty,-\alpha})_\infty. \tag{1.4.29}$$

Indeed (see Theorem A.2.1),

$$\begin{aligned}S_n(W^r_{1,\alpha})_1 &= \sup_{\|\phi\|_1\le 1} \left\|F_r(x,\alpha) * \left(\phi - S_n(\phi)\right)\right\|_1 \\ &= \sup_{\|\phi\|_1\le 1}\ \sup_{\|\psi\|_\infty\le 1} \left|\langle F_r(x,\alpha) * \left(\phi - S_n(\phi)\right), \psi\rangle\right| \\ &= \sup_{\|\phi\|_1\le 1}\ \sup_{\|\psi\|_\infty\le 1} \left|\langle \phi, F_r(x,-\alpha) * \left(\psi - S_n(\psi)\right)\rangle\right| \\ &= S_n(W^r_{\infty,-\alpha})_\infty.\end{aligned}$$

Therefore, to obtain the lower estimate it suffices to consider the case $p = 1$. Let n be given. We consider

$$f(x) := e^{inx}\mathscr{K}_{n-1}(x);$$

then, by the Bernstein inequality,

$$\|D^r_\alpha f\|_1 \le C(r)n^r \|\mathscr{K}_{n-1}\|_1 = C(r)n^r. \tag{1.4.30}$$

Further (see the analogous reasoning in the proof of (1.3.16)),

$$\begin{aligned}\|f - S_n(f)\|_1 &= \left\|\sum_{k=1}^{n}(1-k/n)e^{ikx}\right\|_1 \ge \left\|\sum_{k=1}^{n}(1-k/n)\sin kx\right\|_1 \\ &\ge \left(\sum_{k=1}^{n}(1-k/n)k^{-1}\right)\|\pi - x\|_\infty^{-1} \ge C\ln n. \end{aligned}\tag{1.4.31}$$

Relations (1.4.29)–(1.4.31) imply the theorem. □

1.5 Historical Remarks

In §1.1, along with classical results of Fourier, Du Bois-Reymond, and Weierstrass, which are usually included in a standard course of mathematical analysis, the following papers are cited: Chebyshev (1854), de la Vallée Poussin (1908, 1919), Bernstein (1912, 1914), Jackson (1911), Fredholm (1903), Lebesgue (1910), Kolmogorov (1936, 1985), Favard (1937), Akhiezer and Krein (1937), Ismagilov (1974), Kashin (1977), Tikhomirov (1960b), and Temlyakov (1982a).

Theorem 1.2.1 and its corollary were obtained by Riesz (see Zygmund, 1959, vol. 1). A more detailed treatment of properties of the kernels of Dirichlet, Fejér, de la Vallée Poussin, and Jackson can be found in Dzyadyk (1977). The Rudin–Shapiro polynomials were constructed in Shapiro (1951) and Rudin (1952). The polynomials $G_n(x)$ were considered in Temlyakov (1989b). The proof of relation (1.2.23) is analogous to reasoning from Trigub (1971). Theorem 1.2.5 is a classical result of Gauss. Relation (1.2.31) was obtained by Hardy and Littlewood (1966).

Theorem 1.3.1 was obtained by Kolmogorov (1985), vol. 1, pp. 12–14. Theorem 1.3.2 in the case $p = \infty$, $r = 1$, $\alpha = r$ was proved by Bernstein (1952), vol. 1, pp. 11–104. After this paper appeared, inequalities of this type began to be known as Bernstein inequalities. Today in a number of cases the Bernstein inequalities are known with explicit constants $C(r)$. Theorem 1.3.4 in the case $p = \infty$ was obtained by Jackson (1933) and in the general case by Nikol'skii (1951). Such inequalities are known as Jackson–Nikol'skii or simply Nikol'skii inequalities. Theorem 1.3.6 was obtained by Marciekiewicz (see Zygmund, 1959, vol. 2).

In a number of cases of Theorem 1.4.1 the exact values are known (see the survey Telyakovskii, 1988). Theorem 1.4.2 was proved by Hardy and Littlewood (1928). The classes H^r_q coincide with the Lipschitz classes for $0 < r < 1$ and with the Zygmund classes for $r = 1$. For r non-natural, the classes H^r_q are analogous to the classes $W^{[r]}H_q^{r-[r]}$. This statement follows from both direct and inverse theorems for these classes because these theorems have the same form (see Theorem 1.4.3 and Corollary 1.4.6 as well as the survey Telyakovskii, 1988). Theorem 1.4.3 for $q = \infty$ is a simple consequence of the results of Stechkin (1951). The proof in the general case $1 \leq q \leq \infty$ is carried out in the same way as in the case $q = \infty$. In fact, Theorem 1.4.5 includes both the direct and inverse theorems for the approximation of the classes $H_q^r B$. Theorem 1.4.8 was obtained by Nikol'skii (see his 1969 book). Theorem 1.4.9 is well known but it is not easy to assign priority; the situation is similar for Theorem 1.4.12. Theorem 1.4.13 is due to Lebesgue (1910) for $p = \infty$ and to Nikol'skii for $p = 1$ (see the survey Telyakovskii, 1988).

2

Optimality and Other Properties of the Trigonometric System

2.1 The Widths of the Classes $W^r_{q,\alpha}$ and H^r_q

In this section we show that it is natural to approximate functions in the classes $W^r_{q,\alpha}$ and H^r_q by trigonometric polynomials in $\mathscr{T}(n)$. To do this, we consider three quantities which characterize an optimal performance in approximating a class of functions by means of elements from a subspace with dimension m, when we have various restrictions on the method of constructing an approximating element.

Let $F \subset L_p$. The quantities $(m = 1, 2, \ldots)$

$$d_m(F, L_p) := \inf_{\{u_i\}_{i=1}^m \subset L_p} \sup_{f \in F} \inf_{c_i} \left\| f - \sum_{i=1}^m c_i u_i \right\|_p, \qquad m = 1, 2, \ldots$$

are called the *Kolmogorov widths* of F in L_p. In the definition of the Kolmogorov widths we take for $f \in F$, as an approximating element from $U := \operatorname{span}\{u_i\}_{i=1}^m$ the element of best approximation. This means that in general (i.e., if $p \neq 2$) this method of approximation is not linear. Let us consider quantities in the definitions of which we require the linearity of the approximating method. These quantities

$$\lambda_m(F, L_p) := \inf_{A : \operatorname{rank} A \le m} \sup_{f \in F} \|f - Af\|_p,$$

are called the *linear widths* of F in L_p. Here the infimum is taken over all linear operators A acting from F to L_p such that the dimensions of the ranges of the operators A are not greater than m.

At last, optimizing over linear operators A with $\operatorname{rank} A \le m$ we require an operator A to be an orthogonal projection operator; in other words, A must be the Fourier operator in some orthonormal system. These quantities

$$\varphi_m(F, L_p) := \inf_{\{u_i\}_{i=1}^m} \sup_{f \in F} \left\| f - \sum_{i=1}^m \langle f, u_i \rangle u_i \right\|_p$$

we shall call *orthowidths* or *Fourier widths*. Here the infimum is taken over orthonormal systems of m bounded functions.

We first note some simple properties. From the definition we have

$$d_m(F, L_p) \le \lambda_m(F, L_p) \le \varphi_m(F, L_p). \tag{2.1.1}$$

From the properties of approximation in a Hilbert space it follows that

$$d_m(F, L_2) = \lambda_m(F, L_2) = \varphi_m(F, L_2). \tag{2.1.2}$$

In this chapter we will find the orders of decrease in m of these three widths for classes $W^r_{q,\alpha}$ and H^r_q in the L_p-metric, $1 \le q,\ p \le \infty$. The following three theorems will be proved.

Theorem 2.1.1 *Let* $1 \le q,\ p \le \infty$, $r > r(q,p)$, *then*

$$d_m(W^r_{q,\alpha}, L_p) \asymp d_m(H^r_q, L_p) \asymp m^{-r+\left(1/q-\max(1/2,1/p)\right)_+},$$

where

$$r(q,p) := \begin{cases} (1/q-1/p)_+ & \text{for } 1 \le q \le p \le 2,\ 1 \le p \le q \le \infty, \\ \max(1/2, 1/q) & \text{otherwise}. \end{cases}$$

Theorem 2.1.2 *Let* $1 \le q,\ p \le \infty$; *then*

$$\lambda_m(W^r_{q,\alpha}, L_p) \asymp \lambda_m(H^r_q, L_p) \asymp \begin{cases} m^{-r+(1/q-1/p)_+}, & q \ge 2 \text{ or } p \le 2, \\ & \quad r > (1/q-1/p)_+ \\ m^{-r+\kappa},\ 1 \le q \le 2, & 2 \le p \le \infty, \\ & \quad r > 1+\kappa \end{cases}$$

where $\kappa := \max\big(1/q - 1/2, 1/2 - 1/p\big)$.

Theorem 2.1.3 *Let* $1 \le q,\ p \le \infty$, $r > (1/q-1/p)_+$; *then*

$$\varphi_m(W^r_{q,\alpha}, L_p) \asymp \varphi_m(H^r_q, L_p) \asymp m^{-r+(1/q-1/p)_+}.$$

Before proving these theorems we shall discuss them. Theorems 2.1.1–2.1.3 show that, from the point of view of the widths considered, the classes $W^r_{q,\alpha}$ and H^r_q are the same, although in addition to noting the embedding $W^r_{q,\alpha} \subset H^r_q B$ (see (1.4.20)) it is easy to see that the classes H^r_q are wider than the classes $W^r_{q,\alpha}$.

For the sake of convenience we denote

$$\begin{aligned} D_1 &:= \big\{(q,p) : 1 \le q \le p \le 2 \text{ or } 1 \le p \le q \le \infty\big\}, \\ D_2 &:= \big\{(q,p) : 2 \le q < p \le \infty\big\}, \\ D_3 &:= \big\{(q,p) : 1 \le q < 2 \text{ and } 2 < p \le \infty\big\}, \end{aligned}$$

and let D_3' be a part of D_3 such that $1/q+1/p \geq 1$ and $D_3'' = D_3 \backslash D_3'$. It is convenient to represent the corresponding domains of points $(q,p) \in [1,\infty]^2$ in terms of points $(1/q,1/p) \in [0,1]^2$. For a domain $D \subset [1,\infty]^2$ denote

$$D^* := \{(1/q,1/p) : (q,p) \in D\}.$$

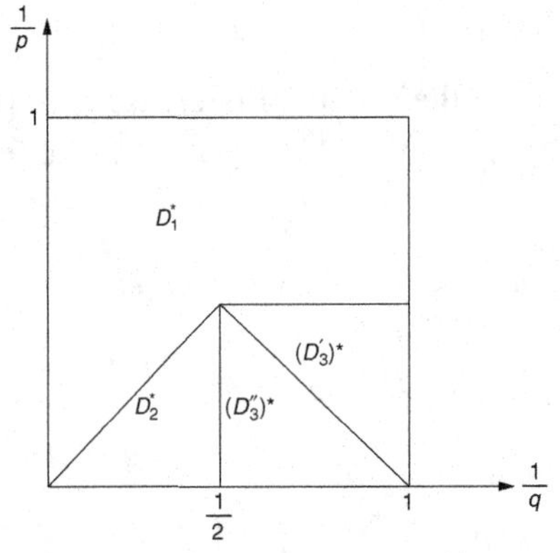

Figure 2.1

A comparison of Theorem 2.1.1 with Theorem 1.4.9 shows that for $(q,p) \in D_1$ approximations by the trigonometric polynomials in $\mathscr{T}(n)$ with $n := [(m-1)/2]$ give the order of decrease of the Kolmogorov widths. But for $(q,p) \notin D_1$ this is not the case; for example, for the H classes we have

$$d_{2n+1}(H_q^r, L_p) = o\big(E_n(H_q^r)_p\big), \qquad (q,p) \notin D_1.$$

Remark 1.4.10 shows that for $(q,p) \in D_1$ the orders of the Kolmogorov widths can be obtained by means of linear operators V_n, $m = 4n-1$. Thus, the conclusion of Theorem 2.1.2 for $(q,p) \in D_1$ is a consequence of Theorem 2.1.1 and Remark 1.4.10. A comparison of Theorem 2.1.2 with Theorem 1.4.9 and Remark 1.4.10 shows that the operators V_n give the orders of linear widths not only in the domain D_1 but also in the domain D_2. In the domain D_3, for the example of H classes, the relation

$$\lambda_{2n+1}(H_q^r, L_p) = o\big(E_n(H_q^r)_p\big)$$

holds, which shows, in particular, that for $(q,p) \in D_3$ the order of linear width cannot be realized by means of the operators V_n.

Theorems 2.1.1 and 2.1.2 show that for $(q,p) \in D_1 \cup D_3'$ the orders of the Kolmogorov widths can be realized by linear methods: in the case $(q,p) \in D_1$ by means of the operators V_n, and in the case $(q,p) \in D'_3$ by means of some other linear operators.

A comparison of Theorems 1.4.12 and 1.4.13 with Theorem 2.1.3 shows that for the classes $W^r_{q,\alpha}$ and H^r_q, for all (q,p) except the cases $(q,p) = (1,1)$, (∞,∞), the operators S_n $(m = 2n+1)$ are optimal Fourier operators in the sense of order.

From Theorems 2.1.2 and 2.1.3 it follows that linear operators A provide the orders of the widths $\lambda_m(H^r_q, L_p)$ for $(q,p) \in D_3$, that is, in the case when A differs from V_n $(m = 4n-1)$, are not orthogonal projections. Moreover, it follows from the proof of Theorem 2.1.5 below that the operators A cannot be bounded uniformly (over m) as operators from L_2 to L_2.

Further, for example for $q = 2$ and $p = \infty$, the Kolmogorov widths decrease faster than the corresponding linear widths:

$$d_m(H^r_2, L_\infty) \asymp m^{-1/2} \lambda_m(H^r_2, L_\infty).$$

However, up to now, no concrete example of a system $\{u_i\}_{i=1}^m$ is known, the best approximations by which would give the order of $d_m(H^r_2, L_\infty)$.

This discussion shows that the sets $\mathscr{T}(n)$ and the operators V_n and S_n are optimal in many cases from the point of view of the Kolmogorov widths, linear widths and orthowidths. In cases when we can approximate in a better way than by means of the operators V_n and S_n, we must sacrifice some useful properties which these operators have.

We proceed to the proof of Theorems 2.1.1–2.1.3.

2.1.1 The Case $1 \le p \le q \le \infty$

In this case the orders of all three kinds of widths coincide. Therefore by (2.1.1) and (1.4.20) it suffices to prove the lower estimates for $d_m(W^r_{q,\alpha}, L_p)$ and the upper estimates for $\varphi_m(H^r_q, L_p)$. The required upper estimates for $\varphi_m(H^r_q, L_p)$ in the cases $(q,p) \ne (1,1)$, (∞,∞) follow from Theorem 1.4.12. From Theorem 1.4.9 and Remark 1.4.10 the required upper estimates for $d_m(H^r_q, L_p)$ and $\lambda_m(H^r_q, L_p)$ follow for all $1 \le p \le q \le \infty$.

Let us prove the upper estimate for orthowidths in the case $q = p = 1, \infty$. By (1.4.20) it suffices to consider H-classes. We first prove the following auxiliary statement.

Lemma 2.1.4 *Let P_a denote the operator giving orthogonal projection of functions $f(x)$, $x \in [0,1]$, $f \in L_1$ onto the subspace of algebraic polynomials of degree $a-1$. Then this operator has the following properties.*

(1) $\|P_a\|_{p \to p} \le C(a)$, $\qquad p = 1, \infty$;

(2) *Let $f(x)$ be a times continuously differentiable; then*

$$\|f-P_a(f)\|_p \le C(a)\|f^{(a)}\|_p, \qquad p=1,\infty.$$

Proof Property (1) is obvious. To prove property (2) we use Taylor's formula with the remainder in integral form:

$$f(x) = \sum_{k=0}^{a-1} f^{(k)}(0)x^k/k! + ((a-1)!)^{-1}\int_0^x (x-t)^{a-1} f^{(a)}(t)dt. \tag{2.1.3}$$

Denoting the second term on the right-hand side of (2.1.3) by $R_a(f)$ and using (1), we get

$$\|(I-P_a)f\|_p = \|(I-P_a)R_a(f)\|_p \le C(a)\|R_a(f)\|_p.$$

From here, using the expression for $R_a(f)$, we obtain

$$\|R_a(f)\|_p \le C(a)\|f^{(a)}\|_p, \qquad p=1,\infty,$$

which concludes the proof of property (2). □

We now define the following orthogonal projection operator on functions $f \in L_1$. For natural numbers n and a the operator $P_{n,a}$ maps a function $f \in L_1$ to a piecewise polynomial function $P_{n,a}(f)$ by the following rule: on each segment $\Delta_j = [(j-1)2\pi/n, j2\pi/n), j = 1,\dots,n$, the operator $P_{n,a}$ maps the function $f(x)$, $x \in \Delta_j$, to its orthogonal projection onto the subspace of algebraic polynomials of degree $a-1$, defined on Δ_j, $j = 1,\dots,n$.

From Lemma 2.1.4 it follows that for an $f(x)$ which is a-times continuously differentiable we have

$$\|f-P_{n,a}(f)\|_p \le C(a)n^{-a}\|f^{(a)}\|_p, \qquad p=1,\infty, \tag{2.1.4}$$

and in addition,

$$\|P_{n,a}\|_{p\to p} \le C(a), \qquad p=1,\infty. \tag{2.1.5}$$

From this, for $f \in H_q^r$, $p=1,\infty$ setting $a := [r]+1$ we get

$$\begin{aligned}\|(I-P_{n,a})f\|_p &\le \sum_{s=0}^{\infty}\|(I-P_{n,a})A_s(f)\|_p \\ &\le C(a)\sum_{2^s\le n} n^{-a}\|A_s(f)^{(a)}\|_p + C(a)\sum_{2^s>n}\|A_s(f)\|_p.\end{aligned} \tag{2.1.6}$$

Using Theorem 1.4.5 and the Bernstein inequality we continue relation (2.1.6) as

$$\le C(a)n^{-a}\sum_{2^s\le n} 2^{(a-r)s} + C(a)\sum_{2^s>n} 2^{-rs} \le C(r)n^{-r}. \tag{2.1.7}$$

To conclude the proof it remains to remark that the dimension of the range of the operator $P_{n,a}$ is equal to na.

The upper estimates for $\varphi_m(H^r_q, L_p)$, $p = 1, \infty$, are now proved.

Let us prove the lower estimates for $d_m(W^r_{q,\alpha}, L_p)$, $1 \le p \le q \le \infty$. Clearly, it suffices to consider the case $p = 1$, $q = \infty$.

Let m and a system of functions $\{u_i\}_{i=1}^m$ be given. We consider the space $\mathscr{T}(m)$ and its subspace

$$\Psi = \{t \in \mathscr{T}(m) : \langle t, u_i \rangle = 0, i = 1, \dots, m\}.$$

Then $\dim \Psi \ge m$ and by Theorem 1.2.4 there is an $f \in \Psi$ such that

$$\|f\|_\infty = 1, \qquad \|f\|_2 \ge C > 0. \tag{2.1.8}$$

Then for any $u \in \operatorname{span}\{u_i\}_{i=1}^m$ we have

$$C^2 \le \langle f, f \rangle = \langle f - u, f \rangle \le \|f - u\|_1 \|f\|_\infty = \|f - u\|_1. \tag{2.1.9}$$

Further, by the Bernstein inequality,

$$\|D^r_\alpha f\|_\infty \le C(r) m^r. \tag{2.1.10}$$

Relations (2.1.9) and (2.1.10) imply that

$$d_m(W^r_{\infty,\alpha}, L_1) \ge C(r) m^{-r}.$$

The proof in the case $1 \le p \le q \le \infty$ is, therefore, complete for all three types of widths.

2.1.2 The Lower Estimates in Theorem 2.1.3 for $1 \le q \le p \le \infty$.

We now prove a slightly stronger statement: we will consider approximating operators from a wider set than the set of orthogonal projection operators.

Let $\mathscr{L}_m(B)_p$, $B \ge 1$, be the set of linear operators A with domains $\mathscr{D}(A)$ containing all trigonometric polynomials and with ranges contained in an m-dimensional subspace of L_p, such that

$$\left\|A(e^{ikx})\right\|_2 \le B. \tag{2.1.11}$$

In particular, $\mathscr{L}_m(1)_p$ contains orthoprojectors in L_p with rank not greater than m and it also contains the operator V_n $(m = 4n - 1)$; $\mathscr{L}_m(B)_p$ contains operators A acting from L_2 to $L_2 \cap L_p$ such that $\operatorname{rank} A \le m$ and

$$\|A\|_{2 \to 2} \le B.$$

For an operator A and a class F of functions we denote

$$A(F,L_p) := \sup_{f\in F\cap \mathscr{D}(A)} \|f-A(f)\|_p.$$

We now prove the following statement.

Theorem 2.1.5 *Let $B\geq 1$ be a fixed number, $1\leq q\leq p\leq\infty$, and $r>1/q-1/p$. Then*

$$\inf_{A\in\mathscr{L}_m(B)_p} A(W^r_{q,\alpha},L_p)\asymp \inf_{A\in\mathscr{L}_m(B)_p} A(H^r_q,L_p)\asymp m^{-r+1/q-1/p}.$$

Proof The upper estimates follow from Theorem 1.4.9 and relation (1.4.26). Let us prove the lower estimates. Clearly, it suffices to consider the classes $W^r_{q,\alpha}$. Let m and $G\in\mathscr{L}_m(B)_p$ be given. We take $N\geq m$ the value of which will be chosen later, and consider the operator

$$A := V_N G\in\mathscr{L}_m(B)_p,$$

the range of which is contained in $\mathscr{T}(2N)$.

Then for any $f\in\mathscr{T}(N)$ and $1\leq p\leq\infty$,

$$\|f-A(f)\|_p = \|V_N(f-G(f))\|_p\leq 3\|f-G(f)\|_p,$$

which implies that it suffices to prove the lower estimate for the class $W^r_{q,\alpha}\cap\mathscr{T}(N)$ and for operators $A\in\mathscr{L}_m(B)_p$ acting in $\mathscr{T}(2N)$.

We first prove an auxiliary statement. For a trigonometric polynomial t, denote by $\hat{t}(k)$, $k\in\mathbb{Z}$, its Fourier coefficients. See (A.3.1) for a general definition.

Lemma 2.1.6 *Let $A\in\mathscr{L}_m(B)_p$ be defined as follows:*

$$A(e^{ikx}) = \sum_{a=1}^{m} b^k_a\psi_a(x),$$

where $\{\psi_a(x)\}_{a=1}^m$ is an orthonormal system of functions.

Then for any trigonometric polynomial $t\in\mathscr{T}(N)$ we have

$$M := \min_{y=x}\operatorname{Re}A\big(t(x-y)\big)\leq B\big(m(2N+1)\big)^{1/2}\max_k|\hat{t}(k)|.$$

Proof We have

$$\begin{aligned} A\big(t(x-y)\big)\,\big|_{y=x} &= A\left(\sum_k \hat{t}(k)e^{ik(x-y)}\right)\Bigg|_{y=x}\\ &= \sum_k \hat{t}(k)e^{-ikx}\sum_{a=1}^m b^k_a\psi_a(x). \end{aligned} \tag{2.1.12}$$

After taking the real part and integrating we get

$$
\begin{aligned}
M &\le \operatorname{Re}\sum_k \sum_{a=1}^m \hat{f}(k) b_a^k \hat{\psi}_a(k) \\
&\le \sum_{a=1}^m \left(\sum_k |\hat{f}(k) b_a^k|^2\right)^{1/2} \left(\sum_k |\hat{\psi}_a(k)|^2\right)^{1/2} \\
&\le \left(m \sum_{a=1}^m \sum_k |\hat{f}(k) b_a^k|^2\right)^{1/2} \le B\big(m(2N+1)\big)^{1/2} \max_k |\hat{f}(k)|.
\end{aligned}
$$

The lemma is proved. □

We consider as an example the function $g(x) := \mathscr{K}_{N-1}(x) \in \mathscr{T}(N-1)$. We have

$$\sigma := \sup_y \big\| g(x-y) - A\big(g(x-y)\big) \big\|_\infty \ge g(0) - \min_{y=x} \operatorname{Re} A\big(g(x-y)\big).$$

Further, using Lemma 2.1.6 and the equality $g(0) = N$, we get

$$\sigma \ge N - B(2mN)^{1/2}.$$

Setting $N = [4B^2 m] + 1$, where $[a]$ means the integer part of a number a, we obtain

$$\sigma \ge N/4. \tag{2.1.13}$$

From (2.1.13) it follows that there is a y^* such that

$$\big\| g(x-y^*) - A\big(g(x-y^*)\big) \big\|_\infty \ge N/4.$$

Taking into account that $A(f) \in \mathscr{T}(2N)$ for all $f \in \mathscr{D}(A)$, this inequality implies using Nikol'skii's inequality that

$$\big\| g(x-y^*) - A\big(g(x-y^*)\big) \big\|_p \ge CN^{1-1/p}, \qquad 1 \le p \le \infty.$$

To conclude the proof it remains to remark that by the Bernstein inequality and the estimate (1.2.11) we have

$$\big\| D^r_\alpha g(x-y^*) \big\|_q \le C(r) N^{r+1-1/q}.$$

Theorem 2.1.5 is proved.

□

Remark 2.1.7 Actually, in the proof of the lower estimates in Theorem 2.1.3 the inequality $(1 \le q \le p \le \infty)$

$$\varphi_m\big(\mathscr{T}(4m) \cap W^r_{q,\alpha}, L_p\big) \gg m^{-r+1/q-1/p}$$

has been proved.

2.1.3 Proof of the Lower Estimate in Theorem 2.1.1 for $1 \le q \le p \le \infty$

Clearly it suffices to consider the case $1 \le q \le p \le 2$. For any system of functions $\{u_i\}_{i=1}^m \subset L_p$ and any $t \in \mathscr{T}(4m)$ we have

$$\left\| t - \sum_{i=1}^m c_i V_{4m}(u_i) \right\|_p = \left\| V_{4m}\left(t - \sum_{i=1}^m c_i u_i \right) \right\|_p \le 3 \left\| t - \sum_{i=1}^m c_i u_i \right\|_p . \tag{2.1.14}$$

Consequently, it suffices to prove the lower estimates for $\mathscr{T}(4m) \cap W^r_{q,\alpha}$, $u_i \in \mathscr{T}(8m)$, $i = 1, \dots, m$.

Let us use Remark 2.1.7 with $p = 2$. Applying (2.1.2) and the Nikol'skii inequality, we have

$$\begin{aligned} m^{-r+1/q-1/2} &\ll \varphi_m\big(\mathscr{T}(4m) \cap W^r_{q,\alpha}, L_2\big) = d_m\big(\mathscr{T}(4m) \cap W^r_{q,\alpha}, L_2\big) \\ &\le d_m\big(\mathscr{T}(4m) \cap W^r_{q,\alpha}, L_2 \cap \mathscr{T}(8m)\big) \\ &\ll d_m\big(\mathscr{T}(4m) \cap W^r_{q,\alpha}, L_p \cap \mathscr{T}(8m)\big) m^{1/p-1/2}. \end{aligned} \tag{2.1.15}$$

From relation (2.1.14) we find that

$$\begin{aligned} & d_m\big(\mathscr{T}(4m) \cap W^r_{q,\alpha}, L_p \cap \mathscr{T}(8m)\big) \\ &\quad \le 3 d_m\big(\mathscr{T}(4m) \cap W^r_{q,\alpha}, L_p\big) \le 3 d_m(W^r_{q,\alpha}, L_p). \end{aligned} \tag{2.1.16}$$

Combining relations (2.1.15) and (2.1.16) we obtain the lower estimates for $1 \le q \le p \le 2$ required in Theorem 2.1.1.

2.1.4 The Lower Estimates in Theorem 2.1.2 for $1 \le q \le p \le \infty$

We first prove the following assertion.

Theorem 2.1.8 *Let $r > 0$, $1 \le q,\ p \le \infty$. Then*

$$\lambda_m(W^r_{q,\alpha}, L_p) = \lambda_m(W^r_{p',-\alpha}, L_{q'}).$$

Proof Let A be a linear operator such that rank $A \le m$, $\|A\|_{q\to p} < \infty$. We mention the following simple properties of conjugate operators:

$$(I^r_\alpha)^* = I^r_{-\alpha},$$
$$\operatorname{rank} G \le m \text{ implies } \operatorname{rank} G^* \le m.$$

Using Theorem A.2.1, we find

$$\begin{aligned} \|(I-A)I^r_\alpha\|_{q\to p} &= \sup_{\|\phi\|_q \le 1,\ \|\psi\|_{p'} \le 1} |\langle (I-A)I^r_\alpha \phi, \psi\rangle| \\ &= \sup_{\|\phi\|_q \le 1,\ \|\psi\|_{p'} \le 1} |\langle \phi, I^r_{-\alpha}(I-A)^*\psi\rangle| \ge \lambda_m(W^r_{p',-\alpha}, L_{q'}) \end{aligned} \tag{2.1.17}$$

Consequently,

$$\lambda_m(W^r_{q,\alpha}, L_p) \geq \lambda_m(W^r_{p',-\alpha}, L_{q'}). \tag{2.1.18}$$

Since $1 \leq q, p \leq \infty$ are arbitrary, it follows from (2.1.18) that the inverse inequality is also true, which proves the theorem. □

The lower estimates for $1 \leq q \leq p \leq \infty$, $1/q + 1/p \geq 1$ follow from the lower estimates for the Kolmogorov widths, which have already been proved. The lower estimates for the remaining $1 \leq q \leq p \leq \infty$ follow from the case already considered and from Theorem 2.1.8.

The lower estimates in Theorem 2.1.2 are proved.

Thus the lower estimates in all these theorems are proved. It remains to prove the upper estimates for $\lambda_m(H^r_q, L_p)$, $(q,p) \in D_3$ and those for $d_m(H^r_q, L_p)$, $(q,p) \in D_2 \cup D_3$.

2.1.5 Completion of the Proof of Theorem 2.1.2

We now prove the upper estimate for $(q,p) \in D_3$. To do this it suffices, by Theorem 2.1.8, to consider the case $(q,p) \in D'_3$.

We first prove two auxiliary statements.

Lemma 2.1.9 *Let the natural numbers s and $n \leq 2^s$ be given. There are a set of integers $G(n,s) \subset (-2^s, -2^{s-2}) \cup (2^{s-2}, 2^s)$ having no more than n elements and a trigonometric polynomial $t_{s,n} \in \mathcal{T}(G(n,s))$ such that*

$$\|\mathscr{A}_s - t_{s,n}\|_\infty \leq C2^{3s/2} n^{-1}, \tag{2.1.19}$$

$$\max_k |\hat{t}_{s,n}(k)| \leq C2^s n^{-1}. \tag{2.1.20}$$

Proof The proof is based on Theorem 1.2.6. Let $b > 2^{s+1}$ be the smallest prime number and let $b = 2a + 1$. Let us consider the function

$$u(x) := \mathscr{D}_a(x) - bt_n(x)/t_n(0),$$

where $t_n(x)$ is the polynomial from Theorem 1.2.6. We set

$$t_{s,n} := \mathscr{A}_s - u * \mathscr{A}_s.$$

Clearly, $\hat{t}_{s,n}(k)$ is nonzero only for those k for which $\hat{\mathscr{A}_s}(k)\hat{t}_n(k) \neq 0$. From this, denoting

$$G(n,s) := \{k : \hat{t}_{s,n}(k) \neq 0\},$$

we get

$$G(n,s) \subset (-2^s, -2^{s-2}) \cup (2^{s-2}, 2^s).$$

Relation (2.1.20) follows from Theorem 1.2.6 since by the Chebyshev theorem the inequality $b \le 2^{s+2}$ is valid.

We now prove relation (2.1.19). By relation (1.2.5) we have

$$u_s(x) := \mathscr{A}_s(x) - t_{s,n}(x) = u(x) * \mathscr{A}_s(x)$$
$$= (2a+1)^{-1} \sum_{\mu=0}^{2a} u(x^\mu)\mathscr{A}_s(x - x^\mu)$$

and

$$\|u_s\|_\infty \le \max_\mu \big|u(x^\mu)\big|(2a+1)^{-1} \max_x \sum_{\mu=0}^{2a} \big|\mathscr{A}_s(x - x^\mu)\big|.$$

Using property (2) of the de la Vallée Poussin kernels we get

$$\|u_s\|_\infty \le C \max_\mu \big|u(x^\mu)\big|.$$

Further, $u(0) = 0$ and, by Theorem 1.2.6 for $\mu = 1, \dots, 2a$,

$$\big|u(x^\mu)\big| \le C 2^{3s/2} n^{-1}.$$

The lemma is proved. □

Lemma 2.1.10 *Let natural numbers s, n be given, and let $t_{s,n}$ be the polynomial from Lemma 2.1.9. For the operator*

$$T_{s,n}(f) := (\mathscr{A}_s - t_{s,n}) * f, \qquad f \in L_1,$$

the following relations hold:

$$T_{s,n} t = 0, \qquad t \in \mathscr{T}(2^{s-2}), \tag{2.1.21}$$
$$\|T_{s,n}\|_{q \to p} \le C 2^{s(1/2+1/q)} n^{-1}, \qquad (q,p) \in D_3'. \tag{2.1.22}$$

Proof Relation (2.1.21) is obvious. We will prove relation (2.1.22). From (2.1.20) we obtain

$$\|T_{s,n}\|_{2 \to 2} \le C 2^s n^{-1}, \tag{2.1.23}$$

and from (2.1.19) we get

$$\|T_{s,n}\|_{1 \to \infty} \le C 2^{3s/2} n^{-1}. \tag{2.1.24}$$

Let $(q,p) \in D_3'$ and $0 \le \upsilon \le 1$ be such that

$$1/q = \upsilon/2 + 1 - \upsilon;$$

then by the Riesz–Thorin theorem, using (2.1.23) and (2.1.24), we get

$$\|T_{s,n}\|_{q\to q'} \le C2^{s(1/2+1/q)}n^{-1}.$$

To complete the proof it remains to observe that for $(q,p) \in D_3'$ we have $p \le q'$ and, consequently,

$$\|T_{s,n}\|_{q\to p} \le \|T_{s,n}\|_{q\to q'}.$$

□

We now proceed to prove Theorem 2.1.2. Clearly, it suffices to consider the case $m \asymp 2^a$. Let $(q,p) \in D_3'$ and $r > 1/2+1/q$. We take $0 < \rho < r-1/2-1/q$ and define the numbers

$$n_s := [2^{a-\rho(s-a)}], \qquad s = a+1,\dots,s_a,$$

where s_a is such that $n_{s_a} > 0$, $n_{s_a+1} = 0$. It is clear that

$$2^{\rho s_a} \asymp 2^{a(1+\rho)}. \tag{2.1.25}$$

We consider the linear operator

$$A := V_{2^{a-1}} + \sum_{s=a+1}^{s_a} (A_s - T_{s,n_s}).$$

Then the dimension of the range of the operator A is not greater than

$$m := 2^{a+1} + \sum_{s=a+1}^{s_a} [2^{a-\rho(s-a)}] \le C(\rho)2^a. \tag{2.1.26}$$

Let $f \in H^r_q$; then

$$f = V_{2^{a-1}}(f) + \sum_{s=a+1}^{\infty} A_s(f)$$

and

$$f - A(f) = \sum_{s=s_a+1}^{\infty} A_s(f) + \sum_{s=a+1}^{s_a} T_{s,n_s}(f) =: \sigma_1 + \sigma_2. \tag{2.1.27}$$

We begin by estimating the first sum. Using Theorem 1.4.5 and the Nikol'skii inequality we get

$$\|\sigma_1\|_p \le \|\sigma_1\|_{q'} \ll \sum_{s=s_a}^{\infty} 2^{-s(r-2/q+1)} \ll 2^{-(r-2/q+1)s_a}.$$

From this, using (2.1.25), it is easy to derive that

$$\|\sigma_1\|_p \ll 2^{-(r-1/q+1/2)a}. \tag{2.1.28}$$

Now we will proceed to estimate the second sum. To do this we use Lemma 2.1.10 which implies that

$$\left\|T_{s,n_s}(f)\right\|_p \leq \|T_{s,n_s}\|_{q\to p} E_{2^{s-2}}(f)_q.$$

From this we find, using Lemma 2.1.10 and Theorem 1.4.3,

$$\left\|T_{s,n_s}(f)\right\|_p \ll 2^{-(r-1/q-1/2)s} n_s^{-1} \ll 2^{-a(1+\rho)} 2^{-s(r-1/q-1/2-\rho)}. \tag{2.1.29}$$

From (2.1.29) we get the estimate

$$\|\sigma_2\|_p \ll 2^{-(r-1/q+1/2)a}. \tag{2.1.30}$$

Combining relations (2.1.27) and (2.1.28), (2.1.30) and using (2.1.26), we obtain

$$\left\|f - A(f)\right\|_p \ll m^{-r+1/q-1/2},$$

as required.

The proof of Theorem 2.1.2 is now complete.

2.1.6 Proof of Theorem 2.1.1

In this subsection we complete the proof of Theorem 2.1.1; that is, we prove the upper estimate in the case $1 \leq q < p \leq \infty$, $p > 2$.

Here we need estimates of the Kolmogorov widths of the Euclidean unit ball $B_2^n \subset \mathbb{R}^n$ in the ℓ_∞^n space.

Theorem 2.1.11 *For any natural numbers n, m, $m < n$ we have*

$$d_m(B_2^n, \ell_\infty^n) \leq Cm^{-1/2}\left(\ln(en/m)\right)^{1/2}.$$

Proof The proof of this theorem is based on the following statement.

Lemma 2.1.12 *For any natural m, n, $m < n$, there exists a subspace $\Gamma \subset \mathbb{R}^n$, $\dim\Gamma \geq n - m$, such that, for $\mathbf{x} \in \Gamma$,*

$$\|\mathbf{x}\|_2 \leq Cm^{-1/2}\left(\ln(en/m)\right)^{1/2}\|\mathbf{x}\|_1.$$

Indeed, let Lemma 2.1.12 be valid and $U := \Gamma^\perp$ be the orthogonal complement of a subspace Γ with respect to $\mathbb{R}^n$. Then, for an arbitrary $\mathbf{x} \in B_2^n$, by the Nikol'skii duality theorem, Theorem A.2.3, we have

$$\begin{aligned}\inf_{\mathbf{u}\in U} \|\mathbf{x} - \mathbf{u}\|_\infty &= \sup_{\mathbf{v}\in\Gamma\cap B_1^n} (\mathbf{x}, \mathbf{v}) \leq \|\mathbf{x}\|_2 \sup_{\mathbf{v}\in\Gamma\cap B_1^n} \|\mathbf{v}\|_2 \\ &\leq Cm^{-1/2}\left(\ln(en/m)\right)^{1/2},\end{aligned}$$

which implies Theorem 2.1.11. □

Proof of Lemma 2.1.12. Let

$$S := S^{n-1} := \{\mathbf{x} \in \mathbb{R}^n : \|\mathbf{x}\|_2 = 1\}$$

be the unit sphere in $\mathbb{R}^n$ and let μ be the normalized Lebesgue measure on S, that is, $\mu(S) = 1$. Denote by P the measure on the product set $Y := (S^{n-1})^m$ of m unit spheres corresponding to the product of the measures μ. We denote

$$\overline{\mathbf{y}} := (\mathbf{y}_1, \dots, \mathbf{y}_m) \in Y, \qquad \mathbf{y}_j \in S, \qquad j = 1, \dots, m.$$

We first prove the following auxiliary statement.

Lemma 2.1.13 *For* $\mathbf{x} \in S$, $\overline{\mathbf{y}} \in Y$, *let*

$$F(\mathbf{x}, \overline{\mathbf{y}}) := m^{-1} \sum_{j=1}^{m} |(\mathbf{x}, \mathbf{y}_j)|.$$

Then, for any $\mathbf{x} \in S$,

$$P\{\overline{\mathbf{y}} \in Y : (0.01)n^{-1/2} \le F(\mathbf{x}, \overline{\mathbf{y}}) \le 3n^{-1/2}\} > 1 - e^{-m/2}.$$

Proof Let

$$E(m,t) := \int_Y e^{tF(\mathbf{x},\overline{\mathbf{y}})} dP.$$

First, we make some simple remarks. The quantity $E(m,t)$ does not depend on $\mathbf{x}$. The following inequality holds: for $t > 0$, $b > 0$,

$$P\{\overline{\mathbf{y}} \in Y : F(\mathbf{x}, \overline{\mathbf{y}}) > b\} \le E(m,t)e^{-bt} \tag{2.1.31}$$

and for $t < 0$, $a > 0$,

$$P\{\overline{\mathbf{y}} \in Y : F(\mathbf{x}, \overline{\mathbf{y}}) < a\} \le E(m,t)e^{-at}. \tag{2.1.32}$$

Since

$$E(m,t) = E(1, t/m)^m, \tag{2.1.33}$$

it suffices to consider the case $m = 1$. We set $E(1,t) =: E(t)$. Taking $\mathbf{x} = (1, 0, \dots, 0)$, we have

$$E(t) = \int_S e^{t|y_1|} d\mu. \tag{2.1.34}$$

Further, let $n \ge 3$. Considering that the $(n-2)$-dimensional volume of the sphere $S^{n-2}(r)$ of radius r is proportional to r^{n-2}, we obtain

$$\mu\{\mathbf{y} \in S : \alpha \le y_1 \le \beta\} = \int_\alpha^\beta |S^{n-2}((1-r^2)^{1/2})|(1-r^2)^{-1/2} dr$$
$$= C(n) \int_\alpha^\beta (1-r^2)^{(n-3)/2} dr.$$

From the normalization condition for the measure μ we get

$$C(n)\int_0^1 (1-r^2)^{(n-3)/2}dr = 1/2. \tag{2.1.35}$$

Thus,

$$E(t) = 2C(n)\int_0^1 e^{rt}(1-r^2)^{(n-3)/2}dr. \tag{2.1.36}$$

To get the upper estimate for $E(t)$ we prove two estimates. First we prove the lower one:

$$\int_0^1 (1-r^2)^{(n-3)/2}dr > \int_0^{n^{-1/2}} (1-r^2)^{(n-3)/2}dr$$
$$> (1-1/n)^{(n-1)/2}n^{-1/2} > (en)^{-1/2}. \tag{2.1.37}$$

Then, using the inequality $1-z \le e^{-z}$, $z \ge 0$, we prove the upper estimate

$$\int_0^1 e^{tr}(1-r^2)^{(n-3)/2}dr \le \int_0^1 e^{tr-nr^2/2+3r^2/2}dr \le e^{3/2}\int_0^1 e^{tr-nr^2/2}dr$$
$$\le e^{3/2+t^2/(2n)}n^{-1/2}\int_{-tn^{-1/2}}^{\infty} e^{-v^2/2}dv, \tag{2.1.38}$$

where $v = n^{1/2}(r-t/n)$. From relations (2.1.35)–(2.1.38), using (2.1.33), we get

$$E(m,t) < e^{2m+t^2/(2mn)}\left(\int_{-t/(mn^{1/2})}^{\infty} e^{-v^2/2}dv\right)^m. \tag{2.1.39}$$

From (2.1.39) and inequality (2.1.31) with $b = 3n^{-1/2}$, $t = 3n^{1/2}m$ we have

$$P\{\bar{\mathbf{y}} \in Y : F(\mathbf{x},\bar{\mathbf{y}}) > 3n^{-1/2}\} < (e^{-5/2}(2\pi)^{1/2})^m. \tag{2.1.40}$$

From (2.1.39) and (2.1.32) with $a = (0.01)n^{-1/2}$ and $t = -100n^{1/2}m$, taking into account the inequality

$$\int_z^{\infty} e^{-v^2/2}dv < e^{-z^2/2}/z, \qquad z > 0,$$

we obtain

$$P\{\bar{\mathbf{y}} \in Y : F(\mathbf{x},\bar{\mathbf{y}}) < 0.01n^{-1/2}\} < (0.01e^3)^m. \tag{2.1.41}$$

The lemma follows from (2.1.40) and (2.1.41). □

We now continue the proof of Lemma 2.1.12. Clearly, we can assume that m is a sufficiently large number. For natural numbers $1 \le l \le n$ let $B^{n,l}$ be the set of all vectors from B_1^n with coordinates of the form k/l, $k \in \mathbb{Z}$. The cardinality of

$B^{n,l}$ does not exceed 2^l times the number of nonnegative integer solutions of the inequality

$$l_1 + \cdots + l_n \leq l.$$

Consequently (here $\binom{n}{k}$ are the binomial coefficients),

$$|B^{n,l}| \leq 2^l \sum_{j=0}^{l} \binom{n-1+j}{n-1} = 2^l \binom{n+l}{l}. \tag{2.1.42}$$

We set

$$l := \left[Am/\ln(en/m)\right] \tag{2.1.43}$$

where A is a sufficiently small number, not depending on n or m, such that there exists $\bar{\mathbf{y}}^* \in Y$ with the property that for all $\mathbf{x} \in B^{n,l}$ we have

$$0.01n^{-1/2}\|\mathbf{x}\|_2 \leq F(\mathbf{x}, \bar{\mathbf{y}}^*) \leq 3n^{-1/2}\|\mathbf{x}\|_2. \tag{2.1.44}$$

Indeed, owing to Lemma 2.1.13 the P-measure of those $\bar{\mathbf{y}}$ for which the relation (2.1.44) does not hold is not greater than

$$|B^{n,l}|e^{-m/2} \leq 2^l(2en/l)^l e^{-m/2}. \tag{2.1.45}$$

The number A is chosen such that the right-hand side of (2.1.45) is less than 1. Let

$$\Gamma := \left\{\mathbf{x} \in \mathbb{R}^n : F(\mathbf{x}, \bar{\mathbf{y}}^*) = 0\right\}.$$

It is clear that $\dim \Gamma \geq n - m$. We will prove that for $\mathbf{x} \in \Gamma \cap B^n_1$,

$$\|\mathbf{x}\|_2 \leq 301 \times l^{-1/2}. \tag{2.1.46}$$

Let $\mathbf{x} \in B^n_1$ and $\mathbf{x}' \in B^{n,l}$ be such that x_j and x'_j have the same sign and $|x'_j| \leq |x_j|$, $|x_j - x'_j| \leq 1/l$, $j = 1, \ldots, n$. We consider $\mathbf{x}'' := \mathbf{x} - \mathbf{x}'$. Then $\mathbf{x}'' \in B^n_1 \cap (1/l)B^n_\infty = \Pi$, and, consequently,

$$\|\mathbf{x}''\|_2 \leq \|\mathbf{x}''\|_1^{1/2}\|\mathbf{x}''\|_\infty^{1/2} \leq l^{-1/2}. \tag{2.1.47}$$

Let us estimate $F(\mathbf{x}'', \bar{\mathbf{y}}^*)$. To do this we prove that $\Pi = \mathrm{conv}(V)$, where V is the set of all vectors having exactly l coordinates that are different from zero and equal to $\pm 1/l$.

The set Π as an intersection of two convex polytopes is a convex polytope. Clearly, the set V belongs to the set of extreme points of Π. We will prove that Π has no other extreme points. Indeed, let $\mathbf{z} \in \Pi \backslash V$ be a boundary point of Π such that $\|\mathbf{z}\|_1 = 1$, $\|\mathbf{z}\|_\infty = 1/l$. Since $\mathbf{z} \notin V$, there are $1 \leq j_1 < j_2 \leq n$ such that $0 < |z_{j_i}| < 1/l$, $i = 1, 2$. Then there is a $\delta > 0$ such that the vectors

$$\mathbf{z}^1 = \mathbf{z} + (0, \ldots, 0, \delta \operatorname{sign} z_{j_1}, 0, \ldots, 0, -\delta \operatorname{sign} z_{j_2}, 0, \ldots, 0),$$
$$\mathbf{z}^2 = \mathbf{z} + (0, \ldots, 0, -\delta \operatorname{sign} z_{j_1}, 0, \ldots, 0, \delta \operatorname{sign} z_{j_2}, 0, \ldots, 0)$$

belong to Π and, obviously, $\mathbf{z} = (\mathbf{z}^1 + \mathbf{z}^2)/2$, which shows that $\mathbf{z}$ is not an extreme point of Π. Thus it is proved that the set of extreme points of Π coincides with V. Consequently $\Pi = \mathrm{conv}(V)$ and, for $\mathbf{x}'' \in \Pi$,

$$F(\mathbf{x}'', \bar{\mathbf{y}}^*) \le \max_{\mathbf{z}\in V} F(\mathbf{z}, \bar{\mathbf{y}}^*)$$
$$\le \max_{\mathbf{z}\in B^{n,l}} \|\mathbf{z}\|_2^{-1} F(\mathbf{z}, \bar{\mathbf{y}}^*) \max_{\mathbf{z}\in V} \|\mathbf{z}\|_2 \le 3(nl)^{-1/2}. \tag{2.1.48}$$

If we now suppose that (2.1.46) does not hold, we get

$$\|\mathbf{x}'\|_2 \ge \|\mathbf{x}\|_2 - \|\mathbf{x}''\|_2 > 300 l^{-1/2},$$

which by (2.1.44), implies for $\mathbf{x}' \in B^{n,l}$ that

$$F(\mathbf{x}', \bar{\mathbf{y}}^*) > 3(nl)^{-1/2}. \tag{2.1.49}$$

From (2.1.48) and (2.1.49) we find

$$F(\mathbf{x}, \bar{\mathbf{y}}^*) \ge F(\mathbf{x}', \bar{\mathbf{y}}^*) - F(\mathbf{x}'', \bar{\mathbf{y}}^*) > 0,$$

which contradicts the condition $\mathbf{x} \in \Gamma$.

Relation (2.1.46) and Lemma 2.1.12 are proved. □

From Theorem 2.1.11 and by means of the Marcinkiewicz theorem (see Theorems 1.3.6 and 1.3.9) we derive the following assertion.

Theorem 2.1.14 *Let $\mathscr{T}(n)_2$ be the unit L_2-ball in $\mathscr{T}(n)$. For any natural numbers n, m, $m < 2n$, we have*

$$d_m(\mathscr{T}(n)_2, L_\infty) \le C(n/m)^{1/2}\big(\ln(en/m)\big)^{1/2}.$$

Proof Clearly, without loss of generality we can consider $\mathscr{RT}(n)$. For each $t \in \mathscr{RT}(n)_2$, let $M(t) := \big(t(x(1)), \dots, t(x(4n))\big) \in \mathbb{R}^{4n}$. Then, by Theorem 1.3.9, we have

$$M(t) \in C n^{1/2} B_2^{4n}. \tag{2.1.50}$$

Using Theorem 2.1.11 we find a subspace $U \subset \mathbb{R}^{4n}$, $\dim U = m$, such that for any $M(t)$ there is a $u(t) \in U$ such that

$$\big\|M(t) - u(t)\big\|_\infty \le C(n/m)^{1/2}\big(\ln(en/m)\big)^{1/2}. \tag{2.1.51}$$

In $\mathscr{RT}(2n)$ we define for $u \in U$ the trigonometric polynomial

$$\psi(u) := (4n)^{-1} \sum_{j=1}^{4n} u_j \mathscr{V}_n\big(x - x(j)\big).$$

Clearly, all these polynomials belong to a subspace Ψ, $\dim\Psi \le m$. From the relation (2.1.51) and Lemma 1.3.10, for any $t \in \mathscr{RT}(n)_2$ it follows that

$$\|t - \psi(u(t))\|_\infty \le C(n/m)^{1/2}(\ln(en/m))^{1/2}.$$

The theorem is proved. □

Remark 2.1.15 We denote

$$d_0(F,X) := \sup_{f\in F} \|f\|_X.$$

From Theorem 2.1.14 and the Nikol'skii inequality it follows that for $m \ge 0$,

$$d_m(\mathscr{T}(n)_2, L_\infty) \le C(n/(m+1))^{1/2}(\ln(en/(m+1)))^{1/2}.$$

Proof of Theorem 2.1.1 We first consider the case $q = 2$, $p = \infty$. Let $0 < \rho < r - 1/2$ and let the numbers m, a, n_s be the same as in subsection 2.1.5 completing the proof of Theorem 2.1.2:

$$n_s = [2^{a-\rho(s-a)}], \qquad s = a+1,\dots,$$
$$m = 2^{a+1} + \sum_s n_s \le C(\rho)2^a.$$

Using the representation

$$f = S_{2^a-1}(f) + \sum_{s=a+1}^{\infty} \delta_s(f),$$

from Theorem 1.4.5 and Corollary 1.4.6 the estimate

$$d_m(H^r_2, L_\infty) \le C(r) \sum_{s=a+1}^{\infty} 2^{-rs} d_{n_s}(\mathscr{T}(2^s)_2, L_\infty) \tag{2.1.52}$$

follows. Applying Theorem 2.1.14 and Remark 2.1.15, we get from (2.1.52)

$$d_m(H^r_2, L_\infty) \le C(r) \sum_{s=a+1}^{\infty} 2^{-rs+(1+\rho)(s-a)/2}\left(\ln\left(e2^s \,/\, (1+[2^{a-\rho(s-a)}])\right)\right)^{1/2}, \tag{2.1.53}$$

choosing ρ such that $r - 1/2 - \rho > 0$, we find from (2.1.53)

$$d_m(H^r_2, L_\infty) \le C(r,\rho)2^{-ra} \le C(r,\rho)m^{-r},$$

as required.

The other cases, $1 \le q \le 2$, $2 < p \le \infty$, follow from the case $q = 2$, $p = \infty$ in view of the inclusion

$$H^r_q \subset H^{r-1/q+1/2}_2 B$$

(see Theorem 1.4.8) and the obvious inequality

$$\|f\|_p \le \|f\|_\infty.$$

The proof of Theorem 2.1.1 is now complete. □

2.2 Further Properties of the Trigonometric System

The trigonometric system and series play a fundamental role in mathematical analysis and have been investigated intensively for about 200 years. The results of §2.1 show that from a modern point of view the trigonometric system takes a special place in the approximation of functions of finite smoothness. In this section we prove that the trigonometric system is optimal for approximating infinitely differentiable functions and analytic functions. These results show the universality of the trigonometric system for approximation and confirm that studying the approximation of arbitrary functions by means of trigonometric polynomials and the representations of functions by Fourier series in the trigonometric system is natural. We present here some characteristic results from this field of investigation.

In the L_p-spaces, $1 \le p \le \infty$, we define for a natural number a the modulus of continuity of order a:

$$\omega_a(f,y)_p := \sup_{|h|\le y} \|\Delta_h^a f\|_p, \qquad 0 \le y \le \pi.$$

Then $\omega_a(f,y)_p$ is a continuous function in y and $\lim_{y\to 0}\omega_a(f,y)_p = 0$ (see Theorem A.1.1). In addition, $\omega_a(f,y)_p$ does not decrease on $[0,\pi]$ and, for a natural number b, we have

$$\omega_a(f,by)_p \le b^a \omega_a(f,y)_p. \tag{2.2.1}$$

We will prove the relation (2.2.1). Let T_h be the operator translating the argument by h, so that $T_h f(x) := f(x+h)$. Clearly, the norm of this operator as an operator from L_p to L_p is equal to one. Further,

$$\Delta_{bh}^a f = (I - T_h^b)^a = \left(\sum_{j=0}^{b-1} T_h^j\right)^a (I - T_h)^a,$$

which implies that

$$\|\Delta_{bh}^a f\|_p \le b^a \|\Delta_h^a f\|_p,$$

and this proves (2.2.1).

We need the following lemma.

Lemma 2.2.1 *For any continuous function $\omega(y) \ge 0$ nondecreasing on $[0,\pi]$*

such that, for an arbitrary natural number $b, \omega(by) \le b^a \omega(y)$, $a > 0$, *we have the estimate* ($\delta > 0$)

$$\int_\delta^\pi \omega(y) y^{-2a} dy \le C(a) \delta^{1-2a} \omega(\delta).$$

Proof We have, for $\delta \le y \le \pi$,

$$\begin{aligned}\omega(y) &= \omega(y\delta/\delta) \le \omega(([y/\delta]+1)\delta) \\ &\le ([y/\delta]+1)^a \omega(\delta) \le (2y/\delta)^a \omega(\delta).\end{aligned}$$

Multiplying this inequality by y^{-2a} and integrating, we obtain the lemma. □

We now prove a statement called the Jackson inequality.

Theorem 2.2.2 *For any* $f \in L_p$, $1 \le p \le \infty$, *we have*

$$E_n(f)_p \le C(a) \omega_a(f, 1/n)_p, \qquad n = 1, 2, \dots$$

Proof We consider (see the proof of Theorem 1.4.3)

$$t(x) := (2\pi)^{-1} \int_{-\pi}^{\pi} (f(x) - \Delta_y^a f(x)) J_n^a(y) dy.$$

Then $t \in \mathscr{T}(an)$ and

$$\begin{aligned}\|f - t\|_p &\le (2\pi)^{-1} \int_{-\pi}^{\pi} \|\Delta_y^a f(x)\|_p J_n^a(y) dy \\ &\le \pi^{-1} \int_0^\pi \omega_a(f, y)_p J_n^a(y) dy \\ &\le \pi^{-1} \omega_a(f, 1/n)_p \int_0^{1/n} J_n^a(y) dy + \pi^{-1} \int_{1/n}^{\pi} \omega_a(f, y)_p J_n^a(y) dy.\end{aligned}$$

To estimate the first term we use (1.2.16), and to estimate the second term we use (1.2.18) and Lemma 2.2.1:

$$\begin{aligned}\|f - t\|_p &\le \omega_a(f, 1/n)_p + C(a) n^{1-2a} \int_{1/n}^{\pi} \omega_a(f, y)_p y^{-2a} dy \\ &\le C(a) \omega_a(f, 1/n)_p.\end{aligned}$$

Thus, it is proved that

$$E_{an}(f)_p \le C(a) \omega_a(f, 1/n)_p,$$

which implies the theorem. □

Corollary 2.2.3 *From Theorem 2.2.2 and the de la Vallée Poussin inequality (1.4.25), the estimate*

$$\left\|f - V_n(f)\right\|_p \le C(a)\omega_a(f,1/n)_p, \qquad 1 \le p \le \infty,$$

follows, this shows, in particular, that for any continuous function, its de la Vallée Poussin sums converge to it uniformly.

Corollary 2.2.4 *From Theorem 2.2.2 and the inequality (1.4.27) it follows that*

$$\left\|f - S_n(f)\right\|_p \le C(a,p)\omega_a(f,1/n)_p, \qquad 1 < p < \infty.$$

Corollary 2.2.5 *From Theorem 2.2.2 and the Lebesgue inequality (1.4.28) it follows that*

$$\left\|f - S_n(f)\right\|_p \le C(a)(\ln n)\omega_a(f,1/n)_p, \qquad n = 2, \ldots, \qquad p = 1, \infty.$$

From this estimate the Dini–Lipschitz criterion for uniform convergence of the Fourier series follows: let f be such that $\omega_a(f,1/n)_\infty \ln n \to 0$ for $n \to \infty$, then the Fourier series of f converges uniformly.

Corollary 2.2.6 (The Weierstrass theorem.) *For any $f \in L_p$, $1 \le p \le \infty$, we have*

$$E_n(f)_p \to 0 \qquad \textit{for } n \to \infty.$$

The corollary follows from Theorem 2.2.2 and from the relation $\omega(f,\delta)_p \to 0$ for $\delta \to 0$ (see Theorem A.1.1).

Corollary 2.2.7 *Let $1 \le p \le \infty$, $f, g \in L_p$ and for all k let $\hat{f}(k) = \hat{g}(k)$. Then the functions f and g are equivalent. Moreover, if f and g are continuous, they coincide.*

Indeed, $V_n(f) = V_n(g)$, and by Corollary 2.2.3,

$$\left\|f - V_n(f)\right\|_p \to 0, \qquad \left\|g - V_n(g)\right\|_p \to 0 \text{ for } n \to \infty,$$

which implies that $\|f - g\|_p = 0$.

We prove an additional statement about the convergence of Fourier series.

Theorem 2.2.8 (The Dirichlet–Jordan criterion.) *Let $f(x)$ be of bounded variation on $[-\pi, \pi]$ and let $f^0(x) := \big(f(x+0) + f(x-0)\big)/2$. Then at each point we have*

$$f^0(x) - S_n(f,x) \to 0 \qquad \textit{for } n \to \infty. \tag{2.2.2}$$

If in addition $f(x)$ is continuous, then

$$\left\|f - S_n(f)\right\|_\infty \to 0 \qquad \textit{for } n \to \infty. \tag{2.2.3}$$

Proof Let $x \in [-\pi, \pi]$ and

$$S_n(f,x) = \pi^{-1} \int_0^{\pi} \big(f(x+y) + f(x-y)\big)\mathscr{D}_n(y)/2dy.$$

Denoting $\psi(y) := \big(f(x+y) + f(x-y)\big)/2$, we get $f^0(x) = \psi(+0)$ and

$$S_n(f,x) - f^0(x) = \pi^{-1} \int_0^{\pi} \big(\psi(y) - \psi(+0)\big)\mathscr{D}_n(y)dy. \tag{2.2.4}$$

We need the following auxiliary statement.

Lemma 2.2.9 *For any* $0 < \delta \le \pi$ *we have*

$$\left|\int_0^{\delta} \mathscr{D}_n(y)dy\right| < C, \qquad \left|\int_{\delta}^{\pi} \mathscr{D}_n(y)dy\right| < C(n\delta)^{-1}.$$

Proof We represent the Dirichlet kernel in the form

$$\mathscr{D}_n(y) = \mathscr{D}_n(y) - \mathscr{K}_{n-1}(y) + \mathscr{K}_{n-1}(y). \tag{2.2.5}$$

From the equality $\|\mathscr{K}_{n-1}\|_1 = 1$ and relation (1.2.10) we get

$$\int_0^{\delta} \mathscr{K}_{n-1}(y)dy \le \pi, \qquad \int_{\delta}^{\pi} \mathscr{K}_{n-1}(y)dy \le C(n\delta)^{-1}. \tag{2.2.6}$$

Further,

$$\begin{aligned}\sigma(\delta) &:= \int_0^{\delta} \big(\mathscr{D}_n(y) - \mathscr{K}_{n-1}(y)\big)dy = 2n^{-1}\sum_{k=1}^{n} \sin k\delta \\ &= 2n^{-1}\big(\cos(\delta/2) - \cos(n+1/2)\delta\big)/\big(2\sin(\delta/2)\big).\end{aligned}$$

Using the equality $\sigma(\pi) = 0$ we obtain

$$|\sigma(\delta)| \le 2; \qquad \left|\int_{\delta}^{\pi} \big(\mathscr{D}_n(y) - K_{n-1}(y)\big)dy\right| \le C(n\delta)^{-1}. \tag{2.2.7}$$

The lemma follows from relations (2.2.5)–(2.2.7). □

We divide the integral in (2.2.4) into two integrals, over $[0,\delta]$ and over $[\delta,\pi]$. Without loss of generality we can assume that $\psi(y)$ is nondecreasing on $[0,\pi]$. Integrating by parts and using Lemma 2.2.9 we find

$$\left|\int_0^{\delta} \big(\psi(y) - \psi(+0)\big)\mathscr{D}_n(y)dy\right| \le C\big(\psi(\delta) - \psi(+0)\big), \tag{2.2.8}$$

$$\left|\int_{\delta}^{\pi} \big(\psi(y) - \psi(+0)\big)\mathscr{D}_n(y)dy\right| \le C\big(\psi(\pi) - \psi(+0)\big)(n\delta)^{-1}. \tag{2.2.9}$$

Relations (2.2.2) and (2.2.3) follow from (2.2.8) and (2.2.9). □

We now prove some assertions about estimates of the Fourier coefficients. We denote by F_q^r one of the classes $W_{q,\alpha}^r$ and H_q^r.

Theorem 2.2.10 *Let $r > 0$; then, for any $1 \le q \le \infty$,*

$$\sup_{f \in F_q^r} |\hat{f}(n)| \asymp |n|^{-r}, \qquad n = \pm 1, \pm 2, \ldots$$

Proof We have

$$\hat{f}(n) = (2\pi)^{-1} \int_0^{2\pi} f(x) e^{-inx} dx = -(2\pi)^{-1} \int_0^{2\pi} f(x + \pi/n) e^{-inx} dx,$$

which implies that

$$\hat{f}(n) = (4\pi)^{-1} \int_0^{2\pi} (f(x) - f(x + \pi/n)) e^{-inx} dx$$
$$= (2\pi)^{-1} \int_0^{2\pi} (1/2) \Delta_{\pi/n} f(x) e^{-inx} dx.$$

Consequently,

$$\hat{f}(n) = (2\pi)^{-1} \int_0^{2\pi} 2^{-a} \Delta_{\pi/n}^a f(x) e^{-inx} dx,$$

and

$$|\hat{f}(n)| \le 2^{-a} \|\Delta_{\pi/n}^a f(x)\|_1 \le 2^{-a} (\pi/|n|)^r.$$

This proves the upper estimate for $q = 1$, which implies the required estimates for all $q \ge 1$.

It suffices to prove the lower estimate for $q = \infty$. We set

$$f(x) = |n|^{-r} e^{inx}.$$

Then $f \in W_{\infty,\alpha}^r B$, which implies the lower estimate. □

In the theory of Fourier series the monotone rearrangement of the Fourier coefficients $\{\hat{f}(\sigma(n))\}$ is used, where $\sigma(n)$ maps the set of natural numbers in a one-to-one manner to the set of integers in such a way that

$$|\hat{f}(\sigma(n))| \ge |\hat{f}(\sigma(n+1))|, \qquad n = 1, 2, \ldots$$

For the sake of brevity we denote $\hat{f}_n = \hat{f}(\sigma(n))$.

Theorem 2.2.11 *Let $r > 0$; then, for $2 \le q \le \infty$,*

$$\sup_{f \in F_q^r} |\hat{f}_n| \asymp n^{-r-1/2}.$$

Proof We will prove the upper estimate for $q=2$ and $F^r_q = H^r_q$. We have

$$n|\hat{f}_{2n}|^2 \leq \sum_{k=n+1}^{2n} |\hat{f}_k|^2 \leq \sum_{k=n+1}^{\infty} |\hat{f}_k|^2 \leq \sum_{|k|\geq n/2} |\hat{f}(k)|^2.$$

By Theorem 1.4.3 we get from this relation

$$n|\hat{f}_{2n}|^2 \leq C(r)n^{-2r},$$

which implies the required upper estimate. The lower estimate follows from the example

$$f(x) := n^{-r-1/2}R_n(x),$$

where $R_n(x)$ is the Rudin–Shapiro polynomial (see §1.2.5).

By the Bernstein inequality and relation (1.2.21) we obtain $f \in W^r_{\infty,\alpha}B$, where B does not depend on n. □

The following condition for the absolute convergence of Fourier series follows directly from Theorem 2.2.11.

Theorem 2.2.12 (The Bernstein theorem) *Let $r > 1/2$; then, for any $f \in H^r_2$, its Fourier series is absolutely convergent:*

$$\sum_k |\hat{f}(k)| < \infty.$$

We now consider the case $1 \leq q < 2$.

Theorem 2.2.13 *Let $r > 0$; then for $1 \leq q < 2$ we have*

$$\sup_{f\in F^r_q} |\hat{f}_n| \asymp n^{-r+1/q-1}.$$

Proof In the case $r > 1/q - 1/2$ the upper estimate follows from Theorem 1.4.5 ($H^r_q \subset H^{r-1/q+1/2}_2 B$) and Theorem 2.2.11. We will prove the upper estimate under the assumption $r > 0$. We use the following simple lemma.

Lemma 2.2.14 *Let $1 \leq p < \infty$ and let a denote a sequence $a_1 \geq a_2 \geq \cdots$ of nonnegative numbers; then*

$$a_n \leq n^{-1/p}\|a\|_p.$$

In the case $q = 1$ the upper estimate in Theorem 2.2.13 follows from Theorem 2.2.10; therefore we assume $1 < q < 2$ and denote $p := q' := q/(q-1)$. By Corollary 1.4.6 we have, for $f \in H^r_q$,

$$\left\|\delta_s(f)\right\|_q \leq C(r,q)2^{-rs}. \tag{2.2.10}$$

Consequently, by the Hausdorff–Young theorem (see Theorem A.3.1) we have

$$\|\{\hat{\delta}_s(f,k)\}\|_p \leq C(r,q)2^{-rs}. \tag{2.2.11}$$

We take a natural number m and set $\kappa := pr$ and

$$n_s := [2^{m-\kappa(s-m)}], \qquad s = m+1,\ldots$$

Then by Lemma 2.2.14 and the relation (2.2.11), all numbers $|\hat{\delta}_s(k)|$, with the possible exception of n_s of them, will not exceed

$$(n_s+1)^{-1/p}\|\{\hat{\delta}_s(f,k)\}\|_p \leq C(r,q)2^{-(r+1/p)m}. \tag{2.2.12}$$

Relation (2.2.12) shows that all $|\hat{f}(k)|$, excepting possibly

$$n := 2^{m+1} + \sum_{s=m+1}^{\infty} n_s \leq C(\kappa)2^m \tag{2.2.13}$$

of them, are not greater than the right-hand side of (2.2.12); that is,

$$|\hat{f}_{n+1}| \leq C(r,q)2^{-(r+1/p)m}.$$

From this relation and from (2.2.13) we get the required upper estimate in the theorem.

The lower estimate follows from the example

$$f(x) = n^{-r+1/q-1}\mathscr{V}_n(x),$$

since $f \in W^r_{q,\alpha}B$ owing to the Bernstein inequality and relation (1.1.14). □

Theorems 2.2.11 and 2.2.13 imply the following assertion.

Theorem 2.2.15 *Let $r > 0$ and $\rho > (r+1/2-(1/q-1/2)_+)^{-1}$, $1 \leq q \leq \infty$. Then for any $f \in H^r_q$ we have*

$$\sum_k |\hat{f}(k)|^\rho < \infty.$$

We present one more statement giving an estimate for the Fourier coefficients of $f \in L_p$, $1 < p \leq 2$, which is a consequence of the more general theorem 3.3.6.

Theorem 2.2.16 *Let $1 < p < 2$, then for $f \in L_p$ we have*

$$\left(\sum_{s=0}^{\infty} \|\delta_s(f)\|_2^p 2^{s(p/2-1)}\right) \leq C(p)\|f\|_p.$$

This theorem implies the following assertion.

Theorem 2.2.17 (The Hardy–Littlewood theorem) *Let $1 < p < 2$; then for $f \in L_p$ we have the inequality*

$$\left(\sum_k |\hat{f}(k)|^p (|k|+1)^{p-2}\right)^{1/p} \leq C(p)\|f\|_p.$$

Proof By the Hölder inequality, for each $s \geq 0$ one has

$$\sum_{k\in\rho(s)} |\hat{f}(k)|^p \leq \left(\sum_{k\in\rho(s)} |\hat{f}(k)|^2\right)^{p/2} 2^{s(1-p/2)}$$

and, consequently,

$$\sum_k |\hat{f}(k)|^p (|k|+1)^{p-2} \leq \sum_{s=0}^{\infty} \|\delta_s(f)\|_2^p 2^{s(p/2-1)}.$$

From this applying Theorem 2.2.16, we obtain the theorem. □

We now prove a criterion for the absolute convergence of the Fourier series of functions $f \in L_2$. This criterion will be formulated in terms of n-term trigonometric approximations in L_2:

$$\sigma_n(f)_2 := \inf_{c_j,k_j \quad j=1,\dots,n} \left\| f(x) - \sum_{j=1}^{n} c_j e^{ik_jx} \right\|_2 .$$

Theorem 2.2.18 (The Stechkin criterion) *For the convergence of the series*

$$\sum_k |\hat{f}(k)| < \infty$$

it is necessary and sufficient that the following series converges:

$$\sum_{n=1}^{\infty} n^{-1/2} \sigma_{n-1}(f)_2 < \infty.$$

Proof Clearly,

$$\sigma_{n-1}(f)_2 = \left(\sum_{k=n}^{\infty} |\hat{f}_k|^2\right)^{1/2}$$

and the series $\sum_k |\hat{f}(k)|$, $\sum_k |\hat{f}_k|$ converge simultaneously. Therefore, to prove Theorem 2.2.18 it suffices to prove the following statement.

Lemma 2.2.19 *Let $a_1 \geq a_2 \geq \cdots$ be a sequence of nonnegative numbers. Then the series*

(I) $\sum_{k=1}^{\infty} a_k$
(II) $\sum_{n=1}^{\infty} n^{-1/2} \left(\sum_{k=n}^{\infty} a_k^2\right)^{1/2}$

converge simultaneously.

Proof On the one hand, by the monotonicity of the sequence we have

$$a_{2n-1}^2 \le n^{-1} \sum_{k=n}^{2n-1} a_k^2,$$

which implies that

$$\sum_{k=1}^{\infty} a_k \le 2 \sum_{n=1}^{\infty} n^{-1/2} \left(\sum_{k=n}^{\infty} a_k^2 \right)^{1/2}.$$

On the other hand,

$$\sigma := \sum_{n=1}^{\infty} n^{-1/2} \left(\sum_{k=n}^{\infty} a_k^2 \right)^{1/2} \le \sum_{s=0}^{\infty} 2^{s/2} \left(\sum_{k=2^s}^{\infty} a_k^2 \right)^{1/2}.$$

Further,

$$\left(\sum_{k=2^s}^{\infty} a_k^2 \right)^{1/2} \le \left(\sum_{j=s}^{\infty} 2^j a_{2^j}^2 \right)^{1/2} \le \sum_{j=s}^{\infty} 2^{j/2} a_{2^j},$$

and consequently

$$\sigma \le \sum_{s=0}^{\infty} 2^{s/2} \sum_{j=s}^{\infty} 2^{j/2} a_{2^j} \le 2 \sum_{s=0}^{\infty} 2^s a_{2^s} \le 4 \sum_{n=1}^{\infty} a_n. \qquad \square$$

This completes the proof of Theorem 2.2.18. □

Remark 2.2.20 (The Stechkin lemma) Under conditions of Lemma 2.2.19, Stechkin 1955 proved the following inequalities:

$$\frac{1}{2} \sum_{n=1}^{\infty} n^{-1/2} \left(\sum_{k=n}^{\infty} a_k^2 \right)^{1/2} \le \sum_{k=1}^{\infty} a_k \le \frac{2}{\sqrt{3}} \sum_{n=1}^{\infty} n^{-1/2} \left(\sum_{k=n}^{\infty} a_k^2 \right)^{1/2}.$$

2.3 Approximation of Functions with Infinite Smoothness

Next we present some results about the approximation of functions whose smoothness is quite different from that of the functions in $W^r_{q,\alpha}$. Letting r and b be positive real numbers, we set

$$g_{r,b}(x) := 1 + 2 \sum_{k=1}^{\infty} e^{-rk^b} \cos kx,$$

and define the class $A_q^{r,b}$, $1 \le q \le \infty$, as the set of functions representable in the form

$$f(x) = (2\pi)^{-1} \int_{-\pi}^{\pi} g_{r,b}(x-y)\phi(y)dy, \qquad \|\phi\|_q \le 1.$$

We note that in the case $b = 1$ the function $g_{r,1}$ coincides with the Poisson kernel

$$(1-\rho^2)(1-2\rho\cos x+\rho^2)^{-1}, \qquad \rho = e^{-r}.$$

We would like to find the orders of the widths of these classes. Let us consider separately two cases: $b \ge 1$ and $0 < b < 1$.

Theorem 2.3.1 *Let $b \ge 1$ and n denote $2m$ or $2m-1$. For all $1 \le q, p \le \infty$ we have the relations*

$$S_{m-1}(A_q^{r,b})_p \asymp \varphi_n(A_q^{r,b}, L_p) \asymp \lambda_n(A_q^{r,b}, L_p)$$
$$\asymp d_n(A_q^{r,b}, L_p) \asymp e^{-rm^b}.$$

Proof We first prove the upper estimates. Let $f \in A_1^{r,b}$; then

$$\left|\hat{f}(k)\right| \le e^{-r|k|^b}$$

and

$$\left\|f - S_{m-1}(f)\right\|_\infty \le 2\sum_{k=m}^{\infty} e^{-rk^b} \le C(r)e^{-rm^b}.$$

To prove the lower estimate it suffices to consider $d_{2m}(A_\infty^{r,b}, L_1)$. We first consider approximations in L_2. Let $2m$ linearly independent functions $\chi_1(x), \dots, \chi_{2m}(x)$ be given. Clearly, we can assume the system of functions $\{\chi_j(x)\}_{j=1}^{2m}$ to be orthonormal. We now estimate approximations of the functions $e_k(x) = e^{ikx}$, $|k| \le m$, in this system.

We set $f_k(x) = e^{-r|k|^b} e_k(x)$. Clearly, $f_k \in A_\infty^{r,b}$. Let

$$S_{2m}(f, \chi) := \sum_{j=1}^{2m} \langle f, \chi_j \rangle \chi_j,$$

then for all $|k| \le m$,

$$\sigma_k := \langle f_k - S_{2m}(f_k, \chi), e_k \rangle = e^{-r|k|^b}\left(1 - \sum_{j=1}^{2m} \left|\hat{\chi}_j(k)\right|^2\right).$$

We will prove that there is a $C(r) > 0$ such that

$$\max_{|k| \le m} \sigma_k \ge C(r)e^{-rm^b}. \qquad (2.3.1)$$

Indeed, assuming the contrary, we have

$$1 \le \sum_{|k|\le m} \left(1 - \sum_{j=1}^{2m} |\hat{\chi}_j(k)|^2\right) = \sum_{|k|\le m} \sigma_k e^{r|k|^b} < C(r)C_1(r),$$

which is impossible for sufficiently small $C(r)$. So (2.3.1) is proved, which implies that there is a k_0 such that $|k_0| \le m$ and

$$\left\| f_{k_0} - S_{2m}(f_{k_0}, \chi)\right\|_2 \ge C(r)e^{-rm^b}.$$

From this relation it follows that, for any $a > 0$,

$$d_{2m}(A_\infty^{a,b} \cap \mathscr{T}(m), L_2) \ge C(a)e^{-am^b}. \tag{2.3.2}$$

Consider the operator

$$G_{\alpha,b}(f) := f(x) * \sum_{|k|\le m} e^{\alpha|k|^b} e^{ikx}.$$

We have

$$\|G_{\alpha,b}\|_{1\to 2} \le \left(\sum_{|k|\le m} e^{2\alpha|k|^b}\right)^{1/2} \le C(\alpha)e^{\alpha m^b}. \tag{2.3.3}$$

From relation (2.3.2) with $a = r/2$ and relation (2.3.3) with $\alpha = r/2$ we find

$$\begin{aligned} C(r)e^{-rm^b/2} &\le d_{2m}(A_\infty^{r/2,b} \cap \mathscr{T}(m), L_2) \\ &\le \|G_{r/2,b}\|_{1\to 2} d_{2m}(A_\infty^{r,b} \cap \mathscr{T}(m), L_1) \\ &\le C(r)e^{rm^b/2} d_{2m}(A_\infty^{r,b}, L_1), \end{aligned}$$

which implies the theorem. □

Theorem 2.3.1 shows that the optimal approximation (in the sense of order) for the classes $A_q^{r,b}$ for all $1 \le q, p \le \infty$ is given by the Fourier sums in the trigonometric system.

We proceed to consider the case $0 < b < 1$, that is, the case of classes of infinitely differentiable functions. The general scheme of studying approximations of the classes $A_q^{r,b}$ for fixed b is analogous to studying the classes $W_{q,\alpha}^r$. We first perform some auxiliary constructions. We construct for $0 < b < 1$ the sequence of numbers $\{N_s\}_{s=1}^\infty$ such that $N_1 = 1$ and

$$N_s^b \le s; \qquad (N_s + 1)^b > s.$$

Denote $n_s := N_{s+1} - N_s$. It follows from the definition that, for all s,

$$s^{1/b} - 1 < N_s \le s^{1/b},$$

which implies the relations

$$n_s \asymp s^{1/b-1}, \qquad n_{s+1}/n_s \asymp 1. \tag{2.3.4}$$

Define the following functions:

$$\mathscr{A}_{b,1}(x) := 1 + 2\cos x + 2\sum_{\nu=1}^{n_1}(1-\nu/n_1)\cos(\nu+1)x,$$

$$\mathscr{A}_{b,s}(x) := 2\sum_{\nu=0}^{n_{s-1}}(1-\nu/n_{s-1})\cos(N_s-\nu)x$$
$$+2\sum_{\nu=1}^{n_s}(1-\nu/n_s)\cos(N_s+\nu)x, \qquad s=2,3,\dots$$

Lemma 2.3.2 *We have*

$$\|\mathscr{A}_{b,s}\|_1 \le C(b), \qquad s=1,2,\dots$$

Proof Clearly it suffices to prove the lemma for large s. Let s be such that $N_s \ge n_s$. We denote $M_s := N_s - n_s$ and represent the function $\mathscr{A}_{b,s+1}(x)$ in the form

$$\mathscr{A}_{b,s+1}(x) = 2\sum_{\nu=0}^{n_s}(1-\nu/n_s)\cos(M_s+2n_s-\nu)x$$
$$+2\sum_{\nu=1}^{n_{s+1}}(1-\nu/n_{s+1})\cos(M_s+2n_s+\nu)x$$
$$= 2(\cos M_s x)\phi_s(x) - 2(\sin M_s x)\tilde{\phi}_s(x),$$

where $\tilde{\phi}_s$ is the function conjugate to ϕ_s, and

$$\phi_s(x) := \sum_{\nu=0}^{n_s}(1-\nu/n_s)\cos(2n_s-\nu)x + \sum_{\nu=1}^{n_{s+1}}(1-\nu/n_{s+1})\cos(2n_s+\nu)x.$$

Applying Theorem 1.3.1 and using relation (2.3.4) we get

$$\|\phi_s\|_1 \le C(b). \tag{2.3.5}$$

Let l_1 and l_2 be such that

$$2^{l_1-1} \le n_s < 2^{l_1}; \qquad 2^{l_2-1} < n_{s+1} + 2n_s \le 2^{l_2}.$$

Then $l_2 - l_1 \le C(b)$ and

$$\tilde{\phi}_s = \phi_s * \left(\sum_{l=l_1}^{l_2} \tilde{\mathscr{A}_l}\right).$$

From this using the relation (2.3.5) and the estimate (1.3.7) we get

$$\|\tilde{\phi}_s\|_1 \le C(b). \tag{2.3.6}$$

The lemma follows from (2.3.5) and (2.3.6). □

We write

$$A_{b,s}(f) := \mathscr{A}_{b,s} * f, \qquad f \in L_1.$$

The following statement can be proved in the same way as Lemma 2.3.2.

Lemma 2.3.3 *We have*

$$\|A_{b,s}(g_{r,b})\|_1 \le C(r,b)e^{-rs}.$$

Lemma 2.3.3 implies the following.

Theorem 2.3.4 *We have*

$$E_n(g_{r,b})_1 \le C(r,b)e^{-rn^b}.$$

Proof Let $n \in [N_l, N_{l+1})$, then

$$E_n(g_{r,b})_1 \le \sum_{s=l}^{\infty} \|A_{b,s}(g_{r,b})\|_1 \le C(r,b)e^{-rl}$$
$$\le C(r,b)e^{-rN_l^b} \le C(r,b)e^{-rn^b}.$$

The theorem is proved. □

We use this theorem to prove the following estimates of the best approximations in the classes $A_q^{r,b}$.

Theorem 2.3.5 *Let* $1 \le q, p \le \infty$, $\beta := (1/q - 1/p)_+$; *then*

$$E_n(A_q^{r,b})_p \asymp e^{-rn^b} n^{(1-b)\beta}.$$

Proof We first prove the upper estimates. It suffices to consider the case $1 \le q \le p \le \infty$. Let $f \in A_q^{r,b}$; then

$$E_n(f)_q \le E_n(g_{r,b})_1,$$

and by Theorem 2.3.4 we obtain the required estimate.

Let $q < p$ and $f \in A_q^{r,b}$; then by Lemma 2.3.3,

$$\|A_{b,s}(f)\|_q \le \|A_{b,s}(g_{r,b})\|_1 \le C(r,b)e^{-rs}. \tag{2.3.7}$$

Using the fact that the function $A_{b,s}(f)$ can have nonzero Fourier coefficients only in $(-N_{s+1}, -N_{s-1}) \cup (N_{s-1}, N_{s+1})$, we obtain, in analogy to Theorem 1.3.4,

$$\|A_{b,s}(f)\|_p \le C(b)\|A_{b,s}(f)\|_q n_s^{\beta}. \tag{2.3.8}$$

Let $n \in [N_l, N_{l+1})$. From relations (2.3.7) and (2.3.8) we find

$$E_n(f)_p \le \sum_{s=l}^{\infty} \|A_{b,s}(f)\|_p \le C(r,b) \sum_{s=l}^{\infty} e^{-rs} s^{(1/b-1)\beta}$$
$$\le C(r,b) e^{-rn^b} n^{(1-b)\beta}.$$

The upper estimates are proved.

Let us prove the lower estimates. We remark that Lemma 2.3.2 and the trivial estimate

$$\|\mathscr{A}_{b,s}\|_\infty \le C(b) s^{1/b-1}$$

imply the estimate

$$\|\mathscr{A}_{b,s}\|_q \le C(b) s^{(1/b-1)(1-1/q)}. \tag{2.3.9}$$

Let $n \in [N_{s-2}, N_{s-1})$. As an example we take

$$f_n(x) := (g_{r,b} * \mathscr{A}_{b,s}) \|\mathscr{A}_{b,s}\|_q^{-1}.$$

Then, on the one hand, for any $t \in \mathscr{T}(n)$ we have

$$\sigma := \langle f_n - t, \mathscr{A}_{b,s} \rangle \le \|f_n - t\|_p \|\mathscr{A}_{b,s}\|_{p'}. \tag{2.3.10}$$

On the other hand it is not difficult to see that

$$\sigma = \langle f_n, \mathscr{A}_{b,s} \rangle \ge C(r,b) e^{-rn^b} n^{(1-b)/q}. \tag{2.3.11}$$

Using that $s \asymp n^b$, we get from relations (2.3.9)–(2.3.11),

$$E_n(f_n) \ge C(r,b) e^{-rn^b} n^{(1-b)\beta},$$

which concludes the proof since $f_n \in A_q^{r,b}$. □

We now prove three theorems about the widths of the classes $A_q^{r,b}$ in the case $0 < b < 1$.

Theorem 2.3.6 *Let $r > 0$ and $1 \le q \le p \le \infty$ or $2 \le p \le q \le \infty$. Then*

$$d_{2m}(A_q^{r,b}, L_p) \asymp e^{-rm^b} m^{(1-b)(1/q-\max(1/2,1/p))_+}.$$

Theorem 2.3.7 *Let $r > 0$ and $1 \le q \le p \le \infty$ or $2 \le p \le q \le \infty$ or $1 \le p \le q \le 2$. Then*

$$\lambda_{2m}(A_q^{r,b}, L_p) \asymp \begin{cases} e^{-rm^b} m^{(1-b)\max(1/q-1/2,1/2-1/p)} & \text{for } 1 \le q < 2 \text{ and } 2 < p \le \infty, \\ e^{-rm^b} m^{(1-b)(1/q-1/p)_+} & \text{otherwise.} \end{cases}$$

Theorem 2.3.8 *Let* $r>0$, $1\le q,p\le\infty$, $(q,p)\neq(1,1),(\infty,\infty)$. *Then*

$$\varphi_{2m}(A_q^{r,b},L_p)\asymp S_m(A_q^{r,b})_p\asymp e^{-rm^b}m^{(1-b)(1/q-1/p)_+}.$$

The proof of the upper estimates in these theorems is analogous to the corresponding proofs of Theorems 2.1.1–2.1.3.

Proof of the upper estimates in Theorem 2.3.8 In the case $1<p<\infty$ the theorem follows from Theorem 2.3.5 and relation (1.4.27). In the case $p=1$, $q>1$ the required estimate follows from the estimate already proved and from the obvious inequality

$$S_m(A_q^{r,b})_1\le S_m(A_{q^*}^{r,b})_{q^*},\qquad q^*=\min(q,2).$$

It remains to consider the case $1\le q<\infty$, $p=\infty$. We define the operators

$$\delta_{b,s}(f):=\sum_{N_{s-1}\le|k|<N_s}\hat f(k)e^{ikx},\qquad s=2,3,\dots$$

Clearly, it suffices to consider the case $m=N_l-1$. Then for $f\in A_q^{r,b}$,

$$\|f-S_m(f)\|_\infty\le\sum_{s>l}\|\delta_{b,s}(f)\|_\infty. \tag{2.3.12}$$

Further,

$$\delta_{b,s}(f)=\left(\sum_{N_{s-1}\le|k|<N_s}e^{ikx}\right)*\left(A_{b,s-1}(f)+A_{b,s}(f)\right), \tag{2.3.13}$$

which by (2.3.7) and (1.2.7) implies that

$$\|\delta_{b,s}(f)\|_\infty\le C(r,b)e^{-rs}\|\mathscr{D}_{n_{s-1}-1}\|_{q'}\ll e^{-rs}s^{(1/b-1)/q}. \tag{2.3.14}$$

From relations (2.3.12) and (2.3.14) we get

$$\|f-S_m(f)\|_\infty\ll e^{-rm^b}m^{(1-b)/q}.$$

The upper estimates in Theorem 2.3.8 are now proved. □

Proof of the upper estimates in Theorem 2.3.6 In the cases $2\le p\le q\le\infty$ and $1\le q\le p\le 2$ the upper estimates follow from Theorem 2.3.5. To prove the estimate in the remaining case it suffices to consider $1\le q\le 2$ and $p=\infty$. Let

$$\delta_{b,s}^+(f)=\sum_{N_{s-1}\le k<N_s}\hat f(k)e^{ikx},$$

$$\delta_{b,s}^-(f)=\delta_{b,s}(f)-\delta_{b,s}^+(f).$$

Then

$$\delta_{b,s}^\pm(f)=\left(e^{\pm iN_{s-1}x}\mathscr{D}_{n_{s-1}-1}(\pm x)\right)*\left(A_{b,s-1}(f)+A_{b,s}(f)\right). \tag{2.3.15}$$

Using for $1 \le q \le 3/2$ the Young inequality and for $3/2 \le q \le 2$ Theorem 1.2.1 and the Nikol'skii inequality, we get from (2.3.7) and (2.3.15),

$$\left\|\delta^{\pm}_{b,s}(f)\right\|_2 \le C(r,b)e^{-rs}s^{(1/b-1)(1/q-1/2)}. \tag{2.3.16}$$

Let $0 < \kappa < r$ and l be fixed and

$$\begin{aligned} m_s &:= \left[n_s e^{-\kappa(s-l)}\right], \qquad s = l+1,\dots, \\ m &:= N_l + \sum_{s>l} m_s \le N_l + C(b,\kappa)N_l^{1-b}. \end{aligned} \tag{2.3.17}$$

Then

$$d_{2m}(A^{r,b}_q, L_\infty) \le C(b,r,\kappa)\sum_{s\ge l} e^{-rs}s^{(1/b-1)(1/q-1/2)}d_{m_s}\left(\mathscr{T}(n_s)_2, L_\infty\right).$$

Applying Theorem 2.1.14, we get

$$\begin{aligned} d_{2m}(A^{r,b}_q, L_\infty) &\ll \sum_{s\ge l} e^{-rs}s^{(1/b-1)(1/q-1/2)}\left(e^{\kappa(s-l)}(s-l+1)\right)^{1/2} \\ &\ll e^{-rl}l^{(1/b-1)(1/q-1/2)} \ll e^{-rm^b}m^{(1-b)(1/q-1/2)}. \end{aligned}$$

The upper estimate for m of the form (2.3.17) is thus proved. Clearly, the general case follows from the case just considered.

The upper estimates in Theorem 2.3.6 are now proved. □

Proof of the upper estimates in Theorem 2.3.7 The following theorem can be proved in the same way as Theorem 2.1.8.

Theorem 2.3.9 *Let $r > 0$ and $1 \le q,p \le \infty$. Then*

$$\lambda_m(A^{r,b}_q, L_p) = \lambda_m(A^{r,b}_{p'}, L_{q'}).$$

Proof From the proof of Theorem 2.3.5 it follows that, for $f \in A^{r,b}_q$,

$$\left\|f - \sum_{s=1}^{l} A_{b,s}(f)\right\|_p \ll e^{-rN_l^b}N_l^{(1-b)(1/q-1/p)_+},$$

which gives the required estimate for all q,p excluding the case $1 \le q < 2$, $2 < p \le \infty$. In this case by virtue of Theorem 2.3.9 it suffices to prove the upper estimate for $(q,p) \in D'_3$, where D'_3 is the same as in the proofs of Theorems 2.1.1–2.1.3. □

The following statement is proved in the same way as Lemmas 2.1.9 and 2.1.10.

Lemma 2.3.10 *Let $b>0$ be given and let*

$$\mathscr{A}^+_{b,s}(x) := \sum_{k>0} \hat{\mathscr{A}}_{b,s}(k)e^{ikx}, \qquad \mathscr{A}^-_{b,s} := \mathscr{A}_{b,s} - \mathscr{A}^+_{b,s}, \qquad s = 2,3,\dots$$

For any natural s and n there are trigonometric polynomials $t^{b,\pm}_{s,n}$ with no more than n terms and such that the operators $Y^{\pm}_{s,n}$, defined by the equality

$$Y^{\pm}_{s,n}(f) := (\mathscr{A}^{\pm}_{b,s} - t^{b,\pm}_{s,n}) * f, \qquad f \in L_1,$$

have the following properties:

$$Y^{\pm}_{s,n}(t) = 0, \qquad t \in \mathscr{T}(N_{s-1})$$

$$\|Y^{\pm}_{s,n}\|_{q\to p} \le C(b)n_s^{1/2+1/q}n^{-1}, \qquad (q,p) \in D'_3.$$

Proof We shall not carry out the complete proof, but merely note some properties of the kernels $\mathscr{A}^{\pm}_{b,s}$. After this the proof will be a repetition of the proofs of Lemmas 2.1.9 and 2.1.10. The properties of the kernels $\mathscr{A}^+_{b,s}$ and $\mathscr{A}^-_{b,s}$ are analogous, therefore we discuss only $\mathscr{A}^+_{b,s}$. We have

$$\mathscr{A}^+_{b,s}(x) = \mathscr{A}_{b,s}(x) * \left(\mathscr{V}_{N_{s-1}}(x)e^{i2N_{s-1}x}\right),$$

from which by Lemma 2.3.2 and the relation (1.2.13) we get

$$\|\mathscr{A}^+_{b,s}\|_1 \le C(b).$$

Further,

$$\mathscr{A}^+_{b,s} = e^{iN_s x}Q_{b,s}(x), \qquad Q_{b,s} \in \mathscr{T}\big(C(b)n_s\big).$$

Consequently,

$$\|Q_{b,s}\|_1 \le C(b).$$

We will use the following assertion (see Remark 1.3.8).

Lemma 2.3.11 *Let $t \in \mathscr{T}(N)$ and $x^{\mu} := 2\pi\mu/(2N+1)$, $\mu = 0,\dots,2N$. Then*

$$\sum_{\mu=0}^{2N} |t(x^{\mu})| \le CN\|t\|_1, \qquad N = 1,2,\dots$$

Proof Let $|t(x^{\mu})| = \varepsilon_{\mu}t(x^{\mu})$, $|\varepsilon_{\mu}| = 1$. We have

$$\sum_{\mu=0}^{2N} |t(x^{\mu})| = (2\pi)^{-1}\int_0^{2\pi} t(y)\sum_{\mu=0}^{2N}\varepsilon_{\mu}\mathscr{V}_N(x^{\mu}-y)dy \le$$

$$\le \|t\|_1 \left\|\sum_{\mu=0}^{2N}\varepsilon_{\mu}\mathscr{V}_N\right\|_{\infty} \le CN\|t\|_1.$$

In the last step we used property (2) of the de la Vallée Poussin polynomials (see §1.2.3). □

Denoting by $2a+1$ the smallest prime which is greater than the order of the polynomial $Q_{b,s}$, further considerations are carried out in the same way as in the proof of Lemmas 2.1.9 and 2.1.10. □

We now conclude the proof of the upper estimate in Theorem 2.3.7. Let κ, l, m_s be the same as in the proof of the upper estimate in Theorem 2.3.6.

Then, setting $Y^{\pm}_{s,m_s} := A^{\pm}_{b,s}$ for $m_s = 0$, we have

$$\lambda_{2m}(A^{r,b}_q, L_p) \ll \sum_{s\geq l} \|Y^{\pm}_{s,m_s}\|_{q\to p} E_{N_{s-1}}(A^{r,b}_q)_q.$$

Applying Lemma 2.3.10 and Theorem 2.3.5, we get

$$\lambda_{2m}(A^{r,b}_q, L_p) \ll \sum_{s\geq l} e^{-rs} s^{(1/b-1)(1/q-1/2)} e^{\kappa(s-1)}$$

$$\ll e^{-rm^b} m^{(1-b)(1/q-1/2)}.$$

The upper estimate in Theorem 2.3.7 is proved. □

We proceed to the proof of the lower estimates in Theorems 2.3.6–2.3.8.

Proof of the lower estimate in Theorem 2.3.8 We first consider the case $1 \leq p \leq q \leq \infty$. Clearly, it suffices to consider $\varphi_{2m}(A^{r,b}_\infty, L_1)$ under the assumption $m = N_n$, $n = 1, 2, \ldots$ Let an orthonormal system of functions $\chi_1, \ldots, \chi_{2N_n}$ be given and let $f_k(x)$, $S_{2m}(f,\chi)$, σ_k be the same as in the proof of Theorem 2.3.1. We will prove that there is a $C(r) > 0$ such that

$$\max_{|k|\leq N_{n+1}} \sigma_k \geq C(r) e^{-rN^b_{n+1}}. \tag{2.3.18}$$

Indeed, otherwise we would have

$$\begin{aligned}2(N_{n+1}-N_n)+1 &\leq \sum_{|k|\leq N_{n+1}} \left(1 - \sum_{j=1}^{2N_n} |\hat{\chi}_j(k)|^2\right) \\ &= \sum_{|k|\leq N_{n+1}} \sigma_k e^{r|k|^b} < C(r)C_1(r)(N_{n+1}-N_n),\end{aligned}$$

which is impossible for sufficiently small $C(r)$. Thus, there is a k_0 such that $|k_0| \leq N_{n+1}$ and

$$\sigma_{k_0} \geq C(r) e^{-rN^b_{n+1}},$$

which implies the inequality

$$\left\|f_{k_0} - S_{2m}(f_{k_0},\chi)\right\|_1 \geq C(r) e^{-rm^b}.$$

To conclude the proof it remains to note that $f_{k_0} \in A^{r,b}_\infty$.

Now let $1 \le q \le p \le \infty$. We consider

$$f_s := g_{r,b} * \mathscr{A}_{b,s},$$
$$g_s := \mathscr{A}_{b,s-1} + \mathscr{A}_{b,s} + \mathscr{A}_{b,s+1}, \qquad s = 2,3,\dots,$$
$$g_1 := \mathscr{A}_{b,1} + \mathscr{A}_{b,2}.$$

We define for $s = 1,\dots,n$ the numbers

$$\sigma_s := (2\pi)^{-1} \int_0^{2\pi} \langle f_s(x+y) - S_{2N_n}(f_s(x+y),\chi), g_s(x+y)\rangle dy$$

and prove that, for some $C(r,b) > 0$,

$$\max_{1\le s\le n} \sigma_s \ge C(r,b) e^{-rN_{n+1}^b}(N_{n+1} - N_n). \tag{2.3.19}$$

Indeed,

$$\sigma_s = \sum_k \hat{f}_s(k)\left(1 - \sum_{j=1}^{2N_n} |\hat{\chi}_j(k)|^2\right)$$

and, if (2.3.19) did not hold, then as above we would obtain

$$N_{n+1} - N_n \le \sum_{s=1}^{n} \sum_{|k|\le N_{n+1}} \hat{f}_s(k) e^{r|k|^b}\left(1 - \sum_{j=1}^{2N_n} |\hat{\chi}_j(k)|^2\right)$$
$$\le \sum_{s=1}^{n} \sigma_s e^{rN_{s+1}^b} \le C(r,b)C_1(r,b)(N_{n+1} - N_n),$$

which is impossible for sufficiently small $C(r,b)$. So, (2.3.19) is proved and, consequently, there are an s^* and a y^* such that

$$\langle f_{s^*}(x+y^*) - S_{2N_n}(f_{s^*}(x+y^*,\chi), g_{s^*}(x+y^*)\rangle$$
$$\ge C(r,b) e^{-rN_{n+1}^b}(N_{n+1} - N_n). \tag{2.3.20}$$

However, using (2.3.9), from (2.3.20) and the inequality $\langle f,g\rangle \le \|f\|_p \|g\|_{p'}$ we get

$$\left\| f_{s^*}(x+y^*) - S_{2N_n}(f_{s^*}(x+y^*,\chi)\right\|_p \ge C(r,b) e^{-rN_{n+1}^b}(N_{n+1} - N_n)^{1-1/p}.$$

To conclude the proof it remains to remark that $f_{s^*}(x+y^*)\|\mathscr{A}_{b,s}\|_q^{-1}$ belongs to $A_q^{r,b}$, and then use (2.3.9).

The proof of Theorem 2.3.8 is now complete. □

We remark that in fact we have proved the slightly stronger statement, for $1 \le q \le p \le \infty$,

$$\varphi_{2N_n}(A_q^{r,b} \cap \mathscr{T}(N_{n+1}), L_p) \gg e^{-rN_n^b}(N_{n+1} - N_n)^{(1/q-1/p)}. \tag{2.3.21}$$

Proof of the lower estimates in Theorem 2.3.6 The required estimates will be derived from Theorem 2.3.8. In the case $p \geq 2$ they follow from Theorem 2.3.8 with $p = 2$ and the relation (2.1.2). Let $1 \leq q \leq p < 2$, $\beta := 1/q - 1/p$. We consider the operator

$$G_{\alpha,b}(f) := f(x) * \left(\sum_{|k| \leq N_{n+1}} e^{\alpha |k|^b} e^{ikx} \right).$$

Then

$$\|G_{\alpha,b}\|_{2\to 2} \leq e^{\alpha N_{n+1}^b},$$

$$\|G_{\alpha,b}\|_{1\to 2} \leq \left(\sum_{|k| \leq N_{n+1}} e^{2\alpha |k|^b} \right)^{1/2} \leq C(\alpha,b) e^{\alpha N_{n+1}^b} (N_{n+1} - N_n)^{1/2}.$$

From these relations by the Riesz–Thorin theorem (see Theorem A.3.2) we find for $1 \leq p \leq 2$,

$$\|G_{\alpha,b}\|_{p\to 2} \leq C(\alpha,b) e^{\alpha N_{n+1}^b} (N_{n+1} - N_n)^{1/p-1/2}. \tag{2.3.22}$$

From (2.3.21), (2.3.22) and the inequality

$$d_{2N_n}(A_q^{r/2,b} \cap \mathscr{T}(N_{n+1}), L_2) \leq \|G_{r/2,b}\|_{p\to 2} d_{2N_n}(A_q^{r,b}, L_p)$$

we obtain the required lower estimate for $1 \leq q \leq p \leq 2$.

The proof of Theorem 2.3.6 is now complete. □

Proof of the lower estimates in Theorem 2.3.7 Owing to relation (2.1.1) the required lower estimates for $2 \leq p \leq q \leq \infty$ and for $1 \leq q \leq p \leq \infty$ and $1/q + 1/p \geq 1$ follow from Theorem 2.3.6. The remaining estimates follow from those already proved and from Theorem 2.3.9. □

2.4 Sampling and Numerical Integration

2.4.1 Numerical Integration

Let f be a continuous and 2π-periodic function. For given m, $\lambda_1, \dots, \lambda_m$, $\xi^1, \dots, \xi^m$ we define the quadrature formula with parameters (m, Λ, ξ):

$$\Lambda_m(f,\xi) := \sum_{j=1}^{m} \lambda_j f(\xi^j), \tag{2.4.1}$$

and for a class of functions F we define the quantities

$$\Lambda_m(F,\xi) := \sup_{f\in F}\left|\Lambda_m(f,\xi) - (2\pi)^{-1}\int_0^{2\pi} f(x)dx\right|,$$

$$\kappa_m(F) := \inf_{\Lambda,\xi}\Lambda_m(F,\xi).$$

The quantity $\kappa_m(F)$ gives the value of the least error of quadrature formulas of the form (2.4.1) with m points over the class F. We prove in this section that the quadrature formulas

$$q_m(f) := m^{-1}\sum_{j=1}^{m} f(2\pi j/m)$$

with equidistant points $\xi^j = 2\pi j/m$ and equal weights $\lambda_j = 1/m$ give the orders of the quantities $\kappa_m(W^r_{p,\alpha})$ and $\kappa_m(H^r_p)$, $1\le p\le\infty$, $r>1/p$. Let us denote

$$q_m(F) := \sup_{f\in F}\left|q_m(f) - (2\pi)^{-1}\int_0^{2\pi} f(x)dx\right|.$$

Theorem 2.4.1 *Let $1\le p\le\infty$ and $r>1/p$. We have*

$$\kappa_m(W^r_{p,\alpha}) \asymp \kappa_m(H^r_p) \asymp q_m(W^r_{p,\alpha}) \asymp q_m(H^r_p) \asymp m^{-r}.$$

Proof We first prove the upper estimate

$$q_m(H^r_p) \ll m^{-r}.$$

By Theorem 1.4.8 $H^r_p \subset H^{r-1/p}_\infty B$. We assume that functions from H^r_p, $r>1/p$, are continuous. By the Dini–Lipschitz criterion (see Corollary 2.2.5) the Fourier series of a function $f\in H^{r-1/p}_\infty$ converges uniformly. Then from the relation

$$\sum_{j=1}^{m} e^{i2\pi jl/m} = \begin{cases} m & \text{for } l\equiv 0 \ (\mathrm{mod}\, m)\\ 0 & \text{for } l\not\equiv 0 \ (\mathrm{mod}\, m)\end{cases}$$

we get

$$q_m(f) = \sum_k \hat f(km),$$

and using Theorem 2.2.10 we obtain, for $r>1$,

$$\left|q_m(f) - \hat f(0)\right| \ll \sum_{|k|\ge 1} |km|^{-r} \ll m^{-r}.$$

Let $1/p<r\le 1$. We now consider the functional q_mV_n and estimate the norm of this linear functional on the space L_p. For $p=\infty$ we have, by Theorem 1.2.3 and the definition of q_m,

$$\|q_mV_n\|_{L_\infty\to\mathbb{C}} \le 3. \tag{2.4.2}$$

For $p = 1$ we have

$$\|q_m V_n\|_{L_1 \to \mathbb{C}} \le \left\| m^{-1} \sum_{j=1}^{m} \mathscr{V}_n(2\pi j/m - y) \right\|_\infty . \tag{2.4.3}$$

Let $n \ge m$. Using again property (1) of the de la Vallée Poussin kernels we get

$$\begin{aligned} \left\| m^{-1} \sum_{j=1}^{m} |\mathscr{V}_n(2\pi j/m - y)| \right\|_\infty &\le C\left(n/m + (m)^{-1} \sum_{l=1}^{\infty} m^2/(nl^2) \right) \\ &\le Cn/m. \end{aligned} \tag{2.4.4}$$

From (2.4.2)–(2.4.4) we find by the Riesz–Thorin theorem A.3.2, for $1 \le p \le \infty$,

$$\|q_m V_n\|_{L_p \to \mathbb{C}} \le C(n/m)^{1/p}, \qquad n \ge m. \tag{2.4.5}$$

Consequently, for a polynomial $t \in \mathscr{T}(n)$, $n \ge m$,

$$|q_m(t)| \le C(n/m)^{1/p} \|t\|_p, \qquad 1 \le p \le \infty. \tag{2.4.6}$$

Let $f \in H^r_p$ and s_m be such that $2^{s_m - 1} \le m < 2^{s_m}$. Then, applying Theorem 1.4.5, we get

$$|q_m(f) - \hat{f}(0)| = \left| \sum_{s=s_m}^{\infty} q_m(A_s(f)) \right| \ll \sum_{s=s_m}^{\infty} (2^s/m)^{1/p} 2^{-rs} \ll m^{-r},$$

which concludes the proof of the upper estimate.

The required lower estimate follows from the Bernstein inequality and the following lemma. □

Lemma 2.4.2 *We have the relation*

$$\kappa_m(\mathscr{RT}(2m)_\infty) \ge C > 0.$$

Proof Let m, $\lambda_1, \dots, \lambda_m$, $\xi^1, \dots, \xi^m$ be given. We consider the subspace $\Psi \subset \mathscr{T}(m)$:

$$\Psi := \{ t \in \mathscr{T}(m) : t(\xi^j) = 0,\ j = 1, \dots, m \}.$$

Clearly, $\dim \Psi \ge m$. By Theorem 1.2.4 we can find a $t \in \Psi$ such that $\|t\|_\infty = 1$ and $\|t\|_2 \ge C > 0$. Then, for $f = |t|^2 \in \mathscr{RT}(2m)_\infty$, we have

$$\Lambda_m(f, \xi) = 0, \qquad (2\pi)^{-1} \int_0^{2\pi} f(x) dx = \|t\|_2^2 \ge C^2.$$

The lemma is now proved. □

2.4.2 Sampling

We proceed to the question of the approximate recovery of functions from values given at m points. Let a function f be 2π-periodic and continuous. For fixed m, $\psi_1(x),\dots,\psi_m(x)$, $\xi^1,\dots,\xi^m$ we define the linear operator

$$\Psi_m(\xi)(f) := \Psi_m(f,\xi) := \sum_{j=1}^{m} f(\xi^j)\psi_j(x) \qquad (2.4.7)$$

and for a class of functions F we define the quantities

$$\Psi_m(F,\xi)_p := \sup_{f\in F}\|\Psi_m(f,\xi)-f\|_p,$$

$$\rho_m(F)_p := \inf_{\psi_1,\dots,\psi_m;\xi^1,\dots,\xi^m} \Psi_m(F,\xi).$$

We consider three special recovery operators:

$$I_m(f) := (2m+1)^{-1}\sum_{l=0}^{2m} f(x^l)\mathscr{D}_m(x-x^l), \qquad x^l := 2\pi l/(2m+1);$$

$$R_m(f) := (4m)^{-1}\sum_{l=1}^{4m} f(x(l))\mathscr{V}_m(x-x(l)), \qquad x(l) := \pi l/(2m);$$

$$B_m(f) := (2m+1)^{-1}\sum_{l=0}^{2m} f(x^l)\mathscr{V}_{m,n}(x-x^l), \qquad n=[3m/2].$$

Clearly, $I_m = S_m B_m$. The operator I_m maps a continuous function $f(x)$ to the trigonometric polynomial $I_m(f)\in T(m)$ in such a way that

$$\big(I_m(f)\big)(x^l) = f(x^l). \qquad l=0,\dots,2m.$$

Therefore, the operator I_m is called an interpolation operator, which is why we use the letter I for this operator. The letter R used for the operator R_m refers to "recovery".

We first study the properties of the operators R_m and B_m. The properties of these operators and their proofs are similar.

Lemma 2.4.3 *Let $1\le p\le\infty$ and $n\ge m$. The following relations hold:*

$$\|R_mV_n\|_{p\to p} \le C(n/m)^{1/p},$$

$$\|B_mV_n\|_{p\to p} \le C(n/m)^{1/p},$$

$$\|I_mV_n\|_{p\to p} \le C(p)(n/m)^{1/p}, \qquad 1<p<\infty.$$

Proof We have

$$R_mV_n(f) = (4m)^{-1}\sum_{l=1}^{4m}\mathscr{V}_m(x-x(l))(2\pi)^{-1}\int_0^{2\pi}\mathscr{V}_n(x(l)-y)f(y)dy.$$

Consequently,

$$\|R_m V_n\|_{1\to 1} \le \max_y \left\| (4m)^{-1} \sum_{l=1}^{4m} \mathscr{V}_m(x - x(l)) \mathscr{V}_n(x(l) - y) \right\|_1$$
$$\le 3 \max_y (4m)^{-1} \sum_{l=1}^{4m} \left| \mathscr{V}_n(x(l) - y) \right|. \tag{2.4.8}$$

Using the estimate (2.4.4) we obtain from (2.4.8)

$$\|R_m V_n\|_{1\to 1} \le Cn/m. \tag{2.4.9}$$

Further, using property (2) of the de la Vallée Poussin kernels, we find

$$\|R_m V_n\|_{\infty\to\infty} \le \|R_m\|_{\infty\to\infty} \|V_n\|_{\infty\to\infty}$$
$$\le 3 \left\| (4m)^{-1} \sum_{l=1}^{4m} \left| \mathscr{V}_m(x - x(l)) \right| \right\|_\infty \le C. \tag{2.4.10}$$

From relations (2.4.9) and (2.4.10) we get by the Riesz–Thorin theorem A.3.2, the required estimate for $\|R_m V_n\|_{p\to p}$, $1 \le p \le \infty$. The norm of the operator $B_m V_n$ can be estimated in just the same way. The estimate of the norm of the operator $I_m V_n$ follows from Theorem 1.2.1 and from the estimate of the norm of $B_m V_n$ since $I_m V_n = S_m B_m V_n$.

The lemma is proved. □

Theorem 2.4.4 *Let $1 \le q, p \le \infty$ and $r > 1/q$. Then*

$$\rho_{4m}(W^r_{q,\alpha})_p \asymp \rho_{4m}(H^r_q)_p \asymp R_m(W^r_{q,\alpha})_p \asymp R_m(H^r_q)_p \asymp m^{-r+(1/q-1/p)_+}.$$

In the case $1 < p < \infty$ the statement of the theorem is valid for the operator I_m instead of the operator R_m.

Proof Using Theorem 1.4.8, we conclude that it suffices to prove the upper estimates in the theorem for $q = p$. The operator R_m has the following simple property:

$$R_m(t) = t, \qquad t \in \mathscr{T}(m). \tag{2.4.11}$$

Consequently, if s_m is such that $2^{s_m} \le m < 2^{s_m+1}$ then since, for $f \in H^r_p$, $r > 1/p$ the series $\sum_{s=0}^\infty A_s(f)$ converges to f uniformly, we have

$$\left\| f - R_m(f) \right\|_p \le \sum_{s > s_m} \left\| A_s(f) - R_m(A_s(f)) \right\|_p$$
$$\le \sum_{s > s_m} \left\| A_s(f) \right\|_p + \sum_{s > s_m} \left\| R_m(A_s(f)) \right\|_p.$$

Using Theorem 1.4.5 and Lemma 2.4.3, we continue the estimate as

$$\ll m^{-r} + \sum_{s > s_m} (2^s/m)^{1/p} 2^{-rs} \ll m^{-r}.$$

This proves the upper estimate for the operator R_m.

In the same way, using the property

$$I_m(t) = t, \qquad t \in \mathcal{T}(m), \tag{2.4.12}$$

we can prove for $1 < p < \infty$ the relation

$$I_m(H^r_q)_p \ll m^{-r}, \qquad r > 1/p.$$

Let us now prove the lower estimates.

Lemma 2.4.5 *Let $1 \le q, p \le \infty$; then*

$$\rho_m\big(\mathcal{T}(2m)_q\big)_p \ge Cm^{(1/q-1/p)_+}.$$

Proof Let t be the polynomial from $\mathcal{T}(m)_\infty$ which was found for fixed m and $\xi^1, \dots, \xi^m$ in the proof of Lemma 2.4.2. Then $\Psi_m(t, \xi) = 0$ and by (1.2.27), $\|t\|_1 \ge C > 0$. This implies the lemma for $1 \le p \le q \le \infty$.

Let $1 \le q < p \le \infty$ and t be the same as above and $\|t\|_\infty = \big|t(x^*)\big|$. We will consider $f(x) = t(x)\mathcal{K}_m(x - x^*)$. Then $f \in \mathcal{T}(2m)$, $\Psi_m(f, \xi) = 0$ and

$$\|f\|_q \le \|t\|_\infty \|\mathcal{K}_m\|_q \le Cm^{1-1/q}. \tag{2.4.13}$$

Further, $\big|f(x^*)\big| = m + 1$; consequently, by the Nikol'skii inequality,

$$\|f\|_p \ge Cm^{1-1/p}. \tag{2.4.14}$$

Relations (2.4.13) and (2.4.14) imply the required lower estimate. □

The lower bounds in Theorem 2.4.4 follow from Lemma 2.4.5 and the Bernstein inequality. □

We note that Theorem 2.4.4 shows that the operator R_m gives the order of the best approximation by trigonometric polynomials of order $2m - 1$ in the classes $W^r_{q,\alpha}$ and H^r_q in the L_p-metric for all $1 \le q, p \le \infty$. A similar statement is valid for the operator I_m in the case $1 < q, p < \infty$.

Remark 2.4.6 Let us estimate the approximation of a continuous function f by means of polynomials $R_m(f)$ and $I_m(f)$ in the uniform metric. By the properties (2.4.11) and (2.4.12) we have

$$\big\|f - R_m(f)\big\|_\infty \le \big(1 + \|R_m\|_{\infty\to\infty}\big) E_m(f)_\infty, \tag{2.4.15}$$

$$\big\|f - I_m(f)\big\|_\infty \le \big(1 + \|I_m\|_{\infty\to\infty}\big) E_m(f)_\infty. \tag{2.4.16}$$

From (2.4.15), by the relation $\|R_m\|_{\infty\to\infty} \le C$ it follows that the recovering operator R_m, which uses for the construction of an approximating polynomial of order

$2m-1$ only the values of a function f at $4m$ points, gives an approximation of the order $E_m(f)_\infty$.

Further,

$$\|I_m\|_{\infty\to\infty} = \max_x (2m+1)^{-1} \sum_{l=0}^{2m} \left|\mathscr{D}_m(x-x^l)\right|.$$

Using the estimate (1.2.2), we get

$$\|I_m\|_{\infty\to\infty} \le C\ln(m+1), \qquad m=1,2,\ldots \tag{2.4.17}$$

From (2.4.16) and (2.4.17) follows an estimate analogous to the Lebesgue inequality:

$$\left\|f - I_m(f)\right\|_\infty \le C\big(\ln(m+1)\big)E_m(f)_\infty.$$

Remark 2.4.7 The recovering operator, defined by relation (2.4.7), is a linear operator with rank $\Psi_m(\xi) \le m$. A comparison of Theorems 2.4.4 and Theorems 2.1.1, 2.1.2 shows that the approximating method R_m is optimal, in the sense of order, from the point of view of the Kolmogorov widths in the cases $1 \le p \le q \le \infty$ and $1 \le q \le p \le 2$, and from the point of view of linear widths for all q, p such that either $q \ge 2$ or $p \le 2$.

2.5 Historical Remarks

The Kolmogorov width was introduced in Kolmogorov (1936), the linear width in Tikhomirov (1960b), and the orthowidth in Temlyakov (1982a). Other widths can be found in Tikhomirov (1976). The results connected with Theorem 2.1.1 were obtained by a number of authors. Exact values of widths have been obtained in some cases but we shall not discuss these results. In the papers cited below the estimates of widths for the W classes were obtained. They imply the corresponding results for the H classes. The first result was obtained by Kolmogorov (1936) (for $q=p=2$; in this case the exact values of the widths were obtained). Rudin (1959) investigated the case $r=1$, $q=1$, $p=2$. Stechkin (1954) generalized Rudin's result to all r and investigated the case $q=p=\infty$. For $1 \le q = p < \infty$ the orders of the widths were obtained by Babadzhanov and Tikhomirov (1967). Makovoz (1972) investigated the case $1 \le p < q \le \infty$. Ismagilov (1974) found the orders of the widths for $1 \le q < p \le 2$ and proved the estimate

$$d_m(W_1^2, L_\infty) \le C(\varepsilon)m^{-6/5+\varepsilon}, \qquad \varepsilon > 0.$$

In particular, this estimate shows that in the case $q=1$, $p=\infty$ the subspace $\mathscr{T}(n)$ of trigonometric polynomials is not optimal from the point of view of the Kolmogorov widths. This result was the first of its kind. The orders of the widths in the case

$q = 1$, $p > 2$, $r \geq 2$ were obtained by Gluskin (1974). For $1 < q < p$, $p > 2$ the orders of the widths were obtained by Kashin (1977). We note that the idea of the proof of the lower estimate for d_m in the case $1 \leq p \leq q \leq \infty$ treated in §2.1.1 is due to Kashin (1980). Theorem 2.1.11 with exponent $3/2$ instead of $1/2$ in the logarithmic factor was proved by Kashin (1977). Later, Garnaev and Gluskin (1984) and Gluskin (1983) improved Kashin's bound to that in Theorem 2.1.11.

The proof of Theorem 2.1.11, which implies the upper estimate for $1 \leq q < p \leq \infty$, $p > 2$ is due to Makovoz. Theorem 2.1.2 in the cases either $p \leq 2$ or $q \geq 2$ was obtained by Ismagilov (1974). His paper Ismagilov (1974) contains Theorem 2.1.8. The first to apply number-theoretical methods to widths was Ismagilov (1974). After that Maiorov used Gaussian sums in Maiorov (1986). Theorems 2.1.3 and 2.1.5 were obtained by Temlyakov (1982a, 1985e). In connection with Theorem 2.1.14 we note that Gluskin (1974) and Maiorov (1975) were the first to apply estimates of the widths of balls B_q^n in ℓ_p^n to estimating the widths of Sobolev classes.

One can find results about properties of the modulus of continuity in Dzyadyk (1977). Theorem 2.2.2 in the case $a = 1$ and $p = \infty$ was proved by Jackson (1911) and in the case $p = \infty$ and arbitrary a by Stechkin (1951). In the case $1 \leq p \leq \infty$, $a = 1$, Theorem 2.2.2 was obtained by Quade (1937). Corollaries 2.2.3–2.2.7 of Theorem 2.2.2 and other results about the convergence of trigonometric Fourier series are contained in Bary (1961) and Zygmund (1959). Theorem 2.2.8 is a classical result due to Dirichlet and Jordan. Theorem 2.2.10 is a simple and well-known result in the theory of trigonometric series. Theorem 2.2.12 was proved by Bernstein (1952), vol. 1, pp. 217–223, and Theorem 2.2.11 is actually contained in the same paper. Theorem 2.2.16 was obtained by Temlyakov (1985d). Theorem 2.2.17 is due to Hardy and Littlewood (see Zygmund, 1959, vol. 2, Chapter 12). Theorem 2.2.18 was obtained by Stechkin (1955).

Many papers are devoted to investigating the approximation of infinitely differentiable, analytic, and harmonic functions. One can find a number of such results in Akhiezer (1965), Dzyadyk (1977), Tikhomirov (1976), and Timan (1960). Theorem 2.3.1 was obtained by Kushpel' (1989) for $1 < q, p < \infty$. Theorem 2.3.5 for $1 < q, p < \infty$ was proved by Stepanets (1987, p. 227). The upper estimates in Theorem 2.3.6 were obtained by Kushpel' (1989, 1990); the lower estimates were proved in Temlyakov (1990a) and announced in Kushpel' (1990).

In connection with Theorem 2.4.1 we note that Nikol'skii (1979) contains exact values for the quantities $\kappa_m(W^r_{p,r})$ for natural r. The conclusion of Theorem 2.4.4 for $q = \infty$ and r a natural number is well known in approximation theory. The upper estimates in Theorem 2.4.4 were obtained in Temlyakov (1985b). For the lower estimates in Theorem 2.4.4 see Temlyakov (1993b).

3
Approximation of Functions from Anisotropic Sobolev and Nikol'skii Classes

3.1 Introduction

The results of Chapter 2 showed that for the approximation of periodic functions of a single variable we can use the trigonometric system $\{e^{ikx}\}$ in the natural order as our instrument of approximation; that is, we can use this system of functions for approximation by forming polynomials. It seems reasonable, therefore, to study approximation by means of polynomials in the trigonometric system $\{e^{i(\mathbf{k},\mathbf{x})}\}$ to approximate periodic functions of several variables. However, in contrast with the univariate case we cannot use any previously determined rule to order the system $\{e^{i(\mathbf{k},\mathbf{x})}\}$ in the multivariate case. As a consequence of this, various partial sums have been studied in the theory of multivariate Fourier series: these with harmonics from squares $(d = 2)$ or cubes $(d = 3)$ or, more generally, from rectangles or parallelepipeds or from disks or balls.

In this chapter we follow the general strategy formulated at the end of §1.1 to choose the kind of trigonometric polynomials, which is best for approximating functions from a given class. The first question is about the classes of functions which we are going to approximate. For the sake of simplicity we discuss this question for functions of two variables. One example of a class of functions of two variables was given in §1.1 in connection with the discussion of the Fredholm result on the eigenvalues of an integral operator. This class is the class of continuous functions $f(x_1,x_2)$ such that for any fixed x_1 the function $f(x_1,x_2)$ as a function of x_2 belongs to the class $\mathrm{Lip}\,r$. In this chapter we consider the Sobolev classes $W^{\mathbf{r}}_{\mathbf{q},\alpha}$ and the Nikol'skii classes $H^{\mathbf{r}}_{\mathbf{q}}$. These classes are defined by a smoothness condition for each variable. In terms of these classes the example considered above can be denoted by $H^{(0,r)}_{(\infty,\infty)}$.

The orders of the Kolmogorov widths, linear widths, and orthowidths of such classes will be obtained. Results of §3.5 show that, for fixed $\mathbf{r}$, for the approximation of the classes $W^{\mathbf{r}}_{\mathbf{q},\alpha}$ and $H^{\mathbf{r}}_{\mathbf{q}}$ the sets $\mathcal{T}^{\mathbf{r}}(n)$ of trigonometric polynomials with harmonics from parallelepipeds whose shape depends on $\mathbf{r}$ and whose size depends on n play the same role as the sets $\mathscr{T}(2^n)$ play for the classes $W^r_{q,\alpha}$, H^r_q in the univariate case.

We emphasize that the optimal subspaces $\mathscr{T}^{\mathbf{r}}(n)$ depend on the vector $\mathbf{r}$. In §5.4 we discuss the problem of finding universal subspaces, i.e., subspaces that do not depend on $\mathbf{r}$ but approximate well the classes $W^{\mathbf{r}}_{\mathbf{q},\alpha}$, $H^{\mathbf{r}}_{\mathbf{q}}$ for various $\mathbf{r}$.

3.2 Trigonometric Polynomials

In this section we define the analogs of the Dirichlet, Fejér, and de la Vallée Poussin kernels and Rudin–Shapiro polynomials for d-dimensional parallelepipeds

$$\Pi(\mathbf{N},d) := \left\{\mathbf{a} \in \mathbb{R}^d : |a_j| \le N_j,\ j = 1,\dots,d\right\},$$

where the N_j are nonnegative integers.

We will formulate properties of these multivariate polynomials which easily follow from the corresponding properties of the univariate polynomials. Then we prove some lower estimates for the volumes of sets of Fourier coefficients of bounded polynomials with harmonics from $\Pi(\mathbf{N},d)$. These estimates allow us to prove the existence of polynomials with properties similar to some properties of the Rudin–Shapiro polynomials, in each subspace $\Psi \subset \mathscr{T}(\mathbf{N},d)$ with sufficiently large dimension. Here $\mathscr{T}(\mathbf{N},d)$ is the set of complex trigonometric polynomials with harmonics from $\Pi(\mathbf{N},d)$. The set of real trigonometric polynomials with harmonics from $\Pi(\mathbf{N},d)$ will be denoted by $\mathscr{RT}(\mathbf{N},d)$.

3.2.1 The Dirichlet Kernels

The multivariate Dirichlet kernels, defined as

$$\mathscr{D}_{\mathbf{N}}(\mathbf{x}) := \prod_{j=1}^{d} \mathscr{D}_{N_j}(x_j), \qquad \mathbf{N} = (N_1,\dots,N_d),$$

have the following properties.

For any trigonometric polynomial $t \in \mathscr{T}(\mathbf{N},d)$,

$$t * \mathscr{D}_{\mathbf{N}} := (2\pi)^{-d} \int_{\mathbb{T}^d} \mathscr{D}_N(\mathbf{x}-\mathbf{y})t(\mathbf{y})d\mathbf{y} = t,\ \mathbb{T}^d = [0,2\pi]^d.$$

For $1 < q \le \infty$,

$$\|\mathscr{D}_{\mathbf{N}}\|_q \asymp \nu(\overline{\mathbf{N}})^{1-1/q}, \tag{3.2.1}$$

where $\overline{N}_j := \max(N_j, 1)$, $\nu(\overline{\mathbf{N}}) = \prod_{j=1}^d \overline{N}_j$, and

$$\|\mathscr{D}_{\mathbf{N}}\|_1 \asymp \prod_{j=1}^{d} \ln(N_j + 2). \tag{3.2.2}$$

We write

$$P(\mathbf{N}) := \{\mathbf{n} = (n_1, \dots, n_d) \text{ for nonnegative integers } n_j, \\ 0 \le n_j \le 2N_j,\ j = 1, \dots, d\},$$

and set

$$\mathbf{x}^{\mathbf{n}} := \left(\frac{2\pi n_1}{2N_1 + 1}, \dots, \frac{2\pi n_d}{2N_d + 1}\right), \qquad \mathbf{n} \in P(\mathbf{N}).$$

Then for any $t \in \mathscr{T}(\mathbf{N}, d)$,

$$t(\mathbf{x}) = \vartheta(\mathbf{N})^{-1} \sum_{\mathbf{n} \in P(\mathbf{N})} t(\mathbf{x}^{\mathbf{n}}) \mathscr{D}_{\mathbf{N}}(\mathbf{x} - \mathbf{x}^{\mathbf{n}}), \tag{3.2.3}$$

where $\vartheta(\mathbf{N}) := \prod_{j=1}^d (2N_j + 1)$ and, for any $t, u \in \mathscr{T}(\mathbf{N}, d)$,

$$\langle t, u \rangle = \vartheta(\mathbf{N})^{-1} \sum_{\mathbf{n} \in P(\mathbf{N})} t(\mathbf{x}^{\mathbf{n}}) \overline{u}(\mathbf{x}^{\mathbf{n}}), \tag{3.2.4}$$

$$\|t\|_2^2 = \vartheta(\mathbf{N})^{-1} \sum_{\mathbf{n} \in P(\mathbf{N})} \left|t(\mathbf{x}^{\mathbf{n}})\right|^2. \tag{3.2.5}$$

3.2.2 The Fejér Kernels

The multivariate Fejér kernels, defined as

$$\mathscr{K}_{\mathbf{N}}(\mathbf{x}) := \prod_{j=1}^{d} \mathscr{K}_{N_j}(x_j), \qquad \mathbf{N} = (N_1, \dots, N_d),$$

are nonnegative trigonometric polynomials from $\mathscr{T}(\mathbf{N}, d)$, which have the following properties:

$$\|\mathscr{K}_{\mathbf{N}}\|_1 = 1, \tag{3.2.6}$$

$$\|\mathscr{K}_{\mathbf{N}}\|_q \asymp \vartheta(\mathbf{N})^{1-1/q}, \qquad 1 \le q \le \infty, \tag{3.2.7}$$

$$\|\mathscr{K}_{\mathbf{N}}\|_{\mathbf{q}} \asymp \prod_{j=1}^{d} \left(\max(1, N_j)\right)^{1-1/q_j}, \qquad \mathbf{1} \le \mathbf{q} \le \infty. \tag{3.2.8}$$

3.2.3 The de la Vallée Poussin Kernels

The multivariate de la Vallée Poussin kernels, defined as

$$\mathcal{V}_{\mathbf{N}}(\mathbf{x}) := \prod_{j=1}^{d} \mathcal{V}_{N_j}(x_j), \qquad \mathbf{N} = (N_1, \dots, N_d),$$

have the following properties:

$$\|\mathcal{V}_{\mathbf{N}}\|_1 \le 3^d, \tag{3.2.9}$$

$$\|\mathcal{V}_{\mathbf{N}}\|_q \asymp \vartheta(\mathbf{N})^{1-1/q}, \qquad 1 \le q \le \infty, \tag{3.2.10}$$

$$\|\mathcal{V}_{\mathbf{N}}\|_{\mathbf{q}} \asymp \prod_{j=1}^{d} \left(\max(1, N_j)\right)^{1-1/q_j}, \qquad \mathbf{1} \le \mathbf{q} \le \infty. \tag{3.2.11}$$

For any $t \in \mathcal{T}(\mathbf{N}, d)$,

$$V_{\mathbf{N}}(t) := t * \mathcal{V}_{\mathbf{N}} = t.$$

We denote

$$P'(\mathbf{N}) := \{\mathbf{n} = (n_1, \dots, n_d) \text{ for natural numbers } n_j, \\ 1 \le n_j \le 4N_j, \ j = 1, \dots, d\}$$

and set

$$\mathbf{x}(\mathbf{n}) := \left(\frac{\pi n_1}{2N_1}, \dots, \frac{\pi n_d}{2N_d}\right), \qquad \mathbf{n} \in P'(\mathbf{N}).$$

In the case $N_j = 0$ we assume $x_j(\mathbf{n}) = 0$. Then for any $t \in \mathcal{T}(\mathbf{N}, d)$ we have the representation

$$t(\mathbf{x}) = \nu(4\bar{\mathbf{N}})^{-1} \sum_{\mathbf{n} \in P'(\mathbf{N})} t(\mathbf{x}(\mathbf{n})) \mathcal{V}_{\mathbf{N}}(\mathbf{x} - \mathbf{x}(\mathbf{n})). \tag{3.2.12}$$

Relation (3.2.9) implies that

$$\|V_{\mathbf{N}}\|_{\mathbf{p} \to \mathbf{p}} \le 3^d, \qquad \mathbf{1} \le \mathbf{p} \le \infty. \tag{3.2.13}$$

3.2.4 The Rudin–Shapiro Polynomials

The multivariate Rudin–Shapiro polynomials are defined as follows:

$$R_{\mathbf{N}}(\mathbf{x}) := \prod_{j=1}^{d} R_{N_j}(x_j), \qquad \mathbf{N} = (N_1, \dots, N_d).$$

They have the following properties:

$$\begin{aligned} &R_{\mathbf{N}} \in \mathscr{T}(\mathbf{N},d),\\ &\hat{R}_{\mathbf{N}}(\mathbf{k}) = \pm 1, \qquad |\mathbf{k}| < \mathbf{N},\\ &\|R_{\mathbf{N}}\|_\infty \le C(d)\vartheta(\mathbf{N})^{1/2}. \end{aligned} \tag{3.2.14}$$

3.2.5 The Fejér Kernels for Dyadic Blocks

Let us define the polynomials $\mathscr{A}_{\mathbf{s}}(\mathbf{x})$ for $\mathbf{s} = (s_1,\dots,s_d)$, where the s_j are nonnegative integers, as follows:

$$\mathscr{A}_{\mathbf{s}}(\mathbf{x}) := \prod_{j=1}^{d} \mathscr{A}_{s_j}(x_j);$$

the $\mathscr{A}_{s_j}(x_j)$ were defined in §1.3.1. Then, by (1.3.2),

$$\left\|\mathscr{A}_{\mathbf{s}}(\mathbf{x})\right\|_1 \le 6^d \tag{3.2.15}$$

and, consequently, we have, for the operator $A_{\mathbf{s}}$ which is the convolution with the kernel $\mathscr{A}_{\mathbf{s}}(\mathbf{x})$, the inequality

$$\|A_{\mathbf{s}}\|_{\mathbf{p}\to\mathbf{p}} \le 6^d, \qquad \mathbf{1} \le \mathbf{p} \le \infty. \tag{3.2.16}$$

We note that in the case $\mathbf{s} \ge \mathbf{2}$, for any $t \in \mathscr{T}(2^{\mathbf{s}-\mathbf{2}},d)$,

$$A_{\mathbf{s}}(t) = 0.$$

3.2.6 Volume Estimates

In this subsection we prove the following assertion.

Theorem 3.2.1 *Let* $\varepsilon \in (0,1]$ *and let a subspace* $\Psi \subset \mathscr{T}(\mathbf{m},d)$ *be such that* $\dim\Psi \ge \varepsilon\vartheta(\mathbf{m})$. *Then there is a* $t \in \Psi$ *such that*

$$\|t\|_\infty = 1, \qquad \|t\|_2 \ge C(\varepsilon,d) > 0.$$

The proof of this theorem is based on the lower estimates of the volumes of the sets of Fourier coefficients for bounded trigonometric polynomials from $\mathscr{T}(\mathbf{m},d)$. We will consider arrays of complex numbers $\bar{y} := \{y^{\mathbf{l}}, \mathbf{l} \in P(\mathbf{m})\}$. Then, since, $y^{\mathbf{l}} = \operatorname{Re} y^{\mathbf{l}} + i \operatorname{Im} y^{\mathbf{l}}$, we can consider $\bar{y}$ as an element of the space $\mathbb{R}^{2\vartheta(\mathbf{m})}$. We define in the space $\mathbb{R}^{2\vartheta(\mathbf{m})}$ the following linear transformation: we map each $\bar{y}$ to the trigonometric polynomial

$$t(\mathbf{x},\bar{y}) := B\bar{y} := \vartheta(\mathbf{m})^{-1} \sum_{\mathbf{l}\in P(\mathbf{m})} y^{\mathbf{l}} \mathscr{D}_{\mathbf{m}}(\mathbf{x}-\mathbf{x}^{\mathbf{l}}),$$

and we map each trigonometric polynomial $t \in \mathscr{T}(\mathbf{m},d)$ to the element $At := \{(\mathrm{Re}\hat{t}(\mathbf{k}), \mathrm{Im}\hat{t}(\mathbf{k}))\}_{|\mathbf{k}|\le\mathbf{m}}$ of the space $\mathbb{R}^{2\vartheta(\mathbf{m})}$. Then

$$\big\|t(\cdot,\bar{y})\big\|_2^2 = \vartheta(\mathbf{m})^{-1} \sum_{\mathbf{l}\in P(\mathbf{m})} |y^{\mathbf{l}}|^2 = \vartheta(\mathbf{m})^{-1}\|\bar{y}\|_2^2$$

and

$$\|t\|_2^2 = \sum_{|\mathbf{k}|\le\mathbf{m}} \big|\hat{t}(\mathbf{k})\big|^2 = \|At\|_2^2. \tag{3.2.17}$$

Thus the operator $AB\vartheta(\mathbf{m})^{1/2}$ gives an orthogonal transformation of $\mathbb{R}^{2\vartheta(\mathbf{m})}$.

We prove the following assertion. For a set $S \subset \mathbb{R}^N$ denote by $\mathrm{vol}(S) := \mathrm{vol}_N(S)$ its volume (Lebesgue measure) in $\mathbb{R}^N$.

Lemma 3.2.2 *Let*

$$S_\infty(\mathbf{m}) := \big\{\bar{y} \in \mathbb{R}^{2\vartheta(\mathbf{m})} : \big\|t(\cdot,\bar{y})\big\|_\infty \le 1\big\}.$$

Then we have the following estimate for the volume of this set in the space $\mathbb{R}^{2\vartheta(\mathbf{m})}$*: there is a* $C(d) > 0$ *such that for all* $\mathbf{m}$*,*

$$\mathrm{vol}\big(S_\infty(\mathbf{m})\big) \ge C(d)^{-\vartheta(\mathbf{m})}.$$

Proof Let $\mathscr{K}_n(t)$ be the Fejér kernel of order n, that is,

$$\mathscr{K}_n(t) = 1 + 2\sum_{k=1}^{n}\left(1 - \frac{k}{n+1}\right)\cos kt$$

and

$$\mathscr{K}_{\mathbf{m}}(\mathbf{x}) := \prod_{j=1}^{d} \mathscr{K}_{m_j}(x_j).$$

We consider the set $G := \{\bar{y} = K\bar{b}\}$, where K is a linear operator defined as follows:

$$y^{\mathbf{l}} = \vartheta(\mathbf{m})^{-1} \sum_{\mathbf{k}\in P(\mathbf{m})} b^{\mathbf{k}} \mathscr{K}_{\mathbf{m}}(\mathbf{x}^{\mathbf{l}} - \mathbf{x}^{\mathbf{k}}), \qquad \mathbf{l} \in P(\mathbf{m}).$$

Here $\bar{b} \in \mathbb{R}^{2\vartheta(\mathbf{m})}$ is such that $|\mathrm{Re}\, b^{\mathbf{k}}|$, $|\mathrm{Im}\, b^{\mathbf{k}}| \le 1$, $\mathbf{k} \in P(\mathbf{m})$. We use the equality

$$(2\pi)^{-d} \int_{\mathbb{T}^d} u(\mathbf{x})w(\mathbf{x})\,d\mathbf{x} = \vartheta(\mathbf{m})^{-1} \sum_{\mathbf{k}\in P(\mathbf{m})} u(\mathbf{x}^{\mathbf{k}})w(\mathbf{x}^{\mathbf{k}}) \tag{3.2.18}$$

which is valid for any pair of trigonometric polynomials $u, w \in \mathscr{T}(\mathbf{m},d)$. Then it is not difficult to see that, for all μ such that $|\mu_j| \le m_j$, $j = 1,\dots,d$, we have

$$\vartheta(\mathbf{m})^{-1} \sum_{\mathbf{k}\in P(\mathbf{m})} e^{i(\mu,\mathbf{x}^{\mathbf{k}})} \mathscr{K}_{\mathbf{m}}(\mathbf{x}^{\mathbf{l}} - \mathbf{x}^{\mathbf{k}}) = \hat{\mathscr{K}}_{\mathbf{m}}(\mu) e^{i(\mu,\mathbf{x}^{l})}. \tag{3.2.19}$$

From relation (3.2.19) it follows that the vectors

$$\varepsilon_\mu := \{(\mathrm{Re}\, e^{i(\mu,\mathbf{x}^\mathbf{k})},\ \mathrm{Im}\, e^{i(\mu,\mathbf{x}^\mathbf{k})})\}_{\mathbf{k}\in P(\mathbf{m})} \in R^{2\vartheta(\mathbf{m})},$$
$$\eta_\mu := \{(\mathrm{Re}\, ie^{i(\mu,\mathbf{x}^\mathbf{k})},\ \mathrm{Im}\, ie^{i(\mu,\mathbf{x}^\mathbf{k})})\}_{\mathbf{k}\in P(\mathbf{m})} \in R^{2\vartheta(\mathbf{m})}$$

are the eigenvectors of the operator K corresponding to the eigenvalues $\hat{\mathscr{K}}_\mathbf{m}(\mu)$, $|\mu| \le \mathbf{m}$. It is not difficult to verify that the vectors ε_μ, η_μ, $|\mu| \le \mathbf{m}$ make up a set of $2\vartheta(\mathbf{m})$ orthogonal vectors from $R^{2\vartheta(\mathbf{m})}$. Consequently, the operator K maps the unit cube of the space $R^{2\vartheta(\mathbf{m})}$ to the set G with volume

$$\mathrm{vol}(G) = 2^{2\vartheta(\mathbf{m})} \prod_{|\mu|\le \mathbf{m}} \hat{\mathscr{K}}_\mathbf{m}(\mu)^2 \ge C_1(d)^{-\vartheta(\mathbf{m})}, \qquad C_1(d) > 0. \tag{3.2.20}$$

Further, let $\bar{y} \in G$; then

$$\begin{aligned} t(\mathbf{x},\bar{y}) &= \vartheta(\mathbf{m})^{-1} \sum_{\mathbf{l}\in P(\mathbf{m})} y^\mathbf{l} \mathscr{D}_\mathbf{m}(\mathbf{x}-\mathbf{x}^\mathbf{l}) \\ &= \vartheta(\mathbf{m})^{-1} \sum_{\mathbf{k}\in P(\mathbf{m})} b^\mathbf{k} \vartheta(\mathbf{m})^{-1} \sum_{\mathbf{l}\in P(\mathbf{m})} \mathscr{D}_\mathbf{m}(\mathbf{x}-\mathbf{x}^\mathbf{l}) \mathscr{K}_\mathbf{m}(\mathbf{x}^\mathbf{l}-\mathbf{x}^\mathbf{k}) \\ &= \vartheta(\mathbf{m})^{-1} \sum_{\mathbf{k}\in P(\mathbf{m})} b^\mathbf{k} \mathscr{K}_\mathbf{m}(\mathbf{x}-\mathbf{x}^\mathbf{k}). \end{aligned} \tag{3.2.21}$$

From the condition $|b^\mathbf{k}| \le 2^{1/2}$, from the estimate

$$\mathscr{K}_n(t) \le C \min\big(n, (nt^2)^{-1}\big),$$

and from representation (3.2.21) it follows that for some $C_2(d)$ we have

$$\|t(\cdot,\bar{y})\|_\infty \le C_2(d). \tag{3.2.22}$$

The lemma now follows from relations (3.2.20) and (3.2.22). □

This lemma and the property of orthogonality of the operator $AB\vartheta(\mathbf{m})^{1/2}$ imply the following statement.

Lemma 3.2.3 *Let*

$$A_\infty(\mathbf{m}) := \{At \in \mathbb{R}^{2\vartheta(\mathbf{m})} : t \in \mathscr{T}(\mathbf{m},d),\ \|t\|_\infty \le 1\}.$$

Then the following estimate holds for the volume of this set in the space $\mathbb{R}^{2\vartheta(\mathbf{m})}$*: there is a* $C(d) > 0$ *such that, for all* $\mathbf{m}$*,*

$$\mathrm{vol}\big(A_\infty(\mathbf{m})\big) \ge \vartheta(\mathbf{m})^{-\vartheta(\mathbf{m})} C(d)^{-\vartheta(\mathbf{m})}.$$

Proof of Theorem 3.2.1 Let $\varepsilon \in (0,1]$ and a subspace $\Psi \in \mathscr{T}(\mathbf{m},d)$ with $\dim\Psi \ge \varepsilon\vartheta(\mathbf{m})$ be given. Let $U \subset \mathbb{R}^{2\vartheta(\mathbf{m})}$ be the image of Ψ under the transformation A defined above. Then $\dim U = 2\dim\Psi \ge 2\varepsilon\vartheta(\mathbf{m})$.

The theorem will follow from Lemma 3.2.3 and the following lemma.

Lemma 3.2.4 *Let a set B be contained in the unit ball B_2^N of the Euclidean space $\mathbb{R}^N$ and let it be a convex and centrally symmetric set with*

$$\mathrm{vol}(B) \geq \mathrm{vol}(B_2^N) C^{-N},$$

where $C > 0$ is a constant independent of N. Then, for any hyperplane $U\,(\mathbf{0} \in U)$ of dimension $n \geq \varepsilon N$, $\varepsilon > 0$, there is an element $b \in B \cap U$ for which

$$\|b\|_2 \geq C(\varepsilon) > 0.$$

Indeed, to prove this it suffices to use the relation

$$\mathrm{vol}(B_2^N) = \pi^{N/2}\Gamma(1+N/2)^{-1} \leq N^{-N/2}C^{-N}, \quad C > 0. \tag{3.2.23}$$

It remains to prove Lemma 3.2.4.

Proof of Lemma 3.2.4 This is based on the Corollary of the Brun theorem (see Theorem A.3.9). Let U be from Lemma 3.2.4 and let $U^\perp$ denote the orthogonal complement of U with respect to $\mathbb{R}^N$ and $B_2(U^\perp)$ be the Euclidean unit ball in $U^\perp$. For any $\mathbf{h} \in U^\perp$ we denote $S_\mathbf{h}(B) := B \cap L_\mathbf{h}$, where $L_\mathbf{h}$ is the linear manifold of the space $\mathbb{R}^N$ such that

$$L_\mathbf{h} = \{\mathbf{l} \in \mathbb{R}^N : \mathbf{l} = \mathbf{h} + \mathbf{u},\ \mathbf{u} \in U\},$$

that is, a hyperplane which is parallel to U and passes through $\mathbf{h}$. Since $B \subset B_2^N$ we have that, for $\mathbf{h} \notin B_2(U^\perp)$, $S_\mathbf{h}(B) = \varnothing$. Along with the set B we consider the set

$$B' := \left\{\mathbf{y} \in \mathbb{R}^N : \mathbf{y} = \mathbf{u} + \mathbf{h},\ \mathbf{u} \in S_\mathbf{0}(B),\ \mathbf{h} \in B_2(U^\perp)\right\}.$$

Then, by Theorem A.3.9, for any $\mathbf{h} \in B_2(U^\perp)$ we have

$$\mathrm{vol}_n(S_\mathbf{h}(B')) = \mathrm{vol}_n(S_\mathbf{0}(B)) \geq \mathrm{vol}_n(S_\mathbf{h}(B)).$$

This implies the estimate

$$\mathrm{vol}_N(B) \leq \mathrm{vol}_N(B'). \tag{3.2.24}$$

Further,

$$\mathrm{vol}_N(B') = \mathrm{vol}_n(S_\mathbf{0}(B))\,\mathrm{vol}_{N-n}(B_2(U^\perp)). \tag{3.2.25}$$

Let

$$a := \max_{\mathbf{u} \in S_0} \|\mathbf{u}\|_2.$$

Then from (3.2.23)–(3.2.25) we get

$$\mathrm{vol}_N(B) \leq a^n n^{-n/2}(N-n)^{-(N-n)/2}C^{-N}. \tag{3.2.26}$$

Owing to the boundedness on $(0,1)$ of the function $x^{-x}(1-x)^{x-1}$ we obtain from (3.2.26),

$$\operatorname{vol}(B) \le a^n N^{-N/2} C^{-N}. \tag{3.2.27}$$

From the hypotheses of Lemma 3.2.4 and the relations (3.2.23), (3.2.27) we get

$$a \ge C(\varepsilon) > 0,$$

which implies the lemma. □

3.3 The Bernstein–Nikol'skii Inequalities and Their Applications. A Generalization of the Marcinkiewicz Theorem

In this section we generalize the results of §1.3 to the multivariate case of trigonometric polynomials from $\mathscr{T}(\mathbf{N},d)$. In addition, by means of the Nikol'skii inequalities we prove inequalities which allow us, in particular, to estimate the L_p-norm of a function in terms of its Fourier coefficients.

3.3.1 The Bernstein Inequality

Let $\mathbf{r} = (r_1,\dots,r_d)$, $r_j \ge 0$, $j = 1,\dots,d$, $\alpha = (\alpha_1,\dots,\alpha_d)$, $\mathbf{N} = (N_1,\dots,N_d)$. We consider the polynomials

$$\mathscr{V}_{\mathbf{N}}^{\mathbf{r}}(\mathbf{x},\alpha) := \prod_{j=1}^{d} \mathscr{V}_{N_j}^{r_j}(x_j,\alpha_j), \tag{3.3.1}$$

where the $\mathscr{V}_{N_j}^{r_j}(x_j,\alpha_j)$ were defined in §1.3.

We define an operator $D_\alpha^{\mathbf{r}}$ on the set of trigonometric polynomials as follows. Let $f \in \mathscr{T}(\mathbf{N},d)$; then

$$D_\alpha^{\mathbf{r}} f := f^{(\mathbf{r})}(\mathbf{x},\alpha) := f(\mathbf{x}) * \mathscr{V}_{\mathbf{N}}^{\mathbf{r}}(\mathbf{x},\alpha),$$

and we call $D_\alpha^{\mathbf{r}} f$ the $(\mathbf{r},\alpha)$-derivative. In the case of identical components $r_j = r$, $j = 1,\dots,d$, we write the scalar r in place of the vector.

Theorem 3.3.1 *Let $\mathbf{r} \ge \mathbf{0}$ and $\alpha \in \mathbb{R}^d$ be such that for $r_j = 0$ we have $\alpha_j = 0$. Then for any $t \in \mathscr{T}(\mathbf{N},d)$, $\mathbf{N} > \mathbf{0}$, the inequality*

$$\left\|t^{(\mathbf{r})}(\cdot,\alpha)\right\|_{\mathbf{p}} \le C(\mathbf{r})\|t\|_{\mathbf{p}} \prod_{j=1}^{d} N_j^{r_j}, \qquad \mathbf{1} \le \mathbf{p} \le \infty,$$

holds.

Proof From the definition of $t^{(\mathbf{r})}(\mathbf{x}, \alpha)$, by the Young inequality for vector norms (see (A.1.16) in the Appendix) it follows that

$$\left\|t^{(\mathbf{r})}(\mathbf{x}, \alpha)\right\|_{\mathbf{p}} \leq \|t\|_{\mathbf{p}} \left\|\mathscr{V}_{\mathbf{N}}^{\mathbf{r}}(\mathbf{x}, \alpha)\right\|_{1}. \tag{3.3.2}$$

From (3.3.1), (1.3.3), (1.2.13), and (3.3.2) we obtain the theorem. □

Theorem 3.3.1 shows that for $\mathbf{r} > \mathbf{0}$ the additional factor in the Bernstein inequality depends on $\mathbf{r}$ but does not depend on α or $\mathbf{p}$.

As an example of the way in which this factor depends on α and p we consider the case $r = 0$, p is scalar. Then, in the same way as in the univariate case for $1 < p < \infty$, we have

$$\left\|t^{(0)}(\mathbf{x}, \alpha)\right\|_{p} \leq C(d, p)\|t\|_{p},$$

where $C(d, p)$ does not depend on α.

Let $p = 1$ or ∞. Since in the univariate case we have

$$t^{(0)}(x, \alpha) = \cos(\alpha\pi/2)\, t^{(0)}(x, 0) + \sin(\alpha\pi/2)\, t^{(0)}(x, 1),$$

in the multivariate case, denoting by $\chi(e)$ for $e \subset [1, d]$ the vector from $\mathbb{R}^d$ such that $(\chi(e))_j = 1$ for $j \in e$ and $(\chi(e))_j = 0$ for $j \notin e$, we get

$$t^{(0)}(\mathbf{x}, \alpha) = \sum_{e \subset [1,d]} t^{(0)}(\mathbf{x}, \chi(e)) \prod_{j \in e} \sin(\alpha_j\pi/2) \prod_{j \notin e} \cos(\alpha_j\pi/2). \tag{3.3.3}$$

Let $e(\alpha)$ denote the set of those j for which $\sin(\alpha_j\pi/2) \neq 0$. From the representation (3.3.3) and the results of §1.3 it is easy to see that for $p = 1, \infty$,

$$\sup_{t \in \mathscr{T}(\mathbf{N}, d)} \left\|t^{(0)}(\mathbf{x}, \alpha)\right\|_{p} / \|t\|_{p} \asymp \prod_{j \in e(\alpha)} \ln(N_j + 2).$$

3.3.2 The Nikol'skii Inequalities.

We now proceed to the Nikol'skii inequalities.

Theorem 3.3.2 *For any $t \in \mathscr{T}(\mathbf{N}, d)$, $\mathbf{N} > \mathbf{0}$ the following inequality holds ($\mathbf{1} \leq \mathbf{q} \leq \mathbf{p} \leq \infty$):*

$$\|t\|_{\mathbf{p}} \leq C(d)\|t\|_{\mathbf{q}} \prod_{j=1}^{d} N_j^{1/q_j - 1/p_j}.$$

Proof Let $t \in \mathscr{T}(\mathbf{N}, d)$; then

$$t = t * \mathscr{V}_{\mathbf{N}}$$

and, by the Young inequality for vector norms (A.1.16)

$$\|t\|_{\mathbf{p}} \le \|t\|_{\mathbf{q}} \|\mathscr{V}_{\mathbf{N}}\|_{\mathbf{a}} \tag{3.3.4}$$

where $1/\mathbf{a} = \mathbf{1} - 1/\mathbf{q} + 1/\mathbf{p}$. Applying relation (3.2.11) to estimate $\|\mathscr{V}_{\mathbf{N}}\|_{\mathbf{a}}$ from (3.3.4) we get

$$\|t\|_{\mathbf{p}} \le \|t\|_{\mathbf{q}} C(d) \prod_{j=1}^{d} N_j^{1/q_j - 1/p_j},$$

as required. □

We now present a generalization of Theorem 3.3.2. It is convenient for us to consider the case of real functions.

Theorem 3.3.3 *Let $X = X_1 \times \cdots \times X_d$ be the tensor product of d finite-dimensional subspaces of the space $C(\mathbb{T})$ of real continuous functions:*

$$X := \operatorname{span}\left(\prod_{i=1}^{d} f_i(x_i),\ f_i \in X_i,\ i = 1, \dots, d \right).$$

Assume that for each X_i, $i = 1, \dots, d$, the following Nikol'skii inequalities hold:

$$\|f\|_\infty \le M_i(q)^{1/q} \|f\|_q, \qquad 1 \le q < \infty, \qquad f \in X_i. \tag{3.3.5}$$

Then for each $f \in X$ and any vectors $\mathbf{p}$, $\mathbf{q}$, such that $\mathbf{1} \le \mathbf{q} \le \mathbf{p} \le \infty$, we have

$$\|f\|_{\mathbf{p}} \le \left(\prod_{i=1}^{d} M_i(q_i)^{\beta_i} \right) \|f\|_{\mathbf{q}}, \qquad \beta_i := 1/q_i - 1/p_i.$$

First we prove a certain auxiliary assertion, which is of interest by itself. Let Φ be a finite-dimensional subspace of the space $C(\mathbb{T})$. For $u \in L_p(\mathbb{T})$ denote

$$E_\Phi^\perp(u)_p := \inf_{v \perp \Phi} \|u - v\|_p.$$

Lemma 3.3.4 *Let Φ be a finite-dimensional subspace of the space $C(\mathbb{T})$ and let $\{\varphi_j\}_{j=1}^N$ be an orthonormal basis of this subspace. Denote by*

$$\mathscr{D}_\Phi(x,y) := \sum_{j=1}^{N} \varphi_j(x)\varphi_j(y)$$

the Dirichlet kernel for the system $\{\varphi_j\}_{j=1}^N$. Then the following equality holds

$$\sup_{\|f\|_q \le 1,\, f \in \Phi} \|f\|_\infty = \sup_x E_\Phi^\perp(\mathscr{D}_\Phi(x,\cdot))_{q'}, \qquad q' := \frac{q}{q-1}.$$

Proof For any $f \in \Phi$ and any $v \perp \Phi$ we have

$$|f(x)| = \left| \frac{1}{2\pi} \int_{\mathbb{T}} (\mathscr{D}_{\Phi}(x,y) - v(y)) f(y) dy \right| \leq \|\mathscr{D}_{\Phi}(x,\cdot) - v(\cdot)\|_{q'} \|f\|_q.$$

Therefore,

$$|f(x)| \leq \|f\|_q \inf_{v \perp \Phi} \|\mathscr{D}_{\Phi}(x,\cdot) - v(\cdot)\|_{q'} = \|f\|_q E^{\perp}_{\Phi}(\mathscr{D}_{\Phi}(x,\cdot))_{q'}$$

and

$$\|f\|_{\infty} \leq \left(\sup_x E^{\perp}_{\Phi}(\mathscr{D}_{\Phi}(x,\cdot))_{q'} \right) \|f\|_q;$$

this proves the inequality $\leq$ in the lemma.

Let us prove the inequality $\geq$. By the Nikol'skii duality theorem (see Theorem A.2.3), we have

$$\begin{aligned} N(q,\infty) &:= \sup_{\|f\|_q \leq 1, f \in \Phi} \|f\|_{\infty} = \sup_{\|f\|_q \leq 1, f \in \Phi} \sup_{\|u\|_1 \leq 1} |\langle f, u \rangle| \\ &= \sup_{\|u\|_1 \leq 1} \sup_{\|f\|_q \leq 1, f \in \Phi} |\langle f, u \rangle| = \sup_{\|u\|_1 \leq 1} E^{\perp}_{\Phi}(u)_{q'}. \end{aligned}$$

Thus, for any $u \in L_1$, $\|u\|_1 \leq 1$, we have

$$E^{\perp}_{\Phi}(u)_{q'} \leq N(q,\infty). \tag{3.3.6}$$

Take the Fejér operator F_m, which acts as a convolution with the Fejér kernel $\mathscr{K}_m$. Then, for any continuous function f,

$$\|f - F_m(f)\|_{\infty} \to 0, \quad m \to \infty.$$

Consider

$$A_m(x,y) := \sum_{j=1}^{N} F_m(\varphi_j)(x) \varphi_j(y) - \mathscr{K}_m(x-y).$$

It is easy to verify that, for any x, we have $A_m(x,\cdot) \perp \Phi$. Furthermore, for

$$u(x,y) := \mathscr{D}_{\Phi}(x,y) - A_m(x,y) = \mathscr{K}_m(x-y) + \sum_{j=1}^{N} (\varphi_j(x) - F_m(\varphi_j)(x)) \varphi_j(y),$$

we have

$$\|u(x,\cdot)\|_1 \leq 1 + \varepsilon_m, \qquad \varepsilon_m \to 0, \qquad m \to \infty.$$

We find from this and (3.3.6) that

$$\sup_x E^{\perp}_{\Phi}(\mathscr{D}_{\Phi}(x,\cdot))_{q'} \leq N(q,\infty).$$

The lemma is proved. □

Corollary 3.3.5 *For any orthonormal system* $\Phi = \{\varphi_j\}_{j=1}^N$,

$$N(2,\infty) \geq N^{1/2}.$$

Proof This inequality follows from the lemma and from the following obvious chain of relations:

$$\sup_x E_\Phi^\perp(\mathscr{D}_\Phi(x,\cdot))_2 = \sup_x \left(\sum_{j=1}^N |\varphi_j(x)|^2\right)^{1/2} \geq \left(\frac{1}{2\pi}\int_\mathbb{T}\sum_{j=1}^N |\varphi_j(x)|^2 dx\right)^{1/2} = N^{1/2}.$$

□

Proof of Theorem 3.3.3 We carry out the proof by induction on the number of variables d. In the case $d=1$ the statement of the theorem follows from (3.3.5) and the inequality

$$\|f\|_p \leq \|f\|_q^{q/p}\|f\|_\infty^{1-q/p}, \qquad q \leq p \leq \infty. \tag{3.3.7}$$

Assume that the theorem is proved for the tensor product of $d-1$ subspaces $X_1,\dots,X_{d-1}$. So, let us prove it for the tensor product of d subspaces $X_1,\dots,X_d$. Consider the following three cases separately:

(1) $p_d = q_d$;
(2) $p_d = \infty$;
(3) $q_d < p_d < \infty$.

In case (1), the assertion follows directly from the supposition of induction. We divide case (2) into the following two subcases:

(2a) $\mathbf{p}^d = \mathbf{q}^d$;
(2b) $\mathbf{q}^d \leq \mathbf{p}^d$,

where we write $\mathbf{y}^d := (y_1,\dots,y_{d-1})$ for the vector $\mathbf{y} = (y_1,\dots,y_d)$, and the vector inequalities are coordinatewise.

(2a). Let $f \in X$. Then, for any $v \perp X_d$,

$$f(\mathbf{x}^d,x_d) = \frac{1}{2\pi}\int_\mathbb{T} f(\mathbf{x}^d,y)\mathscr{D}_{X_d}(x_d,y)dy = \frac{1}{2\pi}\int_\mathbb{T} f(\mathbf{x}^d,y)(\mathscr{D}_{X_d}(x_d,y)-v(y))dy.$$

This implies that, for any x_d,

$$\|f(\cdot,x_d)\|_{\mathbf{p}^d} \leq \|f\|_\mathbf{q} E_{X_d}^\perp(\mathscr{D}_{X_d}(x_d,\cdot))_{q_d'},$$

and, hence, by Lemma 3.3.4, we find that

$$\|f\|_\mathbf{p} = \|f\|_{(\mathbf{p}^d,\infty)} \leq M_d(q_d)^{1/q_d}\|f\|_\mathbf{q}.$$

(2b). By (2a) and (1), we have

$$\|f\|_{\mathbf{p}} \le M_d(q_d)^{\beta_d} \|f\|_{(\mathbf{p}^d, q_d)} \le \left(\prod_{i=1}^{d} M_i(q_i)^{\beta_i} \right) \|f\|_{\mathbf{q}}.$$

It remains to consider case (3). This case follows from the previous two cases and inequality (3.3.7), which is applied for the L_{p_d}- and L_{q_d}-norm with respect to the variable x_d. The theorem is proved. □

3.3.3 Relations Between $\|f\|_p$ and $\{\|\delta_{\mathbf{s}}(f)\|_q\}$.

We proceed to the problem of estimating $\|f\|_p$ in terms of the array $\{\|\delta_{\mathbf{s}}(f)\|_q\}$. Here and below p and q are scalars such that $1 \le q, p \le \infty$. Let an array $\varepsilon = \{\varepsilon_{\mathbf{s}}\}$ be given, where $\varepsilon_{\mathbf{s}} \ge 0$, $\mathbf{s} = (s_1, \dots, s_d)$, and s_j are nonnegative integers, $j = 1, \dots, d$. We denote by $G(\varepsilon, q)$ and $F(\varepsilon, q)$ the following sets of functions $(1 \le q \le \infty)$:

$$G(\varepsilon, q) := \left\{ f \in L_q : \left\| \delta_{\mathbf{s}}(f) \right\|_q \le \varepsilon_{\mathbf{s}} \text{ for all } \mathbf{s} \right\},$$
$$F(\varepsilon, q) := \left\{ f \in L_q : \left\| \delta_{\mathbf{s}}(f) \right\|_q \ge \varepsilon_{\mathbf{s}} \text{ for all } \mathbf{s} \right\}.$$

We now prove the following statement.

Theorem 3.3.6 *The following relations hold:*

$$\sup_{f \in G(\varepsilon, q)} \|f\|_p \asymp \left(\sum_{\mathbf{s}} \varepsilon_{\mathbf{s}}^p 2^{\|\mathbf{s}\|_1 (p/q-1)} \right)^{1/p}, \qquad 1 \le q < p < \infty; \tag{3.3.8}$$

$$\inf_{f \in F(\varepsilon, q)} \|f\|_p \asymp \left(\sum_{\mathbf{s}} \varepsilon_{\mathbf{s}}^p 2^{\|\mathbf{s}\|_1 (p/q-1)} \right)^{1/p}, \qquad 1 < p < q \le \infty, \tag{3.3.9}$$

with constants independent of ε.

Proof We need to prove some auxiliary statements.

Lemma 3.3.7 *Let* $1 \le q < p < \infty$ *and* $f \in L_q$. *Then*

$$\|f\|_p \le C(q, p, d) \left(\sum_{\mathbf{s}} \| \delta_{\mathbf{s}}(f) \|_q^p 2^{\|\mathbf{s}\|_1 (p/q-1)} \right)^{1/p}.$$

Proof We first prove the following assertion.

Lemma 3.3.8 *Let the functions* $g_1, \dots, g_m$ *defined on* $\mathbb{T}^d$ *be given. Then*

$$J := (2\pi)^{-d} \int_{\mathbb{T}^d} \prod_{i=1}^{m} |g_i(\mathbf{x})|^{p/m} d\mathbf{x} \le \left(\prod_{i \ne j} \|g_i\|_{\mathbf{z}p/2}^{p/2} \|g_j\|_{\mathbf{z}'p/2}^{p/2} \right)^{1/(m(m-1))},$$

where $\mathbf{z} = (z_1, \dots, z_d)$, $\mathbf{z}' = (z_1', \dots, z_d')$, $1/z_l + 1/z_l' = 1$, $l = 1, \dots, d$.

Proof We represent the initial product in the form

$$\prod_{i=1}^{m} |g_i|^{p/m} = \prod_{i\neq j} \left(|g_i|^{p/(2m(m-1))} |g_j|^{p/(2m(m-1))} \right). \tag{3.3.10}$$

The product on the right-hand side of (3.3.10) has $\mathscr{P} := m(m-1)$ paired factors. Applying the Hölder inequality (A.1.3) with $\mathscr{P}$ factors $f_{i,j} := |g_i|^{p/(2m(m-1))} |g_j|^{p/(2m(m-1))}$, and $L_{p_{i,j}}$ norms with $p_{i,j} = \mathscr{P}$, we get

$$J \leq \prod_{i\neq j} \left((2\pi)^{-d} \int_{\mathbb{T}^d} |g_i(\mathbf{x})|^{p/2} |g_j(\mathbf{x})|^{p/2} d\mathbf{x} \right)^{1/\mathscr{P}}.$$

For an individual factor, applying the Hölder inequality with exponent z_l and z_l' and integrating with respect to the lth variable, we get

$$(2\pi)^{-d} \int_{\mathbb{T}^d} |g_i(\mathbf{x})|^{p/2} |g_j(\mathbf{x})|^{p/2} d\mathbf{x} \leq \left\| |g_i|^{p/2} \right\|_{\mathbf{z}} \left\| |g_j|^{p/2} \right\|_{\mathbf{z}'} = \|g_i\|_{\mathbf{z}p/2}^{p/2} \|g_j\|_{\mathbf{z}'p/2}^{p/2}.$$

The lemma is proved. □

Remark 3.3.9 Without affecting the validity of Lemma 3.3.8, in any paired factor we can interchange z_l and z_l', $l = 1, \ldots, d$.

We now assume for simplicity that $d = 2$. Of course, it suffices to prove Lemma 3.3.7 for functions of the form

$$f(\mathbf{x}) = \sum_{s_1=0}^{N} \sum_{s_2=0}^{M} \delta_{\mathbf{s}}(f, \mathbf{x}) = \sum_{\mu=0}^{N} \sum_{\nu=0}^{M} f_{\mu\nu}(\mathbf{x}),$$

where

$$f_{\mu\nu}(\mathbf{x}) := \delta_{(\mu,\nu)}(f, \mathbf{x}).$$

For $m = [p] + 1$ we have

$$(2\pi)^2 \|f\|_p^p = \int_{\mathbb{T}^2} \left(\left| \sum_{\mu=0}^{N} \sum_{\nu=0}^{M} f_{\mu\nu}(\mathbf{x}) \right|^{p/m} \right)^m d\mathbf{x}$$

$$\leq \int_{\mathbb{T}^2} \left(\sum_{\mu=0}^{N} \sum_{\nu=0}^{M} |f_{\mu\nu}(\mathbf{x})|^{p/m} \right)^m d\mathbf{x},$$

here we have used the inequality $|a+b|^\kappa \leq |a|^\kappa + |b|^\kappa$, $0 \leq \kappa \leq 1$.

Taking the mth power, we see that the last expression becomes

$$= \int_{\mathbb{T}^2} \sum_{\mu_1=0}^{N} \cdots \sum_{\mu_m=0}^{N} \sum_{\nu_1=0}^{M} \cdots \sum_{\nu_m=0}^{M} \prod_{i=1}^{m} |f_{\mu_i\nu_i}(\mathbf{x})|^{p/m} d\mathbf{x},$$

and Lemma 3.3.8 gives us

$$\leq \sum_{\mu_1=0}^{N} \cdots \sum_{\mu_m=0}^{N} \sum_{\nu_1=0}^{M} \cdots \sum_{\nu_m=0}^{M} (2\pi)^2 \left(\prod_{i \neq j} \|f_{\mu_i \nu_i}\|_{\mathbf{z}p/2}^{p/2} \|f_{\mu_j \nu_j}\|_{\mathbf{z}'p/2}^{p/2} \right)^{1/\mathscr{P}}.$$

Let $2q/p < z_1 < 2$ and $2q/p < z_2 < 2$. Using Remark 3.3.9, we arrange z_1, z_2, z_1', and z_2' so that z_1 goes with (i.e., corresponds to) the function whose first index is larger and z_2 goes with the function with a larger second index.

Suppose for definiteness that $\mu_i \geq \mu_j$ and $\nu_i \geq \nu_j$. Using the Nikol'skii inequality we get

$$\begin{aligned}
\|f_{\mu_i \nu_i}\|_{\mathbf{z}p/2}^{p/2} \|f_{\mu_j \nu_j}\|_{\mathbf{z}'p/2}^{p/2} &\leq C \left(\|f_{\mu_i \nu_i}\|_q 2^{\mu_i \left(1/q - 2/(z_1 p)\right) + \nu_i \left(1/q - 2/(z_2 p)\right)} \right)^{p/2} \\
&\quad \times \left(\|f_{\mu_j \nu_j}\|_q 2^{\mu_j \left(1/q - 2/(z_1' p)\right) + \nu_j \left(1/q - 2/(z_2' p)\right)} \right)^{p/2} \\
&= C \left(\|f_{\mu_j \nu_j}\|_q 2^{\mu_j (1/q - 1/p) + \nu_j (1/q - 1/p)} \right)^{p/2} \\
&\quad \times \left(\|f_{\mu_i \nu_i}\|_q 2^{(\mu_i + \nu_i)(1/q - 1/p)} \right)^{p/2} \\
&\quad \times 2^{-(\mu_i - \mu_j)(1/z_1 - 1/2) - (\nu_i - \nu_j)(1/z_2 - 1/2)}.
\end{aligned}$$

Let $\beta := 1/q - 1/p$, $\theta_1 := 1/z_1 - 1/2$ and $\theta_2 := 1/z_2 - 1/2$. We have

$$\begin{aligned}
\|f\|_p^p &\leq C \sum_{\mu_1=0}^{N} \cdots \sum_{\mu_m=0}^{N} \sum_{\nu_1=0}^{M} \cdots \sum_{\nu_m=0}^{M} \left(\prod_{i \neq j} (\|f_{\mu_i \nu_i}\|_q 2^{\beta(\mu_i + \nu_i)})^{p/2} \right. \\
&\qquad \left. \times (\|f_{\mu_j \nu_j}\|_q 2^{\beta(\mu_j + \nu_j)})^{p/2} 2^{-|\mu_i - \mu_j|\theta_1 - |\nu_i - \nu_j|\theta_2} \right)^{1/\mathscr{P}} \\
&= C \sum_{\mu_1=0}^{N} \cdots \sum_{\mu_m=0}^{N} \sum_{\nu_1=0}^{M} \cdots \sum_{\nu_m=0}^{M} \prod_{i=1}^{m} (\|f_{\mu_i \nu_i}\|_q 2^{\beta(\mu_i + \nu_i)})^{p/m} \\
&\quad \times \prod_{j=1}^{m} 2^{-|\mu_i - \mu_j|\theta_1/\mathscr{P} - |\nu_i - \nu_j|\theta_2/\mathscr{P}}.
\end{aligned}$$

Using the Hölder inequality for sums with exponent m, we see that the above expression becomes

$$\leq C\prod_{i=1}^{m}\Bigg(\sum_{\mu_1=0}^{N}\cdots\sum_{\mu_m=0}^{N}\sum_{\nu_1=0}^{M}\cdots\sum_{\nu_m=0}^{M}\|f_{\mu_i\nu_i}\|_q^p 2^{\beta(\mu_i+\nu_i)p}$$
$$\times\prod_{j=1}^{m}2^{-|\mu_i-\mu_j|\theta_1 m/\mathscr{P}-|\nu_i-\nu_j|\theta_2 m/\mathscr{P}}\Bigg)^{1/m}$$
$$\leq C(m,\theta_1,\theta_2)\sum_{\mu=0}^{N}\sum_{\nu=0}^{M}\|f_{\mu\nu}\|_q^p 2^{\beta(\mu+\nu)p}.$$

The lemma is now proved. □

Remark 3.3.10 In the proof of Lemma 3.3.7 we used only the property $\delta_{\mathbf{s}}(f)\in\mathscr{T}(2^{\mathbf{s}},d)$. That is, if

$$f=\sum_{\mathbf{s}}t_{\mathbf{s}},\qquad t_{\mathbf{s}}\in\mathscr{T}(2^{\mathbf{s}},d),$$

then, for $1\leq q<p<\infty$,

$$\|f\|_p\leq C(q,p,d)\Big(\sum_{\mathbf{s}}\|t_{\mathbf{s}}\|_q^p 2^{\|\mathbf{s}\|_1(p/q-1)}\Big)^{1/p}. \tag{3.3.11}$$

Lemma 3.3.7 implies the upper estimate in relation (3.3.8).

Remark 3.3.11 Further, in the proof of Lemma 3.3.7 the property $\delta_{\mathbf{s}}(f)\in\mathscr{T}(2^{\mathbf{s}},d)$ was used only to apply the Nikol'skii inequalities from Theorem 3.3.2.

Combining Remark 3.3.11 with Theorem 3.3.3 we obtain the following assertion.

Theorem 3.3.12 *Let a system of real functions $\{\varphi_k(x)\}_{k=1}^{\infty}$, $x\in\mathbb{T}$, satisfy the condition*

$$\Big\|\sum_{k=1}^{n}c_k\varphi_k\Big\|_\infty\leq Kn^{1/q}\Big\|\sum_{k=1}^{n}c_k\varphi_k\Big\|_q,\qquad 1\leq q<\infty,\qquad n\in\mathbb{N}.$$

Denote by $\Phi_{\mathbf{s}}$, $\mathbf{s}=(s_1,\dots,s_d)$, a set of polynomials of the form

$$\sum_{k_d=1}^{2^{s_d}}\cdots\sum_{k_1=1}^{2^{s_1}}c_{k_1,\dots,k_d}\varphi_{k_1}(x_1)\cdots\varphi_{k_d}(x_d).$$

Then for any $1\leq q<p<\infty$ and any $f_{\mathbf{s}}\in\Phi_{\mathbf{s}}$, we have

$$\Big\|\sum_{\mathbf{s}}f_{\mathbf{s}}\Big\|_p\leq C(q,p,K,d)\Big(\sum_{\mathbf{s}}\|f_{\mathbf{s}}\|_q^p 2^{\|\mathbf{s}\|_1(p/q-1)}\Big)^{1/p}.$$

Proof We note that, in the formulation of Theorem 3.3.12, only scalar norms L_q and L_p occur. However, we require the Nikol'skii inequalities for vector $L_{\mathbf{q}}$ and $L_{\mathbf{p}}$ norms in order to prove this theorem.

To prove the lower estimate in (3.3.9), we prove a statement which is dual to Lemma 3.3.7.

Lemma 3.3.13 *Let $1 < p < q \le \infty$ and $f \in L_p$. Then*

$$\|f\|_p \ge C(q,p,d)\left(\sum_{\mathbf{s}} \|\delta_{\mathbf{s}}(f)\|_q^p 2^{\|\mathbf{s}\|_1(p/q-1)}\right)^{1/p}.$$

Proof Clearly, it suffices to prove the lemma under the additional restriction that f is a trigonometric polynomial. Let p' and q' denote the exponents dual to p and q, i.e. $1/p+1/p' = 1$ and $1/q+1/q' = 1$.

For an array $\mathbf{a} = \{a_{\mathbf{s}}\}$ we denote

$$\|\mathbf{a}\|_{p,*} := \left(\sum_{\mathbf{s}} |a_{\mathbf{s}}|^p 2^{-\|\mathbf{s}\|_1}\right)^{1/p}.$$

By Theorem A.2.1 we have

$$\|f\|_p = \sup_{\|g\|_{p'}\le 1} |\langle f,g\rangle| = \sup_{\|g\|_{p'}\le 1} \left|\sum_{\mathbf{s}} \langle \delta_{\mathbf{s}}(f), \delta_{\mathbf{s}}(g)\rangle\right|. \tag{3.3.12}$$

Let

$$B(q',p') := \bigcup_{\mathbf{a}} G(\mathbf{a},q'),$$

where the union is taken over all $\mathbf{a}$ such that

$$\sum_{\mathbf{s}} a_{\mathbf{s}}^{p'} 2^{\|\mathbf{s}\|_1(p'/q'-1)} \le 1. \tag{3.3.13}$$

Then by Lemma 3.3.7 for any $g \in B(q',p')$ we have

$$\|g\|_{p'} \le C(q,p,d).$$

By (3.3.12) we then obtain

$$\begin{aligned}
\|f\|_p &\ge C'(q,p,d) \sup_{g\in B(q',p')} \left|\sum_{\mathbf{s}} \langle \delta_{\mathbf{s}}(f), \delta_{\mathbf{s}}(g)\rangle\right| \\
&= \sup_{\mathbf{a}:\|\{a_{\mathbf{s}} 2^{\|\mathbf{s}\|_1/q'}\}\|_{p',*}\le 1} \sum_{\mathbf{s}} a_{\mathbf{s}} \|\delta_{\mathbf{s}}(f)\|_q \\
&= \sup_{\mathbf{a}:\|\{a_{\mathbf{s}} 2^{\|\mathbf{s}\|_1/q'}\}\|_{p',*}\le 1} \sum_{\mathbf{s}} a_{\mathbf{s}} 2^{\|\mathbf{s}\|_1/q'} \|\delta_{\mathbf{s}}(f)\|_q 2^{\|\mathbf{s}\|_1/q} 2^{-\|\mathbf{s}\|_1} \\
&= \left\|\{\|\delta_{\mathbf{s}}(f)\|_q 2^{\|\mathbf{s}\|_1/q}\}\right\|_{p,*},
\end{aligned}$$

which implies the lemma. □

Lemma 3.3.13 implies the lower estimate in the relation (3.3.9).

We now show that the inequalities obtained in Lemmas 3.3.7 and 3.3.13 cannot be improved. Clearly, it suffices to consider the case when ε has only a finite number of nonzero elements.

We define the functions $(1 \le q \le \infty)$

$$f_q^{\varepsilon}(\mathbf{x}) := \sum_{\mathbf{s}} \varepsilon_{\mathbf{s}} \varphi_{\mathbf{s}}(\mathbf{x}),$$

where

$$\varphi_{\mathbf{s}}(\mathbf{x}) := \mathcal{K}^{\mathbf{s}}(\mathbf{x}) \,/\, \|\mathcal{K}^{\mathbf{s}}\|_q,$$

$$\mathcal{K}^{\mathbf{s}}(\mathbf{x}) := e^{i(\mathbf{k}^{\mathbf{s}},\mathbf{x})} \prod_{j=1}^{d} \mathcal{K}_{2^{s_j-2}-1}(x_j),$$

$$k_j^{\mathbf{s}} := \begin{cases} 2^{s_j-1} + 2^{s_j-2} & \text{for } s_j \ge 2, \\ 1 & \text{for } s_j = 1, \\ 0 & \text{for } s_j = 0, \; j = 1,\dots,d, \end{cases}$$

$$\mathcal{K}_m(t) \equiv 1 \qquad \text{for } m < 1.$$

Then for all $\mathbf{s}$,

$$\left\|\delta_{\mathbf{s}}(f_q^{\varepsilon})\right\|_q = \varepsilon_{\mathbf{s}},$$

i.e.,

$$f_q^{\varepsilon} \in G(\varepsilon, q) \cap F(\varepsilon, q).$$

We now prove the lower estimate in (3.3.8). Let $1 \le q < p < \infty$. We apply Lemma 3.3.13 with $q = \infty$ to the function f_q^{ε} and find that

$$\|f_q^{\varepsilon}\|_p \gg \left(\sum_{\mathbf{s}} \varepsilon_{\mathbf{s}}^p \|\varphi_{\mathbf{s}}\|_\infty^p 2^{-\|\mathbf{s}\|_1}\right)^{1/p} \gg \left(\sum_{\mathbf{s}} \varepsilon_{\mathbf{s}}^p 2^{\|\mathbf{s}\|_1(p/q-1)}\right)^{1/p}.$$

The relation (3.3.8) is therefore proved.

Let us prove the upper estimate in (3.3.9). We assume that $1 < p < q \le \infty$, apply Lemma 3.3.7 with $q = 1$ to the function f_q^{ε}, and obtain

$$\begin{aligned} \|f_q^{\varepsilon}\|_p &\le C(p,d) \left(\sum_{\mathbf{s}} \varepsilon_{\mathbf{s}}^p \|\varphi_{\mathbf{s}}\|_1^p 2^{\|\mathbf{s}\|_1(p-1)}\right)^{1/p} \\ &\le C(p,d) \left(\sum_{\mathbf{s}} \varepsilon_{\mathbf{s}}^p 2^{\|\mathbf{s}\|_1(p/q-1)}\right)^{1/p}. \end{aligned}$$

The proof of the theorem is now complete. □

We now prove another assertion analogous to Lemma 3.3.7.

Lemma 3.3.14 *Let $1 \le q < p < \infty$ be given. Assume that a function $f \in L_p$ is represented by a series*

$$f = \sum_{n=0}^{\infty} f_n$$

which converges in L_p, where the functions $f_n \in L_\infty$ have the following property: for all n and $a > q$ the inequalities

$$\|f_n\|_a \le C(a,q,\kappa)\|f_n\|_q 2^{\kappa n(1/q-1/a)}, \qquad \kappa > 0, \tag{3.3.14}$$

hold. Then

$$\|f\|_p \le C(p,q,d,\kappa)\left(\sum_{n=0}^{n} \|f_n\|_q^p 2^{\kappa n(1/q-1/p)p}\right)^{1/p}.$$

Proof The proof of this lemma is analogous to the proof of Lemma 3.3.7 in the univariate case. Set $m := [p]+1$. It clearly suffices to prove the required estimate for an f of the form

$$f = \sum_{n=0}^{N} f_n.$$

We have

$$(2\pi)^d\|f\|_p^p = \int_{\mathbb{T}^d}\left|\sum_{n=0}^{N} f_n\right|^p d\mathbf{x} = \int_{\mathbb{T}^d}\left(\left|\sum_{n=0}^{N} f_n\right|^{p/m}\right)^m d\mathbf{x}$$
$$\le \int_{\mathbb{T}^d}\left(\sum_{n=0}^{N} |f_n|^{p/m}\right)^m d\mathbf{x};$$

here we have used the inequality $|a+b|^\rho \le |a|^\rho + |b|^\rho$, $0 \le \rho \le 1$. This last expression is

$$= \int_{\mathbb{T}^d}\sum_{n_1=0}^{N}\cdots\sum_{n_m=0}^{N}\prod_{i=1}^{m}|f_{n_i}|^{p/m}d\mathbf{x}$$
$$= \sum_{n_1=0}^{N}\cdots\sum_{n_m=0}^{N}\int_{\mathbb{T}^d}\prod_{i=1}^{m}|f_{n_i}|^{p/m}d\mathbf{x}. \tag{3.3.15}$$

Using Lemma 3.3.8 with $\mathbf{z} = (z,\dots,z)$, we continue (3.3.15):

$$\le \sum_{n_1=0}^{N}\cdots\sum_{n_m=0}^{N}(2\pi)^d\left(\prod_{i\neq j}\|f_{n_i}\|_{\mathbf{z}p/2}^{p/2}\|f_{n_j}\|_{\mathbf{z}'p/2}^{p/2}\right)^{1/\mathscr{P}}.$$

Let $2q/p < z < 2$. Taking into account Remark 3.3.9, we apply the $L_{\mathbf{z}p/2}$-norm to which ever of the functions f_{n_i} and f_{n_j} has the larger index. Let $n_i \geq n_j$. Using (3.3.14), we get

$$\begin{aligned}\|f_{n_i}\|_{\mathbf{z}p/2}^{p/2}\|f_{n_j}\|_{\mathbf{z}'p/2}^{p/2} &\leq C(\mathbf{z},p,q,\kappa)\left(\|f_{n_i}\|_q 2^{\kappa n_i\left(1/q-2/(zp)\right)}\right)^{p/2}\\ &\quad\times\left(\|f_{n_j}\|_q 2^{\kappa n_j\left(1/q-2/(z'p)\right)}\right)^{p/2}\\ &=\left(\|f_{n_i}\|_q 2^{\kappa n_i\left(1/q-1/p\right)}\right)^{p/2}\left(\|f_{n_j}\|_q 2^{\kappa n_j\left(1/q-1/p\right)}\right)^{p/2}\\ &\quad\times 2^{-(n_i-n_j)(1/z-1/2)\kappa}.\end{aligned}$$

Let $\beta := 1/q - 1/p$ and $\theta := 1/z - 1/2$. We have

$$\begin{aligned}\|f\|_p^p &\leq C(\mathbf{z},p,q,\kappa)\sum_{n_1=0}^{N}\cdots\sum_{n_m=0}^{N}\Bigg(\prod_{i\neq j}(\|f_{n_i}\|_q 2^{\kappa\beta n_i})^{p/2}\\ &\qquad\times(\|f_{n_j}\|_q 2^{\kappa\beta n_j})^{p/2}\quad 2^{-\kappa|n_i-n_j|\theta}\Bigg)^{1/\mathscr{P}}\\ &= C(\mathbf{z},p,q,\kappa)\sum_{n_1=0}^{N}\cdots\sum_{n_m=0}^{N}\prod_{i=1}^{m}\left(\|f_{n_i}\|_q^{p/m}2^{p\kappa\beta n_i/m}\prod_{i=1}^{m}2^{-\kappa|n_i-n_j|\theta/\mathscr{P}}\right).\end{aligned}$$

Using Hölder's inequality for sums with exponent m, we continue as follows:

$$\begin{aligned}&\leq C(\mathbf{z},p,q,\kappa)\prod_{i=1}^{m}\Bigg(\sum_{n_1=0}^{N}\cdots\sum_{n_m=0}^{N}\|f_{n_i}\|_q^p 2^{\kappa\beta n_i p}\\ &\qquad\times\prod_{j=1}^{m}2^{-\kappa|n_i-n_j|\theta/(m-1)}\Bigg)^{1/m}\ll\sum_{n=0}^{N}\|f_n\|_q^p 2^{\kappa\beta np}.\end{aligned}$$

The lemma is proved. □

3.3.4 The Marcinkiewicz Theorem

In this subsection we prove the equivalence of a mixed norm of a trigonometric polynomial to its mixed lattice norm. We use the notation

$$\ell(\mathbf{N},d) := \left\{\mathbf{a} = \{a_{\mathbf{n}}\}, \mathbf{n} = (n_1,\ldots,n_d), 1 \leq n_j \leq N_j, j = 1,\ldots,d\right\},$$

and for $\mathbf{a} \in \ell(\mathbf{N},d)$ we define the mixed norm (weighted mixed norm)

$$\|\mathbf{a}\|_{\mathbf{p},\mathbf{N}} := \left(\sum_{n_d=1}^{N_d} N_d^{-1} \left(\cdots \left(\sum_{n_1=1}^{N_1} N_1^{-1} |a_{\mathbf{n}}|^{p_1} \right)^{p_2/p_1} \cdots \right)^{p_d/p_{d-1}} \right)^{1/p_d}$$

$$= \|\mathbf{a}\|_{\mathbf{p}} \prod_{j=1}^{d} N_j^{-1/p_j}. \tag{3.3.16}$$

The main result of this subsection is the following proposition:

Theorem 3.3.15 *Let $t \in \mathscr{T}(\mathbf{N},d)$, $\mathbf{N} > \mathbf{0}$. Then for any $\mathbf{1} \le \mathbf{p} \le \infty$,*

$$C_d^{-1} \left\| \{t(\mathbf{x}(\mathbf{n}))\}_{\mathbf{n} \in P'(\mathbf{N})} \right\|_{\mathbf{p},4\mathbf{N}} \le \|t\|_{\mathbf{p}} \le C_d \left\| \{t(\mathbf{x}(\mathbf{n}))\}_{\mathbf{n} \in P'(\mathbf{N})} \right\|_{\mathbf{p},4\mathbf{N}},$$

where C_d is a number depending only on d.

In the case $\mathbf{p} = (p,\dots,p)$ this theorem is an immediate corollary of the corresponding univariate theorem of Marcinkiewicz (see Theorem 1.3.9).

Proof We first prove some auxiliary statements.

Lemma 3.3.16 *Let $\mathbf{a},\mathbf{b} \in \ell(N,1)$ and $\mathbf{b}$ be such that $b_j \ge 0$, $j = 1,\dots,N$, $\sum_{j=1}^N b_j = 1$. Then for any $1 \le p < \infty$,*

$$\left| \sum_{j=1}^N a_j b_j \right| \le \left(\sum_{j=1}^N |a_j|^p b_j \right)^{1/p},$$

and for $p = \infty$

$$\left| \sum_{j=1}^N a_j b_j \right| \le \max_{1 \le j \le N} |a_j|.$$

Proof In the cases $p = 1,\infty$ the lemma is obvious. In the case $1 < p < \infty$ it is a corollary of Hölder's inequality:

$$\left| \sum_{j=1}^N a_j b_j \right| = \left| \sum_{j=1}^N a_j b_j^{1/p} b_j^{1/p'} \right|$$
$$\le \left(\sum_{j=1}^N |a_j|^p b_j \right)^{1/p} \left(\sum_{j=1}^N b_j \right)^{1/p'} = \left(\sum_{j=1}^N |a_j|^p b_j \right)^{1/p}.$$

We note that this lemma is a corollary of the well-known monotonicity of ℓ_p-norms. □

Lemma 3.3.17 *Let* $\mathbf{a} \in \ell(\mathbf{N},d)$ *and suppose that a vector* $\mathbf{b} \in \ell(N_d,1)$ *is such that* $b_j \geq 0$, $j = 1,\dots,N_d$, *and* $b_1 + \cdots + b_{N_d} = 1$. *Then, for any vector* $\mathbf{p} = (p_1,\dots,p_d)$, $1 \leq p_j \leq \infty$, $j = 1,\dots,d$, *the relation*

$$\left\| \sum_{n_d=1}^{N_d} a_{\mathbf{n}} b_{n_d} \right\|_{\mathbf{p}^d} \leq \left(\sum_{n_d=1}^{N_d} \|a_{\mathbf{n}}\|_{\mathbf{p}^d}^{p_d} b_{n_d} \right)^{1/p_d}$$

holds, where $\mathbf{p}^d := (p_1,\dots,p_{d-1})$.

Proof We have

$$\left\| \sum_{n_d=1}^{N_d} a_{\mathbf{n}} b_{n_d} \right\|_{\mathbf{p}^d} \leq \sum_{n_d=1}^{N_d} \|a_{\mathbf{n}}\|_{\mathbf{p}^d} b_{n_d}.$$

Applying Lemma 3.3.16 we obtain the required conclusion. □

With each element $\mathbf{a} \in \ell(4\mathbf{N},d)$, we associate a trigonometric polynomial $t \in \mathscr{T}(2\mathbf{N},d)$ according to the following rule:

$$t(\mathbf{x}) := \nu(4\mathbf{N})^{-1} \sum_{\mathbf{N} \in P'(\mathbf{N})} a_{\mathbf{N}} \mathscr{V}_{\mathbf{N}}(\mathbf{x} - \mathbf{x}(\mathbf{n})). \tag{3.3.17}$$

The following lemma holds.

Lemma 3.3.18 *For a polynomial* $t(\mathbf{x})$ *of the form (3.3.17) and* $\mathbf{1} \leq \mathbf{p} \leq \infty$,

$$\|t(\mathbf{x})\|_{\mathbf{p}} \leq C(d) \|\mathbf{a}\|_{\mathbf{p},4\mathbf{N}}.$$

Proof We carry out the proof by induction on the dimension d of the space. For $d = 1$ the lemma coincides with Lemma 1.3.10. Suppose that it holds for dimension $d-1$. We will deduce it for dimension d. By the induction hypothesis we have that for each x_d

$$\|t(\cdot,x_d)\|_{\mathbf{p}^d} \leq C(d-1) \left\| \sum_{n_d=1}^{N_d} |a_{\cdot,n_d}| |\mathscr{V}_{N_d}(x_d - x(n_d)| (4N_d)^{-1} \right\|_{\mathbf{p}^d,4\mathbf{N}^d},$$

with $\mathbf{N}^d := (N_1,\dots,N_{d-1})$.

Taking into account (3.3.16) and Lemma 3.3.17, we continue this relation:

$$\leq C(d-1) B_{N_d} \left(\sum_{n_d=1}^{N_d} \|a_{\cdot,n_d}\|_{\mathbf{p}^d,4\mathbf{N}^d}^{p_d} |\mathscr{V}_{N_d}(x_d - x(n_d)| (4N_d)^{-1} \right)^{1/p_d}. \tag{3.3.18}$$

Here

$$B_{N_d} := \max \left(\left\| \sum_{n_d=1}^{N_d} |\mathscr{V}_{N_d}(\cdot - x(n_d)| (4N_d)^{-1} \right\|_{\infty}, 1 \right) \leq C,$$

where C is an absolute constant.

From (3.3.18) we find that

$$(2\pi)^{-1}\int_0^{2\pi}\|t(\cdot,x_d)\|_{\mathbf{p}^d}^{p_d}dx_d \le C(d)^{p_d}\left(\sum_{n_d=1}^{N_d}\|a_{\cdot,n_d}\|_{\mathbf{p}^d,4\mathbf{N}^d}^{p_d}\right)(4N_d)^{-1}, \tag{3.3.19}$$

whence we obtain

$$\|t(\mathbf{x})\|_{\mathbf{p}} \le C(d)\|\mathbf{a}\|_{\mathbf{p},4\mathbf{N}},$$

which was to be proved. □

From Lemma 3.3.18 and the representation (3.2.12), the right-hand inequality in Theorem 3.3.15 follows immediately. We now prove the left-hand inequality in Theorem 3.3.15. We have

$$\|\{t(\mathbf{x}(\mathbf{n}))\}_{\mathbf{n}\in P'(\mathbf{N})}\|_{\mathbf{p},4\mathbf{N}} = \sup_{\|\mathbf{a}\|_{\mathbf{p}',4\mathbf{N}}\le 1} \nu(4\mathbf{N})^{-1}\sum_{\mathbf{n}\in P'(\mathbf{N})} t(\mathbf{x}(\mathbf{n}))a_{\mathbf{n}},$$

$$\mathbf{p}' := (p'_1,\dots,p'_d), \qquad p_j^{-1}+(p'_j)^{-1}=1, \qquad j=1,\dots,d.$$

Further, for $t\in\mathscr{T}(\mathbf{N},d)$

$$t(\mathbf{x}(\mathbf{n})) = (2\pi)^{-d}\int_{\mathbb{T}^d} t(\mathbf{y})\mathscr{V}_{\mathbf{N}}(\mathbf{x}(\mathbf{n})-\mathbf{y})d\mathbf{y}$$

and

$$\begin{aligned}\nu(4\mathbf{N})^{-1}\sum_{\mathbf{n}\in P'(\mathbf{N})} t(\mathbf{x}^{\mathbf{n}})a_{\mathbf{n}} &= (2\pi)^{-d}\int_{\mathbb{T}^d} t(\mathbf{y})\nu(4\mathbf{N})^{-1} \\ &\quad\times\sum_{\mathbf{n}\in P'(\mathbf{N})} a_{\mathbf{n}}\mathscr{V}_{\mathbf{N}}(\mathbf{x}(\mathbf{n})-\mathbf{y})d\mathbf{y} \\ &\le \|t\|_p\left\|\nu(4\mathbf{N})^{-1}\sum_{\mathbf{n}\in P'(\mathbf{N})} a_{\mathbf{n}}\mathscr{V}_{\mathbf{N}}(\mathbf{x}(\mathbf{n})-\mathbf{y})\right\|_{\mathbf{p}'}.\end{aligned} \tag{3.3.20}$$

Applying Lemma 3.3.18, we continue (3.3.20):

$$\le C(d)\|t\|_p\|\mathbf{a}\|_{\mathbf{p}',4\mathbf{N}} \le C(d)\|t\|_{\mathbf{p}}.$$

Theorem 3.3.15 is now proved. □

3.4 Approximation of Functions in the Classes $W^{\mathbf{r}}_{\mathbf{q},\alpha}$ and $H^{\mathbf{r}}_{\mathbf{q}}$

In this section we will study the Sobolev and Nikol'skii classes of functions of several variables. The results of this section are similar to those in Chapters 1 and 2 both in their formulations and in the methods of proof. We define the following classes of functions.

The Sobolev class $W^{\mathbf{r}}_{\mathbf{q},\alpha}B$, $\mathbf{r} = (r_1,\dots,r_d)$, $r_j > 0$, $\mathbf{q} = (q_1,\dots,q_d)$, and $\alpha = (\alpha_1,\dots,\alpha_d)$ consists of functions $f(\mathbf{x})$ which have the following integral representation for each $1 \le j \le d$:

$$f(\mathbf{x}) = (2\pi)^{-1} \int_0^{2\pi} \varphi_j(x_1,\dots,x_{j-1},y,x_{j+1},\dots,x_d) F_{r_j}(x_j - y, \alpha_j)dy,$$

$$\|\varphi_j\|_{\mathbf{q}} \le B, \tag{3.4.1}$$

where the functions $F_r(x,\alpha)$ were defined in §1.4.1. Sometimes we write

$$\varphi_j(\mathbf{x}) := f_j^{(r_j)}(\mathbf{x},\alpha_j).$$

The Nikol'skii class $H^{\mathbf{r}}_{\mathbf{q}}B$, $\mathbf{r} = (r_1,\dots,r_d)$ and $\mathbf{q} = (q_1,\dots,q_d)$ is the set of functions $f \in L_{\mathbf{q}}$ such that for each $l_j := [r_j] + 1$, $j = 1,\dots,d$, the following relations hold

$$\|f\|_{\mathbf{q}} \le B, \qquad \|\Delta_h^{l_j,j} f\|_{\mathbf{q}} \le B|h|^{r_j}, \qquad j = 1,\dots,d,$$

where $\Delta_h^{l,j}$ is the lth difference with step h in the variable x_j. In the case $B = 1$ we shall not write it in the notation of the Sobolev and Nikol'skii classes. It is usual to call these classes isotropic in the case $\mathbf{r} = r\mathbf{1}$, and anisotropic in the general case.

We first prove an analog of the Jackson theorem in the case of the approximation of functions of d variables by polynomials from $\mathscr{T}(\mathbf{N},d)$ in the $L_{\mathbf{p}}$-metric, $\mathbf{1} \le \mathbf{p} \le \infty$. For natural numbers $a_1,\dots,a_d$ we define the modulus of smoothness (continuity) in $L_{\mathbf{p}}$,

$$\omega_{a_j,j}(f,y)_{\mathbf{p}} := \sup_{|h| \le y} \|\Delta_h^{a_j,j} f\|_{\mathbf{p}}, \qquad 0 \le y \le \pi, \qquad j = 1,\dots,d.$$

Then, owing to the properties of the space $L_{\mathbf{p}}$, for $\mathbf{1} \le \mathbf{p} < \infty$, $\mathbf{p} = \infty$, all the $\omega_{a_j,j}(f,y)_{\mathbf{p}}$, for $j = 1,\dots,d$ are continuous and nondecreasing functions on y such that $\lim_{y\to 0} \omega_{a_j,j}(f,y)_{\mathbf{p}} = 0$.

In analogy to relation (2.2.1) for a natural number b we have

$$\omega_{a_j,j}(f,by)_{\mathbf{p}} \le b^{a_j} \omega_{a_j,j}(f,y)_{\mathbf{p}}.$$

Theorem 3.4.1 *For any $f \in L_{\mathbf{p}}$, $\mathbf{1} \le \mathbf{p} \le \infty$, we have*

$$E_{\mathbf{N}}(f)_{\mathbf{p}} := \inf_{t \in \mathscr{T}(\mathbf{N},d)} \|f - t\|_{\mathbf{p}} \le C(\mathbf{a}) \sum_{j=1}^{d} \omega_{a_j,j}(f, N_j^{-1})_{\mathbf{p}}.$$

Proof We define recursively the functions $f^0, f^1,\dots,f^d$: $f^0 = f$,

$$f^m(\mathbf{x}) := (2\pi)^{-1} \int_{-\pi}^{\pi} \left(f^{m-1}(\mathbf{x}) - \Delta_y^{a_m,m} f^{m-1}(\mathbf{x})\right) J_{n_m}^{a_m}(y)dy, \tag{3.4.2}$$

where $n_m := [N_m/a_m]$, $m = 1,\dots,d$.

These functions have the following properties: the function $f^m(\mathbf{x})$ is a trigonometric polynomial in the variables $x_1,\dots,x_m$ of orders $N_1,\dots,N_m$ respectively, and

$$\|f^{m-1}-f^m\|_{\mathbf{p}} \le \sum_{i=1}^{m} C(a_i)\omega_{a_i,i}(f,n_i^{-1})_{\mathbf{p}}, \qquad m=1,\dots,d-1, \tag{3.4.3}$$

$$\|\Delta_y^{a_j,j} f^m(\mathbf{x})\|_{\mathbf{p}} \le \omega_{a_j,j}(f,y)_{\mathbf{p}} + \sum_{i=1}^{m} C(a_i)\omega_{a_i,i}(f,n_i^{-1})_{\mathbf{p}}, \tag{3.4.4}$$
$$j=m+1,\dots,d, \qquad m=1,\dots,d-1.$$

We will prove relations (3.4.3) and (3.4.4) by induction. Applying the generalized Minkowski inequality (A.1.9) we obtain from (3.4.2)

$$\|f^{m-1}-f^m\|_{\mathbf{p}} \le (2\pi)^{-1}\int_{-\pi}^{\pi} \|\Delta_y^{a_m,m} f^{m-1}(\mathbf{x})\|_{\mathbf{p}} \mathscr{J}_{n_m}^{a_m}(y)dy$$
$$\le (\pi)^{-1}\int_{0}^{\pi} \omega_{a_m,m}(f^{m-1},y)\mathscr{J}_{n_m}^{a_m}(y)dy.$$

From this we find, in the same way as in the proof of Theorem 2.2.2,

$$\|f^{m-1}-f^m\|_{\mathbf{p}} \le C(a_m)\omega_{a_m,m}(f,n_m^{-1})_{\mathbf{p}}.$$

Relation (3.4.3) in the case $m=1$ is thus proved. Further,

$$\Delta_y^{a_j,j} f^m = \Delta_y^{a_j,j} f^{m-1} + \Delta_y^{a_j,j}(f^m - f^{m-1}),$$

and this, together with (3.4.3), which has just been proved for $m=1$, implies (3.4.4) for f^1. In the same way, assuming the validity of (3.4.4) for f^{m-1}, we get (3.4.3) and (3.4.4) for f^m.

Thus $f^d \in \mathscr{T}(\mathbf{N},d)$ and, by (3.4.3), we have

$$\|f-f^d\| \le \sum_{m=1}^{d} \|f^{m-1}-f^m\|_{\mathbf{p}} \le C(\mathbf{a})\sum_{m=1}^{d} \omega_{a_m,m}(f,n_m^{-1})_{\mathbf{p}}.$$

The theorem is proved. □

Corollary 3.4.2 *From Theorem 3.4.1 and the inequality* (3.2.13) *it follows that*

$$\left\|f-V_{\mathbf{N}}(f)\right\|_{\mathbf{p}} \le C(d)E_{\mathbf{N}}(f)_{\mathbf{p}} \le C(\mathbf{a})\sum_{j=1}^{d}\omega_{a_j,j}(f,N_j^{-1})_{\mathbf{p}}.$$

Corollary 3.4.3 *From Corollary 3.4.2 and the relation*

$$\omega_{a,j}(f,y)_{\mathbf{p}} \to 0 \qquad \textit{for } y\to 0, \qquad \mathbf{1}\le\mathbf{p}<\infty, \qquad \mathbf{p}=\infty,$$

it follows that for any $f\in L_{\mathbf{p}}$, $\mathbf{1}\le\mathbf{p}<\infty$, $\mathbf{p}=\infty$, *we have*

$$\left\|f-V_{\mathbf{N}}(f)\right\|_{\mathbf{p}} \to 0 \qquad \textit{for } \min_{1\le j\le d} N_j\to\infty.$$

Corollary 3.4.4 *Let f and g belong to L_1 and let $\hat{f}(\mathbf{k}) = \hat{g}(\mathbf{k})$ for all $\mathbf{k}$. Then f and g are equivalent.*

Proof Indeed, from the hypothesis of the corollary it follows that $V_{\mathbf{N}}(f) = V_{\mathbf{N}}(g)$. It remains to apply Corollary 3.4.3. □

Corollary 3.4.5 *For any $f \in L_{\mathbf{p}}$, $\mathbf{1} \le \mathbf{p} < \infty$ or $\mathbf{p} = \infty$, we have*

$$E_{\mathbf{N}}(f)_{\mathbf{p}} \to 0 \qquad \textit{for} \min_{1\le j\le d} N_j \to \infty.$$

It will be convenient for us to use the following notations. Let

$$\begin{aligned}
\mathbf{r} &:= (r_1,\dots,r_d), \qquad r_j > 0, \qquad j = 1,\dots,d;\\
g(\mathbf{r}) &:= \left(\sum_{j=1}^{d} r_j^{-1}\right)^{-1},\\
\mathbf{v} := \mathbf{v}(\mathbf{r}) &:= g(\mathbf{r})/\mathbf{r} = \big(g(\mathbf{r})/r_1,\dots,g(\mathbf{r})/r_d\big);\\
2^{\mathbf{v}n} &:= (2^{v_1 n},\dots,2^{v_d n}),\\
[2^{\mathbf{v}n}] &:= \big([2^{v_1 n}],\dots,[2^{v_d n}]\big);\\
\mathscr{T}^{\mathbf{r}}(n) &:= \mathscr{T}\big([2^{\mathbf{v}n}],d\big);\\
E^{\mathbf{r}}_n(f)_{\mathbf{p}} &:= E_{\mathscr{T}^{\mathbf{r}}(n)}(f)_{\mathbf{p}} := \inf_{t\in\mathscr{T}^{\mathbf{r}}(n)} \|f-t\|_{\mathbf{p}}.
\end{aligned}$$

We first prove some theorems for vectors $\mathbf{q} = \mathbf{p}$ and then consider the case of scalar $1 \le q, p \le \infty$.

Theorem 3.4.6 *The inclusion*

$$W^{\mathbf{r}}_{\mathbf{q},\alpha} \subset H^{\mathbf{r}}_{\mathbf{q}} B$$

holds, where B, can depend only on $\mathbf{r}$.

Proof Let $f \in W^{\mathbf{r}}_{\mathbf{q},\alpha}$ and $1 \le j \le d$. From the representation (3.4.1) we find

$$\Delta_h^{l_j,j} f(\mathbf{x}) = (2\pi)^{-1} \int_0^{2\pi} \varphi_j(x_1,\dots,x_{j-1},y,x_{j+1},\dots,x_d) \times \Delta_h^{l_j,j} F_{r_j}(x_j - y,\alpha_j)dy. \tag{3.4.5}$$

We take the L_{q_1}-norm in x_1, the L_{q_2}-norm in $x_2,\dots,$ and the $L_{q_{j-1}}$-norm in x_{j-1} of both sides of the equality (3.4.5) and apply the generalized Minkowski inequality (A.1.9); then we get

$$\begin{aligned}
&\left\|\Delta_h^{l_j,j} f(\cdot,x_j,x_{j+1},\dots,x_d)\right\|_{q_1,\dots,q_{j-1}}\\
&\quad\le (2\pi)^{-1}\int_0^{2\pi} \left\|\varphi_j(\cdot,y,x_{j+1},\dots,x_d)\right\|_{q_1,\dots,q_{j-1}} \left|\Delta_h^{l_j,j} F_{r_j}(x_j-y,\alpha_j)\right| dy.
\end{aligned}$$

Applying the Young inequality (A.1.13) in the jth variable, we obtain

$$\begin{aligned}&\|\Delta_h^{l_j,j} f(\cdot,x_{j+1},\dots,x_d)\|_{q_1,\dots,q_j}\\&\le \|\varphi_j(\cdot,x_{j+1},\dots,x_d)\|_{q_1,\dots,q_j}\|\Delta_h^{l_j,j} F_{r_j}(\cdot,\alpha_j)\|_1 dy.\end{aligned} \tag{3.4.6}$$

Further, $F_{r_j} \in H_1^{r_j} B_j$; see (1.4.19). Consequently, to conclude the proof it remains to take the $L_{q_{j+1}}$-norm in x_{j+1}, the $L_{q_{j+2}}$-norm in $x_{j+2},\dots,$ and the L_{q_d}-norm in x_d of both sides in the inequality (3.4.6). □

For a class F of functions we denote

$$E_n^{\mathbf{r}}(F)_{\mathbf{p}} := \sup_{f\in F} E_n^{\mathbf{r}}(f)_{\mathbf{p}}.$$

Theorem 3.4.7 *Let $1 \le \mathbf{q} \le \infty$; then*

$$E_n^{\mathbf{r}}(W_{\mathbf{q},\alpha}^{\mathbf{r}})_{\mathbf{q}} \asymp E_n^{\mathbf{r}}(H_{\mathbf{q}}^{\mathbf{r}})_{\mathbf{q}} \asymp 2^{-g(\mathbf{r})n}.$$

Proof We first prove the upper estimates. Owing to Theorem 3.4.6 it suffices to prove these estimates for the Nikol'skii classes. Setting

$$N_j := [2^{\nu_j n}], \qquad j = 1,\dots,d,$$

we get the required estimates from Theorem 3.4.1.

To prove the lower estimates, we consider the function

$$f(\mathbf{x}) := N^{-r_1} e^{iNx_1}, \qquad N := [2^{g(\mathbf{r})n/r_1}] + 1.$$

Then $f \in W_{\mathbf{q},\alpha}^{\mathbf{r}}$ and

$$E_n^{\mathbf{r}}(f)_{\mathbf{q}} \ge \hat{f}(N,0,\dots,0) = N^{-r_1} \asymp 2^{-g(\mathbf{r})n}. \tag{3.4.7}$$

The lower estimate for the H-classes follows from (3.4.7) and Theorem 3.4.6.

The theorem is now proved. □

We denote

$$\begin{aligned}V(f,\mathbf{r},n) &:= V(\mathbf{r},n)(f) := f * \mathscr{V}_{[2^{\nu n}]},\\A(f,\mathbf{r},0) &:= V(f,\mathbf{r},0),\\A(f,\mathbf{r},n) &:= V(f,\mathbf{r},n) - V(f,\mathbf{r},n-1), \qquad n = 1,2,\dots\end{aligned}$$

Corollary 3.4.8 *For $f \in H_{\mathbf{q}}^{\mathbf{r}}$ we have*

$$\|f - V(f,\mathbf{r},n)\|_{\mathbf{q}} \ll 2^{-g(\mathbf{r})n}, \qquad n = 0,1,\dots, \tag{3.4.8}$$

$$\|A(f,\mathbf{r},n)\|_{\mathbf{q}} \ll 2^{-g(\mathbf{r})n}, \qquad n = 0,1,\dots \tag{3.4.9}$$

Proof Clearly, it suffices to prove (3.4.8). For any $t \in \mathscr{T}^{\mathbf{r}}(n)$ we have

$$f - V(f,\mathbf{r},n) = f - t - V(f-t,\mathbf{r},n). \tag{3.4.10}$$

By (3.2.9) and the Young inequality (A.1.13),

$$\big\|V(f-t,\mathbf{r},n)\big\|_{\mathbf{q}} \le 3^d \|f-t\|_{\mathbf{q}}. \tag{3.4.11}$$

Relation (3.4.8) follows from (3.4.10), (3.4.11), and Theorem 3.4.7. □

We now prove a statement which, together with Corollary 3.4.8, shows that relation (3.4.9) is a necessary and sufficient condition for f to belong to the class $H^{\mathbf{r}}_{\mathbf{q}}B$.

Theorem 3.4.9 *Let $f \in L_{\mathbf{q}}$ and*

$$\big\|A(f,\mathbf{r},n)\big\|_{\mathbf{q}} \le 2^{-g(\mathbf{r})n}, \qquad n = 0,1,\dots \tag{3.4.12}$$

Then f is equivalent to some function from $H^{\mathbf{r}}_{\mathbf{q}}C(\mathbf{r})$, with some $C(\mathbf{r})$.

Proof Let

$$g = \sum_{n=0}^{\infty} A(f,\mathbf{r},n), \tag{3.4.13}$$

where the series converges in $L_{\mathbf{q}}$. Then by Corollary 3.4.4 the functions f and g are equivalent.

We use the following generalization of Lemma 1.4.4, the proof of which repeats the proof of that lemma.

Lemma 3.4.10 *Let $u(\mathbf{x})$ be a 2π-periodic function, which is continuous and a-times continuously differentiable in x_j.*

Then, for $\mathbf{1} \le \mathbf{q} \le \infty$,

$$\|\Delta_t^{a,j} u\|_{\mathbf{q}} \le |t|^a \left\| \frac{\partial^a u}{\partial x_j^a} \right\|_{\mathbf{q}}.$$

From (3.4.12) we obtain by Lemma 3.4.10 and Theorem 3.3.1

$$\big\|\Delta_h^{l_j,j} A(f,\mathbf{r},n)\big\|_{\mathbf{q}} \le C(\mathbf{r}) 2^{-g(\mathbf{r})n} \min\big(1, \big(|h| 2^{(g(\mathbf{r})/r_j)n}\big)^{l_j}\big). \tag{3.4.14}$$

From relations (3.4.13) and (3.4.14) we get

$$\|\Delta_h^{l_j,j} g\|_{\mathbf{q}} \le C(\mathbf{r}) |h|^{r_j},$$

which proves Theorem 3.4.9. □

We now proceed to the case of scalar q, p. We first prove an embedding theorem. For $1 \le q \le p \le \infty$ denote $\beta := 1/q - 1/p$.

Theorem 3.4.11 *Let $1 \le q \le p \le \infty$ and $g(\mathbf{r}) > \beta$. The following inclusion holds:*

$$H_q^{\mathbf{r}} \subset H_p^{\mathbf{r}'} B, \qquad \mathbf{r}' := \mathbf{r}\big(1 - \beta/g(\mathbf{r})\big).$$

Proof Let $f \in H_q^{\mathbf{r}}$. By Corollary 3.4.8 we have

$$\big\|A(f,\mathbf{r},n)\big\|_q \ll 2^{-g(\mathbf{r})n}, \qquad n = 0,1,\dots;$$

consequently, by the Nikol'skii inequality (Theorem 3.3.2),

$$\big\|A(f,\mathbf{r},n)\big\|_p \ll 2^{-g(\mathbf{r})n+\beta n} = 2^{-g(\mathbf{r}')n}. \tag{3.4.15}$$

Further, $A(f,\mathbf{r},n) = A(f,\mathbf{r}',n)$. From (3.4.15) and Theorem 3.4.9 this theorem follows. □

We now use Theorem 3.4.11 for obtaining estimates of the best approximations.

Theorem 3.4.12 *Let $1 \le q, p \le \infty$ and $g(\mathbf{r}) > \beta := (1/q - 1/p)_+$. Then*

$$E_n^{\mathbf{r}}(W_{q,\alpha}^{\mathbf{r}})_p \asymp E_n^{\mathbf{r}}(H_q^{\mathbf{r}}) \asymp 2^{-(g(\mathbf{r})-\beta)n}.$$

Proof Owing to Theorem 3.4.6 it suffices to prove the upper estimates for the Nikol'skii classes and the lower estimates for the Sobolev classes. We first prove the upper estimates. For $1 \le p \le q \le \infty$ these estimates follow directly from Theorem 3.4.7. Let $1 \le q \le p \le \infty$; then, from Theorems 3.4.11 and 3.4.7, we get that

$$E_n^{\mathbf{r}}(H_q^{\mathbf{r}})_p \ll 2^{-(g(\mathbf{r})-\beta)n},$$

which concludes the proof of the upper estimates.

Let us prove the lower estimates. For $1 \le p \le q \le \infty$ the required lower estimates follow from the proof of the lower estimates in Theorem 3.4.7. Let

$$s_j := \big[g(\mathbf{r})n/r_j\big] + 3, \qquad j = 1,\dots,d,$$
$$f(\mathbf{x}) := 2^{-(g(\mathbf{r})+1-1/q)n} \mathscr{A}_{\mathbf{s}}(\mathbf{x}).$$

Then by the Nikol'skii and Bernstein inequalities it follows from (3.2.15) that

$$\big\|f_j^{(r_j)}(\mathbf{x},\alpha_j)\big\|_q \le C(\mathbf{r}), \qquad j = 1,\dots,d, \tag{3.4.16}$$

i.e., $f \in W_{q,\alpha}^{\mathbf{r}} C(\mathbf{r})$. Further, for any $t \in \mathscr{T}^{\mathbf{r}}(n)$ we have on the one hand

$$\langle f - t, \mathscr{A}_{\mathbf{s}} \rangle = \langle f, \mathscr{A}_{\mathbf{s}} \rangle \gg 2^{-(g(\mathbf{r})-1/q)n}. \tag{3.4.17}$$

On the other hand,

$$\langle f - t, \mathscr{A}_{\mathbf{s}} \rangle \le \|f - t\|_p \|\mathscr{A}_{\mathbf{s}}\|_{p'} \ll \|f - t\|_p 2^{n/p}. \tag{3.4.18}$$

From relations (3.4.17) and (3.4.18) we get

$$E^{\mathbf{r}}_n(f)_p \gg 2^{-(g(\mathbf{r})-\beta)n}.$$

The theorem is proved. □

Theorem 3.4.12 and the properties of the de la Vallée Poussin kernels imply the following assertion.

Theorem 3.4.13 *Let $1 \le q, p \le \infty$ and $g(\mathbf{r}) > \beta := (1/q - 1/p)_+$, and let $F^{\mathbf{r}}_q$ denote one of the classes $W^{\mathbf{r}}_{q,\alpha}$ or $H^{\mathbf{r}}_q$. Then*

$$V(F^{\mathbf{r}}_q, \mathbf{r}, n)_p := \sup_{f \in F^{\mathbf{r}}_q} \|f - V(f, \mathbf{r}, n)\|_p \asymp 2^{-(g(\mathbf{r})-\beta)n}.$$

We now consider approximations by partial sums of Fourier series. For $f \in L_1$, let

$$S(f, \mathbf{r}, n) := f * \mathscr{D}_{[2^{\nu(\mathbf{r})n}]}.$$

Then $S(f, \mathbf{r}, n) \in \mathscr{T}^{\mathbf{r}}(n)$, and Theorems 3.4.12 and 1.2.1 imply an assertion analogous to Theorem 1.4.12.

Theorem 3.4.14 *Let $1 < q, p < \infty$, $g(\mathbf{r}) > \beta := (1/q - 1/p)_+$, and $F^{\mathbf{r}}_q$ be the same as in Theorem 3.4.13. Then*

$$S(F^{\mathbf{r}}_q, \mathbf{r}, n)_p := \sup_{f \in F^{\mathbf{r}}_q} \|f - S(f, \mathbf{r}, n)\|_p \asymp 2^{-(g(\mathbf{r})-\beta)n}.$$

Remark 3.4.15 Theorem 3.4.14 holds for all $(q, p) \ne (1, 1), (\infty, \infty)$.

The proof of this remark is analogous to the corresponding proof in the univariate case (see the proof of Theorem 1.4.12).

Remark 3.4.16 Let $p = 1, \infty$ and let $g(\mathbf{r}) > 0$. Then

$$S(F^{\mathbf{r}}_p, \mathbf{r}, n)_p \asymp n^d 2^{-g(\mathbf{r})n}, \qquad n = 1, 2, \dots$$

Proof The upper estimates follow from Theorem 3.4.12 and relation (3.2.2).

We now prove the lower estimates. It is easy to derive from the proof of Theorem 1.4.13 the following statement: for each $p = 1, \infty$ and $N \ge 2$ there is a $t_N \in \mathscr{T}(2N)$ such that

$$\|t_N\|_p = 1, \qquad \|S_N(t_N)\|_p \gg \ln N. \tag{3.4.19}$$

Let $N_j := [2^{\nu_j n}]$, $j = 1, \dots, d$, where n is sufficiently large, and let

$$t(\mathbf{x}) := \prod_{j=1}^{d} t_{N_j}(x_j).$$

Then

$$\|t\|_p = 1$$

and

$$\|S(t,\mathbf{r},n)\|_p = \left\|\prod_{j=1}^{d} \mathcal{S}_{N_j}(t_{N_j})\right\|_p \gg \prod_{j=1}^{d} \ln N_j \asymp n^d.$$

To conclude the proof it remains to remark that by the Bernstein inequality the function $f := 2^{-g(\mathbf{r})n}t$ belongs to $W^{\mathbf{r}}_{q,\alpha}B$. □

The Nikol'skii inequalities allow us to obtain estimates of the best approximations $E^{\mathbf{r}}_m(f)_p$ in terms of the best approximations $E^{\mathbf{r}}_n(f)_q$, $q < p$ and $n \geq m - C(\mathbf{r})$. We prove the following assertion.

Theorem 3.4.17 *Let $\mathbf{r} > 0$ be given. There is an $n(\mathbf{r}) > 0$ such that for all $m > n(\mathbf{r})$ and $1 \leq q \leq p \leq \infty$ we have, for $f \in L_q$ (in the case $p = \infty$ we assume that f coincides with a continuous function equivalent to it),*

$$E^{\mathbf{r}}_m(f)_p \leq C(\mathbf{r}) \sum_{n=m-n(\mathbf{r})}^{\infty} E^{\mathbf{r}}_n(f)_q 2^{\beta n}. \tag{3.4.20}$$

Moreover, for $1 \leq q < p < \infty$ we have

$$E^{\mathbf{r}}_m(f)_p \leq C(\mathbf{r},q,p) \left(\sum_{n=m-n(\mathbf{r})}^{\infty} (E^{\mathbf{r}}_n(f)_q 2^{\beta n})^p \right)^{1/p}. \tag{3.4.21}$$

Proof We first prove (3.4.20). We will assume that the series on the right-hand side of (3.4.20) converges. From the definition of $A(f,\mathbf{r},n)$ and the properties of the de la Vallée Poussin kernels we find

$$\|A(f,\mathbf{r},n)\|_q \leq C(d) E^{\mathbf{r}}_{n-1}(f)_q. \tag{3.4.22}$$

By the Nikol'skii inequality we have

$$\|A(f,\mathbf{r},n)\|_p \leq C(\mathbf{r}) 2^{n\beta} \|A(f,\mathbf{r},n)\|_q. \tag{3.4.23}$$

Let $n(\mathbf{r})$ be a number such that

$$A(f,\mathbf{r},n) \in \mathcal{T}^{\mathbf{r}}(n+n(\mathbf{r})), \qquad n = 0,1,\ldots$$

Then

$$E^{\mathbf{r}}_m(f)_p \leq \left\| f - \sum_{n=0}^{m-n(\mathbf{r})} A(f,\mathbf{r},n) \right\|_p. \tag{3.4.24}$$

From the representation

$$f = \sum_{n=0}^{\infty} A(f, \mathbf{r}, n) \tag{3.4.25}$$

in L_q, relations (3.4.22)–(3.4.24), and Corollary 3.4.4 it follows that the representation is valid for f in L_p as well, and that for $E_m^{\mathbf{r}}(f)_p$ the inequality (3.4.20) holds.

To prove (3.4.21) we assume that the series on the right-hand side of (3.4.21) converges, and we use Lemma 3.3.14 with $f_n = A(f, \mathbf{r}, n)$, $\kappa = 1$. Then, for any $m_1 < m_2$, we have

$$\left\| \sum_{n=m_1}^{m_2} f_n \right\|_p \leq C(p,q,d) \left(\sum_{n=m_1}^{m_2} \|f_n\|_q^p 2^{\beta np} \right)^{1/p},$$

which implies the convergence of the series (3.4.25) in L_p. Furthermore, applying Lemma 3.3.14 for estimating the right-hand side of (3.4.24), we obtain the inequality (3.4.21). The theorem is now proved. □

3.5 Estimates of the Widths of the Sobolev and Nikol'skii Classes

In this section we consider the Kolmogorov widths, linear widths, and orthowidths of the classes $W_{q,\alpha}^{\mathbf{r}}$ and $H_q^{\mathbf{r}}$ in L_p space with scalar $1 \leq q, p \leq \infty$. It turns out that the orders of each of these widths for the classes $W_{q,\alpha}^{\mathbf{r}}$ and the classes $H_q^{\mathbf{r}}$ are the same. For this reason we formulate theorems for both classes and for the sake of brevity we denote these classes by $F_q^{\mathbf{r}}$.

Theorem 3.5.1 *Let $r(p,q)$ be the same as in Theorem 2.1.1 and let $g(\mathbf{r}) > r(q,p)$. Then*

$$d_m(F_q^{\mathbf{r}}, L_p) \asymp m^{-g(\mathbf{r})+(1/q-\max(1/2,1/p))_+}, \qquad 1 \leq q,p \leq \infty.$$

Theorem 3.5.2 *Let $1 \leq q, p \leq \infty$ and $\kappa := \max(1/q - 1/2; 1/2 - 1/p)$. Then*

$$\lambda_m(F_q^{\mathbf{r}}, L_p) \asymp \begin{cases} m^{-g(\mathbf{r})+(1/q-1/p)_+}, & \text{for } q \geq 2 \text{ or } p \leq 2, \\ & g(\mathbf{r}) > (1/q-1/p)_+, \\ m^{-g(\mathbf{r})+\kappa}, & \text{for } 1 \leq q \leq 2 \text{ and } 2 \leq p \leq \infty, \\ & g(\mathbf{r}) > 1+\kappa. \end{cases}$$

Theorem 3.5.3 *Let $1 \leq q, p \leq \infty$ and $(q,p) \neq (1,1), (\infty,\infty)$. Then*

$$\varphi_m(F_q^{\mathbf{r}}, L_p) \asymp m^{-g(\mathbf{r})+(1/q-1/p)_+}$$

and the operator $S(\cdot, \mathbf{r}, n)$ (see the text after Theorem 3.4.13) with the corresponding n is an orthogonal projection operator giving the order of this quantity.

Theorem 3.5.3 shows that from the point of view of orthowidths it is natural to approximate functions from the classes $F_q^{\mathbf{r}}$ by trigonometric polynomials in $\mathscr{T}^{\mathbf{r}}(n)$, which are defined by the vector $\mathbf{r}$. A comparison of Theorem 3.5.2 with Theorem 3.4.13 shows that the operator $V(\cdot,\mathbf{r},n)$ with appropriate n is an optimal linear operator for all $1 \le q, p \le \infty$ excepting the case $1 \le q < 2, 2 < p \le \infty$. A comparison of Theorem 3.5.1 with Theorem 3.4.13 shows that this operator is optimal from the point of view of the Kolmogorov widths for all $1 \le p \le q \le \infty$ and $1 \le q \le p \le 2$. The above remarks show that the sets $\mathscr{T}^{\mathbf{r}}(n)$ of trigonometric polynomials play the same role for the classes $F_q^{\mathbf{r}}$ in the multivariate case as the sets $\mathscr{T}(2^n)$ play for the classes $W_{q,\alpha}^r$ and H_q^r in the univariate case.

The proofs of Theorems 3.5.1–3.5.3 are quite similar to the corresponding proofs of Theorems 2.1.1–2.1.3.

3.5.1 The Case $1 \le p \le q \le \infty$

The upper estimates in Theorems 3.5.1 and 3.5.2 follow from Theorem 3.4.13, and those in Theorem 3.5.3 follow from Theorem 3.4.14 and Remark 3.4.15. Theorems 3.5.1–3.5.3 state that in this case the orders of all three widths coincide. Therefore, by relation (2.1.1) and Theorem 3.4.6 it suffices to prove the lower estimate for $d_m(W_{\infty,\alpha}^{\mathbf{r}}, L_1)$.

Let m be given and $\{u_i\}_{i=1}^m$ be some system of functions in L_1. We choose n such that

$$\dim \mathscr{T}^{\mathbf{r}}(n) \ge 2m \ge \dim \mathscr{T}^{\mathbf{r}}(n-1).$$

It is clear, that $2^n \asymp m$.

We consider the space $\mathscr{T}^{\mathbf{r}}(n)$ and its subspace

$$\Psi := \{t \in \mathscr{T}^{\mathbf{r}}(n) : \langle t, u_i \rangle = 0,\ i = 1, \dots, m\}.$$

By Theorem 3.2.1 we can find an $f \in \Psi$ such that

$$\|f\|_\infty = 1, \qquad \|f\|_2^2 \ge C(\mathbf{r}) > 0.$$

Then for any polynomial u in the system $\{u_i\}_{i=1}^m$ we have

$$C(\mathbf{r}) \le \langle f, f \rangle = \langle f - u, f \rangle \le \|f - u\|_1. \tag{3.5.1}$$

By Theorem 3.3.1 we get, for $j = 1, \dots, d$,

$$\left\| f_j^{(r_j)}(\mathbf{x}, \alpha_j) \right\|_\infty \le C(\mathbf{r}) 2^{g(\mathbf{r})n}. \tag{3.5.2}$$

The required lower estimate follows from relations (3.5.1) and (3.5.2).

3.5.2 The Lower Estimates in Theorem 3.5.3 for $1 \le q \le p \le \infty$

In this subsection, as in the proof of the lower bounds in Theorem 2.1.3, we will prove a more general statement. Let $\mathscr{L}_m(B)_p$ be the same as in §2.1.2 with a natural modification of (2.1.11):

$$\|Ae^{i(\mathbf{k},\mathbf{x})}\|_2 \le B.$$

Theorem 3.5.4 *Let $B \ge 1$, $1 \le q \le p \le \infty$ and $g(\mathbf{r}) > 1/q - 1/p$. Then*

$$\inf_{A \in \mathscr{L}_m(B)_p} A(F_q^{\mathbf{r}}, L_p) \asymp m^{-g(\mathbf{r})+1/q-1/p}.$$

Proof The upper estimate follows from Theorem 3.4.13. We will prove the lower estimate. Let m be given and $G \in \mathscr{L}_m(B)_p$. For some n, which will be specified later, we consider the operator

$$A := V(\mathbf{r},n)G \in \mathscr{L}_m(B)_p,$$

the range of which belongs to $\mathscr{T}^{\mathbf{r}}\big(n+n(\mathbf{r})\big)$, where

$$n(\mathbf{r}) := \max_j \big[r_j/g(\mathbf{r})\big] + 1. \tag{3.5.3}$$

Then for $f \in \mathscr{T}^{\mathbf{r}}(n)$, $1 \le p \le \infty$, we have

$$\|f - Af\|_p = \big\|V(\mathbf{r},n)(f - Gf)\big\|_p \le 3^d \|f - Gf\|_p.$$

This means it suffices to prove the lower estimate for the class $\mathscr{T}^{\mathbf{r}}(n) \cap W^{\mathbf{r}}_{q,\alpha}$ and operators $A \in \mathscr{L}_m(B)_p$ acting in $\mathscr{T}^{\mathbf{r}}\big(n+n(\mathbf{r})\big)$.

The following assertion is an analog of Lemma 2.1.6.

Lemma 3.5.5 *Let $A \in \mathscr{L}_m(B)_p$ be defined by the relation*

$$Ae^{i(\mathbf{k},\mathbf{x})} = \sum_{a=1}^{m} b_a^{\mathbf{k}} \psi_a(\mathbf{x}),$$

where $\big\{\psi_a(\mathbf{x})\big\}_{a=1}^m$ is an orthonormal system of functions.

Then, for any trigonometric polynomial $t \in \mathscr{T}^{\mathbf{r}}(n)$, we have

$$\min_{\mathbf{y}=\mathbf{x}} \operatorname{Re} A\big(t(\mathbf{x}-\mathbf{y})\big) \le B\big(m \dim \mathscr{T}^{\mathbf{r}}(n)\big)^{1/2} \max_{\mathbf{k}} \big|\hat{t}(\mathbf{k})\big|.$$

We consider as an example the function

$$f(\mathbf{x}) := \mathscr{K}_{\mathbf{N}}(\mathbf{x}), \qquad \mathbf{N} := [2^{(g(\mathbf{r})/\mathbf{r})n}],$$

belonging to $\mathscr{T}^{\mathbf{r}}(n)$. Further, since

$$\mathscr{K}_{\mathbf{N}}(\mathbf{0}) \ge 2^{n-d},$$

choosing n sufficiently large such that $2^n \asymp m$ and using Lemma 3.5.5 we get

$$\sup_{\mathbf{y}} \big\| f(\mathbf{x}-\mathbf{y}) - Af(\mathbf{x}-\mathbf{y}) \big\|_\infty \geq f(\mathbf{0}) - \min_{\mathbf{y}=\mathbf{x}} \operatorname{Re} Af(\mathbf{x}-\mathbf{y})$$
$$\geq 2^{n-d} - B\big(m 2^n C(d)\big)^{1/2} \gg 2^n.$$

Therefore, there is a $\mathbf{y}^*$ such that, for $f^*(\mathbf{x}) := f(\mathbf{x}-\mathbf{y}^*)$, we have

$$\|f^* - Af^*\|_\infty \gg 2^n. \tag{3.5.4}$$

Using that $f^* \in \mathscr{T}^{\mathbf{r}}(n)$ and $Af^* \in \mathscr{T}^{\mathbf{r}}\big(n + n(\mathbf{r})\big)$, we find from (3.5.4), by the Nikol'skii inequality,

$$\|f^* - Af^*\|_p \gg 2^{n(1-1/p)}, \qquad 1 \leq p < \infty. \tag{3.5.5}$$

By Theorem 3.3.1 and by (3.2.7) we have

$$\Big\| f_j^{*(r_j)}(\mathbf{x}, \alpha_j) \Big\|_q \ll 2^{g(\mathbf{r})n} \|f^*\|_q \ll 2^{(g(\mathbf{r})+1-1/q)n}. \tag{3.5.6}$$

The required lower estimates in Theorem 3.5.4 for $1 \leq q \leq p \leq \infty$ follow from relations (3.5.4)–(3.5.6). □

We note that in fact we have proved that

$$\varphi_m\big(\mathscr{T}^{\mathbf{r}}(n) \cap W^{\mathbf{r}}_{q,\alpha}, L_p\big) \gg m^{-g(\mathbf{r})+1/q-1/p}, \qquad 1 \leq q \leq p \leq \infty, \tag{3.5.7}$$

where n is such that $m \asymp 2^n$. □

3.5.3 *Proof of the Lower Estimate in Theorem 3.5.1 for* $1 \leq q \leq p \leq \infty$

Clearly, it suffices to consider the case $1 \leq q \leq p \leq 2$. In the case $p = 2$, owing to (2.1.2) the required estimates follow from Theorem 3.5.3. Moreover, by (3.5.7), for each m there is an n such that $2^n \asymp m$ and

$$d_m\big(\mathscr{T}^{\mathbf{r}}(n) \cap W^{\mathbf{r}}_{q,\alpha}, L_2\big) \gg m^{-g(\mathbf{r})+1/q-1/2}, \qquad 1 \leq q \leq 2. \tag{3.5.8}$$

Further, let $\{u_i\}_{i=1}^m$ be a system of functions in L_p and let

$$E_m(f,U)_p := \inf_{c_i} \left\| f - \sum_{i=1}^m c_i u_i \right\|_p.$$

We consider the system $\mathscr{V}$ of functions $v_i := V(u_i, \mathbf{r}, n)$, $i = 1, \dots, m$. Then, for $f \in \mathscr{T}^{\mathbf{r}}(n)$, we have

$$E_m(f, \mathscr{V})_p \leq 3^d E_m(f,U)_p, \tag{3.5.9}$$

and, by the Nikol'skii inequality,

$$E_m(f, \mathscr{V})_2 \ll 2^{n(1/p-1/2)} E_m(f, \mathscr{V})_p, \qquad 1 \leq p \leq 2. \tag{3.5.10}$$

From relations (3.5.8)–(3.5.10) we get

$$d_m\big(\mathscr{T}^{\mathbf{r}}(n)\cap W^{\mathbf{r}}_{q,\alpha},L_p\big) \gg m^{-g(\mathbf{r})+1/q-1/p}, \qquad 1\le q\le p\le 2, \tag{3.5.11}$$

which implies the required lower estimates in Theorem 3.5.1. □

3.5.4 The Lower Estimates in Theorem 3.5.2 for $1\le q\le p\le\infty$

In this case, under the additional assumption $1/q+1/p\ge 1$, Theorems 3.5.1 and 3.5.2 show that the orders of the Kolmogorov widths and linear widths coincide; therefore, by (2.1.1) the required estimates follow from the estimates in Theorem 3.5.1, which have already been proved. We note that in the proof of the lower estimates of the Kolmogorov widths it was in fact shown that for any m there is an n such that $2^n \asymp m$ and

$$d_m\big(\mathscr{T}^{\mathbf{r}}(n)_q,L_p\big) \gg m^{1/q-1/p}, \qquad 1\le q\le p\le 2, \tag{3.5.12}$$

where $\mathscr{T}^{\mathbf{r}}(n)_q$ is the unit L_q-ball in $\mathscr{T}^{\mathbf{r}}(n)$.

We prove now the following assertion.

Lemma 3.5.6 *For any* $1\le q,p\le\infty$ *we have the inequality*

$$\lambda_m\big(\mathscr{T}(2\mathbf{N},d)_q,L_p\big) \gg \lambda_m\big(\mathscr{T}(\mathbf{N},d)_{p'},L_{q'}\big).$$

Proof Let A be a linear operator of rank m acting from L_q to L_p. We denote

$$A_1 := V_{2\mathbf{N}}AV_{2\mathbf{N}}.$$

Then A_1 is a linear operator of rank not greater than m and it can be considered as an operator in Hilbert space L_2. Further,

$$\begin{aligned}
\sup_{t\in\mathscr{T}(2\mathbf{N},d)_q} \big\|(I-A)t\big\|_p &\ge 3^{-d} \sup_{t\in\mathscr{T}(2\mathbf{N},d)_q} \big\|(I-A_1)t\big\|_p\\
&= 3^{-d} \sup_{t\in\mathscr{T}(2\mathbf{N},d)_q;\|u\|_{p'}\le 1} |\langle (I-A_1)t,u\rangle|\\
&\ge 3^{-d} \sup_{t\in\mathscr{T}(2\mathbf{N},d)_q;u\in\mathscr{T}(\mathbf{N},d)_{p'}} |\langle (I-A_1)t,u\rangle|\\
&= 3^{-d} \sup_{t\in\mathscr{T}(2\mathbf{N},d)_q;u\in\mathscr{T}(\mathbf{N},d)_{p'}} |\langle t,(I-A_1)^*u\rangle|\\
&\ge 3^{-2d} \sup_{\|\varphi\|_q\le 1;u\in\mathscr{T}(\mathbf{N},d)_{p'}} |\langle V_{\mathbf{N}}\varphi,(I-A_1)^*u\rangle|\\
&= 3^{-2d} \sup_{\|\varphi\|_q\le 1;u\in\mathscr{T}(\mathbf{N},d)_{p'}} |\langle \varphi,(I-V_{\mathbf{N}}A_1^*)u\rangle|\\
&\ge 3^{-2d}\lambda_m\big(\mathscr{T}(\mathbf{N},d)_{p'},L_{q'}\big),
\end{aligned}$$

which proves the lemma. □

From relation (3.5.12), using Lemma 3.5.6 and the Bernstein inequality we obtain the required lower estimates in Theorem 3.5.2 for all $1 \le q \le p \le \infty$.

3.5.5 Proof of the Upper Estimates in Theorem 3.5.2

We first prove the following assertion.

Theorem 3.5.7 *Let* $1 \le q \le 2$, $2 \le p \le \infty$; *then*

$$\lambda_m(\mathscr{T}(\mathbf{N},d)_q, L_p) \ll \begin{cases} \vartheta(\mathbf{N})^{1/2+1/q}\big(\max(1,m)\big)^{-1} & \text{for } 1/q+1/p \ge 1, \\ \vartheta(\mathbf{N})^{3/2-1/p}\big(\max(1,m)\big)^{-1} & \text{for } 1/q+1/p \le 1. \end{cases}$$

Proof Owing to Lemma 3.5.6 it suffices to prove the theorem for $1/q+1/p \ge 1$. To begin with, let $d=1$ and b be the least prime greater than $4N$ (clearly, we can assume $N \ge 1$) and set $b = 2a+1$. Let $t_m(x)$ be the polynomial from Theorem 1.2.6. We consider the function

$$u(x) := \mathscr{D}_a(x) - t_m(x)b/t_m(0).$$

Then $u(0)=0$ and, by Theorem 1.2.6, we have

$$\left| u\left(\frac{2\pi l}{2a+1}\right)\right| \le Cb^{3/2}m^{-1}. \tag{3.5.13}$$

We now consider the function

$$w(x) := u(x) * \mathscr{V}_N(x).$$

From (3.5.13), (1.2.5), and property (2) of the de la Vallée Poussin kernels follows the estimate

$$\|w\|_\infty \le Cb^{3/2}m^{-1}.$$

Consequently, for the convolution operator J_w with kernel w we have

$$\|J_w\|_{1\to\infty} \le Cb^{3/2}m^{-1}.$$

From this relation, the inequality

$$\|J_w\|_{2\to 2} \le Cb/m,$$

and the Riesz–Thorin theorem A.3.2, we get, for $1/q+1/p \ge 1$, the relation

$$\|J_w\|_{q\to p} \le Cb^{1/2+1/q}m^{-1},$$

which easily implies the first estimate in the theorem for $d=1$.

The proof in the case $d>1$ is based on the case $d=1$ just considered and on the following lemma.

Lemma 3.5.8 *For any $1 \le q, p \le \infty$ and with $\mathbf{N} \ge \mathbf{1}$, we have*

$$\lambda_m(B_q^{\nu(\mathbf{N})}, \ell_p) \le C(d)\vartheta(\mathbf{N})^{1/p-1/q}\lambda_m(\mathscr{RT}(\mathbf{N},d)_q, L_p),$$

$$\lambda_m(\mathscr{RT}(\mathbf{N},d)_q, L_p) \le C(d)\vartheta(\mathbf{N})^{1/q-1/p}\lambda_m(B_q^{\nu(4\mathbf{N})}, \ell_p);$$

and analogous relations hold for the Kolmogorov widths d_m.

Proof We begin by proving the first relation. Let the coordinates of the elements $\mathbf{y} \in B_q^{\nu(\mathbf{N})}$, $\mathbf{N} \ge 1$, be enumerated by vectors

$$\mathbf{n} = (n_1, \dots, n_d), \qquad 1 \le n_j \le N_j, \qquad j = 1, \dots, d.$$

For $\mathbf{y} \in B_q^{\nu(\mathbf{N})}$ we set

$$t(\mathbf{y})(\mathbf{x}) := t(\mathbf{y},\mathbf{x}) := \sum_{\mathbf{n}} y_{\mathbf{n}}\varphi(\mathbf{x} - \mathbf{z}^{\mathbf{n}}),$$

where

$$\mathbf{z}^{\mathbf{n}} := \left(\frac{2\pi n_1}{N_1}, \dots, \frac{2\pi n_d}{N_d}\right),$$

$$\varphi(\mathbf{x}) := \mathscr{K}_{\mathbf{N}-\mathbf{1}}(\mathbf{x})/\mathscr{K}_{\mathbf{N}-\mathbf{1}}(\mathbf{0}).$$

Then $t(\mathbf{y}, \mathbf{z}^{\mathbf{n}}) = y_{\mathbf{n}}$. It is easy to see that

$$\left\|t(\mathbf{y},\cdot)\right\|_1 \le C(d)\vartheta(\mathbf{N})^{-1}\|\mathbf{y}\|_1.$$

Using a property of the Fejér kernels that is analogous to the property (2) of the de la Vallée Poussin kernels, we get

$$\left\|t(\mathbf{y},\cdot)\right\|_\infty \le C(d)\|\mathbf{y}\|_\infty.$$

From these two inequalities, by the Riesz–Thorin theorem we find

$$\left\|t(\mathbf{y},\cdot)\right\|_q \le C(d)\vartheta(\mathbf{N})^{-1/q}\|\mathbf{y}\|_q, \qquad 1 \le q \le \infty. \tag{3.5.14}$$

Let A be a linear operator with rank $A \le m$ giving the order of $\lambda_m(\mathscr{RT}(\mathbf{N},d)_q, L_p)$. We consider on $B_q^{\nu(\mathbf{N})}$ the operator G such that

$$G\mathbf{y} = \left\{\left(V_{\mathbf{N}}At(\mathbf{y})\right)(\mathbf{z}^{\mathbf{n}})\right\}_{\mathbf{n}},$$

which maps a vector $\mathbf{y}$ to the vector with $\mathbf{n}$th coordinate equal to the value of the function $V_{\mathbf{N}}At(\mathbf{y},\mathbf{x})$ at the point $\mathbf{z}^{\mathbf{n}}$. By Theorem 3.3.15 and relation (3.5.14) we get, for $\mathbf{y} \in B_q^{\nu(\mathbf{N})}$,

$$\begin{aligned}\|\mathbf{y} - G\mathbf{y}\|_p &\le \left\|\left\{t(\mathbf{y},\mathbf{x}(\mathbf{k})) - (V_{\mathbf{N}}At(\mathbf{y}))(\mathbf{x}(\mathbf{k}))\right\}_{\mathbf{k}\in P'(2\mathbf{N})}\right\|_p \\ &\le C(d)\vartheta(\mathbf{N})^{1/p}\|t - V_{\mathbf{N}}At\|_p \le C(d)\vartheta(\mathbf{N})^{1/p}\|t - At\|_p \\ &\le C(d)\vartheta(\mathbf{N})^{1/p-1/q}\lambda_m(\mathscr{RT}(\mathbf{N},d)_q, L_p),\end{aligned}$$

which implies the first relation.

We now prove the second relation. We map each $t \in \mathscr{RT}(\mathbf{N},d)_q$ to the vector $M(t) := \{t(\mathbf{x}(\mathbf{n}))\}_{\mathbf{n}\in P'(\mathbf{N})}$. Then, by Theorem 3.3.15, we have

$$M(t) \in C(d)\vartheta(\mathbf{N})^{1/q} B_q^{\nu(4\mathbf{N})}.$$

Let A be a linear operator with $\operatorname{rank} A \le m$ giving the order of $\lambda_m(B_q^{\nu(4\mathbf{N})}, \ell_p)$ and let $u := A(M(t))$. Then, setting

$$Vu := \nu(4\mathbf{N})^{-1} \sum_{\mathbf{n}\in P'(\mathbf{N})} u_{\mathbf{n}} \mathscr{V}_{\mathbf{N}}(\mathbf{x}-\mathbf{x}(\mathbf{n})),$$

we obtain, using Lemma 3.3.18,

$$\begin{aligned}\|t - Vu\|_p &\le C(d)\nu(4\mathbf{N})^{-1/p} \left\| \{t(\mathbf{x}(\mathbf{n})) - u_{\mathbf{n}}\}_{\mathbf{n}\in P'(\mathbf{N})} \right\|_p \\ &\le C(d)\nu(4\mathbf{N})^{1/q-1/p} \lambda_m(B_q^{\nu(4\mathbf{N})}, \ell_p),\end{aligned}$$

which implies the second relation.

In just the same way the corresponding relations for the Kolmogorov widths can be proved. □

Theorem 3.5.7 for $d > 1$ follows from the conclusion of the theorem for $d = 1$ and the following chain of inequalities which are valid owing to Lemma 3.5.8:

$$\begin{aligned}\lambda_{2m}(\mathscr{T}(\mathbf{N},d)_q, L_p) &\le 2\lambda_m(\mathscr{RT}(\mathbf{N},d)_q, L_p) \\ &\le C(d)\vartheta(\mathbf{N})^{1/q-1/p} \lambda_m(B_q^{\nu(4\mathbf{N})}, \ell_p) \\ &\le C(d)\lambda_m(\mathscr{RT}(\nu(4\mathbf{N}),1)_q, L_p).\end{aligned}$$

Theorem 3.5.7 is now proved. □

We can conclude the proof of Theorem 3.5.2. Let $1 \le q \le 2$, $2 \le p \le \infty$, and $\theta > 0$ be such that

$$g(\mathbf{r}) > 1 + \max(1/q - 1/2, 1/2 - 1/p) + \theta.$$

We take a natural number j and define

$$m_l := [2^{j-(l-j)\theta}], \qquad l = j+1, \dots$$

Then

$$\lambda_m(H_q^{\mathbf{r}}, L_p) \ll \sum_{l>j} \lambda_{m_l}(\mathscr{T}^{\mathbf{r}}(l+n(\mathbf{r})), L_p) \sup_{f\in H_q^{\mathbf{r}}} \|A(f,\mathbf{r},l)\|_q, \tag{3.5.15}$$

where

$$m := \dim \mathscr{T}^{\mathbf{r}}(j+n(\mathbf{r})) + \sum_{l>j} m_l \ll 2^j. \tag{3.5.16}$$

From relation (3.5.15), applying Theorem 3.5.7 and relations (3.4.9) and (3.5.16), we get (for $1/q + 1/p \geq 1$, noting that in the other case the relation is analogous) that

$$\lambda_m(H_q^{\mathbf{r}}, L_p) \ll \sum_{l>j} 2^{l(1/2+1/q)-j+(l-j)\theta-g(\mathbf{r})l}$$

$$\ll m^{-g(\mathbf{r})+1/q-1/2}. \tag{3.5.17}$$

The relation (3.5.17) is proved for m of the form (3.5.16). Clearly, this is sufficient. Theorem 3.5.2 is proved. □

3.5.6 Proof of the Upper Estimates in Theorem 3.5.1

This proof is quite analogous to the proof of Theorem 2.1.1. We present here only some elements of the proof. In the same way as Theorem 2.1.14 was derived from Theorem 2.1.11 we derive the following theorem from Theorem 2.1.11. In this proof instead of Theorem 1.3.9 and Lemma 1.3.10 we use Theorem 3.3.15 and Lemma 3.3.18.

Theorem 3.5.9 *One has the estimate*

$$d_m\big(\mathscr{T}(\mathbf{N},d)_2, L_\infty\big) \ll \big(\vartheta(\mathbf{N})/m\big)^{1/2} \ln\big(e\vartheta(\mathbf{N})/m\big)\big)^{1/2}.$$

Let $\theta > 0$ be such that $g(\mathbf{r}) > 1/2 + \theta$ and let j, m_l, m be the same as above. Then

$$d_m(H_2^{\mathbf{r}}, L_\infty) \ll \sum_{l>j} d_{m_l}(T^{\mathbf{r}}(l+n(\mathbf{r}))_2, L_\infty) 2^{-g(\mathbf{r})l} \ll m^{-g(\mathbf{r})}. \tag{3.5.18}$$

In the remaining cases $1 \leq q < 2$ and $2 < p \leq \infty$ the estimates follow from (3.5.18), the monotonicity of L_p-norms and Theorem 3.4.11. □

3.6 Sampling and Numerical Integration

This section is similar to §2.4 in the formulation of the problems and in the methods of solving them. Here we deal only with continuous functions.

3.6.1 Numerical Integration

For a given m, $\Lambda := (\lambda_1, \ldots, \lambda_m)$, $\xi := (\xi^1, \ldots, \xi^m)$, $\lambda_j \in \mathbb{C}$, $\xi^j \in \mathbb{T}^d$, $j = 1, \ldots, m$, we define the cubature formula (m, Λ, ξ):

$$\Lambda_m(f, \xi) := \sum_{j=1}^{m} \lambda_j f(\xi^j), \tag{3.6.1}$$

and for a class $\mathbf{F}$ of functions we denote

$$\Lambda_m(\mathbf{F},\xi) := \sup_{f\in\mathbf{F}} \left| \Lambda_m(f,\xi) - (2\pi)^{-d} \int_{\mathbb{T}^d} f(\mathbf{x})d\mathbf{x} \right|,$$

$$\kappa_m(\mathbf{F}) := \inf_{\Lambda,\xi} \Lambda_m(\mathbf{F},\xi).$$

We first present some heuristic arguments regarding the choice of a cubature formula for the class $H^{\mathbf{r}}_\infty$. Let $f \in H^{\mathbf{r}}_\infty$. Substituting the multidimensional integral by the iterated integral

$$\int_{\mathbb{T}^d} f(\mathbf{x})d\mathbf{x} = \int_0^{2\pi} \left(\cdots \left(\int_0^{2\pi} f(\mathbf{x})dx_1 \right) \cdots \right) dx_d,$$

we can apply the quadrature formula q_{N_j} (see §2.4.1) instead of integrating in x_j. The function f as a function in x_j, with the other coordinates fixed, belongs to the class $H^{r_j}_\infty$. Therefore, in each step the error of substituting the integral by q_{N_j} will not be greater than $C(r_j)N_j^{-r_j}$. The whole error for the cubature formula

$$q_{\mathbf{N}}(f) := \nu(\mathbf{N})^{-1} \sum_{j_d=1}^{N_d} \cdots \sum_{j_1=1}^{N_1} f(2\pi j_1/N_1,\ldots,2\pi j_d/N_d)$$

will not exceed

$$q_{\mathbf{N}}(H^{\mathbf{r}}_\infty) \le C(\mathbf{r}) \sum_{j=1}^{d} N_j^{-r_j}. \tag{3.6.2}$$

We note that the cubature formula $q_{\mathbf{N}}$ has $\nu(\mathbf{N})$ points. If $\nu(\mathbf{N})$ is fixed then the sum on the right-hand side of (3.6.2) will be smallest in the case

$$r_1 N_1^{-r_1} = r_2 N_2^{-r_2} = \cdots = r_d N_d^{-r_d}.$$

Let a vector $\mathbf{r} = (r_1,\ldots,r_d)$, $r_j > 0$ and a number m be given. We define numbers $N_j := \max([m^{\nu_j}],1)$, $\nu_j := g(\mathbf{r})/r_j$, $j = 1,\ldots,d$, and the cubature formula

$$q_m(\mathbf{r})(f) := q_m(f,\mathbf{r}) := q_{\mathbf{N}}(f), \qquad \mathbf{N} := (N_1,\ldots,N_d). \tag{3.6.3}$$

The number of points of this cubature formula does not exceed m.

We will prove that the orders of $\kappa_m(\mathbf{F})$ are the same for $\mathbf{F} = W^{\mathbf{r}}_{p,\alpha}$ and $\mathbf{F} = H^{\mathbf{r}}_p$ which is similar to the results of §§3.4 and 3.5. Therefore, later we shall formulate theorems for both classes simultaneously and, for the sake of brevity, denote them by $F^{\mathbf{r}}_p$.

Theorem 3.6.1 *Let $1 \le p \le \infty$ and $g(\mathbf{r}) > 1/p$. We have*

$$\kappa_m(F^{\mathbf{r}}_p) \asymp q_m(F^{\mathbf{r}}_p,\mathbf{r}) \asymp m^{-g(\mathbf{r})}.$$

Proof By Theorem 3.4.11, $H^{\mathbf{r}}_p \subset H^{\mathbf{r}'}_\infty B$; therefore we assume that all functions from $H^{\mathbf{r}}_p$ are continuous for $g(\mathbf{r}) > 1/p$. Clearly, it suffices to prove the upper estimate for the classes $H^{\mathbf{r}}_p$. Let $\mathbf{N}$ be the same as in (3.6.3). We consider the linear functional $q_m(\mathbf{r})V(\mathbf{r},n)$ and estimate the norm of this functional on the L_p-space. For $p = \infty$ we have,

$$\left\|q_m(\mathbf{r})V(\mathbf{r},n)\right\|_{L_\infty \to \mathbb{C}} \le C(d). \tag{3.6.4}$$

In the case $p = 1$, using relation (2.4.4) we get, for $2^n \ge m$,

$$\begin{aligned}
&\left\|q_m(\mathbf{r})V(\mathbf{r},n)\right\|_{L_1 \to \mathbb{C}} \\
&\quad \le \left\| m^{-1} \sum_{j_d=1}^{N_d} \cdots \sum_{j_1=1}^{N_1} \mathscr{V}_{[2^{\nu n}]}\big((2\pi j_1/N_1) - y_1, \dots, (2\pi j_d/N_d) - y_d\big) \right\|_\infty \\
&\quad \le C(d)2^n/m.
\end{aligned} \tag{3.6.5}$$

From (3.6.4) and (3.6.5) we obtain, by the Riesz–Thorin theorem A.3.2,

$$\left\|q_m(\mathbf{r})V(\mathbf{r},n)\right\|_{L_p \to \mathbb{C}} \le C(d)(2^n/m)^{1/p}, \qquad 1 \le p \le \infty. \tag{3.6.6}$$

Let $f \in H^{\mathbf{r}}_p$ and $n(\mathbf{r},m)$ denote the largest number n such that

$$A(f,\mathbf{r},n) \in \mathscr{T}(\mathbf{N},d).$$

Clearly $2^{n(\mathbf{r},m)} \ge C(\mathbf{r})m$. Using the fact that the cubature formula $q_m(\mathbf{r})$ is sharp for $t \in \mathscr{T}(\mathbf{N},d)$, i.e.,

$$q_m(t,\mathbf{r}) = \hat{t}(\mathbf{0}), \qquad t \in \mathscr{T}(\mathbf{N},d),$$

and by Corollaries 3.4.8 and 3.4.3, we obtain

$$\begin{aligned}
\left|q_m(f,\mathbf{r}) - \hat{f}(\mathbf{0})\right| &= \left| \sum_{n > n(\mathbf{r},m)} q_m(A(f,\mathbf{r},n),\mathbf{r}) \right| \\
&\ll \sum_{n > n(\mathbf{r},m)} \left\|q_m(\mathbf{r})V(\mathbf{r},n+n(\mathbf{r}))\right\|_{L_p \to \mathbb{C}} 2^{-g(\mathbf{r})n},
\end{aligned} \tag{3.6.7}$$

where $n(\mathbf{r}) \le C(\mathbf{r})$ is the smallest number such that, for all n and f, we have $A(f,\mathbf{r},n) \in \mathscr{T}^{\mathbf{r}}(n+n(\mathbf{r}))$.

Applying (3.6.6) and summing, we obtain from (3.6.7) that

$$\left|q_m(f,\mathbf{r}) - \hat{f}(\mathbf{0})\right| \ll m^{-g(\mathbf{r})}.$$

The upper estimate is proved.

The lower estimate can be obtained from the following lemma by means of the Bernstein inequality.

Lemma 3.6.2 *Let* $\mathbf{N} \geq \mathbf{1}$*; then*

$$\kappa_{\vartheta(\mathbf{N})}\big(\mathscr{RT}(4\mathbf{N},d)_\infty\big) \geq C(d) > 0.$$

Proof Let $\xi^1,\dots,\xi^m$, $\xi^j \in \mathbb{T}^d$, $j = 1,\dots,m$, $m = \vartheta(\mathbf{N})$ be given. We consider the subspace $\Psi \subset \mathscr{T}(2\mathbf{N},d)$:

$$\Psi := \big\{t \in \mathscr{T}(2\mathbf{N},d) : t(\xi^j) = 0,\ j = 1,\dots,m\big\}.$$

Then $\dim\Psi \geq C_1(d)\vartheta(2\mathbf{N})$, $C_1(d) > 0$ and, by Theorem 3.2.1, there is a $t \in \Psi$ such that

$$\|t\|_\infty = 1, \qquad \|t\|_2 \geq C_2(d) > 0.$$

Setting $f := |t|^2 \in \mathscr{RT}(4\mathbf{N},d)_\infty$ we find that, for any cubature formula with points $\xi^1,\dots,\xi^m$, we have $\Lambda_m(f,\xi) = 0$ but

$$\hat{f}(\mathbf{0}) = \|t\|_2^2 \geq C(d) > 0.$$

The lemma is proved since $\xi^1,\dots,\xi^m$ are arbitrary. □

This completes the proof of Theorem 3.6.1 □

3.6.2 Sampling

We consider the approximate recovery of functions from the Sobolev and the Nikol'skii classes. For given m, $\psi_1(\mathbf{x}),\dots,\psi_m(\mathbf{x})$, $\xi := (\xi^1,\dots,\xi^m)$ we define the linear operator

$$\Psi_m(\xi)(f) := \Psi_m(f,\xi) := \sum_{j=1}^{m} f(\xi^j)\psi_j(\mathbf{x}), \tag{3.6.8}$$

and, for a class $\mathbf{F}$ of functions, we define the quantities

$$\Psi_m(\mathbf{F},\xi)_p := \sup_{f\in\mathbf{F}}\big\|\Psi_m(f,\xi) - f\big\|_p,$$

$$\rho_m(\mathbf{F})_p := \inf_{\psi_j,\dots,\psi_m;\xi^1,\dots,\xi^m} \Psi_m(\mathbf{F},\xi).$$

Let $\mathbf{N} = (N_1,\dots,N_d)$, $N_j \geq 1$. We consider the operators $I_\mathbf{N} := \prod_{j=1}^d I_{N_j}$ and $R_\mathbf{N} := \prod_{j=1}^d R_{N_j}$, where the operators I_m and R_m were defined in §2.4.2 and we assume that the operator with index N_j acts on $f(\mathbf{x})$ in the variable x_j.

From the properties (2.4.11) and (2.4.12) of the operators R_m and I_m the following properties of the operators $R_\mathbf{N}$ and $I_\mathbf{N}$ can be derived: for any $t \in \mathscr{T}(\mathbf{N},d)$ we have

$$I_\mathbf{N}(t) = t, \qquad R_\mathbf{N}(t) = t. \tag{3.6.9}$$

Lemma 2.4.3 implies the following assertion.

Lemma 3.6.3 *Let* $1 \le p \le \infty$, $\mathbf{m} \ge \mathbf{N} \ge \mathbf{1}$. *Then*

$$\|R_{\mathbf{N}} V_{\mathbf{m}}\|_{p\to p} \le C(d)\big(\nu(\mathbf{m})/\nu(\mathbf{N})\big)^{1/p}, \qquad 1 \le p \le \infty,$$
$$\|I_{\mathbf{N}} V_{\mathbf{m}}\|_{p\to p} \le C(d)\big(\nu(\mathbf{m})/\nu(\mathbf{N})\big)^{1/p}, \qquad 1 < p < \infty.$$

Let $\mathbf{r} = (r_1,\dots,r_d)$, $r_j > 0$, $j = 1,\dots,d$ and m be given. We define the numbers $N_j := \max\big([m^{g(\mathbf{r})/r_j}], 1\big)$, $j = 1,\dots,d$, and denote

$$I_m(\mathbf{r}) := I_{\mathbf{N}}, \qquad R_m(\mathbf{r}) := R_{\mathbf{N}}.$$

Let us prove the following statement.

Theorem 3.6.4 *Let* $g(\mathbf{r}) > 1/q$; *then*

$$R_m(F_q^{\mathbf{r}},\mathbf{r})_p \asymp \rho_{4^d m}(F_q^{\mathbf{r}})_p \asymp m^{-g(\mathbf{r})+(1/q-1/p)_+}, \qquad 1 \le q,p \le \infty,$$
$$I_m(F_q^{\mathbf{r}},\mathbf{r})_p \asymp \rho_{3^d m}(F_q^{\mathbf{r}})_p \asymp m^{-g(\mathbf{r})+(1/q-1/p)_+}, \qquad 1 < q,p < \infty.$$

Proof Owing to Theorem 3.4.6 it suffices to prove the upper estimate for the class $H_p^{\mathbf{r}}$. Let $f \in H_p^{\mathbf{r}}$. By Theorem 3.4.11 we have, for $1 \le q \le p \le \infty$,

$$H_p^{\mathbf{r}} \subset H_p^{\mathbf{r}'} B, \qquad \mathbf{r}' = \mathbf{r}\big(1-(1/q-1/p)/g(\mathbf{r})\big).$$

From this relation for $p = \infty$ using Corollary 3.4.3 we get that for $g(\mathbf{r}) > 1/q$ the series $\sum_{n=0}^{\infty} A(f,\mathbf{r},n)$ converges uniformly to f and, consequently,

$$\big\|f - R_m(f,\mathbf{r})\big\|_p \le \sum_{n=0}^{\infty} \big\|A(f,\mathbf{r},n) - R_m\big(A(f,\mathbf{r},n),\mathbf{r}\big)\big\|_p.$$

Owing to the embedding theorem 3.4.6 it suffices to consider the case $q = p$.

Let $n(\mathbf{r},m)$ and $n(\mathbf{r})$ be the same as in the proof of Theorem 3.6.1. From the corollary of Theorem 3.4.7 (see (3.4.9)), the relation (3.6.9), and Lemma 3.6.3 we obtain that

$$\begin{aligned}
\big\|f - R_m(f,\mathbf{r})\big\|_p &\le \sum_{n>n(\mathbf{r},m)} \big\|A(f,\mathbf{r},n) - R_m\big(A(f,\mathbf{r},n),\mathbf{r}\big)\big\|_p \\
&= \sum_{n>n(\mathbf{r},m)} \big\|A(f,\mathbf{r},n) - R_m(\mathbf{r})V(\mathbf{r},n+n(\mathbf{r}))A(f,\mathbf{r},n)\big\|_p \\
&\ll \sum_{n>n(\mathbf{r},m)} (2^n/m)^{1/p} 2^{-g(\mathbf{r})n} \ll m^{-g(\mathbf{r})}.
\end{aligned}$$

The upper estimate for the operator $R_m(\mathbf{r})$ is proved.

In just the same way the required estimate for the operator $I_m(\mathbf{r})$ can be proved.

The required lower estimate follows from Theorem 3.3.1 and the following lemma. □

Lemma 3.6.5 *For $1 \le q, p \le \infty$ we have the relation*

$$\rho_{\vartheta(\mathbf{N})}\left(\mathscr{T}(3\mathbf{N},d)_q\right)_p \ge C(d)\vartheta(\mathbf{N})^{(1/q-1/p)_+}, \qquad \mathbf{N} \ge \mathbf{1}.$$

Proof Let the points $\xi^1,\dots,\xi^m$, $m = \vartheta(\mathbf{N})$, be given. In the space $\mathscr{T}(2\mathbf{N},d)$ we consider the subspace

$$L := \left\{t \in \mathscr{T}(2\mathbf{N},d) : t(\xi^j) = 0,\ j = 1,\dots,m\right\}.$$

Then $\dim L \ge C_1(d)\vartheta(2\mathbf{N})$, $C_1(d) > 0$ and, therefore, by Theorem 3.2.1 there is a t in L such that

$$\|t\|_\infty = 1, \qquad \|t\|_2 \ge C_2(d) > 0. \tag{3.6.10}$$

Relation (3.6.10) and the inequality

$$\|t\|_2^2 \le \|t\|_1 \|t\|_\infty$$

imply the estimate $\|t\|_1 \ge C_3(d) > 0$, which gives the required result for $1 \le p \le q \le \infty$. Let $1 \le q < p \le \infty$, t be the same as in (3.6.10), and $\|t\|_\infty = \left|t(\mathbf{x}^*)\right|$. We set $f(\mathbf{x}) = t(\mathbf{x})\mathscr{K}_\mathbf{N}(\mathbf{x}-\mathbf{x}^*)$. Then $f \in \mathscr{T}(3\mathbf{N},d)$, $f(\xi^j) = 0$, $j = 1,\dots,m$, and

$$\|f\|_q \le \|t\|_\infty \|\mathscr{K}_\mathbf{N}\|_q \le C_4(d)\vartheta(\mathbf{N})^{1-1/q}. \tag{3.6.11}$$

In the last step we used relation (3.2.7). However, by (3.2.7) with $q = \infty$ we have

$$\left|f(\mathbf{x}^*)\right| \ge C_5(d)\vartheta(\mathbf{N}).$$

From this inequality and Theorem 3.3.2 we get that

$$\|f\|_p \ge C_6(d)\vartheta(\mathbf{N})^{1-1/p}. \tag{3.6.12}$$

Comparing relations (3.6.11) and (3.6.12) we obtain the lemma. □

Remark 3.6.6 The operators $\Psi_m(\xi)$ defined by relation (3.6.8) are linear operators, and $\operatorname{rank}\Psi_m(\xi) \le m$. Comparing Theorem 3.6.4 with Theorems 3.5.1 and 3.5.2 we see that, for fixed $\mathbf{r}$, $g(\mathbf{r}) > 1/q$, the operator $R_m(\mathbf{r})$ is optimal in the sense of order, from the point of view of the Kolmogorov widths for $1 \le p \le q \le \infty$ and $1 \le q \le p \le 2$ and from the point of view of linear widths for all q, p such that either $q \ge 2$ or $p \le 2$.

3.7 Historical Remarks

Chapter 3 has dealt with multivariate approximation by trigonometric polynomials with frequencies from parallelepipeds. It turns out that this theory goes in parallel to univariate approximation theory. The results of §§3.2.1–3.2.5 are a direct consequence of the corresponding univariate results. Theorem 3.2.1 is a deep and important result. It is contained in Temlyakov (1989d) (see Lemma 1.5 there). We note that the first result (the statement of Lemma 3.2.3 for $d = 1$), about estimating the volumes of sets of Fourier coefficients of bounded trigonometric polynomials, is due to Kashin (1980). Lemma 3.2.4 was also proved by Kashin (1980).

Theorem 3.3.1 is the generalization to the multivariate case of the Bernstein inequality. Theorem 3.3.2 in the case of scalar q, p was obtained by Nikol'skii (1951) and in the case of vector $\mathbf{q}$, $\mathbf{p}$ by Uninskii (1966). Theorem 3.3.3 is from Andrianov and Temlyakov (1997). Theorem 3.3.6 is a very important result on applications of the hyperbolic cross approximation. It was obtained by Temlyakov (see Temlyakov, 1985d, and the monograph Temlyakov, 1986c). We discussed this result in detail. Note that Lemmas 3.3.7, 3.3.8, and 3.3.13 were obtained in Temlyakov (1985d) (see also Temlyakov, 1986c). The first result in the direction of Lemma 3.3.7 was obtained by Ul'yanov (1970). He proved the following inequality in the univariate case for $1 \le q < p < \infty$, $\beta := 1/q - 1/p$:

$$E_n(f)_p \le C(q,p)\left((n+1)^\beta E_n(f)_q + \sum_{k=n+1}^{\infty} k^{-1}(k^\beta E_k(f)_q)^p\right)^{1/p}, \tag{3.7.1}$$

where $E_n(f)_p$ is the best approximation of f in L_p by trigonometric polynomials of order n. We note that an analog of (3.7.1) with the ℓ_p-norm on the right-hand side replaced by the ℓ_1-norm is a straightforward corollary of the corresponding Nikol'skii inequalities (see Konyushkov, 1958). The inequality (3.7.1) captures a very nontrivial effect for $p > 2$. The proof in Ul'yanov (1970) uses the technique of monotone rearrangements of functions. Another proof of (3.7.1), based on the Hardy–Littlewood–Paley method, was given by Timan (1974). Later, the above effect was rediscovered by Franke (1986) and Jawerth (1977) in the case of isotropic multivariate function classes (see a detailed discussion in Dinh Dung *et al.* (2016), Section 3.3). We note that the univariate technique from Timan (1974) works for the isotropic multivariate function classes as well. An important point here is that in both the univariate and isotropic multivariate cases a function f allows a representation in a form of one parametric sum of the corresponding blocks of its Fourier series. In Lemma 3.3.7 we used a multiparametric sum with respect to a vector $\mathbf{s}$. This makes the problem more difficult and requires a new technique. Theorem 3.3.15 was obtained in Temlyakov (1987).

The results in §3.4 are generalizations of the corresponding univariate results. In the case of scalar q, p the results of this section are contained in the book Nikol'skii (1969). One can find there other results in approximation theory and the theory of embedding.

The results in §3.5 were obtained in the same way as the corresponding univariate results from Chapter 2. We note that, in the case of the isotropic classes $W^{\mathbf{r}}_{q,\alpha}$, $\mathbf{r} = r\mathbf{1}$, Theorems 3.5.1 and 3.5.2 were obtained by Höllig (1980). The results in §3.6 are direct generalizations of the the corresponding results from Chapter 2.

4
Hyperbolic Cross Approximation

4.1 Introduction

There is a great variety of ways to measure the smoothness of multivariate functions and of choices for the approximation methods. For example, consider approximation by the trigonometric system $\{e^{i(\mathbf{k},\mathbf{x})}\}_{\mathbf{k}\in\mathbb{Z}^d}$. The multivariate trigonometric system does not have a natural order. Therefore, *a priori*, we have many ways to form finite-dimensional spaces of trigonometric polynomials. Each such choice leads to a problem in approximation. We would like to understand which method to choose for a particular approximation problem. We have already pointed out in Chapters 1 and 3 that in determining a good approximation method it is natural to use an approach based on studying the widths of function classes. We call an approach based on widths the *optimization approach*. In implementation of the optimization approach one needs to prove the upper and the lower bounds for the corresponding widths. In this chapter we consider classes of functions defined by a restriction on a mixed derivative of a function (the classes $\mathbf{W}$) or on the prelimit difference corresponding to it (the classes $\mathbf{H}$). Usually, these classes are called *mixed smoothness classes*. Here, we prove the upper bounds for approximation of these classes by trigonometric polynomials with frequencies from the hyperbolic crosses. In Chapter 5 we show that these bounds are optimal (in the sense of order) in many cases.

We now give a motivation for studying the above-mentioned classes of functions with mixed smoothness. We begin with an example of the classical way of measuring the smoothness of multivariate functions; namely, using the class of functions of bounded variation. At the beginning of the twentieth century Vitali and Hardy generalized the definition of variation to the multivariate case. Roughly speaking, in the univariate case the condition that f is of bounded variation is close to the condition $\|f'\|_1 < \infty$. In the multivariate case the condition that a function has bounded variation in the sense of Hardy and Vitali is close to that requiring

$\|f^{(1,\dots,1)}\|_1 < \infty$, where $f^{(1,\dots,1)}$ is a mixed derivative of f. In our notation the class of functions of bounded variation in the Hardy–Vitali sense is close to $\mathbf{W}^1_{1,1}$.

It turns out that the classes $\mathbf{W}^r_{q,\alpha}$ and $\mathbf{H}^r_q$ are important in applications. At the end of the 1950s N.M. Korobov discovered the importance of these kinds of classes in numerical integration. He constructed a cubature formula with N nodes which guarantees that the accuracy of numerical integration for these classes is of order $N^{-r}(\log N)^{rd}$. Let us compare this estimate with the corresponding estimate for the isotropic Sobolev classes, considered in Chapter 3. It is known that in order to get a numerical integration accuracy of order N^{-r} we should take the class $W_q^{(rd,\dots,rd)}$ of smoothness rd. This means that we impose the restrictions (assume r is integer)

$$\|f^{(r_1,\dots,r_d)}\|_q \le 1 \tag{4.1.1}$$

for all $\mathbf{r} = (r_1,\dots,r_d)$ such that $r_1 + r_2 + \cdots + r_d \le rd$. In the definition of $\mathbf{W}^r_{q,\alpha}$ with $\alpha = (r,\dots,r)$ we impose the restrictions (4.1.1) only for $\mathbf{r}$ such that $r_j \le r$, $j = 1,\dots,d$. It is clear that the class $\mathbf{W}^r_{q,\alpha}$ is much wider than $W_q^{(rd,\dots,rd)}$, but Korobov's result shows that the accuracy of numerical integration of functions in $\mathbf{W}^r_{q,\alpha}$ is very close to that of functions in $W_q^{(rd,\dots,rd)}$. Korobov's discovery pointed out the importance of the classes of functions with dominating mixed derivative in fields such as approximation theory and numerical analysis and boosted the thorough study of these classes, which prove to be important in many areas beyond approximation theory and numerical integration. We refer the reader to Dinh Dung *et al.* (2016) for a detailed discussion of this issue.

We now present an intrinsic motivation in the framework of approximation theory. We show for some examples that the study of the classes $\mathbf{W}^r_{q,\alpha}$ and $\mathbf{H}^r_q$ has revealed new and interesting phenomena and demanded a development of new deep methods in approximation theory.

We define the class $\mathbf{W}^r_{q,\alpha}$ in the following way. Let

$$F_r(\mathbf{x},\alpha) := \prod_{j=1}^{d} F_r(x_j,\alpha_j)$$

be the multivariate analog of the Bernoulli kernel defined in §1.4. We denote by $\mathbf{W}^r_{q,\alpha}$ the class of functions $f(\mathbf{x})$ representable in the form

$$f(\mathbf{x}) = \varphi(\mathbf{x}) * F_r(\mathbf{x},\alpha) := (2\pi)^{-d} \int_{\mathbb{T}^d} \varphi(\mathbf{y}) F_r(\mathbf{x}-\mathbf{y},\alpha) d\mathbf{y},$$

where $\varphi \in L_q$ and $\|\varphi\|_q \le 1$. In this case the function φ is called the (r,α)-derivative of f and is denoted by $\varphi(\mathbf{x}) = f^{(r)}(\mathbf{x},\alpha)$. Note that in the case of integer r the class $\mathbf{W}^r_{q,\alpha}$ with $\alpha_j = r$, $j = 1,\dots,d$, is equivalent to the class defined above by restrictions on mixed derivatives.

In order to get some orientation in the problem of choosing a good approximation method for these classes let us consider first the case of a Hilbert space L_2. The class $\mathbf{W}^r_{2,\alpha}$ is the image of the unit ball for the operator I^r_α defined as follows:

$$I^r_\alpha \varphi := \varphi * F_r(\cdot,\alpha).$$

Then, by Theorem A.3.10, we get

$$d_n(\mathbf{W}^r_{2,\alpha}, L_2) = s_{n+1}(I^r_\alpha). \tag{4.1.2}$$

For the convolution operator I^r_α its singular numbers $s_l(I^r_\alpha)$ coincide with the Fourier coefficients $|\hat{F}_r(\mathbf{k}^l,\alpha)|$ rearranged in decreasing order:

$$|\hat{F}_r(\mathbf{k}^1,\alpha)| \geq |\hat{F}_r(\mathbf{k}^2,\alpha)| \geq \cdots \quad .$$

We have

$$|\hat{F}_r(\mathbf{k},\alpha)| = \left(\prod_{j=1}^{d} \max(|k_j|,1)\right)^{-r}.$$

Thus the optimal subspaces for approximating functions from $\mathbf{W}^r_{2,\alpha}$ in the L_2-norm are the subspaces $\mathcal{T}(\Gamma(N))$ of trigonometric polynomials with frequencies in the hyperbolic crosses (see Figure 4.1):

$$\Gamma(N) := \left\{\mathbf{k} \in \mathbb{Z}^d : \prod_{j=1}^{d} \max(|k_j|,1) \leq N\right\}.$$

Let us introduce some more notation,

$$E_N(f)_p := \inf_{t \in \mathcal{T}(\Gamma(N))} \|f - t\|_p,$$

and, for a function class F,

$$E_N(F)_p := \sup_{f \in F} E_N(f)_p.$$

Using this notation we can formulate the above observation as a theorem.

Theorem 4.1.1 *For each $r > 0$ and $|\Gamma(N)| \leq n < |\Gamma(N+1)|$ we have*

$$d_n(\mathbf{W}^r_{2,\alpha}, L_2) = E_N(\mathbf{W}^r_{2,\alpha})_2 = (N+1)^{-r} \asymp n^{-r}(\log n)^{r(d-1)}.$$

This theorem provides us with a sequence $\{\mathcal{T}(\Gamma(N))\}_{N=1}^{\infty}$ of finite-dimensional spaces of trigonometric polynomials which is suitable for approximation of the classes $\mathbf{W}^r_{2,\alpha}$. The logic of our further development is the following. We will study the approximation of functions from the general classes $\mathbf{W}^r_{q,\alpha}$ and $\mathbf{H}^r_q$ (which will be defined later) by trigonometric polynomials from $\mathcal{T}(\Gamma(N))$ and its modification $\mathcal{T}(Q_n)$, where

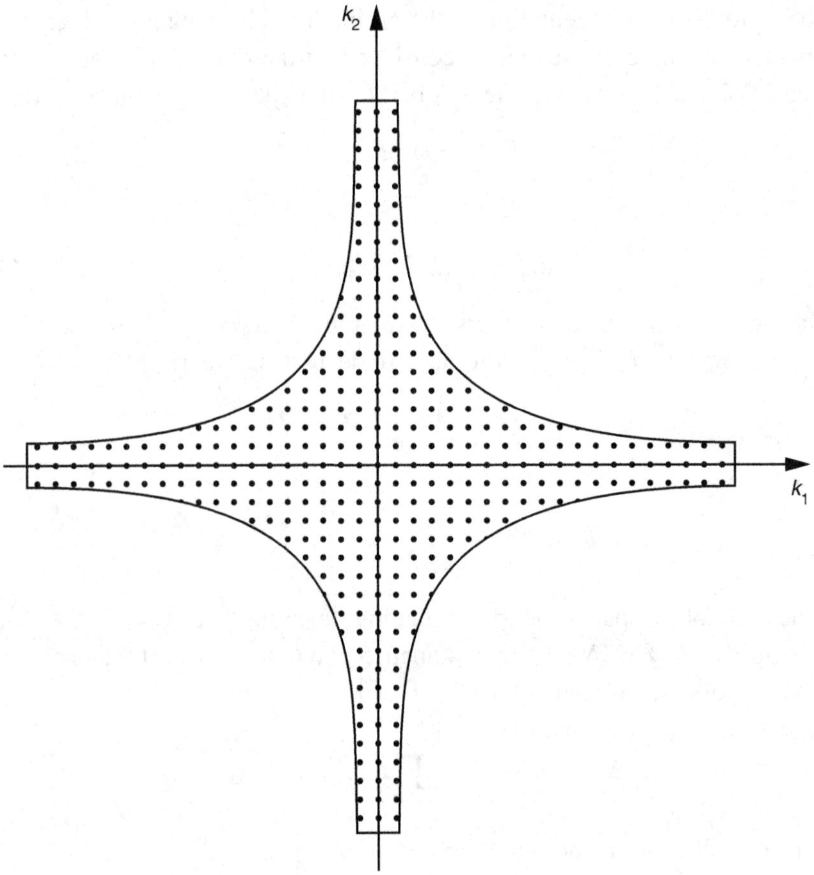

Figure 4.1 Hyperbolic cross for $d = 2$ with $N = 20$.

$$Q_n := \bigcup_{\|\mathbf{s}\|_1 \le n} \{\mathbf{k} \in \mathbb{Z}^d : [2^{s_j-1}] \le |k_j| < 2^{s_j},\ j = 1,\dots,d\},$$

is the step hyperbolic cross (see Figure 4.2).

This investigation will give us the upper estimates for the Kolmogorov widths and the orthowidths. In order to prove the lower estimates for the above-mentioned widths we use some special trigonometric polynomials from $\mathscr{T}(\Gamma(N))$, which will be constructed in §4.2. On the basis of investigations of the Kolmogorov widths and the orthowidths, to be conducted in Chapter 5, it will be established that in questions of the approximation of functions in these classes the sets $\mathscr{T}(Q_n)$ of trigonometric polynomials with harmonics in the step hyperbolic crosses Q_n play the same role as the sets $\mathscr{T}(2^n)$ play in the univariate case.

In order to orient the reader we will highlight some new features which come up in studying approximation by polynomials from $\mathscr{T}(Q_n)$. Let us discuss first the best approximation

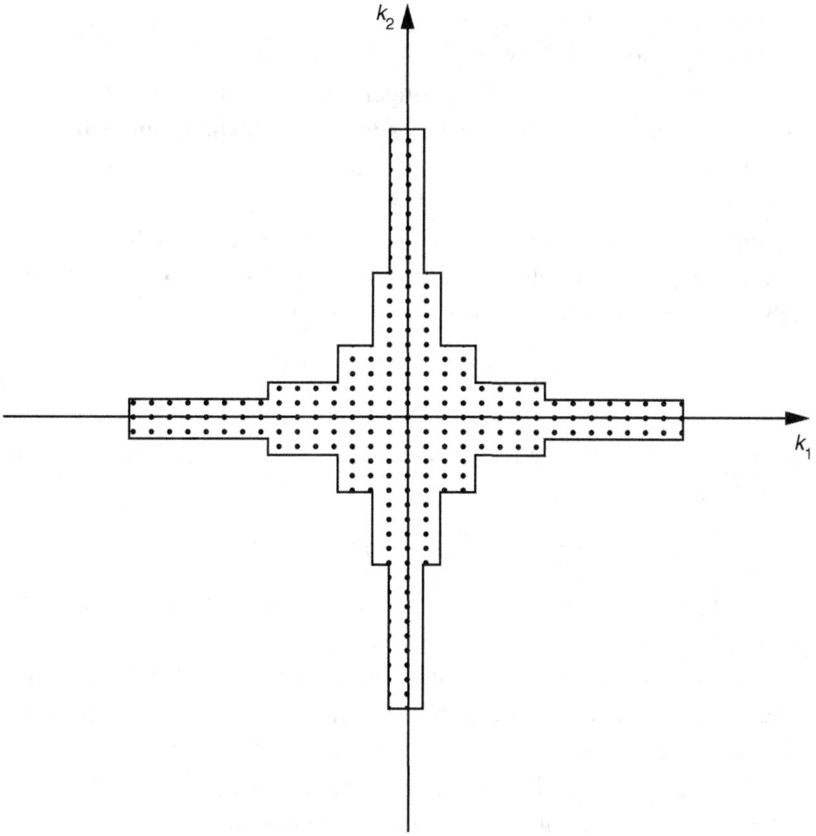

Figure 4.2 Step hyperbolic cross for $d = 2$ with $n = 4$.

$$E_{Q_n}(f)_p := \inf_{t \in \mathscr{T}(Q_n)} \|f - t\|_p.$$

The technique of obtaining estimates for $E_n(f)_p$ in the univariate case was based on studying special kernels: those of Dirichlet, Féjer, Jackson, and de la Vallée Poussin. For instance, the following property of Jackson's kernel

$$|J_n^a(x)| \le C \min(n, n^{1-2a} x^{-2a}), \qquad |x| \le \pi, \tag{4.1.3}$$

allowed us to prove the estimate ($0 \le r < 2a - 1$)

$$\|f - f * J_n^a\|_p \le C n^{-r} \qquad \text{for } f \in W_{p,\alpha}^r$$

simultaneously for all $1 \le p \le \infty$. The boundedness in L_1 of the de la Vallée Poussin kernel, $\|\mathscr{V}_n\|_1 \le 3$, allowed us to prove the inequality

$$\|f - V_n(f)\|_p \le 4 E_n(f)_p, \tag{4.1.4}$$

for all $1 \le p \le \infty$. This inequality is very useful in proving the lower bounds for $E_n(f)_p$. We stress that the approximation technique for univariate-function approximation is the same for all $1 \le p \le \infty$. It turns out that in the case of approximation by polynomials from $\mathscr{T}(Q_n)$ we cannot use the technique from the univariate-function approximation any longer. Moreover, the cases $1 < p < \infty$ and $p = 1, \infty$ require different methods of investigaton and exhibit qualitatively different phenomena. Some methods of harmonic analysis have proved to be useful in studying the case $1 < p < \infty$. Littlewood–Paley theory is of particular importance. The frequent use of this theory pushed us to consider along with $\mathscr{T}(\Gamma(N))$ the subspaces $\mathscr{T}(Q_n)$. In particular, Littlewood–Paley theory (see Corollary A.3.4) implies that the orthogonal projector S_{Q_n} onto $\mathscr{T}(Q_n)$,

$$S_{Q_n}(f) := \sum_{\mathbf{k}\in Q_n} \hat{f}(\mathbf{k})e^{i(\mathbf{k},\mathbf{x})},$$

is bounded as an operator from L_p to L_p for all $1 < p < \infty$. Thus, for each function $f \in L_p$ we have

$$\|f - S_{Q_n}(f)\|_p \le C(p,d)E_{Q_n}(f)_p \tag{4.1.5}$$

and for any function class F the order of best approximation can be realized by the linear operator (on even an orthogonal projector) S_{Q_n}.

The case $p = 1$ or ∞ proved to be much more complicated and intriguing. It is not difficult to understand that there are no simple and convenient representations for the Dirichlet kernel or other kernels for the hyperbolic crosses $\Gamma(N)$ and Q_n. Even worse – there is no analog of the de la Vallée Poussin kernels. Namely, for any number a and polynomial $\mathscr{V} \in \mathscr{T}(Q_{n+a})$ with the property

$$\hat{\mathscr{V}}(\mathbf{k}) = 1 \qquad \text{for } \mathbf{k} \in Q_n$$

we have

$$\|\mathscr{V}\|_1 \ge C(d,a)n^{d-1}. \tag{4.1.6}$$

This means that we do not have an analog of (4.1.4) for the hyperbolic cross approximation in L_1 and L_∞. This property complicates the matter of obtaining the lower estimates but also leads to a surprising phenomenon, as follows.

The relation (4.1.4) guarantees that for each class F of univariate functions with

$$E_n(F)_p \asymp n^{-r}(\log n)^b$$

the rate of approximation in (4.1.4) can be realized by linear methods $\{V_n(\cdot)\}_{n=1}^{\infty}$. It turns out that it is not a case of approximation by polynomials from $\mathscr{T}(\Gamma(N))$. For example, we have the following relations. On the one hand,

$$E_N(\mathbf{W}^r_{2,\alpha})_\infty \asymp N^{-r+1/2} \tag{4.1.7}$$

and on the other hand we have, for any linear bounded operator $A_N : L_\infty \to \mathscr{T}(\Gamma(N))$,

$$\sup_{f \in \mathbf{W}^r_{2,\alpha}} \|f - A_N(f)\|_\infty \geq C(d,r) N^{-r+1/2} (\log N)^{(d-1)/2}. \tag{4.1.8}$$

The above examples show the difficulty of studying approximation by polynomials from $\mathcal{T}(\Gamma(N))$ in the uniform norm. We remark that in most cases the investigation of approximation in the uniform norm has demanded new methods and, in a number of cases, the problems of approximating the classes $\mathbf{W}^r_{q,\alpha}$ and $\mathbf{H}^r_q$ in the uniform norm are still unsolved. We formulate here one of the open problems.

Open Problem 4.1 What is the order of $E_N(\mathbf{W}^r_{\infty,\alpha})_\infty$?

Let us now proceed to a discussion of some other techniques. In the univariate case, when studying the Kolmogorov widths we used a discretization technique which allowed us to treat separately finite-dimensional problems in $\mathcal{T}(2^n)$ and then sum the corresponding geometric progression of errors. We obtain a geometric progression in this case because in the representation of a univariate function by Fourier series we have only one dyadic block $\rho(s)$ of dimension 2^n. In the multivariate case we have difficulties of two different kinds in implementing the discretization idea. In the first place, we can try to represent a function in the form

$$f(\mathbf{x}) = \hat{f}(0) + \sum_{n=0}^{\infty} \sum_{\mathbf{k} \in Q_{n+1} \setminus Q_n} \hat{f}(\mathbf{k}) e^{i(\mathbf{k},\mathbf{x})} =: \hat{f}(0) + \sum_{n=0}^{\infty} f_n(\mathbf{x})$$

and discretize the polynomials $f_n \in \mathcal{T}(Q_{n+1} \setminus Q_n)$. However, we do not have a discretization technique for the subspaces $\mathcal{T}(Q_{n+1} \setminus Q_n)$ or $\mathcal{T}(Q_n)$ of the hyperbolic cross polynomials (see §7.5 for some negative results in this direction). We do have such a technique for $\mathcal{T}(\rho(\mathbf{s}))$, where

$$\rho(\mathbf{s}) := \{\mathbf{k} \in \mathbb{Z}^d : [2^{s_j-1}] \leq |k_j| < 2^{s_j}, \quad j = 1, \dots, d\}.$$

In the second place, when setting

$$f(\mathbf{x}) = \sum_{\mathbf{s}} \sum_{\mathbf{k} \in \rho(\mathbf{s})} \hat{f}(\mathbf{k}) e^{i(\mathbf{k},\mathbf{x})} =: \sum_{\mathbf{s}} \delta_{\mathbf{s}}(f)$$

and discretizing $\mathcal{T}(\rho(\mathbf{s}))$ we need to take into account the "interference "of polynomials from different $\mathcal{T}(\rho(\mathbf{s}))$ and $\mathcal{T}(\rho(\mathbf{s}'))$ having $\|\mathbf{s}\|_1 = \|\mathbf{s}'\|_1$. For fixed n we have $\asymp n^{d-1}$ distinct $\rho(\mathbf{s})$ with $\|\mathbf{s}\|_1 = n$. This causes difficulties and requires a new technique. Littlewood–Paley theory is a convenient and powerful technique here but in some cases it does not work, and then we need something else. We formulate here just one embedding-type inequality which has proved to be useful in investigating approximation in L_p, $1 < p < \infty$ (Lemma 3.3.7):

$$\|\sum_{\mathbf{s}} \delta_{\mathbf{s}}(f)\|_p \leq C(q,p,d) (\sum_{\mathbf{s}} (\|\delta_{\mathbf{s}}(f)\|_q 2^{\|\mathbf{s}\|_1(1/q-1/p)})^p)^{1/p}, \tag{4.1.9}$$

for any $1 \leq q < p < \infty$.

We will illustrate the difficulty with the Littlewood–Paley technique using the following example. Let

$$\mathscr{D}_{Q_n}(\mathbf{x}) := \sum_{\|\mathbf{s}\|_1 \le n} \delta_{\mathbf{s}}(\mathscr{D}_{Q_n}, \mathbf{x}), \qquad \delta_{\mathbf{s}}(\mathscr{D}_{Q_n}, \mathbf{x}) := \sum_{\mathbf{k}\in\rho(\mathbf{s})} e^{i(\mathbf{k},\mathbf{x})}$$

be the Dirichlet kernel for the step hyperbolic cross Q_n. Then, for $2 < p < \infty$, by Corollary A.3.5 one gets

$$\|\mathscr{D}_{Q_n}\|_p \ll \left(\sum_{\|\mathbf{s}\|_1 \le n} \|\delta_{\mathbf{s}}(\mathscr{D}_{Q_n}, \mathbf{x})\|_p^2\right)^{1/2} \ll 2^{(1-1/p)n} n^{(d-1)/2}. \tag{4.1.10}$$

However, the upper bound in (4.1.10) is not sharp. The other technique, see inequality (4.1.9), gives

$$\|\mathscr{D}_{Q_n}\|_p \ll \left(\sum_{|\mathbf{s}|_1 \le n} \left(\|\delta_{\mathbf{s}}(\mathscr{D}_{Q_n}, \mathbf{x})\|_2 2^{\|\mathbf{s}\|_1(1/2-1/p)}\right)^p\right)^{1/p} \ll 2^{(1-1/p)n} n^{(d-1)/p}. \tag{4.1.11}$$

The above example demonstrates the problem, which is related to the fact that in the multivariate Littlewood–Paley formula we have many ($\asymp n^{d-1}$) dyadic blocks of the same size (2^n).

Another important innovation which allows us to attack the "interference"problem is the use of the geometry of volumes. We discuss this new technique in Chapters 5 and 7.

We now turn our discussion to the classes $\mathbf{H}^r_q$. Let $\mathbf{t} = (t_1, \dots, t_d)$ and $\Delta^l_{\mathbf{t}} f(\mathbf{x})$ be the mixed lth difference with step t_j in the variable x_j, that is,

$$\Delta^l_{\mathbf{t}} f(\mathbf{x}) := \Delta^l_{t_d} \cdots \Delta^l_{t_1} f(x_1, \dots, x_d).$$

Let e be a subset of natural numbers in $[1, d]$. We denote

$$\Delta^l_{\mathbf{t}}(e) = \prod_{j\in e} \Delta^l_{t_j}, \qquad \Delta^l_{\mathbf{t}}(\varnothing) = I.$$

We define the class $\mathbf{H}^r_{q,l}B$, $l > r$, as the set of $f \in L_q$ such that for any e

$$\left\|\Delta^l_{\mathbf{t}}(e) f(\mathbf{x})\right\|_q \le B \prod_{j\in e} |t_j|^r. \tag{4.1.12}$$

In the case $B = 1$ we omit B. We will prove in §4.4.2 the following representation theorem for these classes. Let $\mathscr{A}_{\mathbf{s}}$ denote the polynomials defined in §3.2.5. For $f \in L_1$ set

$$A_{\mathbf{s}}(f) := f * \mathscr{A}_{\mathbf{s}}.$$

Theorem 4.1.2 *Let $f \in \mathbf{H}^r_{q,l}$. Then, for* $\mathbf{s} \geq \mathbf{0}$,

$$\|A_{\mathbf{s}}(f)\|_q \leq C(r,d,l)2^{-r\|\mathbf{s}\|_1}, \qquad 1 \leq q \leq \infty, \tag{4.1.13}$$

$$\|\delta_{\mathbf{s}}(f)\|_q \leq C(r,d,q,l)2^{-r\|\mathbf{s}\|_1}, \qquad 1 < q < \infty. \tag{4.1.14}$$

Conversely, from (4.1.13) *or* (4.1.14) *it follows that there exists a $B > 0$ which does not depend on f, such that $f \in \mathbf{H}^r_{q,l}B$.*

This theorem shows that the classes $\mathbf{H}^r_{q,l}$ with different l are equivalent. So, for convenience we fix one $l = [r]+1$ and omit l from the notation. The classes $\mathbf{H}^r_q$ are companions of the classes $\mathbf{W}^r_{q,\alpha}$ in the same sense as the classes H^r_q and $H^{\mathbf{r}}_q$ are companions of the classes $W^r_{q,\alpha}$ and $W^{\mathbf{r}}_{q,\alpha}$ respectively. The results of Chapters 2 and 3 show that, from the point of view of the rate of decrease of their sequences of widths, the classes H^r_q, $H^{\mathbf{r}}_q$ and their companions $W^r_{q,\alpha}$, $W^{\mathbf{r}}_{q,\alpha}$ are the same. It turns out that the classes $\mathbf{W}^r_{q,\alpha}$ and their companions $\mathbf{H}^r_q$ are different, however. We have, for instance,

$$d_n(\mathbf{W}^r_{2,\alpha}, L_2) \asymp n^{-r}(\log n)^{(d-1)r}$$

and

$$d_n(\mathbf{H}^r_2, L_2) \asymp n^{-r}(\log n)^{(d-1)(r+1/2)}.$$

Moreover, in many cases the study of the classes $\mathbf{W}^r_{q,\alpha}$ and $\mathbf{H}^r_q$ requires different methods. This results in the following situation. There are cases (for particular d,q,p) such that the same approximation problem is solved for one class but stands unsolved for the companion class. We present here two examples of this kind.

Open Problem 4.2 What is the order of the quantity $E_N(\mathbf{H}^r_\infty, L_\infty)$ for $d \geq 3$?

For the classes $\mathbf{H}^r_q$ we have the answer to Open Problem 4.2 in the case of functions of two variables ($d = 2$):

$$E_N(\mathbf{H}^r_\infty)_\infty \asymp N^{-r}\log N.$$

Open Problem 4.1 is open for all $d \geq 2$. The opposite situation holds for the Kolmogorov widths in the case $1 \leq p < q < 2$; we know the order of $d_n(\mathbf{W}^r_{q,\alpha}, L_p)$ but we do not know the order of $d_n(\mathbf{H}^r_q, L_p)$.

We hope that this introduction convinces the reader that the classes $\mathbf{W}^r_{q,\alpha}$ and $\mathbf{H}^r_q$ deserve thorough investigation and that this investigation is going to be a journey to a world of new ideas and techniques. We will point out new elements of technique in all sections of this chapter.

4.2 Some Special Polynomials with Harmonics in Hyperbolic Crosses

Let $\mathcal{T}(Q_n)$ and $\mathcal{T}(N) := \mathcal{T}(\Gamma(N))$ denote the sets of trigonometric polynomials with harmonics in Q_n and $\Gamma(N)$ respectively. It is easy to see that

$$Q_n \subset \Gamma(2^n) \subset Q_{n+d}; \tag{4.2.1}$$

therefore it is enough to prove a number of polynomial properties such as the Bernstein and Nikol'skii inequalities for either $\mathcal{T}(Q_n)$ or $\mathcal{T}(N)$.

We consider the following trigonometric polynomials.

4.2.1 The Dirichlet Kernels

Define as above

$$\mathcal{D}_{Q_n}(\mathbf{x}) := \sum_{\mathbf{k}\in Q_n} e^{i(\mathbf{k},\mathbf{x})} = \sum_{\|\mathbf{s}\|_1 \le n} \mathcal{D}_{\rho(\mathbf{s})}(\mathbf{x}),$$

where $\mathcal{D}_{\rho(\mathbf{s})}(\mathbf{x}) := \sum_{\mathbf{k}\in\rho(\mathbf{s})} e^{i(\mathbf{k},\mathbf{x})}$. It is clear that for $t \in \mathcal{T}(Q_n)$,

$$t * \mathcal{D}_{Q_n} = t.$$

Let $r \ge 0$ be a real number and $\alpha = (\alpha_1,\dots,\alpha_d)$ be a vector. We have the following assertion.

Lemma 4.2.1 *Let $1 < p < \infty$. Then*

$$\left\|\mathcal{D}^{(r)}_{Q_n}(\mathbf{x},\alpha)\right\|_p \asymp 2^{(r+1-1/p)n} n^{(d-1)/p}.$$

Proof Suppose that $q = (p+1)/2$; then $1 < q < p$ and, by relation (3.2.1) from Chapter 3,

$$\|\mathcal{D}_{\rho(\mathbf{s})}\|_q \ll 2^{(1-1/q)\|\mathbf{s}\|_1}. \tag{4.2.2}$$

The operator D^0_α is bounded as an operator from L_q to L_q for $1 < q < \infty$. Therefore, by Theorem 3.3.1 we get from (4.2.2)

$$\left\|\mathcal{D}^{(r)}_{\rho(\mathbf{s})}(\mathbf{x},\alpha)\right\|_q \ll 2^{(r+1-1/q)\|\mathbf{s}\|_1}. \tag{4.2.3}$$

From this estimate and Lemma 3.3.7 it follows that

$$\left\|\mathcal{D}^{(r)}_{Q_n}(\mathbf{x},\alpha)\right\|_p \ll \left(\sum_{\|\mathbf{s}\|_1\le n} \left\|\mathcal{D}^{(r)}_{\rho(\mathbf{s})}(\mathbf{x},\alpha)\right\|_q^p 2^{(p/q-1)\|\mathbf{s}\|_1}\right)^{1/p}$$
$$\ll \left(\sum_{\|\mathbf{s}\|_1\le n} 2^{(r+1-1/p)\|\mathbf{s}\|_1 p}\right)^{1/p} \ll 2^{(r+1-1/p)n} n^{(d-1)/p},$$

which gives the required upper estimate.

The lower estimate follows from the upper estimate just proved and the relation

$$2^{2rn}n^{d-1} \ll \left\|\mathscr{D}_{Q_n}^{(r)}(\mathbf{x},\alpha)\right\|_2^2 \le \left\|\mathscr{D}_{Q_n}^{(r)}(\mathbf{x},\alpha)\right\|_p \left\|\mathscr{D}_{Q_n}^{(r)}(\mathbf{x},\alpha)\right\|_{p'}.$$

□

Remark 4.2.2 Similarly to Lemma 4.2.1 we have

$$\left\|\sum_{\|\mathbf{s}\|_1=n} \mathscr{D}_{\rho(\mathbf{s})}(\mathbf{x})\right\|_p \asymp 2^{(r+1-1/p)n} n^{(d-1)/p}.$$

4.2.2 The de la Vallée Poussin Kernels

Let $\mathscr{A}_{\mathbf{s}}(\mathbf{x})$ be the polynomials defined in §3.2.5. These polynomials are from $\mathscr{T}(2^{\mathbf{s}},d)$ and

$$\hat{\mathscr{A}_{\mathbf{s}}}(\mathbf{k}) \neq 0 \qquad \text{only for } \mathbf{k} \text{ such that } 2^{s_j-2} < |k_j| < 2^{s_j},$$
$$s_j \ge 1, \qquad j=1,\dots,d. \tag{4.2.4}$$

We define the polynomials

$$\mathscr{V}_{Q_n}(\mathbf{x}) := \sum_{\|\mathbf{s}\|_1 \le n} \mathscr{A}_{\mathbf{s}}(\mathbf{x}).$$

These are polynomials in $\mathscr{T}(Q_n)$ with the property

$$\hat{\mathscr{V}}_{Q_{n+d}}(\mathbf{k}) = 1 \qquad \text{for } \mathbf{k} \in Q_n. \tag{4.2.5}$$

Indeed, let $\mathbf{k} \in \rho(\mathbf{n})$, $\|\mathbf{n}\|_1 \le n$; then by property (4.2.4) we get

$$\hat{\mathscr{V}}_{Q_{n+d}}(\mathbf{k}) = \sum_{\mathbf{s} \le \mathbf{n}+\mathbf{1}} \hat{\mathscr{A}_{\mathbf{s}}}(\mathbf{k}) = \prod_{j=1}^{d} \left(\sum_{s_j=0}^{n_j+1} \hat{\mathscr{A}}_{s_j}(k_j) \right) = \hat{\mathscr{V}}_{2^{\mathbf{n}}}(\mathbf{k}) = 1.$$

We will use the following notation. Let $f \in L_1$,

$$S_{Q_n}(f) := f * \mathscr{D}_{Q_n},$$

$$V_{Q_n}(f) := f * \mathscr{V}_{Q_n},$$

$$A_{\mathbf{s}}(f) := f * \mathscr{A}_{\mathbf{s}}.$$

From Corollary A.3.4 it follows that for, $1 < p < \infty$,

$$\|S_{Q_n}\|_{p \to p} \le C(p,d). \tag{4.2.6}$$

In Chapters 1 and 3 it was established that the L_1-norms of the de la Vallée Poussin kernels for parallelepipeds are uniformly bounded. This fact played an essential role for approximation problems in the L_1- and L_∞- metrics.

The following lemma shows that unfortunately the kernels $\mathscr{V}_{Q_n}$ have no such property.

Lemma 4.2.3 *Let $1 \le p < \infty$. Then, for $r > 0$ and arbitrary α and for $r = 0$ and $\alpha = 0$, the relation*

$$\left\| \mathscr{V}^{(r)}_{Q_n}(\mathbf{x}, \alpha) \right\|_p \asymp 2^{(r+1-1/p)n} n^{(d-1)/p}$$

holds.

Proof Let $r > 0$. Then, from (3.2.15), by Theorem 3.3.1 it follows that

$$\left\| \mathscr{A}^{(r)}_{\mathbf{s}}(\mathbf{x}, \alpha) \right\|_1 \ll 2^{r\|\mathbf{s}\|_1}. \tag{4.2.7}$$

From this relation we find, for $p = 1$,

$$\left\| \mathscr{V}^{(r)}_{Q_n}(\mathbf{x}, \alpha) \right\|_1 \ll \sum_{\|\mathbf{s}\|_1 \le n} 2^{r\|\mathbf{s}\|_1} \ll 2^{rn} n^{d-1}$$

and, for $p > 1$, using Remark 3.3.10 we get

$$\begin{aligned} \left\| \mathscr{V}^{(r)}_{Q_n}(\mathbf{x}, \alpha) \right\|_p &\ll \left(\sum_{\|\mathbf{s}\|_1 \le n} \left\| \mathscr{A}^{(r)}_{\mathbf{s}}(\mathbf{x}, \alpha) \right\|_1^p 2^{(p-1)\|\mathbf{s}\|_1} \right)^{1/p} \\ &\ll 2^{(r+1-1/p)n} n^{(d-1)/p}. \end{aligned}$$

The lower estimate for $1 < p < \infty$ can be deduced from the upper estimate (which has already been proved) in the same way as in the proof of Lemma 4.2.1. Also this estimate can be obtained from Lemma 4.2.1, the inequality (4.2.6), and the relation

$$S_{Q_n}(\mathscr{V}_{Q_{n+d}}) = \mathscr{D}_{Q_n}.$$

The proof of the lower estimates for $p = 1$ is given at the end of §4.3.3.

We now prove the upper estimate for $r = 0$ and $\alpha = 0$. We denote $\mathbf{s}^d := (s_1, \dots, s_{d-1})$ for $\mathbf{s} = (s_1, \dots, s_d)$ and $\mathbf{x}^d := (x_1, \dots, x_{d-1})$ for $\mathbf{x} = (x_1, \dots, x_d)$. Then

$$\begin{aligned} \mathscr{V}_{Q_n}(\mathbf{x}) &= \sum_{\|\mathbf{s}^d\|_1 \le n} \mathscr{A}_{\mathbf{s}^d}(\mathbf{x}^d) \sum_{s_d=0}^{n-\|\mathbf{s}^d\|_1} \mathscr{A}_{s_d}(x_d) \\ &= \sum_{\|\mathbf{s}^d\|_1 \le n} \mathscr{A}_{\mathbf{s}^d}(\mathbf{x}^d) \mathscr{V}_{2^{n-\|\mathbf{s}^d\|_1-1}}(x_d), \end{aligned}$$

and, therefore,

$$\|\mathscr{V}_{Q_n}\|_1 \ll \sum_{\|\mathbf{s}^d\|_1 \le n} 1 \ll n^{d-1}.$$

4.2.3 The Telyakovskii Polynomials

It is well known that, for any natural number n,

$$\left|\sum_{k=1}^{n} \frac{\sin kx}{k}\right| \le C.$$

It was discovered by S.A. Telyakovskii that there is an analog to this result for the hyperbolic crosses

$$\left|\sum_{\mathbf{k}\in\Gamma(N)} \frac{\sin k_1x_1 \cdots \sin k_dx_d}{k_1 \cdots k_d}\right| \le C(d).$$

He also proved a similar result with the functions $\sin k_jx_j$ replaced by $\cos k_jx_j + \cos(k_j+1)x_j - \cos 2k_jx_j - \cos(2k_j+1)x_j$. We present here the results for slightly more general functions $a_l(k_j,x_j)$ instead of $\sin k_jx_j$. Let Γ be a surface dividing the set $\mathbb{R}^d_+ = \{\mathbf{x} \in \mathbb{R}^d : x_j \ge 0,\ j = 1,\dots,d\}$ into two subsets X and Y with the set X such that if $x \in X$ then the box $\prod_{j=1}^d [0,x_j]$ also belongs to X. We say that X has the property S and Y has the property R. If X_1, X_2 where $X_1 \subset X_2$ have the property S then the set $X_2 \backslash X_1$ will be said to have the property S_1. Denote by $\overline{X}$ the set of points in X whose coordinates are natural numbers. We say that $\overline{X}$ has property $S(S_1)$ if X has the corresponding property. The intersection of two sets having property S or S_1 is a set having property S or S_1, respectively.

Lemma 4.2.4 *Suppose that the even 2π-periodic functions $a_l(n,x)$, $l \ge 1$, satisfy the following conditions:*

(1) $\left|a_l(n,x)\right| \le 2(2^lnx)^\alpha$ *for $0 \le \alpha \le 1$ and $x \ge 0$;*
(2) *for $a > 0$ and $0 < x < \pi$,*

$$\left|\sum_{a\le n\le b} n^{-1}a_l(n,x)\right| \le \frac{C}{ax}$$

where C is an absolute constant.

Then, for any set X having the property S,

$$\left|\chi_l(\mathbf{x},\overline{X})\right| := \left|\sum_{\mathbf{k}\in\overline{X}} \prod_{j=1}^{d} k_j^{-1} a_l(k_j,x_j)\right| \le C(d)l^d.$$

Corollary 4.2.5 *The assertion of Lemma 4.2.4 remains true if X has the property S_1.*

We will carry out the proof of the lemma by induction on the dimension d of the space. Let $d = 1$. Then, using relation (1) with $\alpha = 1/l$ and relation (2), we obtain

$$\left|\sum_{n=1}^{N} n^{-1} a_l(n,x)\right| \leq \sum_{1\leq n\leq 1/x} 4n^{-1}(nx)^{1/l} + \left|\sum_{n>1/x}^{N} n^{-1} a_l(n,x)\right| \ll l. \tag{4.2.8}$$

Assume that the lemma is true for all dimensions $d' \leq d-1$. We take a point $\mathbf{x} = (x_1,\ldots,x_d)$, $x_j > 0, j = 1,\ldots,d$. If some $x_j = 0$, the lemma is obvious. Let $N_1,\ldots,N_d$ be the coordinates of a point where the line $x_1 n_1 = \cdots = x_d n_d$ intersects the surface Γ, and let $N_0 := x_j N_j$, $j = 1,\ldots,d$. We divide the set X by the planes $n_1 = N_1,\ldots,n_d = N_d$ into 2^d disjoint parts X_b in which we have $n_1 < N_1$ or $n_1 \geq N_1,\ldots,n_d < N_d$ or $n_d \geq N_d$.

Let $X_1 = \prod_{j=1}^{d}[1,N_j)$. We find from (4.2.8) that

$$\left|\chi_l(\mathbf{x},\overline{X}_1)\right| \ll l^d. \tag{4.2.9}$$

To conclude the proof it suffices (without loss of generality) to prove that for all $1 \leq i < d$ the required estimate is true for the function

$$\chi_l(\mathbf{x},\overline{X}_b) := \sum_{n_j\leq N_j,\, j=i+1,\ldots,d;\, \mathbf{n}\in\overline{X}} \frac{a_l(n_{i+1},x_{i+1})}{n_{i+1}} \cdots \frac{a_l(n_d,x_d)}{n_d} \times \left\{ \sum_{n_j>N_j,\, j=1,\ldots,i;\, \mathbf{n}\in\overline{X}} \frac{a_l(n_1,x_1)}{n_1} \cdots \frac{a_l(n_i,x_i)}{n_i} \right\}. \tag{4.2.10}$$

Let us estimate the sum σ in the curly brackets. We will prove that

$$\sigma \leq C(d)/N_0. \tag{4.2.11}$$

If $i = 1$ then this estimate follows at once from part (2) of the lemma. For $i > 1$ we divide the domain of summation into parts in which $x_p n_p \leq x_q n_q$ or $x_p n_p > x_q n_q$, $p,q \in \{1,\ldots,i\}$. This partitions the domain of summation into disjoint subdomains in which $x_{m_1} n_{m_1} \leq \cdots \leq x_{m_i} n_{m_i}$, where $m_1,\ldots,m_i$ is some permutation of the numbers $1,\ldots,i$.

It clearly suffices to prove (4.2.11) for a part σ_1 of σ in which the summation is taken over a subdomain. For simplicity we take the subdomain G to be the one in which $x_1 n_1 \leq \cdots \leq x_i n_i$. Then $x_i n_i \geq x_{i-1} n_{i-1}$, and hence

$$\left|\sum_{n_i:\mathbf{n}\in G} n_i^{-1} a_l(n_i,x_i)\right| \leq C(n_{i-1}x_{i-1})^{-1}.$$

Using (1) with $\alpha = 0$, we get

$$\left|a_l(n_j,x_j)\right| \leq 2$$

and

$$\sigma_1 \ll \sum_{n_1,\dots,n_{i-1};\mathbf{n}\in G} (n_1\cdots n_{i-1})^{-1}(n_{i-1}x_{i-1}). \tag{4.2.12}$$

Further, using the inequality

$$\sum_{n>N} n^{-2} \le \frac{2}{N}$$

$i-2$ times along with the fact that $(n_1,\dots,n_i)\in G$, and hence $n_jx_j \le n_{j+1}x_{j+1}$, $j=1,\dots,i-1$, we find from (4.2.12) that

$$\sigma_1 \ll \sum_{n_1>N_1} n_1^{-2}x_1^{-1} \ll \frac{1}{N_1x_1} = N_0^{-1},$$

which proves (4.2.11).

Using the induction hypothesis and corollary, we find another estimate for σ:

$$\sigma \le C(i)l^i. \tag{4.2.13}$$

Let us raise both sides of (4.2.11) to the power $1/l$ and both sides of (4.2.13) to the power $(l-1)/l$ and multiply the resulting inequalities. As a result we get

$$\sigma \le C(d)l^iN_0^{-1/l}. \tag{4.2.14}$$

We now estimate the function $\chi(\mathbf{x},\overline{X}_b)$. Using (4.2.14) and relation (1) with $\alpha = 1/\big(l(d-i)\big)$ we find from (4.2.10) that

$$\begin{aligned}\big|\chi_l(\mathbf{x},\overline{X}_b)\big| \le C(d)l^iN_0^{-1/l} &\sum_{n_{i+1}\le N_{i+1}} n_{i+1}^{-1}(2^ln_{i+1}x_{i+1})^\alpha \times\cdots\\ \times \sum_{n_d\le N_d} n_d^{-1}(2^ln_dx_d)^\alpha &\le C(d)l^iN_0^{-1/l}2^{l\alpha(d-i)}\alpha^{i-d}N_0^{\alpha(d-i)} \le C(d)l^d.\end{aligned}$$

The lemma is proved. □

For the $a_l(n,x)$, take the functions

$$a_l^r(n,x) := 2^{-l}\sum_{j=0}^{2^l-1}\left(\frac{n2^l}{n2^l+j}\right)^r\big(\cos(n+j)x - \cos(2^ln+j)x\big)$$

and denote the corresponding functions χ by χ^r. We will prove that these functions satisfy (1) and (2) in Lemma 4.2.4. Property (1) follows from the obvious inequalities

$$\big|a_l^r(n,x)\big| \le 2, \qquad \big|a_l^r(n,x)\big| \le 2^lnx.$$

Property (2) will be proved first for the functions $a_l^0(n,x)$. We have by the Abel transformation (A.1.18)

$$\left|\sum_{n=a}^{b} n^{-1} a_l^0(n,x)\right| \le \sum_{n=a}^{b-1} \big(n(n+1)\big)^{-1} \left|\sum_{\nu=1}^{n} a_l^0(\nu,x)\right|$$
$$+ b^{-1}\left|\sum_{\nu=1}^{b} a_l^0(\nu,x)\right| + a^{-1}\left|\sum_{\nu=1}^{a-1} a_l^0(\nu,x)\right|.$$

It remains to show that

$$\left|\sum_{k=1}^{s} a_l^0(k,x)\right| \le Cx^{-1}, \qquad 0<x<\pi,$$

where C is an absolute constant. Indeed,

$$\begin{aligned} a_l^0(n,x) &= 2^{-l}\sum_{j=0}^{2^l-1}\big((\cos nx - \cos 2^l nx)\cos jx - (\sin nx - \sin 2^l nx)\sin jx\big) \\ &= 2^{-l}(\cos nx - \cos 2^l nx)\frac{\sin 2^{l-1}x \cos(2^{l-1}-1/2)x}{\sin(x/2)} \\ &\quad - 2^{-l}(\sin nx - \sin 2^l nx)\frac{\sin 2^{l-1}x \sin(2^{l-1}-1/2)x}{\sin(x/2)}. \end{aligned} \tag{4.2.15}$$

Taking into account that

$$\sum_{n=1}^{s} \cos 2^l nx = O\big(|\sin 2^{l-1}x|^{-1}\big), \qquad \sum_{n=1}^{s} \sin 2^l nx = O\big(|\sin 2^{l-1}x|^{-1}\big),$$

we obtain the required estimate from (4.2.15). For the functions $a_l^r(n,x)$ we have that

$$\big|a_l^0(n,x) - a_l^r(n,x)\big| \ll n^{-1}$$

and

$$\left|\sum_{n=a}^{b} n^{-1} a_l^r(n,x)\right| \ll \left|\sum_{n=a}^{b} n^{-1} a_l^0(n,x)\right| + \sum_{n=a}^{b} n^{-2} \ll \frac{1}{ax}.$$

Consequently, the functions $a_l^r(n,x)$ satisfy (2) .

4.2.4 Polynomials with Equivalent L_p-Norms

We know from Chapter 1 that the Rudin–Shapiro polynomials R_N satisfy the inequalities

$$C_1 N^{1/2} \le \|R_N\|_1 \le \|R_N\|_\infty \le C_2 N^{1/2}.$$

This means that all L_p-norms, $1 \le p \le \infty$, of the R_N are equivalent. These polynomials are important in proving the lower estimates. In this subsection we give two examples of polynomials with equivalent L_p-norms which will be used in §§4.3 and 4.4.

Lacunary polynomials We pick any points $\mathbf{k}^{\mathbf{s}} \in \rho(\mathbf{s})$ and consider the polynomial

$$t_n(\mathbf{x}) := \sum_{\|\mathbf{s}\|_1=n} c_{\mathbf{s}} e^{i(\mathbf{k}^{\mathbf{s}},\mathbf{x})}.$$

Then we have, for all $1 < p < \infty$,

$$\|t_n\|_p \asymp \|t_n\|_2.$$

This relation is a direct corollary of the Littlewood–Paley theorem (see Theorem A.3.3).

Polynomials with mixed structure Here we consider polynomials, which are constructed using the idea of lacunary polynomials and the technique of the Fejér kernel. We assume that n is a sufficiently large number. Let

$$\theta_n' := \{\mathbf{s} : \|\mathbf{s}\|_1 = n,\ s_j \geq n/2d,\ j = 1,\dots,d\}.$$

It is easy to see that $|\theta_n'| \asymp n^{d-1}$. We set $v := \left[|\theta_n'|^{1/d}\right]$ and divide the cube $\mathbb{T}^d$ into v^d cubes with edge $2\pi/v$. We set up a one-to-one correspondence between this set of cubes and a subset $\overline{\theta}_n$ of the set θ_n' such that $|\overline{\theta}_n| = v^d$. For $\mathbf{s} \in \overline{\theta}_n$ let the point $\mathbf{x}^{\mathbf{s}} \in \mathbb{T}^d$ be the center of the corresponding cube. We put

$$u := 2^{\left[(1-1/d)\log_2 n\right]},$$

and consider the function

$$\psi(\mathbf{x}) := \sum_{\mathbf{s}\in\overline{\theta}_n} e^{i(\mathbf{k}^{\mathbf{s}},\mathbf{x})} \mathscr{K}_u(\mathbf{x} - \mathbf{x}^{\mathbf{s}}),$$

where $k_j^{\mathbf{s}} := 2^{s_j} + 2^{s_j-1}$, $j = 1,\dots,d$. Then ψ belongs to $\mathscr{T}^{\perp}(Q_n)$ for sufficiently large n. Using property (2) of the de la Vallée Poussin kernels (see §1.2.3), we obtain, by the definition of $\mathbf{x}^{\mathbf{s}}$ and the relation $u \asymp v$,

$$\|\psi\|_\infty \leq \sum_{\mathbf{s}\in\overline{\theta}_n} \mathscr{K}_u(\mathbf{x} - \mathbf{x}^{\mathbf{s}}) \ll u^d \ll n^{d-1}. \tag{4.2.16}$$

Further,

$$n^{2(d-1)} \ll \|\psi\|_2^2 \leq E_{Q_n}(\psi)_1 \|\psi\|_\infty.$$

It follows from this inequality and relation (4.2.16) that

$$\|\psi\|_1 \geq E_{Q_n}(\psi)_1 \gg n^{d-1}. \tag{4.2.17}$$

4.2.5 The Riesz Products

In this subsection we consider one particular example of products, which are multivariate analogs of the Riesz products designed in a way that is convenient for applications in studying approximation by polynomials from $\mathscr{T}(N)$. We have the multivariate analogs of the Riesz products only in the two-dimensional case. We consider the special trigonometric polynomial

$$\Phi_m(\mathbf{x}) := \prod_{k=0}^{m}(1+\cos 4^k x_1 \cos 4^{m-k} x_2). \tag{4.2.18}$$

Clearly, $\Phi_m(\mathbf{x}) \geq 0$. We will prove that

$$\Phi_m(\mathbf{x}) = 1 + \sum_{k=0}^{m} \cos 4^k x_1 \cos 4^{m-k} x_2 + t_m(\mathbf{x}), \qquad t_m \in \mathscr{T}^{\perp}(4^m). \tag{4.2.19}$$

Consider some term obtained after multiplying out the product over k defining Φ_m. This term will have the form

$$\left(\prod_{k\in e} \cos 4^k x_1\right)\left(\prod_{k\in e} \cos 4^{m-k} x_2\right) = w_e(x_1) h_e(x_2),$$

where e is a subset of integers from $[0,m]$. Let

$$a(e) := \max_{k\in e} k, \qquad b(e) := \max_{k\in e}(m-k).$$

Then the Fourier coefficients of the functions $w_e(x_1)$ and $h_e(x_2)$, respectively, up to the indexes

$$4^{a(e)} - \sum_{k\in e\setminus\{a(e)\}} 4^k > (2/3)4^{a(e)}, \qquad (2/3)4^{b(e)},$$

vanish. Further, since for e such that $|e| \geq 2$ we have $a(e)+b(e) \geq m+1$ and $4\,(2/3)^2 > 1$, then

$$w_e(x_1)h_e(x_2) \in \mathscr{T}^{\perp}(4^m), \qquad |e| \geq 2.$$

The relation (4.2.19) is proved. In particular it implies that

$$\|\Phi_m\|_1 = 1.$$

4.2.6 The Small Ball Inequality

Developing the method from the previous subsection we prove here the following inequality, which we call the *Small Ball inequality*. For an even number m denote $Y_m := \{\mathbf{s} = (2n_1, 2n_2), n_1 + n_2 = m/2\}$. We will prove that

$$\left\| \sum_{\mathbf{s}\in Y_m} \delta_{\mathbf{s}}(f) \right\|_\infty \geq C \sum_{\mathbf{s}\in Y_m} \|\delta_{\mathbf{s}}(f)\|_1. \tag{4.2.20}$$

We start with a lemma which looks technical but plays a key role in proving the main inequality. We need some notation. Let $AP(4,b), b = 0,1,2,3$, denote an arithmetic progression of the form $4a+b, a = 0,1,2,\dots$ and let

$$\begin{aligned} \theta_n^3 &:= \{\mathbf{s} = (s_1,s_2) : \mathbf{s} \in Y_n, s_1,s_2 \geq 3\}, \\ A(b) &:= \{s_1 : s_1 \in AP(4,b) \cap [3,n-3]\}; \\ H(b) &:= \{\mathbf{s} \in \theta_n^3 : s_1 \in A(b)\}. \end{aligned}$$

For $\mathbf{k}(\mathbf{s}) = (k_1(\mathbf{s}),k_2(\mathbf{s}))$ such that $2^{s_j-1} \leq k_j(\mathbf{s}) \leq 2^{s_j}, j = 1,2,\ \mathbf{s} \in \theta_n^3$, denote

$$G(\mathbf{s},\mathbf{k}(\mathbf{s})) := \{\mathbf{k} = (k_1,k_2) : -2^{s_j-3} \leq k_j - k_j(\mathbf{s}) \leq 2^{s_j-2}, j = 1,2\}.$$

Lemma 4.2.6 *Consider real trigonometric polynomials $t_{\mathbf{s}} \in \mathscr{T}(G(\mathbf{s},\mathbf{k}(\mathbf{s})))$ such that $\|t_{\mathbf{s}}\|_\infty \leq 1$. Then the function*

$$\Phi(\mathbf{x}) = \prod_{\mathbf{s}\in H(b)} (1 + t_{\mathbf{s}}(\mathbf{x}))$$

has the following properties:

(1) $\Phi(\mathbf{x}) \geq 0$,
(2) $\Phi(\mathbf{x}) = 1 + \Phi_1(\mathbf{x}) + \Phi_2(\mathbf{x})$,

where

$$\Phi_1(\mathbf{x}) = \sum_{\mathbf{s}\in H(b)} t_{\mathbf{s}}(\mathbf{x})$$

and Φ_2 is orthogonal to each $t \in \mathscr{T}(Q_n)$.

Proof The first property is obvious. In order to prove the second we need to prove that, for $\mathbf{k} = (k_1,k_2)$ such that $|k_1k_2| \leq 2^n$ we have $\hat{\Phi}_2(\mathbf{k}) = 0$. Let $w(kt)$ denote either $\cos kt$ or $\sin kt$. Then $\Phi_2(\mathbf{x})$ contains terms of the form

$$h(\mathbf{x}) = c\prod_{i=1}^m w(k_1^i x_1)w(k_2^i x_2), \qquad \mathbf{k}^i \in G(\mathbf{s}^i,\mathbf{k}(\mathbf{s}^i)) \tag{4.2.21}$$

with all $\mathbf{s}^i, i = 1,\dots,m$, $m \geq 2$, distinct. For the sake of simplicity of notations we assume that $s_1^1 > s_1^2 > \dots > s_1^m$. Then, for $h(\mathbf{x})$, the frequencies with respect to x_1 have the form

$$\begin{aligned} k_1 &= k_1^1 \pm k_1^2 \pm \dots \pm k_1^m \\ &\geq 2^{s_1^1-1} - 2^{s_1^1-3} - \sum_{i=1}^{m-1} (2^{s_1^1-4i} + 2^{s_1^1-4i-2}) > 2^{s_1^1-2}. \end{aligned} \tag{4.2.22}$$

Similarly, for the frequencies k_2 of the function $h(\mathbf{x})$ with respect to x_2 we have

$$k_2 > 2^{s_2^m - 2}.$$

Consequently,

$$k_1 k_2 > 2^{s_1^1 + s_2^m - 4}.$$

In order to complete the proof it remains to observe that for all terms $h(\mathbf{x})$ in the function $\Phi_2(\mathbf{x})$ we have $m \geq 2$, which in turn implies $s_1^1 + s_2^m \geq n+4$. The lemma is now proved. □

Theorem 4.2.7 *For an arbitrary real function of two variables $f \in \mathscr{T}(D_n)$ where $D_n := U_{s \in Y_n} \rho(s)$ we have*

$$E_{Q_{n-3}}(f)_\infty \geq C \sum_{\mathbf{s} \in Y_n} \|\delta_{\mathbf{s}}(f)\|_1,$$

where C is a positive absolute constant.

Proof The proof of this theorem is based on Lemma 4.2.6. Let $t \in \mathscr{T}(Q_{n-3})$ be a polynomial of the best approximation of the function f in the uniform norm. We split the sum in the formulation of the theorem into two sums: one over θ_n^3 and the other over the rest of the set, R_n. First, consider the sum over R_n. The number of elements of this set $|R_n| = 6$. Further,

$$\begin{aligned} \sum_{\mathbf{s} \in R_n} \|\delta_{\mathbf{s}}(f)\|_1 &\leq \sum_{\mathbf{s} \in R_n} \|\delta_{\mathbf{s}}(f)\|_2 \leq |R_n|^{1/2} \left(\sum_{\mathbf{s} \in R_n} \|\delta_{\mathbf{s}}(f)\|_2^2 \right)^{1/2} \\ &\leq 3\|f - t\|_2 \leq 3 E_{Q_{n-3}}(f)_\infty. \end{aligned} \tag{4.2.23}$$

We proceed to the remaining sum. First of all, we construct a partition of unity on the plane of frequencies. We start with the univariate case. Let $g(0,0,y) := (1-y)_+$ and for nonnegative integers m and l we define, for $y \geq 0$, continuous functions $g(m,l,y)$ in the following way. Let $y(m,l) = 2^{m-1} + l$ and:

(i) $g(m,l,y(m,l)) = 1$;
(ii) for $1 \leq m \leq 3, l = 0,1,\dots,2^{m-1}-1$, $g(m,l,y) = 0$ for $|y - y(m,l)| \geq 1$ and is linear on $[y(m,l)-1, y(m,l)]$ and $[y(m,l), y(m,l)+1]$;
(iii) for $m > 3$, $l = 0$, $g(m,0,y) = 0$ for $y \leq y(m,0) - 2^{m-4}$ and $y \geq y(m,0) + 2^{m-3}$ and is linear on $[y(m,0) - 2^{m-4}, y(m,0)]$ and $[y(m,0), y(m,0) + 2^{m-3}]$;
(iv) for $m > 3, l = a2^{m-3}, a = 1,2,3$, we have $g(m,l,y) = 0$ for $|y - y(m,l)| \geq 2^{m-3}$ and is linear to the left and to the right of $y(m,l)$.

It is clear that $\{g(m,l,y)\}$ forms a partition of unity for $[0,\infty)$. Let

$$\mathscr{A}(0,0,t) := 1$$

and, for $m \geq 1$,

$$\mathscr{A}(m,l,t) := 2\sum_{k} g(m,l,k)\cos kt.$$

It is well known that

$$\|\mathscr{A}(m,l)\|_1 \leq 2. \tag{4.2.24}$$

We consider functions of two variables

$$g(\mathbf{s},\mathbf{l},\mathbf{y}) := g(s_1,l_1,y_1)g(s_2,l_2,y_2),$$

$$\mathscr{A}(\mathbf{s},\mathbf{l},\mathbf{x}) := \mathscr{A}(s_1,l_1,x_1)\mathscr{A}(s_2,l_2,x_2),$$

where $\mathbf{s} = (s_1,s_2)$ and $\mathbf{l} = (l_1,l_2)$. The family $\{g(\mathbf{s},\mathbf{l},\mathbf{y})\}$ provides a partition of unity on the plane of frequencies. It will be convenient for us to use the notation $\mathscr{A}(m,2^{m-1},t) := \mathscr{A}(m+1,0,t)$. Denote

$$B(m) := \{l : l = a2^{m-3}, a = 0,1,2,3,4\}$$

and

$$B(\mathbf{s}) := \{\mathbf{l} = (l_1,l_2) : l_1 \in B(s_1), l_2 \in B(s_2)\}.$$

For two periodic functions $a(\mathbf{x})$ and $b(\mathbf{x})$, let $(a*b)(\mathbf{x})$ denote their convolution:

$$(a*b)(\mathbf{x}) := (2\pi)^{-2}\int_{\mathbb{T}^2} a(\mathbf{x}-\mathbf{y})b(\mathbf{y})d\mathbf{y}.$$

Consider for $f \in L_1(\pi_2)$ the following trigonometric polynomials:

$$A(f,\mathbf{s},\mathbf{l}) := f * \mathscr{A}(\mathbf{s},\mathbf{l}).$$

It is clear that for $f \in \mathscr{T}(D_n)$ we have

$$\delta_{\mathbf{s}}(f) = \sum_{\mathbf{l}\in B(\mathbf{s})} A(f,\mathbf{s},\mathbf{l}). \tag{4.2.25}$$

We split the set of summation indices $(\mathbf{s},\mathbf{l})$ into a number of sets in order to apply Lemma 4.2.6. The role of $G(\mathbf{s},\mathbf{k}(\mathbf{s}))$ is played now by $\operatorname{supp} g(\mathbf{s},\mathbf{l})$. We can realize this splitting by taking s_1 from $AP(4,b)$ and arbitrary $\mathbf{l} \in B(s_1,n-s_1)$. The number of these new sets does not exceed 100. Let H denote one of these sets. We now prove that

$$\sum_{(\mathbf{s},\mathbf{l})\in H} \|A(f,\mathbf{s},\mathbf{l})\|_1 \leq 4E_{Q_{n-3}}(f)_\infty. \tag{4.2.26}$$

It is clear that (4.2.25) and (4.2.26) together with the previous remark about the number of sets H imply the inequality

$$\sum_{\mathbf{s}\in\theta_n^3} \|\delta_{\mathbf{s}}(f)\|_1 \leq CE_{Q_{n-3}}(f)_\infty.$$

This inequality with the relation (4.2.23) completes the proof of Theorem 4.2.7. Thus, it remains to verify the relation (4.2.26). For each $(\mathbf{s},\mathbf{l}) \in H$ consider

$$t(\mathbf{s},\mathbf{l}) := A(\operatorname{sign} A(f,\mathbf{s},\mathbf{l}),\mathbf{s},\mathbf{l})/\|\mathscr{A}(\mathbf{s},\mathbf{l})\|_1.$$

Then $t(\mathbf{s},\mathbf{l}) \in \mathscr{T}(\operatorname{supp} g(\mathbf{s},\mathbf{l})) \subset \mathscr{T}(G(\mathbf{s},\mathbf{k}(\mathbf{s})))$, where $\mathbf{k}(\mathbf{s}) = (2^{s_1-1}+l_1, 2^{s_2-1}+l_2)$. Further,

$$\|t(\mathbf{s},\mathbf{l})\|_\infty \le 1.$$

This means that the functions $t(\mathbf{s},\mathbf{l})$, $(\mathbf{s},\mathbf{l}) \in H$, meet the conditions of Lemma 4.2.6 and, consequently, that the function

$$\Phi^H := \prod_{(\mathbf{s},\mathbf{l})\in H} (1+t(\mathbf{s},\mathbf{l}))$$

has the properties (1) and (2) from Lemma 4.2.6. Consider now

$$\langle f-t, \Phi^H - 1\rangle = \langle f, \Phi^H_1\rangle = \sum_{(\mathbf{s},\mathbf{l})\in H} \langle f, t(\mathbf{s},\mathbf{l})\rangle. \tag{4.2.27}$$

We have on the one hand

$$\langle f, t(\mathbf{s},\mathbf{l})\rangle = \langle A(f,\mathbf{s},\mathbf{l}), \operatorname{sign} A(f,\mathbf{s},\mathbf{l})\rangle/\|\mathscr{A}(\mathbf{s},\mathbf{l})\|_1 = \|A(f,\mathbf{s},\mathbf{l})\|_1/\|\mathscr{A}(\mathbf{s},\mathbf{l})\|_1. \tag{4.2.28}$$

On the other hand we have

$$|\langle f-t, \Phi^H - 1\rangle| \le \|f-t\|_\infty \|\Phi^H - 1\|_1 \le 2\|f-t\|_\infty = 2E_{Q_{n-3}}(f)_\infty. \tag{4.2.29}$$

The relations (4.2.27)–(4.2.29) and (4.2.24) imply (4.2.26). Theorem 4.2.7 is proved. □

Remark 4.2.8 It is clear that the conclusion of Theorem 4.2.7 is valid for complex functions $f \in \mathscr{T}(D_n)$ as well.

We point out here that there is an analog of Theorem 4.2.7 in which $\delta_{\mathbf{s}}(f)$, is replaced by the "smooth" dyadic block $A_{\mathbf{s}}(f)$.

Theorem 4.2.9 *For an arbitrary trigonometric polynomial $f \in \mathscr{T}(Q_n \setminus Q_{n-1})$ of two variables we have*

$$E_{Q_{n-3}}(f)_\infty \ge C\sum_{\mathbf{s}} \|A_{\mathbf{s}}(f)\|_1,$$

where C is a positive absolute constant.

We discussed above the small ball inequality for polynomials on two variables. The corresponding inequality for polynomials of three and more than three variables is a big open problem. This problem is related to an outstanding problem in probability, which is the small ball problem. This is why the corresponding inequality is called the small ball inequality. We formulate the multivariate small ball inequality as an open problem and refer the reader for further discussion to Temlyakov (2011), Chapter 3, and to Dinh Dung *et al.* (2016).

For even n, set

$$Y_n^d := \{\mathbf{s} = (2l_1, \dots, 2l_d), l_1 + \cdots + l_d = n/2, l_j \in \mathbb{Z}_+, j = 1, \dots, d\}.$$

It is conjectured (see, for instance, Kashin and Temlyakov, 2008) that the following inequality holds for any coefficients $\{c_k\}$:

$$\left\| \sum_{\mathbf{s}\in Y_n^d} \sum_{\mathbf{k}\in\rho(\mathbf{s})} c_{\mathbf{k}} e^{i(\mathbf{k},\mathbf{x})} \right\|_\infty \geq C(d) n^{-(d-2)/2} \sum_{\mathbf{s}\in Y_n^d} \left\| \sum_{\mathbf{k}\in\rho(\mathbf{s})} c_{\mathbf{k}} e^{i(\mathbf{k},\mathbf{x})} \right\|_1 . \tag{4.2.30}$$

We note that a weaker version of (4.2.30) with exponent $(d-2)/2$ replaced by $(d-1)/2$ is a direct corollary of the Parseval identity, the Cauchy inequality, and the monotonicity of the L_p-norms.

Open Problem 4.3 Prove inequality (4.2.30) for $d \geq 3$.

We note that there are interesting results for analogs of inequality (4.2.30) when we replace the L_∞-norm by another norm, namely, by the QC-norm (quasi-continuous norm). The reader can find the corresponding results in Kashin and Temlyakov (2008).

4.3 The Bernstein–Nikol'skii Inequalities

4.3.1 The Bernstein Inequalities

In this subsection we prove

Theorem 4.3.1 *For arbitrary α,*

$$\sup_{t\in\mathscr{T}(N)} \left\|t^{(r)}(\mathbf{x},\alpha)\right\|_p \Big/ \|t\|_p \asymp \begin{cases} N^r & \text{for } 1 < p < \infty,\ r \geq 0, \\ N^r(\log N)^{d-1} & \text{for } p = \infty,\ r > 0. \end{cases}$$

Proof We first analyze the case $1 < p < \infty$. By Theorem A.3.6 the trigonometric conjugation operator is bounded as an operator from L_p to L_p in this case; therefore, it suffices to consider the case $\alpha = \mathbf{0}$.

It is not difficult to verify that in the case $d = 1$ the numbers $\lambda_\nu^r := |\nu|^r 2^{-rs}$, $2^{s-1} \leq |\nu| < 2^s$, $\lambda_0^r := 1$, and the numbers $(\lambda_\nu^r)^{-1}$ are Marcinkiewicz multipliers; that is, for some number M,

$$|\lambda_\nu^r| \leq M, \qquad \sum_{\nu=\pm 2^{s-1}}^{\pm(2^s-1)} |\lambda_\nu^r - \lambda_{\nu+1}^r| \leq M, \qquad s = 1, 2, \dots$$

Then by the Marcinkiewicz theorem (see Theorem A.3.6),

$$\|\Lambda_r f\|_p \asymp \|f\|_p, \qquad \Lambda_r(f) := \sum_{\nu} \lambda_\nu^r \hat{f}(\nu) e^{i\nu x}.$$

Applying the operator Λ_r to a function f as a function of each of the d variables successively, we find that

$$\|\Lambda_r f\|_p \asymp \|f\|_p. \tag{4.3.1}$$

If $t \in \mathscr{T}(N)$, then by the Littlewood–Paley theorem (see Theorem A.3.3),

$$\|t\|_p \asymp \left\| \left(\sum_{\mathbf{s}} |\delta_{\mathbf{s}}(t)|^2 \right)^{1/2} \right\|_p \tag{4.3.2}$$

and, by (4.3.1) and the Littlewood–Paley theorem,

$$\begin{aligned} \left\|t^{(r)}(\mathbf{x}, \mathbf{0})\right\|_p &\asymp \|\Lambda_r^{-1} t^{(r)}\|_p \asymp \left\| \left(\sum_{\mathbf{s}} 2^{2r\|\mathbf{s}\|_1} |\delta_{\mathbf{s}}(t)|^2 \right)^{1/2} \right\|_p \\ &\ll N^r \left\| \left(\sum_{\mathbf{s}} |\delta_{\mathbf{s}}(t)|^2 \right)^{1/2} \right\|_p \asymp N^r \|t\|_p. \end{aligned}$$

The upper estimates are thus proved in the case $1 < p < \infty$.

The lower estimates follow from the example

$$t(\mathbf{x}) := e^{iNx_1} \prod_{j=2}^{d} e^{ix_j}.$$

Consider the case $p = \infty$. We first prove the upper estimates. Let n be such that $2^{n-1} < N \le 2^n$. Then, for $t \in \mathscr{T}(N)$,

$$t^{(r)}(\mathbf{x}, \alpha) = t(\mathbf{x}) * \mathscr{V}_{Q_{n+2d}}^{(r)}(\mathbf{x}, \alpha)$$

and

$$\left\|t^{(r)}(\mathbf{x}, \alpha)\right\|_\infty \le \left\|t(\mathbf{x})\right\|_\infty \left\|\mathscr{V}_{Q_{n+2d}}^{(r)}(\mathbf{x}, \alpha)\right\|_1.$$

Applying Lemma 4.2.3 for $p = 1$, we get

$$\left\|t^{(r)}(\mathbf{x}, \alpha)\right\|_\infty \ll N^r (\log N)^{d-1} \left\|t(\mathbf{x})\right\|_\infty,$$

as required.

Let us proceed to the proof of the lower estimates. We denote

$$\begin{aligned} a(n,x,\alpha) &:= \cos(nx+\alpha\pi/2)+\cos\big((n+1)x+\alpha\pi/2\big) \\ &\quad -\cos(2nx+\alpha\pi/2)-\cos\big((2n+1)x+\alpha\pi/2\big), \qquad n\neq 0, \\ a(0,x,\alpha) &:= 0, \qquad (4.3.3) \\ U_N^r(\mathbf{x},\alpha) &:= \sum_{\mathbf{k}\in\Gamma(N)^+}\prod_{j=1}^{d} a(k_j,x_j,\alpha_j)k_j^r, \quad \Gamma(N)^+ := \Gamma(N)\cap\mathbb{N}^d. \end{aligned}$$

We will use the following lemma for the Telyakovskii polynomials $U_N^{-1}(\mathbf{x},\alpha)$; it is a particular case of Lemma 4.2.4.

Lemma 4.3.2 *We have*

$$\left\|U_N^{-1}(\cdot,\alpha)\right\|_\infty \le C(d).$$

We prove the required lower estimate. Suppose that M and $N \asymp M$ are chosen such that

$$U_N^{-1} \in \mathscr{T}(M).$$

Then, on the one hand, by Lemma 4.3.2 $\|U_N^{-1}\|_\infty \le C(d)$. On the other hand $(r>0)$,

$$(U_N^{-1})^{(r)}(\mathbf{0},\alpha) \asymp \sum_{\mathbf{k}\in\Gamma(N)^+}\prod_{j=1}^{d} k_j^{r-1} \gg N^r(\log N)^{d-1}.$$

The theorem is now proved. □

Remark 4.3.3 The relations in Theorem 4.3.1 do not depend on α.

At the same time, there is an apparent dependence on α for $p=1$ or ∞ and $r=0$. We first find the order of the quantity

$$\sup_{t\in\mathscr{T}(N)} \left\|t^{(0)}(\mathbf{x},\mathbf{1})\right\|_p, \qquad p=1,\infty.$$

Let $\mathbf{N}=(N,\dots,N)$; then it is easy to derive the estimate

$$\sup_{t\in\mathscr{T}(\mathbf{N},d)} \left\|t^{(0)}(\mathbf{x},\mathbf{1})\right\|_p \Big/ \|t\|_p \asymp (\log N)^d, \qquad p=1,\infty, \tag{4.3.4}$$

from the univariate results on the norm of the conjugate polynomial.

Further, since

$$\mathscr{T}\big((N^{1/d},\dots,N^{1/d}),d\big) \subset \mathscr{T}(N) \subset \mathscr{T}(\mathbf{N},d),$$

it follows from (4.3.4) that

$$\sup_{t\in\mathscr{T}(N)} \left\|t^{(0)}(\mathbf{x},\mathbf{1})\right\|_p \Big/ \|t\|_p \asymp (\log N)^d, \qquad p=1,\infty. \tag{4.3.5}$$

This tells us that, for any $e \subset [1,d]$,

$$\sup_{t \in \mathscr{T}(N)} \left\|t^{(0)}(\mathbf{x}, \chi(e))\right\|_p / \|t\|_p \asymp (\log N)^{|e|}, \qquad p = 1, \infty, \tag{4.3.6}$$

where $\left(\chi(e)\right)_j = 1$ for $j \in e$ and $\left(\chi(e)\right)_j = 0$ for $j \notin e$.

Suppose that α is given, and let $0 \le l \le d$ denote the number of indices j such that $\{\alpha_j/2\} \neq 0$; then by (4.3.6) we get

$$\sup_{t \in \mathscr{T}(N)} \left\|t^{(0)}(\mathbf{x}, \alpha)\right\|_p / \|t\|_p \asymp (\log N)^l, \qquad p = 1, \infty.$$

4.3.2 Bernstein Inequalities for $p = 1$

Theorem 4.3.1 provides the correct order for the following quantity $b(n, r, L_p)$ for all $1 < p \le \infty$:

$$b(n, r, L_p) := \sup_{f \in \mathscr{T}(Q_n)} \|f^{(r,\dots,r)}\|_p / \|f\|_p.$$

The order of $b(n, r, L_1)$ is not known. We formulate it as an open problem.

Open Problem 4.4 Find the order in n of the sequence $\{b(n, r, L_1)\}$.

We now present some partial results on the Bernstein inequality in L_1 for polynomials of two variables. In particular, we prove the following assertion.

Theorem 4.3.4 *For any $r > 0$, $\alpha \in \mathbb{R}^2$, and any function $f \in \mathscr{T}(Q_n)$ of two variables we have*

$$\|f^{(r)}(\mathbf{x}, \alpha)\|_1 \le C(r) n^{1/2} 2^{rn} \|f\|_1.$$

We note that Lemma 4.2.3 with $p = 1$ implies the following inequality for all d:

$$\|f^{(r)}(\mathbf{x}, \alpha)\|_1 \le C(r, d) n^{d-1} 2^{rn} \|f\|_1.$$

Thus, Theorem 4.3.4 provides an improvement over the above simple inequality. Also, a comparison of Theorems 4.3.4 and 4.3.1 shows that the behavior of the sequences $\{b(n, r, L_p)\}$ is deferent for $p = \infty$ and $p = 1$. The proof of Theorem 4.3.4 uses the technique of Riesz products.

We introduce some notation. For two integers $a \ge 1$ and $0 \le b < a$, we denote by $AP(a, b)$ the arithmetic progression of the form $al + b$, $l = 0, 1, \dots$ Set

$$H_n(a, b) := \{\mathbf{s} : \mathbf{s} \in \mathbb{Z}^2_+, \|\mathbf{s}\|_1 = n, s_1, s_2 \ge a, s_1 \in AP(a, b)\}.$$

It will be convenient for us to consider subspaces $\mathscr{T}(\rho'(\mathbf{s}))$ of trigonometric polynomials with frequencies (harmonics) from

$$\rho'(\mathbf{s}) := \{\mathbf{k} \in \mathbb{Z}^2 : [2^{s_j - 2}] \le |k_j| < 2^{s_j}, j = 1, 2\}.$$

For a subspace Y in $L_2(\mathbb{T}^d)$ we denote by $Y^\perp$ its orthogonal complement.

Lemma 4.3.5 *Take any trigonometric polynomials $t_\mathbf{s} \in \mathscr{T}(\rho'(\mathbf{s}))$ and form the function*

$$\Phi(\mathbf{x}) := \prod_{\mathbf{s}\in H_n(a,b)} (1+t_\mathbf{s}(\mathbf{x})).$$

Then for any $a \geq 6$ and any $0 \leq b < a$ this function admits the representation

$$\Phi(\mathbf{x}) = 1 + \sum_{\mathbf{s}\in H_n(a,b)} t_\mathbf{s}(\mathbf{x}) + R(\mathbf{x})$$

with $R \in \mathscr{T}(Q_{n+a-6})^\perp$.

The proof of this lemma is similar to the proofs of relation (4.2.19) and of relation (2) in Lemma 4.2.6. We do not give the proof of the above lemma here.

We need the following simple observation, which we formulate without proof.

Remark 4.3.6 For any real numbers $y_l \in [-1,1]$, $l = 1,\dots,N$, we have (with $i^2 = -1$)

$$\left|\prod_{l=1}^{N}\left(1+\frac{iy_l}{\sqrt{N}}\right)\right| \leq C.$$

Lemma 4.3.7 *For any function f of the form*

$$f = \sum_{\mathbf{s}\in H_n(a,b)} t_\mathbf{s}$$

with $a \geq 6$, $b \in [0,a)$, where $t_\mathbf{s} \in \mathscr{T}(\rho'(\mathbf{s}))$ are real trigonometric polynomials such that $\|t_\mathbf{s}\|_\infty \leq 1$, we have

$$E^\perp_{Q_{n+a-6}}(f)_\infty := \inf_{g\in\mathscr{T}(Q_{n+a-6})^\perp} \|f-g\|_\infty \ll (1+n/a)^{1/2}.$$

Proof Let us form the function

$$RP(f) := \operatorname{Im} \prod_{\mathbf{s}\in H_n(a,b)} (1+it_\mathbf{s}(1+n/a)^{-1/2}),$$

which is an analog of the Riesz product associated with f. Then by Remark 4.3.6 we have

$$\|RP(f)\|_\infty \leq C. \tag{4.3.7}$$

Lemma 4.3.5 provides the representation

$$RP(f) = (1+n/a)^{-1/2} \sum_{\mathbf{s}\in H_n(a,b)} t_\mathbf{s} + g, \qquad g \in \mathscr{T}(Q_{n+a-6})^\perp. \tag{4.3.8}$$

Combining (4.3.7) with (4.3.8) we complete the proof of the lemma. □

Remark 4.3.8 It is clear that in Lemma 4.3.7 we can drop the assumption that the $t_{\mathbf{s}}$ are real polynomials.

Lemma 4.3.9 *For any function of the form*

$$f = \sum_{\|\mathbf{s}\|_1 = n} t_{\mathbf{s}}, \quad t_{\mathbf{s}} \in \rho'(\mathbf{s}), \quad \|t_{\mathbf{s}}\|_\infty \le 1,$$

we have, for any $a \ge 6$,

$$E^{\perp}_{Q_{n+a-6}}(f)_\infty \le Ca(1+n/a)^{1/2}.$$

Proof We introduce some more notation. Denote

$$\theta_n := \{\mathbf{s} : \|\mathbf{s}\|_1 = n\}, \qquad \theta_{n,a} := \{\mathbf{s} \in \theta_n : s_1 < a \text{ or } s_2 < a\}.$$

Then

$$f = \sum_{\mathbf{s}\in\theta_n} t_{\mathbf{s}} = \sum_{\mathbf{s}\in\theta_{n,a}} t_{\mathbf{s}} + \sum_{b=0}^{a-1} \sum_{\mathbf{s}\in H_n(a,b)} t_{\mathbf{s}}$$

and

$$E^{\perp}_{Q_{n+a-6}}(f)_\infty \le \sum_{\mathbf{s}\in\theta_{n,a}} \|t_{\mathbf{s}}\|_\infty + \sum_{b=0}^{a-1} E^{\perp}_{Q_{n+a-6}}\left(\sum_{\mathbf{s}\in H_n(a,b)} t_{\mathbf{s}}\right)_\infty.$$

Using the assumption $\|t_{\mathbf{s}}\|_\infty \le 1$, Lemma 4.3.7, and Remark 4.3.8 we obtain the required estimate. □

It proves to be useful in studying approximation of functions with mixed smoothness to consider along with the L_p-norms the Besov-type norms. Set

$$\|f\|_{\mathbf{B}^r_{p,1}} := \sum_{\mathbf{s}} 2^{r\|\mathbf{s}\|_1} \|A_{\mathbf{s}}(f)\|_p.$$

Theorem 4.3.10 *Let $r > 0$ be given. For every polynomial $f \in \mathscr{T}(Q_n)$ of two variables we have*

$$\|f\|_{\mathbf{B}^r_{1,1}} \le C(r) n^{1/2} 2^{rn} \|f\|_1.$$

Proof Take any $k \le n+2$ and consider the sum

$$S_k := \sum_{\|\mathbf{s}\|_1 = k} \|A_{\mathbf{s}}(f)\|_1.$$

Define the polynomials $t_{\mathbf{s}} := A_{\mathbf{s}}(\operatorname{sign}(A_{\mathbf{s}}(f)))$ and

$$g_k := \sum_{\|\mathbf{s}\|_1 = k} \bar{t}_{\mathbf{s}}.$$

It is clear from the definition of $A_\mathbf{s}$ that $\bar{t}_\mathbf{s} \in \mathscr{T}(\rho'(\mathbf{s}))$. Next, on one hand we have

$$\langle f, g_k \rangle = \sum_{\|\mathbf{s}\|_1 = k} \|A_\mathbf{s}(f)\|_1, \tag{4.3.9}$$

and on the other hand

$$\langle f, g_k \rangle \le \|f\|_1 E^{\perp}_{Q_n}(g_k)_\infty. \tag{4.3.10}$$

By Lemma 4.3.9, with $a = \max(n-k+6, 6)$, we get

$$E^{\perp}_{Q_n}(g_k)_\infty \ll (n-k+6)(1+n/(n-k+6))^{1/2}. \tag{4.3.11}$$

Relations (4.3.9)–(4.3.11) imply the inequality

$$\sum_\mathbf{s} 2^{r\|\mathbf{s}\|_1} \|A_\mathbf{s}(f)\|_1 = \sum_{k \le n+2} 2^{rk} S_k \le C(r) n^{1/2} 2^{rn} \|f\|_1,$$

as required. □

Theorem 4.3.10 implies Theorem 4.3.4. Indeed, we have

$$\|f^{(r)}(\mathbf{x}, \alpha)\|_1 \le \sum_\mathbf{s} \|A_\mathbf{s}(f)^{(r)}(\mathbf{x}, \alpha)\|_1 \le C(r) \sum_\mathbf{s} 2^{r\|\mathbf{s}\|_1} \|A_\mathbf{s}(f)\|_1.$$

Using Theorem 4.3.10, we continue this to obtain

$$\le C(r) n^{1/2} 2^{rn} \|f\|_1.$$

This proves Theorem 4.3.4.

Remark 4.3.11 The inequality in Theorem 4.3.10 is sharp. The factor $n^{1/2}2^{rn}$ cannot be replaced by a function that grows more slowly in n even if we replace $\|f\|_1$ by $\|f\|_p$, for some $p < \infty$.

Proof Take

$$f := \sum_{\|\mathbf{s}\|_1 = n} \cos(2^{s_1} x_1) \cos(2^{s_2} x_2).$$

Then

$$\sum_\mathbf{s} 2^{r\|\mathbf{s}\|_1} \|A_\mathbf{s}(f)\|_1 \asymp n 2^{rn},$$

and by the Littlewood–Paley theorem A.3.3 we obtain for $p < \infty$

$$\|f\|_p \ll n^{1/2}.$$ □

The following result from Temlyakov (1998a), which we give without proof, shows that we cannot take $p = \infty$ in Remark 4.3.11.

Theorem 4.3.12 *Let $r > 0$ be given. For every polynomial $f \in \mathscr{T}(Q_n)$ of two variables we have*

$$\|f\|_{\mathbf{B}^r_{1,1}} \le C(r) 2^{rn} \|f\|_\infty.$$

4.3.3 The Nikol'skii Inequalities

The main goal of this subsection is to obtain the Nikol'skii inequalities for polynomials in $\mathscr{T}(N)$ (Theorems 4.3.16 and 4.3.17). We first prove some auxiliary statements.

Lemma 4.3.13 *We have, for the polynomials defined in* (4.3.3),

$$\left\|U_N^r(\cdot,\alpha)\right\|_\infty \ll N^{r+1}, \qquad r \geq -1. \tag{4.3.12}$$

Proof For $r=-1$ the required bound follows from Lemma 4.3.2. We derive inequality (4.3.12) for $r>-1$ from this inequality for $r=-1$. Denote

$$\tilde{g}(\mathbf{k},\alpha) := (2\pi)^{-d}\int_{\mathbb{T}^d} \bar{g}(\mathbf{x})\prod_{j=1}^d a(k_j,x_j,\alpha_j)d\mathbf{x}$$

and

$$\Gamma(N)^+ := \Gamma(N)\cap\mathbb{N}^d.$$

Then, by Theorem A.2.1, we have

$$\begin{aligned}\|U_N^r\|_\infty &= \sup_{\|g\|_1\leq 1}|\langle U_N^r,g\rangle| = \sup_{\|g\|_1\leq 1}\left|\sum_{\mathbf{k}\in\Gamma(N)^+}\tilde{g}(\mathbf{k},\alpha)v(\mathbf{k})^r\right| \\ &= \sup_{\|g\|_1\leq 1}\sum_{l=1}^N l^{r+1}\left|\sum_{v(\mathbf{k})=l}\tilde{g}(\mathbf{k},\alpha)v(\mathbf{k})^{-1}\right|\end{aligned}$$

by the Abel inequality (A.1.17) we continue the above expression, obtaining

$$\begin{aligned}&\leq 2\sup_{\|g\|_1\leq 1}N^{r+1}\max_{1\leq m\leq N}\left|\sum_{\mathbf{k}\in\Gamma(m)^+}\tilde{g}(\mathbf{k},\alpha)v(\mathbf{k})^{-1}\right| \\ &\leq 2N^{r+1}\max_{1\leq m\leq N}\sup_{\|g\|_1\leq 1}\left|\langle U_m^{-1},g\rangle\right| \leq C(d)N^{r+1}.\end{aligned}$$

The lemma is proved. □

Now let

$$\mathscr{D}_N^r(\mathbf{x},\alpha) := \sum_{\mathbf{k}\in\Gamma(N);\mathbf{k}\geq\mathbf{0}}\prod_{j=1}^d(\bar{k}_j)^r\cos(k_jx_j+\alpha_j\pi/2), \qquad \bar{k}_j := \max(k_j,1),$$

$$\mathscr{D}_N^r(\mathbf{x},\alpha)' := \sum_{\mathbf{k}\in\Gamma(N,2)}\prod_{j=1}^d k_j^r\cos(k_jx_j+\alpha_j\pi/2),$$

where

$$\Gamma(N,2) := \{\mathbf{k}\in\Gamma(N) : \mathbf{k}\geq\mathbf{2}\}.$$

Lemma 4.3.14 *We have*

$$E_N^{\perp}(\mathscr{D}_N^r)_\infty \asymp N^{r+1}, \qquad r > -1.$$

Proof Clearly, it suffices to prove this lemma for $\mathscr{D}_N^r(\mathbf{x},\alpha)'$. For simplicity in the notation we treat the case $\alpha = \mathbf{0}$. We begin with a brief description of the idea. For $\mathbf{k} \in \Gamma(N,2)$, in the representation of the function $U_N^r(\mathbf{x}) := U_N^r(\mathbf{x},\mathbf{0})$ each summand of the form $\prod_{j=1}^d \cos k_j x_j$ has a coefficient

$$a(\mathbf{k}) = \sum_F \prod_{j\in F_1} (-(l_j/2)^r) \prod_{j\in F_2} (k_j-1)^r \prod_{j\in F_3} k_j^r, \tag{4.3.13}$$

where

$$l_j := \begin{cases} k_j - 1 & \text{for odd } k_j, \\ k_j & \text{for even } k_j, \end{cases}$$

and $F := (F_1, F_2, F_3)$ is a partition of the interval $[1,d]$ into three sets. If in (4.3.13) we replace all $k_j - 1$ by k_j then we get the new coefficient

$$a(\mathbf{k})' = \left(\prod_{j=1}^d k_j^r\right) \sum_F (-2^{-r})^{|F_1|}.$$

Note that

$$\sum_F (-2^{-r})^{|F_1|} = \sum_{l=0}^d (-2^{-r})^l \binom{d}{l} 2^{d-l} = (2-2^{-r})^d =: A_d^r.$$

Thus, we approximate the function

$$A_d^r \sum_{\mathbf{k}\in\Gamma(N,2)} \prod_{j=1}^d k_j^r \cos k_j x_j$$

by the function

$$\overline{B}(\mathbf{x}) := \left(\prod_{j=1}^d (I - I_j)\right) U_N^r(\mathbf{x}),$$

where I is the identity operator, and

$$I_j f := \left(1/\pi \int_{-\pi}^{\pi} f(\mathbf{x}) \cos x_j \, dx_j\right) \cos x_j.$$

We now give a detailed proof. The technical part of the proof shows how to replace $k_j - 1$ by k_j in a transition from $a(\mathbf{k})$ to $a(\mathbf{k})'$. We prove an assertion, which is more general than the upper estimates in Lemma 4.3.14. Suppose that $G = (G_1, G_2, G_3)$ is a partition of the interval $[1,d]$ into three sets. Denote

$$U(G,d,\mathbf{x}) := \sum_{\mathbf{k}\in\Gamma(N,2)} \left(\prod_{j\in G_1} l_j^r \prod_{j\in G_2} (k_j-1)^r \prod_{j\in G_3} k_j^r \right) \prod_{j=1}^{d} \cos k_j x_j.$$

□

Lemma 4.3.15 *For $r > -1$ the estimate*

$$E_N^{\perp}\big(U(G,d,x)\big)_\infty \ll N^{r+1} \tag{4.3.14}$$

holds for arbitrary G.

Proof The proof is by induction. For $d = 1$ the estimate is true. Assume that it has been proved for all $d' \leq d-1$. We derive from this the statement of the lemma for $d' = d$. We then prove that, for an arbitrary pair of partitions $G := (G_1, G_2, G_3)$ and $F := (F_1, F_2, F_3)$ of $[1,d]$,

$$E_N^{\perp}\big(U(G,d,\mathbf{x}) - U(F,d,\mathbf{x})\big)_\infty \ll N^{r+1}. \tag{4.3.15}$$

Indeed, it clearly suffices to prove (4.3.15) for the case when G and F differ by a single element, that is $j \in F_i$ when $j \in G_i$ for all but one $j \in [1,d]$. We assume for definiteness that $d \in G_1$ and $d \in F_2$. Then, by the induction hypothesis, we have

$$E_N^{\perp}\big(U(G,d,\mathbf{x}) - U(F,d,\mathbf{x})\big)_\infty$$
$$\ll \sum_{k_d=2}^{N} \big(l_d^r - (k_d-1)^r\big)(N/k_d)^{r+1} \ll N^{r+1},$$

which implies (4.3.15). Further, for $g \in \mathscr{T}(N)^{\perp}$,

$$\overline{B}(\mathbf{x}) = \sum_{\mathbf{k}\in\Gamma(N,2)} \sum_{F} ((-2^{-r})^{|F_1|} \prod_{j\in F_1} l_j^r \prod_{j\in F_2} (k_j-1)^r$$
$$\times \prod_{j\in F_3} k_j^r \prod_{j=1}^{d} \cos k_j x_j + g(\mathbf{x}). \tag{4.3.16}$$

Multiplying the difference $U(G,d,\mathbf{x}) - U(F,d,\mathbf{x})$ by $(-2^{-r})^{|F_1|}$ and summing over all partitions F, we get from (4.3.15) and (4.3.16) that

$$E_N^{\perp}\big(A_d^r U(G,d,\mathbf{x}) - \overline{B}(\mathbf{x})\big)_\infty \ll N^{r+1}. \tag{4.3.17}$$

By (4.3.17),

$$E_N^{\perp}\big(A_d^r U(G,d,\mathbf{x})\big)_\infty \leq \|\overline{B}\|_\infty + O\big(N^{r+1}\big). \tag{4.3.18}$$

To finish the proof of the lemma it remains to observe that $A_d^r > 0$ for $r > -1$ and to use Lemma 4.3.13 to estimate $\|\overline{B}\|_\infty \ll N^{r+1}$. □

Lemma 4.3.14 follows from Lemma 4.3.15 for $G = (\varnothing, \varnothing, [1,d])$. The lower estimate in Lemma 4.3.14 follows from the univariate case.

Theorem 4.3.16 *Suppose that* $1 \le p < \infty$ *and* $r \ge 0$. *Then*

$$\sup_{t \in \mathscr{T}(N)} \left\| t^{(r)}(\mathbf{x}, \alpha) \right\|_\infty / \|t\|_p \asymp N^{r+1/p} (\log N)^{(d-1)(1-1/p)}.$$

Proof Let n be such that $2^{n-1} < N \le 2^n$; then

$$t^{(r)}(\mathbf{x}, \alpha) = t(\mathbf{x}) * \mathscr{D}^{(r)}_{Q_{n+d}}(\mathbf{x}, \alpha),$$

$$\left\| t^{(r)}(\mathbf{x}, \alpha) \right\|_\infty \le \|t\|_p \|\mathscr{D}^{(r)}_{Q_{n+d}}\|_{p'}. \tag{4.3.19}$$

From (4.3.19), using Lemma 4.2.1, we obtain the required upper estimate for $1 < p < \infty$.

Let $p = 1$. Then

$$\left\| t^{(r)}(\mathbf{x}, \alpha) \right\|_\infty \le \|t\|_1 E_N^{\perp}\left(\mathscr{D}_N^r(\mathbf{x}, \alpha)\right)_\infty$$

and by applying Lemma 4.3.14 we get the required upper estimate.

We now prove the lower estimates. Let $p = 1$. We consider

$$t(\mathbf{x}) := \mathscr{A}^{(0)}_{n-1}(x_1, \alpha_1) \prod_{j=2}^{d} \cos(x_j + \alpha_j \pi/2).$$

Then

$$\|t\|_1 \ll 1, \qquad t^{(r)}(\mathbf{0}, \alpha) \gg N^{r+1},$$

which implies the required lower estimate for $p = 1$.

Let $1 < p < \infty$. We consider

$$t(\mathbf{x}) := \mathscr{D}^{(0)}_{Q_n}(\mathbf{x}, \alpha).$$

Then, by Lemma 4.2.1, on the one hand we have

$$\|t\|_p \ll N^{1-1/p} (\log N)^{(d-1)/p}. \tag{4.3.20}$$

On the other hand it is easy to see that

$$t^{(r)}(\mathbf{0}, \alpha) \gg N^{r+1} (\log N)^{d-1}. \tag{4.3.21}$$

The required lower estimate follows from (4.3.20) and (4.3.21).

The theorem is now proved. □

Theorem 4.3.17 *Suppose that* $1 \le q \le p < \infty$, $p > 1$, $r \ge 0$. *Then*

$$\sup_{t \in \mathscr{T}(N)} \left\| t^{(r)}(\mathbf{x}, \alpha) \right\|_p / \|t\|_q \asymp N^{r+1/q-1/p}.$$

Proof Let us first consider the case $q = 1$ and estimate $\|t\|_p$ in terms of the norm $\|t\|_1$. We have

$$\|t\|_p \le \|t\|_1^{1/p} \|t\|_\infty^{1-1/p}.$$

By Theorem 4.3.16,

$$\|t\|_\infty \ll N\|t\|_1.$$

Consequently,

$$\|t\|_p \ll N^{1-1/p}\|t\|_1.$$

Using Theorem 4.3.1 for $r \ge 0$ we find that

$$\left\|t^{(r)}(\mathbf{x},\alpha)\right\|_p \ll N^{r+1-1/p}\|t\|_1.$$

We now pass to the case $1 < q \le p < \infty$. Let $\beta = 1/q - 1/p$. Corollary A.3.8 gives us

$$\|A_\beta t\|_p \ll \|t\|_q. \tag{4.3.22}$$

Further,

$$t^{(r)}(\mathbf{x},\alpha) = \left(A_\beta^{-1}(A_\beta t)\right)^{(r)}(\mathbf{x},\alpha) = (A_\beta t)^{(r+\beta)}(\mathbf{x},\alpha). \tag{4.3.23}$$

Here A_β^{-1} denotes the operator inverse to A_β. By (4.3.22), (4.3.23) and Theorem 4.3.1,

$$\left\|t^{(r)}(\mathbf{x},\alpha)\right\|_p \ll N^{r+1/q-1/p}\|t\|_q.$$

For $1 < q = p < \infty$ the theorem follows from Theorem 4.3.1. The upper estimates are thus proved. The lower estimates can be obtained from the univariate case. □

We remark that the lower estimate in Lemma 4.2.3 for $p = 1$ follows from Theorem 4.3.16 with $p = 1$.

4.4 Approximation of Functions in the Classes $\mathbf{W}^r_{q,\alpha}$ and $\mathbf{H}^r_q$

In this section we study the approximation in the L_p-metric of functions in the classes $\mathbf{W}^r_{q,\alpha}$ and $\mathbf{H}^r_q$, $1 \le q, p \le \infty$, by trigonometric polynomials whose harmonics lie in hyperbolic crosses. Certain specific features of the multivariate case are explained in this study.

First and foremost, it will be determined that the least upper bounds of the best approximations by polynomials in $\mathscr{T}(N)$ are different for the classes $\mathbf{W}^r_{q,\alpha}$ and $\mathbf{H}^r_q$ for all $1 < q, p < \infty$. Namely,

$$E_N(\mathbf{W}^r_{q,\alpha})_p = o\left(E_N(\mathbf{H}^r_q)\right)_p, \qquad d \ge 2.$$

As is known, in the univariate case the order of the least upper bounds of the best approximations by trigonometric polynomials are the same for both classes for all $1 \le p, q \le \infty$, even though H^r_q is a wider class than $W^r_{q,\alpha}$.

It turns out that approximation in the uniform metric differs essentially from approximation in the L_p-metric, $1 < p < \infty$, not only in the methods of proof but also in that the results are fundamentally different. For example, in approximation in the L_p-metric, $1 < p < \infty$, the partial Fourier sums $S_{Q_n}(f)$ give the order of the best approximation $E_{Q_n}(f)_p$ and thus, if we are not interested in the dependence of $E_{Q_n}(f)_p$ on p, we can confine ourselves to the study of $S_{Q_n}(f)$.

In the univariate case and the uniform metric the partial sums of the Fourier series give good approximations for the functions in the classes $W^r_{q,\alpha}$ and H^r_q, $1 < q < \infty$ (see Theorems 1.4.9 and 1.4.12):

$$E_n(F)_\infty \asymp \sup_{f \in F} \|f - S_n(f)\|_\infty,$$

where F denotes either $W^r_{q,\alpha}$ or H^r_q.

In the case of the classes $\mathbf{W}^r_{q,\alpha}$ and $\mathbf{H}^r_q$, $1 < q < \infty$, not only do the Fourier sums not give the orders of the least upper bounds of the best approximations in the L_∞-norm, but also no linear method gives the orders of the least upper bounds of the best approximations with respect to the classes $\mathbf{W}^r_{q,\alpha}$ or $\mathbf{H}^r_q$, $d \ge 2$, $1 < q < \infty$. In other words, the operator for the best approximation in the uniform metric by polynomials in $\mathscr{T}(N)$ cannot be replaced by any linear operator without detriment to the order of approximation, for the classes $\mathbf{W}^r_{q,\alpha}$ and $\mathbf{H}^r_q$, $1 < q < \infty$.

An important role in the theory of the approximation of differentiable functions of a single variable is played by Bernoulli kernels, which can be used to give an integral representation of a function in terms of its derivative. In the study of the approximation of functions in the classes $\mathbf{W}^r_{q,\alpha}$ it also turns out to be very useful to approximate the functions $F_r(\mathbf{x}, \alpha)$, which are the multivariate analog of Bernoulli kernels.

4.4.1 Approximation of Bernoulli Kernels

In this subsection we obtain estimates which are sharp (in the sense of order) of the best approximations of the functions $F_r(\mathbf{x}, \alpha)$ by trigonometric polynomials in $\mathscr{T}(N)$ in the L_p-metric, $1 \le p \le \infty$.

We remark that an essential role in obtaining the results of this subsection is played by the Nikol'skii duality theorem (Theorem A.2.3). This theorem plays an important role in the solution of extremal problems in the theory of the approximation of functions of a single variable: in the study of upper estimates for the best approximation of the functions $F_r(\mathbf{x}, \alpha)$ in the uniform metric we use the duality theorem. These estimates are obtained by the following method. Certain functions constructed in the simplest manner are first studied in a special way. The desired estimates are then obtained from results relating to these functions, with the use of the duality theorem.

The use of the duality theorem has the result that we can find the order of the best approximation of $F_r(\mathbf{x}, \alpha)$ in the uniform metric, but we cannot determine a polynomial (for general d and α) giving this approximation. The situation is unusual from the point of view of the approximation of functions of one variable. We prove the following assertion.

Theorem 4.4.1 *For* $1 \le p \le \infty$ *and* $r - 1 + 1/p > 0$ *we have*

$$E_N\big(F_r(\mathbf{x}, \alpha)\big)_p \asymp N^{-r+1-1/p}(\log N)^{(d-1)/p}.$$

Proof We first consider the case $1 < p < \infty$. In the definition of the class H_1^r we considered differences of order $l > r$, and we found in Chapter 1 that the Bernoulli kernels $F_r(y, \alpha)$ of one variable belong to the class $H_1^r B$, with some constant $B > 0$. Consequently, since

$$F_r(\mathbf{x}, \alpha) = \prod_{j=1}^{d} F_r(x_j, \alpha_j),$$

it follows that

$$F_r(\mathbf{x}, \alpha) \in \mathbf{H}_1^r B, \tag{4.4.1}$$

where $B > 0$ is a constant. Relation (4.4.1) and Theorem 4.4.10, which will be proved below, yield the desired upper estimate for $1 < p < \infty$. By Corollary A.3.4 we get from this estimate

$$\big\|F_r - S_{Q_n}(F_r)\big\|_p \ll 2^{-n(r-1+1/p)} n^{(d-1)/p}, \qquad 1 < p < \infty. \tag{4.4.2}$$

We now prove the lower estimates for $1 < p < \infty$. Let $t \in \mathscr{T}(N)$ be a polynomial of the best approximation for F_r in the L_p-metric, and let $n = [\log_2 N] + d + 1$. Consider the quantity

$$J = \langle F_r - t, F_2 - S_{Q_n}(F_2)\rangle = \langle F_r, F_2 - S_{Q_n}(F_2)\rangle.$$

On the one hand

$$J \le E_N(F_r)_p \big\|F_2 - S_{Q_n}(F_2)\big\|_{p'},$$

which, by (4.4.2), implies that

$$J \ll E_N(F_r)_p 2^{-n(2-1/p)} n^{(d-1)(1-1/p)}. \tag{4.4.3}$$

On the other hand, it is not difficult to get a lower estimate for J:

$$J \geq 2^d \sum_{\mathbf{k}>0;\mathbf{k}\notin Q_n} \nu(\mathbf{k})^{-r-2} \asymp \sum_{\|\mathbf{s}\|_1 \geq n} 2^{-(r+1)\|\mathbf{s}\|_1} \asymp 2^{-(r+1)n} n^{d-1}. \tag{4.4.4}$$

Comparing (4.4.3) and (4.4.4), we obtain the required lower estimates. The analysis of the case $1 < p < \infty$ is complete.

Consider now the case $p = 1$. We first prove the upper estimate. It will be more convenient for us to find the required estimate for $E_{Q_n}(F_r)_1$. As an approximating polynomial take

$$t_n(\mathbf{x}) := \sum_{\|\mathbf{s}\|_1 \leq n} A_{\mathbf{s}}(F_r, \mathbf{x}).$$

Then, by (4.4.1) and Theorem 4.4.6 from the next subsection we get

$$\|F_r - t_n\|_1 \leq \sum_{\|\mathbf{s}\|_1 > n} \left\|A_{\mathbf{s}}(F_r)\right\|_1 \ll \sum_{\|\mathbf{s}\|_1 > n} 2^{-r\|\mathbf{s}\|_1} \ll 2^{-rn} n^{d-1}.$$

This proves the upper estimate.

We now prove the lower estimate. Let

$$C\Gamma(N) := \left\{\mathbf{k} : 0 < k_j \leq N,\ j = 1, \dots, d,\ \mathbf{k} \notin \Gamma(N)\right\}$$

$$\varphi_N(\mathbf{x}, \alpha) := \sum_{C\Gamma(N)} \prod_{j=1}^{d} a(k_j, x_j, \alpha_j) k_j^{-1},$$

where, as above,

$$\begin{aligned} a(n, x, \alpha) := {} & \cos(nx + \alpha\pi/2) + \cos\big((n+1)x + \alpha\pi/2\big) \\ & - \cos(2nx + \alpha\pi/2) - \cos\big((2n+1)x + \alpha\pi/2\big). \end{aligned}$$

It is easy to get from Lemma 4.3.2 that

$$\|\varphi_N\|_\infty \ll 1. \tag{4.4.5}$$

Let us estimate $J := \langle F_r(\mathbf{x}, \alpha), \varphi_n(\mathbf{x}, -\alpha)\rangle$. We have on the one hand

$$\begin{aligned} J & \asymp \sum_{\mathbf{k}\in C\Gamma(N)} \prod_{j=1}^{d} k_j^{-1}\left(k_j^{-r} + (k_j+1)^{-r} - (2k_j)^{-r} - (2k_j+1)^{-r}\right) \\ & \asymp \sum_{\mathbf{k}\in C\Gamma(N)} \nu(\mathbf{k})^{-r-1} \gg N^{-r}(\log N)^{d-1}. \end{aligned} \tag{4.4.6}$$

On the other hand,

$$J \leq \|\varphi_N\|_\infty E_N(F_r)_1. \tag{4.4.7}$$

From (4.4.5)–(4.4.7) we get the required lower estimate. The proof for the case $1 \leq p < \infty$ is now complete.

We proceed to the most difficult case $p = \infty$. Here it will be convenient for us to consider $F_r^0(\mathbf{x}, \alpha) := \prod_{j=1}^d \big(F_r(x_j, \alpha_j) - 1\big)$. This is clearly sufficient. We will carry out the proof for $d = 2$ and refer the reader to Temlyakov (1986c) for the proof in the general case.

Let

$$\Gamma_{Nl+b} := \big\{(k,m) \in \mathbb{N}^2 : m \le N,\ km = Nl + b\big\}, \tag{4.4.8}$$

and, for $u = 1, 2, \dots, N$, $l \ge 1$,

$$f_{Nl}^u(x, y, \alpha) := \sum_{b=0}^{u-1} (Nl+b)^{-1} \sum_{(k,m)\in\Gamma_{Nl+b}} \cos(kx - \alpha_1\pi/2)\cos(my - \alpha_2\pi/2).$$

Lemma 4.4.2 *For all $u = 1, 2, \dots, N$*

$$E_N(f_{Nl}^u)_\infty \ll l^{-1}\log(l+1).$$

Proof We represent the function f_{Nl}^u in the form

$$\begin{aligned} f_{Nl}^u(x, y, \alpha) &= \sum_{m=1}^N m^{-1}\cos(my - \alpha_2\pi/2) \sum_{k=k'_m}^{k''_m} k^{-1}\cos(kx - \alpha_1\pi/2) \\ &= \sum_{m=1}^N m^{-1}\cos(my - \alpha_2\pi/2) \sum_{k=0}^{k''_m - k'_m} (k'_m + k)^{-1}\cos((k'_m + k)x - \alpha_1\pi/2), \end{aligned}$$

where $k'_m \ge Nl/m$ and $k''_m - k'_m \le N/m$. If for some m there is no number k with $Nl \le km < Nl + u$ then $\sum_{k=k'_m}^{k''_m}$ is set to zero. Let k_m denote the largest natural number such that $mk_m \le N$.

As an approximating polynomial we take T_{Nl}^u, which is defined by

$$T_{Nl}^u(x, y, \alpha) := \sum_{m=1}^N m^{-1}\cos(my - \alpha_2\pi/2) \sum_{k=0}^{k''_m - k'_m} (k'_m + k)^{-1}\cos\big((k_m - k)x - \alpha_1\pi/2\big).$$

Then

$$\begin{aligned} \big|f_{Nl}^u(x, y, \alpha) - T_{Nl}^u(x, y, \alpha)\big| = \Bigg| &\sum_{m=1}^N m^{-1}\cos(my - \alpha_2\pi/2)\, 2\sin\left(\frac{k'_m + k_m}{2} - \frac{\alpha_1\pi}{2}\right) \\ &\times \sum_{k=0}^{k''_m - k'_m} (k'_m + k)^{-1}\sin\left(\left(\frac{k'_m - k_m}{2} + k\right)x\right)\Bigg|. \end{aligned}$$

To estimate the right-hand side we use the following lemma.

Lemma 4.4.3 *Suppose that L and v are natural numbers and $Q \le L$ is a nonnegative real number. Then*

$$\left|\sum_{k=0}^{v} (L+k)^{-1}\sin(Q+k)x\right| \ll \min\big\{v|x|, v/L, 1/(L|x|)\big\}.$$

Proof The estimate

$$\left|\sum_{k=0}^{v}(L+k)^{-1}\sin(Q+k)x\right| \le (v+1)|x|$$

is obvious. Let us prove the second inequality. We have

$$\left|\sum_{k=0}^{v}(L+k)^{-1}\sin(Q+k)x\right| = \left|\operatorname{Im}\sum_{k=0}^{v}(L+k)^{-1}e^{i(Q+k)x}\right|$$
$$\le \left|\sum_{k=0}^{v}(L+k)^{-1}e^{ikx}\right| \le (1/L)\max_{0\le u\le v}\left|\sum_{k=0}^{u}e^{ikx}\right|$$
$$\ll (1/L)\min\big(v, 1/|x|\big).$$

The lemma is proved. □

Consequently,

$$|f^u_{Nl} - T^u_{Nl}| \ll \sum_{m=1}^{N} m^{-1}\min\left(\frac{N|x|}{m}, \frac{1}{l}, \frac{m}{Nl|x|}\right) \ll \sum_{m=1}^{[N|x|]}(Nl|x|)^{-1}$$
$$+ \sum_{m=\left[N|x|\right]+1}^{\left[Nl|x|\right]}(ml)^{-1} + \sum_{m=\left[Nl|x|\right]+1}^{N} N|x|m^{-2} \ll l^{-1}\log(l+1),$$

and Lemma 4.4.2 is proved. □

We now return to the proof of Theorem 4.4.1. Consider the function

$$h^r_N(x,y,\alpha) = \sum_{l=1}^{\infty}\sum_{b=0}^{N-1}(Nl+b)^{-r}\sum_{(k,m)\in\Gamma_{Nl+b}}\cos(kx-\alpha_1\pi/2)\cos(my-\alpha_2\pi/2).$$

It is not difficult to verify that

$$\left\|F^0_r(x,y,\alpha) - h^r_N(x,y,\alpha) - \sum_{(k,m)\in\Gamma(N)}(km)^{-r}\cos(kx-\alpha_1\pi/2)\cos(my-\alpha_2\pi/2)\right\|_\infty$$
$$\ll N^{1-r}.$$

Therefore,

$$E_N(F^0_r)_\infty \ll E_N(h^r_N)_\infty + N^{1-r}. \tag{4.4.9}$$

We next estimate $E_N(h^r_N)_\infty$. By the Nikol'skii duality theorem we find

$$E_N(h^r_N)_\infty = \sup_{g\in\mathcal{T}(N)^\perp_1}|\langle h^r_N, g\rangle|$$
$$= \sup_{g\in\mathcal{T}(N)^\perp_1}\left|\sum_{l=1}^{\infty}\sum_{b=0}^{N-1}(Nl+b)^{-r}\sum_{(k,m)\in\Gamma_{Nl+b}}\tilde{g}(k,m,\alpha)\right|, \tag{4.4.10}$$

where

$$\tilde{g}(k,m,\alpha) := (2\pi)^{-2}\int_{\pi_2} \bar{g}(x,y)\cos(kx-\alpha_1\pi/2)\cos(my-\alpha_2\pi/2)\,dxdy.$$

Using Lemma 4.4.2 we get, for the functions $g \in \mathscr{T}(N)_1^{\perp}$,

$$\left|\sum_{b=0}^{u-1}(Nl+b)^{-1}\sum_{(k,m)\in\Gamma_{Nl+b}}\tilde{g}(k,m,\alpha)\right| = |\langle f_{Nl}^u, g\rangle|$$
$$\leq \|g\|_1 E_N(f_{Nl}^u)_\infty$$
$$\ll l^{-1}\log(l+1), \qquad u=1,2,\ldots,N.$$

Consequently, for $g \in \mathscr{T}(N)_1^{\perp}$, by the Abel inequality (A.1.17) we have

$$\left|\sum_{b=0}^{N-1}(Nl+b)^{-r}\sum_{(k,m)\in\Gamma_{Nl+b}}\tilde{g}(k,m,\alpha)\right|$$
$$\ll (Nl)^{1-r}\max_{1\leq u\leq N}\left|\sum_{b=0}^{u-1}(Nl+b)^{-1}\sum_{(k,m)\in\Gamma_{Nl+b}}\tilde{g}(k,m,\alpha)\right|$$
$$\ll \frac{\log(l+1)}{N^{r-1}l^r}, \tag{4.4.11}$$

which, by (4.4.10), gives us

$$E_N(h_N^r)_\infty \ll N^{1-r}\sum_{l=1}^{\infty} l^{-r}\log(l+1) \ll N^{1-r}. \tag{4.4.12}$$

The required estimate is obtained by combining (4.4.9) and (4.4.12).

The lower estimates for $p=\infty$ follow from the univariate case. Indeed, let $t \in \mathscr{T}(N)$ be a polynomial of the best approximation of F_r in the L_∞-metric. Set

$$t_1(x_1) := (2\pi)^{1-d}\int_{\mathbb{T}^{d-1}} t(\mathbf{x})dx_2\cdots dx_d;$$

then, for $t_1 \in \mathscr{T}(N)$,

$$\|F_r(x_1,\alpha_1)-t_1(x_1)\|_\infty \leq \|F_r(\mathbf{x},\alpha)-t(\mathbf{x})\|_\infty.$$

We will suppose that n is such that $2^{n-3} < N \leq 2^{n-2}$ and consider

$$\sigma := |\langle F_r(x_1,\alpha_1)-t_1(x_1),\ \mathscr{A}_n^{(r)}(x_1,\alpha_1)\rangle|$$
$$\leq \|F_r(x_1,\alpha_1)-t_1(x_1)\|_\infty \|\mathscr{A}_n^{(r)}(x_1,\alpha_1)\|_1.$$

Using Theorem 1.3.2 we continue the estimate as follows:

$$\ll \|F_r(x_1,\alpha_1)-t_1(x_1)\|_\infty N^r.$$

On the other hand,

$$\sigma = \langle F_r(x_1,\alpha_1),\ \mathscr{A}_n^{(r)}(x_1,\alpha_1)\rangle \gg N.$$

From these estimates we obtain the required lower estimate. This concludes the proof of Theorem 4.4.1. □

We now prove that in the particular case $\alpha = \mathbf{1} = (1,\dots,1)$ the rate of approximation of $F_r(\mathbf{x},\mathbf{1})$ in the L_∞-norm can be realized by Fourier sums. Denote

$$S_N(f) := S_{\Gamma(N)}(f) := \sum_{\mathbf{k}\in\Gamma(N)} \hat{f}(\mathbf{k})e^{i(\mathbf{k},\mathbf{x})}.$$

Proposition 4.4.4 *For $r > 1$ we have*

$$\|F_r(\mathbf{x},\mathbf{1}) - S_N(F_r,\mathbf{x})\|_\infty \ll N^{-r+1}.$$

Proof Clearly it is sufficient to prove the proposition for $F_r^0(\mathbf{x},\mathbf{1})$ instead of $F_r(\mathbf{x},\mathbf{1})$. We estimate the L_∞-norm of the function

$$f_N^r(\mathbf{x}) = F_r^0(\mathbf{x},\mathbf{1}) - S_N\left(F_r^0,\mathbf{x}\right).$$

We denote for $f \in L_1$,

$$\tilde{f}(\mathbf{k}) = \pi^{-d}\int_{\mathbb{T}^d} f(\mathbf{x})\left(\prod_{j=1}^d \sin k_j x_j\right) d\mathbf{x}.$$

Further,

$$\begin{aligned}\|f_N^r\|_\infty &= \sup_{\|g\|_1\le 1} |\langle f_N^r, g\rangle| = \sup_{\|g\|_1\le 1} \sum_{\mathbf{k}} \tilde{f}_N^r(\mathbf{k})\tilde{g}(\mathbf{k})\\ &= \sup_{\|g\|_1\le 1} \sum_{l=N}^\infty l^{1-r} \sum_{\mathbf{k}\in\Gamma_l=\Gamma(l+1)\setminus\Gamma(l)} l^{-1}\tilde{g}(\mathbf{k})\end{aligned}$$

and, applying the inequality (A.1.17) we continue:

$$\begin{aligned}&\le \sup_{\|g\|_1\le 1} N^{1-r} \sup_{L\ge N}\left|\sum_{l=N}^L \sum_{\mathbf{k}\in\Gamma_l} l^{-1}\tilde{g}(\mathbf{x})\right|\\ &= \sup_{\|g\|_1\le 1} N^{1-r}\sup_{L\ge N}\left|\langle f_L^1 - f_{N-1}^1, g\rangle\right| \le N^{1-r}\left(\|f_{N-1}^1\|_\infty + \sup_{L\ge N}\|f_L^1\|_\infty\right).\end{aligned}$$

It is not difficult to derive the estimate

$$\left\|\sum_{\mathbf{k}\in\Gamma(N)} \prod_{j=1}^d k_j^{-1}\sin k_j x_j\right\|_\infty \ll 1$$

from Lemma 4.2.4; therefore

$$\sup_{L\ge N}\|f_L^1\|_\infty \ll 1,$$

which completes the proof of the proposition. □

Remark 4.4.5 In the case $\alpha = \mathbf{0} = (0,\dots,0)$ we have

$$\|F_r(\mathbf{x},\mathbf{0}) - S_N(F_r,\mathbf{x})\|_\infty \asymp N^{-r+1}(\log N)^{d-1}.$$

This statement follows from the estimate

$$\sum_{\mathbf{k}\notin\Gamma(N)} \hat{F}_r(\mathbf{k},\mathbf{0}) = \sum_{\mathbf{k}\notin\Gamma(N)} \left(\prod_{j=1}^{d} \max(|k_j|,1)\right)^{-r} \asymp N^{-r+1}(\log N)^{d-1}.$$

4.4.2 Representation of the Classes $\mathbf{H}_q^r$

In this subsection we prove Theorem 4.1.2. For convenience we repeat the definition from §4.1 and give some further definitions. Let

$$L_q^0 := L_q^0(\mathbb{T}^d) := \left\{ f : f \in L_q,\ \int_{-\pi}^{\pi} f(\mathbf{x})dx_j = 0,\ j = 1,\dots,d \right\}.$$

We denote by $\mathbf{H}_{q,l}^{r,0}$, $l > r$, the class of functions $f \in L_q^0$ such that

$$\left\|\Delta_{\mathbf{t}}^l f(\mathbf{x})\right\|_q \le \prod_{j=1}^{d} |t_j|^r,$$

where $\mathbf{t} = (t_1,\dots,t_d)$ and $\Delta_{\mathbf{t}}^l f(\mathbf{x})$ is the mixed lth difference with step t_j in the variable x_j, that is,

$$\Delta_{\mathbf{t}}^l f(\mathbf{x}) := \Delta_{t_d}^l \cdots \Delta_{t_1}^l f(x_1,\dots,x_d).$$

Let e be a subset of the natural numbers in $[1,d]$. We denote

$$\Delta_{\mathbf{t}}^l(e) := \prod_{j\in e} \Delta_{t_j}^l, \qquad \Delta_{\mathbf{t}}^l(\varnothing) = I.$$

We define the class $\mathbf{H}_{q,l}^r B$, $l > r$, as the set of $f \in L_q$ such that, for any e,

$$\left\|\Delta_{\mathbf{t}}^l(e) f(\mathbf{x})\right\|_q \le B \prod_{j\in e} |t_j|^r. \tag{4.4.13}$$

In the case $B = 1$ omit B. We remark that the class of functions $f(\mathbf{x})$ representable in the form

$$f(\mathbf{x}) = c + \sum_{e\neq\varnothing} f_e(\mathbf{x}), \tag{4.4.14}$$

where $f_e \in \mathbf{H}_{q,l}^{r(e),0}$, with $\mathbf{H}_{q,l}^{r(e),0}$ the class $\mathbf{H}_{q,l}^{r,0}$ of functions of $|e|$ variables x_j, $j \in e$, $|c| \le 1$, is equivalent to the class $\mathbf{H}_{q,l}^r$.

Indeed, let us first prove that if f is represented in the form (4.4.14) then there exists a $B > 0$ such that B does not depend on f and f satisfies the relation (4.4.13).

This assertion follows easily from the following remark for functions of a single variable. Let $f \in L^0_1(\mathbb{T})$, then

$$f(x) = \frac{1}{2\pi} \int_{-\pi}^{\pi} \Delta^l_t f(x) dt.$$

Now, conversely, let $f \in \mathbf{H}^r_{q,l}$. We shall carry out the proof by induction on the dimension d. We define the following functions:

$$g^1(\mathbf{x}) := \frac{1}{2\pi} \int_{-\pi}^{\pi} f(\mathbf{x}) dx_1, \qquad f^1 := f - g^1.$$

Then $\int_{-\pi}^{\pi} f^1(\mathbf{x}) dx_1 = 0$. Defining recursively

$$g^{j+1}(\mathbf{x}) := \frac{1}{2\pi} \int_{-\pi}^{\pi} f^j(\mathbf{x}) dx_{j+1}, \qquad f^{j+1} := f^j - g^{j+1}, \qquad j = 1, \dots, d-1,$$

we obtain

$$f = f^d + \sum_{j=1}^{d} g^j.$$

For each $1 \le j \le d$ we have $\int_{-\pi}^{\pi} f^d(\mathbf{x}) dx_j = 0$, and the functions $g^j(\mathbf{x})$ depend on $d-1$ variables. In addition, it is easy to see that for all $e \in [1,d]$, $|e| \le d-1$ we have

$$\left\|\Delta^l_{\mathbf{t}}(e) g^j(\mathbf{x})\right\|_p \le C(d) \left\|\Delta^l_{\mathbf{t}}(e) f(\mathbf{x})\right\|_p, \qquad j = 1, \dots, d,$$

and $\Delta^l_{\mathbf{t}} f^d(\mathbf{x}) = \Delta^l_{\mathbf{t}} f(\mathbf{x})$. Using the induction hypothesis we find that, for some $\delta > 0$ which does not depend on f, the function δf can be represented in the form (4.4.14).

Let $\mathscr{A}_{\mathbf{s}}$ denote the polynomials defined in §3.2.5 and, for all $f \in L_1$, let

$$A_{\mathbf{s}}(f) := f * \mathscr{A}_{\mathbf{s}}.$$

We now prove the following theorem of representation of the class $\mathbf{H}^r_{q,l}$.

Theorem 4.4.6 *Let $f \in \mathbf{H}^r_{q,l}$. Then, for $\mathbf{s} \ge \mathbf{0}$,*

$$\left\|A_{\mathbf{s}}(f)\right\|_q \le C(r,d,l) 2^{-r\|\mathbf{s}\|_1}, \qquad 1 \le q \le \infty, \tag{4.4.15}$$

$$\left\|\delta_{\mathbf{s}}(f)\right\|_q \le C(r,d,q,l) 2^{-r\|\mathbf{s}\|_1}, \qquad 1 < q < \infty. \tag{4.4.16}$$

Conversely, from (4.4.15) *or* (4.4.16) *it follows that there exists a $B > 0$ which does not depend on f such that $f \in \mathbf{H}^r_{q,l} B$.*

Proof We first prove the relation (4.4.15), which implies (4.4.16) by Corollary A.3.4. Suppose that $\mathbf{n} = (n_1, \dots, n_d)$ and $U(\mathbf{n})$ is the set of functions $u(\mathbf{x})$ of the form

$$u(\mathbf{x}) = \sum_{j=1}^{d} \sum_{\nu=-n_j}^{n_j} e^{i\nu x_j} c_\nu(x_1, \dots, x_{j-1}, x_{j+1}, \dots, x_d).$$

We have the following result.

Lemma 4.4.7 *For every function $f \in \mathbf{H}^r_{q,l}$, $1 \le q \le \infty$, there is a function $u \in U(\mathbf{n})$ such that*

$$\|f - u\|_q \ll \prod_{j=1}^{d} (n_j + 1)^{-r}.$$

Proof Let J_n^a be the Jackson kernel with $a := [r] + 1$. Let

$$u_1(\mathbf{x}) := \frac{1}{2\pi} \int_{-\pi}^{\pi} \left(f(\mathbf{x}) - \Delta^l_{t_1} f(\mathbf{x})\right) J^a_{m_1}(t_1) dt_1, \qquad m_1 := [n_1/a] + 1.$$

It is easy to see that $u_1 \in U(\mathbf{n})$. Further,

$$f(\mathbf{x}) - u_1(\mathbf{x}) = \frac{1}{2\pi} \int_{-\pi}^{\pi} \Delta^l_{t_1} f(\mathbf{x}) J^a_{m_1}(t_1) dt_1.$$

If $f_1 := f - u_1$ then

$$\begin{aligned} \left\|\Delta^l_{t_d} \cdots \Delta^l_{t_2} f_1(\mathbf{x})\right\|_q &= \frac{1}{2\pi} \left\| \int_{-\pi}^{\pi} \Delta^l_{\mathbf{t}} f(\mathbf{x}) J^a_{m_1}(t_1) dt_1 \right\|_q \\ &\le \frac{1}{2\pi} \int_{-\pi}^{\pi} \prod_{j=1}^{d} |t_j|^r J^a_{m_1}(t_1) dt_1 \\ &\ll \left(\prod_{j=2}^{d} |t_j|^r \right) \int_{-\pi}^{\pi} |t_1|^r J^a_{m_1}(t_1) dt_1 \ll m_1^{-r} \prod_{j=2}^{d} |t_j|^r. \end{aligned} \tag{4.4.17}$$

Let

$$u_2(\mathbf{x}) := \frac{1}{2\pi} \int_{-\pi}^{\pi} \left(f_1(\mathbf{x}) - \Delta^l_{t_2} f_1(\mathbf{x})\right) J^a_{m_2}(t_2) dt_2, \qquad m_2 := [n_2/a] + 1,$$

and

$$f_2 := f_1 - u_2.$$

Clearly, $u_2 \in U(\mathbf{n})$. By an argument similar to that for (4.4.17) we find that

$$\|\Delta^l_{t_d} \cdots \Delta^l_{t_3} f_2\|_q \ll (m_1 m_2)^{-r} \prod_{j=3}^{d} |t_j|^r.$$

Continuing this process, we get two sequences of functions $u_j \in U(\mathbf{n})$ and f_j, $j = 1,\dots,d$, such that

$$u_j(\mathbf{x}) = \frac{1}{2\pi}\int_{-\pi}^{\pi} (f_{j-1}(\mathbf{x}) - \Delta^l_{t_j} f_{j-1}(\mathbf{x})) J^a_{m_j}(t_j)dt_j, \qquad m_j = [n_j/a]+1,$$

$$f_j = f_{j-1} - u_j,$$

$$\|\Delta^l_{t_d}\cdots\Delta^l_{t_{j+1}} f_j\|_q \ll \left(\prod_{i=1}^{j} m_i\right)^{-r} \prod_{i=j+1}^{d} |t_i|^r, \qquad j = 1,\dots,d-1,$$

and

$$\|f_d\|_q \ll \left(\prod_{j=1}^{d} m_j\right)^{-r}. \tag{4.4.18}$$

Further, if $u := \sum_{j=1}^d u_j$ then $u \in U(\mathbf{n})$ and $f - u = f_d$. Consequently, we obtain the lemma from (4.4.18). □

Let $f \in \mathbf{H}^r_{q,l}$. By Lemma 4.4.7, there is a function $u \in U(2^{\mathbf{s}-\mathbf{2}})$ such that

$$\|f-u\|_q \ll 2^{-r\|\mathbf{s}\|_1}.$$

Then

$$A_{\mathbf{s}}(f) = A_{\mathbf{s}}(f-u) \qquad \text{and} \qquad \left\|A_{\mathbf{s}}(f)\right\|_q \ll 2^{-r\|\mathbf{s}\|_1}.$$

It is clear that, by virtue of the remarks before Theorem 4.4.6, for proving the second part of the theorem it suffices to show that (4.4.15) implies that, for

$$f^0(\mathbf{x}) = \sum_{\mathbf{s}>\mathbf{0}} A_{\mathbf{s}}(f,\mathbf{x}),$$

the inequality

$$\left\|\Delta^l_{\mathbf{t}} f^0(\mathbf{x})\right\|_q \le C(r,d,l)\prod_{j=1}^{d}|t_j|^r$$

holds.

Let $\mathbf{t}$ be such that $|t_j| > 0$, $j = 1,\dots,d$. Using Lemma 1.4.4 and the Bernstein inequality, we obtain from (4.4.15)

$$\begin{aligned}\left\|\Delta^l_{\mathbf{t}} f^0(\mathbf{x})\right\|_q &\le \sum_{\mathbf{s}>\mathbf{0}} \left\|\Delta^l_{\mathbf{t}} A_{\mathbf{s}}(f,\mathbf{x})\right\|_q \\ &\ll \sum_{\mathbf{s}>\mathbf{0}}\prod_{j=1}^{d}(2^{-rs_j}\min(1,2^{ls_j}|t_j|^l)) \\ &= \prod_{j=1}^{d}\sum_{s_j>0} 2^{-rs_j}\min(1,2^{ls_j}|t_j|^l) \ll \prod_{j=1}^{d}|t_j|^r.\end{aligned}$$

The case when the relation (4.4.16) is satisfied is considered in the same way. □

Theorem 4.4.6 shows that the classes $\mathbf{H}^r_{q,l}$ are equivalent for $l > r$. We are interested only in the orders of the best approximations and the widths of these classes. The classes $\mathbf{H}^r_{q,l}$ with different l turn out to be indistinguishable from this point of view; therefore, we will drop l from the notation.

Corollary 4.4.8 *Let $\beta = 1/q - 1/p$, $r > \beta$. Then*

$$\mathbf{H}^r_q \subset \mathbf{H}^{r-\beta}_p B, \qquad 1 \leq q \leq p \leq \infty.$$

Proof If $f \in \mathbf{H}^r_q$ then, by Theorem 4.4.6,

$$\|A_{\mathbf{s}}(f)\|_q \ll 2^{-r\|\mathbf{s}\|_1}$$

and, by the Niko'lskii inequality,

$$\|A_{\mathbf{s}}(f)\|_p \ll 2^{-(r-\beta)\|\mathbf{s}\|_1}.$$

Using Theorem 4.4.6, we get the required inclusion. □

4.4.3 Approximation in Function Classes

In this subsection we consider approximations in the L_p-metric of functions in the classes $\mathbf{W}^r_{q,\alpha}$ and $\mathbf{H}^r_q$ for all $1 < q, p < \infty$ and for some extreme values (1 or ∞) of q, p.

Since for $1 < q < \infty$ the operator that maps a function to its conjugate function is bounded from L_q to L_q, there is no loss of generality in investigating only the case $\alpha = 0$ when we consider the classes $\mathbf{W}^r_{q,\alpha}$ with $1 < q < \infty$. We denote the class $\mathbf{W}^r_{q,0}$ by $\mathbf{W}^r_q$.

We remark that, by Corollary A.3.4,

$$E_{Q_n}(f)_p \asymp \|f - S_{Q_n}(f)\|_p$$

in the case $1 < p < \infty$; therefore, Theorems 4.4.9 and 4.4.10, which are stated for the least upper bounds of the best approximations, give the order of magnitude of the least upper bounds of the deviations for the Fourier sums $S_{Q_n}(f)$.

Theorem 4.4.9 *Suppose that $1 < q, p < \infty$, $r > (1/q - 1/p)_+$. Then*

$$E_{Q_n}(\mathbf{W}^r_q)_p \asymp 2^{-n\left(r-(1/q-1/p)_+\right)}.$$

Proof We first consider the case $q = p$. Let $f \in \mathbf{W}^r_p$; then

$$f(\cdot) = \varphi(\cdot) * F_r(\cdot, \mathbf{0})$$

and

$$\|\varphi\|_p \le 1.$$

We will estimate $\|f - S_{Q_n}(f)\|_p$. To do this we consider the function

$$\bar{f}(\mathbf{x}) := \sum_{\mathbf{s}} 2^{-r\|\mathbf{s}\|_1} \delta_{\mathbf{s}}(\varphi, \mathbf{x})$$

and prove that

$$\|f\|_p \ll \|\bar{f}\|_p. \tag{4.4.19}$$

In the univariate case ($d = 1$), relation (4.4.19) follows from the Marcinkiewicz theorem A.3.6, since the numbers $\lambda_\nu = 2^{r\nu}/|\bar{\nu}|^r$, $[2^{s-1}] \le |\nu| < 2^s$, are the Marcinkiewicz multipliers; that is, for some number M,

$$|\lambda_\nu| \le M, \qquad \sum_{[2^{s-1}] \le |\nu| < 2^s} |\lambda_{\nu+1} - \lambda_\nu| \le M, \qquad s = 0, 1, \ldots$$

The multivariate case ($d > 1$) is easily derived from the univariate case. Consequently

$$\|f - S_{Q_n}(f)\|_p \ll \|\bar{f} - \bar{S}_{Q_n}\|_p = \left\| \sum_{\|\mathbf{s}\|_1 > n} 2^{-r\|\mathbf{s}\|_1} \delta_{\mathbf{s}}(\varphi, \mathbf{x}) \right\|_p,$$

and, using the Littlewood–Paley theorem, we can continue the inequality:

$$\ll \left\| \left(\sum_{\|\mathbf{s}\|_1 > n} 2^{-2r\|\mathbf{s}\|_1} |\delta_{\mathbf{s}}(\varphi, \mathbf{x})|^2 \right)^{1/2} \right\|_p$$

$$\le 2^{-rn} \left\| \left(\sum_{\|\mathbf{s}\|_1 > n} |\delta_{\mathbf{s}}(\varphi, \mathbf{x})|^2 \right)^{1/2} \right\|_p \ll 2^{-rn} \|\varphi\|_p \ll 2^{-rn}.$$

Thus the theorem is proved for $q = p$. In the case $q > p$ the theorem follows from the case already considered.

Let $q < p$. Then, by Corollary A.3.8, we have

$$\mathbf{W}^r_q \subset \mathbf{W}^{r-\beta}_q B, \qquad \beta = 1/q - 1/p, \qquad r > \beta,$$

and for this case the theorem follows from the case $q = p$ already considered. The upper estimates are therefore proved.

Let us now prove the lower estimates. In the case $q \ge p$ the required estimates follow from the example $f(\mathbf{x}) := e^{i2^n x_1}$. In the case $q < p$ it suffices to consider the function

$$f(\mathbf{x}) = \mathscr{A}_{n+2}(x_1) \,/\, \|\mathscr{A}^{(r)}_{n+2}\|_q. \qquad \square$$

We now proceed to the classes $\mathbf{H}^r_q$.

Theorem 4.4.10 *We have*

$$E_{Q_n}(\mathbf{H}_q^r)_p \asymp \begin{cases} 2^{-n(r-1/q+1/p)}n^{(d-1)/p}, & 1 \le q < p < \infty,\ r > 1/q - 1/p; \\ 2^{-rn}n^{(d-1)/2}, & 1 < p \le q < \infty,\ q \ge 2; \\ & 1 < p < \infty,\ q = \infty; \\ 2^{-rn}n^{(d-1)/q}, & 1 \le p \le q \le 2,\ r > 0. \end{cases}$$

Proof We first consider the case $1 \le q < p < \infty$. From the univariate results (see Theorem 1.4.1 and Corollary 1.4.7) it follows that $F_r \in \mathbf{H}_1^r B$. With Corollary 4.4.8 we find that

$$F_r \in \mathbf{H}_q^{r-1+1/q} B. \tag{4.4.20}$$

From (4.4.20) and Theorem 4.4.1 we obtain the required lower estimates.

We now prove the upper estimates. Let $f \in \mathbf{H}_q^r$. Then, by Theorem 4.4.6,

$$\|A_{\mathbf{s}}(f)\|_q \ll 2^{-r\|\mathbf{s}\|_1}. \tag{4.4.21}$$

Using Remark 3.3.10 we then obtain

$$E_{Q_n}(f)_p \le \left\| \sum_{\|\mathbf{s}\|_1 > n} A_{\mathbf{s}}(f) \right\|_p$$

$$\ll \left(\sum_{\|\mathbf{s}\|_1 > n} 2^{-(r-1/q+1/p)\|\mathbf{s}\|_1 p} \right)^{1/p} \ll 2^{-(r-1/q+1/p)n} n^{(d-1)/p}.$$

The first relation in Theorem 4.4.10 is therefore proved.

Let now $1 < p \le q < \infty$, $q \ge 2$. Clearly, it suffices to prove the upper estimates for $p = q$. We have that

$$\|f - S_{Q_n}(f)\|_q = \left\| \sum_{\|\mathbf{s}\|_1 > n} \delta_{\mathbf{s}}(f) \right\|_q.$$

Using Corollary A.3.5, we continue the above relation (for $2 \le q < \infty$):

$$\ll \left(\sum_{\|\mathbf{s}\|_1 > n} \|\delta_{\mathbf{s}}(f)\|_q^2 \right)^{1/2}. \tag{4.4.22}$$

Using Theorem 4.4.6 to estimate $\|\delta_{\mathbf{s}}(f)\|_q$ we get, from the inequality (4.4.22),

$$E_{Q_n}(f)_q \le \|f - S_{Q_n}(f)\|_q \ll 2^{-rn} n^{(d-1)/2}. \tag{4.4.23}$$

In the case $q = \infty$, $1 < p < \infty$ the upper estimate follows from the case already considered. Let us prove now the lower estimate. We use the lacunary polynomials from §4.2.4. Consider the function

$$\varphi(\mathbf{x}) := \sum_{\|\mathbf{s}\|_1 = n+1} e^{i(2^{\mathbf{s}},\mathbf{x})}.$$

Then, by the Littlewood–Paley theorem A.3.3, we have

$$\|\varphi\|_p \asymp n^{(d-1)/2}, \qquad 1 < p < \infty. \tag{4.4.24}$$

Let $f = 2^{-rn}\varphi$; then, for $1 < p < \infty$,

$$E_{Q_n}(f)_p \gg \left\|f - S_{Q_n}(f)\right\|_p \gg 2^{-rn} n^{(d-1)/2}.$$

In order to conclude the proof of the lower estimate in the second relation in Theorem 4.4.10 we use Theorem 4.4.6, which implies that $f \in \mathbf{H}^r_\infty B$ with $B > 0$ and independent of n.

We proceed to the third relation in Theorem 4.4.10. We first prove the upper estimates. Clearly it suffices to consider the case $1 \le p = q \le 2$. Let $1 < q \le 2$. Then, again by Corollary A.3.5, we have

$$\left\|f - S_{Q_n}(f)\right\|_q \ll \left(\sum_{\|\mathbf{s}\|_1 > n} \|\delta_{\mathbf{s}}(f)\|_q^q\right)^{1/q}. \tag{4.4.25}$$

From (4.4.25), using Theorem 4.4.6 we get the required upper estimate for $1 < p = q \le 2$.

Now let $p = q = 1$; then

$$E_{Q_n}(f)_1 \le \left\|\sum_{\|\mathbf{s}\|_1 > n} \mathscr{A}_{\mathbf{s}}(f)\right\|_1 \le \sum_{\|\mathbf{s}\|_1 > n} \left\|\mathscr{A}_{\mathbf{s}}(f)\right\|_1,$$

and, applying Theorem 4.4.6, we obtain the upper estimate for $p = q = 1$.

Let us now prove the lower estimate. We use the polynomials $\psi(\mathbf{x})$ from §4.2.4. It follows from (4.2.17) that

$$E_{Q_n}(\psi)_1 \gg n^{d-1}. \tag{4.4.26}$$

We consider the function

$$f(\mathbf{x}) := \psi(\mathbf{x}) 2^{-rn} u^{-d(1-1/q)}.$$

By Theorem 4.4.6 and estimate (3.2.7), there is a constant $B > 0$ which does not depend on n such that $f \in \mathbf{H}^r_q B$. From (4.4.26) we find that

$$E_{Q_n}(f)_p \ge E_{Q_n}(f)_1 \gg 2^{-rn} n^{(d-1)/q},$$

which finishes the proof of Theorem 4.4.10. □

4.4.4 Approximation in Function Classes for Extreme Values of q and p.

In this subsection we consider the cases when one or two parameters q, p can take extreme value, 1 or ∞. These results are not as complete as in the case $1 < q, p < \infty$.

Theorem 4.4.11 *We have*

$$E_{Q_n}(\mathbf{W}^r_{q,\alpha})_p \asymp \begin{cases} 2^{-nr}, & r>0;\ q=\infty,\ 1\le p<\infty; \\ & p=1,\ 1<q\le\infty; \\ 2^{-n(r-1+1/p)}n^{(d-1)/p}, & r>1-1/p,\ q=1,\ 1\le p\le\infty; \\ 2^{-n(r-1/q)}, & r>1/q,\ 1\le q\le 2,\ p=\infty. \end{cases}$$

Proof The first relation follows from Theorem 4.4.9 and its proof. Let us prove the second relation. It follows from the definition of the classes $\mathbf{W}^r_{1,\alpha}$ and the generalized Minkowski inequality that

$$E_{Q_n}(\mathbf{W}^r_{1,\alpha})_p \le E_{Q_n}\big(F_r(\mathbf{x},\alpha)\big)_p, \qquad 1\le p\le\infty.$$

For $r > 1-1/p$ the function $F_r(\mathbf{x},\alpha)$ belongs to the closure of the class $\mathbf{W}^r_{1,\alpha}$ in the L_p-metric, and so

$$E_{Q_n}(\mathbf{W}^r_{1,\alpha})_p \ge E_{Q_n}\big(F_r(\mathbf{x},\alpha)\big)_p.$$

Summarizing, we get

$$E_{Q_n}(\mathbf{W}^r_{1,\alpha})_p = E_{Q_n}\big(F_r(\mathbf{x},\alpha)\big)_p.$$

From this relation and Theorem 4.4.1 we obtain the second relation in the theorem.

Let us consider the third relation. The lower estimates follow from consideration of the univariate case. We first prove the upper estimate for $q=2$. Clearly, without loss of generality we can assume that $\alpha = 0$. By the Nikol'skii duality theorem (see Theorem A.2.3),

$$\begin{aligned} E_{Q_n}(\mathbf{W}^r_2)_\infty &= \sup_{\|\varphi\|_2\le 1} E_{Q_n}(\varphi * F_r)_\infty = \sup_{\|\varphi\|_2\le 1}\ \sup_{\psi\in\mathscr{T}(Q_n)_1^\perp} |\langle \varphi * F_r, \psi\rangle| \\ &= \sup_{\psi\in\mathscr{T}(Q_n)_1^\perp}\ \sup_{\|\varphi\|_2\le 1} |\langle \varphi, \psi * F_r\rangle| = \sup_{\psi\in\mathscr{T}(Q_n)_1^\perp} \|\psi * F_r\|_2 \\ &= \sup_{\psi\in\mathscr{T}(Q_n)_1^\perp} |\langle \psi * F_r,\ \psi * F_r\rangle|^{1/2} = \sup_{\psi\in\mathscr{T}(Q_n)_1^\perp} |\langle \psi, \psi * F_{2r}\rangle|^{1/2} \\ &\le E_{Q_n}(F_{2r})_\infty^{1/2} \ll 2^{-n(r-1/2)}. \end{aligned}$$

In the last step we used Theorem 4.4.1.

In the case $q=1$ the required estimate is contained in the second relation. In the case $1<q<2$ the estimate follows from the case $q=2$ in view of the inclusion $\mathbf{W}^r_{q,\alpha} \subset \mathbf{W}^{r-1/q+1/2}_{2,\alpha}B$ (see Corollary A.3.8 of the Hardy–Littlewood inequality).

The theorem 4 is therefore proved. □

We summarize the above results on the behavior of the $E_{Q_n}(\mathbf{W}^r_{q,\alpha})_p$ in Figure 4.3. For $(q,p) \in [1,\infty]^2$, represented by the point $(1/q,1/p) \in [0,1]^2$, we give the order of the $E_{Q_n}(\mathbf{W}^r_{q,\alpha})_p$ and also a reference to the theorem which establishes that relation. In those cases when the order of $E_{Q_n}(\mathbf{W}^r_{q,\alpha})_p$ is not known we refer to the corresponding open problem. As above we use the notation $\beta := 1/q - 1/p$, $\beta_+ := \max(\beta, 0)$.

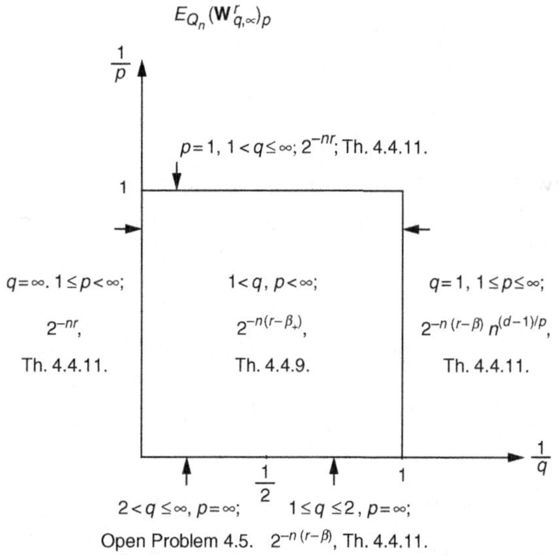

Figure 4.3 The best hyperbolic cross approximation for the $\mathbf{W}$ classes.

We give two theorems on $\mathbf{H}^r_q$ classes which complement Theorem 4.4.10.

Theorem 4.4.12 *Let $2 \le q \le \infty$ and $r > 0$. Then we have*

$$E_{Q_n}(\mathbf{H}^r_q)_1 \asymp 2^{-rn} n^{(d-1)/2}.$$

Proof The upper bounds follow from the upper bounds for $E_{Q_n}(\mathbf{H}^r_2)_2$ from Theorem 4.4.10. The lower bounds follow from Theorem 5.3.17 on Kolmogorov widths, which we prove in Chapter 5. These lower bounds are nontrivial. □

For the classes $\mathbf{H}^r_q$ the following cases were considered in Theorem 4.4.10: $q = 1$, $1 < p < \infty$; $q = \infty$, $1 < p < \infty$; $p = 1$, $1 \le q \le 2$. Here we confine ourselves to the bivariate case.

Theorem 4.4.13 *Let $d = 2$, and $r > 0$. Then*

$$E_{Q_n}(\mathbf{H}^r_\infty)_p \asymp \begin{cases} 2^{-rn} n & \text{for } p = \infty, \\ 2^{-rn} n^{1/2} & \text{for } p = 1. \end{cases}$$

Proof We consider the first relation. Let us prove the upper estimate. We remark that this proof works for all dimensions d. Applying Theorem 4.4.6 we find, for $f \in \mathbf{H}^r_\infty$,

$$E_{Q_n}(f)_\infty \le \left\| \sum_{\|\mathbf{s}\|_1 > n} A_\mathbf{s}(f) \right\|_\infty \le \sum_{\|\mathbf{s}\|_1 > n} \|A_\mathbf{s}(f)\|_\infty$$

$$\ll \sum_{\|\mathbf{s}\|_1 > n} 2^{-r\|\mathbf{s}\|_1} \ll 2^{-rn} n^{d-1}. \tag{4.4.27}$$

To prove the lower estimate we consider the special trigonometric polynomial $\Phi_m(\mathbf{x})$ from §4.2.5. We use the representation (4.2.19) and the property

$$\|\Phi_m\|_1 = 1.$$

Let $n = 2m$ and

$$f(\mathbf{x}) := 2^{-rn} \left(\sum_{k=0}^{m} \cos 4^k x_1 \cos 4^{m-k} x_2 \right).$$

Then f belongs to $\mathbf{H}^r_\infty B$, B does not depend on n, and

$$2^{-rn-2}(m+1) = \langle f, \Phi_m \rangle \le E_{Q_{n-1}}(f)_\infty \|\Phi_m\|_1 = E_{Q_{n-1}}(f)_\infty,$$

which gives the required lower estimate.

Let us now prove the second relation. The upper estimate follows from Theorem 4.4.10. We will prove the lower estimate. Consider the function

$$\varphi(\mathbf{x}) := \sum_{k=0}^{m} e^{i(4^k x_1 + 4^{m-k} x_2)}.$$

Then, for any $t \in \mathscr{T}(Q_{2m-1})$, we have

$$(\varphi - t) * 4(\Phi_m - 1) = \varphi,$$

consequently,

$$\|\varphi\|_1 \le 4\|\Phi_m - 1\|_1 E_{Q_{2m-1}}(\varphi)_1 \le 8 E_{Q_{2m-1}}(\varphi)_1. \tag{4.4.28}$$

Further,

$$(m+1)^{1/2} = \|\varphi\|_2 \le \|\varphi\|_1^{1/3} \|\varphi\|_4^{2/3}.$$

By the Littlewood–Paley theorem A.3.3, $\|\varphi\|_4 \ll m^{1/2}$ and, therefore,

$$\|\varphi\|_1 \gg m^{1/2}.$$

To complete the proof of the theorem it remains to set $f := 2^{-2rm}\varphi$ and apply Theorem 4.4.6.

We note that it is not difficult to derive from the second relation in Theorem 4.4.11 the estimate

$$E_{Q_n}(\mathbf{H}^r_q)_\infty \ll 2^{-n(r-1/q)} n^{(d-1)/2}, \qquad r > 1/q, \qquad 1 \le q \le 2. \tag{4.4.29}$$

Indeed, by Corollary 4.4.8 it suffices to prove this estimate for $q = 2$. Let $f \in \mathbf{H}^r_2$ and

$$f^m := \sum_{\|\mathbf{s}\|_1 = m} \delta_{\mathbf{s}}(f).$$

Using Theorem 4.4.6, we easily find that

$$\|f^m\|_2 \ll 2^{-rm} m^{(d-1)/2}.$$

We choose ρ such that $1/2 < \rho < r$. Then, using Theorem 4.4.11 and the Bernstein inequality, we get

$$\begin{aligned} E_{Q_n}(f)_\infty &\le \sum_{m>n} E_{Q_n}(f^m)_\infty \le \sum_{m>n} E_{Q_n}(\mathbf{W}^\rho_2)_\infty \|D^\rho f^m\|_2 \\ &\ll \sum_{m>n} 2^{-n(\rho-1/2)} 2^{-(r-\rho)m} m^{(d-1)/2} \ll 2^{-n(r-1/2)} n^{(d-1)/2}. \end{aligned}$$

The relation (4.4.29) is proved. □

We summarize the above results on the behavior of the $E_{Q_n}(\mathbf{H}^r_q)_p$ in Figure 4.4. For $(q,p) \in [1,\infty]^2$, represented by the point $(1/q, 1/p) \in [0,1]^2$, we give the order of the $E_{Q_n}(\mathbf{H}^r_q)_p$ and a reference to the theorem which establishes that relation. In those cases when the order of $E_{Q_n}(\mathbf{H}^r_q)_p$ is not known we refer to the corresponding open problem. As above we use the notation $\beta := 1/q - 1/p$.

4.4.5 Linear Methods of Approximation

In this subsection we consider approximations by linear methods of functions in the classes $\mathbf{W}^r_{q,\alpha}$, $\mathbf{H}^r_q$ in the uniform metric.

Theorem 4.4.14 *Let L_{Q_n} be a linear operator assigning to each function $f \in \mathbf{W}^r_{q,\alpha}$ a trigonometric polynomial $L_{Q_n}(f) \in \mathscr{T}(Q_n)$. Then*

$$\sup_{f \in \mathbf{W}^r_{q,\alpha}} \|f - L_{Q_n}(f)\|_\infty \ge E_{Q_n}(F_r(\mathbf{x},\alpha))_{q'}.$$

Proof Let $\theta = (\theta_1, \dots, \theta_d)$, and let T_θ be the operator translating the argument by θ. If, following Marcinkiewicz, we consider the bounded linear operator

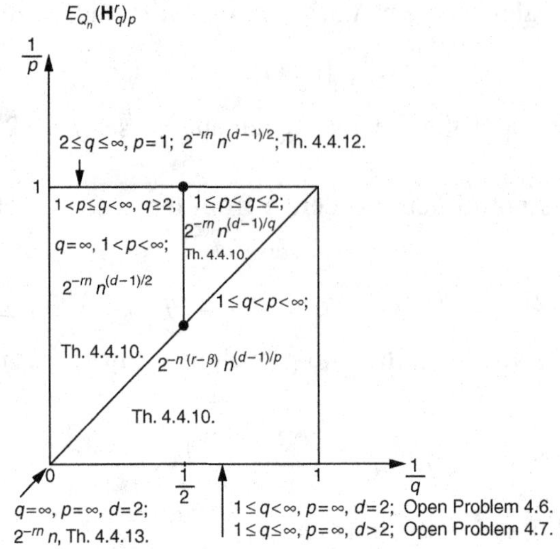

Figure 4.4 The best hyperbolic cross approximation for the **H** classes.

$$L'_{Q_n} := (2\pi)^{-d}\int_{\mathbb{T}^d} T_{-\theta}L_{Q_n}T_\theta d\theta$$

then we find that L'_{Q_n} is a convolution operator, and

$$\sup_{f\in\mathbf{W}^r_{q,\alpha}}\|f-L'_{Q_n}(f)\|_\infty \le \sup_{f\in\mathbf{W}^r_{q,\alpha}}\|f-L_{Q_n}(f)\|_\infty.$$

Consequently, it suffices to prove the theorem for a convolution operator. Let

$$L_{Q_n}(f) = f*\mathscr{L}_n, \qquad \mathscr{L}_n\in\mathscr{T}(Q_n).$$

Then

$$\begin{aligned}\sup_{f\in\mathbf{W}^r_{q,\alpha}}\|f-L_{Q_n}(f)\|_\infty &= \sup_{\|\varphi\|_q\le1}\|\varphi*F_r(\mathbf{x},\alpha)-\varphi*\mathscr{L}_n*F_r(\mathbf{x},\alpha)\|_\infty\\ &= \sup_{\|\varphi\|_q\le1}\|\varphi*(F_r-t_n)\|_\infty = \|F_r-t_n\|_{q'}\ge E_{Q_n}(F_r)_{q'},\end{aligned}$$

where $t_n := \mathscr{L}_n*F_r\in\mathscr{T}(Q_n)$. □

Theorem 4.4.15 *Let $r>1/q$. Then, for $1<q<\infty$*

$$E_{Q_n}(\mathbf{W}^r_{q,\alpha})_\infty = o\left(\sup_{f\in\mathbf{W}^r_{q,\alpha}}\|f-L_{Q_n}(f)\|_\infty\right),$$

for any sequence of bounded linear operators L_{Q_n} acting from $\mathbf{W}^r_{q,\alpha}$ to $\mathscr{T}(Q_n)$.

Theorem 4.4.16 *Let $r > 1/q$. Then for $1 \le q < \infty$*

$$E_{Q_n}(\mathbf{H}^r_q)_\infty = o\left(\sup_{f \in \mathbf{H}^r_q} \|f - L_{Q_n}(f)\|_\infty \right),$$

for any sequence of bounded linear operators L_{Q_n} acting from $\mathbf{H}^r_q$ to $\mathscr{T}(Q_n)$.

Theorems 4.4.15 and 4.4.16 show that, both for the classes $\mathbf{W}^r_{q,\alpha}$ and for the classes $\mathbf{H}^r_q$ with $1 < q < \infty$, approximation by linear methods in the uniform metric does not give the orders of the corresponding least upper bounds of the best approximations. In the case $q = 1$ it turns out that the classes $\mathbf{W}^r_{1,\alpha}$ and $\mathbf{H}^r_1$ are different in this regard. Namely, if for $f \in \mathbf{W}^r_{1,\alpha}$ we take the approximating polynomial to be

$$L_{Q_n}(f) = f * t_n(F_r),$$

where $t_n(F_r)$ is a polynomial of the best approximation of $F_r(\mathbf{x}, \alpha)$ by polynomials in $\mathscr{T}(Q_n)$ in the uniform metric, then we get

$$\sup_{f \in \mathbf{W}^r_{1,\alpha}} \|f - L_{Q_n}(f)\|_\infty = E_{Q_n}(F_r)_\infty = E_{Q_n}(\mathbf{W}^r_{1,\alpha})_\infty,$$

while Theorem 4.4.16 shows that for any sequence of linear operators L_{Q_n} we have

$$E_{Q_n}(\mathbf{H}^r_1)_\infty = o\left(\sup_{f \in \mathbf{H}^r_1} \|f - L_{Q_n}(f)\|_\infty \right).$$

Proof of Theorem 4.4.15 Let $1 < q \le 2$. In view of Theorems 4.4.14 and 4.4.1, for $r > 1/q$ we have

$$\sup_{f \in \mathbf{W}^r_{q,\alpha}} \|f - L_{Q_n}(f)\|_\infty \gg 2^{-(r-1/q)n} n^{(d-1)(1-1/q)}. \tag{4.4.30}$$

By Theorem 4.4.11,

$$E_{Q_n}(\mathbf{W}^r_{q,\alpha})_\infty \asymp 2^{-(r-1/q)n}, \qquad r > 1/q, \tag{4.4.31}$$

for $1 \le q \le 2$. Comparing (4.4.30) and (4.4.31) we get the theorem for $1 < q \le 2$.

Let $2 < q < \infty$. We first prove an auxiliary assertion.

Lemma 4.4.17 *Let $2 \le q \le \infty$; then*

$$E_{Q_n}(\mathbf{W}^r_{q,\alpha}) \ll E_{Q_n}(\mathbf{W}^r_{2,\alpha})^{2/q}_\infty E_{Q_n}(\mathbf{W}^r_{\infty,\alpha})^{1-2/q}_\infty, \qquad 2 \le q \le \infty.$$

Proof Suppose that $2 < q < \infty$ and $f \in \mathbf{W}^r_{q,\alpha}$ is given by

$$f = \varphi * F_r, \qquad \|\varphi\|_q \le 1.$$

Let

$$E := \{\mathbf{x} \in \mathbb{T}^d : |\varphi(\mathbf{x})| \ge D\},$$

where D is a number to be chosen below. We represent φ in the form

$$\varphi = \varphi_D + \varphi^D,$$

where

$$\varphi_D(\mathbf{x}) = \begin{cases} \varphi(\mathbf{x}) & \text{for } \mathbf{x} \notin E, \\ 0 & \text{for } \mathbf{x} \in E. \end{cases}$$

Then $|\varphi_D(\mathbf{x})| \le D$. Let us estimate $\|\varphi^D\|_2$. We have

$$\int_{\mathbb{T}^d} |\varphi^D(\mathbf{x})|^2 d\mathbf{x} = \int_E |\varphi(\mathbf{x})|^2 d\mathbf{x} \le |E|^{1-2/q} \left(\int_E |\varphi(\mathbf{x})|^q d\mathbf{x} \right)^{2/q} \ll |E|^{1-2/q}.$$

Further,

$$1 \ge \|\varphi\|_q \ge \|\varphi^D\|_q \ge D|E|^{1/q}.$$

Therefore

$$\|\varphi^D\|_2 \ll D^{1-q/2}.$$

Let $f_D := \varphi_D * F_r$, $f^D := \varphi^D * F_r$. Then

$$\begin{aligned} E_{Q_n}(f)_\infty &\le E_{Q_n}(f_D)_\infty + E_{Q_n}(f^D)_\infty \\ &\ll D E_{Q_n}(\mathbf{W}^r_{\infty,\alpha})_\infty + D^{1-q/2} E_{Q_n}(\mathbf{W}^r_{2,\alpha}). \end{aligned} \tag{4.4.32}$$

Setting

$$D := \left(E_{Q_n}(\mathbf{W}^r_{2,\alpha})_\infty \,/\, E_{Q_n}(\mathbf{W}^r_{\infty,\alpha})_\infty \right)^{2/q}$$

we get the lemma from (4.4.32). □

Using Theorem 4.4.11 and estimate (4.4.27), we obtain from Lemma 4.4.17

$$E_{Q_n}(\mathbf{W}^r_{q,\alpha})_\infty \ll 2^{-(r-1/q)n} n^{(d-1)(1-2/q)}. \tag{4.4.33}$$

By Theorems 4.4.14 and 4.4.1,

$$\sup_{f \in \mathbf{W}^r_{q,\alpha}} \|f - L_{Q_n}(f)\|_\infty \gg 2^{-(r-1/q)n} n^{(d-1)(1-1/q)}. \tag{4.4.34}$$

Comparing (4.4.33) and (4.4.34) we get Theorem 4.4.15. □

Proof of Theorem 4.4.16 We first prove the following theorem.

Theorem 4.4.18 *Let L_{Q_n} be a bounded linear operator assigning to each function $f \in \mathbf{H}^r_q$ a trigonometric polynomial $L_{Q_n}(f) \in \mathscr{T}(Q_n)$. Then*

$$\sup_{f \in \mathbf{H}^r_q} \|f - L_{Q_n}(f)\|_\infty \gg 2^{-(r-1/q)n} n^{d-1}, \qquad 1 \le q \le \infty. \tag{4.4.35}$$

Proof As in the proof of Theorem 4.4.14, the use of the Marcinkiewicz method shows that it suffices to prove the theorem for a convolution operator

$$L_{Q_n}(f) = f * \mathscr{L}_n, \qquad \mathscr{L}_n \in \mathscr{T}(Q_n).$$

We consider the function

$$f(\mathbf{x}) := 2^{-(r+1-1/q)n} \sum_{\|\mathbf{s}\|_1 = n+2d} \mathscr{A}_{\mathbf{s}}(\mathbf{x});$$

then $f * \mathscr{L}_n = 0$. By Theorem 4.4.6,

$$f \in \mathbf{H}^r_q B, \qquad B > 0.$$

Further,

$$\left\| f - L_{Q_n}(f) \right\|_\infty = \|f\|_\infty \asymp 2^{-(r-1/q)n} n^{d-1}.$$

Consequently,

$$\sup_{f \in \mathbf{H}^r_q} \left\| f - L_{Q_n}(f) \right\|_\infty \gg 2^{-(r-1/q)n} n^{d-1}. \qquad \square$$

For $1 \le q \le 2$, Theorem 4.4.16 follows from a comparison of Theorem 4.4.18 with (4.4.29). Let $2 < q < \infty$. We will obtain an upper estimate for $E_{Q_n}(\mathbf{H}^r_q)_\infty$. If $f \in \mathbf{H}^r_q$ then, by Theorem 4.4.6

$$\left\| \delta_{\mathbf{s}}(f) \right\|_q \ll 2^{-r\|\mathbf{s}\|_1},$$

which implies that

$$\left\| \delta_{\mathbf{s}}(f)^{(r)}(\mathbf{x}, \mathbf{0}) \right\|_q \ll 1. \tag{4.4.36}$$

Let

$$\varphi_m := \sum_{\|\mathbf{s}\|_1 = m} \delta_{\mathbf{s}}(f)^{(r)},$$

then by Corollary A.3.5 and (4.4.36),

$$\|\varphi_m\|_q \ll \left(\sum_{\|\mathbf{s}\|_1 = m} \left\| \delta_{\mathbf{s}}(f)^{(r)} \right\|_q^2 \right)^{1/2} \ll m^{(d-1)/2}.$$

By the Nikol'skii duality theorem (see Theorem A.2.3),

$$\begin{aligned} E_{Q_n}(\mathbf{H}^r_q)_\infty &= \sup_{f \in \mathbf{H}^r_q} \sup_{g \in \mathscr{T}(Q_n)^\perp_1} |\langle f, g \rangle| \\ &= \sup_{g \in \mathscr{T}(Q_n)^\perp_1} \sup_{f \in \mathbf{H}^r_q} \Big| \sum_{\mathbf{s}} \langle \delta_{\mathbf{s}}(f), \delta_{\mathbf{s}}(g) \rangle \Big|. \end{aligned} \tag{4.4.37}$$

Using Theorem 4.4.6 and (4.4.37), we get

$$E_{Q_n}(\mathbf{H}^r_q)_\infty \asymp \sup_{g\in\mathscr{T}(Q_n)^\perp_1} \sum_{\mathbf{s}} 2^{-r\|\mathbf{s}\|_1} \left\|\delta_{\mathbf{s}}(g)\right\|_{q'}$$
$$\asymp \sup_{g\in\mathscr{T}(Q_n)^\perp_1} \sum_{\mathbf{s}} 2^{-r\|\mathbf{s}\|_1} \left\|A_{\mathbf{s}}(g)\right\|_{q'}. \tag{4.4.38}$$

It follows from (4.4.29) and (4.4.38) that, for $q=2$ and $\bar{r}=r/2$,

$$\sum_{\mathbf{s}} \left\|A_{\mathbf{s}}(g)\right\|_2 2^{-r\|\mathbf{s}\|_1/2} \ll 2^{-(r-1)n/2} n^{(d-1)/2}. \tag{4.4.39}$$

Further,

$$\left\|A_{\mathbf{s}}(g)\right\|_{q'} \le \left\|A_{\mathbf{s}}(g)\right\|_1^{2/q'-1} \left\|A_{\mathbf{s}}(g)\right\|_2^{2(1-1/q')}$$

and, by Hölder's inequality,

$$\sum_{\mathbf{s}} \left\|A_{\mathbf{s}}(g)\right\|_{q'} 2^{-r\|\mathbf{s}\|_1} \ll \left(\sum_{\mathbf{s}} \left\|A_{\mathbf{s}}(g)\right\|_1 2^{-r\|\mathbf{s}\|_1/(2-q')}\right)^{(2-q')/q'}$$
$$\times \left(\sum_{\mathbf{s}} \left\|A_{\mathbf{s}}(g)\right\|_2 2^{-r\|\mathbf{s}\|_1/2}\right)^{2(1-1/q')}. \tag{4.4.40}$$

Since $\|g\|_1 \le 1$, it follows that $\left\|A_{\mathbf{s}}(g)\right\|_1 \ll 1$ and, for $g \in \mathscr{T}(Q_n)^\perp_1$,

$$\sum_{\mathbf{s}} \left\|A_{\mathbf{s}}(g)\right\|_1 2^{-r\|\mathbf{s}\|_1/(2-q')} \ll 2^{-rn/(2-q')} n^{d-1}. \tag{4.4.41}$$

Substituting (4.4.39) and (4.4.41) into (4.4.40), we find that, for $g \in \mathscr{T}(Q_n)^\perp_1$,

$$\sum_{\mathbf{s}} \left\|A_{\mathbf{s}}(g)\right\|_{q'} 2^{-r\|\mathbf{s}\|_1} \ll 2^{-(r-1/q)n} n^{(d-1)/q'};$$

and consequently, by (4.4.38),

$$E_{Q_n}(\mathbf{H}^r_q)_\infty \ll 2^{-(r-1/q)n} n^{(d-1)/q'}, \qquad 2 < q < \infty. \tag{4.4.42}$$

Comparing Theorem 4.4.18 and (4.4.42) we get Theorem 4.4.16 for $2 < q < \infty$. The proof of Theorem 4.4.16 is complete. □

4.5 Some Further Remarks

The operators S_{Q_n} and V_{Q_n} play an important role in the hyperbolic cross approximation. These operators can be written in terms of the corresponding univariate operators in the following form. Denote by S^i_l the univariate operator S_l acting on functions of the variable x_i. Then, it follows from the definition of Q_n that

$$S_{Q_n} = \sum_{\mathbf{s}:\|\mathbf{s}\|_1\le n} \prod_{i=1}^{d} (S^i_{2^{s_i}-1} - S^i_{[2^{s_i-1}]-1}), \tag{4.5.1}$$

where $S_{-1} := 0$. A similar formula holds for the V_{Q_n}.

The Smolyak algorithm Operators of the form (4.5.1) with S^i replaced by other univariate operators are used in sampling recovery (see Chapter 6) and other problems. For a discussion we refer to the recent survey Dinh Dung *et al.* (2016). The approximate recovery operators of the form (4.5.1) were first considered by Smolyak (1963); the standard name for operators of the form (4.5.1) is *Smolyak-type algorithms*. Very often the analysis of the operators S_{Q_n}, V_{Q_n}, and other operators of the form (4.5.1) goes along the same lines. The following general framework was suggested in Andrianov and Temlyakov (1997). Let three numbers $a > b \geq 0$ and $1 \leq p \leq \infty$ be given. Consider a family of univariate linear operators $\{Y_s\}_{s=0}^{\infty}$, which are defined on the space $W^a_{p,\alpha}$ and have the following two properties:

(1) for any f from the class $W^a_{p,\alpha}$ we have

$$\|f - Y_s(f)\|_p \leq C_1 2^{-as}, \qquad s = 0,1,2,\dots;$$

(2) for any trigonometric polynomial t of order 2^v, we have

$$\|Y_s(t)\|_p \leq C_2 2^{b(v-s)} \|t\|_p, \qquad v \geq s.$$

As above, let Y^i_s denote the univariate operator Y_s acting on functions of the variable x_i. Consider the d-dimensional operator

$$T_n := \sum_{\mathbf{s}:\|\mathbf{s}\|_1 \leq n} \Delta_{\mathbf{s}}, \qquad \Delta_{\mathbf{s}} := \prod_{i=1}^{d} (Y^i_{s_i} - Y^i_{s_i - 1}),$$

with $Y_{-1} := 0$. We illustrate the above general setting by one result from Andrianov and Temlyakov (1997).

Theorem 4.5.1 *Let operators $\{Y_s\}_{s=0}^{\infty}$ satisfy the above conditions (1) and (2). Then, for any $r \in (b,a)$, we have for $f \in \mathbf{H}^r_p$*

$$\|\Delta_{\mathbf{s}}(f)\|_p \ll 2^{-r\|\mathbf{s}\|_1} \qquad \textit{and} \qquad \|f - T_n(f)\|_p \ll 2^{-rn} n^{d-1}.$$

Proof Let I^a_α denote the operator of convolution with the Bernoulli kernel $F_a(x,\alpha)$ in the univariate case and with $F_a(\mathbf{x},\alpha)$ in the multivariate case (see (1.4.1)). Then, it follows from the definition of the classes $W^a_{p,\alpha}$ that condition (1) is equivalent to the condition

$$\|(I - Y_n) I^a_\alpha\|_{L_p \to L_p} \leq C_1 2^{-an}, \qquad n = 0,1,\dots \tag{4.5.2}$$

This inequality implies the multivariate inequality

$$\|\Delta_{\mathbf{s}} I^a_\alpha\|_{L_p \to L_p} \leq C_1' 2^{-a\|\mathbf{s}\|_1}. \tag{4.5.3}$$

In the same way, condition (2) implies its multivariate analog,

$$\|\Delta_{\mathbf{s}}(t)\|_p \leq C_2' 2^{b(\|\mathbf{v}\|_1 - \|\mathbf{s}\|_1)}, \qquad \mathbf{v} \geq \mathbf{s}, \qquad t \in \mathscr{T}(2^{\mathbf{v}}, d). \tag{4.5.4}$$

Clearly, it is sufficient to prove the bounds required in Theorem 4.5.1 under the assumption that $f \in \mathbf{H}^r_p$ is a trigonometric polynomial. Then, we make the representation

$$f = \sum_{\mathbf{s}} \Delta_{\mathbf{s}}(f) \tag{4.5.5}$$

and estimate

$$\|f - T_n(f)\|_p \le \sum_{\|\mathbf{s}\|_1 > n} \|\Delta_{\mathbf{s}}(f)\|_p. \tag{4.5.6}$$

Let us estimate each summand in (4.5.6). It is known (see Theorem 4.4.6) that for any function $f \in \mathbf{H}^r_p$, $1 \le p \le \infty$, the following expansion holds:

$$f(\mathbf{x}) = \sum_{\mathbf{v} \ge \mathbf{1}} t_{\mathbf{v}}(f)(\mathbf{x}),$$

where $t_{\mathbf{v}} := t_{\mathbf{v}}(f) \in \mathscr{T}(2^{\mathbf{v}}, d)$ and

$$\|t_{\mathbf{v}}\|_p \ll 2^{-r\|\mathbf{v}\|_1}.$$

We now estimate $\|\Delta_{\mathbf{s}}(t_{\mathbf{v}}(f))\|_p$. We use the differentiation operator D^a_α, which is the inverse to the operator I^a_α (see (1.4.8)). Then by (4.5.3) and the Bernstein inequalities from Theorem 3.3.1, on the one hand we obtain for each $i \in [1,d]$

$$\begin{aligned} \|\Delta_{s_i}(t_{\mathbf{v}})\|_p &= \|\Delta_{s_i}(I^a_\alpha D^a_\alpha t_{\mathbf{v}})\|_p \\ &\le \|\Delta_{s_i} I^a_\alpha\|_{L_p \to L_p} \|D^a_\alpha t_{\mathbf{v}}\|_p \ll 2^{-a(s_i - v_i)} \|t_{\mathbf{v}}\|_p. \end{aligned} \tag{4.5.7}$$

On the other hand, by property (2) we have for $v_i \ge s_i$ that

$$\|\Delta_{s_i}(t_{\mathbf{v}})\|_p \ll 2^{b(v_i - s_i)} \|t_{\mathbf{v}}\|_p. \tag{4.5.8}$$

Combining the estimates (4.5.7) and (4.5.8) and the estimate of the norm of $t_{\mathbf{v}}(f)$, we obtain

$$\|\Delta_{\mathbf{s}}(t_{\mathbf{v}}(f))\|_p \ll \prod_{i=1}^d \min(2^{-as_i} 2^{(a-r)v_i}, 2^{b(v_i - s_i)} 2^{-rv_i}),$$

whence, taking into account the fact that $b < r < a$, we deduce that

$$\begin{aligned} \|\Delta_{\mathbf{s}}(f)\|_p &\le \sum_{\mathbf{v} \ge \mathbf{1}} \|\Delta_{\mathbf{s}}(t_{\mathbf{v}}(f))\|_p \\ &\ll \prod_{i=1}^d \left(\sum_{v_i = 1}^{s_i} 2^{-as_i} 2^{(a-r)v_i} + \sum_{v_i > s_i} 2^{b(v_i - s_i)} 2^{-rv_i} \right) \\ &\ll 2^{-r\|\mathbf{s}\|_1}. \end{aligned} \tag{4.5.9}$$

Substituting inequality (4.5.9) into inequality (4.5.6), we obtain

$$\|f - T_n(f)\|_p \ll \sum_{\|\mathbf{s}\|_1 > n} 2^{-r\|\mathbf{s}\|_1} \ll 2^{-rn} n^{d-1}.$$

This proves the theorem. □

4.6 Historical Comments

Babenko (1960a, b) was the first to study the classes $\mathbf{W}^r_q$. Bakhvalov (1963a) and S.M. Nikol'skii were the first to study the classes $\mathbf{H}^r_q$. Many papers have been devoted to the investigation of these and more general classes of mixed smoothness recently. We refer the reader to Temlyakov (1986c, 1993b) and to Dinh Dung *et al.* (2016) for further references.

Theorem 4.1.1 was obtained by (Babenko, 1960a, b). Theorem 4.1.2 is a periodic analog of the corresponding theorem of Nikol'skii (1963). In connection with this theorem see the paper by Nikol'skaya (1974). Lemma 4.2.1 is a direct corollary of Lemma 1.4 in Temlyakov (1980c) and corollary A.3.5 of the Littlewood–Paley theorem. Another proof of Lemma 4.2.1 was given by Galeev (1982). The proof of Lemma 4.2.1 given in §4.2.1 is different from the two proofs mentioned above. It is much easier, and it is based on a deep result, Lemma 3.3.7.

The first Telyakovskii-type polynomials were constructed in Telyakovskii (1963). Lemma 4.2.4 is from Temlyakov (1980c). Lacunary polynomials constitute a classical object for research. The polynomials presented in §4.2.4 are from Temlyakov (1988b). The first use of Riesz products for the hyperbolic cross polynomials, which were described in §4.2.5, was reported in Temlyakov (1980b). The small ball inequality for the bivariate Haar system was proved by Talagrand (1994). The inequality (4.2.20) and all the results of §4.2.6 are from Temlyakov (1995a).

In the case $p = \infty$ the upper estimate in Theorem 4.3.1 was proved by Babenko (1960, a,b) (for certain α) and for arbitrary α it follows easily from results of Telyakovskii (1964). The lower estimate in Theorem 4.3.1 for $p = \infty$ was proved by Telyakovskii (1963). In the case $1 < p < \infty$ this theorem was proved by Mityagin (1962) for r a natural number and by Nikol'skaya (1975) for arbitrary r. Lemma 4.3.2 was proved by Telyakovskii (1963). The results on Bernstein inequalities for $p = 1$ presented in §4.3.2 are from Temlyakov (1998a). The results on the Nikol'skii inequalities presented in §4.3.3 are from Temlyakov (1980c). In the case $1 < q \le p < \infty$ Theorem 4.3.17 was proved by Galeev (1978).

Theorem 4.4.1 was proved by Temlyakov (1980c). The special case $p = 2$ had been considered previously by Bugrov (1964). Theorem 4.4.9 in the case $q = p$ and r natural was proved by Mityagin (1962); in the case $q = p$ and r real, by Nikol'skaya (1975); and in the general case by Galeev (1978) (see also the papers Temlyakov, 1979, 1980c, for $1 < q \le p < \infty$). In the case $q = p = 2$, Theorem 4.4.10 was proved in Bugrov (1964); in the case $1 < q = p < \infty$ in Nikol'skaya (1974); in the case $q = p = 1$ in Temlyakov (1980b) in the case $1 \le q < p \le 2$ in Temlyakov (1982b); in the case $1 \le q < p < \infty$, $p > 2$ in Temlyakov (1982a, 1985d); and in the case $2 \le p < q \le \infty$ the required upper estimates follow from estimates for $1 < q = p < \infty$ and the lower estimates follow from results due to Temlyakov (1980a). In the case $1 < p < 2 \le q < \infty$ Theorem 4.4.10 follows from the results of Galeev (1984) and Dinh Zung (1984); in the case $1 \le p < q \le 2$ it was proved in Temlyakov (1988b). Theorem 4.4.11 was proved in the case $q = 1$, $1 \le p \le \infty$ in Temlyakov (1979, 1980c) and in the case $1 \le q \le 2$, $p = \infty$ in Temlyakov (1980b). Theorem 4.4.13 for $p = \infty$ was proved in Temlyakov (1980b).

The results on linear methods of approximation presented in §4.4.5 are from Temlyakov (1982a, 1985d). The Results in §4.5 are from Andrianov and Temlyakov (1997).

4.7 Open Problems

We have already formulated four of the most acute open problems in the text. We repeat them here and formulate some others.

Open Problem 4.1 What is the order of $E_N(\mathbf{W}^r_{\infty,\alpha})_\infty$?

Open Problem 4.2 What is the order of the quantity $E_N(\mathbf{H}^r_\infty, L_\infty)$ for $d \ge 3$?

Open Problem 4.3 (Small ball inequality) Prove inequality (4.2.30) for $d \ge 3$.

Open Problem 4.4 (Bernstein inequality) Find the order in n of the sequence $\{b(n,r,L_1)\}$.

Open Problem 4.5 Find the order in n of the sequence $E_{Q_n}(\mathbf{W}^r_{q,\alpha})_\infty$ for $2 < q \le \infty$.

Open Problem 4.6 Find the order in n of the sequence $E_{Q_n}(\mathbf{H}^r_q)_\infty$ for $1 \le q < \infty$ in the case $d = 2$.

Open Problem 4.7 Find the order in n of the sequence $E_{Q_n}(\mathbf{H}^r_q)_\infty$ for $1 \le q \le \infty$ in the case $d > 2$.

5

The Widths of Classes of Functions with Mixed Smoothness

5.1 Introduction

This chapter deals with the following important problem regarding the multivariate approximation: what are the natural multivariate trigonometric polynomials? In other words, what is the natural ordering of the multivariate trigonometric system? In this chapter we answer this question. We demonstrate that the natural choice for the multivariate trigonometric polynomials is the hyperbolic cross trigonometric polynomials $\mathcal{T}(Q_n)$. Our argument is twofold. The first approach is based on optimization, which involves a study of the different widths. This approach was discussed in detail in §1.1 and was used in Chapter 3 in the approximation of the anisotropic function classes $W^{\mathbf{r}}_{q,\alpha}$ and $H^{\mathbf{r}}_q$. In this chapter we apply this approach to a study of classes with mixed smoothness, $\mathbf{W}^r_{q,\alpha}$ and $\mathbf{H}^r_q$. We present results only on Kolmogorov widths and on orthowidths (Fourier widths). The second approach is based on the concept of *universality*. We begin our discussion with the optimization approach and then consider the universality approach.

The above problem of finding the natural subspaces of the multivariate trigonometric polynomials is a central problem in the linear approximation of periodic functions. By *linear approximation* we mean approximation using a fixed finite-dimensional subspace. In our study of the approximation of univariate periodic functions in Chapters 1 and 2, the idea of representing a function by its Fourier series is very natural and traditional. It goes back to the work of Fourier from 1807. In this case one can use as a natural tool of approximation the partial sums of the Fourier expansion. In other words this means that we use the subspace $\mathcal{T}(n)$ for a source of approximants and use the orthogonal projection onto $\mathcal{T}(n)$ as the approximation operator. This natural approach is based on a standard ordering of the trigonometric system: 1, e^{ikx}, e^{-ikx}, e^{2ikx}, e^{-2ikx}, … We lose this natural approach when we go from the univariate case to the multivariate case – there is no natural ordering of the $\mathscr{T}^d$ for $d > 1$. The idea of choosing appropriate trigonometric subspaces for the approximation of a given class $\mathbf{F}$ of multivariate functions was suggested by Babenko (1960a). This idea is based on the concept of the Kolmogorov width, introduced in Kolmogorov (1936). Consider a Hilbert space $L_2(\mathbb{T}^d)$ and suppose that the function class $\mathbf{F} = A(B(L_2))$ of interest is an image of the unit ball $B(L_2)$ of $L_2(\mathbb{T}^d)$ under a mapping $A : L_2(\mathbb{T}^d) \to L_2(\mathbb{T}^d)$ by a compact operator A. For instance, in the case $\mathbf{F} = \mathbf{W}^r_2$ the operator $A := A_r$ is the convolution with the kernel $F_r(\mathbf{x})$. It is now well known (see Theorem A.3.10), and was established by Babenko (1960) for a special class of operators A, that

$$d_m(A(B(L_2)), L_2) = s_{m+1}(A), \tag{5.1.1}$$

where $s_j(A)$ are the singular numbers of the operator A: $s_j(A) = (\lambda_j(AA^*))^{1/2}$.

Suppose now that the eigenfunctions of the operator AA^* are the trigonometric functions $e^{i(\mathbf{k}_j,\mathbf{x})}$. Then the m-dimensional subspace that is optimal in the sense of the Kolmogorov width will be the $\operatorname{span}\{e^{i(\mathbf{k}_j,\mathbf{x})}\}_{j=1}^m$. Applying this approach to the class $\mathbf{W}^r_2$ we obtain that for $m = |\Gamma(N)|$ the optimal subspace for approximation in L_2 is the subspace of hyperbolic cross polynomials $\mathscr{T}(\Gamma(N))$ (see Theorem 4.1.1 above). This observation has led to a thorough study of approximation by the hyperbolic cross polynomials. We discussed this in Chapter 4.

Mityagin (1962) used the harmonic analysis technique, in particular, the Marcinkiewicz multipliers (see Theorem A.3.6), to prove that

$$d_m(\mathbf{W}^r_p, L_p) \asymp m^{-r}(\log m)^{r(d-1)}, \qquad 1 < p < \infty.$$

He also proved that the optimal subspaces, in the sense of order, are $\mathscr{T}(Q_n)$ with $|Q_n| \asymp m$ and $|Q_n| \le m$. In addition, the operator $S_{Q_n}(\cdot)$ of orthogonal projection onto $\mathscr{T}(Q_n)$ can be taken as an approximation operator. The use of harmonic analysis techniques for the L_p spaces led to the change from smooth hyperbolic crosses $\Gamma(N)$ to step hyperbolic crosses Q_n. The application of the theory of widths for finding good subspaces for the approximation of classes of functions with mixed smoothness is very natural and has been used in many papers. A typical problem here is to study the approximation of the classes $\mathbf{W}^r_q$ in L_p for all $1 \le p, q \le \infty$. We give a detailed discussion of these results in this chapter. As we mentioned above, in a linear approximation we are interested in approximation using finite-dimensional subspaces. The Kolmogorov width provides a way to determine optimal (usually, in the sense of order) m-dimensional subspaces. The approximation operator used in the Kolmogorov width is the operator of best approximation. Clearly, we would like to use approximation operators that are as simple as possible. As a result we will pay special attention to the orthowidths, where the approximation operator is very simple – an orthogonal projection.

We now discuss the universality principle formulated in Temlyakov (1988c). In

§5.4 we illustrate the following general observation. Methods of approximation which are optimal in the sense of order for the classes with mixed smoothness are universal for the collection of anisotropic smoothness classes. This gives an *a posteriori* justification for the thorough study of classes of functions with mixed smoothness. The phenomenon of saturation is well known in approximation theory; see DeVore and Lorentz (1993), Chapter 11. The classical example of a saturated method is the Fejér operator for the approximation of univariate periodic functions. For a sequence of Fejér operators K_n, saturation means that the approximation order by the operators K_n does not improve over the rate $1/n$ even if we increase the smoothness of functions being approximated. Methods (algorithms) that do not have the saturation property are called unsaturated. The reader can find a detailed discussion of unsaturated algorithms in approximation theory and in numerical analysis in the survey paper Babenko (1985). We point out that the concept of smoothness becomes more complicated in the multivariate case than it is in the univariate case. In the multivariate case a function may have different smoothness properties in different coordinate directions. In other words, functions may belong to different anisotropic smoothness classes (see the anisotropic Sobolev and Nikol'skii classes $W^{\mathbf{r}}_{q,\alpha}$ and $H^{\mathbf{r}}_q$ in Chapter 3). It is known (see Chapter 3 and Temlyakov, 1993b) that the approximation characteristics of anisotropic smoothness classes depend on the average smoothness $g(\mathbf{r})$, and optimal approximation methods depend on the anisotropy of classes and so on the vector $\mathbf{r}$. This motivated a study in Temlyakov (1988c) of the existence of an approximation method that can be used for all anisotropic smoothness classes. This problem concerns the existence of a universal method of approximation. We note that the universality concept in learning theory is very important and it is close to the concepts of adaptation and distribution-free estimation in non-parametric statistics (Györfy *et al.*, 2002; Binev *et al.*, 2005; Temlyakov, 2006).

5.2 The Orthowidths of the Classes $\mathbf{W}^r_{q,\alpha}$ and $\mathbf{H}^r_q$

The main purpose of this chapter is to show that it is natural to approximate functions in the classes $\mathbf{W}^r_{q,\alpha}$ and $\mathbf{H}^r_q$ by trigonometric polynomials in $\mathcal{T}(Q_n)$, that is, by trigonometric polynomials with harmonics in hyperbolic crosses.

In this section we consider the orthowidths. We saw above that the results for the classes $\mathbf{W}^r_{q,\alpha}$ and $\mathbf{H}^r_q$ essentially differ from the results for the univariate case and also for the multivariate classes $W^{\mathbf{r}}_{q,\alpha}$ and $H^{\mathbf{r}}_q$, both in the method of investigation and in the qualitatively different phenomena that are involved. This remark is true for the results of this section as well.

The idea of using an orthoprojector onto a finite-dimensional subspace as an approximating operator is very natural and is widely accepted in approximation theory and numerical analysis. In this section we consider the problem of optimizing over subspaces of fixed dimension the approximation of classes $\mathbf{W}^r_q$ and $\mathbf{H}^r_q$ by orthoprojectors onto these subspaces. The assumption that we can approximate functions by orthoprojectors gives a very convenient way to construct an approximant and also guarantees the stability of approximation in L_2. The stability property is important in numerical analysis. As for the corresponding results in Chapters 2 and 3, we shall see that the stability assumption itself results in the same optimal rates of approximation as the assumption that the operator is an orthoprojector. As in §2.1.2, along with the quantities $\varphi_m(F,L_p)$ we consider the quantities

$$\varphi^B_m(F,L_p) := \inf_{G\in\mathscr{L}_m(B)_p} \sup_{f\in F\cap D(G)} \|f-G(f)\|_p,$$

where $B \geq 1$ is a number and $\mathscr{L}_m(B)_p$ is the set of linear operators G with domains $D(G)$ containing all trigonometric polynomials, and whose ranges are contained in an m-dimensional subspace of L_p, such that $\|Ge^{i(\mathbf{k},\mathbf{x})}\|_2 \leq B$ for all $\mathbf{k}$.

We remark that $\mathscr{L}_m(1)_2$ contains the operators of orthogonal projection onto m-dimensional subspaces as well as operators given by multipliers $\{\lambda_l\}$ with $|\lambda_l| \leq 1$ for all l with respect to an orthonormal system of functions.

For numbers m and r and parameters $1 \leq q,p \leq \infty$ we define the functions

$$w(m,r,q,p) = \left(m^{-1}(\log m)^{d-1}\right)^{r-(1/q-1/p)_+}(\log m)^{(d-1)\xi(q,p)},$$

where $(a)_+ = \max(a,0)$ and

$$\xi(q,p) = \begin{cases} 0 & \text{for } 1<q\leq p<\infty;\ 1\leq p<q\leq\infty; \\ 1-1/q & \text{for } 1<q\leq\infty,\ p=\infty; \\ 1/p & \text{for } q=1,\ 1\leq p<\infty. \end{cases}$$

We also define

$$h(m,r,q,p) = \left(m^{-1}(\log m)^{d-1}\right)^{r-(1/q-1/p)_+}(\log m)^{(d-1)\eta(q,p)},$$

where

$$\eta(q,p) = \begin{cases} 1/p & \text{for } 1<q<p<\infty;\ q=1,\ 1\leq p<\infty; \\ 1 & \text{for } 1\leq q\leq\infty,\ p=\infty; \\ 1/2 & \text{for } 1\leq p\leq q\leq\infty,\ q\geq 2,\ p<\infty; \\ 1/q & \text{for } 1\leq p\leq q\leq 2. \end{cases}$$

We will prove the following assertions.

Theorem 5.2.1 *Suppose that* $r > (1/q-1/p)_+$ *and* $(q,p) \neq (1,1), (\infty,\infty), (1,\infty)$. *Then*

$$\varphi_m(\mathbf{W}^r_{q,\alpha},L_p) \asymp w(m,r,q,p),$$

and orthogonal projections onto the subspaces $\mathscr{T}(Q_n)$ of trigonometric polynomials with harmonics in step hyperbolic crosses with the corresponding n give the order of the quantities $\varphi_m(\mathbf{W}^r_{q,\alpha}, L_p)$.

Theorem 5.2.2 *Suppose that $r > (1/q - 1/p)_+$ and $(q,p) \neq (1,\infty)$. Then*

$$\varphi^B_m(\mathbf{W}^r_{q,\alpha}, L_p) \asymp w(m,r,q,p)$$

and subspaces, optimal in the sense of order, are given by $\mathscr{T}(Q_n)$ with the corresponding n.

Remark 5.2.3 The asymptotic estimate in Theorem 5.2.1 is also true for $(q,p) = (1,1), (\infty,\infty)$.

We shall prove this remark in Chapter 7 using the wavelets technique.

Theorems 5.2.1, 5.2.2 and Remark 5.2.3 show that in the cases under consideration the orders of the quantities $\varphi_m(\mathbf{W}^r_{q,\alpha}, L_p)$ and $\varphi^B_m(\mathbf{W}^r_{q,\alpha}, L_p)$ coincide and do not depend on α. Surprisingly, it turns out that in the case $q = 1$, $p = \infty$ the orders of the quantities $\varphi_m(\mathbf{W}^r_{1,\alpha}, L_\infty)$ depend on α, and for some values of α the orders of $\varphi_m(\mathbf{W}^r_{1,\alpha}, L_\infty)$ and $\varphi^B_m(\mathbf{W}^r_{1,\alpha}, L_\infty)$ coincide while for other values of α they do not coincide. This case has not been thoroughly studied. We present some results.

Theorem 5.2.4 *For $r > 1$ and arbitrary d,*

$$\varphi_m(\mathbf{W}^r_{1,1}, L_\infty) \asymp \varphi^B_m(\mathbf{W}^r_{1,1}, L_\infty) \asymp \left(m^{-1}(\log m)^{d-1}\right)^{r-1}$$

and, for $r > 1$ and $d = 2$,

$$\varphi_m(\mathbf{W}^r_{1,0}, L_\infty) \asymp m^{1-r}(\log m)^r,$$
$$\varphi^B_m(\mathbf{W}^r_{1,\alpha}, L_\infty) \asymp m^{1-r}(\log m)^{r-1} \qquad \text{for } B > B(r).$$

We summarize the above results on the behavior of the $\varphi_m(\mathbf{W}^r_{q,\alpha}, L_p)$ in Figure 5.1. They are in the form

$$\varphi_m(\mathbf{W}^r_{q,\alpha}, L_p) \asymp \left(m^{-1}(\log m)^{d-1}\right)^{r-(1/q-1/p)_+} (\log m)^{(d-1)\xi(q,p)}.$$

For $(q,p) \in [1,\infty]^2$, represented by a point $(1/q, 1/p) \in [0,1]^2$, we give the order of the $\varphi_m(\mathbf{W}^r_{q,\alpha}, L_p)$ by indicating the parameter $\xi(q,p)$ and we also give a reference to the theorem which establishes that relation. If the order of the $\varphi_m(\mathbf{W}^r_{q,\alpha}, L_p)$ is not known then we refer to the corresponding open problem.

We now present results for the $\mathbf{H}$ classes.

Theorem 5.2.5 *Suppose that $r > (1/q - 1/p)_+$ and $(q,p) \neq (1,1), (\infty,\infty)$. Then*

$$\varphi_m(\mathbf{H}^r_q, L_p) \asymp h(m,r,q,p)$$

and the subspaces that are optimal in the sense of order are given by $\mathscr{T}(Q_n)$ with the corresponding n.

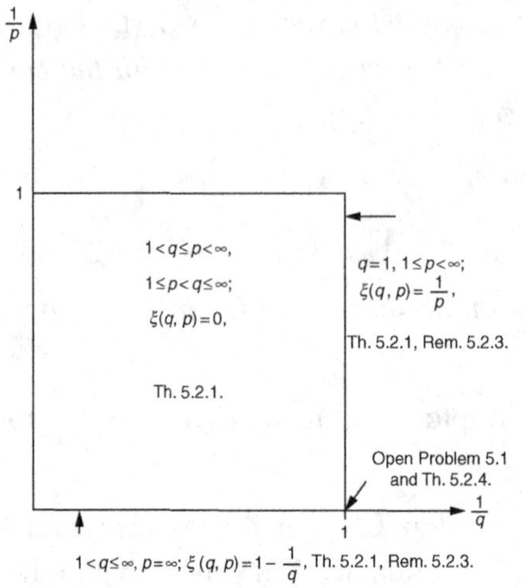

Figure 5.1 The orthowidths of the **W** classes.

Theorem 5.2.6 *Suppose that* $r > (1/q - 1/p)_+$. *Then*

$$\varphi_m^B(\mathbf{H}_q^r, L_p) \asymp h(m, r, q, p)$$

and the subspaces that are optimal in the sense of order are given by $\mathscr{T}(Q_n)$ *with the corresponding n.*

Remark 5.2.7 The asymptotic estimate in Theorem 5.2.5 is also true for $(q, p) = (1,1), (\infty, \infty)$.

We shall prove this remark in Chapter 7 using the wavelets technique.

Theorems 5.2.5, 5.2.6 and Remark 5.2.7 show that the order of the quantities $\varphi_m(\mathbf{H}_q^r, L_p)$ and $\varphi_m^B(\mathbf{H}_q^r, L_p)$ coincide in all cases.

We summarize the above results on the behavior of the $\varphi_m(\mathbf{H}_q^r, L_p)$ in Figure 5.2. They are in the form

$$\varphi_m(\mathbf{H}_q^r, L_p) \asymp \left(m^{-1}(\log m)^{d-1}\right)^{r-(1/q-1/p)_+} (\log m)^{(d-1)\eta(q,p)}.$$

For $(q, p) \in [1, \infty]^2$, represented by the point $(1/q, 1/p) \in [0, 1]^2$, we give the order of the $\varphi_m(\mathbf{H}_q^r, L_p)$ by indicating the parameter $\eta(q, p)$ and also give a reference to the theorem which establishes that relation.

We will proceed to a discussion of the results obtained. We demonstrate in this section that the upper estimates in Theorems 5.2.1, 5.2.2 and 5.2.5, 5.2.6 follow either directly from the results of Chapter 4 or from their proofs. We now discuss the lower estimates in these theorems. The following inequality holds for the quantities $\varphi_m(\mathbf{F}, L_p)$ and $\varphi_m^B(\mathbf{F}, L_p)$:

$$\varphi_m^B(\mathbf{F}, L_p) \le \varphi_m(\mathbf{F}, L_p).$$

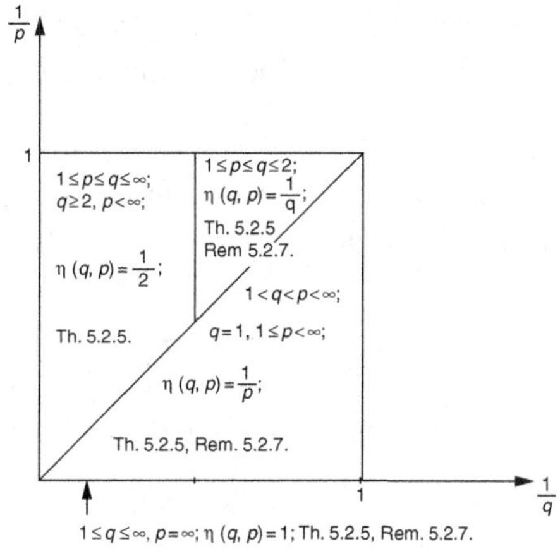

Figure 5.2 The orthowidths of the **H** classes.

Theorems 5.2.1, 5.2.2 and 5.2.5, 5.2.6 show that in all cases considered in these theorems the following order equality holds:

$$\varphi_m(\mathbf{F}^r_q, L_p) \asymp \varphi^B_m(\mathbf{F}^r_q, L_p),$$

where $\mathbf{F}^r_q$ denotes one of the classes $\mathbf{W}^r_{q,\alpha}$ or $\mathbf{H}^r_q$. Thus, the required lower estimates for $\varphi_m(\mathbf{F}^r_q, L_p)$ follow from the lower estimate for $\varphi^B_m(\mathbf{F}^r_q, L_p)$. Theorem 5.2.2 shows that for q, p such that $1 \le p \le q \le \infty$, the orders of the quantities $\varphi^B_m(\mathbf{W}^r_{q,\alpha}, L_p)$ do not depend on q and p. Consequently, to obtain the lower estimate in this case it suffices to prove the lower estimate for $q = \infty$ and $p = 1$.

Let us discuss the classes $\mathbf{H}^r_q$ in the case $1 \le p \le q \le \infty$. It turns out that the orders of the quantities $\varphi_m(\mathbf{H}^r_q, L_p)$ and $\varphi^B_m(\mathbf{H}^r_q, L_p)$ do not depend on q and p for $q \ge 2$, but do depend on q for $q < 2$. These two cases require different methods of investigation. We first discuss the case $q \ge 2$. In this case as for the classes $\mathbf{W}^r_{q,\alpha}$, it is sufficient to prove the required lower estimate for $q = \infty$ and $p = 1$. Theorem 5.2.6 shows that to prove the required lower estimates in the case $q < 2$ it is sufficient to prove for each $1 \le q < 2$ the estimate

$$\varphi^B_m(\mathbf{H}^r_q, L_1) \gg \left(m^{-1}(\log m)^{d-1}\right)^r (\log m)^{(d-1)/q}.$$

We proceed to the proof of the assertions which have been formulated. We first prove a number of auxiliary propositions, which will then be used in the proofs of the lower estimates in Theorems 5.2.2 and 5.2.6. In all these propositions we prove that for an arbitrary operator $G \in \mathscr{L}_m(B)_p$ there exist special trigonometric polynomials which are poorly approximated by the operator G. In the proofs of these propositions we shall assume that the operator G belongs to $\mathscr{L}_m(B)_2$. In the case $p \geq 2$ this follows from $G \in \mathscr{L}_m(B)_p$, but in the case $p < 2$ this is an additional assumption. We now prove that we can accept this assumption without loss of generality.

Remark 5.2.8 Suppose that $G \in \mathscr{L}_m(B)_p$ and consider the operator $A = V_{\mathbf{L}} G$, $\mathbf{L} = (L, \ldots, L)$. Then $A \in \mathscr{L}_m(B)_2$ and for any trigonometric polynomial with harmonics in $[-L, L]^d$ we have

$$\|t - At\|_p \leq 3^d \|t - Gt\|_p.$$

This remark follows from the properties of the de la Vallée Poussin kernels (see §3.2.3).

So, we suppose that $G \in \mathscr{L}_m(B)_2$ and that for all $\mathbf{k}$

$$G(e^{i(\mathbf{k},\mathbf{x})}) = \sum_{\mu=1}^{N} a_\mu^{\mathbf{k}} \psi_\mu(\mathbf{x}), \tag{5.2.1}$$

where N is the dimension of the range of the linear operator G and $\left\{\psi_\mu(\mathbf{x})\right\}_{\mu=1}^{N}$ is an orthonormal basis in this subspace.

It is clear, that for $N \leq m$ and for all $\mathbf{k}$,

$$\sum_{\mu=1}^{N} |a_\mu^{\mathbf{k}}|^2 \leq B^2 \tag{5.2.2}$$

and, for all μ,

$$\sum_{\mathbf{k}} \left|\hat{\psi}_\mu(\mathbf{k})\right|^2 \leq 1. \tag{5.2.3}$$

We first prove two lemmas which will be used in the proofs of Propositions 5.2.12–5.2.18 below.

Lemma 5.2.9 *Let A be a linear operator defined as follows: for any $\mathbf{k} \in \mathbb{Z}^d$,*

$$A\left(e^{i(\mathbf{k},\mathbf{x})}\right) = \sum_{\mu=1}^{N} a_\mu^{\mathbf{k}} \psi_\mu(\mathbf{x}),$$

where $\left\{\psi_\mu(\mathbf{x})\right\}_{\mu=1}^{N}$ is a given system of functions such that

$$\|\psi_\mu\|_2 \leq 1, \qquad \mu = 1, \ldots, N.$$

Then, for any trigonometric polynomial $t(\mathbf{x})$,

$$\min_{\mathbf{y}=\mathbf{x}} \operatorname{Re} At(\mathbf{x}-\mathbf{y}) \leq \operatorname{Re} \sum_{\mathbf{k}} \sum_{\mu=1}^{N} \hat{t}(\mathbf{k}) a_\mu^{\mathbf{k}} \hat{\psi}_\mu(\mathbf{k}) \leq \left(N \sum_{\mu=1}^{N} \sum_{\mathbf{k}} |a_\mu^{\mathbf{k}} \hat{t}(\mathbf{k})|^2\right)^{1/2}.$$

Proof We have

$$A\big(t(\mathbf{x}-\mathbf{y})\big)\Big|_{\mathbf{y}=\mathbf{x}} = A\left(\sum_{\mathbf{k}} \hat{t}(\mathbf{k})e^{i(\mathbf{k},\mathbf{x}-\mathbf{y})}\right)\Bigg|_{\mathbf{y}=\mathbf{x}}$$
$$= \sum_{\mathbf{k}} \hat{t}(\mathbf{k})e^{-i(\mathbf{k},\mathbf{x})} \sum_{\mu=1}^{N} a^{\mathbf{k}}_{\mu} \psi_{\mu}(\mathbf{x}).$$

Integrating both sides of this relation, we get

$$\min_{\mathbf{y}=\mathbf{x}} \operatorname{Re} A\big(t(\mathbf{x}-\mathbf{y})\big)$$
$$\le (2\pi)^{-d} \int_{\mathbb{T}^d} \operatorname{Re} A\big(t(\mathbf{x}-\mathbf{y})\big)\Big|_{\mathbf{y}=\mathbf{x}} d\mathbf{x}$$
$$= \operatorname{Re} \sum_{\mathbf{k}} \sum_{\mu=1}^{N} \hat{t}(\mathbf{k}) a^{\mathbf{k}}_{\mu} \hat{\psi}_{\mu}(\mathbf{k}) \le \sum_{\mu=1}^{N} \left(\sum_{\mathbf{k}} \big|\hat{t}(\mathbf{k}) a^{\mathbf{k}}_{\mu}\big|^2\right)^{1/2} \left(\sum_{\mathbf{k}} \big|\hat{\psi}_{\mu}(\mathbf{k})\big|^2\right)^{1/2}$$
$$\le \sum_{\mu=1}^{N} \left(\sum_{\mathbf{k}} \big|\hat{t}(\mathbf{k}) a^{\mathbf{k}}_{\mu}\big|^2\right)^{1/2} \le \left(N \sum_{\mu=1}^{N} \sum_{\mathbf{k}} \big|\hat{t}(\mathbf{k}) a^{\mathbf{k}}_{\mu}\big|^2\right)^{1/2}.$$

The lemma is proved. □

Lemma 5.2.10 *Let $G \in \mathscr{L}_m(B)_2$ be defined by* (5.2.1). *Then, for any two trigonometric polynomials $t(\mathbf{x})$ and $u(\mathbf{x})$, we have*

$$\min_{\mathbf{y}} \operatorname{Re}\langle G\big(t(\mathbf{x}+\mathbf{y})\big), u(\mathbf{x}+\mathbf{y})\rangle$$
$$\le B\left(\max_{\mathbf{k}\in\Phi} \big|\hat{t}(\mathbf{x})\big|\right)\left(\max_{\mathbf{k}\in\Phi} \big|\hat{u}(\mathbf{k})\big|\right) |\Phi|^{1/2} \left(\sum_{\mathbf{k}\in\Phi} \sum_{\mu=1}^{N} \big|\hat{\psi}_{\mu}(\mathbf{k})\big|^2\right)^{1/2},$$

where $\Phi := \big\{\mathbf{k} : \hat{t}(\mathbf{k})\hat{u}(\mathbf{k}) \ne 0\big\}$.

Proof Denote $\beta(\mathbf{y}) := \langle G\big(t(\mathbf{x}+\mathbf{y})\big), u(\mathbf{x}+\mathbf{y})\rangle$ and consider

$$J := (2\pi)^{-d} \int_{\mathbb{T}^d} \beta(\mathbf{y}) d\mathbf{y}.$$

We have

$$J = \sum_{\mu=1}^{N} \sum_{\mathbf{k}\in\Phi} \hat{t}(\mathbf{k}) a^{\mathbf{k}}_{\mu} \hat{\psi}_{\mu}(\mathbf{k}) \overline{\hat{u}(\mathbf{k})}.$$

Further,

$$
\begin{aligned}
|J| &\le \left(\max_{\mathbf{k}\in\Phi}|\hat{u}(\mathbf{k})|\right)\sum_{\mu=1}^{N}\sum_{\mathbf{k}\in\Phi}|\hat{t}(\mathbf{k})a_{\mu}^{\mathbf{k}}\hat{\psi}_{\mu}(\mathbf{k})| \\
&\le \left(\max_{\mathbf{k}\in\Phi}|\hat{u}(\mathbf{k})|\right)\sum_{\mathbf{k}\in\Phi}|\hat{t}(\mathbf{k})|\left(\sum_{\mu=1}^{N}|a_{\mu}^{\mathbf{k}}|^2\right)^{1/2}\left(\sum_{\mu=1}^{N}|\hat{\psi}_{\mu}(\mathbf{k})|^2\right)^{1/2} \\
&\le B\left(\max_{\mathbf{k}\in\Phi}|\hat{u}(\mathbf{k})|\right)\sum_{\mathbf{k}\in\Phi}|\hat{t}(\mathbf{k})|\left(\sum_{\mu=1}^{N}|\hat{\psi}_{\mu}(\mathbf{k})|^2\right)^{1/2} \\
&\le B\left(\max_{\mathbf{k}\in\Phi}|\hat{u}(\mathbf{k})|\right)\max_{\mathbf{k}\in\Phi}|\hat{t}(\mathbf{k})|\,|\Phi|^{1/2}\left(\sum_{\mathbf{k}\in\Phi}\sum_{\mu=1}^{N}|\hat{\psi}_{\mu}(\mathbf{k})|^2\right)^{1/2},
\end{aligned}
$$

which gives the lemma. □

Remark 5.2.11 The following estimate can be proved in the same way as Lemma 5.2.10:

$$
\min_{\mathbf{y}}\operatorname{Re}\langle G(t(\mathbf{x}+\mathbf{y})),u(\mathbf{x}+\mathbf{y})\rangle \le B\left(\max_{\mathbf{k}\in\Phi}|\hat{t}(\mathbf{k})|\right)\|u\|_2\left(\sum_{\mathbf{k}\in\Phi}\sum_{\mu=1}^{N}|\hat{\psi}_{\mu}(\mathbf{k})|^2\right)^{1/2}.
$$

Proof Indeed, we have

$$
\begin{aligned}
|J| &\le \left(\max_{\mathbf{k}\in\Phi}|\hat{t}(\mathbf{k})|\right)\left(\sum_{\mathbf{k}\in\Phi}|\hat{u}(\mathbf{k})|\sum_{\mu=1}^{N}|a_{\mu}^{\mathbf{k}}\hat{\psi}_{\mu}(\mathbf{k})|\right) \\
&\le \left(\max_{\mathbf{k}\in\Phi}|\hat{t}(\mathbf{k})|\right)\|u\|_2\left(\sum_{\mathbf{k}\in\Phi}\left(\sum_{\mu=1}^{N}|a_{\mu}^{\mathbf{k}}\hat{\psi}_{\mu}(\mathbf{k})|\right)^2\right)^{1/2} \\
&\le B\left(\max_{\mathbf{k}\in\Phi}|\hat{t}(\mathbf{k})|\right)\|u\|_2\left(\sum_{\mathbf{k}\in\Phi}\sum_{\mu=1}^{N}|\hat{\psi}_{\mu}(\mathbf{k})|^2\right)^{1/2}.
\end{aligned}
$$

This proves the remark. □

We proceed to the construction of examples. Let

$$
\Omega_n := \bigcup_{\mathbf{s}\in\theta_n}\rho(\mathbf{s}), \qquad \Omega_n^+ := \bigcup_{\mathbf{s}\in\theta_n}\rho(\mathbf{s})^+, \qquad \rho(\mathbf{s})^+ := \rho(\mathbf{s})\cap\mathbb{N}^d,
$$

where $\theta_n = \{\mathbf{s} : \|\mathbf{s}\|_1 = n;\ s_j \text{ are natural numbers},\ j = 1,\dots,d\}$. Then

$$
|\theta_n| \asymp n^{d-1}, \qquad |\Omega_n| \asymp 2^n n^{d-1}, \qquad |\Omega_n^+| = 2^{-d}|\Omega_n|. \tag{5.2.4}
$$

Proposition 5.2.12 *Let $G \in \mathscr{L}_m(B)_1$ and n be such that*

$$|\Omega_{n-1}| < 4B^2 m \leq |\Omega_n|.$$

Then there exists a $\mathbf{k}^0 \in \Omega_n$ such that

$$\left\| e^{i(\mathbf{k}^0,\mathbf{x})} - G(e^{i(\mathbf{k}^0,\mathbf{x})}) \right\|_1 \geq 3^{-d}/2.$$

Proof For $G \in \mathscr{L}_m(B)_2$ we consider approximations of the functions $e^{i(\mathbf{k},\mathbf{x})}$, $\mathbf{k} \in \Omega_n$. Denote

$$\beta_{\mathbf{k}} := \langle G(e^{i(\mathbf{k},\mathbf{x})}), e^{i(\mathbf{k},\mathbf{x})} \rangle.$$

We have

$$\beta_{\mathbf{k}} = \sum_{\mu=1}^{N} a_\mu^{\mathbf{k}} \hat{\psi}_\mu(\mathbf{k})$$

and

$$|\beta_{\mathbf{k}}|^2 \leq B^2 \sum_{\mu=1}^{N} |\hat{\psi}_\mu(\mathbf{k})|^2.$$

Consequently, by (5.2.3)

$$\sum_{\mathbf{k} \in \Omega_n} |\beta_{\mathbf{k}}|^2 \leq B^2 N.$$

Taking into account the definition of n we find that there is a $\mathbf{k}^0 \in \Omega_n$ such that $|\beta_{\mathbf{k}^0}| \leq 1/2$. Then

$$\begin{aligned} 1/2 \leq |1 - \beta_{\mathbf{k}^0}| &= \left| \langle e^{i(\mathbf{k}^0,\mathbf{x})} - G(e^{i(\mathbf{k}^0,\mathbf{x})}), e^{i(\mathbf{k}^0,\mathbf{x})} \rangle \right| \\ &\leq \left\| e^{i(\mathbf{k}^0,\mathbf{x})} - G(e^{i(\mathbf{k}^0,\mathbf{x})}) \right\|_1, \end{aligned}$$

which, by Remark 5.2.3, gives the required estimate for $G \in \mathscr{L}_m(B)_1$. □

Proposition 5.2.13 *Suppose that $G \in \mathscr{L}_m(B)_p$, $1 < p \leq \infty$. Then there exists an n such that $|\Omega_n| \leq C(B,d)m$ and a $\mathbf{y}^*$ such that*

$$\left\| h(\mathbf{x} - \mathbf{y}^*) - G\big(h(\mathbf{x} - \mathbf{y}^*)\big) \right\|_p \gg \begin{cases} 2^{n(1-1/p)} n^{(d-1)/p}, & 1 < p < \infty, \\ 2^n n^{d-1}, & p = \infty, \end{cases}$$

where

$$h(\mathbf{x}) := \sum_{\mathbf{s}\in\theta_n} h_{\mathbf{s}}(\mathbf{x}),$$

$$h_{\mathbf{s}}(\mathbf{x}) := e^{i(\mathbf{k}^s,\mathbf{x})} \prod_{j=1}^{d} \mathcal{K}_{2^{s_j-2}-1}(x_j),$$

$$k_j^{\mathbf{s}} := \begin{cases} 2^{s_j-1} + 2^{s_j-2} & \text{for } s_j \geq 2, \\ 1 & \text{for } s_j = 1,\ j = 1,\dots,d, \end{cases}$$

and where $\mathcal{K}_\mu(t)$ *is the Fejér kernel of order* μ *and* $\mathcal{K}_l(t) \equiv 1$ *for* $l < 1$.

Proof We first consider the case $p = \infty$. Let us estimate the quantity

$$J := \sup_{\mathbf{y}} \big\| h(\mathbf{x}-\mathbf{y}) - G\big(h(\mathbf{x}-\mathbf{y})\big) \big\|_\infty \geq h(\mathbf{0}) - \min_{\mathbf{y}=\mathbf{x}} \operatorname{Re} G\big(h(\mathbf{x}-\mathbf{y})\big).$$

Applying Lemma 5.2.9 with $A = G$, $t(\mathbf{x}) = h(\mathbf{x})$ and using the bound $|\hat{h}(\mathbf{k})| \leq 1$ we obtain on the one hand

$$\min_{\mathbf{y}=\mathbf{x}} \operatorname{Re} G\big(h(\mathbf{x}-\mathbf{y})\big) \leq B\big(m|\Omega_n|\big)^{1/2}.$$

On the other hand,

$$h(\mathbf{0}) \geq C(d)|\Omega_n|, \qquad C(d) > 0.$$

We may choose n such that

$$C(d)|\Omega_{n-1}|^{1/2} \leq 2Bm^{1/2} < C(d)|\Omega_n|^{1/2}.$$

Then for J we get

$$J \geq B\big(m|\Omega_n|\big)^{1/2} \gg 2^n n^{d-1},$$

which gives the proposition for $p = \infty$.

The case $1 < p < \infty$ can be derived from the above with $A = (S_{Q_n} - S_{Q_{n-1}})G$, by means of the inequalities (see Corollary A.3.5)

$$\|S_{Q_l}\|_{p\to p} \leq C(d,p), \qquad 1 < p < \infty;$$

and the Nikol'skii inequality: for any $t \in \mathcal{T}(Q_n)$ (see Theorem 4.3.16),

$$\|t\|_\infty \ll 2^{n/p} n^{(d-1)(1-1/p)} \|t\|_p, \qquad 1 \leq p < \infty.$$

□

We proceed to the next propositions and use the Telyakovskii polynomials. As in §4.3.1, let

$$a(n,x,\alpha) := \cos(nx+\alpha\pi/2) + \cos\big((n+1)x+\alpha\pi/2\big) - \cos(2nx+\alpha\pi/2) - \cos\big((2n+1)x+\alpha\pi/2\big),$$
$$a(0,x,\alpha) := 0,$$

and

$$\varphi_L(\mathbf{x},\alpha) := U_{2L}^{-1}(\mathbf{x},\alpha) - U_L^{-1}(\mathbf{x},\alpha) \tag{5.2.5}$$

with $U_N^r(\mathbf{x},\alpha)$ defined by (4.3.3).

By Lemma 4.3.2

$$\|\varphi_L\|_\infty \le C(d). \tag{5.2.6}$$

Proposition 5.2.14 *Let $G \in \mathscr{L}_m(B)_p$, $1 < p < \infty$. Then there exists an n with $|\Omega_n| < C(B,d)m$ and $\mathbf{y}^*, \mathbf{s}^*, \mathbf{s}^* \in \theta_n$ such that*

$$\big\|h_{\mathbf{s}^*}(\mathbf{x}-\mathbf{y}^*) - G\big(h_{\mathbf{s}^*}(\mathbf{x}-\mathbf{y}^*)\big)\big\|_p \gg 2^{n(1-1/p)},$$

where $h_{\mathbf{s}}(\mathbf{x})$ is the same as in Proposition 5.2.13 .

Proof Let m and n be the same as in Proposition 5.2.13. For a given operator $G \in \mathscr{L}_m(B)_p$ and for $\mathbf{s} \in \theta_n$ we consider the operators

$$G_{\mathbf{s}}(e^{i(\mathbf{k},\mathbf{x})}) := \sum_{\mu=1}^{N} a_\mu^{\mathbf{k}} \delta_{\mathbf{s}}(\psi_\mu, \mathbf{x}),$$

where ψ_μ is from (5.2.1), and the quantities

$$J_{\mathbf{s}} := \sup_{\mathbf{y}} \big\|h_{\mathbf{s}}(\mathbf{x}-\mathbf{y}) - G_{\mathbf{s}}\big(h_{\mathbf{s}}(\mathbf{x}-\mathbf{y})\big)\big\|_\infty.$$

Denote

$$b_{\mathbf{s}} := \min_{\mathbf{y}=\mathbf{x}} \operatorname{Re} G_{\mathbf{s}}\big(h_{\mathbf{s}}(\mathbf{x}-\mathbf{y})\big);$$

then

$$J_{\mathbf{s}} \ge h_{\mathbf{s}}(\mathbf{0}) - b_{\mathbf{s}}.$$

By Lemma 5.2.9, for $b_{\mathbf{s}}$ we obtain the estimates

$$b_{\mathbf{s}} \le \operatorname{Re} \sum_{\mathbf{k}\in\rho(\mathbf{s})} \sum_{\mu=1}^{N} \hat{h}_{\mathbf{s}}(\mathbf{k}) a_\mu^{\mathbf{k}} \hat{\psi}_\mu(\mathbf{k}), \qquad \mathbf{s} \in \theta_n,$$

which give

$$\sum_{\mathbf{s}\in\theta_n} b_{\mathbf{s}} \le \operatorname{Re} \sum_{\mathbf{k}\in\Omega_n} \sum_{\mu=1}^{N} \hat{h}(\mathbf{k}) a_\mu^{\mathbf{k}} \hat{\psi}_\mu(\mathbf{k}) \le B\big(m|\Omega_n|\big)^{1/2}.$$

Let n be the smallest number satisfying

$$h(\mathbf{0}) > 2B\big(m|\Omega_n|\big)^{1/2}.$$

Using $h(\mathbf{0}) = \sum_{\mathbf{s}\in\theta_n} h_{\mathbf{s}}(\mathbf{0})$, we obtain, for some $\mathbf{s}^* \in \theta_n$,

$$h_{\mathbf{s}^*}(\mathbf{0}) - b_{\mathbf{s}^*} \ge B\big(m|\Omega_n|\big)^{1/2} \,/\, |\theta_n| \asymp 2^n.$$

Thus, for some $\mathbf{y}^*$,

$$\big\| h_{\mathbf{s}^*}(\mathbf{x}-\mathbf{y}^*) - G_{\mathbf{s}^*}\big(h_{\mathbf{s}^*}(\mathbf{x}-\mathbf{y}^*)\big)\big\|_\infty \gg 2^n.$$

Using the Nikol'skii inequality we get, for $1 \le p < \infty$,

$$\big\| h_{\mathbf{s}^*}(\mathbf{x}-\mathbf{y}^*) - G_{\mathbf{s}^*}\big(h_{\mathbf{s}^*}(\mathbf{x}-\mathbf{y}^*)\big)\big\|_p \gg 2^{n(1-1/p)}.$$

To conclude the proof, it remains to remark that, for $1 < p < \infty$,

$$\big\| h_{\mathbf{s}}(\mathbf{x}-\mathbf{y}) - G_s\big(h_{\mathbf{s}}(\mathbf{x}-\mathbf{y})\big)\big\|_p \ll \big\| h_{\mathbf{s}}(\mathbf{x}-\mathbf{y}) - G\big(h_{\mathbf{s}}(\mathbf{x}-\mathbf{y})\big)\big\|_p.$$

□

Proposition 5.2.15 *Let $G \in \mathscr{L}_m(B)_\infty$ and*

$$f_r(\mathbf{x}, L, \alpha) := F_r(\mathbf{x}, \alpha) * \varphi_L(\mathbf{x}, \alpha),$$

where $F_r(\mathbf{x},\alpha)$ is the multivariate Bernoulli kernel and $\varphi_L(\mathbf{x},\alpha)$ is defined by (5.2.5). Then there exist L and $\mathbf{y}^*$ such that

$$\big\| f_r(\mathbf{x}-\mathbf{y}^*, L, \alpha) - G\big(f_r(\mathbf{x}-\mathbf{y}^*, L, \alpha)\big)\big\|_\infty \gg m^{-r}(\log m)^{(d-1)(r+1)}.$$

Proof It is easy to see that

$$f_r(\mathbf{0}, L, \alpha) \ge C(r,d) L^{-r} (\log L)^{d-1}, \qquad C(r,d) > 0. \tag{5.2.7}$$

Applying Lemma 5.2.9 with $A = G$ and $t(\mathbf{x}) = f_r(\mathbf{x}, L, \alpha)$ we get

$$\min_{\mathbf{y}=\mathbf{x}} \operatorname{Re} G\big(f_r(\mathbf{x}-\mathbf{y}, L, \alpha)\big) \le BC(d)\big(N\big|\Gamma(c(d)L)\big|\big)^{1/2} L^{-r-1}. \tag{5.2.8}$$

We choose L such that

$$2BC(d)\big(N\big|\Gamma(c(d)L)\big|\big)^{1/2} L^{-1} \le C(r,d)(\log L)^{d-1},$$
$$L(\log L)^{d-1} \ll N.$$

Then from (5.2.7) and (5.2.8) it follows that there exists a $\mathbf{y}^*$ such that

$$\big\|f_r(\mathbf{x}-\mathbf{y}^*,L,\alpha)-G\big(f_r(\mathbf{x}-\mathbf{y}^*,L,\alpha)\big)\big\|_\infty \gg m^{-r}(\log m)^{(d-1)(r+1)}.$$

This completes the proof of the proposition. □

We note that in the proof of Propositions 5.2.13 and 5.2.14 only Lemma 5.2.9 is used. We proceed to construct examples in which Lemma 5.2.10 is used.

Proposition 5.2.16 *Let $G \in \mathscr{L}_m(B)_p$, $1 \le p < \infty$. Then there exist $\mathbf{y}^*$ and $\mathbf{M} = (M,\dots,M)$ such that, for $r > 1-1/p$,*

$$\big\|V_{\mathbf{M}}\big(F_r(\mathbf{x}+\mathbf{y}^*,\alpha)\big)-G\big(V_{\mathbf{M}}\big(F_r(\mathbf{x}+\mathbf{y}^*,\alpha)\big)\big)\big\|_p$$
$$\gg m^{-r+1-1/p}(\log m)^{(d-1)(r-1+2/p)}.$$

Proof We first consider the case $1 < p < \infty$. Let

$$D_n(\mathbf{x},\alpha) := 2^d \sum_{\mathbf{k}\in\Omega_n^+} \prod_{j=1}^d \cos(k_j x_j - \alpha_j\pi/2).$$

Then, by Lemma 4.2.1, for $1 < q < \infty$,

$$\big\|D_n(\mathbf{x},\alpha)\big\|_q \ll 2^{n(1-1/q)} n^{(d-1)/q}. \tag{5.2.9}$$

For $G \in \mathscr{L}_m(B)_2$ we consider the function

$$\beta(\mathbf{y}) := \big\langle G\big(V_{\mathbf{M}}\big(F_r(\mathbf{x}+\mathbf{y},\alpha)\big)\big), D_n(\mathbf{x}+\mathbf{y},\alpha)\big\rangle.$$

By Lemma 5.2.10 we have on the one hand

$$\min_{\mathbf{y}} \operatorname{Re}\beta(\mathbf{y}) \le B2^{r(d-n)}\big(|\Omega_n|N\big)^{1/2}.$$

On the other hand, for $M \ge 2^n$,

$$\big\langle V_{\mathbf{M}}\big(F_r(\mathbf{x}+\mathbf{y},\alpha)\big), D_n(\mathbf{x}+\mathbf{y},\alpha)\big\rangle = \sum_{\mathbf{k}\in\Omega_n} \nu(\mathbf{k})^{-r} \ge 2^{-rn}|\Omega_n|.$$

Choosing n such that

$$|\Omega_{n-1}|^{1/2} \le 2B2^{rd}N^{1/2} < |\Omega_n|^{1/2},$$

we obtain for some $\mathbf{y}^*$

$$J = \operatorname{Re}\big\langle V_{\mathbf{M}}\big(F_r(\mathbf{x}+\mathbf{y}^*,\alpha)\big) - G\big(V_{\mathbf{M}}\big(F_r(\mathbf{x}+\mathbf{y}^*,\alpha)\big)\big), D_n(\mathbf{x}+\mathbf{y}^*,\alpha)\big\rangle$$
$$\gg 2^{-n(r-1)} n^{d-1}. \tag{5.2.10}$$

Further,

$$J \le \big\|V_{\mathbf{M}}\big(F_r(\mathbf{x}+\mathbf{y}^*,\alpha)\big) - G\big(V_{\mathbf{M}}\big(F_r(\mathbf{x}+\mathbf{y}^*,\alpha)\big)\big)\big\|_p \big\|D_n(\mathbf{x}+\mathbf{y}^*,\alpha)\big\|_{p'}, \tag{5.2.11}$$

where $p' = p/(p-1)$. The required estimate follows from (5.2.9)–(5.2.11).

We now consider the case $p = 1$. We use the function $\varphi_L(\mathbf{x}, -\alpha)$ defined by (5.2.5). By Lemma 5.2.10 we have, for $G \in \mathscr{L}_m(B)_2$,

$$\min_{\mathbf{y}} \operatorname{Re}\langle G\big(V_{\mathbf{M}}\big(F_r(\mathbf{x}+\mathbf{y},\alpha)\big)\big), \varphi_L(\mathbf{x}+\mathbf{y},-\alpha)\rangle$$
$$\le BC(d)L^{-r-1}\big(|\Gamma(c(d)L)|N\big)^{1/2}. \tag{5.2.12}$$

However, it is easy to see that, for $M \ge L$,

$$\langle V_{\mathbf{M}}\big(F_r(\mathbf{x}+\mathbf{y},\alpha)\big), \varphi_L(\mathbf{x}+\mathbf{y},-\alpha)\rangle \ge C(d,r)L^{-r}(\log L)^{d-1}. \tag{5.2.13}$$

Choosing L such that

$$2BC(d)\big|\Gamma(c(d)L)\big|^{1/2}N^{1/2} \le C(d,r)L(\log L)^{d-1},$$
$$L(\log L)^{d-1} \ll N,$$

we obtain, for some $\mathbf{y}^*$,

$$\operatorname{Re}\langle V_M\big(F_r(\mathbf{x}+\mathbf{y}^*,\alpha)\big) - G\big(V_{\mathbf{M}}\big(F_r(\mathbf{x}+\mathbf{y}^*,\alpha)\big)\big), \varphi_L(\mathbf{x}+\mathbf{y}^*,-\alpha)\rangle$$
$$\gg m^{-r}(\log m)^{(d-1)(r+1)}. \tag{5.2.14}$$

Further,

$$\big|\langle V_{\mathbf{M}}\big(F_r(\mathbf{x}+\mathbf{y}^*,\alpha)\big) - G\big(V_{\mathbf{M}}\big(F_r(\mathbf{x}+\mathbf{y}^*,\alpha)\big)\big), \varphi_L(\mathbf{x}+\mathbf{y}^*,-\alpha)\rangle\big|$$
$$\le \big\|V_{\mathbf{M}}\big(F_r(\mathbf{x}+\mathbf{y}^*,\alpha)\big) - G\big(V_{\mathbf{M}}\big(F_r(\mathbf{x}+\mathbf{y}^*,\alpha)\big)\big)\big\|_1 \big\|\varphi_L(\mathbf{x}+\mathbf{y}^*,-\alpha)\big\|_\infty. \tag{5.2.15}$$

From (5.2.14) and (5.2.15), using the estimate (5.2.6), we get the required inequality.

Applying Remark 5.2.8 concludes the proof of the proposition. □

Proposition 5.2.17 *Let $G \in \mathscr{L}_m(B)_1$. Then there exist an n and $\theta_n^1 \subset \theta_n$ such that $|\Omega_n| < C(B,d)m$, $|\theta_n^1| \ge |\theta_n|/2$ and in each $\rho(\mathbf{s})$ with $\mathbf{s} \in \theta_n^1$ there is a vector $\mathbf{k}^{\mathbf{s}} \in \rho(\mathbf{s})$ for which, for the function*

$$g(\mathbf{x}) := \sum_{\mathbf{s}\in\theta_n^1} e^{i(\mathbf{k}^s,\mathbf{x})},$$

there exists a $\mathbf{y}^*$ such that

$$\big\|g(\mathbf{x}+\mathbf{y}^*) - G\big(g(\mathbf{x}+\mathbf{y}^*)\big)\big\|_1 \gg (\log m)^{(d-1)/2}.$$

Proof For $G \in \mathscr{L}_m(B)_2$ we consider the quantities

$$\rho_{\mathbf{s}} := \sum_{\mathbf{k}\in\rho(\mathbf{s})} \sum_{\mu=1}^{N} |\hat{\psi}_\mu(\mathbf{k})|^2.$$

It follows from (5.2.3) that

$$\sum_{\mathbf{s}\in\theta_n} \rho_{\mathbf{s}} \le N. \tag{5.2.16}$$

Let $\theta_n^1 := \{\mathbf{s}\in\theta_n : \rho_{\mathbf{s}} \le 2N/|\theta_n|\}$. From (5.2.16) we have

$$|\theta_n^1| \ge |\theta_n|/2.$$

In each $\rho(\mathbf{s})$ with $\mathbf{s}\in\theta_n^1$ we choose $\mathbf{k}^{\mathbf{s}}$ such that

$$\sum_{\mu=1}^{N} |\hat{\psi}_\mu(\mathbf{k}^{\mathbf{s}})|^2 \le \rho_{\mathbf{s}} \,/\, |\rho(\mathbf{s})| \le 2N \,/\, |\Omega_n|. \tag{5.2.17}$$

Further, we consider

$$\beta(\mathbf{y}) := \langle G(g(\mathbf{x}+\mathbf{y})), \overline{\operatorname{sign} g(\mathbf{x}+\mathbf{y})}\rangle.$$

By Remark 5.2.11,

$$\min_{\mathbf{y}} \operatorname{Re}\beta(\mathbf{y}) \le B\left(\sum_{\mathbf{s}\in\theta_n^1}\sum_{\mu=1}^{N} |\hat{\psi}_\mu(\mathbf{k}^{\mathbf{s}})|^2\right)^{1/2};$$

taking into account (5.2.17) we continue the estimate:

$$\le B|\theta_n^1|^{1/2}\left(2N|\Omega_n|^{-1}\right)^{1/2}. \tag{5.2.18}$$

We now estimate $\|g\|_1$ from below. We have

$$\|g\|_2 = |\theta_n^1|^{1/2}, \qquad \|g\|_4 \le C_d|\theta_n^1|^{1/2}. \tag{5.2.19}$$

Further,

$$\|g\|_2 \le \|g\|_1^{1/3}\|g\|_4^{2/3}. \tag{5.2.20}$$

From (5.2.19) and (5.2.20) we get

$$\|g\|_1 \ge C(d)|\theta_n^1|^{1/2}. \tag{5.2.21}$$

Let n be such that

$$4B^2N|\Omega_n|^{-1} < C(d)^2 \le 4B^2N|\Omega_{n-1}|^{-1}.$$

Then from (5.2.21) and (5.2.18) we obtain, for some $\mathbf{y}^*$, on the one hand

$$J = \left|\langle (g(\mathbf{x}+\mathbf{y}^*) - G(g(\mathbf{x}+\mathbf{y}^*)), \overline{\operatorname{sign} g(\mathbf{x}+\mathbf{y}^*)}\rangle\right| \gg |\theta_n^1|^{1/2}. \tag{5.2.22}$$

On the other hand,

$$J \le \left\| g(\mathbf{x}+\mathbf{y}^*) - G\big(g(\mathbf{x}+\mathbf{y}^*)\big) \right\|_1. \tag{5.2.23}$$

Comparing (5.2.22) and (5.2.23) we obtain the proposition for $G \in \mathscr{L}_m(B)_2$. To conclude the proof, it remains to use Remark 5.2.8. □

We proceed to the last proposition, using the construction and notation from §4.2.4.

Proposition 5.2.18 *Let $G \in \mathscr{L}_m(B)_1$. There exist an n and a set $\theta_n^2 \subset \overline{\theta}_n$ such that $|\Omega_n| < C(B,d)m$, $|\theta_n^2| \ge |\overline{\theta}_n|/2$ and in each $\rho(\mathbf{s})$ with $\mathbf{s} \in \theta_n^2$ there is a cube $\Delta_\mathbf{s} \subset \rho(\mathbf{s})^+$ with center $\mathbf{k}^\mathbf{s}$ and edge $2u$ such that, for the function,*

$$g(\mathbf{x}) := \sum_{\mathbf{s}\in\theta_n^2} e^{i(\mathbf{k}^s,\mathbf{x})} \mathscr{K}_\mathbf{u}(\mathbf{x}-\mathbf{x}^\mathbf{s}), \qquad \mathbf{u} = (u,\dots,u),$$

there exists a $\mathbf{y}^*$ for which

$$\left\| g(\mathbf{x}+\mathbf{y}^*) - G\big(g(\mathbf{x}+\mathbf{y}^*)\big) \right\|_1 \gg (\log m)^{d-1}.$$

Proof There exists a constant $0 < C_d \le 1$ such that

$$\|\mathscr{K}_\mathbf{u}\|_2^2 \ge C_d u^d.$$

Let

$$\rho_\mathbf{s}^+ := \sum_{\mathbf{k}\in\rho(\mathbf{s})^+} \sum_{\mu=1}^{N} |\hat{\psi}_\mu(\mathbf{k})|^2$$

and let

$$\theta_n^2 := \left\{ \mathbf{s} \in \overline{\theta}_n : \rho_\mathbf{s} \le 2N/|\overline{\theta}_n| \right\}.$$

From (5.2.16) it follows that

$$|\theta_n^2| \ge |\overline{\theta}_n|/2$$

and, in each $\rho(\mathbf{s})^+$ with $\mathbf{s} \in \theta_n^2$, there is a cube $\Delta_\mathbf{s}$ with edge $2u$ such that

$$\sum_{\mathbf{k}\in\Delta_\mathbf{s}} \sum_{\mu=1}^{N} |\hat{\psi}_\mu(\mathbf{k})|^2 \le \frac{|\Delta_\mathbf{s}|}{|\rho(\mathbf{s})^+|} \rho_\mathbf{s}^+ \le \frac{|\Delta_\mathbf{s}|}{|\rho(\mathbf{s})^+|} \frac{2N}{|\overline{\theta}_n|}. \tag{5.2.24}$$

We consider the function

$$\beta(\mathbf{y}) := \langle G\big(g(\mathbf{x}+\mathbf{y})\big), g(\mathbf{x}+\mathbf{y}) \rangle.$$

Let us denote

$$W_n = \bigcup_{\mathbf{s}\in\theta_n^2} \Delta_\mathbf{s}.$$

Then, by Lemma 5.2.10, we have on the one hand

$$\min_{\mathbf{y}} \operatorname{Re}\beta(\mathbf{y}) \leq B|W_n|^{1/2}\left(|\theta_n^2|\frac{|\Delta_{\mathbf{s}}|}{|\rho(\mathbf{s})^+|}\frac{2N}{|\overline{\theta}_n|}\right)^{1/2}. \tag{5.2.25}$$

On the other hand,

$$\langle g(\mathbf{x}+\mathbf{y}), g(\mathbf{x}+\mathbf{y})\rangle = \sum_{\mathbf{s}\in\theta_n^2} \|\mathscr{K}_{\mathbf{u}}\|_2^2 \geq C_d|\theta_n^2|u^d. \tag{5.2.26}$$

We choose n such that $2^n n^{d-1} \asymp N$ and such that the right-hand side of (5.2.26) is greater than the right-hand side of (5.2.25) multiplied by two. Then for this n there is a $\mathbf{y}^*$ such that

$$J = \left|\langle g(\mathbf{x}+\mathbf{y}^*) - G\big(g(\mathbf{x}+\mathbf{y}^*)\big), g(\mathbf{x}+\mathbf{y})\rangle\right| \gg n^{2(d-1)}. \tag{5.2.27}$$

However, we also have,

$$J \leq \left\|g(\mathbf{x}+\mathbf{y}^*) - G\big(g(\mathbf{x}+\mathbf{y}^*)\big)\right\|_1 \|g\|_\infty. \tag{5.2.28}$$

In the same way as in (4.2.16) we deduce that

$$\|g\|_\infty \leq C(d)u^d \ll n^{d-1}. \tag{5.2.29}$$

Combining (5.2.27)–(5.2.29) we get the required estimate. □

We proceed to the proof of the theorems.

Proof of Theorems 5.2.1 and 5.2.2 We first prove the upper estimates. In the cases $1 < q, p < \infty$, $q = \infty$, $1 \leq p < \infty$, and $p = 1$, $1 < q < \infty$ the required upper estimates follow from Theorems 4.4.9 and 4.4.11, the boundedness of the operator S_{Q_n} as an operator from L_p to L_p for $1 < p < \infty$, and the inequality $\|f\|_p \leq \|f\|_q$ for $1 \leq p \leq q \leq \infty$. Regarding the upper estimates in Theorem 5.2.1, it remains to consider the case $1 < q < \infty$ and $p = \infty$.

Let $f \in \mathbf{W}^r_{q,\alpha}$; then, by Theorem 4.4.9,

$$\left\|S_{Q_{l+1}}(f) - S_{Q_l}(f)\right\|_q \ll 2^{-rl}. \tag{5.2.30}$$

Further,

$$\left\|f - S_{Q_n}(f)\right\|_\infty \leq \sum_{l=n}^{\infty}\left\|S_{Q_{l+1}}(f) - S_{Q_l}(f)\right\|_\infty.$$

Applying Theorem 4.3.16 and (5.2.30) we continue the estimate:

$$\ll \sum_{l=n}^{\infty} 2^{l/q} l^{(d-1)(1-1/q)} 2^{-rl} \ll 2^{-n(r-1/q)} n^{(d-1)(1-1/q)}.$$

The upper estimates in Theorem 5.2.1 are now proved. In the case $(q,p) \neq$

$(1,1),(\infty,\infty)$ these estimates imply the upper estimates in Theorem 5.2.2. In the case $(q,p)=(1,1),(\infty,\infty)$ the upper estimates in Theorem 5.2.2 follow from Theorem 5.2.6 and from the inclusion $\mathbf{W}^r_{q,\alpha} \subset \mathbf{H}^r_q C(r,q)$. Thus, it is sufficient to prove the lower estimates in Theorem 5.2.2. To prove these estimates, the functions constructed in the Propositions 5.2.12–5.2.18 will be applied as follows. If the function g in the proposition does not belong to the class involved, we consider the function

$$f(\mathbf{x}) = g(\mathbf{x}) \,/\, \|g^{(r)}(\mathbf{x},\alpha)\|_q,$$

which belongs to the class $\mathbf{W}^r_{q,\alpha}$. Then, to obtain the required estimate we need to estimate $\|g^{(r)}(\mathbf{x},\alpha)\|_q$ from above.

We now indicate which propositions are used for specific q,p.

Proposition 5.2.12 is used for $1 \le p < q \le \infty$ and $1 < q = p < \infty$. The upper estimates for $\|g^{(r)}(\mathbf{x},\alpha)\|_q$ in this example are trivial.

Proposition 5.2.13 is used for $1 < q < \infty$ and $p = \infty$. The upper estimates for $\|g^{(r)}(\mathbf{x},\alpha)\|_q$ in this example follow from Theorem 4.3.1 and Lemma 3.3.7:

$$\begin{aligned}\|g^{(r)}(\mathbf{x},\alpha)\|_q &= \|h^{(r)}(\mathbf{x},\alpha)\|_q \ll 2^{nr}\|h\|_q \\ &\ll 2^{nr}\Big(\sum_{\mathbf{s}\in\theta_n}\|h_{\mathbf{s}}\|_1^q 2^{n(q-1)}\Big)^{1/q} \ll 2^{n(r+1-1/q)} n^{(d-1)/q}.\end{aligned}$$

Proposition 5.2.14 is used for $1 < q \le p < \infty$. The upper estimates for $\|g^{(r)}(\mathbf{x},\alpha)\|_q$ in this example follow from the Bernstein inequality; see §3.3.1.

Proposition 5.2.15 is used for $q = p = \infty$. The function $f_r(\mathbf{x},L,\alpha)C(d)^{-1}$ in this example belongs to the class $\mathbf{W}^r_{q,\alpha}$, where $C(d)$ is the constant from (5.2.6).

Proposition 5.2.16 is used for $q = 1$ and $1 \le p < \infty$. The function $g(\mathbf{x}) = V_{\mathbf{M}}\big(F_r(\mathbf{x}+\mathbf{y}^*,\alpha)\big)$ constructed in this example belongs to the class $\mathbf{W}^r_{1,\alpha}3^d$.

Thus, the Propositions 5.2.12–5.2.16 give the required lower estimates for all $1 \le q,p \le \infty$ with the exception of the case $q = 1$ and $p = \infty$.

The latter case requires special consideration both for obtaining lower estimates and for obtaining upper estimates. The investigation of this case is not complete.

Proof of Theorem 5.2.4 We first prove the following relation for all d:

$$\varphi_m(\mathbf{W}^r_{1,1},L_\infty) \asymp \varphi^B_m(\mathbf{W}^r_{1,1},L_\infty) \asymp \big(m^{-1}(\log m)^{d-1}\big)^{r-1}. \tag{5.2.31}$$

Let us prove the upper estimates. We consider the orthogonal projection operator onto $\mathscr{T}(N)$:

$$S_N(f,\mathbf{x}) := S_{\Gamma(N)}(f,\mathbf{x}) := \sum_{\mathbf{k}\in\Gamma(N)} \hat f(\mathbf{k}) e^{i(\mathbf{k},\mathbf{x})}.$$

Write

$$f^r_N(\mathbf{x}) = F_r(\mathbf{x},\mathbf{1}) - S_N\big(F_r,\mathbf{x}\big).$$

Then it is easy to see that

$$\sup_{f\in\mathbf{W}^r_{1,1}} \big\|f - S_N(f)\big\|_\infty \le \|f^r_N\|_\infty. \tag{5.2.32}$$

The upper estimate in (5.2.31) follows from (5.2.32) and Proposition 4.4.4.

We now prove the lower estimate in (5.2.31). Let $h(\mathbf{x}-\mathbf{y}^*)$ be from Proposition 5.2.13 and

$$f(\mathbf{x}) := h(\mathbf{x}-\mathbf{y}^*) \;/\; \big\|h^{(r)}(\mathbf{x}-\mathbf{y}^*,\mathbf{1})\big\|_1.$$

Then $f\in\mathbf{W}^r_{1,1}$. It is easy to see that, by the Bernstein inequality,

$$\big\|h^{(r)}_{\mathbf{s}}(\mathbf{x},\mathbf{1})\big\|_1 \ll 2^{r\|\mathbf{s}\|_1}$$

and, consequently,

$$\big\|h^{(r)}(\mathbf{x}-\mathbf{y}^*,\mathbf{1})\big\|_1 \ll 2^{rn}n^{d-1}.$$

This inequality and Proposition 5.2.13 for $p=\infty$ give

$$\big\|f - G(f)\big\|_\infty \gg 2^{-n(r-1)} \gg \big(m^{-1}(\log m)^{d-1}\big)^{r-1}.$$

The lower estimate is proved.

Let us consider now the case $d=2$ and $\alpha=0$. It turns out that in this case the quantities $\varphi_m(\mathbf{W}^r_{1,0},L_\infty)$ and $\varphi^B_m(\mathbf{W}^r_{1,0},L_\infty)$, $B>B(r)$, have different orders. We first find the order of $\varphi_m(\mathbf{W}^r_{1,0},L_\infty)$. Let us prove that

$$\varphi_m(\mathbf{W}^r_{1,0},L_\infty) \asymp m^{1-r}(\log m)^r. \tag{5.2.33}$$

The upper estimate follows from the estimate

$$\big\|F_r(\mathbf{x},0) - S_N(F_r,\mathbf{x})\big\|_\infty \ll N^{1-r}\log N,$$

which can be easily verified.

We now prove the lower estimate. Let $\Psi=\{\psi_\mu\}_{\mu=1}^m$ be an orthonormal system of functions and let $S(\cdot,\Psi)$ denote the orthogonal projector onto $\operatorname{span}(\Psi)$. We will consider the approximation of the functions $f(\mathbf{x}-\mathbf{y})$, $f(\mathbf{x}) := F_r(\mathbf{x},\mathbf{0})$. We have

$$S\big(f(\mathbf{x}-\mathbf{y}),\Psi\big) = \sum_{\mathbf{k}} \hat f(\mathbf{k})e^{-i(\mathbf{k},\mathbf{y})}\sum_{\mu=1}^m \overline{\hat\psi_\mu(\mathbf{k})}\psi_\mu(\mathbf{x}).$$

Further,

$$\inf_{\mathbf{y}=\mathbf{x}} \operatorname{Re} S\big(f(\mathbf{x}-\mathbf{y}),\Psi\big) \le \operatorname{Re}(2\pi)^{-2} \int_{\mathbb{T}^2} S\big(f(\mathbf{x}-\mathbf{y}),\Psi\big)\Big|_{\mathbf{y}=\mathbf{x}} d\mathbf{x}$$
$$= \operatorname{Re} \sum_{\mathbf{k}} \hat{f}(\mathbf{k}) \sum_{\mu=1}^{m} \overline{\hat{\psi}_\mu(\mathbf{k})} \hat{\psi}_\mu(\mathbf{k})$$
$$= \sum_{\mathbf{k}} \hat{f}(\mathbf{k}) \sum_{\mu=1}^{m} \big|\hat{\psi}_\mu(\mathbf{k})\big|^2.$$

Therefore,

$$\sup_{\mathbf{y}} \big\|f(\mathbf{x}-\mathbf{y}) - S\big(f(\mathbf{x}-\mathbf{y}),\Psi\big)\big\|_\infty \ge f(\mathbf{0}) - \inf_{\mathbf{y}=\mathbf{x}} \operatorname{Re} S\big(f(\mathbf{x}-\mathbf{y}),\Psi\big) \tag{5.2.34}$$
$$\ge \sum_{\mathbf{k}} \hat{f}(\mathbf{k}) \left(1 - \sum_{\mu=1}^{m} \big|\hat{\psi}_\mu(\mathbf{k})\big|^2\right). \tag{5.2.35}$$

Let L be such that

$$\big|\Gamma(L-1)\big| < 2m \le \big|\Gamma(L)\big|.$$

Then we continue the relation (5.2.35):

$$\ge \sum_{\mathbf{k}\in\Gamma(L)} \hat{f}(\mathbf{k}) \left(1 - \sum_{\mu=1}^{m} \big|\hat{\psi}_\mu(\mathbf{k})\big|^2\right)$$
$$\ge \left(\min_{\mathbf{k}\in\Gamma(L)} \hat{f}(\mathbf{k})\right) \sum_{\mathbf{k}\in\Gamma(L)} \left(1 - \sum_{\mu=1}^{m} \big|\hat{\psi}_\mu(\mathbf{k})\big|^2\right)$$
$$\ge \left(\min_{\mathbf{k}\in\Gamma(L)} \hat{f}(\mathbf{k})\right) \big(\big|\Gamma(L)\big| - m\big) \gg m^{1-r}(\log m)^r.$$

The relation (5.2.33) is proved.

We now prove the following relation for $B > B(r)$:

$$\varphi_m^B(\mathbf{W}^r_{1,\alpha}, L_\infty) \asymp (m^{-1}\log m)^{r-1}. \tag{5.2.36}$$

We first prove the upper estimate. To do this, we construct an approximating polynomial $t \in \mathscr{T}(N)$ such that

$$\big\|F_r(\cdot,\alpha) - t(\cdot)\big\|_\infty \ll N^{1-r}$$

and

$$\big|\hat{t}(\mathbf{k})\big| \le B\big(\nu(\bar{\mathbf{k}})\big)^{-r}. \tag{5.2.37}$$

We construct $t(x,y)$ in the same way as in the proof of Lemma 4.4.2. Clearly, it suffices to approximate the $F_r^0(\mathbf{x},\alpha) := \prod_{j=1}^d (F_r(x_j,\alpha_j)-1)$. Let Γ_{Nl+b} be as in (4.4.8). For $r>1$, set

$$f_{Nl}(x,y,\alpha) := \sum_{b=0}^{N-1} (Nl+b)^{-r} \sum_{(k,\mu)\in\Gamma_{Nl+b}} \cos(kx-\alpha_1\pi/2)\cos(\mu y-\alpha_2\pi/2)$$

and represent it in the form

$$\begin{aligned} f_{Nl}(x,y,\alpha) &= \sum_{\mu=1}^{N} \mu^{-r}\cos(\mu y-\alpha_2\pi/2) \sum_{k=k'_m}^{k''_m} k^{-r}\cos(kx-\alpha_1\pi/2) \\ &= \sum_{\mu=1}^{N} \mu^{-r}\cos(\mu y-\alpha_2\pi/2) \sum_{k=0}^{k''_\mu-k'_\mu} (k'_\mu+k)^{-r}\cos((k'_\mu+k)x-\alpha_1\pi/2), \end{aligned}$$

where $k'_\mu \geq Nl/\mu$ and $k''_\mu - k'_\mu \leq N/\mu$. If for some μ there is no number k with $Nl \leq k\mu < Nl+N$ then $\sum_{k=k'_\mu}^{k''_\mu}$ is set to zero. Let k_μ denote the largest natural number such that $\mu k_\mu \leq N$.

As an approximating polynomial we take T_{Nl}, which is defined by

$$T_{Nl}(x,y,\alpha) := \sum_{\mu=1}^{N} \mu^{-r}\cos(\mu y-\alpha_2\pi/2) \sum_{k=0}^{k''_\mu-k'_\mu} (k'_\mu+k)^{-r}\cos\big((k_\mu-k)x-\alpha_1\pi/2\big).$$

Then

$$\begin{aligned} &\big|f_{Nl}(x,y,\alpha)-T_{Nl}(x,y,\alpha)\big| \\ &\quad= \Bigg|\sum_{\mu=1}^{N} \mu^{-r}\cos(\mu y-\alpha_2\pi/2)\times 2\sin\big(1/2(k'_\mu+k_\mu)-\alpha_1\pi/2\big) \\ &\qquad\times \sum_{k=0}^{k''_\mu-k'_\mu} (k'_\mu+k)^{-r}\sin\left(\frac{k'_\mu-k_\mu}{2}+k\right)x\Bigg|. \end{aligned}$$

To estimate the right-hand side of this relation, we use the following lemma which is a corollary of Lemma 4.4.3.

Lemma 5.2.19 *Let L and ϑ be natural numbers and $Q \leq L$ be a nonnegative real number. Then, for $x \in (0,\pi)$ and $r>1$, we have*

$$\left|\sum_{k=0}^{\vartheta}(L+k)^{-r}\sin(Q+k)x\right| \ll L^{1-r}\min\{\vartheta x, \vartheta L^{-1}, (Lx)^{-1}\}.$$

Proof By the Abel inequality we obtain

$$\left|\sum_{k=0}^{\vartheta}(L+k)^{-r}\sin(Q+k)x\right| \leq L^{-r+1}\max_{0\leq\nu\leq\vartheta}\left|\sum_{k=0}^{\nu}(L+k)^{-1}\sin(Q+k)x\right|.$$

An application of Lemma 4.4.3 completes the proof. □

Using the inequalities $k'_\mu \geq Nl/\mu$ and $k''_\mu - k'_\mu \leq N/\mu$, we obtain by Lemma 5.2.19

$$\begin{aligned}
&\left|f_{Nl}(x,y,\alpha) - T_{Nl}(x,y,\alpha)\right| \\
&\quad \ll \left|\sum_{\mu=1}^{m} \mu^{-r} \left(\frac{Nl}{\mu}\right)^{1-r} \min\left\{\frac{Nx}{\mu}, \frac{1}{l}, \frac{\mu}{Nlx}\right\}\right| \\
&\quad \ll (Nl)^{1-r} \sum_{\mu=1}^{m} \mu^{-1} \min\left\{\frac{Nx}{\mu}, \frac{1}{l}, \frac{\mu}{Nlx}\right\} \\
&\quad \ll (Nl)^{1-r} \left(\sum_{\mu=1}^{[Nx]} (Nlx)^{-1} + \sum_{\mu=[Nx]+1}^{[Nlx]} (\mu l)^{-1} + \sum_{\mu=[Nlx]+1}^{\infty} Nx\mu^{-2}\right) \\
&\quad \ll N^{1-r} l^{-r} \ln(l+1).
\end{aligned}$$

We set

$$t_1(\mathbf{x},\mathbf{y}) := \sum_{l=1}^{\infty} T_{Nl}(x,y,\alpha), \qquad t := t_1 + S_N(F_r).$$

Then

$$\left\|\sum_{l=1}^{\infty} f_{Nl} - t_1\right\|_\infty \ll N^{1-r}, \qquad r > 1, \tag{5.2.38}$$

and it is easy to see that

$$\left\|F_r - \sum_{l=1}^{\infty} f_{Nl} - S_N(F_r)\right\|_\infty \ll N^{1-r}. \tag{5.2.39}$$

From (5.2.38) and (5.2.39) it follows that

$$\|F_r - t\|_\infty \ll N^{1-r}.$$

From the construction of the polynomial t_1 we have

$$\left|\hat{t}_1(\mathbf{k})\right| \leq \sum_{l=1}^{\infty} \left|\hat{T}_{Nl}(\mathbf{k})\right| \leq N^{-r} \sum_{l=1}^{\infty} l^{-r} \leq B'(r) N^{-r},$$

which implies the required estimate (5.2.37). Thus the upper estimate in (5.2.36) is proved. The lower estimate can be proved in the same way as in (5.2.31) This concludes the proof of Theorem 5.2.4. □

Proof of Theorems 5.2.5 and 5.2.6 We first prove the upper estimates. Theorem 4.4.10, the boundedness of the operator S_{Q_n} as an operator from L_p to L_p for $1 < p < \infty$, and the monotonicity of L_p-norms imply the upper estimates in Theorem 5.2.5 for all $(q,p) \neq (1,1), (\infty,\infty)$ with the exception of the case $1 \leq q < \infty$ and $p = \infty$. We now consider this case. First, remark that the norm of the operator $\delta_\mathbf{s}$ as an operator from L_q, $1 \leq q < \infty$, to L_∞ is less in order than $2^{\|\mathbf{s}\|_1/q}$. In the case $q = 1$

this is trivial and in the case $1 < q < \infty$ it follows from the Nikol'skii inequality and from the boundedness of the operator $\delta_{\mathbf{s}}$ as an operator from L_q to L_q for $1 < q < \infty$. Using Theorem 4.4.6 we obtain

$$\|f - S_{Q_n}(f)\|_\infty \leq \sum_{\|\mathbf{s}\|_1 > n} \|\delta_{\mathbf{s}}(f)\|_\infty = \sum_{\|\mathbf{s}\|_1 > n} \left\| \delta_{\mathbf{s}} \left(\sum_{\|\mathbf{s}'-\mathbf{s}\|_\infty \leq 1} A_{\mathbf{s}'}(f) \right) \right\|_\infty$$
$$\ll \sum_{\|\mathbf{s}\|_1 > n; \|\mathbf{s}'-\mathbf{s}\|_\infty \leq 1} \|A_{\mathbf{s}'}(f)\|_q 2^{\|\mathbf{s}\|_1/q} \ll 2^{-n(r-1/q)} n^{d-1}.$$

In the case $(q,p) \neq (1,1), (\infty,\infty)$ the upper estimates in Theorem 5.2.5 imply the upper estimate in Theorem 5.2.6. Let $q = p$ and let p take one of the values 1 or ∞. Then, using Theorem 4.4.6, we find that

$$\|f - V_{Q_n}(f)\|_p \leq \sum_{\|\mathbf{s}\|_1 > n} \|A_{\mathbf{s}}(f)\|_p \ll 2^{-rn} n^{d-1}.$$

It remains to remark that $V_{Q_n} \in \mathscr{L}_{2^d|Q_n|}(1)_p$, $1 \leq p \leq \infty$.

For the lower estimates it is sufficient to carry out the proof of Theorem 5.2.6. This proof is based on Propositions 5.2.13, 5.2.17, and 5.2.18, and for all cases it goes as follows. Let $G \in \mathscr{L}_m(B)_p$, and $g(\mathbf{x})$ be a function from the Propositions, then we consider the function

$$f(\mathbf{x}) = g(\mathbf{x}) / \max_{\mathbf{s}} \|A_{\mathbf{s}}(g)\|_q 2^{r\|\mathbf{s}\|_1},$$

which by Theorem 4.4.6 belongs to the class $\mathbf{H}^r_q B$ with some constant $B > 0$ which does not depend on m. The required lower estimates follow from the propositions and from the upper estimate for $\max_{\mathbf{s}} \|A_{\mathbf{s}}(g)\|_q 2^{r\|\mathbf{s}\|_1}$.

We now indicate which propositions are used for specific q, p.

Proposition 5.2.13 is used for the case $1 < q < p \leq \infty$; $q = 1, 1 < p \leq \infty$; $1 < q = p \leq 2$; $q = p = \infty$. In these cases we can easily estimate the quantities $\|A_{\mathbf{s}}(g)\|_q$, $1 \leq q \leq \infty$:

$$\|A_{\mathbf{s}}(g)\|_q \ll 2^{\|\mathbf{s}\|_1(1-1/q)}.$$

Proposition 5.2.17 is used for the case $q \geq 2$, $1 \leq p \leq q \leq \infty$, $p < \infty$. In this case the estimates of the quantities $\|A_{\mathbf{s}}(g)\|_q$ are trivial.

Proposition 5.2.18 is used for $1 \leq p \leq q \leq 2$. In this case we can easily estimate $\|A_{\mathbf{s}}(g)\|_q$ for $1 \leq q \leq 2$:

$$\|A_{\mathbf{s}}(g)\|_q \ll n^{(d-1)(1-1/q)}.$$

We remark that in the case $q = p = 1$ the required estimate can be obtained from Proposition 5.2.16.

The results of this section show that the operators V_{Q_n} and the corresponding operators for partial sums S_{Q_n} are the best (in the sense of order) approximation operators of among a sufficiently wide class $\mathscr{L}_m(B)_p$ of operators.

5.3 The Kolmogorov Widths of the Classes $\mathbf{W}^r_{q,\alpha}$ and $\mathbf{H}^r_q$

In this section we study the Kolmogorov widths of the classes $\mathbf{W}^r_q$ and $\mathbf{H}^r_q$ in the L_p-norm for a number of values of the parameters q and p. We do not know the orders of the Kolmogorov widths for certain values of q and p, but for those q and p for which we do know the answer we observe the following phenomenon. For values of q and p (i.e., $1 \le q \le p \le 2$ and $1 \le p \le q \le \infty$) for which approximation by trigonometric polynomials from $\mathscr{T}(n)$ gives the order of the width in the univariate case, approximation by polynomials in $\mathscr{T}(N)$ (i.e., trigonometric polynomials with harmonics in the hyperbolic cross) gives the order of the width in the multivariate case ($d > 1$).

As in Chapter 4 and §5.2 the investigation of cases when one of the parameters q or p takes the value 1 or ∞ has required in most cases a new technique. In Chapters 2 and 3 we saw that two results from finite-dimensional geometry, namely, Theorem 2.1.11 and Lemma 2.1.12, played a very important role in proving the upper bounds for the Kolmogorov widths in the case $2 < p \le \infty$. It turns out that methods from finite-dimensional geometry also play a key role in proving the lower estimates for classes with mixed smoothness as well. We illustrate the use of some of these methods in the proof of Lemma 5.3.7. The further use of these methods is implicit in the following scheme of proof. We use geometrical methods to prove the lower estimates for entropy numbers (see Chapter 7) and then apply a general inequality between entropy numbers and Kolmogorov widths. In this section we will use Lemma 5.3.14, which is a corollary of the above-mentioned general inequality.

Theorem 5.3.1 *Let $r(q,p)$ be the same as in Theorem 2.1.1. Then*

$$d_m(\mathbf{W}^r_{q,\alpha}, L_p) \asymp \begin{cases} \left((\log m)^{d-1}/m\right)^{r-\left(1/q-\max(1/2,1/p)\right)_+} & \text{for} \quad 1 < q,p < \infty, \\ & \qquad r > r(q,p) \\ \left((\log m)^{d-1}/m\right)^{r-1/2} (\log m)^{(d-1)/2} & \text{for} \quad q = 1, \\ & \qquad 2 \le p < \infty, \\ & \qquad r > 1. \end{cases}$$

Theorem 5.3.1 covers the case $1 < q,p < \infty$ and, partially, the case $q = 1$ under certain conditions on r. We illustrate the above results and also some results which are proved later in this chapter on the behavior of the $d_m(\mathbf{W}^r_{q,\alpha}, L_p)$ in Figure 5.3.

They are in the form

$$d_m(\mathbf{W}^r_{q,\alpha}, L_p) \asymp \left(m^{-1}(\log m)^{d-1}\right)^{a^r(q,p)} (\log m)^{(d-1)b(q,p)}$$

under certain conditions on r, which we do not specify. For $(q,p) \in [1,\infty]^2$, represented by the point $(1/q, 1/p) \in [0,1]^2$, we give the order of the $d_m(\mathbf{W}^r_{q,\alpha}, L_p)$ by indicating the parameters $a^r(q,p)$ and $b(q,p)$ and a reference to the theorem which establishes that relation. In those cases when the order of the $d_m(\mathbf{W}^r_{q,\alpha}, L_p)$ is not known we refer to the corresponding open problem.

Theorem 5.3.2 *Let $r(q,p)$ be the same as in Theorem 2.1.1. Then, for $1 < p < q \le \infty$, $q \ge 2$ and $1 \le q \le p < \infty$, $p \ge 2$, with $r > r(q,p)$, we have*

$$d_m(\mathbf{H}^r_q, L_p) \asymp \left(\frac{(\log m)^{d-1}}{m}\right)^{r-\left(1/q-\max(1/2,1/p)\right)_+} (\log m)^{(d-1)/2}.$$

We illustrate the results of Theorem 5.3.2 and results proved later in this chapter on the behavior of the $d_m(\mathbf{H}^r_q, L_p)$ in Figure 5.4. They are in the form

$$d_m(\mathbf{H}^r_q, L_p) \asymp \left(m^{-1}(\log m)^{d-1}\right)^{u^r(q,p)} (\log m)^{(d-1)v(q,p)}$$

under certain conditions on r, which we do not specify. For $(q,p) \in [1,\infty]^2$, represented by the point $(1/q, 1/p) \in [0,1]^2$, we give the order of the $d_m(\mathbf{H}^r_q, L_p)$ by indicating the parameters $u^r(q,p)$ and $v(q,p)$ and a reference to the theorem which establishes that relation. In those cases when the order of $d_m(\mathbf{H}^r_q, L_p)$ is not known we refer to the corresponding open problem.

Proof of Theorems 5.3.1 and 5.3.2 We first consider the upper estimates. In all the cases $1 \le q \le p \le 2$ and $1 \le p \le q \le \infty$ included in the theorems, the upper

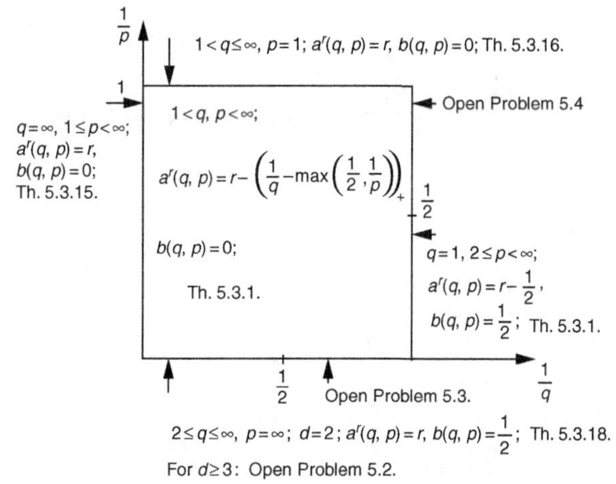

Figure 5.3 The Kolmogorov widths of the **W** classes.

estimates follow from Theorems 4.4.9 and 4.4.10 (they also follow from Theorems 5.2.1 and 5.2.5). It remains to obtain the upper estimates for the case $2 < p < \infty$. We prove the following auxiliary assertion. Let

$$\theta_n := \{\mathbf{s} : \|\mathbf{s}\|_1 = n\}, \qquad \Delta Q_n := \bigcup_{\mathbf{s}\in\theta_n} \rho(\mathbf{s}) = Q_n \backslash Q_{n-1}.$$

Lemma 5.3.3 *We have*

$$d_m\big(\mathscr{T}(\Delta Q_n)_2, L_\infty\big) \ll \big(|\Delta Q_n|/m\big)^{1/2} n^{(d-1)/2} \big(\ln\big(e|\Delta Q_n|/m\big)\big)^{1/2},$$
$$d_m\big(\mathscr{T}(\Delta Q_n)_2, L_p\big) \ll \big(|\Delta Q_n|/m\big)^{1/2} \big(\ln\big(e|\Delta Q_n|/m\big)\big)^{1/2}, \qquad 2 < p < \infty.$$

Proof We start with the first relation. Let $f \in \mathscr{T}(\Delta Q_n)_2$, then

$$f = \sum_{\|\mathbf{s}\|_1 = n} \delta_{\mathbf{s}}(f) \qquad \text{and} \qquad \sum_{\|\mathbf{s}\|_1 = n} \big\|\delta_{\mathbf{s}}(f)\big\|_2^2 \le 1. \tag{5.3.1}$$

Further, let $\overline{m} := \big[m/|\theta_n|\big]$ and $\varepsilon > 0$. We suppose that there are functions $u_i^{\mathbf{s}}(\mathbf{x})$, $i = 1, \dots, \overline{m}$, such that for any function $g \in \mathscr{T}\big(\rho(\mathbf{s})\big)$ with $\|\mathbf{s}\|_1 = n$ there are numbers $a_i^{\mathbf{s}}$ for which

$$\left\| g - \sum_{i=1}^{\overline{m}} a_i^{\mathbf{s}} u_i^{\mathbf{s}} \right\|_\infty \le (1+\varepsilon) d_{\overline{m}}\big(\mathscr{T}\big(\rho(\mathbf{s})\big)_2, L_\infty\big) \, \|g\|_2.$$

Then, for a function $f \in \mathscr{T}(\Delta Q_n)$, there are numbers $v_i^{\mathbf{s}}$ such that

$$\left\| f - \sum_{\|\mathbf{s}\|_1 = n} \sum_{i=1}^{\overline{m}} v_i^{\mathbf{s}} u_i^{\mathbf{s}} \right\|_\infty \le (1+\varepsilon) \sum_{\|\mathbf{s}\|_1 = n} d_{\overline{m}}\big(\mathscr{T}\big(\rho(\mathbf{s})\big)_2, L_\infty\big) \big\|\delta_{\mathbf{s}}(f)\big\|_2.$$

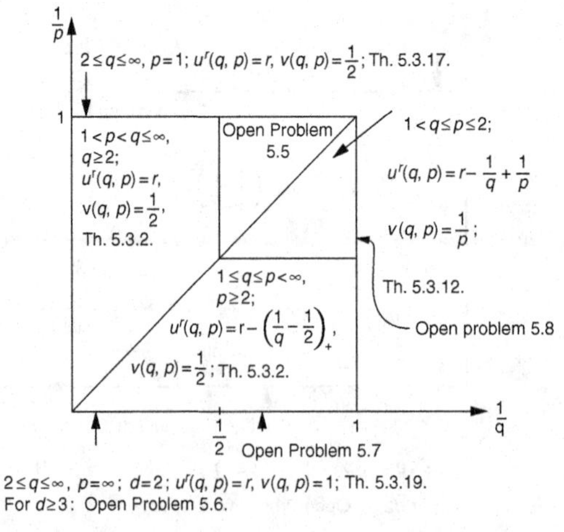

Figure 5.4 The Kolmogorov widths of the **H** classes.

Consequently,

$$d_m(\mathscr{T}(\Delta Q_n)_2, L_\infty) \le \left(\sum_{\|\mathbf{s}\|_1=n} d_{\overline{m}}\big(\mathscr{T}(\rho(\mathbf{s}))_2, L_\infty\big)^2 \right)^{1/2}.$$

Applying Theorem 3.5.9 we get the required estimate.

We prove the second relation. Let $\overline{m}$ be the same as above and the functions $u_i^{\mathbf{s}}(\mathbf{x})$, $i=1,\dots,m$, be such that for any function $g \in \mathscr{T}\big(\rho(\mathbf{s})\big)$ with $\|\mathbf{s}\|_1 = n$ there are numbers $a_i^{\mathbf{s}}$ such that

$$\left\| g - \sum_{i=1}^{\overline{m}} v_i^{\mathbf{s}} u_i^{\mathbf{s}} \right\|_p \le C(d,p) d_{\overline{m}}\big(\mathscr{T}(\rho(\mathbf{s}))_2, L_p\big) \|g\|_2. \tag{5.3.2}$$

Clearly, by Corollary A.3.4 to the Littlewood–Paley theorem, we may assume without affecting (5.3.2) that $u_i^{\mathbf{s}} \in \mathscr{T}\big(\rho(\mathbf{s})\big)$. Then, using Corollary A.3.5, we obtain from (5.3.2) that for $f \in \mathscr{T}(\Delta Q_n)$ there are numbers $v_i^{\mathbf{s}}$ such that

$$\left\| f - \sum_{\|\mathbf{s}\|_1=n} \sum_{i=1}^{\overline{m}} v_i^{\mathbf{s}} u_i^{\mathbf{s}} \right\|_p \ll \left(\sum_{\|\mathbf{s}\|_1=n} \left\| \delta_{\mathbf{s}}(f) - \sum_{i=1}^{\overline{m}} v_i^{\mathbf{s}} u_i^{\mathbf{s}} \right\|_p^2 \right)^{1/2}$$
$$\ll \left(\sum_{\|\mathbf{s}\|_1=n} d_{\overline{m}}\big(\mathscr{T}(\rho(\mathbf{s}))_2, L_p\big)^2 \|\delta_{\mathbf{s}}(f)\|_2^2 \right)^{1/2}.$$

Therefore,

$$d_m\big(\mathscr{T}(\Delta Q_n)_2, L_p\big) \ll \max_{\|\mathbf{s}\|_1=n} d_{\bar{m}}\big(\mathscr{T}(\rho(\mathbf{s}))_2, L_p\big).$$

Applying Theorem 3.5.9 we get the required estimate. □

Remark 5.3.4 In the case $m=0$ it follows from Theorems 4.3.16 and 4.3.17 that

$$d_0\big(\mathscr{T}(\Delta Q_n)_2, L_\infty\big) \ll |\Delta Q_n|^{1/2},$$
$$d_0\big(\mathscr{T}(\Delta Q_n)_2, L_p\big) \ll 2^{n(1/2-1/p)}.$$

We proceed to prove the upper estimates in Theorem 5.3.1 for $2 < p < \infty$. We first consider the case $q=2$. Let $0 < \kappa < r - 1/2$ and let a be a natural number and

$$m_n := \big[|\Delta Q_a| 2^{-\kappa(n-a)}\big], \qquad n = a+1, \dots,$$
$$m := |Q_a| + \sum_{n>a} m_n \le C(\kappa, d) 2^a a^{d-1}.$$

From the representation

$$f = S_{Q_a}(f) + \sum_{n=a+1}^{\infty} \left(\sum_{\|\mathbf{s}\|_1=n} \delta_{\mathbf{s}}(f) \right),$$

using Theorem 4.4.9, we obtain

$$d_m(\mathbf{W}^r_{2,\alpha}, L_p) \ll \sum_{n=a+1}^{\infty} d_{m_n}(\mathscr{T}(\Delta Q_n)_2, L_p) 2^{-rn}. \tag{5.3.3}$$

Applying Lemma 5.3.3 and Remark 5.3.4 we get from (5.3.3),

$$\begin{aligned} d_m(\mathbf{W}^r_{2,\alpha}, L_p) &\ll \left(|\Delta Q_a| 2^{\kappa a}\right)^{-1/2} \sum_{n>a} 2^{-n(r-1/2-\kappa)} n^{(d-1)/2} \\ &\times \left(\ln \frac{e|\Delta Q_n|}{m_n+1} \right)^{1/2} \ll 2^{-ra} \\ &\ll \left(m^{-1}(\log m)^{d-1}\right)^r. \end{aligned}$$

In the above proof we used the inequality

$$\left\| \sum_{\|\mathbf{s}\|_1=n} \delta_{\mathbf{s}}(f) \right\|_2 \ll 2^{-rn}, \tag{5.3.4}$$

which is valid for $f \in \mathbf{W}^r_{2,\alpha}$. For functions from $\mathbf{H}^r_2$, by Theorem 4.4.6 we have the inequality

$$\left\| \sum_{\|\mathbf{s}\|_1=n} \delta_{\mathbf{s}}(f) \right\|_2 \ll 2^{-rn} n^{(d-1)/2}. \tag{5.3.5}$$

Clearly, if in the previous arguments we use (5.3.5) instead of (5.3.4) then we get

$$d_m(\mathbf{H}^r_2, L_p) \ll \left(m^{-1}(\log m)^{d-1}\right)^r (\log m)^{(d-1)/2}.$$

So, the case $q = 2$, $2 < p < \infty$ in Theorems 5.3.1 and 5.3.2 has been investigated. The remaining cases follow from this one. Indeed, for $1 < q < 2$ we have

$$\mathbf{W}^r_{q,\alpha} \subset \mathbf{W}^{r-1/q+1/2}_{2,\alpha} B, \qquad \mathbf{H}^r_q \subset \mathbf{H}^{r-1/q+1/2}_2 B,$$

and, for $q = 1$,

$$\mathbf{W}^r_{1,\alpha} \subset \mathbf{H}^r_1 B \subset \mathbf{H}^{r-1/2}_2 B.$$

Thus the upper estimates in Theorems 5.3.1 and 5.3.2 are proved.

We now proceed to prove the lower estimates. We see that in the case $2 \le p < \infty$ the estimates in Theorems 5.3.1 and 5.3.2 do not depend on p. Therefore, it suffices to prove the lower estimates for $p = 2$. Using the relation (2.1.2), which gives the coincidence of the Kolmogorov widths and the orthowidths in a Hilbert space, we get from Theorems 5.2.1 and 5.2.5 the required estimates for $2 \le p < \infty$.

We now prove the lower estimates for the $\mathbf{W}$ classes in the case $1 < q \le p \le 2$. In the proof of the lower estimates for $\varphi_m(\mathbf{W}^r_{q,\alpha}, L_2)$ it has been proved that there is an n such that $|Q_n| \le C(d)m$ and

$$d_m\big(\mathbf{W}^r_{q,\alpha} \cap \mathscr{T}(Q_n), L_2\big) \gg 2^{-n(r-1/q+1/2)}. \tag{5.3.6}$$

Let $1 < q \le p < 2$. Then by Corollary A.3.4 and by the Nikol'skii inequality (see Theorem 4.3.17) we obtain from (5.3.6)

$$\begin{aligned} d_m(\mathbf{W}^r_{q,\alpha}, L_p) &\ge d_m\big(\mathbf{W}^r_{q,\alpha} \cap \mathscr{T}(Q_n), L_p\big) \\ &\gg d_m\big(\mathbf{W}^r_{q,\alpha} \cap \mathscr{T}(Q_n), L_p \cap \mathscr{T}(Q_n)\big) \\ &\gg 2^{n(1/2-1/p)} d_m\big(\mathbf{W}^r_{q,\alpha} \cap \mathscr{T}(Q_n), L_2 \cap \mathscr{T}(Q_n)\big) \\ &= 2^{n(1/2-1/p)} d_m\big(\mathbf{W}^r_{q,\alpha} \cap \mathscr{T}(Q_n), L_2\big) \gg 2^{-n(r-1/q+1/p)}, \end{aligned}$$

which gives the required lower estimate.

We now demonstrate the use of an important technique for proving the lower bounds which is based on finite-dimensional geometry results. More specifically, it is based on the volume estimates of the sets of Fourier coefficients of trigonometric polynomials with harmonics in hyperbolic crosses. We have already used this technique to prove the fundamental theorem 3.2.1. Further development of the technique is presented in Chapter 7. Here, we use it to prove lower bounds for the Kolmogorov widths of classes with mixed smoothness. We need some modifications of $\rho(\mathbf{s})$, θ_n, and ΔQ_n. As above, for $\mathbf{s} \in \mathbb{N}^d$ set

$$\rho(\mathbf{s})^+ := \big\{\mathbf{k} = (k_1, \dots, k_d) : 2^{s_j-1} \le k_j < 2^{s_j},\ j = 1, \dots, d\big\}.$$

For an even number n define the sets

$$\begin{aligned} \theta'_n &:= \big\{\mathbf{s} : \|\mathbf{s}\|_1 = n,\ s_j \quad \text{are even natural numbers},\ j = 1, \dots, d\big\}, \\ \Delta Q'_n &:= \bigcup_{\mathbf{s} \in \theta'_n} \rho(\mathbf{s})^+. \end{aligned}$$

Let us prove the following analog of Theorem 3.2.1.

Theorem 5.3.5 *Let $\varepsilon > 0$ and a subspace $\Psi \subset \mathscr{T}(\Delta Q'_n)$ be such that* $\dim \Psi \ge$ $\varepsilon |\Delta Q'_n|$. *Then there is a $t \in \Psi$ such that, for all $\mathbf{s} \in \theta'_n$,*

$$\Big\| \sum_{\mathbf{k} \in \rho(\mathbf{s})^+} \hat{t}(\mathbf{k}) e^{i(\mathbf{k},\mathbf{x})} \Big\|_\infty \le |\theta'_n|^{-1/2}$$

and

$$\|t\|_2 \ge C(\varepsilon, d) > 0.$$

Proof We will use some notation from Chapter 7. For a set $\Lambda \subset \mathbb{Z}^d$ denote

$$\mathscr{T}(\Lambda) := \{f \in L_1 : \hat{f}(\mathbf{k}) = 0, \mathbf{k} \in \mathbb{Z}^d \setminus \Lambda\}, \quad \mathscr{T}(\Lambda)_p := \{f \in \mathscr{T}(\Lambda) : \|f\|_p \le 1\}.$$

For a finite set Λ we assign to each $f = \sum_{\mathbf{k}\in\Lambda} \hat{f}(\mathbf{k}) e^{i(\mathbf{k},\mathbf{x})} \in \mathscr{T}(\Lambda)$ a vector

$$A(f) := \{(\mathrm{Re}(\hat{f}(\mathbf{k})), \mathrm{Im}(\hat{f}(\mathbf{k}))),\ \mathbf{k} \in \Lambda\} \in \mathbb{R}^{2|\Lambda|},$$

where $|\Lambda|$ denotes the cardinality of Λ, and define

$$B_\Lambda(L_p) := \{A(f) : f \in \mathscr{T}(\Lambda)_p\}.$$

For Λ of the special form $\Lambda = \bigcup_{\mathbf{s}\in\theta} \rho(\mathbf{s})$ we write

$$B_\Lambda(\mathbf{H}_q) := \{A(f) : f \in \mathscr{T}(\Lambda),\ \|\delta_{\mathbf{s}}(f)\|_q \le 1,\ \mathbf{s} \in \theta\}.$$

Lemma 5.3.6 *We have*

$$\mathrm{vol}(B_{\Delta Q'_n}(\mathbf{H}_\infty))^{(2|\Delta Q'_n|)^{-1}} \gg 2^{-n/2}.$$

Proof From the definition of $B_{\Delta Q'_n}(\mathbf{H}_\infty)$ we have

$$\mathrm{vol}(B_{\Delta Q'_n}(\mathbf{H}_\infty)) = \prod_{\mathbf{s}\in\theta'_n} \mathrm{vol}(B_{\Pi(2^{\mathbf{s}-1}, 2^{\mathbf{s}}-1, d)}(L_\infty)), \tag{5.3.7}$$

where for $\mathbf{a} = (a_1, \dots, a_d)$ and $\mathbf{b} = (b_1, \dots, b_d)$ we denote

$$\Pi(\mathbf{a}, \mathbf{b}, d) := [\mathbf{a}, \mathbf{b}] \cap \mathbb{Z}^d := [a_1, b_1] \times \cdots \times [a_d, b_d] \cap \mathbb{Z}^d.$$

The lemma then follows from (5.3.7) and Theorem 7.5.2. □

We can now complete the proof of Theorem 5.3.5. Lemma 5.3.6 implies that

$$\mathrm{vol}(|\theta'_n|^{-1/2} B_{\Delta Q'_n}(\mathbf{H}_\infty))^{(2|\Delta Q'_n|)^{-1}} \gg 2^{-n/2} |\theta'_n|^{-1/2} \asymp (2|\Delta Q'_n|)^{-1/2}.$$

It remains to apply Lemma 3.2.4.

□

We now prove the lower bounds in Theorems 5.3.1 and 5.3.2 in the case $1 \le p \le q \le \infty$. Here it will be convenient for us to obtain the bounds in some other norm. We suppose that $1 \le q \le \infty$, $1 \le \theta \le \infty$ and consider the following norms for $f \in L_q$:

$$\|f\|_{B_{q,\theta}} := \left(\sum_{\mathbf{s}} \|A_{\mathbf{s}}(f)\|_q^\theta \right)^{1/\theta}.$$

We remark that, in the case $1 < q < \infty$,

$$\|f\|_{B_{q,\theta}} \asymp \left(\sum_{\mathbf{s}} \|\delta_{\mathbf{s}}(f)\|_q^\theta \right)^{1/\theta}.$$

Lemma 5.3.7 *Let $r > 0$; then*

$$d_m(\mathbf{H}^r_\infty, B_{1,1}) \gg m^{-r}(\log m)^{(d-1)(r+1)}.$$

Proof Let m be given. We choose an even n such that

$$|\Delta Q'_{n-2}| < 2m \leq |\Delta Q'_n|, \tag{5.3.8}$$

where $\Delta Q'_n$ is defined above.

Let $\Psi \subset L_1$ be a subspace of dimension not exceeding m. By Theorem 5.3.5 we can find a function $g \in \mathscr{T}(\Delta Q'_n)$ such that

$$\big\|\delta_{\mathbf{s}}(g)\big\|_\infty \leq |\theta'_n|^{-1/2}, \qquad \mathbf{s} \in \theta'_n, \tag{5.3.9}$$

$$\|g\|_2 \geq C(d) > 0, \tag{5.3.10}$$

$$\langle g, \psi \rangle = 0, \qquad \psi \in \Psi. \tag{5.3.11}$$

We set

$$f := g 2^{-rn} |\theta'_n|^{1/2}.$$

Then by Theorem 4.4.6 we find that $f \in \mathbf{H}^r_\infty B$ with some B which does not depend on n.

Let $\psi \in \Psi$. We will estimate the quantity $\|f - \psi\|_{B_{1,1}}$. Let us consider $J := \langle f - \psi, g \rangle$. On the one hand by (5.3.10) and (5.3.11) we have

$$J = \langle f, g \rangle \geq 2^{-rn} |\theta'_n|^{1/2} C(d)^2. \tag{5.3.12}$$

On the other hand,

$$J = \sum_{\mathbf{s} \in \theta'_n} \langle \delta_{\mathbf{s}}(f - \psi), \delta_{\mathbf{s}}(g) \rangle = \sum_{\mathbf{s} \in \theta'_n} \left\langle \sum_{\|\mathbf{s}' - \mathbf{s}\|_\infty \leq 1} A_{\mathbf{s}'}(f - \psi), \delta_{\mathbf{s}}(g) \right\rangle$$

$$\ll \left(\max_{\mathbf{s} \in \theta'_n} \big\|\delta_{\mathbf{s}}(g)\big\|_\infty \right) \sum_{\mathbf{s}} \big\|A_{\mathbf{s}}(f - \psi)\big\|_1.$$

From this relation, taking into account (5.3.12) and (5.3.9), we get

$$\|f - \psi\|_{B_{1,1}} \gg 2^{-rn} |\theta'_n|,$$

which implies the conclusion of the lemma. □

Remark 5.3.8 In fact, it has been proved that for m and n satisfying the condition (5.3.8), we have

$$d_m\big(\mathbf{H}^r_\infty \cap \mathscr{T}(\Delta Q'_n), B_{1,1} \cap \mathscr{T}(Q_{n+d} \backslash Q_{n-d})\big) \gg m^{-r}(\log m)^{(d-1)(r+1)}.$$

We now present a number of corollaries to Lemma 5.3.7 and Remark 5.3.8.

Corollary 5.3.9 *Let $r > 0$; then*

$$d_m(\mathbf{H}^r_\infty, B_{1,2}) \gg m^{-r}(\log m)^{(d-1)(r+1/2)}.$$

Proof This corollary follows from Remark 5.3.8 and the following simple inequality for $f \in \mathscr{T}(Q_{n+d} \backslash Q_{n-d})$:

$$\|f\|_{B_{1,1}} \le C(d)|\theta'_n|^{1/2}\|f\|_{B_{1,2}}.$$ □

Corollary 5.3.10 *For any $1 < p < \infty$ and $r > 0$ the inequality*

$$d_m(\mathbf{H}^r_\infty, L_p) \gg m^{-r}(\log m)^{(d-1)(r+1/2)}$$

holds.

Proof The estimate follows from Corollary 5.3.9 and the inequality

$$\|f\|_{B_{1,2}} \ll \left(\sum_{\mathbf{s}} \|\delta_{\mathbf{s}}(f)\|_1^2\right)^{1/2} \ll \left(\sum_{\mathbf{s}} \|\delta_{\mathbf{s}}(f)\|_p^2\right)^{1/2} \ll \|f\|_p$$

which is valid for $f \in \mathscr{T}(Q_{n+d} \backslash Q_{n-d})$, $1 < p \le 2$.

The last inequality can be easily derived from the following inequality for $2 \le q < \infty$ (see Corollary A.3.5):

$$\|f\|_q \ll \left(\sum_{\mathbf{s}} \|\delta_{\mathbf{s}}(f)\|_q^2\right)^{1/2}.$$ □

Corollary 5.3.11 *Let $1 \le q < \infty$; then*

$$d_m(\mathbf{W}^r_{q,\alpha}, B_{1,1}) \gg m^{-r}(\log m)^{(d-1)(r+1/2)}.$$

Proof This corollary follows from Remark 5.3.8 and the following inclusion: let $f \in \mathbf{H}^r_\infty \cap \mathscr{T}(Q_{n+d} \backslash Q_{n-d})$; then

$$f|\theta_n|^{-1/2} \in \mathbf{W}^r_{q,\alpha}B$$

for some $B > 0$ which does not depend on n. In the same way as that in which Corollaries 5.3.9 and 5.3.10 were derived from Lemma 5.3.7 we can derive from Corollary 5.3.11 and its proof the following relations: for any $1 \le q < \infty$,

$$d_m(\mathbf{W}^r_{q,\alpha}, B_{1,2}) \gg m^{-r}(\log m)^{r(d-1)}$$

and, for any $1 \le q < \infty$, $1 < p < \infty$,

$$d_m(\mathbf{W}^r_{q,\alpha}, L_p) \gg m^{-r}(\log m)^{r(d-1)}.$$

Thus the proofs of Theorems 5.3.1 and 5.3.2 are complete. □

We note that Lemma 5.3.3 and the Parseval identity imply the following upper bounds for the Kolmogorov widths in the case $p = \infty$, $r > 1/2$:

$$d_m(\mathbf{W}^r_{2,\alpha}, L_\infty) \ll m^{-r}(\log m)^{(d-1)(r+1/2)}, \tag{5.3.13}$$

$$d_m(\mathbf{H}^r_2, L_\infty) \ll m^{-r}(\log m)^{(d-1)(r+1)}. \tag{5.3.14}$$

In the case $d \geq 3$, better upper bounds are known. E. Belinsky proved the following bounds:

$$d_m(\mathbf{W}^r_{2,\alpha}, L_\infty) \ll m^{-r}(\log m)^{(d-1)r+1/2}, \tag{5.3.15}$$

$$d_m(\mathbf{H}^r_2, L_\infty) \ll m^{-r}(\log m)^{(d-1)r+d/2)}. \tag{5.3.16}$$

We will not prove these bounds here but refer the reader to Trigub and Belinsky (2004).

Theorem 5.3.12 *For any $1 < q \leq p \leq 2$ and $r > 1/q - 1/p$ we have*

$$d_m(\mathbf{H}^r_q, L_p) \asymp \left(\frac{(\log m)^{d-1}}{m}\right)^{r-1/q+1/p} (\log m)^{(d-1)/p}.$$

Proof The upper estimate follows from Theorem 4.4.10 (see also Theorem 5.2.5). We will prove the lower estimate by studying the approximation of functions of a special form. For a fixed n consider the set $\theta_n := \{\mathbf{s} \in \mathbb{N}^d : \|\mathbf{s}\|_1 = n\}$ and enumerate its elements by $j(\mathbf{s}) \in [1, M]$, $M := |\theta_n|$. Note that $M \asymp n^{d-1}$. Let $r_j(t)$ denote the Rademacher functions. Denote by T the set of all selections

$$S := (\mathbf{k}^1, \dots, \mathbf{k}^M), \qquad \mathbf{k}^j \in \rho(\mathbf{s}^j)^+, \qquad j = 1, \dots, M,$$

where $\rho(\mathbf{s})^+$ is defined above (see, for instance, before Theorem 5.3.5). In our construction we use the orthonormal system $\mathscr{U}^d = \{u_{\mathbf{k}}(\mathbf{x})\}_{\mathbf{k} \in \mathbb{Z}^d}$ studied in §7.6.3. For $S = (\mathbf{k}^1, \dots, \mathbf{k}^M) \in T$ consider the functions

$$f_S(\mathbf{x}, t) := \sum_{j=1}^{M} r_j(t) u_{\mathbf{k}^j}(\mathbf{x}).$$

It is clear that

$$f_S(\cdot, t) \in \mathscr{T}(\Delta Q_n^+) := \mathscr{T}\left(\bigcup_{\mathbf{s} \in \theta_n} \rho(\mathbf{s})^+\right)$$

and, for $\mathbf{u} \in \theta_n$, we have

$$\delta_{\mathbf{u}}(f_S(\mathbf{x}, t)) = r_{j(\mathbf{u})}(t) u_{\mathbf{k}^{j(\mathbf{u})}}(\mathbf{x}).$$

Thus, it follows from the definition of $u_{\mathbf{k}}$ that

$$\|\delta_{\mathbf{u}}(f_S(\cdot, t))\|_q = \|u_{\mathbf{k}^{j(\mathbf{u})}}(\cdot)\|_q \asymp 2^{n(1/2-1/q)}. \tag{5.3.17}$$

Let m be given. Find the smallest n such that

$$\dim \mathscr{T}(\Delta Q_n^+) \geq 2m.$$

Then $m \asymp 2^n n^{d-1}$. Denote $N := \dim \mathscr{T}(\Delta Q_n^+)$. Let an approximating subspace Ψ, $\dim \Psi = m$, be given. Assuming that $1 < p \leq 2$ we get from Corollary A.3.4 that an orthogonal projector onto $\mathscr{T}(\Delta Q_n^+)$ is bounded as an operator from L_p to L_p. Therefore, keeping in mind that we are studying approximation of functions $f_S(\mathbf{x},t)$ which are in $\mathscr{T}(\Delta Q_n^+)$, we can assume without loss of generality that $\Psi \subset \mathscr{T}(\Delta Q_n^+)$. Denote by $\Psi^\perp$ the orthogonal complement of Ψ to $\mathscr{T}(\Delta Q_n^+)$ and by P the orthoprojector onto $\Psi^\perp$. It is clear that $\dim \Psi^\perp := N_1 = N - m \geq m$. Then, for any $f \in \mathscr{T}(\Delta Q_n^+)$ and any $\psi \in \Psi$, we have

$$|\langle f, P(f)\rangle| = |\langle f-\psi, P(f)\rangle| \leq \|f-\psi\|_p \|P(f)\|_{p'}.$$

So

$$\|f-\psi\|_p \geq \frac{|\langle f, P(f)\rangle|}{\|P(f)\|_{p'}}$$

and

$$\|f-\psi\|_p^{p'} \geq \frac{|\langle f, P(f)\rangle|^{p'}}{\|P(f)\|_{p'}^{p'}}. \tag{5.3.18}$$

Using the inequality $\max(a/b, c/d) \geq (a+c)/(b+d)$, which is valid for any positive numbers, we get from (5.3.18)

$$\max_{S,t} \inf_{\psi \in \Psi} \|f_S(\mathbf{x},t) - \psi(\mathbf{x})\|_p^{p'} \geq \frac{A}{B}, \tag{5.3.19}$$

with

$$A := \frac{1}{|T|} \sum_{S \in T} \int_0^1 |\langle f_S(\cdot,t), P(f_S(\cdot,t))\rangle|^{p'} dt$$

and

$$B := \frac{1}{|T|} \sum_{S \in T} \int_0^1 \|P(f_S(\cdot,t))\|_{p'}^{p'} dt.$$

We denote for convenience $a := p'$ and first estimate A from below. By the monotonicity of weighted ℓ_p-norms and L_p-norms we get

$$A^{1/a} \geq \frac{1}{|T|} \sum_{S \in T} \int_0^1 |\langle f_S, P(f_S)\rangle| dt \geq \frac{1}{|T|} \sum_{S \in T} \int_0^1 \langle f_S, P(f_S)\rangle dt.$$

Let $\{\psi_i\}_{i=1}^{N_1}$ be an orthonormal basis for $\Psi^\perp$. Then

$$P(f_S) = \sum_{j=1}^{M} \sum_{i=1}^{N_1} r_j(t) \langle u_{\mathbf{k}^j}, \psi_i\rangle \psi_i,$$

and

$$\langle f_S, P(f_S)\rangle = \sum_{l=1}^{M}\sum_{j=1}^{M}\sum_{i=1}^{N_1} r_j(t)r_l(t)\overline{\langle u_{\mathbf{k}^j}, \psi_i\rangle}\langle u_{\mathbf{k}^l}, \psi_i\rangle.$$

Therefore,

$$\int_0^1 \langle f_S, P(f_S)\rangle dt = \sum_{j=1}^{M}\sum_{i=1}^{N_1} |\langle u_{\mathbf{k}^j}, \psi_i\rangle|^2. \tag{5.3.20}$$

The system $\{u_{\mathbf{k}}\}_{\mathbf{k}\in\Delta Q_n^+}$ forms an orthonormal basis of $\mathscr{T}(\Delta Q_n^+)$. By the Parseval identity we obtain, for each i,

$$\sum_{j=1}^{M}\sum_{\mathbf{k}\in\rho(\mathbf{s}^j)^+} |\langle \psi_i, u_{\mathbf{k}}\rangle|^2 = 1. \tag{5.3.21}$$

We note that the sum over $S=(\mathbf{k}^1,\dots,\mathbf{k}^M)\in T$ contains $|T|=(2^{n-d})^M$ summands, where $2^{n-d}=|\rho(\mathbf{s})^+|$ for $\mathbf{s}\in\theta_n$ and

$$\sum_{S\in T} |\langle \psi_i, u_{\mathbf{k}^j}\rangle|^2 = 2^{(n-d)(M-1)} \sum_{\mathbf{k}^j\in\rho(\mathbf{s}^j)^+} |\langle \psi_i, u_{\mathbf{k}^j}\rangle|^2.$$

Thus, taking into account (5.3.21), we get from (5.3.20)

$$\frac{1}{|T|}\sum_{S\in T}\int_0^1 \langle f_S, P(f_S)\rangle dt \ge \frac{1}{|T|}N_1\frac{|T|}{N/M} \ge M/2.$$

This implies the following estimate for A:

$$A \gg M^a. \tag{5.3.22}$$

We now proceed to the more difficult part of the proof – the upper estimate for B. The key role in this proof is played by the inequality (see Lemma 3.3.7)

$$\|f\|_a^a \ll \sum_{\mathbf{u}} \left(2^{\|\mathbf{u}\|_1(1/2-1/a)}\|\delta_{\mathbf{u}}(f)\|_2\right)^a, \qquad a\ge 2. \tag{5.3.23}$$

We can estimate $\|P(f_S)\|_a^a$ using this inequality. Denoting

$$g_{\mathbf{u},\mathbf{k}^j} := \delta_{\mathbf{u}}(P(u_{\mathbf{k}^j})),$$

we get from (5.3.23)

$$\|P(f_S)\|_a^a \ll 2^{n(a/2-1)} \sum_{\mathbf{u}\in\theta_n} \left\| \sum_{j=1}^{M} r_j(t) g_{\mathbf{u},\mathbf{k}^j}(\cdot)\right\|_2^a.$$

Let $\mathbf{u}$ be fixed, and estimate

$$A_{\mathbf{u},S} := \int_0^1 \left\| \sum_{j=1}^{M} r_j(t) g_{\mathbf{u},\mathbf{k}^j}(\cdot)\right\|_2^a dt.$$

Using the inequality (see (A.1.10))

$$\| \|g(\cdot,t\|_2 \|_a \le \| \|g(\mathbf{x},\cdot)\|_a \|_2, \qquad \text{for } a \ge 2,$$

and the Khinchin inequality, we get

$$A_{\mathbf{u},S}^{1/a} \ll \left(\sum_{j=1}^{M} \|g_{\mathbf{u},\mathbf{k}^j}\|_2^2 \right)^{1/2}.$$

We need the following simple inequalities for $y_{\mathbf{k}^j} := \|g_{\mathbf{u},\mathbf{k}^j}\|_2^2$:

$$0 \le y_{\mathbf{k}^j} \le 1, \tag{5.3.24}$$

$$\sum_{j=1}^{M} \sum_{\mathbf{k}\in\rho(\mathbf{s}^j)^+} y_{\mathbf{k}} = \sum_{\mathbf{k}} \|\delta_{\mathbf{u}}(P(u_{\mathbf{k}}))\|_2^2 = \sum_{\mathbf{m}\in\rho(\mathbf{u})^+} \sum_{\mathbf{k}} |\langle P(u_{\mathbf{k}}), u_{\mathbf{m}}\rangle|^2$$
$$\sum_{\mathbf{m}\in\rho(\mathbf{u})^+} \sum_{\mathbf{k}} |\langle u_{\mathbf{k}}, P(u_{\mathbf{m}})\rangle|^2 \le |\rho(\mathbf{u})^+| = 2^{n-d}. \tag{5.3.25}$$

Thus

$$\frac{1}{|T|} \sum_{S\in T} \int_0^1 \|P(f_S)\|_a^a dt \ll 2^{n(a/2-1)} \sum_{\mathbf{u}\in\theta_n} \frac{1}{|T|} \sum_{S\in T} \left(\sum_{j=1}^{M} \|g_{\mathbf{u},\mathbf{k}^j}\|_2^2 \right)^{a/2}. \tag{5.3.26}$$

The rest of the proof is based on a combinatorial lemma, which we formulate in a form that is convenient for us.

Lemma 5.3.13 *Assume that the $\{y_{\mathbf{k}}\}_{\mathbf{k}\in\Delta Q_n^+}$ satisfy the inequalities (5.3.24) and (5.3.25). Then, for any $1 \le b < \infty$, we have*

$$Y := \frac{1}{|T|} \sum_{S\in T} \left(\sum_{j=1}^{M} y_{\mathbf{k}^j} \right)^b \le C(b).$$

Proof By the monotonicity of weighted ℓ_p-norms we can assume that b is an integer. Then

$$Y = \frac{1}{|T|} \sum_{S\in T} \sum_{(b_1,\dots,b_M)} \frac{b!}{b_1!\cdots b_M!} y_{\mathbf{k}^1}^{b_1} \cdots y_{\mathbf{k}^M}^{b_M} \le \frac{b!}{|T|} \sum_{S\in T} \sum_{(b_1,\dots,b_M)} y_{\mathbf{k}^1}^{b_1} \cdots y_{\mathbf{k}^M}^{b_M},$$

where $b_1,\dots,b_M$ satisfy the equality $b_1 + \cdots + b_M = b$. It is clear that the number of nonzero b_j is at most b. Let $B(\Lambda)$ be the set of $(b_1,\dots,b_M)$ such that $b_j \ge 1$ for $j \in \Lambda$ and $b_j = 0$ otherwise. Denote by $\Sigma(k)$ the set of all subsets $\Lambda \subset \{1,\dots,M\}$ with $|\Lambda| = k$. For all $\Lambda \in \Sigma(k)$, $k \le b$, we have

$$|B(\Lambda)| \le C(b)$$

and, by (5.3.24), for $(b_1,\dots,b_M) \in B(\Lambda)$,

$$y_{\mathbf{k}^1}^{b_1} \cdots y_{\mathbf{k}^M}^{b_M} \le \prod_{l\in\Lambda} y_{\mathbf{k}^l}.$$

Thus

$$\begin{aligned}Y &\le \frac{C(b)}{|T|}\sum_{S\in T}\sum_{k=1}^{b}\sum_{\Lambda\in\Sigma(k)}\prod_{l\in\Lambda} y_{\mathbf{k}^l} \ll \sum_{k=1}^{b}\sum_{\Lambda\in\Sigma(k)}\frac{1}{|T|}\sum_{S\in T}\prod_{l\in\Lambda} y_{\mathbf{k}^l}\\ &\le \sum_{k=1}^{b}\sum_{\Lambda\in\Sigma(k)} 2^{-nk}\prod_{l\in\Lambda}\Bigg(\sum_{\mathbf{k}\in\rho(\mathbf{s}^l)^+} y_{\mathbf{k}^l}\Bigg) \le \sum_{k=1}^{b} 2^{-nk}\sum_{\Lambda\in\Sigma(k)}\prod_{l\in\Lambda}\Bigg(\sum_{\mathbf{k}\in\rho(\mathbf{s}^l)^+} y_{\mathbf{k}^l}\Bigg)\\ &\le \sum_{k=1}^{b} 2^{-nk}\Big(\sum_{\mathbf{k}} y_{\mathbf{k}}\Big)^k \ll 1.\end{aligned}$$

The lemma is proved. □

Using Lemma 5.3.13 we get from (5.3.26)

$$B \ll 2^{n(a/2-1)}M. \tag{5.3.27}$$

Combining (5.3.22) and (5.3.27) we obtain

$$(A/B)^{1/a} \gg 2^{-n(1/2-1/a)}M^{1-1/a}.$$

From this and (5.3.17) we find that

$$d_m(\mathbf{H}^r_q, L_p) \gg 2^{-rn+n(1/q-1/p)}n^{(d-1)/p} \asymp \left(\frac{(\log m)^{d-1}}{m}\right)^{r-1/q+1/p}(\log m)^{(d-1)/p}.$$

Theorem 5.3.12 is proved. □

We now illustrate a powerful method of proving lower bounds for Kolmogorov widths. This method is based on the lower bounds for entropy numbers. We will discuss entropy numbers in Chapter 7: here we just use some results from there. For further discussion of related results on the entropy numbers we refer the reader to Temlyakov (2011), Chapter 3, and Dinh Dung *et al.* (2016). The following lemma plays a fundamental role in that method.

Lemma 5.3.14 *Let A be centrally symmetric compact in a separable Banach space X such that for two real numbers $r > 0$ and $a \in \mathbb{R}$ we have*

$$d_m(A,X) \ll m^{-r}(\log m)^a$$

and

$$\varepsilon_m(A,X) \gg m^{-r}(\log m)^a.$$

Then the relations

$$d_m(A,X) \asymp \varepsilon_m(A,X) \asymp m^{-r}(\log m)^a$$

hold.

Proof The key role in the proof is played by Theorem 7.4.1. We begin with the upper bounds for $\varepsilon_m(A,X)$. Using the inequality

$$n^{-r}(\log n)^a \le C(a)n^{-r-1}m(\log m)^a, \qquad 2 \le n \le m,$$

and applying Theorem 7.4.1 with $r+1$ we obtain, for $k=m$,

$$\varepsilon_m(A,X) \le C(r,a)m^{-r-1}m(\log m)^a = C(r,a)m^{-r}(\log m)^a.$$

This proves the required upper bound.

We now prove the lower bound for $d_m(A,X)$. Our assumption on $\{\varepsilon_m(A,X)\}$ implies that

$$\varepsilon_m(A,X) \ge C_1 m^{-r-1}m(\log m)^a.$$

Therefore, by Theorem 7.4.1

$$\max_{n\le m} d_{n-1}(A,X)n^{r+1} \ge C_2 m(\log m)^a. \tag{5.3.28}$$

The assumptions on $d_m(A,X)$ and (5.3.28) guarantee that there is a $c>0$ such that, for some $n \in [cm,m]$, we have

$$d_{n-1}(A,X) \ge C_2 n^{-r}(m/n)(\log m)^a \ge C_4 n^{-r}(\log n)^a.$$

This implies the required lower bound. □

Next we formulate now some results about the Kolmogorov widths of classes $\mathbf{W}^r_{q,\alpha}$ and $\mathbf{H}^r_q$. Theorems 5.3.1 and 7.7.2 together with Lemma 5.3.14 imply the following result.

Theorem 5.3.15 *For all $r>0$ and $1 \le p < \infty$ we have*

$$d_m(\mathbf{W}^r_{\infty,\alpha}, L_p) \asymp m^{-r}(\log m)^{r(d-1)}.$$

Proof The lower bound follows from Theorem 7.7.2. The corresponding upper bound follows from the case $1 < q = p < \infty$ covered by Theorem 5.3.1. □

As a corollary of the lower bound in Theorem 5.3.15 and the upper bound from Theorem 5.3.1 we obtain the following result.

Theorem 5.3.16 *For all $r>0$ and $1 < q \le \infty$ we have*

$$d_m(\mathbf{W}^r_{q,\alpha}, L_1) \asymp m^{-r}(\log m)^{r(d-1)}.$$

Theorem 5.3.17 *For all $r > 0$ and $2 \le q \le \infty$ we have*

$$d_m(\mathbf{H}^r_q, L_1) \asymp m^{-r}(\log m)^{(r+1/2)(d-1)}.$$

Proof Theorems 5.3.2 and 7.8.1 together with Lemma 5.3.14 imply the lower bound for $q = \infty$. The corresponding upper bound follows from the upper bound for $d_m(\mathbf{H}^r_2, L_2)$ covered by Theorem 5.3.2. □

Lemma 5.3.14 and the results on entropy numbers from Chapter 7 give the following two theorems for the classes of functions of two variables.

Theorem 5.3.18 *In the case $d = 2$ we have, for $2 \le q \le \infty$ and $r > 1/2$,*

$$d_m(\mathbf{W}^r_{q,\alpha}, L_\infty) \asymp m^{-r}(\log m)^{r+1/2}.$$

Proof The lower bound for $q = \infty$ follows from Theorem 7.7.13 and Lemma 5.3.14. The corresponding upper bound for $q = 2$ is easily derived from Lemma 5.3.3. □

Theorem 5.3.19 *In the case $d = 2$ we have, for $2 \le q \le \infty$ and $r > 1/2$,*

$$d_m(\mathbf{H}^r_q, L_\infty) \asymp m^{-r}(\log m)^{r+1}.$$

Proof The lower bound for $q = \infty$, $r > 0$, follows from Theorem 7.8.4 and Lemma 5.3.14. The corresponding upper bound for $q = 2$ is easily derived from Theorem 3.5.9. □

Remark 5.3.20 In the case $q = \infty$ Theorem 5.3.19 holds for $r > 0$.

Proof As we pointed out in the proof of Theorem 5.3.19 the lower bound holds for $r > 0$. The corresponding upper bound follows from Theorem 4.4.13. □

5.4 Universality of Approximation by Trigonometric Polynomials from the Hyperbolic Crosses

The results of the previous sections of this chapter show that the sets $\mathscr{T}(Q_n)$ of trigonometric polynomials and the operators S_{Q_n} for the approximation of the classes $\mathbf{W}^r_{q,\alpha}$ and $\mathbf{H}^r_q$ play the same role as the sets $\mathscr{T}(2^n)$ and the operators S_{2^n} do in the univariate case. The results of Chapter 3 and of §§5.1–5.3 give, for the classes under consideration, those sets of trigonometric polynomials or, in other words, those ways of ordering the trigonometric system, which are best for the approximation of functions in these classes. We emphasize that this optimal way of ordering depends on the class involved. For the Sobolev classes $W^{\mathbf{r}}_{q,\alpha}$ and the Nikol'skii classes $H^{\mathbf{r}}_q$ the optimal ordering of the system $\{e^{i(\mathbf{k},\mathbf{x})}\}$ is an ordering in such a manner that for their partial sums one can take the Fourier sums $S(\cdot,\mathbf{r},n)$

for all n. This ordering essentially depends on the anisotropy of the class, defined by the vector $\mathbf{r}$. For this reason, in this section we study the following problem. Let a collection of function classes $F_q^{\mathbf{r}}$, $\mathbf{r} \in P$, be given. For a fixed m we define the number $N(m,q,p)$ as the smallest among those numbers N, for which there is an orthonormal system $\Psi = \{\psi_i\}_{i=1}^N$ such that, for all $\mathbf{r} \in P$,

$$\sup_{f \in F_q^{\mathbf{r}}} \left\| f - \sum_{i=1}^{N} \langle f, \psi_i \rangle \psi_i \right\|_p \leq \varphi_m(F_q^{\mathbf{r}}, L_p).$$

The number

$$u(m,q,p) := N(m,q,p)/m,$$

is called the *index of universality* for the collection of classes $\{F_q^{\mathbf{r}}\}_{\mathbf{r} \in P}$, with respect to the orthowidth in L_p.

In the same way we define the *index of universality* $\kappa(m,q,p)$ for the collection of classes $\{F_q^{\mathbf{r}}\}_{\mathbf{r} \in P}$ with respect to the Kolmogorov widths in L_p:

$$\kappa(m,q,p) := L(m,q,p)/m,$$

where $L(m,q,p)$ is the smallest number among those L for which there is a system Ψ of functions $\{\psi_i\}_{i=1}^L$ such that, for all $\mathbf{r} \in P$,

$$\sup_{f \in F_q^{\mathbf{r}}} \inf_{c_i} \left\| f - \sum_{i=1}^{L} c_i \psi_i \right\|_p \leq d_m(F_q^{\mathbf{r}}, L_p).$$

The problem of finding universal systems of functions can be raised in the following way. Assume that we know that the function f belongs to, for example, the Nikol'skii class $H_q^{\mathbf{r}}$ but the vector $\mathbf{r}$ is not known exactly; we know only that $\mathbf{r} \in P := \prod_{j=1}^d [A_j, B_j]$. What is the most natural form of the partial Fourier series sums for the approximation of the function $f(\mathbf{x})$?

5.4.1 Universality with respect to the Orthowidth

In this subsection we prove two theorems which imply that the universal system for the collections of anisotropic Sobolev or Nikol'skii classes is the trigonometric system ordered in such a manner that for their partial sums one can take $S_{Q_n}(\cdot)$ for all n. It is interesting to compare these results with Theorems 5.2.1 and 5.2.5, which imply that the same system is optimal for the classes $\mathbf{W}^r_{q,\alpha}$ and $\mathbf{H}^r_q$.

Theorem 5.4.1 *Let $F_q^{\mathbf{r}}$ denote one of the classes of functions of d variables $W^{\mathbf{r}}_{q,\alpha}$, $H^{\mathbf{r}}_q$. Then, for $1 \leq q,p \leq \infty$, $(q,p) \neq (1,1), (\infty,\infty)$ and for any nonsingular parallelepiped $P := \prod_{j=1}^d [A_j, B_j]$ with $g(\mathbf{A}) > (1/q - 1/p)_+$, $\mathbf{A} = (A_1, \dots, A_d)$, we* *have the relation*

$$u(m,q,p) \asymp (\log m)^{d-1}.$$

The subspace onto which the orthogonal projections give approximations of order $\varphi_m(F_q^{\mathbf{r}}, L_p)$, *for all* $\mathbf{r} \in P$, *is the subspace of trigonometric polynomials* $\mathscr{T}(Q_n)$ *with* $n = [\log m]$.

We first prove an auxiliary assertion.

Lemma 5.4.2 *Assume that a number* $\theta > 0$ *is given, with* $\theta d \le 1$. *Then, for any numbers* $\beta_1, \dots, \beta_d$ *satisfying the conditions*

$$\sum_{j=1}^{d} \beta_j = 1, \qquad \beta_j \ge \theta, \qquad j = 1, \dots, d,$$

there exists an $n(\theta,d)$ *such that, for* $n > n(\theta,d)$, *the number* $N(n,\theta,d)$ *of solutions of the inequalities*

$$|s_j - \beta_j n| \le \theta n$$

in natural numbers s_j, $j = 1, \dots, d$, *under the condition*

$$\sum_{j=1}^{d} s_j = n,$$

satisfies the estimate

$$N(n,\theta,d) \ge C(\theta,d) n^{d-1}, \qquad C(\theta,d) > 0.$$

Proof The proof will be carried out by induction on the dimension d. For $d = 1$ the conclusion of the lemma is valid with $C(\theta,d) = 1$. Now assume that the lemma holds for $d - 1 \ge 1$. From this we can derive the conclusion of the lemma for d. We set

$$\gamma_j = \beta_j (1 - \beta_d)^{-1}, \qquad j = 1, \dots, d-1.$$

Then

$$\sum_{j=1}^{d-1} \gamma_j = 1$$

and

$$\gamma_j > \theta, \qquad j = 1, \dots, d-1.$$

Let m be an arbitrary natural number such that

$$\bigl|m - n(1 - \beta_d)\bigr| \le \kappa n, \tag{5.4.1}$$
$$\kappa := (d-1)\theta^3 \bigl(1 + (d-1)\theta^2\bigr)^{-1}.$$

We set

$$n(\theta,d) := \left[\frac{2n(\theta,d-1)}{(d-1)\theta}\right] + 1.$$

Then for $n > n(\theta,d)$ we have $m > n(\theta,d-1)$. Applying the lemma for $d-1$ with $\gamma_1,\dots,\gamma_{d-1}$ and with m satisfying (5.4.1), we find that the number $N(m,\theta,d-1)$ of natural numbers s_j, $j=1,\dots,d-1$, such that

$$|s_j - \gamma_j m| \le \theta m \tag{5.4.2}$$

under the condition

$$\sum_{j=1}^{d-1} s_j = m$$

is estimated from below by the number

$$C(\theta,d-1)m^{d-2}. \tag{5.4.3}$$

Now, for each m satisfying (5.4.1), we set $s_d = n - m$. Then, by (5.4.1) and the definition of κ, we obtain

$$|s_d - \beta_d n| \le \theta n.$$

Next we prove the relations

$$|s_j - \beta_j n| \le \theta n, \qquad j = 1,\dots,d-1. \tag{5.4.4}$$

By (5.4.1), (5.4.2), and the definitions of γ_j and κ we have

$$s_j \le \beta_j n + \theta(1-\beta_d)n + \kappa n\left(\theta + \frac{\beta_j}{1-\beta_d}\right) \le \beta_j n + \theta n.$$

Similarly,

$$s_j \ge \beta_j n - \theta n.$$

Thus relations (5.4.4) are proved.

For the number of vectors $\mathbf{s} = (s_1,\dots,s_d)$ satisfying the condition of the lemma with $n > n(\theta,d)$ we obtain the estimate

$$N(n,\theta,d) \ge \sum_{m:\left|m-(1-\beta_d)n\right| \le \kappa n} N(m,\theta,d-1) \ge C(\theta,d)n^{d-1}, \qquad C(\theta,d) > 0.$$

The lemma is proved. □

Proof of Theorem 5.4.1 The upper estimates for $\left\|f - S_{Q_n}(f)\right\|_p$ are easily derived from the results of Chapter 3. Indeed, assume first that $1 < p < \infty$. For $f \in H_p^{\mathbf{r}}$ we have the relation (see Theorem 3.4.14)

$$\left\|f - S(f,\mathbf{r},n)\right\|_p \ll 2^{-g(\mathbf{r})n}, \qquad S(f,\mathbf{r},n) \in T^{\mathbf{r}}(n),$$

and for each polynomial $t \in T^{\mathbf{r}}(n)$ one has $S_{Q_{n+d}}(t) = t$. From this and from the boundedness of $S_{Q_{n+d}}$ as an operator from L_p to L_p, we obtain

$$\|f - S_{Q_{n+d}}(f)\|_p \ll m^{-g(\mathbf{r})}, \qquad m = 2^n.$$

For $1 \le q < p < \infty$, by the inclusion $H_q^{\mathbf{r}} \subset H_p^{\mathbf{r}'}$, $\mathbf{r}' = \big(1 - (1/q - 1/p)g(\mathbf{r})^{-1}\big)\mathbf{r}$ we obtain for $f \in H_q^{\mathbf{r}}$,

$$\|f - S_{Q_{n+d}}(f)\|_p \ll m^{-g(\mathbf{r})+1/q-1/p}.$$

Let us consider the case $p = \infty$, $1 \le q < \infty$. We have, for $f \in H_q^{\mathbf{r}}$, $1 < q < \infty$,

$$\begin{aligned}\|f - S_{Q_{n+d}}(f)\|_\infty &\le \|f - S(f,\mathbf{r},n)\|_\infty \\ &\quad + \sum_{j=n}^{\infty} \|S_{Q_{n+d}}(S(f,\mathbf{r},j+1) - S(f,\mathbf{r},j)\|_\infty.\end{aligned}$$

By the Nikols'kii inequality and by Theorem 3.4.14 and Remark 3.4.15, we continue the inequality as

$$\begin{aligned}&\ll 2^{-n\left(g(\mathbf{r})-1/q\right)} + \sum_{j=n}^{\infty} \|S_{Q_{n+d}}(S(f,\mathbf{r},j+1) - S(f,\mathbf{r},j)\|_q 2^{j/q} \\ &\ll 2^{-n\left(g(\mathbf{r})-1/q\right)} = m^{-g(\mathbf{r})+1/q}.\end{aligned}$$

In the case $q = 1$ we use $V(f,\mathbf{r},j)$ instead of $S(f,\mathbf{r},j)$ and make the estimation

$$\begin{aligned}&\|S_{Q_{n+2d}}\big(V(f,\mathbf{r},j+1) - V(f,\mathbf{r},j)\big)\|_\infty \\ &\quad \ll 2^{j/2} \|S_{Q_{n+2d}}\big(V(f,\mathbf{r},j+1) - V(f,\mathbf{r},j)\big)\|_2 \\ &\quad \ll 2^j \|V(f,\mathbf{r},j+1) - V(f,\mathbf{r},j)\|_1 \ll 2^{-j\left(g(\mathbf{r})-1\right)}.\end{aligned}$$

Summing over j we obtain the required estimates. The upper estimates for $p < q$ follow from the estimates for $p = q$.

Let us now prove the lower estimates for Theorem 5.4.1. We first prove a somewhat stronger statement.

Theorem 5.4.3 *Let $1 \le q, p \le \infty$ and $G \in \mathscr{L}_N(B)_p$ and assume that, for all $\mathbf{r} \in P = \prod_{j=1}^d [A_j, B_j]$, $A_j < B_j$, $j = 1, \dots, d$, $g(\mathbf{A}) > (1/q - 1/p)_+$, one has the estimate*

$$\sup_{f \in F_q^{\mathbf{r}}} \|f - G(f)\|_p \le K m^{-g(\mathbf{r}) + (1/q-1/p)_+},$$

with a constant K which does not depend on $\mathbf{r}$ or m. Then, for N, one has the lower estimate

$$N \gg m(\log m)^{d-1}.$$

Proof Let n be a natural number, whose value will be chosen later. Let

$$w_{\mathbf{s}}(\mathbf{x}) := e^{i(\mathbf{k}^{\mathbf{s}},\mathbf{x})} \prod_{j=1}^{d} \mathscr{K}_{2^{s_j-3}-1}(x_j),$$

$$k_j^{\mathbf{s}} := \begin{cases} 2^{s_j-1} + 2^{s_j-2}, & s_j \geq 2 \\ 1, & s_j = 1,\ j = 1,\dots,d, \end{cases}$$

where $\mathscr{K}_l(t)$ is the Fejér kernel of order l and $\mathscr{K}_l(t) \equiv 1$ for $l < 1$. Then

$$\|w_{\mathbf{s}}\|_1 = 1.$$

We define the numbers

$$r_j := \frac{2A_jB_j}{A_j + B_j}, \qquad \beta_j := g(\mathbf{r})r_j^{-1}, \qquad j = 1,\dots,d,$$

$$\theta := \min\Big\{g(\mathbf{A});\ 1/2(1/A_j - 1/B_j)g(\mathbf{r}),\ j = 1,\dots,d;\ 1/d\Big\}.$$

Denote by $h_n(\beta,\theta)$ the set of vectors $\mathbf{s}$ with positive integer components such that

$$|s_j - \beta_j n| \leq \theta n, \qquad j = 1,\dots,d, \qquad \sum_{j=1}^{d} s_j = n.$$

Then, by Lemma 5.4.2,

$$\big|h_n(\beta,\theta)\big| \geq C(\theta,d)n^{d-1}.$$

We now consider the functions $w_{\mathbf{s}}$ with $\mathbf{s} \in h_n(\beta,\theta)$ and the operators

$$T_{\mathbf{s}} := U_{\mathbf{s}}G, \qquad T := \sum_{\mathbf{s}\in h_n(\beta,\theta)} T_{\mathbf{s}},$$

$$U_{\mathbf{s}}(f) := f * \left(e^{i(\mathbf{k}^{\mathbf{s}},\mathbf{x})} \prod_{j=1}^{d} \mathscr{V}_{2^{s_j-3}}(x_j) \right).$$

The range of the operator T is a subspace Ψ_N of the space $\mathscr{T}(Q_n)$ of dimension $\dim \Psi_N =: \overline{N} \leq N$.

Let $\{\psi_\mu\}_{\mu=1}^{\overline{N}}$ be an orthonormal basis of Ψ_N and let

$$T(e^{i(\mathbf{k},\mathbf{x})}) = \sum_{\mu=1}^{\overline{N}} a_\mu^{\mathbf{k}} \psi_\mu(\mathbf{x}).$$

Then

$$\sum_{\mu=1}^{\overline{N}} |a_\mu^{\mathbf{k}}|^2 \leq B^2.$$

It is easy to see that for the operator $T_{\mathbf{s}}$ we have

$$T_{\mathbf{s}}(e^{i(\mathbf{k},\mathbf{x})}) = \sum_{\mu=1}^{\overline{N}} a_{\mu}^{\mathbf{k}} \delta_{\mathbf{s}}(\psi_{\mu}, \mathbf{x}).$$

We consider the quantities

$$J_{\mathbf{s}} := \sup_{\mathbf{y}} \left\| w_{\mathbf{s}}(\mathbf{x}-\mathbf{y}) - T_{\mathbf{s}}\big(w_{\mathbf{s}}(\mathbf{x}-\mathbf{y})\big) \right\|_{\infty}, \qquad \mathbf{s} \in h_n(\beta,\theta).$$

Denoting

$$b_{\mathbf{s}} := \min_{\mathbf{y}=\mathbf{x}} \operatorname{Re} T_{\mathbf{s}}\big(w_{\mathbf{s}}(\mathbf{x}-\mathbf{y})\big),$$

since $w(\mathbf{0})$ is a real number we have

$$J_{\mathbf{s}} \geq w_{\mathbf{s}}(\mathbf{0}) - b_{\mathbf{s}}.$$

By Lemma 5.2.9 for $b_{\mathbf{s}}$ we obtain the estimates

$$b_{\mathbf{s}} \leq \operatorname{Re} \sum_{\mathbf{k}\in\rho(\mathbf{s})^+} \sum_{\mu=1}^{\overline{N}} \hat{w}_{\mathbf{s}}(\mathbf{k}) a_{\mu}^{\mathbf{k}} \hat{\psi}_{\mu}(\mathbf{k}),$$

$$\sum_{\mathbf{s}\in h_n(\beta,\theta)} b_{\mathbf{s}} \leq \sum_{\mu=1}^{\overline{N}} \sum_{\mathbf{s}\in h_n(\beta,\theta)} \sum_{\mathbf{k}\in\rho(\mathbf{s})^+} \hat{w}_{\mathbf{s}}(\mathbf{k}) a_{\mu}^{\mathbf{k}} \hat{\psi}_{\mu}(\mathbf{k}).$$

Applying Cauchy's inequality and using the inequalities $|\hat{w}_{\mathbf{s}}(\mathbf{k})| \leq 1$, we continue the above estimate:

$$\leq B\big(N2^n |h_n(\beta,\theta)|\big)^{1/2}.$$

However, $w_{\mathbf{s}}(\mathbf{0}) \geq C_d 2^{\|\mathbf{s}\|_1}$, $C_d > 0$. Therefore

$$\sum_{\mathbf{s}\in h_n(\beta,\theta)} w_{\mathbf{s}}(\mathbf{0}) \geq C_d 2^n |h_n(\beta,\theta)|.$$

We choose n such that

$$C_d 2^n |h_n(\beta,\theta)| - B\big(N2^n |h_n(\beta,\theta)|\big)^{1/2} > C_1(\theta,d) 2^n n^{d-1},$$
$$2^n n^{d-1} < C(\theta,d,B,\beta) N. \tag{5.4.5}$$

Clearly, this can be done. Then we find that for some $\mathbf{s}^*$ we have $J_{\mathbf{s}^*} \gg 2^n$ with a constant independent of n. From this we find that for some $\mathbf{y}^*$ we have

$$\left\| w_{\mathbf{s}^*}(\mathbf{x}-\mathbf{y}^*) - T_{\mathbf{s}^*}\big(w_{\mathbf{s}^*}(\mathbf{x}-\mathbf{y}^*)\big) \right\|_{\infty} \gg 2^n$$

and, by the Nikol'skii inequality,

$$\left\| w_{\mathbf{s}^*}(\mathbf{x}-\mathbf{y}^*) - T_{\mathbf{s}^*}\big(w_{\mathbf{s}^*}(\mathbf{x}-\mathbf{y}^*)\big) \right\|_{p} \gg 2^{n(1-1/p)}, \qquad 1 \leq p < \infty.$$

Further, we denote $v(\mathbf{x}) := w_{\mathbf{s}^*}(\mathbf{x}-\mathbf{y}^*)$. Then

$$\|v - T_{\mathbf{s}^*}v\|_p = \left\|U_{\mathbf{s}^*}\left(v - G(v)\right)\right\|_p \le C(d)\left\|v - G(v)\right\|_p.$$

Next we define $\mathbf{r}^*$ as follows:

$$r_j^* := \frac{g(\mathbf{r})n}{s_j^*}.$$

Then it is easy to verify that, from the definition of $\mathbf{r}, \beta$, and θ, it follows that $\mathbf{r}^* \in P$. Also, $g(\mathbf{r}^*) = g(\mathbf{r})$. The function

$$f(\mathbf{x}) := 2^{-n\left(g(\mathbf{r})+1-1/q\right)}v(\mathbf{x})$$

belongs to the class $W^{\mathbf{r}^*}_{q,\alpha}R$ for some constant $R > 0$ that is independent of n and $\mathbf{r}$. Thus we obtain

$$\left\|f - G(f)\right\|_p \gg 2^{-n\left(g(\mathbf{r})-1/q+1/p\right)}, \qquad f \in W^{\mathbf{r}^*}_{q,\alpha}R.$$

Taking into account the hypothesis of Theorem 5.4.3 we obtain

$$2^{-n\left(g(\mathbf{r}^*)-1/q+1/p\right)} \ll m^{-\left(g(\mathbf{r}^*)-1/q+1/p\right)},$$

which implies that

$$2^n \gg m$$

and, by (5.4.5), we obtain the required estimate.

Let us now consider the case $1 \le p < q \le \infty$. Assume that m and an operator $G \in \mathscr{L}_m(B)_1$ are given such that, for an arbitrary $f \in F^{\mathbf{r}}_\infty$, $\mathbf{r} \in P$,

$$\left\|f - G(f)\right\|_1 \le Km^{-g(\mathbf{r})}.$$

We can show that $N \gg m(\log m)^{d-1}$. Let $\mathbf{r}, \beta, \theta, h_n(\beta,\theta)$ be as above and let

$$\overline{Q}_n = \bigcup_{\mathbf{s}\in h_n(\beta,\theta)} \rho(\mathbf{s}).$$

We choose n such that

$$2^d|\overline{Q}_n| \ge 4B^2N, \qquad |\overline{Q}_n| \le B^2C(d)N. \tag{5.4.6}$$

Now we consider approximations of the functions $e^{i(\mathbf{k},\mathbf{x})}$, $|\mathbf{k}| \in \overline{Q}_n$. If $\{\psi_\mu\}_{\mu=1}^{\overline{N}}$ ($\overline{N} \le N$) is an orthonormal basis in $G\left(F^{\mathbf{r}}_\infty \cap T(Q_n)\right)$ and

$$G(e^{i(\mathbf{k},\mathbf{x})}) = \sum_{\mu=1}^{\overline{N}} a_\mu^{\mathbf{k}}\psi_\mu(\mathbf{x}),$$

$$b_\mu^{\mathbf{k}} := \langle\psi_\mu(\mathbf{x}), e^{i(\mathbf{k},\mathbf{x})}\rangle,$$

then

$$\sum_{\mu=1}^{\overline{N}} |a_\mu^{\mathbf{k}}|^2 \le B^2, \qquad \sum_{\mu=1}^{\overline{N}} |b_\mu^{\mathbf{k}}|^2 \le 1, \qquad \sum_{\mathbf{k}} |b_\mu^{\mathbf{k}}|^2 \le 1. \tag{5.4.7}$$

Write

$$\rho_{\mathbf{k}} := \langle G(e^{i(\mathbf{k},\mathbf{x})}), e^{i(\mathbf{k},\mathbf{x})}\rangle.$$

We have

$$\rho_{\mathbf{k}} = \sum_{\mu=1}^{\overline{N}} a_\mu^{\mathbf{k}} b_\mu^{\mathbf{k}}$$

and

$$|\rho_{\mathbf{k}}|^2 \le B^2 \sum_{\mu=1}^{\overline{N}} |b_\mu^{\mathbf{k}}|^2.$$

From this and from (5.4.7) we find

$$\sum_{|\mathbf{k}|\in\overline{Q}_n} |\rho_{\mathbf{k}}|^2 \le B^2 \overline{N}. \tag{5.4.8}$$

Consequently, taking into account (5.4.6) we obtain from (5.4.8) that there exist $\mathbf{s}^* \in h_n(\beta,\theta)$ and $\mathbf{k}^* : |\mathbf{k}^*| \in \rho(\mathbf{s}^*)$ such that

$$|\rho_{\mathbf{k}^*}| \le 1/2.$$

Then

$$\begin{aligned} |1-\rho_{\mathbf{k}^*}| &= \left|\langle e^{i(\mathbf{k}^*,\mathbf{x})} - G(e^{i(\mathbf{k}^*,\mathbf{x})}), e^{i(\mathbf{k}^*,\mathbf{x})}\rangle\right| \\ &\le \left\| e^{i(\mathbf{k}^*,\mathbf{x})} - G(e^{i(\mathbf{k}^*,\mathbf{x})})\right\|_1, \end{aligned}$$

and

$$\left\| e^{i(\mathbf{k}^*,\mathbf{x})} - G(e^{i(\mathbf{k}^*,\mathbf{x})})\right\|_1 \ge 1/2. \tag{5.4.9}$$

It is easy to verify that $\mathbf{r}^* \in P$, where $r_j^* = g(\mathbf{r})n/s_j^*$, and that the function

$$f(\mathbf{x}) = 2^{-ng(\mathbf{r})} e^{i(\mathbf{k}^*,\mathbf{x})}$$

belongs to the class $W^{\mathbf{r}^*}_{\infty,\alpha}$. Then from (5.4.9) we find for $f(\mathbf{x})$

$$\|f - G(f)\|_1 \ge 2^{-ng(\mathbf{r})}.$$

From this and from the initial assumption on the operator G we obtain

$$2^n \gg m.$$

Taking into account (5.4.6) and applying Lemma 5.4.2, we obtain the required estimate for N.

Theorem 5.4.3 is now proved. □

In order to conclude the proof of Theorem 5.4.1 it remains to note that, for $\mathbf{r} \in P$, one has the following estimate for $\varphi_m(F_q^{\mathbf{r}}, L_p)$:

$$\varphi_m(F_q^{\mathbf{r}}, L_p) \ll m^{-g(\mathbf{r})+(1/q-1/p)_+}$$

uniformly with respect to $\mathbf{r} \in P$. □

5.4.2 Universality with respect to the Kolmogorov Width

Here we prove the following theorem.

Theorem 5.4.4 *Let $F_q^{\mathbf{r}}$ denote one of the classes of functions of d variables $W_{q,\alpha}^{\mathbf{r}}$, $H_q^{\mathbf{r}}$ for the following cases: $1 < q \le p \le 2$; $q = 1$, $1 < p \le 2$; $2 \le p \le q \le \infty$. Let $P = \prod_{j=1}^{d}[A_j, B_j]$ be an arbitrary nonsingular parallelepiped such that $g(\mathbf{A}) > (1/q - 1/p)_+$, $\mathbf{A} = (A_1, \dots, A_d)$. Then we have the relation*

$$\kappa(m, q, p) \asymp (\log m)^{d-1}.$$

Proof The upper estimates follow from the corresponding upper estimates for $E_{Q_n}(f)_p$. For $f \in F_q^{\mathbf{r}}$ we have

$$E_{Q_{n+2d}}(f)_p \le \left\|f - V(f, \mathbf{r}, n)\right\|_p \ll 2^{-n\left(g(\mathbf{r})-(1/q-1/p)_+\right)}.$$

We now prove the lower estimates. Let a subspace U be such that, for all $\mathbf{r} \in P$ and for all $f \in F_q^{\mathbf{r}}$,

$$\inf_{u \in U} \|f - u\|_p \le K d_m(F_q^{\mathbf{r}}, L_p), \qquad 1 \le q \le p \le 2.$$

We consider the subspace $\Psi := \{\psi = S_{Q_n}(u),\ u \in U\}$. Then $\dim \Psi \le \dim U$. From the proof of Theorem 5.4.3 with $G(f) = S(f, \Psi)$, where $S(f, \Psi)$ is the orthogonal projection onto Ψ, we select an $f \in W_{q,\alpha}^{\mathbf{r}^*}$ such that

$$\left\|f - S(f, \Psi)\right\|_2 \gg 2^{-n\left(g(\mathbf{r})-(1/q-1/2)\right)}, \qquad 2^n n^{d-1} \ll \dim \Psi. \tag{5.4.10}$$

In addition, we note that $S_{Q_n}(f) = f$. Let $u \in U$ be such that

$$\|f - u\|_p \le 2K d_m(W_q^{\mathbf{r}^*}, L_p). \tag{5.4.11}$$

Then

$$\|f - S_{Q_n}(u)\|_p \le C(d, p) K \|f - u\|_p, \qquad 1 < p \le 2, \tag{5.4.12}$$

and, using Theorem 4.3.17, we get

$$\|f - S_{Q_n}(u)\|_2 \le C(d, p) K \|f - u\|_p 2^{n(1/p-1/2)}, \qquad 1 < p \le 2. \tag{5.4.13}$$

Further,

$$\|f - S_{Q_n}(u)\|_2 \geq \|f - S(f, \Psi)\|_2. \tag{5.4.14}$$

Comparing the relations (5.4.10)–(5.4.14) and taking into account that for $\mathbf{r} \in P$, we obtain

$$d_m(F_q^{\mathbf{r}}, L_p) \ll m^{-g(\mathbf{r})+1/q-1/p}, \qquad 1 \leq q \leq p \leq 2,$$

with a constant, which does not depend on m or $\mathbf{r} \in P$, and we get

$$2^n \gg m,$$

which implies the required estimate.

In the case $2 \leq p \leq q \leq \infty$ the theorem follows from Theorem 5.4.1 for the case $p = 2$, $q = \infty$.

The theorem is now proved. □

Let $\mathbf{r}^0$, such that $g(\mathbf{r}^0) > (1/q - 1/p)_+$, be given. For a number $\eta \in (0, \min_j 1/r_j^0)$ define

$$A_j := \left(\frac{1}{r_j^0} + \eta\right)^{-1}, \qquad B_j := \left(\frac{1}{r_j^0} - \eta\right)^{-1}.$$

Suppose that η is small enough to guarantee that $g(\mathbf{A}) > (1/q - 1/p)_+$. Denote

$$\mathcal{H}(\mathbf{r}^0, \eta) := \{\mathbf{r} \in [\mathbf{A}, \mathbf{B}] : g(\mathbf{r}) = g(\mathbf{r}^0)\}.$$

The following remark follows from the proofs of Theorems 5.4.3 and 5.4.4 if we construct $\mathbf{r}^*$ such that $g(\mathbf{r}^*) = g(\mathbf{r})$.

Remark 5.4.5 Theorems 5.4.3 and 5.4.4 hold with P replaced by $\mathcal{H}(\mathbf{r}^0, \eta)$.

5.5 Historical Remarks

Theorem 5.2.1, in the special case $1 < q \leq p < \infty$, was obtained in Temlyakov (1982a, 1985d) and in the general case it was proved in Temlyakov (1988a, 1989d). Theorem 5.2.2 was proved in Temlyakov (1988a, 1989d). (See also the book Temlyakov, 1993b). Theorem 5.2.4 was obtained in Temlyakov (1988a, 1989d). Theorem 5.2.5 in the cases $1 < q < p < \infty$, $q = 1$, $1 < p < \infty$, and $1 < q = p \leq 2$ was proved in Temlyakov (1982a, 1985d), in the case $1 < p < q < 2$ by Galeev (1988), and the in general case in Temlyakov (1988a, 1989d). Theorem 5.2.6 for $1 < q < p < \infty$, $q = 1$, $1 < p < \infty$, and $1 < q = p \leq 2$ was proved in the book Temlyakov (1993b) and in the remaining cases in Temlyakov (1988a, 1989d).

Theorem 5.3.1 for $q = p = 2$ was proved by (Babenko, 1960a, b), for $1 < q = p <$

∞ by Mityagin (1962) (for r a natural number), and by Galeev (1978) (for arbitrary r). It was proved for $1 < q < p \le 2$ by Temlyakov (1980a, 1982b), for $1 \le q < p < \infty$, $2 \le p < \infty$ by Temlyakov (1982a, 1985d), and for $1 < p < q < \infty$ by Galeev (1985). Theorem 5.3.2 for $2 \le p \le q \le \infty$, $p < \infty$ was proved in Temlyakov (1980a), for $1 \le q < 2$, $p = 2$ in Temlyakov (1982b), for $1 \le q < p < \infty$, $p \ge 2$ in Temlyakov (1982a, 1985d), for $1 < p < 2 \le q < \infty$ in Galeev (1985, 1984) and in Dinh Zung (1984), and for $q = \infty$, $1 < p < \infty$ in Temlyakov (1988a, 1989d). Lemma 5.3.7 is from Temlyakov (1989d). Theorem 5.3.12 was obtained by Galeev (1990); the proof of lower bounds in this theorem required new techniques. The upper bounds in Theorem 5.3.15 follow from the case $1 < q = p < \infty$. The corresponding lower bounds in Theorem 5.3.15 were proved by Kashin and Temlyakov (1994) in the case $1 < p < \infty$. The lower bounds in the case $p = 1$ were obtained in Kashin and Temlyakov (1995). Theorem 5.3.17 holds for the class $\mathbf{H}^r_q$, $2 \le q \le \infty$. The upper bounds follow from the corresponding upper bounds for $E_{Q_n}(\mathbf{H}^r_2)_2$. The nontrivial part of this theorem is the lower bound for the class $\mathbf{H}^r_\infty$. It was observed by Belinskii that the lower bounds follow from the corresponding lower bounds for the entropy numbers $\varepsilon_k(\mathbf{H}^r_\infty, L_1)$ obtained in Temlyakov (1988a, 1989d). The most difficult part of Theorem 5.3.18 is the lower bounds. The proof of this part is based on the small ball inequality. This limits us to the case $d = 2$, where this inequality is known. In the case $2 \le q < \infty$ Theorem 5.3.18 was proved in Temlyakov (1996) and in the case $q = \infty$ in Temlyakov (1998a). As for Theorem 5.3.18, the most difficult part of Theorem 5.3.19 – the lower bounds – is based on the small ball inequality. It was proved in Temlyakov (1996).

The settings of the problem of finding the universal system of functions and the results of §5.4 are due to Temlyakov (1988c).

5.6 Open Problems

Two asymptotic characteristics of classes of functions with mixed smoothness are discussed in this section: Kolmogorov widths and orthowidths (Fourier widths). It seems that the most complete results are obtained for orthowidths (see §5.2). However, even in the case of orthowidths there are still unresolved problems. We mention one relating to the **W** classes.

Open Problem 5.1 Find the order of $\varphi_m(\mathbf{W}^r_{1,\alpha}, L_\infty)$ for all $r > 1$, α, and d.

The results in §5.3 show that the correct orders of the Kolmogorov widths $d_m(\mathbf{W}^r_q, L_p)$ are known for all $1 < q, p < \infty$ and $r > r(q,p)$. However, in the case of extreme values of p or q (p or q takes the value 1 or ∞) not much is known. Here are some relevant open problems.

Open Problem 5.2 Find the order of $d_m(\mathbf{W}^r_{q,\alpha}, L_\infty)$ for $2 \le q \le \infty$ in the case $d \ge 3$.

Open Problem 5.3 Find the order of $d_m(\mathbf{W}^r_{q,\alpha}, L_\infty)$ for $1 \le q < 2$.

Open Problem 5.4 Find the order of $d_m(\mathbf{W}^r_{1,\alpha}, L_p)$ for $1 \le p < 2$.

It turns out that the problem of finding the correct orders of the Kolmogorov widths for the **H** classes is more difficult than for the **W** classes. In addition to some open problems in the case of extreme values of p or q, the following case is not settled.

Open Problem 5.5 Find the order of $d_m(\mathbf{H}^r_q, L_p)$ for $1 \le p < q < 2$.

We now formulate some open problems in the case of extreme values of p or q.

Open Problem 5.6 Find the order of $d_m(\mathbf{H}^r_q, L_\infty)$ for $2 \le q \le \infty$ in the case $d \ge 3$.

Open Problem 5.7 Find the order of $d_m(\mathbf{H}^r_q, L_\infty)$ for $1 \le q < 2$.

Open Problem 5.8 Find the order of $d_m(\mathbf{H}^r_1, L_p)$ for $1 \le p < 2$.

6

Numerical Integration and Approximate Recovery

6.1 Introduction

Numerical integration is a challenging multivariate problem in which approximation theory methods are very useful. As we explained earlier, for a given function class $\mathbf{F}$ we want to find m points $\xi^1, \dots, \xi^m$ in Ω such that $\sum_{j=1}^{m} m^{-1} f(\xi^j)$ approximates well the integral $\int_\Omega f d\mu$, where μ is the normalized Lebesgue measure on Ω. Classical discrepancy theory provides constructions of point sets that can be used for the numerical integration of characteristic functions of parallelepipeds of the form $[\mathbf{a}, \mathbf{b}] := \prod_{j=1}^{d} [a_j, b_j]$. The typical error bound is of the form $m^{-1}(\log m)^{d-1}$. Note that a regular grid for $m = n^d$, which is used in the cubature formula $q_m(\mathbf{1})$ (see (3.6.3)), provides an error of the order $m^{-1/d}$.

The above-mentioned results of discrepancy theory are closely related to the numerical integration of functions with bounded mixed derivative (for the case of the mixed derivatives of order one). This example shows that in the multivariate numerical integration, even for a natural class of functions, we need nontrivial cubature formulas for satisfactory numerical integration. Thus, as in §5.1 we consider an optimization problem consisting of finding cubature formulas which provide numerical integration errors for the class $\mathbf{F}$ that are close to optimal, i.e., $\kappa_m(\mathbf{F})$ (see, for instance, §3.6). In §3.6 we discussed this problem in detail for the anisotropic classes $W^{\mathbf{r}}_{q,\alpha}$ and $H^{\mathbf{r}}_q$. We established there (see Theorem 3.6.1) that the simple cubature formulas $q_m(\mathbf{r})$ which use regular grids as a set $\{\xi^j\}$ are optimal in the sense of order for the numerical integration of functions from the classes $W^{\mathbf{r}}_{q,\alpha}$ and $H^{\mathbf{r}}_q$.

In this chapter we study the above optimization problem of numerical integration in classes of functions with mixed smoothness, i.e., the classes $\mathbf{W}^r_{q,\alpha}$ and $\mathbf{H}^r_q$. It is easy to verify that the cubature formulas $q_m(\mathbf{r})$ based on regular grids are far from being optimal for these classes. The problem of numerical integration in the classes $\mathbf{W}^r_{q,\alpha}$ and $\mathbf{H}^r_q$ turns out to be very challenging and very important in applications. It has been known for a long time that this problem is closely related to the discrepancy problem. We discuss this connection in §6.2. Recently, it has been understood that this problem is equivalent to the special problem of nonlinear m-term approximation. In particular, this connects the above numerical integration problem with the recently developed theory of greedy approximation with respect to redundant dictionaries. We discuss this issue in §6.3.

Our main goal in this chapter is to find the correct order of decay of the sequences $\{\kappa_m(\mathbf{W}^r_{p,\alpha})\}$ and $\{\kappa_m(\mathbf{H}^r_p)\}$. For this purpose we need both the lower and the upper bounds for $\kappa_m(\cdot)$. In §6.4 we consider the lower bounds. We discuss there a deep method which is based on nontrivial results on volume estimates from Chapter 3. The upper bounds are discussed in three sections: §6.5–6.7. Mostly, we present there classical constructions based on number-theoretical methods. In §6.5 we study in detail the numerical integration of bivariate functions. We prove there that simple and natural cubature formulas, namely the Fibonacci cubature formulas, are optimal in many cases. Some cases, for instance the case of small smoothness, are still open. However, the study of the Fibonacci cubature formulas in §6.5 is complete. It reveals a nontrivial phenomenon for small smoothness. In §6.6 we study the Korobov cubature formulas, which represent lattice rules and quasi-Monte-Carlo methods. These results are not as complete as the results in §6.5. Also, only a few cases are known for which the Korobov cubature formulas provide an optimal numerical integration error for the classes of mixed smoothness. The very important Frolov cubature formulas are studied in §6.7. These cubature formulas provide the correct upper bounds for $\kappa_m(\mathbf{W}^r_{p,\alpha})$ and $\kappa_m(\mathbf{H}^r_p)$ for many values of the parameters r, p. The theoretical study of these cubature formulas requires some concepts and results from algebraic number theory. We present such an introduction in §6.7.

In Chapter 5 we addressed, along with the optimality problem, the universality problem. In the same way, in this chapter we study the universal cubature formulas. Similarly to results in Chapter 5 we can state that methods of numerical integration which are optimal in the sense of order for classes with mixed smoothness are universal for the collection of anisotropic smoothness classes. We study this issue in §6.8. The results on universal methods of approximation from §5.4 show that we have to pay a price for this universality – we must increase the dimension of the approximating subspace by a factor $(\log m)^{d-1}$. Surprisingly, in contrast with these results, we show in §6.8 that we do not need to pay any price for universality in numerical integration. In this sense the problem of numerical integration is easier than the problem of approximation.

In §6.9 we present some results on the approximate recovery of functions from classes with mixed smoothness. The critical role there is played by recovery methods based on sparse grids. Recently, there has been substantial progress in the study of recovery in classes with mixed smoothness. For these results we refer the reader to the recent survey Dinh Dung *et al.* (2016).

6.2 Cubature Formulas and Discrepancy

The main goal of this section is to demonstrate connections between two large areas of research: the theory of cubature formulas (numerical integration) and discrepancy theory. We discuss the relation between the results on cubature formulas and on discrepancy. In particular, we show here how settings that are standard in the theory of cubature formulas can be translated into the discrepancy problem and into a natural generalization of the discrepancy problem. This leads to the concept of the r-discrepancy. One of the important messages of this section is that the theory of discrepancy is closely connected with the theory of cubature formulas for classes of functions with bounded mixed derivatives.

Numerical integration seeks effective ways of approximating an integral

$$\int_\Omega f(\mathbf{x})d\mu$$

by an expression of the form

$$\Lambda_m(f,\xi) := \sum_{j=1}^m \lambda_j f(\xi^j), \qquad \xi = (\xi^1,\dots,\xi^m), \qquad \xi^j \in \Omega, \qquad j = 1,\dots,m. \tag{6.2.1}$$

Clearly we must assume that f is integrable and defined at the points $\xi^1,\dots,\xi^m$. Expression (6.2.1) is called a *cubature formula* (Λ,ξ), if $\Omega \subset \mathbb{R}^d$, $d \ge 2$, or a *quadrature formula* (Λ,ξ), if $\Omega \subset \mathbb{R}$, with knots $\xi = (\xi^1,\dots,\xi^m)$ and weights $\Lambda := (\lambda_1,\dots,\lambda_m)$. For a function class $\mathbf{W}$ we introduce the concept of error of the cubature formula $\Lambda_m(\cdot,\xi)$:

$$\Lambda_m(\mathbf{W},\xi) := \sup_{f\in\mathbf{W}} \left| \int_\Omega f d\mu - \Lambda_m(f,\xi) \right|. \tag{6.2.2}$$

In order to orient the reader we begin with univariate periodic functions. For $r > 0$, let $F_r(x) := F_r(x,r)$, that is,

$$F_r(x) := 1 + 2\sum_{k=1}^\infty k^{-r}\cos(kx - r\pi/2), \tag{6.2.3}$$

and let $W_p^r := W_{p,r}^r$, defined in §1.4:

$$W_p^r := \{f : f = \varphi * F_r,\ \|\varphi\|_p \le 1\}. \tag{6.2.4}$$

We consider here this special case of the classes $W_{p,\alpha}^r$ for convenience and note that in the case $1 < p < \infty$ classes with different α are equivalent and, therefore, it is sufficient to study one of them. It is well known that for $r > 1/p$ the class W_p^r is embedded into the space of continuous functions $C(\mathbb{T})$. In the particular case of W_1^1 we also have embedding into $C(\mathbb{T})$. From the definitions (6.2.1), (6.2.2), and

(6.2.4) we see that, for the normalized measure $d\mu := (2\pi)^{-1}dx$

$$\Lambda_m(W^r_p,\xi) = \sup_{\|\varphi\|_p\le 1}\left|\frac{1}{2\pi}\int_{\mathbb{T}}\left(\int_{\mathbb{T}}F_r(x-y)d\mu - \sum_{j=1}^m \lambda_j F_r(\xi^j - y)\right)\varphi(y)dy\right|$$
$$= \left\|1-\sum_{j=1}^m \lambda_j F_r(\xi^j-\cdot)\right\|_{p'}, \qquad p' := \frac{p}{p-1}. \tag{6.2.5}$$

Thus the quality of the quadrature formula $\Lambda_m(\cdot,\xi)$ for the function class W^r_p is controlled by the quality of $\Lambda_m(\cdot,\xi)$ for the representing kernel $F_r(x-y)$. In the particular case of W^1_1 we have

$$\Lambda_m(W^1_1,\xi) = \max_y\left|1-\sum_{j=1}^m \lambda_j F_1(\xi^j - y)\right|. \tag{6.2.6}$$

In this case the function

$$F_1(x) = 1+2\sum_{k=1}^\infty \frac{\sin kx}{k} =: 1+S(x)$$

has a simple form: $S(x)=0$ for $x=l\pi$ and $S(x)=\pi-x$ for $x\in(0,2\pi)$. This allows us to associate the quantity $\Lambda_m(W^1_1,\xi)$ with one that has a simple geometrical interpretation. Denote by χ the class of all characteristic functions $\chi_{[0,a]}(x)$, $a\in[0,2\pi)$. Then we have the following property.

Proposition 6.2.1 *There exist two positive absolute constants C_1 and C_2 such that, for any $\Lambda_m(\cdot,\xi)$ with the property $\sum_j\lambda_j=1$, we have*

$$C_1\Lambda_m(\chi,\xi)\le\Lambda_m(W^1_1,\xi)\le C_2\Lambda_m(\chi,\xi). \tag{6.2.7}$$

Proof We have, for any $a\in[0,2\pi)$,

$$\chi_{[0,a]}(x) = \frac{1}{2\pi}(a+F_1(x)-F_1(x-a)). \tag{6.2.8}$$

Thus using (6.2.6) we get

$$\Lambda_m(\chi,\xi)\le\frac{1}{\pi}\Lambda_m(W^1_1,\xi),$$

which proves the left-hand inequality in (6.2.7).

Let us prove the right-hand inequality in (6.2.7). Denote $\varepsilon := \Lambda_m(\chi,\xi)$. Then by (6.2.8) we get, for any $a\in[0,2\pi)$,

$$-2\pi\varepsilon\le\int_{\mathbb{T}}(F_1(x)-F_1(x-a))d\mu - \Lambda_m(F_1(x)-F_1(x-a),\xi)\le 2\pi\varepsilon. \tag{6.2.9}$$

Integrating these inequalities against a over $\mathbb{T}$ we get

$$\left|\int_{\mathbb{T}}F_1(x)d\mu - \Lambda_m(F_1,\xi)\right|\le 2\pi\varepsilon.$$

This inequality combined with (6.2.9) implies that

$$\Lambda_m(W_1^1,\xi) \le 4\pi\Lambda_m(\chi,\xi). \qquad \square$$

We now proceed to the multivariate case. For $\mathbf{x} = (x_1,\dots,x_d)$ denote

$$F_r(\mathbf{x}) := \prod_{j=1}^{d} F_r(x_j)$$

and

$$\mathbf{W}_p^r := \{f : f = \varphi * F_r, \|\varphi\|_p \le 1\}.$$

For $f \in \mathbf{W}_p^r$ we write $f^{(r)} := \varphi$, where φ is such that $f = \varphi * F_r$. In the case of integer r the class $\mathbf{W}_p^r$ is very close to the class of functions f satisfying $\|f^{(r,\dots,r)}\|_p \le 1$, where $f^{(r,\dots,r)}$ is the mixed derivative of f of order rd. A multivariate analog of the class χ is the class

$$\chi^d := \{\chi_{[\mathbf{0},\mathbf{a}]}(\mathbf{x}) := \prod_{j=1}^{d} \chi_{[0,a_j]}(x_j),\ a_j \in [0,2\pi),\ j = 1,\dots,d\}.$$

As in the univariate case one obtains analogs of (6.2.5), (6.2.6), and Proposition 6.2.1:

$$\Lambda_m(\mathbf{W}_p^r,\xi) = \left\| 1 - \sum_{j=1}^{m} \lambda_j F_r(\xi^j - \cdot) \right\|_{p'}, \tag{6.2.10}$$

$$\Lambda_m(\mathbf{W}_1^1,\xi) = \max_{\mathbf{y}} \left| 1 - \sum_{j=1}^{m} \lambda_j F_1(\xi^j - \mathbf{y}) \right|. \tag{6.2.11}$$

Proposition 6.2.2 *There exist two positive constants $C_1(d)$ and $C_2(d)$ such that for any $\Lambda_m(\cdot,\xi)$ with the property $\sum_j \lambda_j = 1$ we have*

$$C_1(d)\Lambda_m(\chi^d,\xi) \le \Lambda_m(\mathbf{W}_1^1,\xi) \le C_2(d)\Lambda_m(\chi^d,\xi). \tag{6.2.12}$$

The reader can find the proof of Proposition 6.2.2 in Temlyakov (2003b).

The classical definition of the discrepancy of a set X of points $\mathbf{x}^1,\dots,\mathbf{x}^m \subset [0,1]^d$ is as follows:

$$D(X,m,d)_\infty := \max_{\mathbf{a}\in[0,1]^d} \left| \prod_{j=1}^{d} a_j - \frac{1}{m}\sum_{\mu=1}^{m} \chi_{[\mathbf{0},\mathbf{a}]}(\mathbf{x}^\mu) \right|.$$

It is clear that

$$D(X,m,d)_\infty = \Lambda_m(\chi^d,2\pi X) \qquad \text{with } \lambda_1 = \cdots = \lambda_m = 1/m.$$

Thus, Proposition 6.2.2 shows that the classical concept of discrepancy is directly related to the efficiency of the corresponding cubature formulas for a special function class $\mathbf{W}_1^1$. It is well known that W_1^1 is very close to the class of functions of bounded variation and also that $\mathbf{W}_1^1$ is very close to the class of functions with bounded variation in the sense of Hardy–Vitali. For historical comments we refer the reader to §4.1.

In addition to the classes of 2π-periodic functions it will be convenient for us to consider the classes of nonperiodic functions defined on $\Omega_d := [0,1]^d$. Let r be a natural number and let $\mathbf{W}_p^r(\Omega_d)$, $1 \le p \le \infty$, denote the closure in the uniform metric of the set of rd-times continuously differentiable functions $f(\mathbf{x})$ such that

$$\|f\|_{\mathbf{W}_p^r} := \sum_{0\le n_j\le r; j=1,\dots,d} \left\| \frac{\partial^{n_1+\cdots+n_d} f}{\partial x_1^{n_1}\dots\partial x_d^{n_d}} \right\|_p \le 1, \qquad (6.2.13)$$

where

$$\|g\|_p := \left(\int_{\Omega_d} |g(\mathbf{x})|^p d\mathbf{x} \right)^{1/p}.$$

It will be convenient for us to consider the class $\dot{\mathbf{W}}_p^r(\Omega_d)$ consisting of the functions $f(\mathbf{x})$ representable in the form

$$f(\mathbf{x}) = \int_{\Omega_d} B_r(\mathbf{t},\mathbf{x})\varphi(\mathbf{t})d\mathbf{t}, \qquad \|\varphi\|_p \le 1,$$

where

$$B_r(\mathbf{t},\mathbf{x}) := \prod_{j=1}^{d} \left((r-1)!\right)^{-1} (t_j - x_j)_+^{r-1},$$

$$\mathbf{t},\mathbf{x} \in \Omega_d, \qquad (a)_+ := \max(a,0).$$

In connection with the definition of the class $\dot{\mathbf{W}}_p^r(\Omega_d)$ we remark here that the following relation (6.2.14) holds for the error of the cubature formula (Λ,ξ) with weights $\Lambda = (\lambda_1,\dots,\lambda_m)$ and knots $\xi = (\xi^1,\dots,\xi^m)$. Let

$$\left| \Lambda_m(f,\xi) - \int_{\Omega_d} f(\mathbf{x})d\mathbf{x} \right| =: R_m(\Lambda,\xi,f);$$

then one obtains an expression similar to (6.2.5) and (6.2.10):

$$\begin{aligned}
\Lambda_m\left(\dot{\mathbf{W}}_p^r(\Omega_d),\xi\right) &:= \sup_{f\in\dot{\mathbf{W}}_p^r(\Omega_d)} R_m(\Lambda,\xi,f) \\
&= \left\| \sum_{\mu=1}^{m} \lambda_\mu B_r(\mathbf{t},\xi^\mu) - \prod_{j=1}^{d} (t_j^r/r!) \right\|_{p'} \\
&=: D_r(\xi,\Lambda,m,d)_{p'}, \qquad p' = p/(p-1). \qquad (6.2.14)
\end{aligned}$$

The quantity $D_r(\xi,\Lambda,m,d)_q$ in the case $r=1$, $\Lambda=(1/m,\dots,1/m)$ is the classical discrepancy of the set of points $\{\xi^\mu\}$. In the case $\Lambda=(1/m,\dots,1/m)$ we write $D_r(\xi,m,d)_q := D_r(\xi,(1/m,\dots,1/m),m,d)_q$ and call it the r-discrepancy (see Temlyakov, 1994, 2003b). Thus, the quantity $D_r(\xi,\Lambda,m,d)_q$ defined in (6.2.14) is a natural generalization of the discrepancy concept

$$D(\xi,m,d)_q := D_1(\xi,m,d)_q. \tag{6.2.15}$$

This generalization contains two ingredients: general weights Λ instead of the special case of equal weights $(1/m,\dots,1/m)$ and any natural number r instead of $r=1$. We note that in approximation theory we usually study the whole range of smoothness classes rather than an individual smoothness class. The above generalization of the discrepancy to arbitrary positive integer r allows us to study the question: how does the smoothness r affect the rate of decay of the generalized discrepancy?

Remark 6.2.3 Let $r\in\mathbb{N}$, then the class $\mathbf{W}^r_p$ of 2π-periodic functions, defined on $\mathbb{T}^d$, turns into a subclass of the class $\mathbf{W}^r_p(\Omega_d)B := \{f : f/B \in \mathbf{W}^r_p(\Omega_d)\}$, after a linear change of variables

$$x_j = 2\pi t_j, \qquad j=1,\dots,d.$$

We are interested in the dependence on m of the quantities

$$\kappa_m(\mathbf{W}) := \inf_{\lambda_1,\dots,\lambda_m;\xi^1,\dots,\xi^m} \Lambda_m(\mathbf{W},\xi)$$

for the classes $\mathbf{W}$ defined above. Remark 6.2.3 shows that

$$\kappa_m(\mathbf{W}^r_p) \ll \kappa_m\big(\mathbf{W}^r_p(\Omega_d)\big). \tag{6.2.16}$$

Let $\bar{\mathbf{W}}^r_p(\Omega_d)$ denote the subset of functions in $\mathbf{W}^r_p(\Omega_d)$ which is the closure in the uniform metric of the set of rd-times continuously differentiable functions f, satisfying the condition (6.2.13), such that

$$\operatorname{supp}(f) := \{\mathbf{x} : f(\mathbf{x}) \neq 0\} \subset (0,1)^d.$$

Theorem 6.2.4 *Let $1\le p\le\infty$. Then, for $r\in\mathbb{N}$,*

$$\kappa_m\big(\bar{\mathbf{W}}^r_p(\Omega_d)\big) \asymp \kappa_m\big(\dot{\mathbf{W}}^r_p(\Omega_d)\big) \asymp \kappa_m\big(\mathbf{W}^r_p(\Omega_d)\big). \tag{6.2.17}$$

Proof Let Λ and ξ be given. We construct a function $g\in\bar{\mathbf{W}}^r_p(\Omega_d)$ such that

$$\left|\Lambda_m(g,\xi) - \int_{\Omega_d} g(\mathbf{t})d\mathbf{t}\right| \gg \kappa_m\big(\mathbf{W}^r_p(\Omega_d)\big). \tag{6.2.18}$$

Suppose that an infinitely differentiable function $\psi(x)$ is such that $\psi(x)=0$ for $x\le 0$, $\psi(x)=1$ for $x\ge 1$, and $\psi(x)$ strictly increases on $[0,1]$. For the cubature formula (Λ, ξ) defined on the class $\bar{\mathbf{W}}^r_p(\Omega_d)$, we define a cubature formula (Λ', η), whose error will be investigated for the class $\mathbf{W}^r_p(\Omega_d)$ as follows:

$$\eta^\mu_j := \psi(\xi^\mu_j), \qquad j = 1, \dots, d,$$
$$\lambda'_\mu := \lambda_\mu \prod_{j=1}^{d} \psi'(\xi^\mu_j), \qquad \mu = 1, \dots, m.$$

Then for functions f and g related by

$$g(\mathbf{t}) = f(\psi(t_1), \dots, \psi(t_d)) \prod_{j=1}^{d} \psi'(t_j) \tag{6.2.19}$$

we have

$$\int_{\Omega_d} f(\mathbf{x})\, d\mathbf{x} = \int_{\Omega_d} g(\mathbf{t})\, d\mathbf{t},$$
$$\sum_{\mu=1}^{m} \lambda_\mu g(\xi^\mu) = \sum_{\mu=1}^{m} \lambda'_\mu f(\eta^\mu).$$

It remains to check that there exists a number $\delta > 0$, which does not depend on m, such that $\delta g \in \bar{\mathbf{W}}^r_p(\Omega_d)$ provided that $f \in \mathbf{W}^r_p(\Omega_d)$.

Differentiating (6.2.19) we see that the expression for $g^{(\mathbf{s})}(\mathbf{t})$, $\mathbf{s} := (s_1, \dots, s_d)$, $0 \le s_j \le r$, $j = 1, \dots, d$, will contain terms of the form

$$\omega(\mathbf{t}, \mathbf{k}) := f^{(\mathbf{k})}(\psi(\mathbf{t})) \prod_{j=1}^{d} \prod_{i} \left(\psi^{(l^i_j)}(t_j)\right)^{m^i_j}, \qquad \psi(\mathbf{t}) := \left(\psi(t_1), \dots, \psi(t_d)\right),$$
$$\mathbf{k} := (k_1, \dots, k_d), \qquad 0 \le k_j \le s_j, \qquad \sum_i l^i_j m^i_j = s_j + 1, \qquad j = 1, \dots, d. \tag{6.2.20}$$

The number of terms $\omega(\mathbf{t}, \mathbf{k})$ depends on the vector $\mathbf{s}$.

It is obvious that in the case $p = \infty$ we have

$$\|\omega(\mathbf{t}, \mathbf{k})\|_\infty \le C(r) \|f\|_{\mathbf{W}^r_\infty(\Omega_d)}, \qquad C(r) = C(r, \psi, d).$$

To estimate $\|\omega(\cdot, \mathbf{k})\|_p$, $1 \le p < \infty$, we use the following simple lemma.

Lemma 6.2.5 *Suppose that $f \in \mathbf{W}^r_p(\Omega_d)$, and the vector $\mathbf{k} \in \mathbb{Z}^d_+$ is such that $k_j = r$ for $j \in e_1$ and $k_j \le r - 1$ for $j \in e_2 = [1, d] \setminus e_1$. Then*

$$\sup_{x_j, j \in e_2} \int_{[0,1]^{|e_1|}} |f^{(k_1, \dots, k_d)}(\mathbf{x})|^p \left(\prod_{j \in e_1} dx_j\right) \le C(p, r, d) \|f\|^p_{\mathbf{W}^r_p(\Omega_d)}, \qquad 1 \le p < \infty.$$

Proof We first prove the following statement. Let f be such that f, $(\partial f)/(\partial x_j)$ are continuous. Then

$$\sup_{x_j} \int_{[0,1]^{d-1}} |f(\mathbf{x})|^p \Big(\prod_{i\neq j} dx_i\Big) \leq 2^p \Big(\|f\|_p^p + \Big\|\frac{\partial f}{\partial x_j}\Big\|_p^p\Big). \tag{6.2.21}$$

Indeed, there is an $a \in [0,1]$ such that

$$\int_{[0,1]^{d-1}} |f(x_1,\dots,x_{j-1},a,x_{j+1},\dots,x_d)| \Big(\prod_{i\neq j} dx_i\Big) \leq \|f\|_1 \leq \|f\|_p.$$

We now represent the function $f(\mathbf{x})$ in the form

$$\begin{aligned} f(\mathbf{x}) &= f(x_1,\dots,x_{j-1},a,x_{j+1},\dots,x_d) \\ &\quad + \int_a^{x_j} \frac{\partial f}{\partial x_j}(x_1,\dots,x_{j-1},u,x_{j+1},\dots,x_d)\, du. \end{aligned}$$

Then, for any $x_j \in [0,1]$, we have

$$\int_{[0,1]^{d-1}} |f(\mathbf{x})|^p \Big(\prod_{i\neq j} dx_i\Big) \leq 2^p \Big(\|f\|_p^p + \Big\|\frac{\partial f}{\partial x_j}\Big\|_p^p\Big).$$

This proves (6.2.21).

Applying relation (6.2.21) successively, we obtain the lemma. □

We return to estimating $\|\omega(\cdot,\mathbf{k})\|_p$. By Lemma 6.2.5 and by the uniform boundedness of the functions $|\psi^{(l)}(t_j)| \leq C(r,\psi)$, $l \leq r+1$, we obtain, for $\mathbf{k}$ such that $k_j < r$, $j = 1,\dots,d$,

$$\|\omega(\cdot,\mathbf{k})\|_p \leq \|\omega(\cdot,\mathbf{k})\|_\infty \ll \|f^{(k_1,\dots,k_d)}\|_\infty \leq \|f\|_{\mathbf{W}^r_\infty(\Omega_d)}.$$

Thus, it remains to consider only those $\mathbf{k}$ for which there is a j such that $k_j = r$. Then, with respect to the jth variable, $(\psi'(t_j))^{r+1}$ participates as an additional cofactor in expression (6.2.20). Taking into account that

$$(\psi'(t_j))^{p(r+1)} \leq C(p,r,\psi)\psi'(t_j),$$

we obtain

$$\int_0^1 |\omega(\mathbf{t},\mathbf{k})|^p\,dt_j$$
$$\leq C(p,r_j)\left(\int_0^1 \left|f^{(\mathbf{k})}(\psi(t_1),\ldots,\psi(t_{j-1}),x_j,\psi(t_{j+1}),\ldots,\psi(t_d))\right|^p dx_j\right)$$
$$\times \prod_{\nu\neq j}\prod_i \left(\psi^{(\theta_i^\nu)}(t_\nu)\right)^{m_i^\nu p}.$$

Reasoning in this way for all j such that $k_j = r$ and applying Lemma 6.2.5, we find that

$$\|\omega(\cdot,\mathbf{k})\|_p \leq C(p,r,d)\|f\|_{\mathbf{W}_p^r(\Omega_d)} \tag{6.2.22}$$

for all $\mathbf{k}$. This implies that $c(p,r,d)g \in \bar{W}_p^r(\Omega_d)$ for some positive $c(p,r,d)$.

The arguments presented show that

$$\sup_{g\in\bar{\mathbf{W}}_p^r(\Omega_d)} \left|\int_{Q_d} g(\mathbf{t})\,d\mathbf{t} - \sum\nolimits_{\mu=1}^N \lambda_\mu g(\xi^\mu)\right|$$
$$\geq c(p,r,d)^{-1} \sup_{f\in\mathbf{W}_p^r(\Omega_d)} \left|\int_{Q_d} f(\mathbf{x})\,d\mathbf{x} - \sum_{\mu=1}^N \lambda'_\mu f(\eta^\mu)\right|. \tag{6.2.23}$$

Relation (6.2.23) and the embeddings $\bar{\mathbf{W}}_p^r(\Omega_d) \hookrightarrow \dot{\mathbf{W}}_p^r(\Omega_d) \hookrightarrow \mathbf{W}_p^r(\Omega_d)$ yield Theorem 6.2.4. □

Remark 6.2.6 Let $1 \leq p \leq \infty$ and $r > 1/p$. Then

$$\kappa_m(\mathbf{W}_p^r) \asymp \kappa_m\big(\mathbf{W}_p^r(\Omega_d)\big).$$

The upper estimate follows from (6.2.16). The lower estimate follows from Theorem 6.2.4.

6.3 Optimal Cubature Formulas and Nonlinear Approximation

In this section we present results on the relation between the construction of an optimal cubature formula with m knots for a given function class and the best nonlinear m-term approximation of a special function determined by the function class. The nonlinear m-term approximation is taken with regard to a redundant dictionary also determined by the function class.

Relations (6.2.10) and (6.2.11) can be interpreted as giving a connection between the error of the cubature formula (Λ,ξ) on the class $\mathbf{W}_p^r$ and the approximation error of a special function $1 = \int_{\mathbb{T}^d} F_r(\mathbf{x})d\mu$ by the m-term linear combination of functions $F_r(\xi^j - \cdot)$, $j = 1,\ldots,m$. The latter problem is one of nonlinear m-term approximation with regard to a given system of functions, in the above case with regard to the system $\{F_r(\mathbf{x}-\cdot), \mathbf{x} \in \mathbb{T}^d\}$. The problem of nonlinear m-term approximation has attracted considerable attention during the last 20 years because of its importance in numerical applications (see the surveys DeVore, 1998, and Temlyakov, 2003a). In this section we use some known results from m-term approximation in Banach spaces in order to estimate the error of optimal cubature formulas. We present these ideas in a general setting.

Let $1 \le q \le \infty$. We define a set $\mathscr{K}_q$ of kernels possessing the following properties. Let $K(\mathbf{x},\mathbf{y})$ be a measurable function on $\Omega^1 \times \Omega^2$. We assume that for any $\mathbf{x} \in \Omega^1$ we have $K(\mathbf{x},\cdot) \in L_q(\Omega^2)$ and, for any $\mathbf{y} \in \Omega^2$, $K(\cdot,\mathbf{y})$ is integrable over Ω^1 and $\int_{\Omega^1} K(\mathbf{x},\cdot)d\mathbf{x} \in L_q(\Omega^2)$. For a kernel $K \in \mathscr{K}_{p'}$ we define the class

$$\mathbf{W}_p^K := \left\{f : f = \int_{\Omega^2} K(\mathbf{x},\mathbf{y})\varphi(\mathbf{y})d\mathbf{y},\ \|\varphi\|_{L_p(\Omega^2)} \le 1\right\}. \tag{6.3.1}$$

Then each $f \in \mathbf{W}_p^K$ is integrable on Ω^1 (by Fubini's theorem) and defined at each point of Ω^1. We denote for convenience

$$J(\mathbf{y}) := J_K(\mathbf{y}) := \int_{\Omega^1} K(\mathbf{x},\mathbf{y})d\mathbf{x}.$$

For a cubature formula $\Lambda_m(\cdot,\xi)$ we have

$$\begin{aligned}\Lambda_m(\mathbf{W}_p^K,\xi) &= \sup_{\|\varphi\|_{L_p(\Omega^2)}\le 1} \left|\int_{\Omega^2}\left(J(\mathbf{y}) - \sum_{\mu=1}^m \lambda_\mu K(\xi^\mu,\mathbf{y})\right)\varphi(\mathbf{y})d\mathbf{y}\right| \\ &= \left\|J(\cdot) - \sum_{\mu=1}^m \lambda_\mu K(\xi^\mu,\cdot)\right\|_{L_{p'}(\Omega^2)}.\end{aligned} \tag{6.3.2}$$

We use the same definition as above of the error of the optimal cubature formula with m knots for a class $\mathbf{W}$:

$$\kappa_m(\mathbf{W}) := \inf_{\lambda_1,\dots,\lambda_m;\xi^1,\dots,\xi^m} \Lambda_m(\mathbf{W},\xi).$$

Thus, by (6.3.2),

$$\kappa_m(\mathbf{W}_p^K) = \inf_{\lambda_1,\dots,\lambda_m;\xi^1,\dots,\xi^m} \left\|J(\cdot) - \sum_{\mu=1}^m \lambda_\mu K(\xi^\mu,\cdot)\right\|_{L_{p'}(\Omega^2)}. \tag{6.3.3}$$

Let us apply some results on greedy approximation, which we will obtain in Chapter 8, to numerical integration. Consider a dictionary

$$\mathscr{D} := \mathscr{D}(K,p') := \{g : g(\mathbf{x},\mathbf{y}) = \pm K(\mathbf{x},\mathbf{y})/\|K(\mathbf{x},\cdot)\|_{L_{p'}(\Omega^2)}, \mathbf{x} \in \Omega^1\}$$

(in the case $\|K(\mathbf{x},\cdot)\|_{L_{p'}(\Omega^2)} = 0$ we set $g(\mathbf{x},\cdot) = 0$), and define a Banach space $X := X(K,p')$ as the $L_{p'}(\Omega^2)$-closure of the span of $\mathscr{D}$. Assume now that $J_K \in X$.

We use the convergence result for the weak Chebyshev greedy algorithm (WCGA); see Corollary 8.6.5. Then for $1 < p' < \infty$ the WCGA satisfying (8.6.3) with $q := \min(2, p')$ provides a deterministic algorithm for constructing a sequence of cubature formulas $\Lambda_m^c(\cdot, \xi)$ such that

$$\Lambda_m^c(\mathbf{W}_p^K, \xi) \to 0 \qquad \text{as } m \to \infty.$$

We will discuss in more detail the question of the rate of convergence. We illustrate it using the WCGA. For further discussion, in particular on the use of the relaxed greedy algorithm, we refer the reader to Temlyakov (2003b). The following theorem is proved in Chapter 8 (see Theorem 8.6.6).

Theorem 6.3.1 *Let X be a uniformly smooth Banach space with modulus of smoothness $\rho(u) \le \gamma u^q$, $1 < q \le 2$. Then, for a sequence $\tau := \{t_k\}_{k=1}^{\infty}$, $t_k \in [0,1]$, $k = 1, 2, \dots$, we have for any $f \in A_1(\mathscr{D}^{\pm})$ that*

$$\|f_m^{c,\tau}\| \le C_1(q,\gamma)\left(1 + \sum_{k=1}^{m} t_k^p\right)^{-1/p}, \qquad p := \frac{q}{q-1},$$

where the constant $C_1(q,\gamma)$ may depend only on q and γ.

Corollary 6.3.2 *In a particular case $\tau = \{t_k\}_{k=1}^{\infty}$, $t_k = t$, $k = 1, 2, \dots$, with some $t \in (0,1]$, we have under the assumptions of Theorem 6.3.1 that*

$$\|f_m^{c,\tau}\| \le C_1(q,\gamma,t) m^{-1/p}, \qquad p := \frac{q}{q-1}.$$

In order to apply Theorem 6.3.1 to the numerical integration of the $\mathbf{W}_p^K$ we need to check that $J_K \in A_1(\mathscr{D}(K,p'))$ (or there exists a positive constant c such that $cJ_K \in A_1(\mathscr{D}(K,p'))$. It could be a difficult problem. An inspection of the proof of Theorem 6.3.1 shows that it is sufficient to check that

$$\int_{\Omega^1} \|K(\mathbf{x},\cdot)\|_{L_{p'}(\Omega^2)} d\mathbf{x} < \infty.$$

We formulate this as a theorem.

Theorem 6.3.3 *Let $\mathbf{W}_p^K$ be a class of functions defined by* (6.3.1). *Assume that $K \in \mathscr{K}_{p'}$ satisfies the condition*

$$\int_{\Omega^1} \|K(\mathbf{x},\cdot)\|_{L_{p'}(\Omega^2)} d\mathbf{x} \le M$$

and that $J_K \in X(K,p')$. Then for any m there exists a cubature formula $\Lambda_m(\cdot,\xi)$ (provided by the WCGA with $\tau = \{t\}$) such that

$$\Lambda_m(\mathbf{W}_p^K, \xi) \le MC(p,t) \begin{cases} m^{-1/2}, & 1 \le p \le 2, \\ m^{-1/p}, & 2 \le p < \infty. \end{cases}$$

Let us consider a particular example, $K(\mathbf{x},\mathbf{y}) := (2\pi)^{-d}F(\mathbf{x}-\mathbf{y})$, $\Omega^1 := \Omega^2 := \mathbb{T}^d$. We denote the corresponding class $\mathbf{W}^K_p$ by $\mathbf{W}^F_p$.

Proposition 6.3.4 *Let $1 < p < \infty$ and $\|F\|_{p'} \le M$. Then the kernel $K(\mathbf{x},\mathbf{y}) := (2\pi)^{-d}F(\mathbf{x}-\mathbf{y})$ satisfies the assumptions of Theorem 6.3.3.*

Proof It is obvious that $K \in \mathscr{K}_{p'}$. Next,

$$(2\pi)^{-d}\int_{\mathbb{T}^d} \|F(\mathbf{x}-\cdot)\|_{p'} d\mathbf{x} = \|F\|_{p'} \le M.$$

It remains to check that $J_K \in X(K,p')$. We have

$$J_K(\mathbf{y}) = (2\pi)^{-d}\int_{\mathbb{T}^d} F(\mathbf{x}-\mathbf{y})d\mathbf{x} = \hat{F}(0).$$

Denote

$$S_N(F,\mathbf{x}) := \sum_{|k_j|\le N,\, j=1,\dots,d} \hat{F}(\mathbf{k})e^{i(\mathbf{k},\mathbf{x})}.$$

Then it is well known (by the Riesz theorem) that

$$\|F - S_N(F)\|_{p'} \to 0 \qquad \text{as } N \to \infty.$$

For a given N we consider the cubature formula

$$q_N(f) := N^{-d}\sum_{\mu_d=1}^{N}\cdots\sum_{\mu_1=1}^{N} f(2\pi\mu_1/N,\dots,2\pi\mu_d/N).$$

Then we have

$$\begin{aligned}
&\|\hat{F}(0) - q_N(F(\cdot-\mathbf{y}))\|_{p'}\\
&\quad = \|\hat{F}(0) - q_N(S_{N-1}(F,\cdot-\mathbf{y})) + q_N(S_{N-1}(F,\cdot-\mathbf{y}) - F(\cdot-\mathbf{y}))\|_{p'}\\
&\quad = \|q_N(S_{N-1}(F,\cdot-\mathbf{y}) - F(\cdot-\mathbf{y}))\|_{p'} \le \|S_{N-1}(F) - F\|_{p'} \to 0
\end{aligned}$$

as $N \to \infty$. This proves the proposition. □

Theorem 6.3.3 and Proposition 6.3.4 yield the following result.

Theorem 6.3.5 *Let $1 < p < \infty$. Assume that $F \in L_{p'}(\mathbb{T}^d)$, $p' = p/(p-1)$. Consider the class*

$$\mathbf{W}^F_p := \{f : f = F * \varphi,\ \|\varphi\|_p \le 1\}.$$

Then for any m there exists a cubature formula $\Lambda_m(\cdot,\xi)$ such that

$$\Lambda_m(W^F_p,\xi) \le C(p)\|F\|_{p'}\begin{cases} m^{-1/2}, & 1 < p \le 2,\\ m^{-1/p}, & 2 \le p < \infty.\end{cases}$$

A sequence of $\{\Lambda_m(\cdot,\xi)\}$ from Theorem 6.3.5 can be obtained by applying the WCGA with a fixed $\tau = \{t\}$.

We now return to relation (6.3.3). For convenience we formulate it as a proposition.

Proposition 6.3.6

$$\kappa_m(\mathbf{W}^K_p) = \inf_{\lambda_1,\dots,\lambda_m;\xi^1,\dots,\xi^m} \left\| J(\cdot) - \sum_{\mu=1}^{m} \lambda_\mu K(\xi^\mu,\cdot) \right\|_{L_{p'}(\Omega^2)}.$$

Thus, the problem of finding the optimal error of a cubature formula with m knots for the class $\mathbf{W}^K_p$ is equivalent to the problem of the best m-term approximation of a special function J with respect to the dictionary $\mathscr{D} = \{K(\mathbf{x},\cdot), \mathbf{x} \in \Omega^1\}$. We saw in §6.2 how an analog of Proposition 6.3.6, namely, relations (6.2.10) and (6.2.11), connected the error of a cubature formula with the discrepancy given in formula (6.2.14).

Consider now the numerical integration of the functions $K(\mathbf{x},\mathbf{y})$, $\mathbf{y} \in \Omega^2$, with respect to $\mathbf{x}$, $K \in \mathscr{K}_q$:

$$\int_{\Omega^1} K(\mathbf{x},\mathbf{y})d\mathbf{x} - \sum_{\mu=1}^{m} \lambda_\mu K(\xi^\mu,\mathbf{y}).$$

Definition 6.3.7 The (K,q)-discrepancy of a cubature formula Λ_m with knots $\xi^1,\dots,\xi^m$ and weights $\lambda_1,\dots,\lambda_\mu$ is

$$D(\Lambda_m,K,q) := \left\| \int_{\Omega^1} K(\mathbf{x},\mathbf{y})d\mathbf{x} - \sum_{\mu=1}^{m} \lambda_\mu K(\xi^\mu,\mathbf{y}) \right\|_{L_q(\Omega^2)}.$$

The above definition of the (K,q)-discrepancy implies straight away the following relation.

Proposition 6.3.8

$$\inf_{\lambda_1,\dots,\lambda_m;\xi^1,\dots,\xi^m} D(\Lambda_m,K,q) = \inf_{\lambda_1,\dots,\lambda_m;\xi^1,\dots,\xi^m} \left\| J(\cdot) - \sum_{\mu=1}^{m} \lambda_\mu K(\xi^\mu,\cdot) \right\|_{L_q(\Omega^2)}.$$

Therefore, the problem of finding the minimal (K,q)-discrepancy is equivalent to the problem of the best m-term approximation of a special function J with respect to the dictionary $\mathscr{D} = \{K(\mathbf{x},\cdot), \mathbf{x} \in \Omega^1\}$.

The particular case $K(\mathbf{x},\mathbf{y}) := \chi_{[0,\mathbf{y}]}(\mathbf{x}) := \prod_{j=1}^{d} \chi_{[0,y_j]}(x_j), y_j \in [0,1)$, $j = 1,\dots,d$, where $\chi_{[0,y]}(x)$, $y \in [0,1)$, is the characteristic function of an interval $[0,y)$, leads to the classical concept of the L_q-discrepancy.

We illustrated above the use of the WCGA in numerical integration. There are other greedy-type algorithms that can be successfully used in numerical integration (see Temlyakov, 2003b, 2016a). We now discuss an application of the incremental algorithm, IA(ε), to be studied in §9.2. An important feature of the IA(ε) is that its application in numerical integration gives a cubature formula with equal weights $\lambda_\mu = 1/m$, $\mu = 1,\dots,m$.

Theorem 6.3.9 *Let $\mathbf{W}^K_p$ be the class of functions defined above. Assume that $K \in \mathscr{K}_{p'}$ satisfies the condition*

$$\|K(\mathbf{x},\cdot)\|_{L_{p'}(\Omega^2)} \le 1, \qquad \mathbf{x} \in \Omega^1, \qquad |\Omega^1| = 1$$

and that $J_K \in X(K,p')$. Then for any m there exists (provided by the Incremental Algorithm) a cubature formula $\Lambda_m(\cdot,\xi)$ with $\lambda_\mu = 1/m$, $\mu = 1,\dots,m$, and

$$\Lambda_m(\mathbf{W}^K_p,\xi) \le C(p-1)^{-1/2}m^{-1/2}, \qquad 1 < p \le 2.$$

Theorem 6.3.9 provides a constructive way of finding, for a wide variety of classes $\mathbf{W}^K_p$, cubature formulas that give the error bounds similar to those of the Monte Carlo method. We stress that in Theorem 6.3.9 we do not assume any smoothness of the kernel $K(\mathbf{x},\mathbf{y})$.

Proof By (6.3.2),

$$\Lambda_m(\mathbf{W}^K_p,\xi) = \left\| J(\cdot) - \sum_{\mu=1}^{m} \lambda_\mu K(\xi^\mu,\cdot) \right\|_{L_{p'}(\Omega^2)}.$$

We will apply Theorem 9.2.3 to the incremental algorithm with $X = X(K,p') \subset L_{p'}(\Omega^2)$, $f = J_K$. We need to check Condition B from §9.2. Let F be a bounded linear functional on $L_{p'}$. Then, by the Riesz representation theorem, there exists an $h \in L_p$ such that, for any $\phi \in L_{p'}$,

$$F(\phi) = \int_{\Omega^2} h(\mathbf{y})\phi(\mathbf{y})d\mathbf{y}.$$

By the Hölder inequality we have for any $\mathbf{x} \in \Omega^1$

$$\int_{\Omega^2} |h(\mathbf{y})K(\mathbf{x},\mathbf{y})|d\mathbf{y} \le \|h\|_p.$$

Therefore, the functions $|h(\mathbf{y})K(\mathbf{x},\mathbf{y})|$ and $h(\mathbf{y})K(\mathbf{x},\mathbf{y})$ are integrable on $\Omega^1 \times \Omega^2$ and, by Fubini's theorem,

$$\begin{aligned} F(J_K) &= \int_{\Omega^2} h(\mathbf{y}) \int_{\Omega^1} K(\mathbf{x},\mathbf{y})d\mathbf{x} = \int_{\Omega^1} \left(\int_{\Omega^2} h(\mathbf{y})K(\mathbf{x},\mathbf{y})d\mathbf{y} \right) d\mathbf{x} \\ &= \int_{\Omega^1} F(K(\mathbf{x},\mathbf{y}))d\mathbf{x} \le \sup_{\mathbf{x}\in\Omega^1} F(K(\mathbf{x},\mathbf{y})), \end{aligned}$$

which proves Condition B. Applying Theorem 9.2.3 and taking into account (9.2.1) completes the proof. □

Proposition 6.3.8 and the above proof imply the following theorem on the (K,q)-discrepancy.

Theorem 6.3.10 *Assume that* $K \in \mathscr{K}_q$ *satisfies the condition*

$$\|K(\mathbf{x},\cdot)\|_{L_q(\Omega^2)} \le 1, \quad \mathbf{x} \in \Omega^1, \quad |\Omega^1| = 1$$

and that $J_K \in X(K,q)$. *Then for any* m *there exists (provided by the Incremental Algorithm) a cubature formula* $\Lambda_m(\cdot,\xi)$ *with* $\lambda_\mu = 1/m$, $\mu = 1,\dots,m$, *and*

$$D(\Lambda_m, K, q) \le C q^{1/2} m^{-1/2}, \qquad 2 \le q < \infty.$$

We now proceed to applications of greedy approximation to discrepancy estimates in high dimensions. It will be convenient for us to study a slight modification of $D(\xi,m,d)_q$ (see (6.2.15)). For $a,t \in [0,1]$ denote

$$H(a,t) := \chi_{[0,a]}(t) - \chi_{[a,1]}(t)$$

and, for $\mathbf{x},\mathbf{y} \in \Omega_d$,

$$H(\mathbf{x},\mathbf{y}) := \prod_{j=1}^{d} H(x_j,y_j).$$

We define a symmetrized L_q-discrepancy by

$$D^s(\xi,m,d)_q := \left\| \int_{\Omega_d} H(\mathbf{x},\mathbf{y})d\mathbf{y} - \frac{1}{m}\sum_{\mu=1}^{m} H(\mathbf{x},\xi^\mu) \right\|_{L_q(\Omega_d)}.$$

Using the identity

$$\chi_{[0,x_j]}(y_j) = \frac{1}{2}(H(1,y_j) + H(x_j,y_j))$$

we get the simple inequality

$$D(\xi,m,d)_\infty \le D^s(\xi,m,d)_\infty. \tag{6.3.4}$$

We are interested in sets ξ having a small discrepancy. Consider

$$D^s(m,d)_q := \inf_{\xi} D^s(\xi,m,d)_q.$$

The following relation is known:

$$D(m,d)_q \asymp m^{-1}(\ln m)^{(d-1)/2}, \qquad 1 < q < \infty, \tag{6.3.5}$$

with constants in $\asymp$ depending on q and d. The correct order of $D(m,d)_q$, $q =$

$1, \infty$, for $d \geq 3$ is unknown. The following estimate was obtained in Heinrich *et al.* (2001).

$$D(m,d)_\infty \leq Cd^{1/2}m^{-1/2}. \tag{6.3.6}$$

It was pointed out in Heinrich *et al.* (2001) that (6.3.6) is only an existence theorem and even the constant C in (6.3.6) is unknown. Their proof is a probabilistic one. There are also some other estimates in Heinrich *et al.* (2001) with explicit constants. We mention one of them:

$$D(m,d)_\infty \leq C(d\ln d)^{1/2}((\ln m)/m)^{1/2} \tag{6.3.7}$$

with an explicit constant C. The proof of (6.3.7) is also probabilistic.

In this section we apply greedy-type algorithms to obtain upper estimates of $D(m,d)_q$, $1 \leq q \leq \infty$ in the style of (6.3.6) and (6.3.7). An important feature of our proof is that it is deterministic and, moreover, constructive. Formally the optimization problem

$$D(m,d)_q = \inf_\xi D(\xi,m,d)_q$$

is deterministic: one needs to minimize over $\{\xi^1,\dots,\xi^m\} \subset \Omega_d$. However, minimization by itself does not provide any upper estimate. It is known (see Davis *et al.*, 1997) that simultaneous optimization over many parameters ($\{\xi^1,\dots,\xi^m\}$ in our case) is a very difficult problem. We note that

$$D(m,d)_q = \sigma_m^e(J,\mathscr{B})_q := \inf_{g_1,\dots,g_m \in \mathscr{B}} \left\| J(\cdot) - \frac{1}{m}\sum_{\mu=1}^m g_\mu \right\|_{L_q(\Omega_d)}$$

where

$$J(\mathbf{x}) = \int_{\Omega_d} B(\mathbf{x},\mathbf{y})d\mathbf{y}, \qquad B(\mathbf{x},\mathbf{y}) := B_1(\mathbf{y},\mathbf{x}),$$

and

$$\mathscr{B} := \{B(\mathbf{x},\mathbf{y}), \qquad \mathbf{y} \in \Omega_d\}.$$

It was proved in Davis *et al.* (1997) that if an algorithm finds the best m-term approximation for each $f \in \mathbb{R}^N$ and for every dictionary $\mathscr{D}$ whose number of elements is of order N^k, $k \geq 1$, then this algorithm solves an NP-hard problem. Thus, in nonlinear m-term approximation we look for methods (algorithms) which provide approximation close to the best m-term approximation and at each step solve an optimization problem over only one parameter (ξ^μ in our case). In this section we provide such an algorithm for estimating $\sigma_m^e(J,\mathscr{B})_q$. We call this algorithm *constructive* because it provides an explicit construction with feasible one-parameter optimization steps.

We now proceed to the construction. We use in our construction the IA(ε), which is defined in §9.2. We use the following corollaries of Theorem 9.2.3.

Corollary 6.3.11 *We apply Theorem 9.2.3 for* $X = L_q(\Omega_d)$, $q \in [2,\infty)$, $\mathscr{D} = \{H(\mathbf{x},\mathbf{y}), \mathbf{y} \in \Omega_d\}$, $f = J^s(\mathbf{x})$, *where*

$$J^s(\mathbf{x}) := \int_{\Omega_d} H(\mathbf{x},\mathbf{y})d\mathbf{y} \in A_1(\mathscr{D}).$$

Using (9.2.1) we obtain by Theorem 9.2.3 a constructive set $\xi^1,\dots,\xi^m$ *such that*

$$D^s(\xi,m,d)_q = \|(J^s)_m^{i,\varepsilon}\|_{L_q(\Omega_d)} \le Cq^{1/2}m^{-1/2},$$

with absolute constant C.

Corollary 6.3.12 *We apply Theorem 9.2.3 for* $X = L_q(\Omega_d)$, $q \in [2,\infty)$, $\mathscr{D} = \{B(\mathbf{x},\mathbf{y}), \mathbf{y} \in \Omega_d\}$, $f = J(\mathbf{x})$, *where*

$$J(\mathbf{x}) := \int_{\Omega_d} B(\mathbf{x},\mathbf{y})d\mathbf{y} \in A_1(\mathscr{D}).$$

Using (9.2.1) we obtain by Theorem 9.2.3 a constructive set $\xi^1,\dots,\xi^m$ *such that*

$$D(\xi,m,d)_q = \|J_m^{i,\varepsilon}\|_{L_q(\Omega_d)} \le Cq^{1/2}m^{-1/2},$$

with absolute constant C.

Corollary 6.3.13 *We apply Theorem 9.2.3 for* $X = L_q(\Omega_d)$, $q \in [2,\infty)$, $\mathscr{D} = \{B(\mathbf{x},\mathbf{y})/\|B(\cdot,\mathbf{y})\|_{L_q(\Omega_d)},\ \ \mathbf{y} \in \Omega_d\}$, $f = J(\mathbf{x})$. *Using (9.2.1) we obtain by Theorem 9.2.3 a constructive set* $\xi^1,\dots,\xi^m$ *such that*

$$\left\| \int_{\Omega_d} B(\mathbf{x},\mathbf{y})d\mathbf{y} - \frac{1}{m}\sum_{\mu=1}^{m}\left(\frac{q}{q+1}\right)^d \left(\prod_{j=1}^{d}(1-\xi_j^\mu)^{-1/q}\right) B(\mathbf{x},\xi^\mu)\right\|_{L_q(\Omega_d)}$$
$$\le C\left(\frac{q}{q+1}\right)^d q^{1/2}m^{-1/2}$$

with absolute constant C.

We note that in the case $X = L_q(\Omega_d)$, $q \in [2,\infty)$, $\mathscr{D} = \{H(\mathbf{x},\mathbf{y}), \mathbf{x} \in \Omega_d\}$, $f = J^s(\mathbf{y})$ the implementation of the IA(ε) involves a sequence of maximization steps, in which we maximize functions of d variables. An important advantage of L_q-spaces is the simple and explicit form of the norming functional F_f of a function $f \in L_q(\Omega_d)$. For real L_q-spaces the functional F_f acts as follows:

$$F_f(g) = \int_{\Omega_d} \|f\|_q^{1-q}|f(\mathbf{x})|^{q-2}f(\mathbf{x})g(\mathbf{x})d\mathbf{x}.$$

Thus the IA(ε) should find at step m an approximate solution to the following optimization problem (over $\mathbf{y} \in \Omega_d$):

$$\int_{\Omega_d} |f_{m-1}^{i,\varepsilon}(\mathbf{x})|^{q-2} f_{m-1}^{i,\varepsilon}(\mathbf{x})H(\mathbf{x},\mathbf{y})d\mathbf{x} \quad \to \quad \max.$$

We now derive an estimate for $D(m,d)_\infty$ from Corollary 6.3.12.

Proposition 6.3.14 *For any m there exists a constructive set $\xi = \{\xi^1,\dots,\xi^m\} \subset \Omega_d$ such that*

$$D(\xi,m,d)_\infty \le Cd^{3/2}(\max(\ln d,\ln m))^{1/2}m^{-1/2}, \qquad d,m \ge 2 \tag{6.3.8}$$

with effective absolute constant C.

Proof We use an inequality from Niederreiter *et al.* (1990),

$$D(\xi,m,d)_\infty \le c(d,q)d(3d+4)D(\xi,m,d)_q^{q/(q+d)}, \tag{6.3.9}$$

and an estimate for $c(d,q)$ from Heinrich *et al.* (2001),

$$c(d,q) \le 3^{1/3}d^{-1+2/(1+q/d)}. \tag{6.3.10}$$

Specifying $q = d\max(\ln d,\ln m)$ and using Corollary 6.3.12 we get (6.3.8) from (6.3.9) and (6.3.10). □

6.4 Lower Estimates

It is convenient for us to assume that the functions are 1-periodic and to keep the notation $\Omega_d := [0,1]^d$. We begin with the following result.

Theorem 6.4.1 *For any cubature formula (Λ,ξ) with m knots the following relation holds ($r > 1/2$):*

$$\Lambda_m(\mathbf{W}^r_2,\xi) \ge C(r,d)m^{-r}(\log m)^{(d-1)/2}.$$

Proof Let us denote

$$\Lambda(\mathbf{k}) := \Lambda_m(e^{i2\pi(\mathbf{k},\mathbf{x})},\xi) = \sum_{\mu=1}^{m} \lambda_\mu e^{i2\pi(\mathbf{k},\xi^\mu)}$$

for the cubature formula (Λ,ξ) and for $\mathbf{k} \in \mathbb{Z}^d$.

Lemma 6.4.2 *The following inequality is valid for any $r > 1/2$:*

$$\sum_{\mathbf{k}\ne 0} |\Lambda(\mathbf{k})|^2 \nu(\bar{\mathbf{k}})^{-2r} \ge C(r,d)|\Lambda(0)|^2 m^{-2r}(\log m)^{d-1},$$

where $\bar{k}_j := \max(|k_j|,1)$ and $\nu(\bar{\mathbf{k}}) := \prod_{j=1}^d \bar{k}_j$.

First we deduce Theorem 6.4.1 from Lemma 6.4.2 and then prove this lemma. We assume that $|\Lambda(\mathbf{0})| \ge 1/2$ because otherwise it is sufficient to take as an example the function $f(\mathbf{x}) \equiv 1$. Let us consider the function

$$f(\mathbf{x}) := \sum_{\mathbf{k}\ne\mathbf{0}} \overline{\Lambda(\mathbf{k})}\nu(\bar{\mathbf{k}})^{-2r}e^{i2\pi(\mathbf{k},\mathbf{x})}$$

where $\overline{\Lambda(\mathbf{k})}$ is the complex conjugate of $\Lambda(\mathbf{k})$. Then

$$\|f^{(r)}\|_2 = \left(\sum_{\mathbf{k} \neq 0} \left|\Lambda(\mathbf{k})\right|^2 \nu(\bar{\mathbf{k}})^{-2r} \right)^{1/2} \tag{6.4.1}$$

and, taking into account that $\hat{f}(\mathbf{0}) = 0$, we get

$$\Lambda_m(f,\xi) - \hat{f}(\mathbf{0}) = \sum_{\mathbf{k} \neq \mathbf{0}} \left|\Lambda(\mathbf{k})\right|^2 \nu(\bar{\mathbf{k}})^{-2r}. \tag{6.4.2}$$

By (6.4.1) and (6.4.2),

$$\Lambda_m(\mathbf{W}_2^r,\xi) \geq \left(\sum_{\mathbf{k} \neq 0} \left|\Lambda(\mathbf{k})\right|^2 \nu(\bar{\mathbf{k}})^{-2r} \right)^{1/2};$$

using Lemma 6.4.2 we get

$$\Lambda_m(\mathbf{W}_2^r,\xi) \geq C(r,d) m^{-r} (\log m)^{(d-1)/2},$$

which proves Theorem 6.4.1. □

Proof We return to the proof of Lemma 6.4.2. Let $b(x)$ be an infinitely differentiable function such that $b(x) = 0$ for $x \notin (0,1)$ and $b(x) > 0$ for $x \in (0,1)$. Let m be given and choose $n \in \mathbb{N}$ such that

$$2m \leq 2^n < 4m. \tag{6.4.3}$$

Denote for $\mathbf{s} := (s_1, \ldots, s_d)$, where the s_j are nonnegative integers,

$$b_{\mathbf{s}}(\mathbf{x}) := \prod_{j=1}^{d} b(2^{s_j+2} x_j),$$

and

$$Y_{\mathbf{s}} := \left\{ \mathbf{y} \in \Omega_d \text{ such that } \Lambda_m\big(b_{\mathbf{s}}(\mathbf{x}-\mathbf{y}),\xi\big) = 0 \right\}.$$

It is easy to verify that, for all $\mathbf{s}$ such that $\|\mathbf{s}\|_1 = n$, the estimate

$$|Y_{\mathbf{s}}| \geq C(d) > 0$$

is valid for the measure $|Y_{\mathbf{s}}|$ of the set $Y_{\mathbf{s}}$. Further,

$$\begin{aligned}
\left|\hat{b}_{\mathbf{s}}(\mathbf{0})\Lambda(\mathbf{0})\right|^2 |Y_{\mathbf{s}}| &= \int_{Y_{\mathbf{s}}} \left|\Lambda_m\big(b_{\mathbf{s}}(\mathbf{x}-\mathbf{y}),\xi\big) - \hat{b}_{\mathbf{s}}(\mathbf{0})\Lambda(\mathbf{0})\right|^2 d\mathbf{y} \\
&\leq \int_{\Omega_d} \left|\Lambda_m\big(b_{\mathbf{s}}(\mathbf{x}-\mathbf{y}),\xi\big) - \hat{b}_{\mathbf{s}}(\mathbf{0})\Lambda(\mathbf{0})\right|^2 d\mathbf{y} \\
&= \sum_{\mathbf{k} \neq \mathbf{0}} \left|\Lambda(\mathbf{k})\right|^2 \left|\hat{b}_{\mathbf{s}}(\mathbf{k})\right|^2.
\end{aligned} \tag{6.4.4}$$

Let $a := [r] + 1$. Then, for $\mathbf{s}$ such that $\|\mathbf{s}\|_1 = n$, we have

$$\left|\hat{b}_{\mathbf{s}}(\mathbf{k})\right| \le C(d,a) \prod_{j=1}^{d} (2^{-s_j} \min(1, 2^{as_j} (\bar{k}_j)^{-a})$$
$$= C(d,a) 2^{(r-1)n} \prod_{j=1}^{d} 2^{-rs_j} \min\left(1, 2^{as_j} (\bar{k}_j)^{-a}\right). \tag{6.4.5}$$

By summing relation (6.4.5) over all $\mathbf{s}$ such that $\|\mathbf{s}\|_1 = n$ and using the inequalities

$$\sum_{\|\mathbf{s}\|_1 = n} \prod_{j=1}^{d} 2^{-rs_j} \min\left(1, 2^{as_j} (\bar{k}_j)^{-a}\right) \le \prod_{j=1}^{d} \sum_{s_j=0}^{\infty} 2^{-rs_j} \min\left(1, 2^{as_j} (\bar{k}_j)^{-a}\right)$$
$$\ll \prod_{j=1}^{d} (\bar{k}_j)^{-r}$$

we obtain from (6.4.4)

$$n^{d-1} 2^{-2n} \left|\Lambda(\mathbf{0})\right|^2 C(d) \le 2^{2(r-1)n} C(d,r) \sum_{\mathbf{k} \ne \mathbf{0}} \left|\Lambda(\mathbf{k})\right|^2 \nu(\bar{\mathbf{k}})^{-2r},$$

which proves the lemma. □

Next, we proceed to a stronger result than Theorem 6.4.1: we replace the class $\mathbf{W}^r_2$ in Theorem 6.4.1 by $\mathbf{W}^r_p$ with $2 \le p < \infty$ and prove the following proposition.

Theorem 6.4.3 *The following lower estimate is valid for any cubature formula (Λ, ξ) with m knots $(r > 1/p)$:*

$$\Lambda_m(\mathbf{W}^r_p, \xi) \ge C(r,d,p) m^{-r} (\log m)^{(d-1)/2}, \qquad 1 \le p < \infty.$$

Proof The proof of this theorem is based on Theorem 3.2.1. For convenience we recall the corresponding notation and reformulate Theorem 3.2.1 below as Theorem 6.4.4. Let

$$\Pi(\mathbf{N}, d) := \left\{(a_1, \dots, a_d) \in R^d : |a_j| \le N_j,\ j = 1, \dots, d\right\},$$

where N_j are nonnegative integers and $\mathbf{N} := (N_1, \dots, N_d)$. We denote

$$\mathscr{T}(\mathbf{N}, d) := \left\{t : t = \sum_{\mathbf{k} \in \Pi(\mathbf{N}, d)} c_{\mathbf{k}} e^{i(\mathbf{k}, \mathbf{x})}\right\}.$$

Then

$$\dim \mathscr{T}(\mathbf{N}, d) = \prod_{j=1}^{d} (2N_j + 1) =: \vartheta(\mathbf{N}).$$

Theorem 6.4.4 *Let $\varepsilon \in (0,1]$ and let a subspace $\Psi \subset \mathscr{T}(\mathbf{N}, d)$ be such that $\dim \Psi \ge \varepsilon \vartheta(\mathbf{N})$. Then there is a $t \in \Psi$ such that*

$$\|t\|_\infty = 1, \qquad \|t\|_2 \ge C(\varepsilon, d) > 0.$$

First, we prove the following assertion.

Lemma 6.4.5 *Let the coordinates of the vector* $\mathbf{s}$ *be natural numbers and* $\|\mathbf{s}\|_1 = n$*. Then, for any* $N \leq 2^{n-1}$ *and an arbitrary cubature formula* (Λ, ξ) *with* N *knots, there is a* $t_\mathbf{s} \in \mathscr{T}(2^\mathbf{s}, d)$ *such that* $\|t_\mathbf{s}\|_\infty \leq 1$ *and*

$$\hat{t}_\mathbf{s}(\mathbf{0}) - \Lambda_N(t_\mathbf{s}, \xi) \geq C(d) > 0. \tag{6.4.6}$$

Proof Let N and (Λ, ξ) be given. Let us consider in $\mathscr{T}((2^{s_1-1}, \dots, 2^{s_d-1}), d)$ the linear subspace Ψ of polynomials t satisfying the conditions

$$t(\xi^j) = 0, \qquad j = 1, \dots, N. \tag{6.4.7}$$

Then

$$\dim \Psi \geq 2^n - N \geq 2^{n-1}.$$

Consequently, by Theorem 6.4.4 there is a $t^1 \in \Psi$ such that $\|t^1\|_\infty = 1$ and

$$\|t^1\|_2 \geq C(d) > 0. \tag{6.4.8}$$

We define

$$t_\mathbf{s}(x) := \begin{cases} 1 & \text{if } \sum_{j=1}^N \lambda_j \leq 1/2, \\ |t^1(x)|^2 & \text{otherwise.} \end{cases}$$

The relations (6.4.7) and (6.4.8) prove the lemma. □

We can now complete the proof of Theorem 6.4.3. Let m be given. We choose n such that

$$m \leq 2^{n-1} < 2m.$$

Consider the polynomial

$$t(\mathbf{x}) := \sum_{\|\mathbf{s}\|_1 = n} t_\mathbf{s}(\mathbf{x}),$$

where the $t_\mathbf{s}$ are polynomials from Lemma 6.4.5 with $N = m$. Then

$$\hat{t}(\mathbf{0}) - \Lambda_m(t, \xi) \geq C(d) n^{d-1}. \tag{6.4.9}$$

Let us estimate $\|t^{(r)}\|_p$, $2 \leq p < \infty$. We will use Corollary A.3.5. We have

$$\|t^{(r)}\|_p \ll \left(\sum_{\|\mathbf{u}\|_1 \leq n+d} \|\delta_\mathbf{u}(t^{(r)})\|_p^2 \right)^{1/2}.$$

Using the Bernstein inequality (see Theorem 3.3.1) we continue the above estimate:

$$\ll \left(\sum_{\|\mathbf{u}\|_1 \le n+d} 2^{2r\|\mathbf{u}\|_1} \left\| \delta_{\mathbf{u}}(t) \right\|_p^2 \right)^{1/2}. \tag{6.4.10}$$

Next we have

$$\delta_{\mathbf{u}}(t) = \sum_{\|\mathbf{s}\|_1 = n} \delta_{\mathbf{u}}(t_{\mathbf{s}}) = \sum_{\|\mathbf{s}\|_1 = n; \mathbf{s}+\mathbf{1} \ge \mathbf{u}} \delta_{\mathbf{u}}(t_{\mathbf{s}}).$$

By the inequality $\|t_{\mathbf{s}}\|_\infty \le 1$ we find

$$\left\| \delta_{\mathbf{u}}(t) \right\|_p \ll (n+d+1-\|\mathbf{u}\|_1)^{d-1}. \tag{6.4.11}$$

Estimates (6.4.10), (6.4.11) result in

$$\left\| t^{(r)} \right\|_p \ll \left(\sum_{\|\mathbf{u}\|_1 \le n+d} 2^{2r\|\mathbf{u}\|_1} (n+d+1-\|\mathbf{u}\|_1)^{2(d-1)} \right)^{1/2} \ll 2^{rn} n^{(d-1)/2}. \tag{6.4.12}$$

Comparing (6.4.9) and (6.4.12) we obtain Theorem 6.4.3 for $2 \le p < \infty$. Clearly, the lower estimate for $1 \le p < 2$ follows from the estimate which we have just proved. □

We now show how the technique developed above can be used for proving the lower bounds for the error of numerical integration for other classes and for special cubature formulas. We begin with obtaining lower bounds for numerical integration with respect to a special class of knots for the $\mathbf{W}^r_p$ classes. Let $\mathbf{s} = (s_1, \dots, s_d)$, $s_j \in \mathbb{N}_0$, $j = 1, \dots, d$. We associate with $\mathbf{s}$ a web $W(\mathbf{s})$ as follows: denote

$$w(\mathbf{s}, \mathbf{x}) := \prod_{j=1}^{d} \sin(2^{s_j} x_j)$$

and define

$$W(\mathbf{s}) := \{\mathbf{x} : w(\mathbf{s}, \mathbf{x}) = 0\}.$$

Definition 6.4.6 We say that a set of knots $X_m := \{\xi^i\}_{i=1}^m$ is an (n,l)-net if $|X_m \setminus W(\mathbf{s})| \le 2^l$ for all $\mathbf{s}$ such that $\|\mathbf{s}\|_1 = n$.

Theorem 6.4.7 *For any cubature formula (Λ, X_m) with respect to an $(n, n-1)$-net X_m we have*

$$\Lambda(\mathbf{W}^r_p, X_m) \gg 2^{-rn} n^{(d-1)/2}, \qquad 1 \le p < \infty.$$

Proof The proof is similar to that of Theorem 6.4.3. Take $\mathbf{s}$ such that $\|\mathbf{s}\|_1 = n$ and consider $\mathscr{T}(\mathbf{N}, d)$ with $N_j := 2^{s_j - 1}$, $j = 1, \dots, d$. Then

$$\dim \mathscr{T}(\mathbf{N}, d) \ge 2^{\|\mathbf{s}\|_1} = 2^n.$$

Let $I(\mathbf{s})$ be a set of indices such that

$$X_m \setminus W(\mathbf{s}) = \{\xi^i\}_{i\in I(\mathbf{s})}.$$

Then by our assumption $|I(\mathbf{s})| \le 2^{n-1}$. Consider

$$\Psi(\mathbf{s}) := \{t \in \mathscr{T}(\mathbf{N},d) : t(\xi^i) = 0, i \in I(\mathbf{s})\}.$$

Then $\dim \Psi(\mathbf{s}) \ge 2^{n-1}$. By Theorem 6.4.4 we find $t_{\mathbf{s}}^1 \in \Psi(\mathbf{s})$ such that $\|t_{\mathbf{s}}^1\|_\infty = 1$ and $\|t_{\mathbf{s}}^1\|_2 \ge c(d) > 0$. Consider

$$t(\mathbf{x}) := \sum_{\|\mathbf{s}\|_1 = n} t_{\mathbf{s}}(\mathbf{x}), \qquad t_{\mathbf{s}}(\mathbf{x}) := |t_{\mathbf{s}}^1(\mathbf{x})|^2 w(\mathbf{s},\mathbf{x})^2.$$

We have

$$\begin{aligned} &t_{\mathbf{s}}^1(\mathbf{x}) w(s_1,x_1) \\ &= (2i)^{-1} \left(\sum_{\mathbf{k}:|k_j|\le 2^{s_j-1}} \hat{t}_{\mathbf{s}}^1(\mathbf{k}) \exp\left(i(\mathbf{k},\mathbf{x}) + i2^{s_1}x_1\right) \right. \\ &\quad \left. - \sum_{\mathbf{k}:|k_j|\le 2^{s_j-1}} \hat{t}_{\mathbf{s}}^1(\mathbf{k}) \exp\left(i(\mathbf{k},\mathbf{x}) - i2^{s_1}x_1\right) \right), \end{aligned}$$

and

$$\|t_{\mathbf{s}}^1(\mathbf{x}) w(s_1,x_1)\|_2^2 = 2^{-1}\|t_{\mathbf{s}}^1\|_2^2.$$

Therefore,

$$\|t_{\mathbf{s}}^1(\mathbf{x}) w(\mathbf{s},\mathbf{x})\|_2^2 = 2^{-d}\|t_{\mathbf{s}}^1\|_2^2 \ge c_1(d) > 0.$$

Then relation (6.4.9) is obviously satisfied for the t defined above. The relation

$$\|t\|_{\mathbf{W}^r_p} \ll \|t\|_{\mathbf{B}^r_{p,2}} \ll 2^{rn} n^{(d-1)/2} \tag{6.4.13}$$

is proved in the same way as relation (6.4.12). □

We proceed to the classes $\mathbf{H}^r_q$ and $\mathbf{B}^r_{q,\theta}$. Define

$$\|f\|_{\mathbf{H}^r_q} := \sup_{\mathbf{s}} \|\delta_{\mathbf{s}}(f)\|_q 2^{r\|\mathbf{s}\|_1},$$

and for $1 \le \theta < \infty$ define

$$\|f\|_{\mathbf{B}^r_{q,\theta}} := \left(\sum_{\mathbf{s}} \left(\|\delta_{\mathbf{s}}(f)\|_q 2^{r\|\mathbf{s}\|_1} \right)^\theta \right)^{1/\theta}.$$

We will write $\mathbf{B}^r_{q,\infty} := \mathbf{H}^r_q$. With a little abuse of notation, we denote the corresponding unit ball as

$$\mathbf{B}^r_{q,\theta} := \{f : \|f\|_{\mathbf{B}^r_{q,\theta}} \le 1\}.$$

The example that was constructed in the proof of Theorem 6.4.3 (see above) provides the lower bound for the Besov-type classes.

Theorem 6.4.8 *The following lower estimate is valid for any cubature formula* (Λ, X_m) *with* m *knots* $(r > 1/p)$*:*

$$\Lambda(\mathbf{B}^r_{p,\theta}, X_m) \geq C(r,d,p) m^{-r} (\log m)^{(d-1)(1-1/\theta)}, \qquad 1 \leq p \leq \infty, \qquad 1 \leq \theta \leq \infty.$$

Indeed, the proof of (6.4.12) implies that

$$\|t\|_{\mathbf{B}^r_{p,\theta}} \ll 2^{rn} n^{(d-1)/\theta}. \tag{6.4.14}$$

In the same way the proof of Theorem 6.4.7 gives the following result.

Theorem 6.4.9 *For any cubature formula* (Λ, X_m) *with respect to an* $(n, n-1)$*-net* X_m *we have*

$$\Lambda(\mathbf{B}^r_{p,\theta}, X_m) \gg 2^{-rn} n^{(d-1)(1-1/\theta)}, \quad 1 \leq p \leq \infty.$$

We note that Theorems 6.4.7 and 6.4.9 provide lower bounds for numerical integration with respect to sparse grids and their modifications. For $n \in \mathbb{N}$ we define the sparse grid $\mathrm{SG}(n)$ as follows:

$$\mathrm{SG}(n) := \big\{\xi(\mathbf{n},\mathbf{k}) = (\pi k_1 2^{-n_1}, \dots, \pi k_d 2^{-n_d}), \\ 0 \leq k_j < 2^{n_j}, j = 1, \dots, d, \ \|\mathbf{n}\|_1 = n\big\}.$$

Then it is easy to check that $\mathrm{SG}(n) \subset W(\mathbf{s})$ for any $\mathbf{s}$ such that $\|\mathbf{s}\|_1 = n$. Indeed, let $\xi(\mathbf{n},\mathbf{k}) \in \mathrm{SG}(n)$. Take any $\mathbf{s}$ with $\|\mathbf{s}\|_1 = n$. Then $\|\mathbf{s}\|_1 = \|\mathbf{n}\|_1$ and there exists a j such that $s_j \geq n_j$. For this j we have

$$\sin 2^{s_j} \xi(\mathbf{n},\mathbf{k})_j = \sin 2^{s_j} \pi k_j 2^{-n_j} = 0 \qquad \text{and} \qquad w(\mathbf{s}, \xi(\mathbf{n},\mathbf{k}) = 0.$$

This means that $\mathrm{SG}(n)$ is an (n,l)-net for any l. We note that $|\mathrm{SG}(n)| \asymp 2^n n^{d-1}$. It is known (see (6.9.1) and Theorem 6.9.2 below) that there exists a cubature formula $(\Lambda, \mathrm{SG}(n))$ such that

$$\Lambda(\mathbf{H}^r_p, \mathrm{SG}(n)) \ll 2^{-rn} n^{d-1}, \quad 1 \leq p \leq \infty, \quad r > 1/p. \tag{6.4.15}$$

Theorem 6.4.9 with $\theta = \infty$ shows that the bound (6.4.15) is sharp. Moreover, it shows that even an addition of extra 2^{n-1} arbitrary knots to $\mathrm{SG}(n)$ will not improve the bound in (6.4.15). In the case $X_m = \mathrm{SG}(n)$ other proof of Theorem 6.4.9 is given in Dinh Dung and Ullrich (2014).

Theorem 6.4.3 gives the same lower estimate for different parameters $1 \leq p < \infty$. It is clear that the larger is p, the stronger is the statement. We now discuss an improvement of Theorem 6.4.3 in the particular case $p = 1$. We will improve the lower estimate by replacing the exponent $(d-1)/2$ by $d-1$. However, this revised estimate will be proved under some (mild) assumptions on the weights of a cubature formula (Λ, ξ) and also under a slight modification of the classes $\mathbf{W}^r_1$, namely, the classes $\mathbf{W}^r_{1,0}$. For convenience we recall their definition here. Denote

$$F_{r,0}(x) := F_r(x,0) := 1 + 2\sum_{k=1}^{\infty} k^{-r}\cos kx, \qquad x \in \mathbb{T},$$

$$F_{r,0}(\mathbf{x}) := \prod_{j=1}^{d} F_{r,0}(x_j), \qquad \mathbf{x} = (x_1, \dots, x_d) \in \mathbb{T}^d,$$

$$\mathbf{W}^r_{1,0} := \{f : f = \varphi * F_{r,0}, \qquad \|\varphi\|_1 \le 1\}.$$

It is clear that if r is an even integer then we have $\mathbf{W}^r_{1,0} = \mathbf{W}^r_1$. Let B be a positive number and $Q(B,m)$ be the set of cubature formulas $\Lambda_m(\cdot, \xi)$ satisfying the additional condition

$$\sum_{\mu=1}^{m} |\lambda_\mu| \le B.$$

We will now obtain the lower estimates for the quantities

$$\kappa^B_m(\mathbf{W}) := \inf_{\Lambda_m(\cdot,\xi)\in Q(B,m)} \Lambda_m(\mathbf{W}, \xi).$$

First, we prove the following relation.

Theorem 6.4.10 *Let $r > 1$. Then*

$$\kappa^B_m(\mathbf{W}^r_{1,0}) \ge C(r,B,d)m^{-r}(\log m)^{d-1}, \qquad C(r,B,d) > 0.$$

Proof We use a notation similar to the above:

$$\Lambda(\mathbf{k}) := \Lambda_m(e^{i(\mathbf{k},\mathbf{x})}, \xi) = \sum_{\mu=1}^{m} \lambda_\mu e^{i(\mathbf{k},\xi^\mu)}.$$

In the case $|\Lambda(\mathbf{0})| < 1/2$ it is sufficient to consider a function $f(\mathbf{x}) \equiv 1$ as an example, and therefore we consider only $|\Lambda(\mathbf{0})| \ge 1/2$. From the cubature formula $\tilde{\Lambda}_m(\xi) := \Lambda_m(\xi)/\Lambda(\mathbf{0})$ we see that

$$\Lambda_m(\mathbf{W}, \xi) \ge \tilde{\Lambda}_m(\mathbf{W}, \xi)/4$$

for $\mathbf{W}$ such that $\frac{1}{2}(f - \hat{f}(\mathbf{0})) \in \mathbf{W}$, provided that $f \in \mathbf{W}$ and that $\tilde{\Lambda}_m(\cdot, \xi)$ is exact on the function $f(\mathbf{x}) \equiv 1$, i.e. $\tilde{\Lambda}(\mathbf{0}) = 1$. Thus it is sufficient for our purpose to consider the cubature formulas $\Lambda_m(\cdot, \xi)$ satisfying the additional condition $\Lambda(\mathbf{0}) = 1$. Let us consider the cubature formula Λ' constructed with the use of $\Lambda_m(\cdot, \xi)$ as follows:

$$\Lambda'(f) := \sum_{\nu=1}^{m} \bar{\lambda}_\nu \Lambda_m(f(\mathbf{x} - \xi^\nu), \xi)$$

where $\bar{\lambda}_\nu$ is the complex conjugate of λ_ν. Then

$$\Lambda'(\mathbf{k}) = \Lambda'(e^{i(\mathbf{k},\mathbf{x})}) = \sum_{\nu=1}^{m} \bar{\lambda}_\nu \sum_{\mu=1}^{m} \lambda_\mu e^{i(\mathbf{k},(\xi^\mu-\xi^\nu))} = |\Lambda(\mathbf{k})|^2. \tag{6.4.16}$$

The function $F_{r,0}$ belongs to the closure in the uniform norm $(r > 1)$ of the class $\mathbf{W}^r_{1,0}$. Therefore, by (6.4.16) and Lemma 6.4.2, we obtain on the one hand

$$\begin{aligned}\Lambda'(\mathbf{W}^r_{1,0}) &\geq \Lambda'(F_{r,0}) - \hat{F}_{r,0}(\mathbf{0}) = \sum_{\mathbf{k}\neq\mathbf{0}} \Lambda'(\mathbf{k})\hat{F}_{r,0}(\mathbf{k}) = \sum_{\mathbf{k}\neq\mathbf{0}} |\Lambda(\mathbf{k})|^2 \hat{F}_{r,0}(\mathbf{k})\\ &\geq C(r,d)m^{-r}(\log m)^{d-1}.\end{aligned} \tag{6.4.17}$$

On the other hand, for the cubature formula Λ' we have

$$\Lambda'(f) - \hat{f}(\mathbf{0}) = \sum_{\nu=1}^{m} \bar{\lambda}_\nu (\Lambda_m(f(\mathbf{x}-\xi^\nu),\xi) - \hat{f}(\mathbf{0}))$$

which, for $\Lambda_m(\cdot,\xi) \in Q(B,m)$, implies the inequality

$$\Lambda'(\mathbf{W}^r_{1,0}) \leq B\Lambda_m(\mathbf{W}^r_{1,0},\xi). \tag{6.4.18}$$

Relations (6.4.17) and (6.4.18) yield the required lower estimate for $\kappa^B_m(\mathbf{W}^r_{1,0})$. $\square$

Let us discuss how Theorems 6.4.3 and 6.4.10 can be used to estimate from below the generalized discrepancy $D_r(\xi,\Lambda,m,d)_q$. Theorem 6.4.3 combined with Theorem 6.2.4 and Remark 6.2.6 implies the following result.

Theorem 6.4.11 *Let $1 < q < \infty$ and r be a positive integer. Then, for any points $\xi = (\xi^1,\dots,\xi^m) \subset \Omega_d$ and any weights $\Lambda = (\lambda_1,\dots,\lambda_m)$, we have*

$$D_r(\xi,\Lambda,m,d)_q \geq C(d,r)m^{-r}(\log m)^{(d-1)/2},$$

with a positive constant $C(d,r)$.

We now turn to an application of Theorem 6.4.10. Let r be an even integer. Then $\mathbf{W}^r_{1,0} = \mathbf{W}^r_1$. Assume a given cubature formula $\Lambda_m(\cdot,\xi) \in Q(B,m)$. Then, using the definition of $D_r(\xi,\Lambda,m,d)_\infty$ (see (6.2.14)) and the embedding $\bar{\mathbf{W}}^r_p(\Omega_d) \hookrightarrow \dot{\mathbf{W}}^r_p(\Omega_d)$ we get

$$D_r(\xi,\Lambda,m,d)_\infty \gg \Lambda_m(\bar{\mathbf{W}}^r_1(\Omega_d),\xi). \tag{6.4.19}$$

By (6.2.23) and the embedding $\mathbf{W}^r_1 \hookrightarrow \mathbf{W}^r_1(\Omega_d)$ we obtain

$$\Lambda_m(\bar{\mathbf{W}}^r_1(\Omega_d),\xi) \gg \Lambda'_m(\mathbf{W}^r_1,\theta), \tag{6.4.20}$$

where $\theta = \{\theta^1,\dots,\theta^m\}$, $\theta^\mu = 2\pi\eta^\mu$, $\eta^\mu_j = \psi(\xi^\mu_j)$, $j = 1,\dots,d$;

$$\lambda'_\mu = \lambda_\mu \prod_{j=1}^{d} \psi'(\xi_j^\mu), \qquad \mu = 1, \ldots, m.$$

Next, it is clear that $\Lambda_m(\cdot,\xi) \in Q(B,m)$ implies that $\Lambda'_m(\cdot,\theta) \in Q(C(d)B,m)$. Therefore, by Theorem 6.4.10 we get

$$\Lambda'_m(\mathbf{W}_1^r,\theta) \gg m^{-r}(\log m)^{d-1}. \tag{6.4.21}$$

Combining (6.4.19)–(6.4.21) we obtain the following statement.

Theorem 6.4.12 *Let B be a positive number. For any points $\xi^1,\ldots,\xi^m \subset \Omega_d$ and any weights $\Lambda = (\lambda_1,\ldots,\lambda_m)$ satisfying the condition*

$$\sum_{\mu=1}^{m} |\lambda_\mu| \le B$$

we have, for even integers r,

$$D_r(\xi,\Lambda,m,d)_\infty \ge C(d,B,r)m^{-r}(\log m)^{d-1}$$

with a positive constant $C(d,B,r)$.

Corollary 6.4.13 *Let r be an even integer. Then we have for the r-discrepancy*

$$D_r(\xi,m,d)_\infty := D_r(\xi,(1/m,\ldots,1/m),m,d)_\infty \gg m^{-r}(\log m)^{d-1}.$$

The case $p=\infty$ is excluded in Theorem 6.4.3. There is no nontrivial general lower estimate in this case. We give one conditional result in this direction.

Theorem 6.4.14 *Let the cubature formula (Λ,ξ) be such that the inequality*

$$\Lambda_m(\mathbf{W}_p^r,\xi) \ll m^{-r}(\log m)^{(d-1)/2}, \qquad r > 1/p,$$

holds for some $1 < p < \infty$. Then

$$\Lambda_m(\mathbf{W}_\infty^r,\xi) \gg m^{-r}(\log m)^{(d-1)/2}.$$

Proof We denote as above

$$\Lambda(\mathbf{k}) := \sum_{j=1}^{m} \lambda_j e^{i(\mathbf{k},\xi^j)}.$$

Let us consider the function

$$g_{\Lambda,\xi,r}(\mathbf{x}) := \sum_{\mathbf{k}} \Lambda(\mathbf{k})\hat{F}_r(\mathbf{k})e^{i(\mathbf{k},\mathbf{x})} - 1.$$

Then, for the quantity $\Lambda_m(\mathbf{W}^r_p,\xi)$, writing $p' = p/(p-1)$ we have

$$\Lambda_m(\mathbf{W}^r_p,\xi) = \sup_{f\in\mathbf{W}^r_p} \left|\Lambda_m(f,\xi) - \hat{f}(\mathbf{0})\right| = \sup_{\|\varphi\|_p\le 1} \left|\Lambda_m\left(F_r(\mathbf{x}) * \varphi(\mathbf{x}),\xi\right) - \hat{\varphi}(\mathbf{0})\right|$$
$$= \sup_{\|\varphi\|_p\le 1} \left|\langle g_{\Lambda,\xi,r}(-\mathbf{y}), \overline{\varphi(\mathbf{y})}\rangle\right| = \|g_{\Lambda,\xi,r}\|_{p'}. \tag{6.4.22}$$

Therefore, by the hypothesis of the theorem, for some $1 < q < \infty$ with $q = p'$ we have

$$\|g_{\Lambda,\xi,r}\|_q \ll m^{-r}(\log m)^{(d-1)/2}. \tag{6.4.23}$$

Further, for arbitrary $1 < a < b$ and $f \in L_b$ the following inequality holds:

$$\|f\|_a \le \|f\|_1^{\kappa}\|f\|_b^{1-\kappa}, \qquad \kappa := \left(\frac{1}{a} - \frac{1}{b}\right)\left(1 - \frac{1}{b}\right)^{-1}. \tag{6.4.24}$$

By Theorem 6.4.3 we have, for any $1 \le z < \infty$,

$$\Lambda_m(\mathbf{W}^r_{z,\alpha},\xi) \gg m^{-r}(\log m)^{(d-1)/2}. \tag{6.4.25}$$

Therefore, by (6.4.22)

$$\|g_{\Lambda,\xi,r}\|_{z'} \gg m^{-r}(\log m)^{(d-1)/2}. \tag{6.4.26}$$

Now setting $b = q$, $a = \frac{1}{2}(b+1)$, $z' = a$ we obtain from the relations (6.4.24), (6.4.23), (6.4.26)

$$\|g_{\Lambda,\xi,r}\|_1 \gg m^{-r}(\log m)^{(d-1)/2}.$$

It suffices to apply (6.4.22) to complete the proof.

The theorem is proved. □

Remark 6.4.15 We have actually proved the following inequality. Let $1 \le p_1 < p_2 < \infty$; then, for any (Λ,ξ),

$$\Lambda_m(\mathbf{W}^r_{p_2},\xi) \le \Lambda_m(\mathbf{W}^r_{p_1},\xi)^{p_1/p_2}\Lambda_m(\mathbf{W}^r_{\infty},\xi)^{1-p_1/p_2}, \qquad \text{with } r > 1/p_1.$$

6.5 The Fibonacci Cubature Formulas

In this section we provide the first results on the upper bounds for $\kappa_m(\mathbf{W}^r_p)$ and $\kappa_m(\mathbf{H}^r_p)$, which match the corresponding lower bounds from §6.4. We prove these upper bounds in a constructive way by presenting an effective cubature formula. Here, we discuss the case $d = 2$. It turns out that in this case the optimal cubature formulas (in the sense of order) are given by simple expressions.

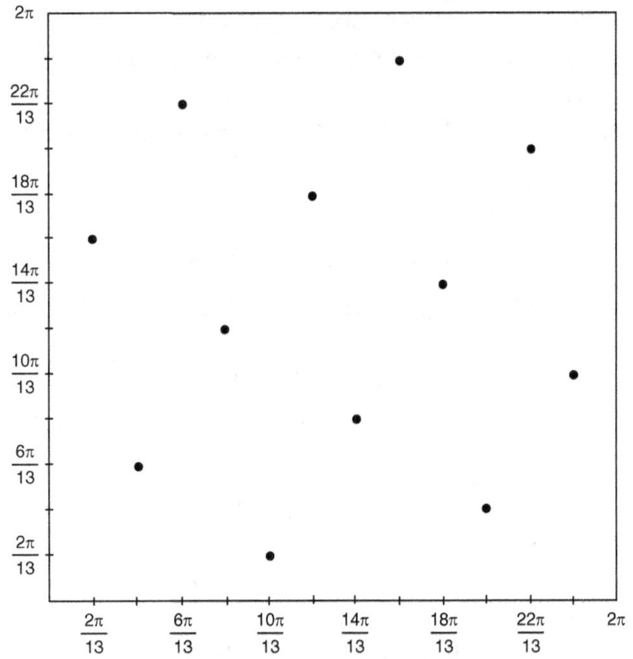

Figure 6.1 For $n = 6$, $b_6 = 13$, $b_5 = 8$ the Fibonacci set is given by $(x_1^\mu, x_2^\mu) = \left(\frac{2\pi\mu}{13}, 2\pi\left\{\frac{8\mu}{13}\right\}\right)$, $\mu = 1, \ldots, 13$.

Let $\{b_n\}_{n=0}^{\infty}$, $b_0 = b_1 = 1$, $b_n = b_{n-1} + b_{n-2}$, $n \geq 2$, be the Fibonacci numbers. For the continuous functions of two variables which are 2π-periodic in each variable we define the *Fibonacci cubature formulas*:

$$\Phi_n(f) = b_n^{-1} \sum_{\mu=1}^{b_n} f\left(\frac{2\pi\mu}{b_n}, 2\pi\left\{\frac{\mu b_{n-1}}{b_n}\right\}\right).$$

In this definition $\{a\}$ is the fractional part of the number a. In Figure 6.1 we illustrate the Fibonacci set for $n = 6$, $b_6 = 13$, $b_5 = 8$:

$$(x_1^\mu, x_2^\mu) = \left(\frac{2\pi\mu}{13}, 2\pi\left\{\frac{8\mu}{13}\right\}\right), \quad \mu = 1, \ldots, 13.$$

For a function class $\mathbf{F}$ denote

$$\Phi_n(\mathbf{F}) := \sup_{f \in \mathbf{F}} |\Phi_n(f) - \hat{f}(\mathbf{0})|.$$

The following lower estimate (see Theorems 6.4.1 and 6.4.3) was proved in §6.4 above: for all $r > 1/p$, $1 \leq p < \infty$,

$$\Phi_n(\mathbf{W}^r_{p,\alpha}) \geq \kappa_m(\mathbf{W}^r_{p,\alpha}) \gg b_n^{-r}(\log b_n)^{1/2}.$$

In §6.5.1 we will prove a similar upper estimate for $r > 1/2$, $2 \le p < \infty$ and for $r > 1/p$, $1 < p \le 2$ and find the orders of decrease of the quantities $\Phi_n(\mathbf{W}^r_{\infty,\alpha})$, $\Phi_n(\mathbf{W}^r_{1,0})$ respectively for $r > 1/2$ and $r > 1$. In §6.5.2 we consider the case of small smoothness $1/p < r < 1/2$, $2 \le p \le \infty$, which is not considered in §6.5.1 and in addition we find there the order of decrease of $\Phi_n(\mathbf{H}^r_p)$.

6.5.1 Large Smoothness

First, we denote $\mathbf{y}^\mu := \big(2\pi\mu/b_n, 2\pi\{\mu b_{n-1}/b_n\}\big)$, $\mu = 1,\dots,b_n$, and

$$\Phi(\mathbf{k}) := b_n^{-1}\sum_{\mu=1}^{b_n} e^{i(\mathbf{k},\mathbf{y}^\mu)}.$$

Note that

$$\Phi_n(f) = \sum_{\mathbf{k}} \hat{f}(\mathbf{k})\Phi(\mathbf{k}), \tag{6.5.1}$$

where for the sake of simplicity we may assume that f is a trigonometric polynomial.

It is easy to see that the following relation holds

$$\Phi(\mathbf{k}) = \begin{cases} 1 & \text{for } \mathbf{k} \in L(n) \\ 0 & \text{for } \mathbf{k} \notin L(n), \end{cases} \tag{6.5.2}$$

where

$$L(n) := \big\{\mathbf{k} = (k_1,k_2) : k_1 + b_{n-1}k_2 \equiv 0 \pmod{b_n}\big\}.$$

We now prove the following assertion.

Lemma 6.5.1 *There exists an absolute constant $\gamma > 0$ such that, for any $n > 2$,*

$$\Gamma(\gamma b_n) \cap \big(L(n)\backslash \mathbf{0}\big) = \varnothing.$$

Proof Let $\mathbf{k} \in L(n)$, $k_1k_2 \ne 0$, $|k_1|, |k_2| \le b_n/2$. Then, for some integer l,

$$k_1k_2 = b_nk_2^2\left(\frac{l}{k_2} - \frac{b_{n-1}}{b_n}\right). \tag{6.5.3}$$

We denote $\Delta := l/k_2 - b_{n-1}/b_n$, $P(x) := x^2 + x - 1$ and consider the value $P(l/k_2)$. By Taylor's formula,

$$P\left(\frac{l}{k_2}\right) = P\left(\frac{b_{n-1}}{b_n}\right) + P'\left(\frac{b_{n-1}}{b_n}\right)\Delta + \Delta^2. \tag{6.5.4}$$

Further, $P(l/k_2)$ is nonzero; consequently $|P(l/k_2)| \ge 1/k_2^2$. We have

$$\begin{aligned}\left|P\left(\frac{b_{n-1}}{b_n}\right)\right| &= b_n^{-2}|b_{n-1}^2 + b_nb_{n-1} - b_n^2| \\ &= b_n^{-2}|b_{n-2}^2 + b_{n-1}b_{n-2} - b_{n-1}^2| = \cdots = b_n^{-2}.\end{aligned}$$

Clearly, it is sufficient to consider the case $|\Delta| \le 1$. Then by (6.5.4) we find that $|\Delta| \ge Ck_2^{-2}$ and by (6.5.3) we obtain the estimate $|k_1k_2| \ge Cb_n$ for $\mathbf{k} \in L(n)$, $k_1k_2 \neq 0$ $|k_1|, |k_2| \le b_n/2$. These inequalities prove the lemma. □

Theorem 6.5.2 *For $1 < p \le \infty$, $r > \max(1/p, 1/2)$ we have*

$$\Phi_n(\mathbf{W}^r_{p,\alpha}) \asymp b_n^{-r}(\log b_n)^{1/2}.$$

Proof We first prove the upper estimates. Clearly, it suffices to consider the case $1 < p \le 2$. Owing to the restriction on r the class $\mathbf{W}^r_{p,\alpha}$ is embedded in the space of continuous functions. By Theorem 4.4.11 with $q = p$ and $p = \infty$ it is sufficient to consider trigonometric polynomials. Let

$$Z_l := \big(\Gamma(2^{l+1}\gamma b_n)\backslash\Gamma(2^l\gamma b_n)\big) \cap L(n), \qquad l = 0, 1, \dots$$

By using Lemma 6.5.1 it is easy to get the estimate

$$|Z_l| \ll 2^l(l + \log b_n), \qquad l = 0, 1, \dots \tag{6.5.5}$$

Further, by (6.5.1), (6.5.2) and Lemma 6.5.1,

$$\Phi_n(f) - \hat{f}(\mathbf{0}) = \sum_{l\ge 0}\sum_{\mathbf{k}\in Z_l} \hat{f}(\mathbf{k}). \tag{6.5.6}$$

Let

$$\psi_l(\mathbf{x}) := \sum_{\mathbf{k}\in Z_l} e^{i(\mathbf{k},\mathbf{x})}.$$

Then

$$\left|\Phi_n(f) - \hat{f}(\mathbf{0})\right| = \left|\sum_{l\ge 0}\langle f, \psi_l\rangle\right| \le \sum_{l\ge 0} E_{2^l\gamma b_n}(f)_p \|\psi_l\|_{p'}. \tag{6.5.7}$$

For $\|\psi_l\|_{p'}$ we have, by Corollary A.3.5,

$$\|\psi_l\|_{p'} \ll \left(\sum_{\mathbf{s}} \|\delta_{\mathbf{s}}(\psi_l)\|_{p'}^2\right)^{1/2}.$$

It is not difficult to see that for ψ_l only those $\delta_{\mathbf{s}}(\psi_l)$ can be nonzero for which

$$\big|\,\|\mathbf{s}\|_1 - \log_2(2^l\gamma b_n)\big| \le C,$$

where C is an absolute constant. In addition, by Lemma 6.5.1 the number of nonzero terms of $\delta_{\mathbf{s}}(\psi_l)$ is not greater than $C2^l$. Consequently,

$$\left\|\delta_{\mathbf{s}}(\psi_l)\right\|_{p'} \le \left\|\delta_{\mathbf{s}}(\psi_l)\right\|_2^{2/p'} \left\|\delta_{\mathbf{s}}(\psi_l)\right\|_\infty^{1-2/p'} \ll 2^{l/p}$$

and

$$\|\psi_l\|_{p'} \ll (l + \log b_n)^{1/2} 2^{l/p}. \tag{6.5.8}$$

The required upper estimate follows from (6.5.8), (6.5.7), and Theorem 4.4.9.

Let us proceed to the proof of the lower estimates. In the case $p < \infty$ the lower estimates follow from the more general results of §6.4 (see Theorem 6.4.3). It remains to consider the case $p = \infty$. In this case the required lower bound follows from Theorem 6.4.14 and from upper bounds that have been proved already. □

We now consider the case $p = 1$.

Theorem 6.5.3 *Let $r > 1$. Then*

$$\Phi_n(\mathbf{W}^r_{1,0}) \asymp b_n^{-r} \log b_n.$$

Proof The upper estimate holds for all $\mathbf{W}^r_{1,\alpha}$; this estimate follows from (6.5.6), (6.5.5), and the inequality

$$|\hat{f}(\mathbf{k})| \ll \nu(\bar{\mathbf{k}})^{-r},$$

which holds for $f \in \mathbf{W}^r_{1,\alpha}$. The required lower bound follows from Theorem 6.4.10. □

Remark 6.5.4 Theorem 6.5.3 holds for the classes $\mathbf{W}^r_{1,\alpha}$ for any α.

As we pointed out in the proof of Theorem 6.5.3 the upper bound holds for $\mathbf{W}^r_{1,\alpha}$ with arbitrary α. The lower bound in Theorem 6.5.3 was derived from Theorem 6.4.10, which holds for cubature formulas that are much more general than the Fibonacci cubature formulas. However, we proved Theorem 6.4.10 only for $\alpha = \mathbf{0}$. We do not know whether it holds for all α. It turns out, however, that one can prove the required lower bound for the classes $\mathbf{W}^r_{1,\alpha}$ with arbitrary α by a method which uses specific features of the Fibonacci cubature formulas. This proof is somewhat technically involved and we do not present it here. Instead, we refer the reader to the paper Temlyakov (1994).

6.5.2 Small Smoothness

The results in §6.5.1 give the orders of the errors of the Fibonacci cubature formulas for the classes $\mathbf{W}^r_{p,\alpha}$ for $1 \le p \le \infty$ and $r > \max(1/p, 1/2)$. A natural restriction on r is the condition for the embedding of the class $\mathbf{W}^r_{p,\alpha}$ into the space of continuous functions, namely $r > 1/p$, $1 \le p \le \infty$. Thus, it remains to investigate the case $2 < p \le \infty$, $1/p < r \le 1/2$. It turns out that in this case the character of the dependence of $\Phi_n(\mathbf{W}^r_{p,\alpha})$ on r and n is different from the case $r > 1/2$.

Theorem 6.5.5 *Let $2 < p \le \infty$, $1/p < r < 1/2$. Then*

$$\Phi_n(\mathbf{W}^r_{p,\alpha}) \asymp b_n^{-r} (\log b_n)^{1-r}.$$

Proof We first prove the upper estimates. Clearly, it suffices to consider the case $2 < p < \infty$, $\alpha = \mathbf{0}$ and f a trigonometric polynomial. For the quantity $R_n(f) := \Phi_n(f) - \hat{f}(\mathbf{0})$ the relations (6.5.1), (6.5.2) imply that

$$R_n(f) = \sum_{\mathbf{k}\in L(n);\mathbf{k}\neq\mathbf{0}} \hat{f}(\mathbf{k}). \tag{6.5.9}$$

It is not difficult to prove the following assertion.

Lemma 6.5.6 *The set $L(n)$ may be represented in the form*

$$L(n) = \big\{(ub_{n-2} - vb_{n-3},\ u + 2v),\ u, v \textit{ are integers}\ \big\}.$$

For the class $\mathbf{W}^r_{p,0}$ we have, by (6.5.9),

$$\sup_{f\in\mathbf{W}^r_{p,0}} \big|R_n(f)\big| = \Bigg\|\sum_{\mathbf{k}\in L(n);\mathbf{k}\neq\mathbf{0}} \hat{F}_r(\mathbf{k})e^{i(\mathbf{k},\mathbf{x})}\Bigg\|_q, \tag{6.5.10}$$

where $\hat{F}_r(\mathbf{k}) = \hat{F}_r(\mathbf{k},\mathbf{0})$, $q := p/(p-1)$. Denote

$$g(\mathbf{x}) := \sum_{\mathbf{k}\in L(n);\mathbf{k}\neq\mathbf{0}} \hat{F}_r(\mathbf{k})e^{i(\mathbf{k},\mathbf{x})}.$$

We write $m := [\log \gamma b_n]$, where γ is from Lemma 6.5.1. Then

$$g(\mathbf{x}) = \sum_{\|\mathbf{s}\|_1\geq m} \delta_{\mathbf{s}}(g,\mathbf{x}).$$

We will use the following lemma to estimate $\big\|\delta_{\mathbf{s}}(g,\mathbf{x})\big\|_q$.

Lemma 6.5.7 *For $1 \leq q \leq 2$ we have*

$$\big\|A_{\mathbf{s}}(g)\big\|_q \ll 2^{-r\|\mathbf{s}\|_1+\left(\|\mathbf{s}\|_1-m\right)(1-1/q)}.$$

Proof It is easy to deduce from Lemma 6.5.1 that the following estimate holds for the number of elements of the set $G^n_{\mathbf{s}} := L(n)\cap\rho(\mathbf{s})$ ($\|\mathbf{s}\|_1 \geq m$):

$$|G^n_{\mathbf{s}}| \ll 2^{\|\mathbf{s}\|_1-m}. \tag{6.5.11}$$

By (6.5.11) we get

$$\big\|A_{\mathbf{s}}(g)\big\|_2 \ll 2^{-r\|\mathbf{s}\|_1+\left(\|\mathbf{s}\|_1-m\right)/2}. \tag{6.5.12}$$

Let us estimate $\big\|A_{\mathbf{s}}(g)\big\|_1$. Let $\mathbf{s}$ be given and let N be sufficiently large. We consider the function

$$G_N(\mathbf{x}) := \mathscr{V}_N(b_{n-2}x_1 + x_2)\mathscr{V}_N(-b_{n-3}x_1 + 2x_2). \tag{6.5.13}$$

Then, using Lemma 6.5.6, we get

$$A_{\mathbf{s}}(g,\mathbf{x}) = \big(A_{\mathbf{s}}(\mathbf{x}) * G_N(\mathbf{x})\big) * F_r(\mathbf{x}). \tag{6.5.14}$$

We will prove that

$$\|G_N\|_1 \ll 1. \tag{6.5.15}$$

Performing the change of variables $\mathbf{y} = B\mathbf{x}$, with

$$B = \begin{bmatrix} b_{n-2} & 1 \\ -b_{n-3} & 2 \end{bmatrix}, \qquad \det B = b_n,$$

we get

$$\int_{\mathbb{T}^2} |G_N(\mathbf{x})|d\mathbf{x} = \int_{B(\mathbb{T}^2)} |\mathscr{V}_N(y_1)\mathscr{V}_N(y_2)|b_n^{-1}d\mathbf{y} \ll 1.$$

By (6.5.14), (6.5.15), and the well-known estimate (see Theorem 1.4.1)

$$E_m(F_r)_1 \ll m^{-r},$$

we get

$$\|A_{\mathbf{s}}(g)\|_1 \ll 2^{-r\|\mathbf{s}\|_1}. \tag{6.5.16}$$

The relations (6.5.12), (6.5.16) imply the lemma. □

Now let $l := [\log m]$. Then $(1 < q \le 2)$

$$\|g\|_q \le \left\| \sum_{m \le \|\mathbf{s}\|_1 \le m+l} \delta_{\mathbf{s}}(g) \right\|_2 + \left\| \sum_{\|\mathbf{s}\|_1 > m+l} \delta_{\mathbf{s}}(g) \right\|_q = \sigma_1 + \sigma_2. \tag{6.5.17}$$

Further, by (6.5.12) and the above choice of l we obtain

$$\sigma_1^2 \ll 2^{-m} \sum_{m \le \|\mathbf{s}\|_1 \le m+l} 2^{2(1/2-r)\|\mathbf{s}\|_1} \ll 2^{-2rm} m^{2(1-r)}. \tag{6.5.18}$$

The operator $\delta_{\mathbf{s}}$ is bounded as an operator from L_q to L_q for $1 < q < \infty$. Therefore, by Lemma 6.5.7 and Corollary A.3.5 we get

$$\begin{aligned}\sigma_2^q &\ll \sum_{\|\mathbf{s}\|_1 > m+l} \|\delta_{\mathbf{s}}(g)\|_q^q \ll 2^{-mq(1-1/q)} \times \sum_{\|\mathbf{s}\|_1 > m+l} 2^{-(r-1+1/q)\|\mathbf{s}\|_1 q} \\ &\ll 2^{-qrm} m^{q(1-r)}. \end{aligned} \tag{6.5.19}$$

From (6.5.17)–(6.5.19) we find

$$\|g\|_q \ll 2^{-rm} m^{1-r},$$

which proves the upper estimate in Theorem 6.5.5.

Let us proceed to the proof of the lower estimates. We first consider the case $2 \le$

$p < \infty$. Let m, l be the same as in the proof of the upper estimates. The Littlewood–Paley theorem and (6.5.10) give

$$\Phi_n(\mathbf{W}^r_{p,0}) \gg \left\| \left(\sum_{\|\mathbf{s}\|_1 = m+l} |\delta_\mathbf{s}(g)|^2 \right)^{1/2} \right\|_q . \tag{6.5.20}$$

It is easy to see that for an arbitrary $\mathbf{k} \in L(n)$ we have

$$e^{i(\mathbf{k},\xi^\mu)} = 1, \qquad \xi^\mu = (2\pi\mu/b_n, 2\pi\{\mu b_{n-1}/b_n\}), \qquad \mu = 1,\dots,b_n.$$

Let us estimate from below the number of elements of the set $G^n_\mathbf{s} = L(n) \cap \rho(\mathbf{s})$ for $\mathbf{s}$ such that $\|\mathbf{s}\|_1 = m+l$. We assume that $2^{s_j} \le b_n$, $j = 1,2$. By Lemma 6.5.6 the number of elements of $G^n_\mathbf{s}$ will be equal to the number of integer u, v such that

$$2^{s_2-1} \le |u+2v| < 2^{s_2}, \qquad 2^{s_1-1} \le |ub_{n-2} - vb_{n-3}| < 2^{s_1}.$$

We denote $u+2v =: \mu$. Then, to obtain a lower estimate for $|G^n_\mathbf{s}|$ it suffices to get a lower estimate for the number of points of the form $\xi^\mu = (2\pi\mu/b_n, 2\pi\{\mu b_{n-1}/b_n\})$ which belong to the rectangle

$$P := \left(\frac{2\pi 2^{s_2-1}}{b_n}, \frac{2\pi 2^{s_2}}{b_n} \right) \times \left(\frac{2\pi 2^{s_1-1}}{b_n}, \frac{2\pi 2^{s_1}}{b_n} \right).$$

Take a rectangle $Q := (a,b) \times (c,d) \subset \mathbb{T}^2$ and consider the function

$$f(\mathbf{x}) := \left(1 - 2(b-a)^{-1} \left| x_1 - \frac{a+b}{2} \right| \right)_+ \left(1 - 2(d-c)^{-1} \left| x_2 - \frac{c+d}{2} \right| \right)_+ .$$

Let $|Q| := (b-a)(d-c)$ be the area of the rectangle Q. It is not difficult to see that the function $f|Q|/8$ belongs to the closure of the class $\mathbf{W}^2_{1,0}$ in the L_∞-metric. Furthermore,

$$\sum_{\xi^\mu \in Q} 1 \ge \sum_{\xi^\mu \in Q} f(\xi^\mu) \ge b_n \left((2\pi)^{-2} \int_{\mathbb{T}^2} f(\mathbf{x}) d\mathbf{x} - \Phi_n(\mathbf{W}^2_{1,0}) \frac{8}{|Q|} \right).$$

By Theorem 6.5.3, for sufficiently large n this inequality with $Q = P$ implies that

$$|G^n_\mathbf{s}| \gg 2^{\|\mathbf{s}\|_1}/b_n \asymp 2^l.$$

Consequently, at each point ξ^μ we have for $\mathbf{s}$ such that $\|\mathbf{s}\|_1 = m+l$

$$\delta_\mathbf{s}(g, \xi^\mu) \gg 2^{l - r\|\mathbf{s}\|_1}. \tag{6.5.21}$$

Furthermore, by the Bernstein inequality we have

$$\left\| \frac{\partial \delta_\mathbf{s}(g,\mathbf{x})}{\partial x_1} \right\|_\infty \ll 2^{s_1 + l - r\|\mathbf{s}\|_1}, \qquad \left\| \frac{\partial \delta_\mathbf{s}(g,\mathbf{x})}{\partial x_2} \right\|_\infty \ll 2^{s_2 + l - r\|\mathbf{s}\|_1},$$

and it follows from (6.5.21) that for $\mathbf{x}$ belonging to the rectangle

$$P_{\mathbf{s},\mu} := \{\mathbf{x} : a2^{-s_j} \le x_j - \xi_j^\mu < a2^{-s_j+1},\ j = 1,2,\\ 0 < a \le a_1 < 1,\ a_1 \text{ does not depend on } \mathbf{s}, l\}$$

the inequality

$$\delta_{\mathbf{s}}(g,\mathbf{x}) \gg 2^{l-r\|\mathbf{s}\|_1} \tag{6.5.22}$$

holds. The number a will be chosen later.

Let us consider the collection $\mathscr{P}$ of rectangles $P_{\mathbf{s},\mu}$ such that

$$s_1, s_2 \ge m/3, \qquad \|\mathbf{s}\|_1 = m + l, \qquad \xi^\mu \in [0,\pi] \times [0,\pi]$$

and the sets

$$G_\mu = \{\mathbf{x} : (x_1 - \xi_1^\mu)(x_2 - \xi_2^\mu) \le 4a^2 2^{-(m+l)},\ 0 \le x_j - \xi_j^\mu \le 2^{-m/3+1},\ j = 1,2\}.$$

Then for all $P_{\mathbf{s},\mu} \in \mathscr{P}$ the inclusion $P_{\mathbf{s},\mu} \subset G_\mu$ holds. Next, for fixed μ the rectangles $P_{\mathbf{s},\mu}$ are mutually disjoint and

$$\left|\bigcup_{\mathbf{s}} P_{\mathbf{s},\mu}\right| \ge a^2 2^{-(m+l)} m/3. \tag{6.5.23}$$

We estimate the measure of intersection of the sets G_μ and $G_{\mu'}$. Clearly, it is sufficient to consider the case $\xi_1^{\mu'} < \xi_1^\mu$, $\xi_2^{\mu'} > \xi_2^\mu$. Then we have

$$|G_\mu \cap G_{\mu'}| \le (4a^2 2^{-(m+l)})^2 ((\xi_1^\mu - \xi_1^{\mu'})(\xi_2^{\mu'} - \xi_2^\mu))^{-1}. \tag{6.5.24}$$

Further, we write $k_2 := b_n(\xi_1^\mu - \xi_1^{\mu'})/(2\pi)$ and $k_1 := b_n(\xi_2^{\mu'} - \xi_2^\mu)/2\pi$. From the definition of ξ^μ it follows that k_1, k_2 are natural numbers satisfying the condition

$$k_1 + k_2 b_{n-1} \equiv 0 \pmod{b_n},$$

that is, $\mathbf{k} \in L(n)$. Thus, using (6.5.5) we get

$$\sum_{\mu' \ne \mu} |G_\mu \cap G_{\mu'}| \le (a^2 b_n 2^{-(m+l)})^2 \sum_{\mathbf{k} \in L(n) \cap \mathbb{N}^2; k_j \le b_n} (k_1 k_2)^{-1} \\ \le C(a^2 2^{-(m+l)})^2 b_n (\log b_n)^2. \tag{6.5.25}$$

We now choose the number a in such a way that the right-hand side of the relation (6.5.25) is less than half the right-hand side of the relation (6.5.23). In addition, we will have $a \ge a_0 > 0$, where a_0 is an absolute constant. Further, we denote

$$P'_{\mathbf{s},\mu} := P_{\mathbf{s},\mu} \Big\backslash \bigcup_{(\mathbf{s}',\mu') \ne (\mathbf{s},\mu)} P_{\mathbf{s}',\mu'},$$

where all the $P_{\mathbf{s},\mu}$ and $P_{\mathbf{s}',\mu'} \in \mathscr{P}$. Then the $P'_{\mathbf{s},\mu}$ are mutually disjoint and by the choice of a,

$$\left|\bigcup_{\mathbf{s}} P'_{\mathbf{s},\mu}\right| \gg m2^{-(m+l)}. \tag{6.5.26}$$

By (6.5.20) and (6.5.22) we get

$$\Phi_n(\mathbf{W}^r_{p,0})^q \gg \sum_{\mathbf{s},\mu} 2^{lq}|P'_{\mathbf{s},\mu}|2^{-rq\|\mathbf{s}\|_1},$$

which implies that

$$\Phi_n(\mathbf{W}^r_{p,0}) \gg b_n^{-r}(\log b_n)^{1-r}.$$

The case $2 \le p < \infty$ is now complete.

The required lower estimate for $p = \infty$ is deduced from the case already considered in the same way as for the case $r > 1/2$ (see Remark 6.4.15). □

6.5.3 H Classes

In this subsection we obtain estimates for $\Phi_n(\mathbf{H}^r_p)$.

Theorem 6.5.8 *Let $r > 1/p$, $1 \le p \le \infty$. Then*

$$\Phi_n(\mathbf{H}^r_p) \asymp b_n^{-r} \log b_n.$$

Proof The lower estimate follows from the more general lower estimates for arbitrary cubature formulas (see Theorem 6.4.8).

Let us prove the upper estimates. The required estimate in the case $r > 1$ follows from (6.5.6), (6.5.5), and the inequality

$$|\hat{f}(\mathbf{k})| \ll \nu(\bar{\mathbf{k}})^{-r},$$

which is valid for $f \in \mathbf{H}^r_p$, $1 \le p \le \infty$. Clearly, it suffices now to consider the case $1 < p < \infty$. Let

$$\chi_{\mathbf{s}}(\mathbf{x}) := \sum_{\mathbf{k} \in L(n) \cap \rho(\mathbf{s})} e^{i(\mathbf{k},\mathbf{x})}.$$

Then by (6.5.1), (6.5.2), and Lemma 6.5.1

$$|\Phi_n(f) - \hat{f}(\mathbf{0})| = \left|\sum_{\mathbf{s}} \langle \delta_{\mathbf{s}}(f), \chi_{\mathbf{s}}\rangle\right| \le \sum_{\|\mathbf{s}\|_1 \ge \log b_n - C} \|\delta_{\mathbf{s}}(f)\|_p \|\chi_{\mathbf{s}}\|_{p'}. \tag{6.5.27}$$

From Lemma 6.5.1 it follows that

$$\|\chi_{\mathbf{s}}\|_\infty \le |L(n) \cap \rho(\mathbf{s})| \ll 2^{\|\mathbf{s}\|_1} b_n^{-1}. \tag{6.5.28}$$

Using Lemma 6.5.7, the boundedness of the operator $\delta_{\mathbf{s}}$ from L_q to L_q for $1 < q < \infty$, and the Bernstein inequality we get

$$\|\chi_{\mathbf{s}}\|_q \ll (2^{\|\mathbf{s}\|_1}/b_n)^{(1-1/q)}, \qquad 1 < q \le 2. \tag{6.5.29}$$

By (6.5.28) and (6.5.29),

$$\|\chi_{\mathbf{s}}\|_{p'} \ll (2^{\|\mathbf{s}\|_1}/b_n)^{1/p}, \qquad 1 < p' \le \infty. \tag{6.5.30}$$

Further, by Theorem 4.4.6,

$$\left\|\delta_{\mathbf{s}}(f)\right\|_p \ll 2^{-r\|\mathbf{s}\|_1}. \tag{6.5.31}$$

The required upper estimate for $r > 1/p$ is obtained from (6.5.30), (6.5.31), and (6.5.27). □

6.5.4 Some Further Results

In §§6.5.2 and 6.5.3 we have found the correct order of $\Phi_n(\mathbf{W}^r_{p,\alpha})$ for different parameters r, p, α. As we have already pointed out above, in the case $1 < p < \infty$ the classes $\Phi_n(\mathbf{W}^r_{p,\alpha})$ with different α are equivalent and, therefore, it is sufficient to find the correct order of the $\Phi_n(\mathbf{W}^r_{p,\alpha})$ for any specific α. In the case $p = 1$ or $p = \infty$ these classes are no longer equivalent for different α. Moreover, as we saw in Chapter 5, the behavior of the orthowidths $\varphi_m(\mathbf{W}^r_{1,\alpha}, L_\infty)$ depends on α. The natural restriction on r is $r > 1/p$, which guarantees the embedding of the class $\mathbf{W}^r_{p,\alpha}$ into the space of continuous functions. Thus, the problem of the correct orders of the $\Phi_n(\mathbf{W}^r_{p,\alpha})$ will be completely solved if we solve it for all $1 \le p \le \infty$, all $r > 1/p$, and all α. Theorem 6.5.2 covers the case $1 < p \le \infty$, large $r > \max(1/p, 1/2)$, and all α. Theorem 6.5.3 and Remark 6.5.4 cover the case $p = 1$, $r > 1$, and all α. Theorem 6.5.5 covers the case $2 < p \le \infty$, small $r \in (1/p, 1/2)$, and all α. Thus, the only case which has not been covered yet is $2 < p \le \infty$, $r = 1/2$. This case was settled in Temlyakov (1994). We do not present its proof here.

Theorem 6.5.9 *Let $2 < p \le \infty$. For any α*

$$\Phi_n(\mathbf{W}^{1/2}_{p,\alpha}) \asymp b_n^{-1/2}\left((\log b_n)(\log\log b_n)\right)^{1/2}.$$

6.5.5 Some Applications

In this subsection we demonstrate how the above results can be applied for obtaining the lower estimates for the quantities $E_N(\mathbf{W}^r_{\infty,\alpha})_\infty$ and $E_N(\mathbf{H}^r_\infty)_\infty$. The correct order of $E_N(\mathbf{W}^r_{\infty,\alpha})_\infty$ is not known for all $d \ge 2$. The correct order of $E_N(\mathbf{H}^r_\infty)_\infty \asymp N^{-r}\log N$ in the case $d = 2$ is given by Theorem 4.4.13. Theorem 6.5.10 gives another proof of that lower bound.

Theorem 6.5.10 *Let $d = 2$. Then*

$$E_N(\mathbf{W}^r_{\infty,\alpha})_\infty \gg N^{-r}(\log N)^{\max(1-r,1/2)}, \qquad r > 0, \qquad r \neq 1/2,$$
$$E_N(\mathbf{H}^r_\infty)_\infty \gg N^{-r}\log N, \qquad r > 0.$$

Proof Let us define the following linear functional on the space of continuous functions

$$R_n(f) := \Phi_n(f) - \hat{f}(\mathbf{0}).$$

Then $\|R_n\| \le 2$. It follows from Lemma 6.5.1 that, for any $t \in \mathscr{T}(\gamma b_n)$,

$$R_n(f) = 0.$$

Consequently, for any continuous f we have

$$\big|R_n(f)\big| = \big|R_n(f-t)\big| \le 2E_{\gamma b_n}(f)_\infty. \tag{6.5.32}$$

The first inequality in the theorem follows from (6.5.32) and Theorems 6.5.2 and 6.5.5. The second inequality follows from (6.5.32) and Theorem 6.5.8. □

6.5.6 Optimality of the Fibonacci Cubature Formulas

Theorem 6.4.3, which provides the lower bounds for any cubature formula, when combined with Theorem 6.5.2, which gives the upper bounds for the Fibonacci cubature formulas, leads to the following result.

Theorem 6.5.11 *In the case $d = 2$ for $1 < p < \infty$, $r > \max(1/p, 1/2)$ we have*

$$\kappa_m(\mathbf{W}^r_{p,\alpha}) \asymp \Phi_{b_n}(\mathbf{W}^r_{p,\alpha}) \asymp m^{-r}(\log m)^{1/2}, \qquad m \in [b_n, b_{n+1}).$$

Thus, in the case of large smoothness $r > \max(1/p, 1/2)$ the Fibonacci cubature formulas are optimal (in the sense of order) for all $1 < p < \infty$. It is likely that these formulas are also optimal in certain other cases, but we do not have the corresponding lower bounds for $\kappa_m(\mathbf{W}^r_{p,\alpha})$ in those cases.

Theorem 6.4.8, which provides the lower bounds for any cubature formula, when combined with Theorem 6.5.8, which gives the upper bounds for the Fibonacci cubature formulas, leads to the following result.

Theorem 6.5.12 *In the case $d = 2$ for all $1 \le p \le \infty$ and $r > 1/p$ we have*

$$\kappa_m(\mathbf{H}^r_p) \asymp \Phi_{b_n}(\mathbf{H}^r_p) \asymp m^{-r}\log m, \qquad m \in [b_n, b_{n+1}).$$

Theorem 6.5.8 provides the correct order of $\kappa_m(\mathbf{H}^r_p)$ for all values of the parameters p and r. This solves the optimization problem of numerical integration for the classes $\mathbf{H}^r_p$ completely and shows that the Fibonacci cubature formulas are optimal in this case.

We do not know the correct order of the $\kappa_m(\mathbf{W}^r_{1,\alpha})$. However, Theorems 6.4.10 and 6.5.3 give the correct order of the $\kappa^B_m(\mathbf{W}^r_{1,\mathbf{0}})$.

Theorem 6.5.13 *In the case $d = 2$ for $r > 1$ we have*

$$\kappa^B_m(\mathbf{W}^r_{1,\mathbf{0}}) \asymp \Phi_{b_n}(\mathbf{W}^r_{p,\mathbf{0}}) \asymp m^{-r}\log m, \qquad m \in [b_n, b_{n+1}).$$

6.6 The Korobov Cubature Formulas

The Fibonacci cubature formulas, studied above, are very simple and powerful numerical integration rules. They apply to the bivariate functions. It would be very nice to be able to construct numerical integration rules as simple and as good as the Fibonacci cubature formulas in higher dimensions ($d \geq 3$) as well. Unfortunately, in the case $d \geq 3$ the problem of finding concrete cubature formulas as good as the Fibonacci cubature formulas in the case $d = 2$ is unsolved. The results of this section deal with the case $d \geq 3$ and are not as complete as those of §6.5.

Let $m \in \mathbb{N}$, $\mathbf{a} := (a_1, \dots, a_d)$, $a_1, \dots, a_d \in \mathbb{Z}$. We consider the *Korobov cubature formulas*

$$P_m(f, \mathbf{a}) := m^{-1} \sum_{\mu=1}^{m} f\left(2\pi\left\{\frac{\mu a_1}{m}\right\}, \dots, 2\pi\left\{\frac{\mu a_d}{m}\right\}\right).$$

For a function class $\mathbf{F}$ denote

$$P_m(\mathbf{F}, \mathbf{a}) := \sup_{f \in \mathbf{F}} |P_m(f, \mathbf{a}) - \hat{f}(\mathbf{0})|.$$

In the case $d = 2$, $m = b_n$, $\mathbf{a} = (1, b_{n-1})$ we have

$$P_m(f, \mathbf{a}) = \Phi_n(f).$$

We now introduce some notations:

$$S(\mathbf{k}, \mathbf{a}) := P_m(e^{i(\mathbf{k},\mathbf{x})}, \mathbf{a}),$$

$$\chi_{\mathbf{s}}(\mathbf{x}, \mathbf{a}) := \sum_{\mathbf{k} \in \rho(\mathbf{s}) \cap L(m,\mathbf{a})} e^{i(\mathbf{k},\mathbf{x})} = \sum_{\mathbf{k} \in \rho(\mathbf{s})} S(\mathbf{k}, \mathbf{a}) e^{i(\mathbf{k},\mathbf{x})},$$

where

$$L(m, \mathbf{a}) := \{\mathbf{k} : (\mathbf{a}, \mathbf{k}) \equiv 0 \pmod{m}\}.$$

In the special case $\mathbf{a} = (1, a, a^2, \dots, a^{d-1})$ we can work in terms of the scalar a instead of the vector $\mathbf{a}$.

In this section it will be convenient for us to assume that m is a prime number and denote it by p. We may accept this assumption without loss of generality because we will obtain power-type order estimates.

6.6.1 Large Smoothness

It is convenient for us to consider, besides the classes $\mathbf{W}^r_{q,\alpha}$ and $\mathbf{H}^r_q$, the class $\mathbf{E}^rB$ of continuous functions f that are 2π-periodic in each variable and such that

$$|\hat{f}(\mathbf{k})| \le B\nu(\bar{\mathbf{k}})^{-r}.$$

Clearly, $\mathbf{W}^r_{q,\alpha}$ and $\mathbf{H}^r_q$ are embedded into $\mathbf{E}^rB$ for all $1 \le q \le \infty$, α. As above, when $B = 1$, we simply write $\mathbf{E}^r$.

We first prove the following auxiliary assertion.

Lemma 6.6.1 *Let p, κ, L be a prime, a positive real, and a natural number, respectively, such that*

$$\big|\Gamma(L)\big| < (p-1)(1-2^{-\kappa})/d. \tag{6.6.1}$$

Then there is a natural number $a \in I_p := [1,p)$ such that for all $\mathbf{m} \in \Gamma(L)$, $\mathbf{m} \ne \mathbf{0}$, we have

$$m_1 + am_2 + \cdots + a^{d-1}m_d \not\equiv 0 \pmod{p}, \tag{6.6.2}$$

and relation (6.6.2) *holds for all vectors $\mathbf{m} \in F_l(L) := \Gamma(L2^l)\backslash\Gamma(L2^{l-1})$, $\mathbf{m} \ne p\mathbf{m}'$, with the exception of no more than*

$$A_l^L := \big|F_l(L)\big| d2^{(l+1)\kappa}(2^\kappa - 1)^{-1}(p-1)^{-1}, \qquad l = 1,2,\ldots$$

such vectors.

Proof Let $a \in I_p$ be a natural number. We consider the congruence

$$m_1 + am_2 + \cdots + a^{d-1}m_d \equiv 0 \pmod{p}. \tag{6.6.3}$$

For the fixed vector $\mathbf{m} = (m_1,\ldots,m_d)$ we denote by $A_p(\mathbf{m})$ the set of natural numbers $a \in I_p$ which are solutions of the congruence (6.6.3). It is well known that, for $\mathbf{m} \ne \mathbf{0}$, $|m_j| < p$, $j = 1,\ldots,d$, the number $\big|A_p(\mathbf{m})\big|$ of elements of the set $A_p(\mathbf{m})$ satisfies the inequality

$$\big|A_p(\mathbf{m})\big| \le d - 1 < d. \tag{6.6.4}$$

We denote by G_1 the set of the numbers a for which there is a nontrivial solution $\mathbf{m} \in \Gamma(L)$ of the congruence (6.6.3), that is,

$$G_1 = \bigcup_{\mathbf{m} \in \Gamma(L)\backslash \mathbf{0}} A_p(\mathbf{m}).$$

Let us estimate the number $|G_1|$ of elements of the set G_1. By (6.6.4) and (6.6.1) we have

$$|G_1| \le \sum_{\mathbf{m} \in \Gamma(L)\backslash \mathbf{0}} \big|A_p(\mathbf{m})\big| < d\big|\Gamma(L)\big| < (p-1)(1-2^{-\kappa}). \tag{6.6.5}$$

For any $a \in I_p\backslash G_1$ and for all $\mathbf{m} \in \Gamma(L) \setminus \{\mathbf{0}\}$ we have

$$m_1 + am_2 + \cdots + a^{d-1}m_d \not\equiv 0 \pmod{p}.$$

Let G_{l+1}, $l = 1, 2, \ldots$ denote the set of those a for which the number of elements of the set

$$M_a^l := \{\mathbf{m} : \mathbf{m} \in F_l(L),\ \mathbf{m} \neq p\mathbf{m}',\ m_1 + am_2 + \cdots + a^{d-1}m_d \equiv 0 \pmod{p}\}$$

satisfies the inequality

$$|M_a^l| > A_l^L. \tag{6.6.6}$$

Then, by (6.6.6), we have on the one hand

$$\sum_{a \in G_{l+1}} |M_a^l| > A_l^L |G_{l+1}|. \tag{6.6.7}$$

On the other hand, by (6.6.4) each $\mathbf{m}$ can belong to at most $d-1$ different sets M_a^l and consequently

$$\sum_{a \in G_{l+1}} |M_a^l| < d\big|F_l(L)\big|. \tag{6.6.8}$$

Comparing (6.6.7) and (6.6.8) we find that

$$|G_{l+1}| < d\big|F_l(L)\big|/A_l^L = (p-1)(2^\kappa - 1)2^{-\kappa(l+1)}. \tag{6.6.9}$$

From relations (6.6.5) and (6.6.9) it follows that

$$\sum_{l=1}^{\infty} |G_l| < p - 1.$$

This means that there exists a number $a \in I_p$ which does not belong to any set G_l, $l = 1, \ldots$ This a is the required number by the definition of the sets G_l. The lemma is proved. □

Theorem 6.6.2 *For any $r > 1$ and any prime number p there is a number $a \in [1, p)$ such that*

$$P_p(\mathbf{E}^r, a) \ll p^{-r}(\log p)^{r(d-1)}.$$

Proof Let $0 < \kappa \le (r-1)/2$ and let p be given. Suppose that L satisfies (6.6.1) and, in addition, $\big|\Gamma(L)\big| \asymp p$. Let a be the number from Lemma 6.6.1, determined by p, κ, L and

$$Z_l := \{\mathbf{m} : \mathbf{m} \in F_l(L), \mathbf{m} \neq p\mathbf{m}', m_1 + am_2 + \cdots + a^{d-1}m_d \equiv 0 \pmod{p}\}.$$

Then, by Lemma 6.6.1,

$$|Z_l| \le A_l^L, \tag{6.6.10}$$

and the error of the cubature formula can be estimated as follows:

$$\big|P_p(f, a) - \hat{f}(\mathbf{0})\big| \le \left|\sum_{l=1}^{\infty} \sum_{\mathbf{m} \in Z_l} \hat{f}(\mathbf{m})\right| + \left|\sum_{\mathbf{m} = p\mathbf{m}'} \hat{f}(\mathbf{m})\right| = \sigma_1 + \sigma_2. \tag{6.6.11}$$

First, we estimate σ_1. By (6.6.10) we have, for $f \in \mathbf{E}^r$,

$$\sigma_1 \ll \sum_{l=1}^{\infty} 2^{-lr} L^{-r} 2^{l(1+\kappa)} L(\log L2^l)^{d-1} p^{-1}$$
$$\ll L^{-r} \asymp p^{-r}(\log p)^{r(d-1)}. \tag{6.6.12}$$

Second, for σ_2 we have

$$\sigma_2 \ll \sum_{e \subset [1,d]; |e| \geq 1} \frac{1}{p^{|e|r}} \left(\sum_{m'_j \neq 0; j \in e} \prod_{j \in e} (m'_j)^{-r} \right) \ll p^{-r}. \tag{6.6.13}$$

The theorem follows from (6.6.12) and (6.6.13). □

For large r ($r > 1$) Theorem 6.6.2 implies the following error estimates for the Korobov cubature formulas for the classes $\mathbf{W}^r_{q,\alpha}$ and $\mathbf{H}^r_q$, $1 \leq q \leq \infty$.

Corollary 6.6.3 *For any $r > 1$ and any prime number p there is a number $a \in [1, p)$ such that, for all $1 \leq q \leq \infty$,*

$$P_p(\mathbf{W}^r_{q,\alpha}, a) \ll p^{-r}(\log p)^{r(d-1)},$$
$$P_p(\mathbf{H}^r_q, a) \ll p^{-r}(\log p)^{r(d-1)}.$$

6.6.2 Small Smoothness

For the classes $\mathbf{W}^r_{q,\alpha}$ and $\mathbf{H}^r_q$ the natural restriction on r is the condition for the embedding of these classes into the space of continuous functions, namely, $r > 1/q$, $1 \leq q \leq \infty$. We now consider the case $\max(1/q, 1/2) < r \leq 1$.

Theorem 6.6.4 *For any r such that $\max(1/q, 1/2) < r \leq 1$ and any prime number p there is a natural number $a \in [1, p)$ for which*

$$P_p(\mathbf{W}^r_{q,\alpha}, a) \ll p^{-r}(\log p)^{r(d-1)}, \qquad 1 < q \leq \infty.$$

Proof The proof of this theorem is similar to that of Theorem 6.6.2. Clearly, it suffices to consider the case $1 < q \leq 2$. Let $0 < \kappa < r - 1/q$ and let the numbers p, L, a be the same as in the proof of Theorem 6.6.2. Let us estimate σ_1, σ_2 from (6.6.11) for $f \in \mathbf{W}^r_{q,\alpha}$. Denote

$$\psi_l(\mathbf{x}) := \sum_{\mathbf{m} \in Z_l} e^{i(\mathbf{k},\mathbf{x})}.$$

We have

$$\sigma_1 = \left| \sum_{l=1}^{\infty} \langle f, \psi_l \rangle \right| \leq \sum_{l=1}^{\infty} E_{L2^{l-1}}(\mathbf{W}^r_{q,\alpha})_q \|\psi_l\|_{q'}. \tag{6.6.14}$$

Further

$$\|\psi_l\|_{q'} \le \|\psi_l\|_2^{2/q'} \|\psi_l\|_\infty^{1-2/q'} \le |Z_l|^{1/q}. \tag{6.6.15}$$

From (6.6.14) and (6.6.15), using Lemma 6.6.1 and Theorem 4.4.9 we get

$$\sigma_1 \ll L^{-r} \ll p^{-r}(\log p)^{r(d-1)}.$$

For σ_2 we have

$$\sigma_2 \le \sum_{\mathbf{m}' \neq \mathbf{0}} |\hat{f}(p\mathbf{m}')| \ll p^{-r} \sum_{\mathbf{m}'} \nu(\bar{\mathbf{m}}')^{-r} |\hat{\varphi}(p\mathbf{m}')|, \tag{6.6.16}$$

where φ is such that $f = F_r * \varphi$, $\|\varphi\|_q \le 1$. From (6.6.16), applying the Hölder inequality and the Hausdorff–Young theorem (see Theorem A.3.1), we get

$$\sigma_2 \ll p^{-r} \left(\sum_{\mathbf{m}'} \nu(\bar{\mathbf{m}}')^{-rq} \right)^{1/q} \left(\sum_{\mathbf{m}'} |\hat{\varphi}(p\mathbf{m}')|^{q'} \right)^{1/q'} \ll p^{-r} \|\varphi\|_q \le p^{-r}.$$

The conclusion of the theorem follows from the estimates for σ_1 and σ_2. □

We now prove the following assertion, which shows that the Korobov cubature formulas are optimal in the sense of order in some cases, when $d \ge 3$.

Theorem 6.6.5 *Let* $0 < r < 1$. *There is a vector* $\mathbf{a}$ *such that*

$$P_m(\mathbf{H}^r_\infty, \mathbf{a}) \ll m^{-r}(\log m)^{d-1}.$$

Proof Denote $I(m,d) := [1, m-1]^d$ and set $n := [\log m]$. Take an $\mathbf{a} \in I(m,d)$. The error of the cubature formula $P_m(f, \mathbf{a})$ for $f \in \mathbf{H}^r_\infty$ can be estimated as follows:

$$\begin{aligned} |P_m(f,\mathbf{a}) - \hat{f}(\mathbf{0})| &\le |P_m(V_{Q_n}(f),\mathbf{a}) - \hat{f}(\mathbf{0})| + |P_m(f - V_{Q_n}(f),\mathbf{a})| \\ &\le |P_m(V_{Q_n}(f),\mathbf{a}) - \hat{f}(\mathbf{0})| + \|f - V_{Q_n}(f)\|_\infty. \end{aligned} \tag{6.6.17}$$

By relation (4.4.27) for $f \in \mathbf{H}^r_\infty$.

$$\|f - V_{Q_n}(f)\|_\infty \ll m^{-r}(\log m)^{d-1} \tag{6.6.18}$$

and, by Theorem 4.4.6 $V_{Q_n}(f) \in \mathbf{H}^r_\infty B$. Therefore, it is sufficient to assume that $f \in \mathbf{H}^r_\infty \cap \mathscr{T}(Q_l)$, $l := n + d$. Then, using the functions $\chi_\mathbf{s}(\mathbf{a})$, defined at the beginning of the section and Theorem 4.4.6, we get

$$|P_m(f,\mathbf{a}) - \hat{f}(\mathbf{0})| = \left| \sum_{\|\mathbf{s}\|_1 \le l} \langle f(\mathbf{x}), \chi_\mathbf{s}(\mathbf{x},\mathbf{a}) \rangle \right| \ll \sum_{\|\mathbf{s}\|_1 \le l} 2^{-r\|\mathbf{s}\|_1} \|\chi_\mathbf{s}(\mathbf{x},\mathbf{a})\|_1. \tag{6.6.19}$$

Let σ be the minimum over all $\mathbf{a} \in I(m,d)$ of the expression on the right-hand side of (6.6.19), which is attained at some point $\mathbf{a}'$. We estimate σ by the average value over all $\mathbf{a} \in I(m,d)$:

$$\sigma \le (m-1)^{-d} \left(\sum_{\mathbf{a} \in I(m,d)} \sum_{\|\mathbf{s}\|_1 \le l} 2^{-r\|\mathbf{s}\|_1} \left\| \chi_{\mathbf{s}}(\mathbf{x}, \mathbf{a}) \right\|_1 \right).$$

We remark that, for all $\mathbf{k} \neq \mathbf{0}$, $|k_j| < m$, the following estimate holds:

$$(m-1)^{-d} \sum_{\mathbf{a} \in I(m,d)} S(\mathbf{k}, \mathbf{a}) \ll m^{-1}. \tag{6.6.20}$$

We have

$$\sigma \le (m-1)^{-d} \left(\sum_{\|\mathbf{s}\|_1 \le l} 2^{-r\|\mathbf{s}\|_1} \sum_{\mathbf{k} \in \rho(\mathbf{s})} \sum_{\mathbf{a} \in I(m,d)} S(\mathbf{k}, \mathbf{a}) \right).$$

Applying (6.6.20) we get

$$\sigma \ll 2^{(1-r)l} l^{d-1} m^{-1} \ll m^{-r} (\log m)^{d-1},$$

which completes the proof of the theorem. □

Theorem 6.6.5 together with Theorem 6.4.8 shows that the Korobov cubature formulas are optimal for the classes $\mathbf{H}^r_\infty$ for small r, $0 < r < 1$.

6.7 The Frolov Cubature Formulas

The main purpose of this section is to obtain precise orders of decrease of the optimal errors of cubature formulas for classes of functions with bounded mixed derivatives or differences. Here, for $d > 2$, cubature formulas that are optimal in the sense of order will be constructed. The construction of such cubature formulas is based on some results in algebraic number theory.

6.7.1 Nonperiodic Case

In this subsection we construct cubature formulas that are optimal (in the sense of order) for the classes $\bar{\mathbf{W}}^r_p(\Omega_d)$, $2 \le p < \infty$. These cubature formulas will then be used for investigating similar classes of functions with bounded mixed differences. The following lemma plays a fundamental role in the construction of such cubature formulas.

Lemma 6.7.1 *There exists a matrix A such that the lattice $L(\mathbf{m}) = A\mathbf{m}$, where $\mathbf{m}$*

is a (column) vector with integer coordinates, has the following properties

$$\left(L(\mathbf{m}) = \begin{bmatrix} L_1(\mathbf{m}) \\ \vdots \\ L_d(\mathbf{m}) \end{bmatrix} \right);$$

(1) $\left|\prod_{j=1}^{d} L_j(\mathbf{m})\right| \geq 1$ *for all* $\mathbf{m} \neq \mathbf{0}$;

(2) *each parallelepiped* P *with volume* $|P|$ *whose edges are parallel to the coordinate axes contains no more than* $|P|+1$ *lattice points.*

Proof The proof is based on the following two auxiliary propositions.

Lemma 6.7.2 *Let* $P_n(x) := \prod_{j=1}^{n}(x-a_j) = \sum_{j=0}^{n}(-1)^{n-j}\sigma_{n-j}(\mathbf{a})x^j$ *be a polynomial with integer coefficients and* $\varphi(x_1,\ldots,x_n)$ *be a symmetric polynomial in* n *variables (*φ *does not change upon any rearrangement of the variables* $x_1,\ldots,x_n$*), with integer coefficients. Then* $\varphi(a_1,\ldots,a_n)$ *is an integer.*

Proof We denote $\mathbf{x} := (x_1,\ldots,x_n)$ and prove the following well-known proposition (see, for example, the book Cassels, 1971, p. 124): $\varphi(\mathbf{x})$ may be represented as the polynomial $\psi(\sigma_1(\mathbf{x}),\ldots,\sigma_n(\mathbf{x}))$ with integer coefficients. From this proposition the conclusion of the lemma follows immediately.

So, let $\varphi(\mathbf{x})$ be a symmetric polynomial and

$$\begin{aligned}
\sigma_1(\mathbf{x}) &= x_1 + x_2 + \cdots + x_n, \\
\sigma_2(\mathbf{x}) &= x_1x_2 + x_1x_3 + \cdots + x_2x_3 + \cdots + x_{n-1}x_n, \\
&\vdots \\
\sigma_n(\mathbf{x}) &= x_1x_2\cdots x_n.
\end{aligned}$$

The function $\varphi(\mathbf{x})$ has a finite number of terms of the form

$$c(\mathbf{r})x_1^{r_1}\ldots x_n^{r_n}, \qquad c(\mathbf{r}) \in \mathbb{Z}.$$

Let the vector $\mathbf{r}$ have the form $r_1 \geq r_2 \geq \cdots \geq r_n$ and

$$u_{\mathbf{r}}(\mathbf{x}) := \sum_{\rho \in A(\mathbf{r})} x_1^{\rho_1}\cdots x_n^{\rho_n},$$

where $A(\mathbf{r})$ is the set of all vectors obtained by any rearrangement of the coordinates of the vector $\mathbf{r}$. Clearly, the function $\varphi(\mathbf{x})$ can be represented in the form of the finite sum

$$\varphi(\mathbf{x}) = \sum_{\mathbf{r}} c(\mathbf{r})u_{\mathbf{r}}(\mathbf{x}), \qquad c(\mathbf{r}) \in \mathbb{Z}.$$

Therefore, it suffices to consider the functions $u_{\mathbf{r}}(\mathbf{x})$. For two vectors $\mathbf{a} = (a_1,\ldots,a_n)$,

$a_1 \ge a_2 \ge \cdots \ge a_n$ and $\mathbf{b} = (b_1, \dots, b_n)$, $b_1 \ge b_2 \ge \cdots \ge b_n$ with integer nonnegative coordinates a_j, b_j, $j = 1, \dots, n$ we write $\mathbf{b} < (l)\mathbf{a}$ and say that $\mathbf{b}$ is less than $\mathbf{a}$ in the lexicographical order if there is an m such that $a_1 = b_1, \dots, a_{m-1} = b_{m-1}$, $a_m > b_m$. We have $(r_1 \ge r_2 \ge \cdots \ge r_n)$

$$u_{\mathbf{r}}(\mathbf{x}) = \sigma_1(\mathbf{x})^{r_1 - r_2} \sigma_2(\mathbf{x})^{r_2 - r_3} \cdots \sigma_n(\mathbf{x})^{r_n} + \sum_{\mathbf{r}' : \mathbf{r}' < (l)\mathbf{r}} a(\mathbf{r}') u_{\mathbf{r}'}(\mathbf{x}). \tag{6.7.1}$$

From the representation (6.7.1) the assertion of the lemma follows by induction. □

Lemma 6.7.3 *Let $P_n(x) = \prod_{j=1}^n (x - a_j)$ be an irreducible algebraic polynomial with real roots $a_1, \dots, a_n$. Then, for any integers $m_1, \dots, m_n$ such that at least one of them is nonzero, we have*

$$\prod_{i=1}^n \left| \sum_{j=1}^n m_j a_i^{j-1} \right| \ge 1.$$

Proof Denote

$$Q(\mathbf{m}, x) := \sum_{j=1}^n m_j x^{j-1}, \qquad \mathbf{m} := (m_1, \dots, m_n).$$

By Lemma 6.7.2 it is sufficient to prove that, for any $\mathbf{m} \ne \mathbf{0}$,

$$Q(\mathbf{m}, a_i) \ne 0, \qquad i = 1, \dots, n.$$

Assume the contrary, that is, for some i we have $Q(\mathbf{m}, a_i) = 0$; then

$$P_n(x) = S(x) Q(\mathbf{m}, x) + R(x),$$

where $S(x)$, $R(x)$ are polynomials with rational coefficients and $\deg R < \deg Q$. From the hypothesis of the lemma it follows that $R(a_i) = 0$. In the case $R(x) \equiv 0$ we get a contradiction to the irreducibility of $P_n(x)$, otherwise we continue the process by dividing P_n by R and so on. As a result we get a contradiction to the irreducibility of P_n. □

To complete the proof of Lemma 6.7.1 it remains to exhibit for each n an example of an irreducible algebraic polynomial P_n of degree n with integer coefficients and leading coefficient 1 such that P_n has n real roots. As an example we can choose the polynomial

$$P_n(x) = \prod_{k=1}^n \bigl(x - (2k-1)\bigr) - 1, \qquad n > 2.$$

It is not difficult to see that this polynomial has n real roots. For example in the case of even n the polynomial $P_n(x)$ at the points $2k - 1$, $k = 1, \dots, n$, obtains a negative value equal to -1 and at the points $4k$, $k = 0, 1, \dots, n/2$ it attains positive values.

Thus, $P_n(x)$ has at least one root per segment for $[0,1]$, $[2n-1,2n]$ and two roots per segment for $[4k-1,4k+1]$, $k=1,2,\ldots,n/2-1$. This implies that $P_n(x)$ has n real roots. The case of odd n is quite similar.

The following assertion gives the irreducibility of the polynomial $P_n(x)$.

Lemma 6.7.4 *The polynomial $\prod_{j=1}^{n}(x-b_j)-1$, where b_j, $j=1,\ldots,n$, are distinct integers, is irreducible.*

Proof First we prove a proposition about reducible polynomials.

Lemma 6.7.5 *Let $\varphi(x)$ be a reducible polynomial of degree n with integer coefficients, the highest of which is equal to* 1. *Then there are two polynomials $f(x)$, $g(x)$ each with nonzero degree and integer coefficients such that $\varphi(x)=f(x)g(x)$.*

Proof An algebraic polynomial with integer coefficients the greatest common divisor of which equals 1 is called primitive. Clearly, $\varphi(x)$ is primitive. By the hypothesis of the lemma $\varphi(x)$ is reducible. Let $\varphi(x)=f(x)u(x)$, where $f(x)$ is a primitive polynomial and $u(x)$ is a polynomial with rational coefficients. Then $u(x)=(a/b)g(x)$ where a, b are integers and $g(x)$ is a primitive polynomial. We have

$$b\varphi(x)=af(x)g(x). \tag{6.7.2}$$

We now prove a theorem of Gauss that says that the product of two primitive polynomials is a primitive polynomial. Let us assume the contrary and let p be a prime number which divides $f(x)g(x)$ for all integer x, that is,

$$f(x)g(x)\equiv 0 \pmod p. \tag{6.7.3}$$

We prove that either $f(x)\equiv 0 \pmod p$ or $g(x)\equiv 0 \pmod p$. Indeed, otherwise both $f(x)$ and $g(x)$ have coefficients which are not divisible by p. Let f_k, g_l be the corresponding coefficients with largest indices. Then the coefficient of x^{k+l} of the polynomial $f(x)g(x)$ will not be divisible by p, which is in contradiction to (6.7.3). Thus, from (6.7.3) it follows that either $f(x)$ or $g(x)$ is not a primitive polynomial. This contradiction proves Gauss's theorem.

Gauss's theorem shows that in (6.7.2) we have $a=b$, which completes the proof of the lemma. □

We now continue the proof of Lemma 6.7.4. Let us assume the contrary, that is, that $\prod_{j=1}^{n}(x-b_j)-1$ is a reducible polynomial. Then, by Lemma 6.7.5,

$$\varphi(x)=f(x)g(x),$$

where $f(x)$, $g(x)$ are polynomials with integer coefficients. This implies that

$$f(b_j)g(b_j)=-1, \qquad j=1,\ldots,n,$$

which results in

$$f(b_j) + g(b_j) = 0, \qquad j = 1, \ldots, n.$$

Consequently, $f(x) \equiv -g(x)$ and

$$\varphi(x) = -f(x)^2,$$

which is impossible because of the difference in signs of the coefficients of x^n.

This proves Lemma 6.7.4. □

Property (1) of Lemma 6.7.1 follows from Lemma 6.7.3, and in addition we can take the matrix $A = [a_i^{j-1}]_{i,j=1}^d$, where a_i, $i = 1, \ldots, n$ are the roots of an irreducible polynomial of degree d with integer coefficients and leading coefficient 1.

Property (2) follows from conclusion (1), which has already been proved. Indeed, first let $|P| < 1$. If P contains two different points $L(\mathbf{m}')$, $L(\mathbf{m}'')$ on the lattice then

$$|P| \geq \prod_{j=1}^{d} \left| L_j(\mathbf{m}') - L_j(\mathbf{m}'') \right| = \prod_{j=1}^{d} \left| L_j(\mathbf{m}' - \mathbf{m}'') \right| \geq 1,$$

which leads to a contradiction.

The case $|P| \geq 1$ is easily reduced to the previous case by dividing P into $\big[|P|\big] + 1$ identical parallelepipeds whose edges are parallel to the coordinate axes.

The proof of Lemma 6.7.1 is now complete. □

Certainly it would be desirable to have as simple a matrix A as possible in Lemma 6.7.1. We give one example of such a matrix for the case $d = 2^n$, $n = 2, 3, \ldots$ We define recurrently the sequence $\{Q_{2^k}(x)\}_{k=1}^{\infty}$ of polynomials

$$Q_{2^k}(x) := Q_{2^{k-1}}(x)^2 - 2, \qquad Q_2(x) := x^2 - 2.$$

Then it is not difficult to see that

$$Q_{2^n}(x) = 2T_{2^n}(x/2), \qquad n = 1, 2, \ldots,$$

where $T_{2^n}(y) = \cos 2^n \arccos y$ is a Chebyshev polynomial. Consequently, the roots of the polynomial $Q_{2^n}(x)$ are

$$a_l = 2\cos\frac{\pi(2l-1)}{2^{n+1}}, \qquad l = 1, 2, \ldots, 2^n.$$

We now prove that $Q_{2^n}(x)$ is irreducible. Let us assume the contrary. Then by Lemma 6.7.5

$$Q_{2^n}(x) = \left(\sum_{\mu=0}^{l} p_\mu x^\mu\right)\left(\sum_{\nu=0}^{m} q_\nu x^\nu\right),$$

where all the p_μ, and q_ν are integers. Clearly, we may assume that $p_l = q_m = 1$. Further, it is easy to see that $\left|Q_{2^n}(0)\right| = 2$. Consequently, from the equality $p_0 q_0 = Q_{2^n}(0)$ it follows that one of the numbers p_0, q_0 is even and the other is odd. Without loss of generality let us assume that p_0 is even. By the definition of $Q_{2^n}(x)$ it is clear that all its coefficients excluding the highest are even. Considering step by step the coefficients of $x, x^2, x^3, \dots$ we verify that all p_μ, $\mu = 0, \dots, l$ must be even numbers, but $p_l = 1$. This contradiction proves the irreducibility of $Q_{2^n}(x)$.

We now state the Poisson formula and prove it in a form that will be convenient for us. Write, for $f \in L_1(\mathbb{R}^d)$,

$$\hat{f}(\mathbf{y}) := \int_{\mathbb{R}^d} f(\mathbf{x}) e^{-2\pi i(\mathbf{y},\mathbf{x})} d\mathbf{x}.$$

Lemma 6.7.6 (Poisson formula) *Let $f(\mathbf{x})$ be continuous and have compact support and let the series $\sum_{\mathbf{k}\in\mathbb{Z}^d} \hat{f}(\mathbf{k})$ converge. Then*

$$\sum_{\mathbf{k}\in\mathbb{Z}^d} \hat{f}(\mathbf{k}) = \sum_{\mathbf{n}\in\mathbb{Z}^d} f(\mathbf{n}).$$

Proof We consider the auxiliary function

$$\varphi(\mathbf{y}) := \sum_{\mathbf{n}} f(\mathbf{y}+\mathbf{n}). \tag{6.7.4}$$

From the hypothesis of the lemma it follows that for each $\mathbf{y}$ we have a finite number of summands in (6.7.4) and $\varphi(\mathbf{y})$ is the periodic function whose domain of periodicity is Ω_d. Therefore

$$\begin{aligned} c_{\mathbf{k}} &= \int_{\Omega_d} \varphi(\mathbf{y}) e^{-2\pi i(\mathbf{k},\mathbf{y})} d\mathbf{y} = \sum_{\mathbf{n}} \int_{\Omega_d} f(\mathbf{y}+\mathbf{n}) e^{-2\pi i(\mathbf{k},\mathbf{y})} d\mathbf{y} \\ &= \int_{\mathbb{R}^d} f(\mathbf{y}) e^{-2\pi i(\mathbf{k},\mathbf{y})} d\mathbf{y} \\ &= \hat{f}(\mathbf{k}). \end{aligned}$$

Further,

$$\varphi(\mathbf{0}) = \sum_{\mathbf{n}} f(\mathbf{n}).$$

However, we have

$$\varphi(\mathbf{x}) \sim \sum_{\mathbf{k}} \hat{f}(\mathbf{k}) e^{2\pi i(\mathbf{k},\mathbf{x})}.$$

We know that the series $\sum_{\mathbf{k}} \hat{f}(\mathbf{k})$ converges and the function $\varphi(\mathbf{x})$ is continuons. This implies that

$$\varphi(\mathbf{0}) = \sum_{\mathbf{k}} \hat{f}(\mathbf{k}), \tag{6.7.5}$$

which gives the conclusion of the lemma. □

Remark 6.7.7 It is clear from the proof of Lemma 6.7.6 that the convergence of the series $\sum_{\mathbf{k}} \hat{f}(\mathbf{k})$ can be understood as convergence with respect to any sequence of parallelepipeds $R_l := \prod_{j=1}^{d}[-b_j^l, b_j^l]$ such that $\lim_{l\to\infty} \min b_j^l = \infty$. Also, the assumption of convergence of the series $\sum_{\mathbf{k}} \hat{f}(\mathbf{k})$ can be replaced by an assumption of the summability of this series, using a method which provides (6.7.5).

Let $\mathbf{E}^r(\Omega_d)$ denote a set of continuous functions $f(\mathbf{x})$ whose supports are in Ω_d and are such that

$$|\hat{f}(\mathbf{y})| \le \prod_{j=1}^{d} (\max(1,|y_j|))^{-r}.$$

Let $a > 1$ and A be the matrix from Lemma 6.7.1. We consider the cubature formula

$$\Phi(a,A)(f) := \left(a^d |\det A|\right)^{-1} \sum_{\mathbf{m}\in\mathbb{Z}^d} f\left(\frac{(A^{-1})^{\mathrm{T}}\mathbf{m}}{a}\right),$$

for $f \in \mathbf{E}^r(\Omega_d)$. Clearly, the number N of points of this cubature formula does not exceed $C(A)a^d|\det A|$.

Theorem 6.7.8 *Let $r > 1$ be a real number. Then*

$$\Phi(a,A)(\mathbf{E}^r(\Omega_d)) := \sup_{f\in\mathbf{E}^r(\Omega_d)} \left|\Phi(a,A)(f) - \int_{\Omega_d} f(\mathbf{x})d\mathbf{x}\right| \ll a^{-rd}(\log a)^{d-1}.$$

Proof Let $f \in \mathbf{E}^r(\Omega_d)$, $r > 1$. By Lemma 6.7.6 the identity

$$\begin{aligned}\Phi(a,A)(f) &= \left(a^d |\det A|\right)^{-1} \sum_{\mathbf{m}} \int_{\mathbb{R}^d} f\left(\frac{(A^{-1})^{\mathrm{T}}\mathbf{x}}{a}\right) e^{-2\pi i(\mathbf{m},\mathbf{x})} d\mathbf{x} \\ &= \sum_{\mathbf{m}} \int_{\mathbb{R}^d} f(\mathbf{y}) e^{-2\pi i(aA\mathbf{m},\mathbf{y})} d\mathbf{y} = \sum_{\mathbf{m}} \hat{f}(aA\mathbf{m}) \end{aligned} \tag{6.7.6}$$

holds under the assumption that the series on the right-hand side of (6.7.6) converges. The convergence of this series will follow from further considerations. In the relation (6.7.6) we carried out the linear change of variables $\mathbf{y} = ((A^{-1})^{\mathrm{T}}\mathbf{x})/a$.

We have for the error of this cubature formula

$$\delta := \Phi(a,A)(f) - \hat{f}(\mathbf{0}) = \sum_{\mathbf{m}\neq\mathbf{0}} \hat{f}(aA\mathbf{m}). \tag{6.7.7}$$

Let l be such that

$$a^{-d}2^{l-1} < 1, \qquad a^{-d}2^l \ge 1.$$

Then by property (1) of Lemma 6.7.1 the inequality $\|\mathbf{s}\|_1 \ge l$ holds for $\mathbf{s}$ such that $\rho(\mathbf{s})$ contains the point $aA\mathbf{m}$ with $\mathbf{m} \neq \mathbf{0}$. From (6.7.7), using property (2) of Lemma 6.7.1, we get

$$\delta \leq \sum_{\|\mathbf{s}\|_1 \geq l} \sum_{aL(\mathbf{m}) \in \rho(\mathbf{s})} \prod_{j=1}^{d} |aL_j(\mathbf{m})|^{-r} \ll \sum_{\|\mathbf{s}\|_1 \geq l} 2^{-r\|\mathbf{s}\|_1} \left(\frac{|\rho(\mathbf{s})|}{a^d} + 1 \right)$$
$$\ll 2^{-rl} l^{d-1} \ll a^{-rd} (\log a)^{d-1}. \tag{6.7.8}$$

The theorem is proved. □

Theorem 6.7.9 *Let r be a natural number. Then*

$$\Phi(a,A)\left(\bar{\mathbf{W}}_2^r(\Omega_d)\right) \ll a^{-rd} (\log a)^{(d-1)/2}.$$

Proof We begin the proof with the following simple assertion .

Lemma 6.7.10 *Let $\|\varphi\|_2 \leq 1$ and let the support of φ be contained in Ω_d. Then, for any $a > 1$ and nonsingular matrix A, we have*

$$\sum_{\mathbf{m}} |\hat{\varphi}(aA\mathbf{m})|^2 \leq C(A).$$

Proof Similarly to (6.7.6) we have

$$\hat{\varphi}(aA\mathbf{m}) = \left(a^d |\det A|\right)^{-1} \int_{\mathbb{R}^d} \varphi\left(\frac{(A^{-1})^{\mathrm{T}}\mathbf{x}}{a}\right) e^{-2\pi i(\mathbf{m},\mathbf{x})} d\mathbf{x}. \tag{6.7.9}$$

Let

$$\Omega_d(\mathbf{n}) := \{\mathbf{x} = \mathbf{y} + \mathbf{n},\ \mathbf{y} \in \Omega_d\}$$

and

$$G := \left\{\mathbf{n} : \left(\operatorname{supp} \varphi\left(\frac{(A^{-1})^{\mathrm{T}}\mathbf{x}}{a}\right)\right) \cap \Omega_d(\mathbf{n}) \neq \varnothing\right\}.$$

From the hypothesis of the lemma it follows that

$$|G| \leq C_1(A) a^d. \tag{6.7.10}$$

Denote

$$c_{\mathbf{m}}(\mathbf{n}) := \int_{\Omega_d(\mathbf{n})} \varphi\left(\frac{(A^{-1})^{\mathrm{T}}\mathbf{x}}{a}\right) e^{-2\pi i(\mathbf{m},\mathbf{x})} d\mathbf{x}.$$

By the Parseval identity,

$$\sum_{\mathbf{m}} |c_{\mathbf{m}}(\mathbf{n})|^2 = \int_{\Omega_d(\mathbf{n})} \left|\varphi\left(\frac{(A^{-1})^{\mathrm{T}}\mathbf{x}}{a}\right)\right|^2 d\mathbf{x}. \tag{6.7.11}$$

From the relation (6.7.9), using the Cauchy inequality and the inequality (6.7.10), we get

$$\left|\hat{\varphi}(aA\mathbf{m})\right|^2 = \left(a^d|\det A|\right)^{-2}\left|\sum_{\mathbf{n}\in G} c_{\mathbf{m}}(\mathbf{n})\right|^2 \le \left(a^d|\det A|\right)^{-2}|G|\sum_{\mathbf{n}\in G}\left|c_{\mathbf{m}}(\mathbf{n})\right|^2$$
$$\le \left(a^d|\det A|\right)^{-1}C_2(A)\sum_{\mathbf{n}\in G}\left|c_{\mathbf{m}}(\mathbf{n})\right|^2.$$

Performing the summation over $\mathbf{m}$ and taking into account the relation (6.7.11) gives

$$\sum_{\mathbf{m}}\left|\hat{\varphi}(aA\mathbf{m})\right|^2 \le C_2(A)\left(a^d|\det A|\right)^{-1}\int_{\mathbb{R}^d}\left|\varphi\left(\frac{(A^{-1})^{\mathrm{T}}\mathbf{x}}{a}\right)\right|^2 d\mathbf{x}$$
$$= C_2(A)\int_{\mathbb{R}^d}|\varphi(\mathbf{y})|^2 d\mathbf{y}$$
$$\le C_2(A).$$

The lemma is proved. □

We continue the proof of Theorem 6.7.9. By Lemma 6.7.6, according to (6.7.7) we get

$$\delta := \Phi(a,A)(f) - \hat{f}(\mathbf{0}) = \sum_{\mathbf{m}\neq\mathbf{0}} \hat{f}(aA\mathbf{m}),$$

under an assumption about the convergence of the series, which follows from further considerations.

Let $f \in \bar{\mathbf{W}}_2^r(\Omega_d)$. We denote

$$\varphi(\mathbf{x}) = \frac{\partial^{rd} f}{\partial x_1^r \dots \partial x_d^r}.$$

Then, for $\mathbf{m}\neq\mathbf{0}$,

$$\hat{f}(aA\mathbf{m}) = \hat{\varphi}(aA\mathbf{m})\prod_{j=1}^{d}\left(2\pi i a L_j(\mathbf{m})\right)^{-r}.$$

Let l be the same as in the proof of Theorem 6.7.8. Then

$$\delta \ll \sum_{\|\mathbf{s}\|_1\ge l}\sum_{aL(\mathbf{m})\in\rho(\mathbf{s})}|\hat{\varphi}(aA\mathbf{m})|\prod_{j=1}^{d}|aL_j(\mathbf{m})|^{-r}$$
$$\le \left(\sum_{\|\mathbf{s}\|_1\ge l}\sum_{aL(\mathbf{m})\in\rho(\mathbf{s})}\prod_{j=1}^{d}|aL_j(\mathbf{m})|^{-2r}\right)^{1/2}\left(\sum_{\mathbf{m}}|\hat{\varphi}(aA\mathbf{m})|^2\right)^{1/2}. \quad (6.7.12)$$

Applying Lemma 6.7.10 and using the relation (6.7.8), in which a similar estimate was carried out, we get

$$\delta \ll a^{-rd}(\log a)^{(d-1)/2}.$$

Theorem 6.7.9 is proved. □

6.7.2 Periodic Case

In this subsection we apply the results of the previous subsection to obtain precise estimates, in the sense of order of errors, of cubature formulas for the classes $\mathbf{W}^r_{p,\alpha}$ and $\mathbf{H}^r_p$.

Theorem 6.7.11 *Let $r \geq 1$ be a real number and $2 \leq p < \infty$. Then*

$$\kappa_m(\mathbf{W}^r_{p,\alpha}) \asymp m^{-r}(\log m)^{(d-1)/2}.$$

Proof The lower estimate follows from Theorem 6.4.3. Clearly, it is sufficient to prove the upper estimate for $p = 2$. For natural numbers r the upper estimate follows from Theorem 6.2.4, Remark 6.2.6, and Theorem 6.7.9. As an optimal cubature formula in this case we may take

$$\tilde{\Phi}(a,A)(f(2\pi\mathbf{x})) := \Phi(a,A)\left(f(2\pi\psi(\mathbf{x}))\prod_{j=1}^{d}\psi'(x_j)\right), \qquad \mathbf{x} \in \Omega_d,$$

where $\psi(u)$ was defined in the proof of Theorem 6.2.4. Moreover, from the proof of Theorem 6.2.4 it follows that for all natural numbers b the following relation holds:

$$\tilde{\Phi}(a,A)(\mathbf{W}^b_2) \ll \Phi(a,A)(\bar{\mathbf{W}}^b_2(\Omega_d)).$$

Consequently, by Theorem 6.7.9,

$$\tilde{\Phi}(a,A)(\mathbf{W}^b_2) \ll a^{-bd}(\log a)^{(d-1)/2}. \tag{6.7.13}$$

Let $r > 1$ be different from a natural number. We denote $b := [r]$. Let $f \in \mathbf{W}^r_2$; we define the linear functional

$$e_a(f) := \tilde{\Phi}(a,A)(f) - \hat{f}(\mathbf{0}).$$

Denote

$$f_\mu(\mathbf{x}) := \sum_{\mathbf{k}\in\Delta Q_\mu} \hat{f}(\mathbf{k})e^{i(\mathbf{k},\mathbf{x})}, \qquad \Delta Q_\mu := Q_\mu \setminus Q_{\mu-1}.$$

Then

$$\|f_\mu\|_2 \ll 2^{-r\mu} \tag{6.7.14}$$

and

$$f(\mathbf{x}) = \sum_{\mu=1}^{\infty} f_\mu(\mathbf{x}),$$

where the series converges uniformly. Then

$$|e_a(f)| \leq \sum_{\mu=1}^{\infty} |e_a(f_\mu)|.$$

Let l be such that $2^{l-1} \le a^d < 2^l$. We have

$$|e_a(f)| \le \sum_{\mu=1}^{l} \tilde{\Phi}(a,A)(\mathbf{W}_2^{b+1})\|f_\mu\|_{\mathbf{W}_2^{b+1}} + \sum_{\mu>l} \tilde{\Phi}(a,A)(\mathbf{W}_2^{b})\|f_\mu\|_{\mathbf{W}_2^{b}}. \quad (6.7.15)$$

Applying (6.7.13), (6.7.14) and the Bernstein inequality, we get

$$|e_a(f)| \ll a^{-(b+1)d}(\log a)^{(d-1)/2} \sum_{\mu=1}^{l} 2^{(b+1-r)\mu} + a^{-bd}(\log a)^{(d-1)/2} \sum_{\mu>l} 2^{(b-r)\mu}$$
$$\ll a^{-rd}(\log a)^{(d-1)/2}.$$

From this the conclusion of the theorem follows because the number m of points of the cubature formula $\tilde{\Phi}(a,A)$ does not exceed a^d in order. □

From Theorems 6.7.11, 6.2.4 and relation (6.2.14) we obtain the following result concerning discrepancies.

Theorem 6.7.12 *Let r be a natural number and $1 < q \le 2$. Then*

$$\inf_{\lambda_1,\dots,\lambda_N \xi^\mu,\dots,\xi^N} D_r(\xi,\Lambda,N,d) \asymp N^{-r}(\log N)^{(d-1)/2}.$$

Let us proceed to the **H**-classes.

Theorem 6.7.13 *Let $r > 1$ be a real number, $1 \le p \le \infty$. Then*

$$\kappa_m(\mathbf{H}_p^r) \asymp m^{-r}(\log m)^{d-1}.$$

Proof The lower estimate follows from Theorem 6.4.8. Let us prove the upper estimate. For $p = 2$ the upper estimate may be proved in the same way as in Theorem 6.7.11 if instead of relation (6.7.14) we use the inequality

$$\|f_\mu\|_2 \ll 2^{-r\mu}\mu^{(d-1)/2}, \quad (6.7.16)$$

which is valid for f_μ, constructed for $f \in \mathbf{H}_2^r$. The estimate (6.7.16) is a simple corollary of Theorem 4.4.6. Now let $1 < p < 2$, $r > 1$. We consider the auxiliary class $\mathbf{H}_p^r(\Omega_d)$ of functions $f(\mathbf{x})$ with supports in Ω_d. Let $l := [r] + 1$. For any set $G \subset [1,d]$ of natural numbers the following condition holds for the functions $f(\mathbf{x})$:

$$\left\| \left(\prod_{j\in G} \Delta_{t_j}^{l,j} \right) f \right\|_p \le \prod_{j\in G} |t_j|^r. \quad (6.7.17)$$

We prove one sufficient condition for the function f to belong to the class $\mathbf{H}_p^r(\Omega_d)$.

Lemma 6.7.14 *Let the functions $h_{\mathbf{s}}(\mathbf{x})$ have continuous derivatives of the order $\mathbf{r} \le l\mathbf{1}$. The supports of the $h_{\mathbf{s}}(\mathbf{x})$ are contained in Ω_d and, for all* $\mathbf{s}$,

$$\|h_{\mathbf{s}}\|_p \ll 2^{-r\|\mathbf{s}\|_1}. \tag{6.7.18}$$

Moreover, for any vector $\mathbf{l}(G)$, $G \subset [1,d]$ whose coordinates with indices from G are equal to l while the other coordinates are equal to zero, the following inequalities hold:

$$\|D^{\mathbf{l}(G)} h_{\mathbf{s}}\|_p \ll 2^{-r\|\mathbf{s}\|_1 + \left(\mathbf{l}(G),\mathbf{s}\right)}. \tag{6.7.19}$$

Then there is a $\delta > 0$ which may depend on r, d, and constants from the relations (6.7.18), (6.7.19), *such that*

$$\delta \sum_{\mathbf{s}} h_{\mathbf{s}} \in \mathbf{H}^r_p(\Omega_d).$$

Proof Let $G \subset [1,d]$. We have

$$\left(\prod_{j\in G} \Delta^{l,j}_{t_j}\right)\left(\sum_{\mathbf{s}} h_{\mathbf{s}}\right) =: \Delta^l_t(G)\left(\sum_{\mathbf{s}} h_{\mathbf{s}}\right) = \sum_{\mathbf{s}} \Delta^l_t(G) h_{\mathbf{s}}. \tag{6.7.20}$$

Applying Lemma 1.4.4 and the relation (6.7.19) we get

$$\left\|\Delta^l_t(G) h_{\mathbf{s}}\right\|_p \ll 2^{-r\|\mathbf{s}\|_1} \prod_{j\in G} \min\left(1, 2^{ls_j}|t_j|^l\right). \tag{6.7.21}$$

Performing the summation over $\mathbf{s}$ from (6.7.20), (6.7.21) in the same way as in the proof of Theorem 4.4.6, we get

$$\left\|\Delta^l_t(G)\sum_{\mathbf{s}} h_{\mathbf{s}}\right\|_p \ll \prod_{j\in G} |t_j|^r,$$

which proves the lemma. □

Lemma 6.7.15 *Let $f \in \mathbf{H}^r_p(\Omega_d)$, $1 \le p \le \infty$. Then*

$$\left|\hat{f}(\mathbf{y})\right| \ll \prod_{j=1}^{d} (\max(1, |y_j|))^{-r}.$$

Proof It is easy to see that

$$\mathscr{F}(f)(\mathbf{y}) := \hat{f}(\mathbf{y}) = \mathscr{F}\left(\frac{1}{2}\Delta^{1,j}_{1/(2y_j)} f\right)(\mathbf{y}).$$

Let $G := \{j : |y_j| \ge 1\}$; then

$$\mathscr{F}(f)(\mathbf{y}) = \mathscr{F}\left(\left(\prod_{j\in G} \frac{1}{2^l}\Delta^{l,j}_{1/(2y_j)}\right) f\right)(\mathbf{y}),$$

which implies that

$$|\mathscr{F}(f)(\mathbf{y})| \le \left\| \left(\left(\prod_{j\in G} \frac{1}{2^l} \Delta^{l,j}_{1/(2y_j)} \right) f \right)(\mathbf{x}) \right\|_1 \ll \prod_{j\in G} |y_j|^{-r}.$$

The lemma is proved. □

Let $f \in \mathbf{H}^r_p$, $r > 1$, and

$$g(\mathbf{x}) := f(2\pi\psi(\mathbf{x})) \prod_{j=1}^{d} \psi'(x_j), \qquad \mathbf{x} \in \Omega_d.$$

By Theorem 4.4.6 for $1 \le p \le \infty$,

$$f(\mathbf{x}) = \sum_{\mathbf{s}} A_{\mathbf{s}}(f)(\mathbf{x}), \qquad \|A_{\mathbf{s}}(f)\|_p \ll 2^{-r\|\mathbf{s}\|_1}.$$

Consequently, by the Bernstein inequality for any $\mathbf{l}(G)$ from Lemma 6.7.14, we have

$$\left\|D^{\mathbf{l}(G)} A_{\mathbf{s}}(f)\right\|_p \ll 2^{-r\|\mathbf{s}\|_1 + (\mathbf{l}(G),\mathbf{s})}. \tag{6.7.22}$$

Further,

$$g(\mathbf{x}) = \sum_{\mathbf{s}} h_{\mathbf{s}}(\mathbf{x}),$$

where

$$h_{\mathbf{s}}(\mathbf{x}) := A_{\mathbf{s}}(f)(2\pi\psi(\mathbf{x})) \prod_{j=1}^{d} \psi'(x_j).$$

From the relation (6.7.22) and the proof of Theorem 6.2.4 (see (6.2.22)) we find that $h_{\mathbf{s}}(\mathbf{x})$ satisfies Lemma 6.7.14. Consequently, $\delta g \in \mathbf{H}^r_p(\Omega_d)$. Applying Lemma 6.7.15 and Theorem 6.7.8 we get

$$\left|\Phi(a,A)(g) - \hat{g}(\mathbf{0})\right| \ll a^{-rd}(\log a)^{d-1}. \tag{6.7.23}$$

Let $e_a(f)$ be the same as in the proof of Theorem 6.7.11. From (6.7.23) it follows that

$$\left|e_a(f)\right| \ll a^{-rd}(\log a)^{d-1},$$

which implies Theorem 6.7.13. □

6.8 Universal Cubature Formulas

In Chapter 3 and in this chapter we have considered various cubature formulas and found the optimal cubature formulas (in the sense of order) either for the Sobolev and Nikol'skii classes or for the classes of functions with bounded mixed derivatives or differences. We emphasize that, for instance, for the anisotropic Nikol'skii classes $H^{\mathbf{r}}_p$ the cubature formula $q_m(f,\mathbf{r})$ giving the optimal order of the error for this class is determined by the vector $\mathbf{r}$. In this section we discuss the question of finding cubature formulas that give an acceptable error for all the classes $H^{\mathbf{r}}_p$. As in §5.4 we observe that methods which are optimal for the classes $\mathbf{W}^r_{p,\alpha}$, $\mathbf{H}^r_p$ are universal for the collection of classes $W^{\mathbf{r}}_{p,\alpha}$, $H^{\mathbf{r}}_p$.

6.8.1 Bivariate Functions

We first consider the case $d=2$ and prove that the Fibonacci cubature formulas are universal.

Theorem 6.8.1 *Let $1\le p\le\infty$, $g(\mathbf{r})>1/p$. Then*

$$\Phi_n(H^{\mathbf{r}}_p)\asymp\kappa_{b_n}(H^{\mathbf{r}}_p)\asymp b_n^{-g(\mathbf{r})}.$$

Proof The lower estimate follows from Theorem 3.6.1. Let us prove the upper estimate. Let $f\in H^{\mathbf{r}}_p$, $g(\mathbf{r})>1/p$. Clearly, it is sufficient to prove the conclusion of the theorem under the assumption that f is a trigonometric polynomial. Then using the above notation we have

$$\Phi_n(f)=\sum_{\mathbf{k}\in L(n)}\hat f(\mathbf{k}). \tag{6.8.1}$$

Let

$$\mathscr{A}(\mathbf{r},l,\mathbf{x}):=\mathscr{V}_{[2^{\mathbf{v}l}]}(\mathbf{x})-\mathscr{V}_{[2^{\mathbf{v}(l-1)}]}(\mathbf{x}),\qquad l=1,2,\dots,$$
$$\mathscr{A}(\mathbf{r},0,\mathbf{x}):=\mathscr{V}_1(\mathbf{x}),\qquad \text{where } \mathbf{v}:=\mathbf{v}(\mathbf{r}):=g(\mathbf{r})/\mathbf{r},$$

and

$$\mathscr{A}(\mathbf{r},l,\mathbf{x})_{L(n)}:=\sum_{\mathbf{k}\in L(n)}\hat{\mathscr{A}}(\mathbf{r},l,\mathbf{k})e^{i(\mathbf{k},\mathbf{x})}.$$

Lemma 6.8.2 *Let*

$$l_0:=\max\big(l:\mathscr{A}(\mathbf{r},l)\in\mathscr{T}(\gamma b_n)\big),$$

where γ is from Lemma 6.5.1. Then

$$\mathscr{A}(\mathbf{r},0)_{L(n)}=1,\qquad \mathscr{A}(\mathbf{r},l)_{L(n)}=0,\qquad l=1,2,\dots,l_0,$$

and, for $l > l_0$,

$$\left\|\mathscr{A}(\mathbf{r},l)_{L(n)}\right\|_q \ll (2^l/b_n)^{1-1/q}, \qquad 1 \le q \le \infty.$$

Proof The first conclusion of the lemma follows directly from Lemma 6.5.1. Lemma 6.5.1 implies that the number of nonzero terms in $\mathscr{A}(\mathbf{r},l)$ is less (in the sense of order) than $2^l/b_n$; this gives the second inequality for $q=\infty$. Let us prove this inequality for $q=1$. We use the function $G_N(\mathbf{x})$, which was defined in (6.5.13). Let l be fixed. We assume that N is so large that, for all $\mathbf{k} \in L(n)$ such that $\hat{\mathscr{A}}(\mathbf{r},l,\mathbf{k}) \neq 0$, we have $\hat{G}_N(\mathbf{k}) = 1$. Then

$$\mathscr{A}(\mathbf{r},l)_{L(n)} = \mathscr{A}(\mathbf{r},l) * G_N$$

and, by (6.5.15),

$$\left\|\mathscr{A}(\mathbf{r},l)_{L(n)}\right\|_1 \ll 1,$$

proving the inequality in the lemma for $q=1$. The general case, $1 \le q \le \infty$, follows from the previous cases by the inequality

$$\|f\|_q \le \|f\|_1^{1/q}\|f\|_\infty^{1-1/q}. \qquad \square$$

Remark 6.8.3 The number l_0 is such that

$$2^{l_0} \asymp b_n.$$

From (6.8.1) and Lemma 6.8.2 we obtain

$$\Phi_n(f) - \hat{f}(\mathbf{0}) = \sum_{l>l_0} \langle f, \mathscr{A}(\mathbf{r},l)_{L(n)}\rangle. \tag{6.8.2}$$

Theorem 4.4.9 and Lemma 6.8.2 give

$$\left|\langle f, \mathscr{A}(\mathbf{r},l)_{L(n)}\rangle\right| \ll 2^{-g(\mathbf{r})l}\left\|\mathscr{A}(\mathbf{r},l)_{L(n)}\right\|_{p'} \ll 2^{-(g(\mathbf{r})-1/p)l} b_n^{-1/p}.$$

Substituting this estimate into (6.8.2) and performing the summation we find that

$$\left|\Phi_n(f) - \hat{f}(\mathbf{0})\right| \ll b_n^{-g(\mathbf{r})}.$$

Theorem 6.8.1 is proved. $\square$

6.8.2 General Case

Now let $d > 2$. We consider the cubature formulas

$$\tilde{\Phi}(a,A)(f(2\pi\mathbf{x})) := \Phi(a,A)\left(f(2\pi\psi(\mathbf{x}))\prod_{j=1}^{d}\psi'(x_j)\right), \qquad \mathbf{x} \in \Omega_d,$$

which were used in the proof of Theorems 6.7.11 and 6.7.13.

Theorem 6.8.4 *Let $1 \le p \le \infty$, $g(\mathbf{r}) > 1$. Then for all $\mathbf{r}$ the following relation holds:*

$$\tilde{\Phi}(a,A)(H_p^{\mathbf{r}}) \asymp a^{-g(\mathbf{r})d}.$$

Proof The lower estimate follows from Theorem 3.6.1. Let us prove the upper estimate. In the same way as in the proofs of Theorems 6.7.11 and 6.7.13 we reduce the problem to estimating the error of the cubature formula $\Phi(a,A)$ for functions with supports in Ω_d.

Let $f \in H_p^{\mathbf{r}}$, $g(\mathbf{r}) > 1$. Then we can represent f by the uniformly convergent series

$$f = \sum_{m=0}^{\infty} A(f,\mathbf{r},m),$$

and (see Corollary 3.4.8) we have

$$\|A(f,\mathbf{r},m)\|_p \ll 2^{-g(\mathbf{r})m}. \tag{6.8.3}$$

By Theorem 3.3.1 with $l_j = [r_j]+1$ we obtain

$$\|A_j^{(r)}(f,\mathbf{r},m)\|_p \ll 2^{-g(\mathbf{r})m+g(\mathbf{r})mr/r_j}, \qquad 0 \le r \le l_j. \tag{6.8.4}$$

In the same way as in Lemma 6.7.14 it follows from (6.8.3), (6.8.4), and (6.2.22) that the function

$$\varphi(\mathbf{x}) := f(2\pi\psi(\mathbf{x}))\prod_{j=1}^{d}\psi'(x_j) = \sum_{m=0}^{\infty} A(f,\mathbf{r},m)(2\pi\psi(\mathbf{x}))\prod_{j=1}^{d}\psi'(x_j)$$

has the following properties: the support of $\varphi(\mathbf{x})$ is contained in Ω_d, $\varphi(\mathbf{x})$ is continuous, and, for each $j = 1,\dots,d$,

$$\|\Delta_{t_j}^{l_j,j}\varphi\|_p \ll |t_j|^{r_j}. \tag{6.8.5}$$

From (6.8.5) we obtain in the same way as in Lemma 6.7.15 the estimate

$$|\hat{\varphi}(\mathbf{y})| \ll \min_{1\le j\le d}\left(\max(1,|y_j|)\right)^{-r_j}. \tag{6.8.6}$$

We denote by $\mathbf{E}^{\mathbf{r}}(\Omega_d)$ the set of continuous functions with supports in Ω_d which satisfies (6.8.6) with a constant 1 in the inequality. Clearly, in order to complete the proof of Theorem 6.8.4 it is sufficient to prove the following assertion.

Theorem 6.8.5 *Let $g(\mathbf{r}) > 1$. Then*

$$\Phi(a,A)(\mathbf{E}^{\mathbf{r}}(\Omega_d)) \ll a^{-g(\mathbf{r})d}.$$

Proof Let $f \in \mathbf{E}^{\mathbf{r}}(\Omega_d)$, $g(\mathbf{r}) > 1$. Applying the Poisson formula (Lemma 6.7.6) in the same way as in (6.7.6), we obtain the relation

$$\Phi(a,A)(f) = \sum_{\mathbf{m}} \hat{f}(aA\mathbf{m}). \tag{6.8.7}$$

This is correct under the assumption of convergence of the series, which will follow from further estimates. Consider the following sets:

$$G_l(\mathbf{r}) := \{\mathbf{k} \in \mathbb{R}^d : |k_j| \le 2^{g(\mathbf{r})l/r_j},\ j = 1,\dots,d\},$$
$$H_l(\mathbf{r}) := G_{l+1}(\mathbf{r}) \setminus G_l(\mathbf{r}), \qquad l = 0,1,\dots$$

Then, using Lemma 6.7.1, we find that

$$\Phi(a,A)(f) - \hat{f}(\mathbf{0}) = \sum_{l \ge l_0} \sum_{aA\mathbf{m} \in H_l(\mathbf{r})} \hat{f}(aA\mathbf{m}),$$

where l_0 is such that

$$a^{-d}2^{l_0-1} < 1, \qquad a^{-d}2^{l_0} \ge 1.$$

Using Lemma 6.7.1 and the definition of the class $\mathbf{E}^{\mathbf{r}}(\Omega_d)$ we get

$$\Phi(a,A)(f) - \hat{f}(\mathbf{0}) \ll \sum_{l \ge l_0} 2^{-g(\mathbf{r})l} \left(\frac{|H_l(\mathbf{r})|}{a^d} + 1 \right) \ll a^{-d} \sum_{l \ge l_0} 2^{-\left(g(\mathbf{r})-1\right)l}$$
$$\ll a^{-g(\mathbf{r})d}.$$

The theorem 6.8.5 is now proved. □

This completes the proof of Theorem 6.8.4. □

6.9 Recovery of Functions

In this section we study the approximation of functions from the classes $\mathbf{W}^r_{q,\alpha}$ and $\mathbf{H}^r_q$ in the L_p-norm by polynomials constructed from values of the function at m points. We have already discussed the problem of approximate recovery, for functions of a single variable in §2.4.2 and for multivariate functions from the anisotropic smoothness classes in §3.6.2. It was established in Theorems 2.4.4 and 3.6.4 that sampling operators which provide approximation by the appropriate trigonometric polynomials are optimal in the sense of order for all $1 \le q, p \le \infty$ under a natural smoothness assumption: $r > 1/q$ for the univariate case and $g(\mathbf{r}) > 1/q$ for the multivariate case. Thus, in the multivariate case of the classes $W^{\mathbf{r}}_{q,\alpha}$ and $H^{\mathbf{r}}_q$, the problem of approximate recovery is completely solved. In contrast with that it turns out that the problem of optimal recovery for classes of functions with mixed smoothness is wide open. In this section we present some known results in this direction. We first recall the setting of optimal recovery. For fixed m, $\xi := \{\xi^j\}_{j=1}^m$ (sometimes we use the notation X_m instead of ξ), and $\psi_1(\mathbf{x}), \dots, \psi_m(\mathbf{x})$ we define the linear operator

$$\Psi(f,\xi) := \sum_{j=1}^{m} f(\xi^j)\psi_j(\mathbf{x}).$$

For a function class $\mathbf{F}$ define

$$\Psi(\mathbf{F},\xi)_p := \sup_{f\in\mathbf{F}} \|f - \Psi(f,X_m)\|_p.$$

Denote, for the class $\mathbf{F}$,

$$\rho_m(\mathbf{F})_p := \inf_{\xi;\psi_1,\dots,\psi_m} \Psi(\mathbf{F},\xi)_p.$$

Let us make a simple well-known observation on the relation between recovery and numerical integration. Associate with the recovery operator $\Psi(\cdot,\xi)$ the cubature formula $\Lambda_m(\cdot,\xi)$ with

$$\lambda_j := \int_\Omega \psi_j(\mathbf{x})d\mu.$$

Then

$$\left|\Lambda_m(f,\xi) - \int_\Omega f(\mathbf{x})d\mu\right| = \left|\int_\Omega (\Psi(f,\xi)-f)d\mu\right| \le \|\Psi(f,\xi)-f\|_1$$
$$\le \|\Psi(f,\xi)-f\|_p, \qquad p\ge 1.$$

Therefore, for any function class $\mathbf{F}$ and each $p \ge 1$ we have

$$\kappa_m(\mathbf{F}) \le \rho_m(\mathbf{F})_p. \tag{6.9.1}$$

6.9.1 Upper Bounds in the Case $p = q$

We begin our presentation with the use of the Smolyak algorithm for recovery. In §4.5 we gave a general way of constructing good approximation methods for functions with mixed smoothness using univariate methods of approximation. We now discuss a direct corollary of Theorem 4.5.1. As the operators Y_s from Theorem 4.5.1 we will use two univariate recovery operators from §2.4.2:

$$I_{2^s}(f) := (2^{s+1}+1)^{-1}\sum_{l=0}^{2^{s+1}} f(x^l)\mathscr{D}_{2^s}(x-x^l), \qquad x^l := 2\pi l/(2^{s+1}+1),$$

$$R_{2^s}(f) := 2^{-s-2}\sum_{l=1}^{2^{s+2}} f\big(x(l)\big)\mathscr{V}_{2^s}\big(x-x(l)\big), \qquad x(l) := \pi l 2^{-s-1}.$$

The operator I_{2^s} maps a continuous function $f(x)$ to the trigonometric polynomial $I_{2^s}(f) \in \mathcal{T}(2^s)$. Theorem 2.4.4 states that for any $f \in W^r_{p,\alpha}$, $1 < p < \infty$, $r > 1/p$,

$$\|f - I_{2^s}(f)\|_p \le C_1(p,r)2^{-rs}. \tag{6.9.2}$$

Lemma 2.4.3 gives that for $t \in \mathcal{T}(2^v)$, $v \ge s$,

$$\|I_{2^s}(t)\|_p \le C_2(p)2^{(v-s)/p}\|t\|_p, \qquad 1 < p < \infty. \tag{6.9.3}$$

Similarly, the operator R_{2^s} maps a continuous function $f(x)$ to the trigonometric polynomial $R_{2^s}(f) \in \mathcal{T}(2^{s+1}-1)$. In this case Theorem 2.4.4 states that for any $f \in W^r_{p,\alpha}$, $1 \le p \le \infty$, $r > 1/p$,

$$\|f - R_{2^s}(f)\|_p \le C_1(r)2^{-rs}. \tag{6.9.4}$$

Lemma 2.4.3 now gives that for $t \in \mathcal{T}(2^v)$, $v \ge s$,

$$\|R_{2^s}(t)\|_p \le C_2 2^{(v-s)/p}\|t\|_p, \qquad 1 \le p \le \infty. \tag{6.9.5}$$

Thus the operators I_{2^s} and R_{2^s} satisfy conditions (1) and (2) for the operators Y_s (see §4.5) in the cases $1 < p < \infty$ for I_{2^s}, $1 \le p \le \infty$ for R_{2^s}, $a > 1/p$, $b = 1/p$. The corresponding d-dimensional operators are

$$T^I_n := \sum_{\mathbf{s}:\|\mathbf{s}\|_1 \le n} \Delta^I_{\mathbf{s}}, \qquad \Delta^I_{\mathbf{s}} := \prod_{i=1}^d (I^i_{2^{s_i}} - I^i_{2^{s_i-1}}),$$

with $I_{1/2} := 0$, and

$$T_n := T^R_n := \sum_{\mathbf{s}:\|\mathbf{s}\|_1 \le n} \Delta_{\mathbf{s}}, \qquad \Delta_{\mathbf{s}} := \Delta^R_{\mathbf{s}} := \prod_{i=1}^d (R^i_{2^{s_i}} - R^i_{2^{s_i-1}}),$$

with $R_{1/2} := 0$. Then Theorem 4.5.1 gives the following two theorems.

Theorem 6.9.1 *Let $1 < p < \infty$ and $r > 1/p$. Then we have for $f \in \mathbf{H}^r_p$*

$$\|\Delta^I_{\mathbf{s}}(f)\|_p \ll 2^{-r\|\mathbf{s}\|_1} \qquad \text{and} \qquad \|f - T^I_n(f)\|_p \ll 2^{-rn}n^{d-1}.$$

Theorem 6.9.2 *Let $1 \le p \le \infty$ and $r > 1/p$. Then we have for $f \in \mathbf{H}^r_p$*

$$\|\Delta^R_{\mathbf{s}}(f)\|_p \ll 2^{-r\|\mathbf{s}\|_1} \qquad \text{and} \qquad \|f - T^R_n(f)\|_p \ll 2^{-rn}n^{d-1}.$$

We will mention two simple properties of the operators $\Delta_{\mathbf{s}}$ which are used in further subsections. These properties follow from the corresponding properties of operators R_m (see §2.4.2).

Property 1 For any continuous function f we have

$$\|\Delta_{\mathbf{s}}(f)\|_\infty \le C(d)\|f\|_\infty.$$

Property 2 Let vectors $\mathbf{s} = (s_1,\dots,s_d)$ and $\mathbf{k} = (k_1,\dots,k_d)$ be such that, for some j, we have $|k_j| \le 2^{s_j-1}$. Then

$$\Delta_{\mathbf{s}}(e^{i(\mathbf{k},\mathbf{x})}) = 0.$$

This property follows from (2.4.11).

6.9.2 Upper Bounds in the Case $1 \le q < p < \infty$

Here we use the fundamental inequality from Lemma 3.3.7 in the form of Remark 3.3.10 to deduce the following two theorems from Theorems 6.9.1 and 6.9.2.

Theorem 6.9.3 *For any* $f \in \mathbf{H}^r_q$, $1 < q < p < \infty$, $r > 1/q$,

$$\|f - T^I_n(f)\|_p \ll 2^{-n(r-\beta)} n^{(d-1)/p}, \qquad \beta := 1/q - 1/p. \tag{6.9.6}$$

Theorem 6.9.4 *For any* $f \in \mathbf{H}^r_q$, $1 \le q < p < \infty$, $r > 1/q$,

$$\|f - T^R_n(f)\|_p \ll 2^{-n(r-\beta)} n^{(d-1)/p}, \qquad \beta := 1/q - 1/p. \tag{6.9.7}$$

We note that the operator $T_n := T^R_n$ uses the sparse grid $\mathrm{SG}(n+d)$ and $\psi_j \in \mathscr{T}(Q_{n+d})$. Theorems 6.9.4 and 4.4.10 show that the recovering operator T_n provides an optimal rate of approximation by the hyperbolic cross polynomials from $\mathscr{T}(Q_{n+d})$.

6.9.3 Optimal Rates for Recovery

It easily follows from the definition of $\rho_m(\mathbf{F})_p$ that $\rho_m(\mathbf{F})_p \ge d_m(\mathbf{F}, L_p)$. The upper bound from Theorem 6.9.2 and the lower bound for the Kolmogorov width from Theorem 5.3.19 for $d = 2$,

$$d_m(\mathbf{H}^r_\infty, L_\infty) \asymp m^{-r}(\log m)^{r+1}$$

imply that, for $d = 2$,

$$\rho_m(\mathbf{H}^r_\infty)_\infty \asymp m^{-r}(\log m)^{r+1}. \tag{6.9.8}$$

The upper bound (6.9.7) and the known bounds for the Kolmogorov width (see Theorem 5.3.2): for $1 \le q \le 2$, $r > 1/q$,

$$d_m(\mathbf{H}^r_q, L_2) \asymp m^{-r+\eta}(\log m)^{(d-1)(r+1-1/q)}, \qquad \eta := 1/q - 1/2,$$

imply that, for $1 \le q < 2$, $r > 1/q$,

$$\rho_m(\mathbf{H}^r_q)_2 \asymp m^{-r+\eta}(\log m)^{(d-1)(r+1-1/q)}. \tag{6.9.9}$$

Further, the upper bound (6.9.7) and Theorem 5.3.12 imply that, for $1 < q < p \le 2$, $r > \beta$,

$$\rho_m(\mathbf{H}^r_q)_p \asymp m^{-r+\beta}(\log m)^{(d-1)(r-\beta+1/p)}. \tag{6.9.10}$$

Thus, the above upper bounds for the operator T_n and the lower bounds for the Kolmogorov widths give the correct order for the quantities $\rho_m(\mathbf{H}^r_q)_p$ in the following cases: for all d, when $1 < q < p \le 2$ or $q = 1$, $p = 2$ with $r > 1/q$; for $d = 2$, when $q = p = \infty$.

We now prove a result on the optimal rate of recovery of the $\mathbf{W}$ classes.

Theorem 6.9.5 *The relation*

$$\rho_m(\mathbf{W}^r_2)_\infty \asymp m^{-r+1/2}(\log m)^{r(d-1)}$$

holds for all $r > 1/2$.

Proof First, we prove the upper estimate. For this purpose we study the errors involved in using the recovery operator T_n defined above.

Theorem 6.9.6 *Let* $1 < q < \infty$ *and* $r > 1/q$. *Then*

$$T_n(\mathbf{W}^r_q)_\infty := \sup_{f\in\mathbf{W}^r_q} \|f - T_n(f)\|_\infty \asymp 2^{-(r-1/q)n} n^{(d-1)(1-1/q)}.$$

Proof We have

$$\|f - T_n(f)\|_\infty \le \sum_{\|\mathbf{s}\|_1 > n} \|\Delta_{\mathbf{s}}(f)\|_\infty. \tag{6.9.11}$$

Now we bound each $\|\Delta_{\mathbf{s}}(f)\|_\infty$:

$$\|\Delta_{\mathbf{s}}(f)\|_\infty \le \sum_{\mathbf{u}} \|\Delta_{\mathbf{s}}(\delta_{\mathbf{u}}(f))\|_\infty = \sum_{\mathbf{u}\ge\mathbf{s}} \|\Delta_{\mathbf{s}}(\delta_{\mathbf{u}}(f))\|_\infty. \tag{6.9.12}$$

Further, we find by Property 1 above that

$$\|\Delta_{\mathbf{s}}(\delta_{\mathbf{u}}(f))\|_\infty \le C(d)\|\delta_{\mathbf{u}}(f)\|_\infty. \tag{6.9.13}$$

Let $f = F_r * \varphi$, $\|\varphi\|_2 \le 1$. Then,

$$\|\delta_{\mathbf{u}}(f)\|_\infty \ll 2^{-r\|\mathbf{u}\|_1}\|\delta_{\mathbf{u}}(\varphi)\|_\infty. \tag{6.9.14}$$

Relations (6.9.11)–(6.9.14) imply that

$$\begin{aligned}\|f - T_n(f)\|_\infty &\ll \sum_{\|\mathbf{s}\|_1 > n}\sum_{\mathbf{u}\ge\mathbf{s}} 2^{-r\|\mathbf{u}\|_1}\|\delta_{\mathbf{u}}(\varphi)\|_\infty \\ &= \sum_{\|\mathbf{u}\|_1 > n} \nu_n(\mathbf{u})2^{-r\|\mathbf{u}\|_1}\|\delta_{\mathbf{u}}(\varphi)\|_\infty,\end{aligned} \tag{6.9.15}$$

where $\nu_n(\mathbf{u})$ is the number of $\mathbf{s} \in \mathbb{N}^d_0$ such that $\mathbf{s} \le \mathbf{u}$ and $\|\mathbf{s}\|_1 > n$. It is easy to see that $v_n(\mathbf{u})$ does not exceed the number of $\mathbf{s}' := \mathbf{u} - \mathbf{s} \in \mathbb{N}_0^d$ such that $\|\mathbf{s}'\|_1 \le \|\mathbf{u}\|_1 - n$. Therefore,

$$v_n(\mathbf{u}) \ll (\|\mathbf{u}\|_1 - n + 1)^d. \tag{6.9.16}$$

By the Hölder inequality we obtain from (6.9.15) that

$$\|f - T_n(f)\|_\infty \ll \Big(\sum_{\mathbf{u}} 2^{-\|\mathbf{u}\|_1} \|\delta_{\mathbf{u}}(\varphi)\|_\infty^q\Big)^{1/q} \Big(\sum_{\|\mathbf{u}\|_1 > n} \big(v_n(\mathbf{u}) 2^{-(r-1/q)\|\mathbf{u}\|_1}\big)^{q'}\Big)^{1/q'}. \tag{6.9.17}$$

By Lemma 3.3.13 we obtain

$$\Big(\sum_{\mathbf{u}} 2^{-\|\mathbf{u}\|_1} \|\delta_{\mathbf{u}}(\varphi)\|_\infty^q\Big)^{1/q} \ll \|\varphi\|_q. \tag{6.9.18}$$

Inequality (6.9.16) implies

$$\Big(\sum_{\|\mathbf{u}\|_1 > n} \big(v_n(\mathbf{u}) 2^{-(r-1/q)\|\mathbf{u}\|_1}\big)^{q'}\Big)^{1/q'} \ll 2^{-(r-1/q)} n^{(d-1)(1-1/q)}, \tag{6.9.19}$$

which proves the upper estimates in Theorem 6.9.6.

We now proceed to consider the lower estimates in Theorem 6.9.6. We will prove these lower estimates in the following way. First, we prove the lower estimate in Theorem 6.9.5. This estimate implies the lower estimate in Theorem 6.9.6 for $q = 2$. Next, using the upper estimate that has been proved for all $1 < q < \infty$ and the lower estimate obtained for $q = 2$, we can prove the lower estimate for all $1 < q < \infty$.

Proof of lower bound in Theorem 6.9.5 We derive this lower bound from the known results about the Kolmogorov and the linear widths. It follows from the definitions of ρ_m and the linear width that

$$\rho_m(\mathbf{W}_2^r)_\infty \ge \lambda_m(\mathbf{W}_2^r, L_\infty). \tag{6.9.20}$$

By the multivariate analog of Theorem 2.1.8 (the proof for the multivariate case is the same as in the univariate case) we get

$$\lambda_m(\mathbf{W}_2^r, L_\infty) = \lambda_m(\mathbf{W}_{1,-r}^r, L_2). \tag{6.9.21}$$

Relations (6.9.20) and (6.9.21) imply that

$$\rho_m(\mathbf{W}_2^r)_\infty \ge d_m(\mathbf{W}_{1,-r}^r, L_2). \tag{6.9.22}$$

Using Theorem 5.3.1 we derive

$$\rho_m(\mathbf{W}_2^r)_\infty \ge m^{-r+1/2} (\log m)^{r(d-1)}, \tag{6.9.23}$$

which proves the lower bound in Theorem 6.9.5. The corresponding upper bound follows from upper bound already proved in Theorem 6.9.6. This completes the proof of Theorem 6.9.5. □

We now continue the proof of Theorem 6.9.6. We will prove the inequality

$$T_n(\mathbf{W}_2^r)_\infty^2 \leq 4T_n(\mathbf{W}_q^r)_\infty T_n(\mathbf{W}_{q'}^r)_\infty \tag{6.9.24}$$

for all $1 < q < \infty$.

Suppose that $f \in \mathbf{W}_2^r$ that is

$$f = \varphi * F_r, \qquad \|\varphi\|_2 \leq 1.$$

Let

$$E := \{\mathbf{x} \in \mathbb{T}^d : |\varphi(\mathbf{x})| \geq D\},$$

where D is a number to be chosen below. We represent φ in the form

$$\varphi = \varphi_D + \varphi^D,$$

where

$$\varphi_D(\mathbf{x}) = \begin{cases} \varphi(\mathbf{x}) & \text{for } \mathbf{x} \notin E, \\ 0 & \text{for } \mathbf{x} \in E. \end{cases}$$

Without loss of generality we can assume that $1 < q < 2$. It is not difficult to verify that

$$\|\varphi^D\|_q \leq D^{1-2/q} =: A; \qquad \|\varphi\|_{q'} \leq D^{2/q-1} = A^{-1}.$$

Thus, we have

$$\begin{aligned} T_n(\mathbf{W}_2^r)_\infty &\leq \min_A \left(AT_n(\mathbf{W}_q^r)_\infty + A^{-1}T_n(\mathbf{W}_{q'}^r)_\infty\right) \\ &\leq 2T_n(\mathbf{W}_q^r)_\infty^{1/2} T_n(\mathbf{W}_{q'}^r)_\infty^{1/2}, \end{aligned}$$

which proves (6.9.24).

Using (6.9.24), the lower bound for $\rho_m(\mathbf{W}_2^r)_\infty$, and the upper bound for $T_n(\mathbf{W}_{q'}^r)_\infty$ we obtain

$$2^{-2n(r-1/2)} n^{d-1} \ll T_n(\mathbf{W}_2^r)_\infty^2 \ll T_n(\mathbf{W}_q^r)_\infty 2^{-(r-1+1/q)n} n^{(d-1)/q}.$$

This gives the required lower estimate for $T_n(\mathbf{W}_q^r)_\infty$. □

6.9.4 Some Other Lower Bounds

The main result of this subsection is the following theorem.

Theorem 6.9.7 *For any recovering operator* $\Psi(\cdot, X_m)$ *with respect to an* $(n, n-1)$ *-net* X_m *we have for* $1 \le q < p < \infty$

$$\Psi(\mathbf{H}^r_q, X_m)_p \gg 2^{-n(r-\beta)} n^{(d-1)/p}, \qquad \beta := 1/q - 1/p.$$

Proof We use the polynomials $t^1_{\mathbf{s}}$ constructed in the proof of Theorem 6.4.7. We also need some more constructions. Let

$$\mathcal{K}_N(x) := \sum_{|k| \le N} (1 - |k|/N) e^{ikx} = (\sin(Nx/2))^2 \,/\, (N(\sin(x/2))^2)$$

be a univariate Fejér kernel. The Fejér kernel $\mathcal{K}_N$ is an even nonnegative trigonometric polynomial in $\mathcal{T}(N-1)$. From the obvious relations

$$\|\mathcal{K}_N\|_1 = 1, \qquad \|\mathcal{K}_N\|_\infty = N,$$

together with the inequality

$$\|f\|_q \le \|f\|_1^{1/q} \|f\|_\infty^{1-1/q}$$

and the duality argument, we get

$$CN^{1-1/q} \le \|\mathcal{K}_N\|_q \le N^{1-1/q}, \qquad 1 \le q \le \infty. \tag{6.9.25}$$

In the multivariate case, define

$$\mathcal{K}_{\mathbf{N}}(\mathbf{x}) := \prod_{j=1}^{d} \mathcal{K}_{N_j}(x_j), \qquad \mathbf{N} = (N_1, \dots, N_d).$$

Then the $\mathcal{K}_{\mathbf{N}}$ are nonnegative trigonometric polynomials from $\mathcal{T}(\mathbf{N} - \mathbf{1}, d)$ which have the following properties:

$$\|\mathcal{K}_{\mathbf{N}}\|_1 = 1, \tag{6.9.26}$$

$$\|\mathcal{K}_{\mathbf{N}}\|_q \asymp \vartheta(\mathbf{N})^{1-1/q}, \qquad 1 \le q \le \infty. \tag{6.9.27}$$

For n of the form $n = 4l$, $l \in \mathbb{N}$, define

$$Y(n,d) := \{\mathbf{s} : \mathbf{s} = (4l_1, \dots, 4l_d),\ l_1 + \cdots + l_d = n/4,\ l_j \in \mathbb{N},\ j = 1, \dots, d\}.$$

Define for $\mathbf{s} \in Y(n,d)$

$$t_{\mathbf{s}}(\mathbf{x}) := t^1_{\mathbf{s}}(\mathbf{x}) \mathcal{K}_{2^{\mathbf{s}-2}}(\mathbf{x} - \mathbf{x}^*),$$

where $\mathbf{x}^*$ is a maximum point of $|t^1_{\mathbf{s}}(\mathbf{x})|$. Finally, define

$$t(\mathbf{x}) := \sum_{\mathbf{s} \in Y(n,d)} t_{\mathbf{s}}(\mathbf{x}) w(\mathbf{s}, \mathbf{x}).$$

Then we have

$$|t_{\mathbf{s}}(\mathbf{x}^*| \gg 2^n$$

and, therefore, by Nikol'skii's inequality,

$$\|t_{\mathbf{s}}\|_2 \gg 2^{n/2}.$$

It follows from our definition of $Y(n,d)$ that the polynomials $t_{\mathbf{s}}(\mathbf{x})w(\mathbf{s},\mathbf{x})$, $\mathbf{s} \in Y(n,d)$, form an orthogonal system. This implies that

$$\|t\|_2^2 \gg 2^n n^{d-1}. \tag{6.9.28}$$

Take any $p \in (1,\infty)$ and, by Theorem 3.3.6, estimate as follows:

$$\|t\|_{p'} \ll \left(\sum_{\mathbf{s}\in Y(n,d)} \|t_{\mathbf{s}}\|_1^{p'} 2^{\|\mathbf{s}\|_1(p'-1)} \right)^{1/p'} \ll 2^{n/p} n^{(d-1)/p'}. \tag{6.9.29}$$

Relations (6.9.28) and (6.9.29) imply that

$$\|t\|_p \gg 2^{n(1-1/p)} n^{(d-1)/p}. \tag{6.9.30}$$

It is now clear that

$$\|t\|_{\mathbf{H}_q^r} \ll 2^{n(r+1-1/q)}. \tag{6.9.31}$$

The bounds (6.9.30), (6.9.31) and the fact that $\Psi(t,X_m) = 0$ imply the required bound in Theorem 6.9.7. Theorem 6.9.7 is proved. □

The inequality

$$\|t\|_{\mathbf{B}_{q,\theta}^r} \ll 2^{n(r+1-1/q)} n^{(d-1)/\theta}$$

and (6.9.30) imply the following result.

Theorem 6.9.8 *For any recovering operator $\Psi(\cdot,X_m)$ with respect to an $(n,n-1)$-net X_m we have, for $1 \le q < p < \infty$, $r > \beta$,*

$$\Psi(\mathbf{B}_{q,\theta}^r, X_m)_p \gg 2^{-n(r-\beta)} n^{(d-1)(1/p-1/\theta)}, \qquad \beta := 1/q - 1/p.$$

6.9.5 Universality in Recovering

In §5.4 we discussed the universality of approximation with respect to orthowidths and with respect to the Kolmogorov widths. We established there that approximation by the hyperbolic cross trigonometric polynomials is a universal method of approximation in many cases. For instance, Theorem 5.4.1 implies that the operator S_{Q_n} of orthogonal projection onto the $\mathscr{T}(Q_n)$ is universal for collections $\{W_{q,\alpha}^{\mathbf{r}}\}_{\mathbf{r}\in P}$ and $\{H_q^{\mathbf{r}}\}_{\mathbf{r}\in P}$ with respect to the orthowidths $\varphi_m(\cdot,L_p)$ for all $1 \le q,p \le \infty$ except $(q,p) = (1,1),(\infty,\infty)$. In this subsection we show that the recovery operator T_n is universal. We consider the recovery of functions from the classes $H_p^{\mathbf{r}}$ by the operator T_n defined above.

Theorem 6.9.9 *Let $1 \leq q, p \leq \infty$, $g(\mathbf{r}) > 1/q$. Then*

$$\sup_{f \in H_q^{\mathbf{r}}} \|f - T_n(f)\|_p \ll 2^{-n\left(g(\mathbf{r}) - (1/q - 1/p)_+\right)}.$$

Proof Clearly, it suffices to carry out the proof of the theorem under the assumption that $f(\mathbf{x})$ is a trigonometric polynomial. We first consider the case $q = p$. We have

$$f - T_n(f) = \sum_{\|\mathbf{s}\|_1 > n} \Delta_{\mathbf{s}}(f) = \sum_{\|\mathbf{s}\|_1 > n} \sum_{l} \Delta_{\mathbf{s}}(A(f, \mathbf{r}, l)), \tag{6.9.32}$$

where the $A(f, \mathbf{r}, l)$ are defined in §3.4. We will estimate the quantities

$$\sigma_m := \sum_{\|\mathbf{s}\|_1 = m} \sum_{l} \|\Delta_{\mathbf{s}}(A(f, \mathbf{r}, l))\|_p \tag{6.9.33}$$

using Property 2 of the operators $\Delta_{\mathbf{s}}$, which shows that, for l satisfying one of the inequalities

$$2^{lg(\mathbf{r})/r_j + 1} \leq 2^{s_j - 1}, \qquad j = 1, \dots, d, \tag{6.9.34}$$

we have

$$\Delta_{\mathbf{s}}(A(f, \mathbf{r}, l)) = 0. \tag{6.9.35}$$

Let $m > 2d$. It is clear that if $l \leq l_m := m - 2d$ then, for $\mathbf{s}$ such that $\|\mathbf{s}\|_1 = m$, the relation (6.9.34) holds for some j and, consequently, (6.9.35) also holds. Therefore, it suffices to carry out the summation with respect to l in (6.9.33) for $l \geq l_m$. For these l the number of nonzero summands in the sum

$$\sum_{\|\mathbf{s}\|_1 = m} \|\Delta_{\mathbf{s}}(A(f, \mathbf{r}, l))\|_p$$

does not exceed $(l - l_m + 1)^{d-1}$ in order. Using Lemma 3.6.3 and Theorem 4.4.6 for fixed $\mathbf{s}$, $\|\mathbf{s}\|_1 = m$, and $l > l_m$ we get

$$\|\Delta_{\mathbf{s}}(A(f, \mathbf{r}, l))\|_p \ll 2^{(l-m)/p - g(\mathbf{r})l}. \tag{6.9.36}$$

From (6.9.33) and (6.9.36) we find that

$$\begin{aligned} \sigma_m \leq \sum_{l \geq l_m} \sum_{\|\mathbf{s}\|_1 = m} \|\Delta_{\mathbf{s}}(A(f, \mathbf{r}, l))\|_p &\ll \sum_{l \geq l_m} (l - l_m + 1)^{d-1} 2^{(l-m)/p - g(\mathbf{r})l} \\ &\ll 2^{-g(\mathbf{r})m}. \end{aligned} \tag{6.9.37}$$

Relations (6.9.32) and (6.9.37) give, for $f \in H_p^{\mathbf{r}}$,

$$\|f - T_n(f)\|_p \ll 2^{-g(\mathbf{r})n},$$

which proves the theorem for the case $q = p$.

Let $p < q$; then $\|\varphi\|_p \le \|\varphi\|_q$ and the theorem follows from the case $q = p$.
Let $1 \le q < p \le \infty$. Then, by Theorem 3.4.11,

$$H_q^{\mathbf{r}} \subset H_p^{\mathbf{r}'}B, \qquad \mathbf{r}' = \big(1 - (1/q - 1/p)/g(\mathbf{r})\big)\mathbf{r}.$$

Applying the conclusion of the theorem already proved for $q = p$ to the class $H_p^{\mathbf{r}'}B$ we obtain the required estimate.

The theorem is now proved. □

The recovery operator T_n is a linear operator with the dimension D_n of its range satisfying the inequality $D_n \ll 2^n n^{d-1}$. Therefore, Theorem 5.4.4 on the universality with respect to the Kolmogorov widths shows that the operator T_n is universal, even in the sense of the Kolmogorov widths, in the following cases: $1 < q \le p \le 2$; $q = 1$, $1 < p \le 2$; $2 \le p \le q \le \infty$.

6.10 Historical Notes, Comments, and Some Open Problems

First we will give a brief historical survey on discrepancy. For a complete survey we refer the reader to the following books on discrepancy and related topics: Kuipers and Niederreiter (1974), Beck and Chen (1987), Matoušek (1999), and Chazelle (2000). We will formulate all results in the notation of this book and in a form convenient for us. As above, we use the following notation for discrepancy:

$$D(\xi,m,d)_q := \left(\int_{\Omega_d}\left|\prod_{j=1}^{d} a_j - \frac{1}{m}\sum_{\mu=1}^{m}\chi_{[\mathbf{0},\mathbf{a}]}(\xi^\mu)\right|^q d\mathbf{a}\right)^{1/q}, \quad 1 \le q < \infty,$$

$$D(\xi,m,d)_\infty := \max_{\mathbf{a}\in[0,1]^d}\left|\prod_{j=1}^{d} a_j - \frac{1}{m}\sum_{\mu=1}^{m}\chi_{[\mathbf{0},\mathbf{a}]}(\xi^\mu)\right|,$$

where $\xi = (\xi^1,\dots,\xi^m)$. The first result in this area was a conjecture of van der Corput 1935a, b. Let $\xi^j \in [0,1]$, $j = 1,2,\dots$; then we have

$$\limsup_{m\to\infty} mD((\xi^1,\dots,\xi^m),m,1)_\infty = \infty.$$

The above conjecture was proved by van Aardenne-Ehrenfest (1945); who obtained

$$\limsup_{m\to\infty}\frac{\log\log\log m}{\log\log m} mD((\xi^1,\dots,\xi^m),m,1)_\infty > 0.$$

Let us denote

$$D(m,d)_q := \inf_{\xi} D(\xi,m,d)_q, \qquad 1 \le q \le \infty.$$

In 1954 Roth proved that

$$D(m,d)_2 \ge C(d)m^{-1}(\log m)^{(d-1)/2}. \tag{6.10.1}$$

Schmidt (1972) proved that

$$D(m,2)_\infty \geq Cm^{-1}\log m. \tag{6.10.2}$$

The same author found, in Schmidt (1977a), that

$$D(m,d)_q \geq C(d,q)m^{-1}(\log m)^{(d-1)/2}, \qquad 1 < q \leq \infty. \tag{6.10.3}$$

Furthermore, in 1981 Halász obtained

$$D(m,d)_1 \geq C(d)m^{-1}(\log m)^{1/2}. \tag{6.10.4}$$

The following conjecture was formulated in Beck and Chen (1987) as an "excruciatingly difficult" and important open problem.

Conjecture 6.10.1 *We have, for $d \geq 3$,*

$$D(m,d)_\infty \geq C(d)m^{-1}(\log m)^{d-1}.$$

This problem is still open. Bilyk and Lacey (2008) and Bilyk *et al.* (2008) proved that

$$D(m,d)_\infty \geq C(d)m^{-1}(\log m)^{(d-1)/2+\delta(d)},$$

for some positive $\delta(d)$.

We now present results on the lower estimates for the r-discrepancy. We denote

$$D_r(m,d)_q := \inf_\xi D_r(\xi,(1/m,\dots,1/m),m,d)_q,$$

where $D_r(\xi,\Lambda,m,d)_q$ is defined in (6.2.14), and we also denote the optimal discrepancy by

$$D_r^o(m,d)_q := \inf_{\xi,\Lambda} D_r(\xi,\Lambda,m,d)_q.$$

It is clear that

$$D_r^o(m,d)_q \leq D_r(m,d)_q.$$

The first result in estimating the generalized discrepancy was obtained in 1985 by Bykovskii:

$$D_r^o(m,d)_2 \geq C(r,d)m^{-r}(\log m)^{(d-1)/2}. \tag{6.10.5}$$

This result is a generalization of Roth's result (6.10.1). The generalization of Schmidt's result (6.10.3) was obtained by Temlyakov (1990b, see Theorem 6.4.11 above):

$$D_r^o(m,d)_q \geq C(r,d,q)m^{-r}(\log m)^{(d-1)/2}, \qquad 1 < q \leq \infty. \tag{6.10.6}$$

In 1994, Temlyakov (1994) proved that for even integers r we have for the r-discrepancy (see Theorem 6.4.12 and Corollary 6.4.13)

$$D_r(m,d)_\infty \geq C(r,d)m^{-r}(\log m)^{d-1}. \tag{6.10.7}$$

This result encourages us to formulate the following generalization of Conjecture 6.10.1.

Conjecture 6.10.2 *For all $d, r \in \mathbb{N}$ we have*

$$D_r^o(m,d)_\infty \geq C(r,d)m^{-r}(\log m)^{d-1}.$$

The above lower estimates are formally stronger for $D_1^o(m,d)_q$ than the corresponding estimates for $D(m,d)_q$ because in $D_1^o(m,d)_q$ we are optimizing over the weights Λ. However, the proofs for $D(m,d)_q$ could be adjusted to give the estimates for $D_1^o(m,d)_q$. The results (6.10.5)–(6.10.7) for the generalized discrepancy were obtained as a corollary of the corresponding results on cubature formulas (see Theorems 6.2.4, 6.4.11 and 6.4.12). We do not know whether existing methods for $D(m,d)_q$ could be modified to get the estimates for $D_r^o(m,d)_q$, $r \geq 2$.

We now proceed to consider the lower estimates for cubature formulas. Theorem 6.4.1 and Lemma 6.4.2 were established in Bykovskii (1985). Here the proofs were taken from Temlyakov (1993b). Theorem 6.4.3 and Lemma 6.4.5 were proved in Temlyakov (1990b). Theorems 6.4.7–6.4.9 are from Temlyakov (2015a). Theorems 6.4.10–6.4.12 were proved in Temlyakov (1994). Theorem 6.4.14 is from Temlyakov (1991b). There are two big open problems in this area. We formulate them as conjectures.

Conjecture 6.10.3 *For any $d \geq 2$ and any $r \geq 1$ we have*

$$\kappa_m(\mathbf{W}_1^r) \geq C(r,d)m^{-r}(\log m)^{d-1}.$$

Conjecture 6.10.4 *For any $d \geq 2$ and any $r > 0$ we have*

$$\kappa_m(\mathbf{W}_\infty^r) \geq C(r,d)m^{-r}(\log m)^{(d-1)/2}.$$

We note that by Proposition 6.2.2, Theorem 6.2.4 and (6.2.14), Conjecture 6.10.3 implies Conjecture 6.10.2, and Conjecture 6.10.4 implies that

$$D_r^o(m,d)_1 \geq C(r,d)m^{-r}(\log m)^{(d-1)/2}. \tag{6.10.8}$$

We also note that, in the case $d=2$ and $r=1$, Conjecture 6.10.3 holds. This follows from the Schmidt result (6.10.2) and Proposition 6.2.2.

We turn to the upper estimates and begin with the cubature formulas. Here, we will discuss only classes of functions with mixed smoothness. For results on cubature formulas for Sobolev-type classes we refer the reader to the books Sobolev (1994), Novak (1988), and Temlyakov (1993b, Chapter 2), and Chapter 3 above.

The first result in this direction was obtained by Korobov in 1959 (see also his book Korobov, 1963). He used the cubature formulas $P_m(f,\mathbf{a})$ defined in §6.6. We note that similar cubature formulas were also used by Hlawka (1962). Korobov's results lead to the following estimate:

$$\kappa_m(\mathbf{W}_1^r) \ll \kappa_m(\mathbf{H}_1^r) \leq C(r,d)m^{-r}(\log m)^{rd}, \qquad r>1. \tag{6.10.9}$$

In 1959 Bakhvalov improved (6.10.9) to

$$\kappa_m(\mathbf{H}_1^r) \leq C(r,d)m^{-r}(\log m)^{r(d-1)}, \qquad r>1.$$

The results on the Korobov cubature formulas found in Lemma 6.6.1 and Theorems 6.6.2 and 6.6.4 are from Temlyakov (1986a). Theorem 6.6.4 addresses the case of small smoothness: $r \in (\max(1/q,1/2),1]$ instead of $r>1$. For Theorem 6.6.5 see Temlyakov (1993b), Chapter 4.

The first best-possible upper estimate for the classes $\mathbf{W}_p^r$ was obtained in Bakhvalov (1963b). He proved in the case $d=2$ that

$$\kappa_m(\mathbf{W}_2^r) \leq C(r)m^{-r}(\log m)^{1/2}, \qquad r \in \mathbb{N}. \tag{6.10.10}$$

Bakhvalov used the Fibonacci cubature formulas defined in §6.5. The error bound (6.10.10) was proved for real $r>1/2$ in Temlyakov (1989a). The results on the Fibonacci cubature formulas in §6.5 are from Temlyakov (1991b) and Temlyakov (1994).

In 1976 Frolov used the cubature formulas defined in §6.7 to extend (6.10.10) to the case $d>2$:

$$\kappa_m(\mathbf{W}_2^r) \leq C(r,d)m^{-r}(\log m)^{(d-1)/2}, \qquad r \in \mathbb{N}. \tag{6.10.11}$$

In 1985 this estimate was further generalized by Bykovskii to $r \in \mathbb{R}$, $r \geq 1$. Bykovskii also used the Frolov cubature formulas. One can find these results in §6.7 above. Lemma 6.7.1 is a well-known result in algebraic number theory (see Cassels, 1971). Theorem 6.7.9 was obtained in Frolov (1979). The proof given here was taken from Temlyakov (1993b).

The Frolov cubature formulas (Frolov, 1979) give the following estimate (see, for instance, Theorem 6.7.13):

$$\kappa_m(\mathbf{W}_1^r) \leq C(r,d)m^{-r}(\log m)^{d-1}, \qquad r>1. \tag{6.10.12}$$

Thus the lower estimate in Conjecture 6.10.3 is the best possible.

In 1994 Skriganov proved the following estimate:

$$\kappa_m(\mathbf{W}_p^r) \leq C(r,d,p)m^{-r}(\log m)^{(d-1)/2}, \qquad 1<p\leq\infty, \qquad r \in \mathbb{N}. \tag{6.10.13}$$

This estimate combined with Theorem 6.4.3 implies that

$$\kappa_m(\mathbf{W}_p^r) \asymp m^{-r}(\log m)^{(d-1)/2}, \qquad 1<p<\infty, \qquad r \in \mathbb{N}. \tag{6.10.14}$$

Other proofs of (6.10.13) and Theorem 6.4.3 were given in 1995 by Bykovskii.

We now present upper estimates for the discrepancy. In 1956 Davenport proved that

$$D(m,2)_2 \leq Cm^{-1}(\log m)^{1/2}.$$

Other proofs of this estimate were later given by Vilenkin (1967), Halton and Zaremba (1969), and Roth (1976). In 1979 Roth proved that

$$D(m,3)_2 \leq Cm^{-1}\log m$$

and in 1980 Roth and also Frolov proved that

$$D(m,d)_2 \leq C(d)m^{-1}(\log m)^{(d-1)/2}.$$

Chen (1980) obtained

$$D(m,d)_q \leq C(d)m^{-1}(\log m)^{(d-1)/2}, \qquad q < \infty.$$

The estimate (6.10.12) and Theorem 6.2.4 imply

$$D_r^o(m,d)_\infty \leq C(r,d)m^{-r}(\log m)^{d-1}, \qquad r \geq 2.$$

We note that the upper estimates for $D(m,d)_q$ are stronger than the same upper estimates for $D_1^o(m,d)_q$.

Let us now make some remarks on the **H** classes. The lower bound in Theorem 6.4.8 in the case $p = \infty$ and $\theta = \infty$ was obtained by Bakhvalov (1972). In the case $p = \infty$ and $\theta = 2$ it was proved in Temlyakov (1990b). It was pointed out in Temlyakov (2015a) that the proof from Temlyakov (1990b) works for all θ and provides the lower bound in Theorem 6.4.8. Another proof is given in Dinh Dung and Ullrich (2014). For Theorem 6.5.8 see the book Temlyakov (1993b), Chapter 4. Lemma 6.6.1, Theorem 6.6.2, and Theorem 6.6.4 are from Temlyakov (1986a) (see also Temlyakov, 1985a, c). For Theorem 6.6.5 see the book Temlyakov (1993b), Chapter 4. The upper estimates in Theorem 6.7.13 were proved by Dubinin (1992).

We also mention the classic book of Nikol'skii (1979) on quadrature formulas and the books by Schmidt (1977b) and Hua Loo Keng and Wang Yuan (1981) on discrepancy and related topics.

The results on the universal cubature formulas in §6.8 are from Temlyakov (1989a, 1990b, 1991a).

Theorem 6.9.2 was proved in Temlyakov (1985b). Theorem 6.9.4 is a direct corollary of Theorem 6.9.2 together with a version (see Remark 3.3.10) of the fundamental inequality from Lemma 3.3.7 (see Dinh Zung, 1991, and Temlyakov, 1993b). Theorems 6.9.5 and 6.9.6 are from Temlyakov (1993a). Theorems 6.9.7 and 6.9.8 are from Temlyakov (2015a). Theorem 6.9.9 on the universality of the recovery operator T_n is from Temlyakov (1989b). We refer the reader for further discussion of recent results on recovery to the survey Dinh Dung *et al.* (2016).

The results in §6.2 are from Temlyakov (2003b). For the results in §6.3 see Temlyakov (2003b, 2011) and Temlyakov (2016a).

6.11 Open Problems

In addition to the four conjectures formulated above we list here some more open problems.

Open Problem 6.1 Find the correct order of $\kappa_m(\mathbf{W}^r_{p,\alpha})$ in the case of small smoothness, $r \in (1/p, 1/2]$, $2 < p < \infty$.

Open Problem 6.2 Let $d \geq 3$. For which r and p are the Korobov cubature formulas optimal in the sense of order for the classes $\mathbf{W}^r_{p,\alpha}$?

Open Problem 6.3 Let $d \geq 3$. For which r and p are the Korobov cubature formulas optimal in the sense of order for the classes $\mathbf{H}^r_p$?

7
Entropy

7.1 Introduction. Definitions and Some Simple Properties

The concept of entropy is also known as the *Kolmogorov entropy* or the *metric entropy*. The entropy concept allows us to measure the size of a compact set. In the case of finite-dimensional compact sets it is convenient to compare them by their volumes. For infinite-dimensional Banach spaces this does not work and in this case the entropy concept is a good replacement for that of volume. We present here some classical basic results and some recently developed techniques. In a certain sense this chapter complements Chapter 3 of Temlyakov (2011). However, the reader does not need to be familiar with the latter in order to understand the present chapter, in which we emphasise the new techniques based on m-term approximation.

We discuss a technique for proving upper bounds which is based on the following two-step strategy. At the first step we obtain bounds for the best m-term approximation with respect to a dictionary. At the second step we use general inequalities relating the entropy numbers to the best m-term approximations. We prove the corresponding general inequalities in §7.4. This technique can be applied to study classes of functions with mixed smoothness. For the lower bounds we use the volume-estimate method, which is a well-known and powerful method for proving the lower bounds for entropy numbers. Estimating the volumes of sets related to the hyperbolic cross trigonometric polynomials is highly nontrivial. We present the corresponding results in §7.5.

Let X be a Banach space and let B_X denote the unit ball of X whose center is at 0. Denote by $B_X(y,r)$ a ball with center y and radius r: $\{x \in X : \|x-y\| \le r\}$. For a compact set A and a positive number ε we define the *covering number* $N_\varepsilon(A)$ as follows:

$$N_\varepsilon(A) := N_\varepsilon(A,X) := \min\left\{ n : \exists y^1,\dots,y^n : A \subseteq \bigcup_{j=1}^{n} B_X(y^j,\varepsilon) \right\}.$$

Let us list three obvious properties of covering numbers:

$$N_\varepsilon(A) \le N_\varepsilon(B), \qquad \text{provided } A \subseteq B; \tag{7.1.1}$$

$$N_{\varepsilon_1\varepsilon_2}(A,X) \le N_{\varepsilon_1}(A,X)N_{\varepsilon_2}(B_X,X); \tag{7.1.2}$$

for $C = A \oplus B := \{c : c = a+b, a \in A, b \in B\}$,

$$N_{\varepsilon_1+\varepsilon_2}(C) \le N_{\varepsilon_1}(A)N_{\varepsilon_2}(B). \tag{7.1.3}$$

For a compact A we define an ε-distinguishable set $\{x^1,\dots,x^m\} \subseteq A$ as a set with the property

$$\|x^i - x^j\| > \varepsilon, \qquad \text{for all } i,j : i \ne j.$$

Denote by $M_\varepsilon(A) := M_\varepsilon(A,X)$ the maximal cardinality of the ε-distinguishable sets of a compact set A.

Theorem 7.1.1 *For any compact set A we have*

$$M_{2\varepsilon}(A) \le N_\varepsilon(A) \le M_\varepsilon(A). \tag{7.1.4}$$

Proof We first prove the second inequality. Let an ε-distinguishable set F realize $M_\varepsilon(A)$, i.e.,

$$F = \{x^1,\dots,x^{M_\varepsilon(A)}\}.$$

By the definition of $M_\varepsilon(A)$ as the maximal cardinality of ε-distinguishable sets of a compact set A, we obtain for any $x \in A$ an index $j := j(x) \in [1, M_\varepsilon(A)]$ such that $\|x - x^j\| \le \varepsilon$. Thus we have

$$A \subseteq \bigcup_{j=1}^{M_\varepsilon(A)} B_X(x^j,\varepsilon),$$

and the inequality $N_\varepsilon(A) \le M_\varepsilon(A)$ follows.

A proof by contradiction gives the first inequality in (7.1.4). Let $\{y^1,\dots,y^{N_\varepsilon(A)}\}$ be a set such that

$$A \subseteq \bigcup_{j=1}^{N_\varepsilon(A)} B_X(y^j,\varepsilon).$$

Assume that $M_{2\varepsilon}(A) > N_\varepsilon(A)$. Then the corresponding 2ε-distinguishable set F contains two points that are in the same ball $B_X(y^j,\varepsilon)$, for some $j \in [1, N_\varepsilon(A)]$. This clearly amounts to a contradiction. □

Proposition 7.1.2 *Let $A \subset Y$, and let Y be a subspace of X. Then*

$$N_\varepsilon(A,X) \ge N_{2\varepsilon}(A,Y).$$

Indeed, by Theorem 7.1.1 we have

$$N_{2\varepsilon}(A,Y) \le M_{2\varepsilon}(A,Y) = M_{2\varepsilon}(A,X) \le N_\varepsilon(A,X).$$

It is convenient to consider, along with the entropy $H_\varepsilon(A,X) := \log_2 N_\varepsilon(A,X)$, the *entropy numbers* $\varepsilon_k(A,X)$:

$$\varepsilon_k(A,X) := \inf\left\{\varepsilon : \exists y^1,\dots,y^{2^k} \in X : A \subseteq \bigcup_{j=1}^{2^k} B_X(y^j,\varepsilon)\right\}.$$

Properties (7.1.1)–(7.1.3) give the following inequalities for the entropy numbers:

$$\varepsilon_k(A,X) \le \varepsilon_k(B,X), \qquad \text{provided } A \subseteq B; \tag{7.1.5}$$

$$\varepsilon_{k+m}(A,X) \le \varepsilon_k(A,X)\varepsilon_m(B_X,X); \tag{7.1.6}$$

for $C = A \oplus B := \{c : c = a+b, a \in A, b \in B\}$,

$$\varepsilon_{k+m}(C,X) \le \varepsilon_k(A,X) + \varepsilon_m(B,X). \tag{7.1.7}$$

7.2 Finite-Dimensional Spaces. Volume Estimates

Let us consider the space $\mathbb{R}^n$ equipped with different norms, say $\|\cdot\|_X$ and $\|\cdot\|_Y$. For a Lebesgue measurable set $E \in \mathbb{R}^n$ we denote its Lebesgue measure by $\mathrm{vol}(E) := \mathrm{vol}_n(E)$.

Theorem 7.2.1 *For any two norms X and Y and any $\varepsilon > 0$ we have*

$$\frac{1}{\varepsilon^n}\frac{\mathrm{vol}(B_Y)}{\mathrm{vol}(B_X)} \le N_\varepsilon(B_Y,X) \le \frac{\mathrm{vol}(B_Y(0,2/\varepsilon)\oplus B_X)}{\mathrm{vol}(B_X)}. \tag{7.2.1}$$

Proof We begin with the first inequality in (7.2.1). We have

$$B_Y \subseteq \bigcup_{j=1}^{N_\varepsilon(B_Y,X)} B_X(y^j,\varepsilon)$$

and, therefore,

$$\mathrm{vol}(B_Y) \le \sum_{j=1}^{N_\varepsilon(B_Y,X)} \mathrm{vol}(B_X(y^j,\varepsilon)) \le N_\varepsilon(B_Y,X)\varepsilon^n \mathrm{vol}(B_X).$$

This gives the required inequality.

We proceed to the second inequality in (7.2.1). Let $\{x^1,\dots,x^{M_\varepsilon}\}$, $M_\varepsilon := M_\varepsilon(A)$, be an ε-distinguishable set of B_Y. Consider the set

$$C := \bigcup_{j=1}^{M_\varepsilon} B_X(x^j,\varepsilon/2).$$

Note that the balls $B_X(x^j, \varepsilon/2)$ are disjoint. Then

$$C \subseteq B_Y \oplus B_X(0,\varepsilon/2) \qquad \text{and} \qquad M_\varepsilon(\varepsilon/2)^n \operatorname{vol}(B_X) \le \operatorname{vol}(B_Y \oplus B_X(0,\varepsilon/2))$$

and the second inequality in (7.2.1), with N_ε replaced by M_ε, follows. It remains to use Theorem 7.1.1. □

Let us formulate one immediate corollary of Theorem 7.2.1.

Corollary 7.2.2 *For any n-dimensional Banach space X we have*

$$\varepsilon^{-n} \le N_\varepsilon(B_X, X) \le (1+2/\varepsilon)^n,$$

and, therefore,

$$\varepsilon_k(B_X, X) \le 3(2^{-k/n}).$$

Let us consider some typical n-dimensional Banach spaces. These are the spaces ℓ_p^n, i.e., the linear space $\mathbb{R}^n$ equipped with the norms

$$\|\mathbf{x}\|_p := \|\mathbf{x}\|_{\ell_p^n} := \left(\sum_{j=1}^n |x_j|^p\right)^{1/p}, \quad 1 \le p < \infty, \qquad \|\mathbf{x}\|_\infty := \|\mathbf{x}\|_{\ell_\infty^n} := \max_j |x_j|.$$

Denote $B_p^n := B_{\ell_p^n}$. It is obvious that

$$\operatorname{vol}(B_\infty^n) = 2^n.$$

It is also not difficult to see that

$$\operatorname{vol}(B_1^n) = 2^n/n!. \tag{7.2.2}$$

Indeed, consider for $r > 0$

$$O_n(r) := \left\{\mathbf{x} \in \mathbb{R}^n : x_j \ge 0, j = 1,\dots,n, \sum_{j=1}^n x_j \le r\right\}, \qquad O_n := O_n(1).$$

Then $\operatorname{vol}(B_1^n) = 2^n \operatorname{vol}(O_n)$ and

$$\begin{aligned}\operatorname{vol}(O_n) &= \int_0^1 \operatorname{vol}_{n-1}(O_{n-1}(1-t))dt = \operatorname{vol}_{n-1}(O_{n-1}) \int_0^1 (1-t)^{n-1} dt \\ &= \operatorname{vol}_{n-1}(O_{n-1})/n.\end{aligned}$$

Taking into account that $\operatorname{vol}_1(O_1) = 1$ we deduce that $\operatorname{vol}_n(O_n) = 1/n!$.

Let us proceed to the Euclidean case $p = 2$. We will prove the following estimates: there exist two positive absolute constants C_1 and C_2 such that

$$C_1^n n^{-n/2} \le \operatorname{vol}(B_2^n) \le C_2^n n^{-n/2}. \tag{7.2.3}$$

We have

$$\mathrm{vol}_n(B_2^n) = \int_{-1}^{1} \mathrm{vol}_{n-1}(B_2^{n-1}(0,(1-t^2)^{1/2}))dt$$
$$= 2\int_0^1 (1-t^2)^{(n-1)/2}\,\mathrm{vol}_{n-1}(B_2^{n-1})dt. \tag{7.2.4}$$

We will estimate the integrals $\int_0^1 (1-t^2)^{(n-1)/2}dt$ and prove that

$$C_3 n^{-1/2} \le \int_0^1 (1-t^2)^{(n-1)/2}dt \le C_4 n^{-1/2}. \tag{7.2.5}$$

It is clear that the identity (7.2.4) and the inequalities (7.2.5) and $(n/e)^n \le n! \le n^n$ imply (7.2.3). We begin by proving the first inequality in (7.2.5). We have

$$\int_0^1 (1-t^2)^{(n-1)/2}dt \ge \int_0^{n^{-1/2}} (1-t^2)^{(n-1)/2}dt \ge C_3 n^{-1/2}.$$

We now proceed to the second inequality in (7.2.5). Using the inequality $1-x \le e^{-x}$ we get

$$\int_0^1 (1-t^2)^{(n-1)/2}dt \le \int_0^1 e^{-t^2(n-1)/2}dt \le e^{1/2}\int_0^1 e^{-nt^2/2}dt \le e^{1/2}\int_0^\infty e^{-nt^2/2}dt$$
$$\le n^{-1/2}e^{1/2}\int_0^\infty e^{-y^2/2}dy;$$

here, we have made the substitution $t = n^{-1/2}y$. This completes the proof of (7.2.5) and (7.2.3).

7.3 Some Simple General Inequalities

We now proceed to two multiplicative inequalities for L_p-spaces. Let Ω be a domain in $\mathbb{R}^d$ and let $L_p := L_p(\Omega)$ denote the corresponding L_p-space, $1 \le p \le \infty$, with respect to the Lebesgue measure. We note that the inequalities below hold for any measure μ on Ω.

Theorem 7.3.1 *Let $A \subset L_1 \cap L_\infty$. Then, for any $1 \le p \le \infty$, we have $A \subset L_p$ and*

$$\varepsilon_{n+m}(A,L_p) \le 2\varepsilon_n(A,L_1)^{1/p}\varepsilon_m(A,L_\infty)^{1-1/p}.$$

Proof The simple inequality $\|f\|_p \le \|f\|_1^{1/p}\|f\|_\infty^{1-1/p}$ implies that $A \subset L_p$. Let a and b be any positive numbers satisfying $a > \varepsilon_n(A,L_1)$ and $b > \varepsilon_m(A,L_\infty)$. By the definition of the entropy number (see §7.1) there exist $g_1,\dots,g_{2^n}$ in L_1 and $h_1,\dots,h_{2^m}$ in L_∞ such that

$$A \subset \bigcup_{k=1}^{2^n} B_{L_1}(g_k,a), \qquad A \subset \bigcup_{l=1}^{2^m} B_{L_\infty}(h_l,b).$$

We now set $\varepsilon := 2a^{1/p}b^{1-1/p}$ and bound the $M_\varepsilon(A)$ from above. We want to prove that $M_\varepsilon(A) \le 2^{n+m}$. We take any set $f_1,\dots,f_N$ of elements of A with $N > 2^{n+m}$ and prove that for some i and j we have $\|f_i - f_j\|_p \le \varepsilon$. Indeed, the total number of sets $G_{k,l} := B_{L_1}(g_k,a) \cap B_{L_\infty}(h_l,b)$ is less than or equal to $2^{n+m} < N$. Therefore, there exist two indices i and j such that for some k and l we have $f_i \in G_{k,l}$ and $f_j \in G_{k,l}$. This means that

$$\|f_i - f_j\|_1 \le 2a, \qquad \|f_i - f_j\|_\infty \le 2b$$

and

$$\|f_i - f_j\|_p \le \|f_i - f_j\|_1^{1/p}\|f_i - f_j\|_\infty^{1-1/p} \le \varepsilon.$$

This implies $M_\varepsilon(A) \le 2^{n+m}$ and, by Theorem 7.1.1, $N_\varepsilon(A) \le 2^{n+m}$, which completes the proof. □

Remark 7.3.2 The above proof of Theorem 7.3.1 gives a more general result. Let $1 \le u \le p \le v \le \infty$ and let $A \subset L_u \cap L_v$. Then $A \subset L_p$ and

$$\varepsilon_{n+m}(A, L_p) \le 2\varepsilon_n(A, L_u)^\alpha \varepsilon_m(A, L_v)^{1-\alpha}$$

with $\alpha := (1/p - 1/v)(1/u - 1/v)^{-1}$.

Proof Instead of the inequality $\|f\|_p \le \|f\|_1^{1/p}\|f\|_\infty^{1-1/p}$, which we used in the proof of Theorem 7.3.1, we use the inequality $\|f\|_p \le \|f\|_u^\alpha \|f\|_v^{1-\alpha}$ (see (A.1.6)). □

It will be convenient for us to formulate one more inequality, in terms of the entropy numbers of operators. Let S be a linear operator from X to Y. We define the nth entropy number of S as

$$\varepsilon_n(S : X \to Y) := \varepsilon_n(S(B_X), Y),$$

where $S(B_X)$ is the image of B_X under the mapping S.

Theorem 7.3.3 *For any $1 \le u < q < p \le \infty$ and any Banach space Y we have*

$$\varepsilon_{n+m}(S : L_q \to Y) \le 2\varepsilon_n(S : L_u \to Y)^{1-\theta}\varepsilon_m(S : L_p \to Y)^\theta,$$

with $\theta := (1/u - 1/q)(1/u - 1/p)^{-1}$.

Proof We begin with a simple well-known lemma.

Lemma 7.3.4 *Let $1 \le u < q < p \le \infty$. For any $f \in L_q(\Omega)$, $\|f\|_q \le 1$, and any positive numbers a, b there exists a representation $f = f_1 + f_2$ such that*

$$a\|f_1\|_u \le a^{1-\theta}b^\theta, \qquad b\|f_2\|_p \le a^{1-\theta}b^\theta, \qquad \theta := (1/u - 1/q)(1/u - 1/p)^{-1}. \tag{7.3.1}$$

Proof Let f_T denote the T cutoff of f. Thus $f_T(x) = f(x)$ if $|f(x)| \le T$ and $f_T(x) = 0$ otherwise. Clearly, $\|f_T\|_\infty \le T$. Set $f^T := f - f_T$. We now estimate the L_u-norm of the f^T. Let $E := \{x : f^T(x) \ne 0\}$. First, we bound from above the measure $|E|$ of E. We have

$$1 \ge \int_\Omega |f(\mathbf{x})|^q d\mathbf{x} \ge \int_E T^q d\mathbf{x} = T^q |E|.$$

Second, we bound the $\|f^T\|_u$:

$$\|f^T\|_u^u = \int_E |f^T(\mathbf{x})|^u d\mathbf{x} \le \left(\int_E |f^T(\mathbf{x})|^q d\mathbf{x} \right)^{u/q} |E|^{1-u/q} \le T^{u-q}.$$

Third, we bound the $\|f_T\|_p$. Using the inequality

$$\|g\|_p \le \|g\|_q^{q/p} \|g\|_\infty^{1-q/p}$$

we obtain

$$\|f_T\|_p \le T^{1-q/p}.$$

Specifying $T = (a/b)^{(q(1/u-1/p))^{-1}}$ we get

$$a\|f^T\|_u \le a^{1-\theta} b^\theta, \qquad b\|f_T\|_p \le a^{1-\theta} b^\theta.$$

This proves the lemma. □

We now continue the proof of Theorem 7.3.3. Let a and b be such that

$$a > \varepsilon_n(S : L_u \to Y), \qquad b > \varepsilon_m(S : L_p \to Y).$$

Find $y_1, \dots, y_{2^n}$ and $z_1, \dots, z_{2^m}$ such that

$$S(B_{L_u}) \subset \bigcup_{k=1}^{2^n} B_Y(y_k, a), \qquad S(B_{L_p}) \subset \bigcup_{l=1}^{2^m} B_Y(z_l, b).$$

Take any $f \in L_q$, $\|f\|_q \le 1$. Set $\varepsilon := a^{1-\theta} b^\theta$ and, by Lemma 7.3.4, find f_1 and f_2 such that $f = f_1 + f_2$ and

$$a\|f_1\|_u \le \varepsilon, \qquad b\|f_2\|_p \le \varepsilon.$$

Clearly, for some k,

$$S(af_1/\varepsilon) \in B_Y(y_k, a) \quad \Rightarrow \quad S(f_1) \in B_Y(\varepsilon y_k/a, \varepsilon) \tag{7.3.2}$$

and, for some l,

$$S(bf_2/\varepsilon) \in B_Y(z_l, b) \quad \Rightarrow \quad S(f_2) \in B_Y(\varepsilon z_l/b, \varepsilon). \tag{7.3.3}$$

Consider the sets $G_{i,j} := B_Y(\epsilon y_i/a + \epsilon z_j/b, 2\epsilon)$, $i = 1,\dots,2^n$, $j = 1,\dots,2^m$. Relations (7.3.2) and (7.3.3) imply that $S(f) \in G_{k,l}$. Thus

$$\varepsilon_{n+m}(S : L_q \to Y) \le 2\varepsilon. \qquad \square$$

7.4 An Inequality Between Entropy Numbers and Best m-Term Approximations

There has been an increasing interest in the last few decades in nonlinear m-term approximation with regard to different systems. In Temlyakov (1998b) we generalized the concept of the classical Kolmogorov width in order to use it in estimating best m-term approximations. For this purpose we introduced the nonlinear Kolmogorov (N,m)-width:

$$d_m(F,X,N) := \inf_{\Lambda_N, \#\Lambda_N \le N} \sup_{f\in F} \inf_{L\in\Lambda_N} \inf_{g\in L} \|f-g\|_X,$$

where Λ_N is a set of at most N m-dimensional subspaces L. It is clear that

$$d_m(F,X,1) = d_m(F,X).$$

The new feature of $d_m(F,X,N)$ is that it allows us to choose a subspace $L \in \Lambda_N$ depending on $f \in F$. It is clear that the larger N is, the more flexibility we have in approximating f. The following theorem from Temlyakov (2013) plays a fundamental role in our new technique for proving the upper bounds for entropy numbers.

Theorem 7.4.1 *Let a compact $F \subset X$ and a number $r > 0$ be such that, for some $n \in \mathbb{N}$,*

$$d_{m-1}(F,X,(Kn/m)^m) \le m^{-r}, \qquad m \le n.$$

Then, for $k \le n$,

$$\varepsilon_k(F,X) \le C(r,K)\left(\frac{\log(2n/k)}{k}\right)^r.$$

Proof Let $X(N,m)$ denote the union of not more than N subspaces L, with $\dim L \le m$. Consider a collection $\mathscr{K}(l) := \{X((Kn2^{-s-1})^{2^{s+1}}, 2^{s+1})\}_{s=1}^{l}$, $2^{l+1} \le n$, and write

$$H^r(\mathscr{K}(l)) := \Big\{ f \in X : \exists L_1(f),\dots,L_l(f) : L_s(f) \in X((Kn2^{-s-1})^{2^{s+1}}, 2^{s+1}),$$
$$\text{and } \exists t_s(f) \in L_s(f) \text{ such that } \|t_s(f)\|_X \le 2^{-r(s-1)},$$
$$s = 1,\dots,l;\ \Big\|f - \textstyle\sum_{s=1}^{l} t_s(f)\Big\|_X \le 2^{-rl}\Big\}.$$

Lemma 7.4.2 *We have for $r > 0$*

$$\varepsilon_{2^l}(H^r(\mathscr{K}(l)),X) \le C(r,K)2^{-rl}(\log(Kn2^{-l}))^r, \qquad 2^{l+1} \le n.$$

Proof Clearly, it is sufficient to prove this lemma for large enough $l \geq c(r,K)$. We use Corollary 7.2.2 to estimate the entropy numbers $\varepsilon_n(B_X, X)$ for the unit ball B_X in a d-dimensional space X:

$$\varepsilon_n(B_X, X) \leq 3(2^{-n/d}). \tag{7.4.1}$$

Take any sequence $\{n_s\}_{s=1}^{l(r)}$ of $l(r) \leq l-2$ nonnegative integers. We will specify $l(r)$ later. Construct ε_{n_s}-nets each consisting of 2^{n_s} points for all unit balls of the spaces in $X((Kn2^{-s-1})^{2^{s+1}}, 2^{s+1})$. Then the total number of the elements y_j^s in these ε_{n_s}-nets does not exceed

$$M_s := (Kn2^{-s-1})^{2^{s+1}} 2^{n_s}.$$

We now consider the set A of elements of the form

$$y_{j_1}^1 + 2^{-r} y_{j_2}^2 + \cdots + 2^{-r(l(r)-1)} y_{j_{l(r)}}^{l(r)}, \qquad j_s \in [1, M_s], \qquad s = 1, \ldots, l(r).$$

The total number of these elements does not exceed

$$M = \prod_{s=1}^{l(r)} M_s, \qquad \log M \leq \sum_{s=1}^{l(r)} 2^{s+1} \log(Kn2^{-s-1}) + \sum_{s=1}^{l(r)} n_s.$$

It is easy to see that

$$\sum_{s=1}^{l(r)} 2^{s+1} \log(Kn2^{-s-1}) \leq C_1 2^{l(r)} \log(Kn2^{-l(r)}).$$

We now set

$$n_s := [(r+1)(l-s)2^{s+1}], \qquad s = 1, \ldots, l(r),$$

where as before $[x]$ denotes the integer part of a number x. We choose $l(r) \leq l-2$ as a maximal natural number satisfying

$$\sum_{s=1}^{l(r)} n_s \leq 2^{l-1}$$

(such a number exists if $l \geq c(r,K)$ is large enough) and

$$C_1 2^{l(r)} \log(Kn2^{-l(r)}) \leq 2^{l-1}.$$

It is clear that

$$2^{l(r)} \geq C_2 2^l (\log(Kn2^{-l}))^{-1}. \tag{7.4.2}$$

Then we have

$$M \leq 2^{2^l}.$$

For the error $\varepsilon(f)$ of approximation of $f \in H^r(\mathscr{K}(l))$ by elements of A we have

$$\varepsilon(f) \le 2^{-rl} + \sum_{s=1}^{l(r)} \left\| t_s(f) - 2^{-r(s-1)} y^s_{j_s} \right\|_X + \sum_{s=l(r)+1}^{l} \|t_s(f)\|_X$$

$$\le C(r)2^{-rl(r)} + \sum_{s=1}^{l(r)} 2^{-r(s-1)} \varepsilon_{n_s}(B_{L_s(f)}, X)$$

$$\le C(r)2^{-rl(r)} + 3\sum_{s=1}^{l(r)} 2^{-r(s-1)} 2^{-n_s/2^{s+1}} \le C'(r)2^{-rl(r)}.$$

Taking into account (7.4.2) completes the proof of the lemma. □

We continue the proof of Theorem 7.4.1. Without loss of generality assume that

$$\max_{1 \le m \le n} m^r d_{m-1}(F, X, (Kn/m)^m) < 1/2.$$

Then for $s = 1, 2, \dots, l$, $l \le [\log(n-1)]$, we have

$$d_{2^s}(F, X, (Kn2^{-s})^{2^s}) < 2^{-rs-1}.$$

This means that, for each $s = 1, 2, \dots, l$, there is a collection Λ_s of $(Kn2^{-s})^{2^s}$ 2^s-dimensional spaces L^s_j, $j = 1, \dots, [(Kn2^{-s})^{2^s}]$, such that for each $f \in F$ there exists a subspace $L^s_{j_s}(f)$ and an approximant $a_s(f) \in L^s_{j_s}(f)$ such that

$$\|f - a_s(f)\| \le 2^{-rs-1}.$$

Consider

$$t_s(f) := a_s(f) - a_{s-1}(f), \qquad s = 2, \dots, l. \tag{7.4.3}$$

Then we have

$$t_s(f) \in L^s_{j_s}(f) \oplus L^{s-1}_{j_{s-1}}(f), \qquad \dim\left(L^s_{j_s}(f) \oplus L^{s-1}_{j_{s-1}}(f)\right) \le 2^s + 2^{s-1} < 2^{s+1}.$$

Note that for K large enough (i.e., $K \ge 8$) we have

$$(Kn2^{-s})^{2^s}(Kn2^{-s+1})^{2^{s-1}} \le (Kn2^{-s-1})^{2^{s+1}}.$$

We denote by $X((Kn2^{-s-1})^{2^{s+1}}, 2^{s+1})$ the collection of all $L^s_{j_s} \oplus L^{s-1}_{j_{s-1}}$ over various $1 \le j_s \le (Kn2^{-s})^{2^s}$, $1 \le j_{s-1} \le (Kn2^{-s+1})^{2^{s-1}}$. For $t_s(f)$ defined by (7.4.3) we have

$$\|t_s(f)\| \le 2^{-rs-1} + 2^{-r(s-1)-1} \le 2^{-r(s-1)}.$$

Next, for $a_1(f) \in L^1(f)$ we have

$$\|f - a_1(f)\| \le 1/2,$$

and from $d_0(F,X) \le 1/2$ we get

$$\|a_1(f)\| \le 1.$$

Take $t_1(f) = a_1(f)$. Then we have $F \subset H^r(\mathscr{K}(l))$ and Lemma 7.4.2 gives the required bound,

$$\varepsilon_{2^l}(F) \le C(r,K)2^{-rl}(\log(Kn2^{-l}))^r, \qquad 1 \le l \le [\log(n-1)].$$

It is clear that these inequalities imply Theorem 7.4.1. □

Applications of Theorem 7.4.1 We begin with an application which motivated a study of $d_m(F,X,N)$ with $N = (Kn/m)^m$. Let $\mathscr{D} = \{g_j\}_{j=1}^n$ be a system of elements of cardinality $|\mathscr{D}| = n$ in a Banach space X. Consider the best m-term approximations of f with respect to $\mathscr{D}$:

$$\sigma_m(f,\mathscr{D})_X := \inf_{\{c_j\};\Lambda:|\Lambda|=m} \left\| f - \sum_{j\in\Lambda} c_j g_j \right\|.$$

For a function class F set

$$\sigma_m(F,\mathscr{D})_X := \sup_{f\in F} \sigma_m(f,\mathscr{D})_X.$$

Then it is clear that, for any system $\mathscr{D}$, $|\mathscr{D}| = n$,

$$d_m\left(F,X,\binom{n}{m}\right) \le \sigma_m(F,\mathscr{D})_X.$$

We also have

$$\binom{n}{m} \le (en/m)^m.$$

Thus Theorem 7.4.1 implies the following theorem.

Theorem 7.4.3 *Let a compact $F \subset X$ be such that there exists a system $\mathscr{D}$, $|\mathscr{D}| = n$, and a number $r > 0$ such that*

$$\sigma_m(F,\mathscr{D})_X \le m^{-r}, \qquad m \le n.$$

Then, for $k \le n$,

$$\varepsilon_k(F,X) \le C(r)\left(\frac{\log(2n/k)}{k}\right)^r. \tag{7.4.4}$$

Remark 7.4.4 Suppose that a compact F from Theorem 7.4.3 belongs to an n-dimensional subspace $X_n := \operatorname{span}(\mathscr{D})$. Then, in addition to (7.4.4), we have for $k \ge n$

$$\varepsilon_k(F,X) \le C(r)n^{-r}2^{-k/n}. \tag{7.4.5}$$

Proof Inequality (7.4.5) follows from Theorem 7.4.3 with $X = X_n$, $k = n$, inequality (7.4.1) and the simple well-known inequality (see (7.1.6))

$$\varepsilon_{k_1+k_2}(A, X_n) \leq \varepsilon_{k_1}(A, X_n)\varepsilon_{k_2}(B_{X_n}, X_n),$$

where A is compact and B_{X_n} is a unit ball of X_n. □

We point out that Remark 7.4.4 is formulated for a real Banach space X. In the case of a complex Banach space X we have $2^{-k/(2n)}$ instead of $2^{-k/n}$ in (7.4.5). We formulate this as a remark, which will be used later.

Remark 7.4.5 Suppose that a compact F from Theorem 7.4.3 belongs to an n-dimensional subspace $X_n := \operatorname{span}(\mathscr{D})$. Then, in addition to (7.4.4), we have for $k \geq n$

$$\varepsilon_k(F, X) \leq C(r)n^{-r}2^{-k/(2n)}.$$

As a corollary of Theorem 7.4.3 and Remark 7.4.4 we obtain the following classical bound.

Corollary 7.4.6 *For any* $0 < q \leq \infty$ *and* $\max(1, q) \leq p \leq \infty$ *we have*

$$\varepsilon_k(B^n_q, \ell^n_p) \leq C(q,p)\begin{cases} ((\log(2n/k))/(k))^{1/q-1/p}, & k \leq n, \\ 2^{-k/n}n^{1/p-1/q}, & k \geq n. \end{cases}$$

Proof Indeed, it is well known and easy to check (see Lemma 7.6.6 below) that, for a sequence of nonnegative numbers $x_1 \geq x_2 \geq \cdots \geq x_n$, we have for $0 < q \leq p$

$$\left(\sum_{j=m}^{n} x_j^p\right)^{1/p} \leq m^{1/p-1/q}\left(\sum_{j=1}^{n} x_j^q\right)^{1/q}. \tag{7.4.6}$$

Therefore, for $0 < q \leq p$,

$$\sigma_m(B^n_q, \{e_j\}_{j=1}^n)_{\ell^n_p} \leq m^{1/p-1/q}, \quad m \leq n,$$

where $\{e_j\}_{j=1}^n$ is a canonical basis for $\mathbb{R}^n$. Applying Theorem 7.4.3 and Remark 7.4.4 we obtain the corollary. □

For a normalized system $\mathscr{D}$, define $A_q(\mathscr{D})$, $q > 0$, as a closure in X of the set

$$\left\{x : x = \sum_j c_j g_j,\ g_j \in \mathscr{D},\ \sum_j |c_j|^q \leq 1\right\}.$$

Corollary 7.4.7 *Let* $1 < p < \infty$. *For a normalized system* $\mathscr{D}$ *of cardinality* $|\mathscr{D}| = n$ *we have*

$$\varepsilon_k(A_1(\mathscr{D}), L_p) \leq C(p)\left(\frac{\log(2n/k)}{k}\right)^{\max(1/2,1/p)-1}, \qquad k \leq n.$$

Proof It is known (see Donahue *et al.*, 1997, and Temlyakov, 2011) that

$$\sigma_m(A_1(\mathscr{D}),\mathscr{D})_{L_p} \leq C(p)m^{\max(1/2,1/p)-1}. \tag{7.4.7}$$

It remains to apply Theorem 7.4.3. □

Corollary 7.4.8 *Let $\mathscr{D}$ be a normalized system of cardinality $|\mathscr{D}| = n$. Then, for $0 < q \leq 1$ and $1 < p < \infty$, we have*

$$\varepsilon_k(A_q(\mathscr{D}),L_p) \leq C(q,p)\begin{cases} ((\log(2n/k))/(k))^{1/q-\max(1/2,1/p)}, & k \leq n, \\ 2^{-k/n}n^{\max(1/2,1/p)-1/q}, & k \geq n. \end{cases}$$

Proof Let us estimate $\sigma_m(A_q(\mathscr{D}),\mathscr{D})_{L_p}$. If $q = 1$ then the bound is given by (7.4.7). If $q < 1$ then we use (7.4.6) with $p = 1$ and, by (7.4.7), we get

$$\sigma_{2m}(A_q(\mathscr{D}),\mathscr{D})_{L_p} \leq C(q,p)m^{\max(1/2,1/p)-1/q}.$$

Applying Theorem 7.4.3 and Remark 7.4.4 we obtain the corollary. □

We note that Corollary 7.4.8 gives the same upper bounds as in Theorem 1 of Gao *et al.* (2013), where it was proved that these bounds are best possible up to a constant.

7.5 Volume Estimates for Balls of Trigonometric Polynomials

We now consider finite-dimensional subspaces of trigonometric polynomials. For a set $\Lambda \subset \mathbb{Z}^d$, denote

$$\mathscr{T}(\Lambda) := \{f \in L_1 : \hat{f}(\mathbf{k}) = 0, \mathbf{k} \in \mathbb{Z}^d \setminus \Lambda\}, \quad \mathscr{T}(\Lambda)_p := \{f \in \mathscr{T}(\Lambda) : \|f\|_p \leq 1\}.$$

For a finite set Λ we assign to each $f = \sum_{\mathbf{k}\in\Lambda} \hat{f}(\mathbf{k})e^{i(\mathbf{k},\mathbf{x})} \in \mathscr{T}(\Lambda)$ a vector

$$A(f) := \{(\mathrm{Re}(\hat{f}(\mathbf{k})), \mathrm{Im}(\hat{f}(\mathbf{k}))),\ \mathbf{k} \in \Lambda\} \in \mathbb{R}^{2|\Lambda|},$$

where $|\Lambda|$ denotes the cardinality of Λ, and we define

$$B_\Lambda(L_p) := \{A(f) : f \in \mathscr{T}(\Lambda)_p\}.$$

The volume estimates of the sets $B_\Lambda(L_p)$ and related questions have been studied in a number of papers: the case $\Lambda = [-n,n]$, $p = \infty$ in Kashin (1980); the case $\Lambda = [-N_1,N_1] \times \cdots \times [-N_d,N_d]$, $p = \infty$ in Temlyakov (1989d, 1992a). In the case $\Lambda = \Pi(\mathbf{N},d) := [-N_1,N_1] \times \cdots \times [-N_d,N_d]$, $\mathbf{N} := (N_1,\ldots,N_d)$, the following estimates are known.

Theorem 7.5.1 *For any $1 \leq p \leq \infty$ we have*

$$(\mathrm{vol}(B_{\Pi(\mathbf{N},d)}(L_p)))^{(2|\Pi(\mathbf{N},d)|)^{-1}} \asymp |\Pi(\mathbf{N},d)|^{-1/2},$$

with constants in $\asymp$ that may depend only on d.

Proof We note that the most difficult part of Theorem 7.5.1 is the lower estimate for $p = \infty$. The corresponding estimate was proved in the case $d = 1$ in Kashin (1980) and in the general case in Temlyakov (1989d, 1992a) (see Lemma 3.2.3). The upper estimate for $p = 1$ in Theorem 7.5.1 can be easily reduced to the volume estimate for an octahedron (see for instance Temlyakov, 1992b). Indeed, denote

$$P(\mathbf{N}) := \{\mathbf{l} = \{l_1,\dots,l_d\} : \; l_j \in \mathbb{Z}_+, \; 0 \le l_j \le 2N_j, \; j = 1,\dots,d\},$$

$$\mathbf{x}^{\mathbf{l}} := \left(\frac{2\pi l_1}{2N_1+1},\dots,\frac{2\pi l_d}{2N_d+1}\right), \qquad \mathbf{l} \in P(\mathbf{N}).$$

With a polynomial $t \in \mathscr{T}(\mathbf{N},d) := \mathscr{T}(\Pi(\mathbf{N},d))$ associate two vectors: $A(t) \in \mathbb{R}^D$, $D = \vartheta(\mathbf{N}) := \prod_{j=1}^d (2N_j+1)$, and

$$S(t) := \{(\mathrm{Re}(t(\mathbf{x}^{\mathbf{l}})), \mathrm{Im}(t(\mathbf{x}^{\mathbf{l}})), \qquad \mathbf{l} \in P(\mathbf{N})\}.$$

Consider a linear operator mapping $S(t)$ into $A(t)$. It follows easily from the identity (3.2.5) that

$$\|S(t)\|_2 = \vartheta(\mathbf{N})^{1/2}\|t\|_2 = \vartheta(\mathbf{N})^{1/2}\|A(t)\|_2.$$

Therefore, this linear operator reduces the length of the vector $\vartheta(\mathbf{N})^{1/2}$ times and reduces the volume $\vartheta(\mathbf{N})^{D/2}$ times. Further, by Remark 1.3.8, for any $t \in \mathscr{T}(\mathbf{N},d)$ we have

$$\sum_{\mathbf{l} \in P(\mathbf{N})} |t(\mathbf{x}^{\mathbf{l}})| \le C(d)\vartheta(\mathbf{N})\|t\|_1.$$

This implies that

$$\mathrm{vol}(B_{\Pi(\mathbf{N},d)}(L_1)) \le (2C(d)\vartheta(\mathbf{N}))^{\vartheta(\mathbf{N})}\vartheta(\mathbf{N})^{-\vartheta(\mathbf{N})/2}\mathrm{vol}(B_1^{\vartheta(\mathbf{N})}). \tag{7.5.1}$$

Relation (7.2.2) for $\mathrm{vol}(B_1^D)$ and relation (7.5.1) imply the upper bound in the theorem. □

A slight modification in the above proof allows us to obtain the following general result for any parallelepiped. Let $\mathbf{a} = (a_1,\dots,a_d)$ and $\mathbf{b} = (b_1,\dots,b_d)$. In the case $\Lambda = \Pi(\mathbf{a},\mathbf{b},d) := [\mathbf{a},\mathbf{b}] \cap \mathbb{Z}^d := [a_1,b_1] \times \cdots \times [a_d,b_d] \cap \mathbb{Z}^d$ the following estimates hold.

Theorem 7.5.2 *For any* $1 \le p \le \infty$ *we have*

$$(\mathrm{vol}(B_{\Pi(\mathbf{a},\mathbf{b},d)}(L_p)))^{(2|\Pi(\mathbf{a},\mathbf{b},d)|)^{-1}} \asymp |\Pi(\mathbf{a},\mathbf{b},d)|^{-1/2},$$

with constants in $\asymp$ *that may depend only on* d.

The case of arbitrary Λ and $p = 1$ was studied in Kashin and Temlyakov (1994). The results there imply the following estimate.

Theorem 7.5.3 *For any finite set $\Lambda \subset \mathbb{Z}^d$ and any $1 \le p \le 2$ we have*

$$\mathrm{vol}(B_\Lambda(L_p))^{(2|\Lambda|)^{-1}} \asymp |\Lambda|^{-1/2}.$$

Proof We begin with the following lemma.

Lemma 7.5.4 *The unit ball $B_2^{2|\Lambda|}$ is an ellipsoid of maximal volume lying in $B_\Lambda(L_p)$, $1 \le p \le 2$.*

Proof First, it is clear from the obvious relations

$$\|t\|_1 \le \|t\|_p \le \|t\|_2, \qquad \|t\|_2 = \|A(t)\|_2 := \|A(t)\|_{\ell_2^D}, \quad D = 2|\Lambda|, \quad t \in \mathscr{T}(\Lambda),$$

that $B_2^D \subseteq B_\Lambda(L_p)$, $1 \le p \le 2$, and so it suffices to prove the lemma for $p = 1$. Second, the proof is based on the following well-known classical result from the theory of convex bodies.

Theorem 7.5.5 *Let B be a centrally symmetric convex body in $\mathbb{R}^D$. Then there exists a unique ellipsoid of maximal volume contained in B.*

Let

$$\mathscr{E} := \mathscr{E}_\Lambda := \left\{ \mathbf{y} \in \mathbb{R}^D : \textstyle\sum_{j=1}^D (\mathbf{y}, \mathbf{h}_j)^2/\mu_j^2 \le 1 \right\}$$

be the ellipsoid of maximal volume inscribed in $B_\Lambda(L_1)$, let $\{\mathbf{h}_j\}$, $\|\mathbf{h}_j\|_2 = 1$, $j = 1,\dots,D$, be the directions of the semi-axes of $\mathscr{E}_\Lambda$, and let μ_j, $j = 1,\dots,D$, be their lengths. We define the operator

$$J : \mathbb{R}^D \to \mathbb{R}^D$$

by

$$J(\{a_\mathbf{k}, b_\mathbf{k}\}_{\mathbf{k}\in\Lambda}) = (\{-b_\mathbf{k}, a_\mathbf{k}\}_{\mathbf{k}\in\Lambda}).$$

It is clear that J is an orthogonal operator in $\mathbb{R}^D$, $J^2 = -Id$, and, for $t \in \mathscr{T}(\Lambda)$,

$$J(A(t)) = A(it), \qquad \|it\|_1 = \|t\|_1.$$

By the last relation and Theorem 7.5.5, we have

$$J(\mathscr{E}_\Lambda) = \mathscr{E}_\Lambda.$$

Moreover, $(\mathbf{y}, J(\mathbf{y})) = 0$ for any $\mathbf{y} \in \mathbb{R}^D$, and the two-dimensional subspace generated by the vectors $\mathbf{y}$ and $J(\mathbf{y})$ is J-invariant. Therefore, it is easy to see that the ellipsoid $\mathscr{E}_\Lambda$ can be written in the form

$$\mathscr{E}_\Lambda := \left\{ \mathbf{y} \in \mathbb{R}^D : \textstyle\sum_{j=1}^{D/2} ((\mathbf{y}, \mathbf{u}_j)^2 + (\mathbf{y}, J(\mathbf{u}_j))^2)/\lambda_j^2 \le 1 \right\},$$

where $\|\mathbf{u}_j\|_2 = \|J(\mathbf{u}_j\|_2 = 1$ for $j = 1,\ldots,D/2$. With each pair $\{\mathbf{u}_j, J(\mathbf{u}_j)\}$ of vectors we associate a polynomial $t_j \in \mathscr{T}(\Lambda)$ such that $\mathbf{u}_j = A(t_j)$. Then, by the orthogonality of the system $\{\mathbf{u}_j, J(\mathbf{u}_j)\}$ of semi-axes, we see that the polynomials t_j are orthogonal and

$$\mathscr{E}_\Lambda := \left\{ \mathbf{y} \in \mathbb{R}^D : \sum_{j=1}^{D/2} |\langle A^{-1}(\mathbf{y}), t_j\rangle|^2/\lambda_j^2 \le 1 \right\}. \tag{7.5.2}$$

Let $T_\mathbf{h}$ be an $\mathbf{h}$-translation operator, $T_\mathbf{h}(f(\mathbf{x})) = f(\mathbf{x}-\mathbf{h})$, and

$$F_\mathbf{h}(\mathbf{y}) := A(T_\mathbf{h}(A^{-1}(\mathbf{y}))), \qquad \mathbf{y} \in \mathbb{R}^D.$$

It is clear that $F_\mathbf{h}$ is an orthogonal operator in $\mathbb{R}^D$. Since the space $\mathscr{T}(\Lambda)$ and the norms $\|\cdot\|_1$ are translation invariant for any $\mathbf{h} \in \mathbb{R}^d$, we have

$$F_\mathbf{h}(\mathscr{E}_\Lambda) = \mathscr{E}_\Lambda, \qquad \mathbf{h} \in \mathbb{R}^d. \tag{7.5.3}$$

In turn, property (7.5.3) implies that:

TI For any λ the linear hull $E_\lambda := \operatorname{span}(\{t_j\}, j \in S_\lambda)$, where $S_\lambda := \{j : \lambda_j = \lambda\}$, is translation invariant. Note that the span is taken over the field of complex numbers.

We now need the following simple well-known lemma.

Lemma 7.5.6 *Let E be a finite-dimensional translation-invariant subspace of $L_1(\mathbb{T}^d)$. Then there exists a basis of exponents in E given by*

$$E = \operatorname{span}\{e^{i(\mathbf{k}^1,\mathbf{x})}, \ldots, e^{i(\mathbf{k}^s,\mathbf{x})}\}, \qquad s := \dim E. \tag{7.5.4}$$

Proof Let $f \in E$ be such that $f \neq 0$. Then there exists $\mathbf{k}^1$ such that $\hat{f}(\mathbf{k}^1) \neq 0$. By assumption E is a closed and translation invariant subspace. Therefore

$$\hat{f}(\mathbf{k}^1)e^{i(\mathbf{k}^1,\mathbf{x})} = (2\pi)^{-d}\int_{\mathbb{T}^d} f(\mathbf{x}-\mathbf{y})e^{i(\mathbf{k}^1,\mathbf{y})}d\mathbf{y} \in E.$$

Thus, we have found an exponent $e^{i(\mathbf{k}^1,\mathbf{x})}$ which belongs to E. Consider now

$$E^1 := \{f(\mathbf{x}) - \hat{f}(\mathbf{k}^1)e^{i(\mathbf{k}^1,\mathbf{x})}, f \in E\}$$

and repeat the above argument with E^1 instead of E. After s iterations we have found a basis of exponents for E. □

By (7.5.4) with $E = E_\lambda$ and (7.5.2) we conclude that

$$\mathscr{E}_\Lambda := \left\{ \mathbf{y} \in \mathbb{R}^D : \sum_{\lambda: E_\lambda \neq \emptyset} \sum_{\mathbf{k}: e^{i(\mathbf{k},\mathbf{x})} \in E_\lambda} |\langle A^{-1}(\mathbf{y}), e^{i(\mathbf{k},\mathbf{x})}\rangle|^2/\lambda_j^2 \le 1 \right\},$$

and therefore the ellipsoid $\mathscr{E}_\Lambda$ can be represented in the form

$$\mathscr{E}_\Lambda = \left\{ \mathbf{y} : \sum_{\mathbf{k}\in\Lambda} |\widehat{A^{-1}(\mathbf{y})}(\mathbf{k})|^2 / v_{\mathbf{k}}^2 \leq 1 \right\}.$$

In other words, $\| \sum_{\mathbf{k}\in\Lambda} c_{\mathbf{k}} e^{i(\mathbf{k},\mathbf{x})} \|_1 \leq 1$ for any numbers $c_{\mathbf{k}}$ such that

$$\sum_{\mathbf{k}\in\Lambda} |c_{\mathbf{k}}|^2 / v_{\mathbf{k}}^2 \leq 1.$$

Thus, $v_{\mathbf{k}}^2 \leq 1$ for any $\mathbf{k} \in \Lambda$ and, therefore, $\mathscr{E}_\Lambda \subseteq B_2^D$. This completes the proof of Lemma 7.5.4. □

We now continue the proof of Theorem 7.5.3. The lower bounds follow from Lemma 7.5.4 and relation (7.2.3) with $n = D$. The upper bounds follow from (7.2.3) with $n = D$ and Lemma 7.5.7 below.

Lemma 7.5.7 *There exists an absolute constant C such that, for any finite set $\Lambda \subset \mathbb{Z}^d$,*

$$\mathrm{vol}(B_\Lambda(L_1)) \leq C^D \mathrm{vol}(B_2^D), \qquad D = 2|\Lambda|.$$

Proof We need one more classical result from the theory of convex bodies. Let us recall the definition of the cotype constant. For a normed space X, a constant $C_2(X)$ of cotype 2 is called the least constant C such that, for any finite set $\{x_1, \ldots, x_m\} \subset X$,

$$C \int_0^1 \left\| \sum_{j=1}^m r_j(t) x_j \right\|_X dt \geq \left(\sum_{j=1}^m \|x_j\|_X^2 \right)^{1/2};$$

here $r_j(t)$, $j = 1, \ldots, m$, are Rademacher functions.

A constant of cotype 2 is well defined for any finite-dimensional space X and, what is important for us, for the spaces L_p, $1 \leq p \leq 2$. Moreover, by the Khinchin inequality we have

$$C_2(X) \leq 10, \qquad X = L_1(\mathbb{R}^d). \tag{7.5.5}$$

The following theorem is from Bourgain and Milman (1987).

Theorem 7.5.8 *Let X be a D-dimensional real normed space with unit ball B, and let $\mathscr{E}$ be an ellipsoid of maximal volume contained in B. Then*

$$\left(\frac{\mathrm{vol}(B)}{\mathrm{vol}(\mathscr{E})} \right)^{1/D} \leq K C_2(X) (\log C_2(X))^4,$$

with K an absolute constant.

Let X be a normed space for which $B_\Lambda(L_1)$ is the unit ball. Then, by (7.5.5), we have $C_2(X) \le 10$. Applying Theorem 7.5.8 and Lemma 7.5.4, we arrive at the conclusion of Lemma 7.5.7. □

This completes the proof of Theorem 7.5.3. □

Lemma 7.5.9 *Let $\Lambda \subseteq [-2^n, 2^n]^d$ and $D := 2|\Lambda|$. Then*

$$(\mathrm{vol}(B_\Lambda(L_\infty)))^{1/D} \ge C(d)(Dn)^{-1/2}.$$

Proof We will use the following result of Gluskin (1989).

Theorem 7.5.10 *Let $Y = \{\mathbf{y}_1, \dots, \mathbf{y}_M\} \subset \mathbb{R}^D$, $\|\mathbf{y}_i\| = 1$, $i = 1, \dots, M$, and*

$$W(Y) := \{\mathbf{x} \in \mathbb{R}^D : |(\mathbf{x}, \mathbf{y}_i)| \le 1,\ i = 1, \dots, M\}.$$

Then

$$(\mathrm{vol}(W(Y)))^{1/D} \ge C(1 + \log(M/D))^{-1/2}.$$

Consider the following lattice on the $\mathbb{T}^d$:

$$G_n := \left\{\mathbf{x}(l) = (l_1, \dots, l_d)\pi 2^{-n-1},\ 1 \le l_j \le 2^{n+2},\ l_j \in \mathbb{N},\ j = 1, \dots, d\right\}.$$

Clearly $|G_n| = 2^{d(n+2)}$. It is well known that for any $f \in \mathscr{T}([-2^n, 2^n]^d)$ one has

$$\|f\|_\infty \le C_1(d) \max_{\mathbf{x} \in G_n} |f(\mathbf{x})|.$$

Thus, for any $\Lambda \subseteq [-2^n, 2^n]^d$ we have

$$\left\{A(f) : f \in \mathscr{T}(\Lambda),\ |f(\mathbf{x})| \le C_1(d)^{-1},\ \mathbf{x} \in G_n\right\} \subseteq B_\Lambda(L_\infty). \tag{7.5.6}$$

Further,

$$\begin{aligned} |f(\mathbf{x})|^2 &= |\sum_{\mathbf{k} \in \Lambda} \hat{f}(\mathbf{k}) e^{i(\mathbf{k},\mathbf{x})}|^2 \\ &= \left(\sum_{\mathbf{k} \in \Lambda} \mathrm{Re}\hat{f}(\mathbf{k}) \cos(\mathbf{k},\mathbf{x}) - \mathrm{Im}\hat{f}(\mathbf{k}) \sin(\mathbf{k},\mathbf{x})\right)^2 \\ &\quad + \left(\sum_{\mathbf{k} \in \Lambda} \mathrm{Re}\hat{f}(\mathbf{k}) \sin(\mathbf{k},\mathbf{x}) + \mathrm{Im}\hat{f}(\mathbf{k}) \cos(\mathbf{k},\mathbf{x})\right)^2. \end{aligned}$$

With each point $\mathbf{x} \in G_n$ we associate two vectors $\mathbf{y}^1(\mathbf{x})$ and $\mathbf{y}^2(\mathbf{x})$ from $\mathbb{R}^D$:

$$\begin{aligned} \mathbf{y}^1(\mathbf{x}) &:= \{(\cos(\mathbf{k},\mathbf{x}), -\sin(\mathbf{k},\mathbf{x})),\ \mathbf{k} \in \Lambda\}, \\ \mathbf{y}^2(\mathbf{x}) &:= \{(\sin(\mathbf{k},\mathbf{x}), \cos(\mathbf{k},\mathbf{x})),\ \mathbf{k} \in \Lambda\}. \end{aligned}$$

Then

$$\|\mathbf{y}^1(\mathbf{x})\|^2 = \|\mathbf{y}^2(\mathbf{x})\|^2 = |\Lambda|$$

and

$$|f(\mathbf{x})|^2 = (A(f),\mathbf{y}^1(\mathbf{x}))^2 + (A(f),\mathbf{y}^2(\mathbf{x}))^2.$$

It is clear that the condition $|f(\mathbf{x})| \le C_1(d)^{-1}$ is satisfied if

$$|(A(f),\mathbf{y}^i(\mathbf{x}))| \le 2^{-1/2}C_1(d)^{-1}, \qquad i = 1,2.$$

Now let

$$Y := \{\mathbf{y}^i(\mathbf{x})/\|\mathbf{y}^i(\mathbf{x})\|,\ \mathbf{x} \in G_n,\ i = 1,2\}.$$

Then $M = 2^{d(n+2)+1}$ and, by Theorem 7.5.10,

$$(\mathrm{vol}(W(Y)))^{1/D} \gg (1+\log(M/D))^{-1/2} \gg n^{-1/2}. \tag{7.5.7}$$

Using the fact that the condition

$$|(A(f),\mathbf{y}^i(\mathbf{x}))| \le 1$$

is equivalent to the condition

$$\left|(A(f),\mathbf{y}^i(\mathbf{x})/\|\mathbf{y}^i(\mathbf{x})\|)\right| \le (D/2)^{-1/2},$$

we get from (7.5.6) and (7.5.7)

$$(\mathrm{vol}(B_\Lambda(L_\infty)))^{1/D} \gg (Dn)^{-1/2}.$$

This completes the proof of the lemma. □

We now present some results on a specific set Λ related to the hyperbolic crosses. Denote for a natural number n

$$Q_n := \bigcup_{\|\mathbf{s}\|_1 \le n} \rho(\mathbf{s}), \qquad \Delta Q_n := Q_n \setminus Q_{n-1} = \bigcup_{\|\mathbf{s}\|_1 = n} \rho(\mathbf{s}),$$

with $\|\mathbf{s}\|_1 = s_1 + \cdots + s_d$ for $\mathbf{s} \in \mathbb{Z}^d_+$. We call a set ΔQ_n a *hyperbolic layer*. As a direct corollary of Lemma 7.5.9 we obtain the following result.

Lemma 7.5.11 *Let Λ be either Q_n or ΔQ_n and $D := 2|\Lambda|$. Then*

$$(\mathrm{vol}(B_\Lambda(L_\infty)))^{1/D} \ge C(d)(|Q_n|n)^{-1/2} \asymp (2^n n^d)^{-1/2}.$$

We complement the lower bound in Lemma 7.5.11 by the corresponding upper bound. However, we have a corresponding bound only in dimension $d = 2$. We need some more notation. Let

$$E^\perp_\Lambda(f)_p := \inf_{g \perp \mathscr{T}(\Lambda)} \|f - g\|_p,$$

and

$$B^{\perp}_{\Lambda}(L_p) := \{A(f) : f \in \mathscr{T}(\Lambda),\ E^{\perp}_{\Lambda}(f)_p \leq 1\}.$$

Theorem 7.5.12 *In the case $d = 2$ we have with $D := 2|\Delta Q_n|$,*

$$(\mathrm{vol}(B_{\Delta Q_n}(L_\infty)))^{1/D} \asymp (2^n n^2)^{-1/2}, \tag{7.5.8}$$

$$(\mathrm{vol}(B^{\perp}_{\Delta Q_n}(L_1)))^{1/D} \asymp 2^{-n/2}. \tag{7.5.9}$$

Proof The lower bound in (7.5.8) follows from Lemma 7.5.11. We now proceed to the proof of the upper estimate in (7.5.8). This proof uses the geometry of ΔQ_n. Comparing the estimate (7.5.8) with Theorem 7.5.1 we conclude that the upper estimate in (7.5.8) cannot be generalized for all $\Lambda \subseteq [-2^n, 2^n]^2$ with $|\Lambda| \asymp |\Delta Q_n|$. First we prove the lower estimate in (7.5.9). We will use the following lemma, which follows from Lemma 4.2.6 (see also Lemma 7.6.18 below).

Lemma 7.5.13 *Let $d = 2$. For any $f \in \mathscr{T}(\Delta Q_n)$ satisfying*

$$\|\delta_{\mathbf{s}}(f)\|_\infty \leq 1, \qquad \|\mathbf{s}\|_1 = n,$$

we have

$$E^{\perp}_{Q_n}(f)_1 \leq C.$$

Let

$$H_\infty(\Delta Q_n) := \{f \in \mathscr{T}(\Delta Q_n) : \|\delta_{\mathbf{s}}(f)\|_\infty \leq 1\}$$

and

$$A(H_\infty(\Delta Q_n)) := \{A(f) : f \in H_\infty(\Delta Q_n)\}.$$

Writing $D = 2|\Delta Q_n|$, Lemma 7.5.13 implies that

$$(\mathrm{vol}(B^{\perp}_{\Delta Q_n}(L_1)))^{1/D} \gg (\mathrm{vol}(A(H_\infty(\Delta Q_n))))^{1/D}. \tag{7.5.10}$$

Using Theorem 7.5.1 we get

$$(\mathrm{vol}(A(H_\infty(\Delta Q_n))))^{1/D} = \left(\prod_{\|\mathbf{s}\|_1 = n} \mathrm{vol}(A(\mathscr{T}(\rho(\mathbf{s}))_\infty))\right)^{1/D} \gg 2^{-n/2}, \tag{7.5.11}$$

where

$$\mathscr{T}(\rho(\mathbf{s}))_\infty := \{t \in \mathscr{T}(\rho(\mathbf{s})) : \|t\|_\infty \leq 1\}.$$

The lower bound in (7.5.9) follows from (7.5.10) and (7.5.11). The upper bound in (7.5.8) is derived from the lower bound in (7.5.9). To prove this we use the following result of Bourgain and Milman (1987), which plays an important role in the volume estimates of finite-dimensional bodies.

Theorem 7.5.14 *For any convex centrally symmetric body $K \subset \mathbb{R}^D$ we have*

$$(\mathrm{vol}(K)\,\mathrm{vol}(K^o))^{1/D} \asymp (\mathrm{vol}(B_2^D))^{2/D} \asymp 1/D,$$

where K^o is a polar for K; that is,

$$K^o := \left\{ x \in \mathbb{R}^D : \sup_{y \in K} (x,y) \le 1 \right\}.$$

Having in mind applications of Theorem 7.5.14 we define some sets other than $B_\Lambda(L_p)$. Let

$$E^{\perp}_{\Lambda,R}(f)_p := \inf_{g \perp \mathscr{T}(\Lambda), \hat{g}(\mathbf{k}) \in \mathbb{R}} \|f - g\|_p.$$

Consider

$$\mathscr{T}_R(\Lambda) := \{f \in \mathscr{T}(\Lambda) : \hat{f}(\mathbf{k}) \in \mathbb{R}\}, \qquad B_{\Lambda,R}(L_p) := \{A(f) : f \in \mathscr{T}_R(\Lambda),\ \|f\|_p \le 1\}$$

and

$$B^{\perp}_{\Lambda,R}(L_p) := \{A(f) : f \in \mathscr{T}_R(\Lambda),\ E^{\perp}_{\Lambda,R}(f)_p \le 1\}.$$

We note that

$$B_\Lambda(L_p) \subseteq B^{\perp}_{\Lambda}(L_p).$$

Moreover, if the orthogonal projector P_Λ onto $\mathscr{T}(\Lambda)$ is bounded as an operator from L_p to L_p then we have

$$\mathrm{vol}(B_\Lambda(L_p))^{(2|\Lambda|)^{-1}} \asymp \mathrm{vol}(B^{\perp}_{\Lambda}(L_p))^{(2|\Lambda|)^{-1}}; \tag{7.5.12}$$

for example, this is the case when $\Lambda = \bigcup_{\mathbf{s} \in A} \rho(\mathbf{s})$ and $1 < p < \infty$.

Using the Nikol'skii duality theorem one can prove the following relation:

$$B^{\perp}_{\Lambda,R}(L_{p'}) \subseteq B_{\Lambda,R}(L_p)^o \subseteq 2B^{\perp}_{\Lambda,R}(L_{p'}), \qquad p' = \frac{p}{p-1}. \tag{7.5.13}$$

Proof of (7.5.13) This proof uses standard ideas from the duality arguments. First, we note that the relation

$$B^{\perp}_{\Lambda,R}(L_{p'}) \subseteq B_{\Lambda,R}(L_p)^o$$

follows immediately from an inequality that holds for any $f, g \in \mathscr{T}_R(\Lambda)$:

$$\left| \sum_{\mathbf{k} \in \Lambda} \hat{f}(\mathbf{k}) \hat{g}(\mathbf{k}) \right| = \frac{1}{(2\pi)^d} \left| \int_{\mathbb{T}^d} f \bar{g} \right| \le E^{\perp}_{\Lambda}(f)_{p'} \|g\|_p.$$

Second, we will prove the inverse inclusion

$$B_{\Lambda,R}(L_p)^o \subseteq 2B^{\perp}_{\Lambda,R}(L_{p'}). \tag{7.5.14}$$

By the Nikol'skii duality theorem we get

$$E_\Lambda^\perp(t)_{p'} = \sup_{g\in\mathscr{T}(\Lambda),\|g\|_p\le 1} |\langle t,g\rangle|. \tag{7.5.15}$$

Let us represent an arbitrary polynomial $g \in \mathscr{T}(\Lambda)$, $\|g\|_p \le 1$, in the form

$$g = g' + ig'', \qquad g' := \sum_{\mathbf{k}\in\Lambda} \operatorname{Re}\hat{g}(\mathbf{k})e^{i(\mathbf{k},\mathbf{x})}, \qquad g'' := \sum_{\mathbf{k}\in\Lambda} \operatorname{Im}\hat{g}(\mathbf{k})e^{i(\mathbf{k},\mathbf{x})}.$$

Then $g', g'' \in \mathscr{T}_R(\Lambda)$ and

$$\|g'\|_p = \left\|\frac{1}{2}(g(\mathbf{x}) + \bar{g}(-\mathbf{x}))\right\|_p \le 1, \qquad \|g''\|_p = \left\|\frac{1}{2}(g(\mathbf{x}) - \bar{g}(-\mathbf{x}))\right\|_p \le 1.$$

Since

$$|\langle t,g\rangle| \le 2\max\{|\langle t,g'\rangle|, |\langle t,g''\rangle|\},$$

it follows from (7.5.15) that

$$E_\Lambda^\perp(t)_{p'} \le 2 \sup_{g\in\mathscr{T}_R(\Lambda),\|g\|_p\le 1} |\langle t,g\rangle|.$$

If $t = A_R^{-1}(\mathbf{a})$ and $\mathbf{a}$ is an arbitrary vector of $\mathbb{R}^{|\Lambda|}$ then the last inequality implies the inclusion (7.5.14). □

We will now show that the volumes of $B_\Lambda(L_p) \subset \mathbb{R}^{2|\Lambda|}$ and $B_{\Lambda,R}(L_p) \subset \mathbb{R}^{|\Lambda|}$ are closely related. First, if

$$\left\|\sum_{\mathbf{k}\in\Lambda} a_{\mathbf{k}}e^{i(\mathbf{k},\mathbf{x})}\right\|_p \le 1/2, \qquad \left\|\sum_{\mathbf{k}\in\Lambda} b_{\mathbf{k}}e^{i(\mathbf{k},\mathbf{x})}\right\|_p \le 1/2$$

then

$$\left\|\sum_{\mathbf{k}\in\Lambda} (a_{\mathbf{k}} + ib_{\mathbf{k}})e^{i(\mathbf{k},\mathbf{x})}\right\|_p \le 1.$$

Therefore

$$\left(\frac{1}{2}B_{\Lambda,R}(L_p)\right) \otimes \left(\frac{1}{2}B_{\Lambda,R}(L_p)\right) \subseteq B_\Lambda(L_p). \tag{7.5.16}$$

Next, let

$$f(\mathbf{x}) = \sum_{\mathbf{k}\in\Lambda} (a_{\mathbf{k}} + ib_{\mathbf{k}})e^{i(\mathbf{k},\mathbf{x})}.$$

Then

$$\sum_{\mathbf{k}\in\Lambda} a_{\mathbf{k}}e^{i(\mathbf{k},\mathbf{x})} = \frac{1}{2}(f(\mathbf{x}) + \bar{f}(-\mathbf{x}))$$

and

$$i\sum_{\mathbf{k}\in\Lambda} b_{\mathbf{k}}e^{i(\mathbf{k},\mathbf{x})} = \frac{1}{2}(f(\mathbf{x}) - \bar{f}(-\mathbf{x})).$$

This implies that

$$B_\Lambda(L_p) \subseteq B_{\Lambda,R}(L_p) \otimes B_{\Lambda,R}(L_p). \tag{7.5.17}$$

We get from (7.5.16) and (7.5.17) that

$$(\mathrm{vol}(B_\Lambda(L_p))^{(2|\Lambda|)^{-1}} \asymp (\mathrm{vol}(B_{\Lambda,R}(L_p))^{(|\Lambda|)^{-1}}. \tag{7.5.18}$$

Similarly, we obtain

$$(\mathrm{vol}(B_\Lambda^\perp(L_p))^{(2|\Lambda|)^{-1}} \asymp (\mathrm{vol}(B_{\Lambda,R}^\perp(L_p))^{(|\Lambda|)^{-1}}. \tag{7.5.19}$$

This observation and Theorems 7.5.3 and 7.5.14 combined with (7.5.12) imply the following statement.

Theorem 7.5.15 *Let Λ have the form $\Lambda = \bigcup_{\mathbf{s}\in S} \rho(\mathbf{s})$, where $S \subset \mathbb{Z}^d_+$ is a finite set. Then, for any $1 \le p < \infty$, we have*

$$\mathrm{vol}(B_\Lambda(L_p))^{(2|\Lambda|)^{-1}} \asymp |\Lambda|^{-1/2}.$$

In particular, Theorem 7.5.15 implies for $d = 2$ and $1 \le p < \infty$ that

$$(\mathrm{vol}(B_{\Delta Q_n}(L_p)))^{(2|\Delta Q_n|)^{-1}} \asymp |\Delta Q_n|^{-1/2} \asymp (2^n n)^{-1/2}. \tag{7.5.20}$$

We now derive the upper bound in (7.5.8) from the lower bound in (7.5.9). The lower bounds (7.5.10), (7.5.11), and (7.5.19) imply that

$$(\mathrm{vol}(B_{\Delta Q_n,R}^\perp(L_1))^{2/D} \gg 2^{-n/2}.$$

By (7.5.13) we then obtain

$$(\mathrm{vol}(B_{\Delta Q_n,R}(L_1)^o)^{2/D} \gg 2^{-n/2}.$$

Theorem 7.5.14 gives

$$(\mathrm{vol}(B_{\Delta Q_n,R}(L_\infty))^{2/D} \ll 2^{n/2}/D \ll (n2^{n/2})^{-1}.$$

Finally, by (7.5.18) we get

$$(\mathrm{vol}(B_{\Delta Q_n}(L_\infty))^{1/D} \ll 2^{n/2}/D \ll (n^2 2^n)^{-1/2}.$$

In the same way we can derive the upper bound in (7.5.9) from the lower bound in (7.5.8). □

The discrete L_∞-norm for polynomials from $\mathscr{T}(\Delta Q_n)$ Theorem 7.5.12 implies an interesting and surprising result on discretization for polynomials from $\mathscr{T}(\Delta Q_n)$. We can derive from Theorem 7.5.12 that there is no analog of the Marcinkiewicz theorem (see Theorem 3.3.15) in L_∞ for polynomials from $\mathscr{T}(\Delta Q_n)$. We present here some results from Kashin and Temlyakov (2003) and begin with the following conditional statement.

Theorem 7.5.16 *Assume that a finite set $\Lambda \subset \mathbb{Z}^d$ has the properties*

$$(\mathrm{vol}(B_\Lambda(L_\infty)))^{1/D} \le K_1 D^{-1/2}, \qquad D := 2|\Lambda|, \tag{7.5.21}$$

and that a set $\Omega = \{x^1, \dots, x^M\}$ satisfies the condition

$$\forall f \in \mathscr{T}(\Lambda), \qquad \|f\|_\infty \le K_2 \|f\|_{\infty,\Omega}, \quad \|f\|_{\infty,\Omega} := \max_{x\in\Omega} |f(x)|. \tag{7.5.22}$$

Then there exists an absolute constant $C > 0$ such that

$$M \ge D \exp(C(K_1K_2)^{-2}).$$

Proof Using the assumption (7.5.22) we can derive from Theorem 7.5.10, in the same way as we proved Lemma 7.5.9, the following volume estimate:

$$(\mathrm{vol}(B_\Lambda(L_\infty)))^{1/D} \ge C_1 K_2^{-1}(D \log(M/D))^{-1/2} \tag{7.5.23}$$

with $C_1 > 0$ an absolute constant. Comparing (7.5.23) with the assumption (7.5.21) we get

$$M \ge D \exp(C(K_1K_2)^{-2}), \qquad C = C_1^2.$$

The theorem is proved. □

We now give some corollaries of Theorem 7.5.16.

Theorem 7.5.17 *Assume that a finite set $\Omega \subset \mathbb{T}^2$ has the following property:*

$$\forall t \in \mathscr{T}(\Delta Q_n), \qquad \|t\|_\infty \le K_2 \|t\|_{\infty,\Omega}. \tag{7.5.24}$$

Then

$$|\Omega| \ge 2|\Delta Q_n| \exp(Cn/K_2^2),$$

with $C > 0$ an absolute constant.

Proof By Theorem 7.5.12 (see (7.5.8)) we have

$$(\mathrm{vol}(B_{\Delta Q_n}(L_\infty)))^{1/D} \le C(2^n n^2)^{-1/2} \le Cn^{-1/2} D^{-1/2},$$

with $C > 0$ an absolute constant. Using Theorem 7.5.16 we obtain

$$|\Omega| \ge 2|\Delta Q_n| \exp(Cn/K_2^2).$$

This proves the theorem. □

Remark 7.5.18 In the particular case $K_2 = bn^\alpha$, $0 \le \alpha \le 1/2$, Theorem 7.5.17 gives

$$|\Omega| \ge 2|\Delta Q_n| \exp\left(Cb^{-2}n^{1-2\alpha}\right).$$

Corollary 7.5.19 *Let a set $\Omega \subset \mathbb{T}^d$ have the property*

$$\forall t \in \mathscr{T}(\Delta Q_n), \qquad \|t\|_\infty \le bn^\alpha \|t\|_{\infty,\Omega}$$

with some $0 \le \alpha < 1/2$. Then

$$|\Omega| \ge C_3 2^n n \exp\left(Cb^{-2}n^{1-2\alpha}\right) \ge C_1(b,d,\alpha)|Q_n| \exp\left(C_2(b,d,\alpha)n^{1-2\alpha}\right).$$

7.6 Entropy Numbers of the Balls of Trigonometric Polynomials

7.6.1 Bounds for $\mathscr{T}(\Pi(\mathbf{N},d))$.

We begin with the simplest case, where Λ is the parallelepiped $\Pi(\mathbf{N},d)$, and use Theorem 7.5.1. First, we prove the lower bounds.

Lemma 7.6.1 *Let $D = 2|\Pi(\mathbf{N},d)|$. Then*

$$\varepsilon_D(\mathscr{T}(\mathbf{N},d)_\infty, L_1) \gg 1.$$

Proof By Theorem 7.2.1,

$$N_\varepsilon(B_Y, X) \ge \varepsilon^{-D} \frac{\mathrm{vol}(B_Y)}{\mathrm{vol}(B_X)} \tag{7.6.1}$$

with $B_Y := B_{\Pi(\mathbf{N},d)}(L_\infty)$ and $B_X := B_{\Pi(\mathbf{N},d)}(L_1)$. Inequality (7.6.1) and Theorem 7.5.1 imply that

$$\varepsilon_D(\mathscr{T}(\mathbf{N},d)_\infty, L_1 \cap \mathscr{T}(\mathbf{N},d)) \gg 1. \tag{7.6.2}$$

Proposition 7.1.2 and inequality (7.6.2) complete the proof. □

Second, we prove the upper bounds.

Lemma 7.6.2 *We have, for $1 \le q \le p \le \infty$, $\beta := 1/q - 1/p$,*

$$\varepsilon_k(\mathscr{T}(\mathbf{N},d)_q, L_p) \ll \begin{cases} \left((|\Pi(\mathbf{N},d)|/k)\log(4|\Pi(\mathbf{N},d)|/k)\right)^\beta, & k \le 2|\Pi(\mathbf{N},d)|, \\ 2^{-k/(2|\Pi(\mathbf{N},d)|)}, & k \ge 2|\Pi(\mathbf{N},d)|. \end{cases}$$

Proof This lemma follows from the finite-dimensional result in Corollary 7.4.6 with the help of the Marcinkiewicz theorem (see Theorem 3.3.15). □

7.6.2 Hyperbolic Crosses. Lower Bounds

We now proceed to the hyperbolic cross case. The results here are not as complete as in the above case of the parallelepiped $\Pi(\mathbf{N},d)$. First, we prove the lower bounds.

Lemma 7.6.3 *Let Λ be either Q_n or ΔQ_n and $D = 2|\Lambda|$. Then, for any $q < \infty$,*

$$\varepsilon_D(\mathscr{T}(\Lambda)_q, L_1) \gg 1.$$

Proof This proof is similar to the proof of lower bounds in Lemma 7.6.1. We use the inequality from Theorem 7.2.1, i.e.,

$$N_\varepsilon(B_Y, X) \geq \varepsilon^{-D} \frac{\mathrm{vol}(B_Y)}{\mathrm{vol}(B_X)}, \tag{7.6.3}$$

with $B_Y := B_\Lambda(L_q)$ and $B_X := B_\Lambda(L_1)$. Relation (7.6.3) and Theorem 7.5.15 imply that

$$\varepsilon_k(\mathscr{T}(\Lambda)_q, L_1 \cap \mathscr{T}(\Lambda)) \gg 1. \tag{7.6.4}$$

Proposition 7.1.2 and inequality (7.6.4) give

$$\varepsilon_k(\mathscr{T}(\Lambda)_q, L_1) \gg 1.$$ □

In the case $q = \infty$ we have a weaker lower bound.

Lemma 7.6.4 *Let Λ be either Q_n or ΔQ_n and $D = 2|\Lambda|$. Then*

$$\varepsilon_D(\mathscr{T}(\Lambda)_\infty, L_1) \gg n^{-1/2}.$$

Proof This proof repeats the above proof for lower bounds in Lemma 7.6.3. We again use the inequality from Theorem 7.2.1,

$$N_\varepsilon(B_Y, X) \geq \varepsilon^{-D} \frac{\mathrm{vol}(B_Y)}{\mathrm{vol}(B_X)}, \tag{7.6.5}$$

with $B_Y := B_\Lambda(L_\infty)$ and $B_X := B_\Lambda(L_1)$. Relation (7.6.5) and Lemma 7.5.11 imply that

$$\varepsilon_D(\mathscr{T}(\Lambda)_\infty, L_1 \cap \mathscr{T}(\Lambda)) \gg n^{-1/2}. \tag{7.6.6}$$

Proposition 7.1.2 and inequality (7.6.6) give

$$\varepsilon_D(\mathscr{T}(\Lambda)_\infty, L_1) \gg n^{-1/2},$$

which completes the proof. □

7.6.3 Hyperbolic Crosses. Upper Bounds for $1 < q < p < \infty$

First, we discuss the entropy numbers $\varepsilon_k(\mathscr{T}(Q_n)_q, L_p)$ in the case $1 < q \le 2 \le p < \infty$. For the m-term approximation we use the following system, described and studied in Temlyakov (2000b). We define a system of orthogonal trigonometric polynomials which is optimal in a certain sense (see Temlyakov, 2000b) for m-term approximations. Variants of this system are well known and very useful in the interpolation of functions by trigonometric polynomials. We begin with a construction of the system $\mathscr{U}$ in the univariate case. Denote

$$U_n^+(x) := \sum_{k=0}^{2^n-1} e^{ikx} = \frac{e^{i2^n x} - 1}{e^{ix} - 1}, \qquad n = 0, 1, 2, \dots,$$

$$U_{n,j}^+(x) := e^{i2^n x} U_n^+(x - 2\pi j 2^{-n}), \qquad j = 0, 1, \dots, 2^n - 1,$$

$$U_{n,j}^-(x) := e^{-i2^n x} U_n^+(-x + 2\pi j 2^{-n}), \qquad j = 0, 1, \dots, 2^n - 1.$$

It will be more convenient for us to normalize the system of functions $\{U_{n,j}^+, U_{n,j}^-\}$ in L_2. We write

$$u_{n,j}^+(x) := 2^{-n/2} U_{n,j}^+(x), \qquad u_{n,j}^-(x) := 2^{-n/2} U_{n,j}^-(x).$$

For $k = 2^n + j$, $n = 0, 1, 2, \dots$, and $j = 0, 1, \dots, 2^n - 1$, define

$$u_k(x) := u_{n,j}^+(x), \qquad u_{-k}(x) := u_{n,j}^-(x).$$

The above formulas define u_k for all $k \in \mathbb{Z} \setminus \{0\}$. Finally, define $u_0(x) = 1$. Set $\mathscr{U} := \{u_k\}_{k \in \mathbb{Z}}$. In the multivariate case of $\mathbf{x} = (x_1, \dots, x_d)$ we define the system $\mathscr{U}^d$ as the tensor product of the univariate systems $\mathscr{U}$. Namely, $\mathscr{U}^d := \{u_{\mathbf{k}}(\mathbf{x})\}_{\mathbf{k} \in \mathbb{Z}^d}$, where

$$u_{\mathbf{k}}(\mathbf{x}) := \prod_{i=1}^{d} u_{k_i}(x_i), \qquad \mathbf{k} = (k_1, \dots, k_d).$$

For $s \in \mathbb{N}$ denote

$$\rho^+(s) := \{k : 2^{s-1} \le k < 2^s\}, \qquad \rho^-(s) := \{-k : 2^{s-1} \le k < 2^s\}$$

and, for $s = 0$, denote

$$\rho^+(0) = \rho^-(0) = \rho(0) := \{0\}.$$

Then, for $\varepsilon = +$ or $\varepsilon = -$, we have

$$\mathscr{T}(\rho^\varepsilon(s)) = \operatorname{span}\{u_k,\ k \in \rho^\varepsilon(s)\} = \operatorname{span}\{e^{ikx},\ k \in \rho^\varepsilon(s)\}.$$

In the multivariate case, for $\mathbf{s} = (s_1, \dots, s_d)$ and $\varepsilon = (\varepsilon_1, \dots, \varepsilon_d)$, define a Cartesian product:

$$\rho^\varepsilon(\mathbf{s}) := \rho^{\varepsilon_1}(s_1) \times \cdots \times \rho^{\varepsilon_d}(s_d).$$

Then

$$\mathscr{T}(\rho^\varepsilon(\mathbf{s})) = \operatorname{span}\{u_{\mathbf{k}},\ \mathbf{k} \in \rho^\varepsilon(\mathbf{s})\} = \operatorname{span}\{e^{i(\mathbf{k},\mathbf{x})},\ \mathbf{k} \in \rho^\varepsilon(\mathbf{s})\}.$$

It is easy to check that for any $\mathbf{k} \neq \mathbf{m}$ we have

$$\langle u_{\mathbf{k}}, u_{\mathbf{m}} \rangle = (2\pi)^{-d} \int_{\mathbb{T}^d} u_{\mathbf{k}}(\mathbf{x}) \bar{u}_{\mathbf{m}}(\mathbf{x}) d\mathbf{x} = 0$$

and

$$\|u_{\mathbf{k}}\|_2 = 1.$$

We use the following notation for $f \in L_1$:

$$f_{\mathbf{k}} := \langle f, u_{\mathbf{k}} \rangle := (2\pi)^{-d} \int_{\mathbb{T}^d} f(\mathbf{x}) \bar{u}_{\mathbf{k}}(\mathbf{x}) d\mathbf{x}, \qquad \hat{f}(\mathbf{k}) := (2\pi)^{-d} \int_{\mathbb{T}^d} f(\mathbf{x}) e^{-i(\mathbf{k},\mathbf{x})} d\mathbf{x}$$

and

$$\delta_{\mathbf{s}}^\varepsilon(f) := \sum_{\mathbf{k} \in \rho^\varepsilon(\mathbf{s})} \hat{f}(\mathbf{k}) e^{i(\mathbf{k},\mathbf{x})}.$$

An analog, important for us, of the Marcinkiewicz theorem holds:

$$\|\delta_{\mathbf{s}}^\varepsilon(f)\|_p^p \asymp \sum_{\mathbf{k} \in \rho^\varepsilon(\mathbf{s})} \|f_{\mathbf{k}} u_{\mathbf{k}}\|_p^p, \qquad 1 < p < \infty, \tag{7.6.7}$$

with constants depending on p and d.

We will often use the inequalities

$$\left(\sum_{\mathbf{s},\varepsilon} \|\delta_{\mathbf{s}}^\varepsilon(f)\|_p^p \right)^{1/p} \ll \|f\|_p \ll \left(\sum_{\mathbf{s},\varepsilon} \|\delta_{\mathbf{s}}^\varepsilon(f)\|_p^2 \right)^{1/2}, \qquad 2 \le p < \infty, \tag{7.6.8}$$

$$\left(\sum_{\mathbf{s},\varepsilon} \|\delta_{\mathbf{s}}^\varepsilon(f)\|_p^2 \right)^{1/2} \ll \|f\|_p \ll \left(\sum_{\mathbf{s},\varepsilon} \|\delta_{\mathbf{s}}^\varepsilon(f)\|_p^p \right)^{1/p}, \qquad 1 < p \le 2, \tag{7.6.9}$$

which are corollaries of the well-known Littlewood–Paley inequalities

$$\|f\|_p \asymp \left\| \left(\sum_{\mathbf{s}} \left| \sum_{\varepsilon} \delta_{\mathbf{s}}^\varepsilon(f) \right|^2 \right)^{1/2} \right\|_p.$$

Lemma 7.6.5 *Let $1 < q \le 2 \le p < \infty$. Let $\mathscr{D}_n^1 := \{u_{\mathbf{k}} : \mathbf{k} \in \Delta Q_n\}$. Then*

$$\sigma_m(\mathscr{T}(\Delta Q_n)_q, \mathscr{D}_n^1)_p \ll (|\Delta Q_n|/m)^\beta, \qquad \beta = 1/q - 1/p.$$

Proof Theorem 4.3.17 implies the lemma for $m \le m_n := [|Q_n| 2^{-n}]$. Let $m \ge m_n$ and take $f \in \mathscr{T}(\Delta Q_n)$. Then

$$f = \sum_{\|\mathbf{s}\|_1 = n} \sum_{\varepsilon} \delta_{\mathbf{s}}^\varepsilon(f).$$

We use the following representation:

$$\delta_{\mathbf{s}}^{\varepsilon}(f) = \sum_{\mathbf{k}\in\rho^{\varepsilon}(\mathbf{s})} f_{\mathbf{k}} u_{\mathbf{k}}.$$

For convenience we will omit ε in the notation $\delta_{\mathbf{s}}^{\varepsilon}(f)$, $\rho^{\varepsilon}(\mathbf{s})$, meaning that we are estimating a quantity $\delta_{\mathbf{s}}^{\varepsilon}(f)$ for a fixed ε and all estimates we are going to make are the same for all ε. We now need the following well-known simple lemma (see, for instance, Temlyakov, 1986c, p. 92).

Lemma 7.6.6 *Let $a_1 \geq a_2 \geq \cdots \geq a_M \geq 0$ and $1 \leq q \leq p \leq \infty$. Then, for all $m < M$, one has*

$$\left(\sum_{k=m}^{M} a_k^p\right)^{1/p} \leq m^{-\beta}\left(\sum_{k=1}^{M} a_k^q\right)^{1/q}, \qquad \beta := 1/q - 1/p.$$

Proof Denote

$$A := \left(\sum_{k=1}^{M} a_k^q\right)^{1/q}.$$

The monotonicity of $\{a_k\}$ implies that

$$ma_m^q \leq A^q \qquad \text{and} \qquad a_m \leq Am^{-1/q}.$$

Then

$$\sum_{k=m}^{M} a_k^p \leq a_m^{p-q} \sum_{k=m}^{M} a_k^q \leq a_m^{p-q} A^q.$$

The above two inequalities imply the lemma. □

We will now apply Lemma 7.6.6 to each set of $f_{\mathbf{k}}$, $\mathbf{k} \in \rho(\mathbf{s})$, $\|\mathbf{s}\|_1 = n$, with $m_{\mathbf{s}} := [m/m_n]$. Denote by $G_{\mathbf{s}}$ the set, with cardinality $|G_{\mathbf{s}}| = m_{\mathbf{s}}$, of $\mathbf{k}$ from $\rho(\mathbf{s})$ with the largest $|f_{\mathbf{k}}|$. Then by (7.6.7) we obtain

$$\left\|\sum_{\mathbf{k}\in\rho(\mathbf{s})\setminus G_{\mathbf{s}}} f_{\mathbf{k}} u_{\mathbf{k}}\right\|_p \asymp 2^{n(1/2-1/p)}\left(\sum_{\mathbf{k}\in\rho(\mathbf{s})\setminus G_{\mathbf{s}}} |f_{\mathbf{k}}|^p\right)^{1/p}.$$

Applying Lemma 7.6.6 we continue the right-hand side of the above relation:

$$\ll 2^{n(1/2-1/p)}(m_{\mathbf{s}}+1)^{-\beta}\left(\sum_{\mathbf{k}\in\rho(\mathbf{s})} |f_{\mathbf{k}}|^q\right)^{1/q}.$$

Using (7.6.7) again we obtain

$$\left\|\sum_{\mathbf{k}\in\rho(\mathbf{s})\setminus G_{\mathbf{s}}} f_{\mathbf{k}} u_{\mathbf{k}}\right\|_p \ll (|\Delta Q_n|/m)^{\beta} \|\delta_{\mathbf{s}}(f)\|_q.$$

Estimating the norm $\|\cdot\|_p$ from above by (7.6.8) and the norm $\|\cdot\|_q$ from below by (7.6.9) completes the proof of Lemma 7.6.5. □

It is easy to see that Lemma 7.6.5 implies the corresponding result for $\mathscr{T}(Q_n)$.

Lemma 7.6.7 *Let $1 < q \le 2 \le p < \infty$. Let $\mathscr{D}_n' := \{u_{\mathbf{k}} : \mathbf{k} \in Q_n\}$. Then*

$$\sigma_m(\mathscr{T}(Q_n)_q, \mathscr{D}_n')_p \ll (|Q_n|/m)^\beta, \qquad \beta = 1/q - 1/p.$$

We now apply the second step of the strategy described in the beginning of this chapter. Theorem 7.4.3, Remark 7.4.5 and Lemma 7.6.7 imply the following lemma.

Lemma 7.6.8 *Let $1 < q \le 2 \le p < \infty$ and $\beta := 1/q - 1/p$. Then*

$$\varepsilon_k(\mathscr{T}(Q_n)_q, L_p) \ll \begin{cases} (|Q_n|/k)^\beta (\log(4|Q_n|/k))^\beta, & k \le 2|Q_n|, \\ 2^{-k/(2|Q_n|)}, & k \ge 2|Q_n|. \end{cases}$$

We can extend Lemma 7.6.8 to the case $1 < q < p < \infty$. We will use Lemma 7.3.4 and derive from lemma the following inequality for entropy numbers.

Lemma 7.6.9 *For $1 < u < q < p < \infty$ we have, for $\theta := (1/u - 1/q)(1/u - 1/p)^{-1}$,*

$$\varepsilon_{k+l}(\mathscr{T}(Q_n)_q, L_p) \le C(u,p,d) \left(\varepsilon_k(\mathscr{T}(Q_n)_u, L_p)\right)^{1-\theta} \left(\varepsilon_l(\mathscr{T}(Q_n)_p, L_p)\right)^\theta.$$

Proof This proof goes along the lines of the proof of Theorem 7.3.3. Let $t \in \mathscr{T}(Q_n)_q$. Applying Lemma 7.3.4 we split the polynomial t into a sum $t = f_1 + f_2$ satisfying (7.3.1). Consider

$$t_1 := S_{Q_n}(f_1), \qquad t_2 := S_{Q_n}(f_2).$$

Then

$$at_1 \in \mathscr{T}(Q_n)_u C(u,d) a^{1-\theta} b^\theta \qquad \text{and} \qquad bt_2 \in \mathscr{T}(Q_n)_p C(p,d) a^{1-\theta} b^\theta.$$

Let a and b be such that

$$a > \varepsilon_k(\mathscr{T}(Q_n)_u, L_p), \qquad b > \varepsilon_l(\mathscr{T}(Q_n)_p, L_p).$$

Now we find $y_1, \dots, y_{2^k}$ and $z_1, \dots, z_{2^l}$ such that

$$\mathscr{T}(Q_n)_u \subset \bigcup_{i=1}^{2^k} B_{L_p}(y_i, a), \qquad \mathscr{T}(Q_n)_p \subset \bigcup_{j=1}^{2^l} B_{L_p}(z_j, b).$$

Take any $f \in \mathscr{T}(Q_n)_q$. Set $\varepsilon := a^{1-\theta} b^\theta \max\{C(u,d), C(p,d)\}$ and as above find t_1 and t_2 from $\mathscr{T}(Q_n)$ such that $t = t_1 + t_2$ and

$$a\|t_1\|_u \le \varepsilon, \qquad b\|t_2\|_p \le \varepsilon.$$

Clearly, for some i,

$$at_1/\varepsilon \in B_{L_p}(y_i, a) \quad \Rightarrow \quad t_1 \in B_{L_p}(\varepsilon y_i/a, \varepsilon) \tag{7.6.10}$$

and, for some j,

$$bt_2/\varepsilon \in B_{L_p}(z_j, b) \quad \Rightarrow \quad t_2 \in B_{L_p}(\varepsilon z_j/b, \varepsilon). \tag{7.6.11}$$

Consider the sets $G_{i,j} := B_{L_p}(\varepsilon y_i/a + \varepsilon z_j/b, 2\varepsilon)$, $i = 1, \dots, 2^k$, $j = 1, \dots, 2^l$. Relations (7.6.10) and (7.6.11) that imply $t \in G_{i,j}$. Thus

$$\varepsilon_{k+l}(\mathscr{T}(Q_n)_q) \le 2\varepsilon. \qquad \square$$

Now let $1 \le q < p \le v \le \infty$. The simple inequality

$$\|g\|_p \le \|g\|_q^{1-\mu} \|g\|_v^{\mu}, \qquad \mu := (1/q - 1/p)(1/q - 1/v)^{-1}, \tag{7.6.12}$$

implies that

$$\varepsilon_k(\mathscr{T}(Q_n)_q, L_p) \le 2\left(\varepsilon_k(\mathscr{T}(Q_n)_q, L_v)\right)^{\mu}. \tag{7.6.13}$$

Indeed, denote $\varepsilon_k := \varepsilon_k(\mathscr{T}(Q_n)_q, L_v)$ and, by Theorem 7.1.1, obtain

$$M_{2\varepsilon_k}(\mathscr{T}(Q_n)_q, L_v) \le N_{\varepsilon_k}(\mathscr{T}(Q_n)_q, L_v) \le 2^k. \tag{7.6.14}$$

By inequality (7.6.12), for any two $g_1, g_2 \in \mathscr{T}(Q_n)_q$ such that $\|g_1 - g_2\|_v \le 2\varepsilon_k$ we get the bound

$$\|g_1 - g_2\|_p \le 2^{1-\mu}(2\varepsilon_k)^{\mu} = 2\varepsilon_k^{\mu}.$$

This and inequality (7.6.14) imply that

$$M_{\varepsilon}(\mathscr{T}(Q_n)_q, L_p) \le 2^k, \qquad \varepsilon := 2\varepsilon_k^{\mu}.$$

It remains to apply Theorem 7.1.1.

Lemma 7.6.10 *Let $1 < q < p < \infty$ and $\beta := 1/q - 1/p$. Then*

$$\varepsilon_k(\mathscr{T}(Q_n)_q, L_p) \ll \begin{cases} (|Q_n|/k)^{\beta}(\log(4|Q_n|/k))^{\beta}, & k \le 2|Q_n|, \\ 2^{-k/(2|Q_n|)}, & k \ge 2|Q_n|. \end{cases}$$

Proof In the case $1 < q \le 2 \le p < \infty$, the lemma follows directly from Lemma 7.6.8. Consider the case $2 < q < p < \infty$. Applying Lemma 7.6.9 with $u = 2$ and, using Lemma 7.6.8 with $q = 2$, we obtain the required bound. In the case $1 < q < p < 2$ the required bound follows from inequality (7.6.13) with $v = 2$ and Lemma 7.6.8 with $p = 2$. $\square$

7.6.4 Hyperbolic Crosses. Upper Bounds for $d = 2$, $q = 1$, and $q = \infty$

The construction of the orthonormal basis in the previous subsection uses classical methods and classical building blocks – an analog of the Dirichlet kernels. That construction works very well for L_q-spaces with $q \in (1, \infty)$. However, because of use of Dirichlet kernels it does not work well in the cases $q = 1$ and $q = \infty$. We present here another construction based on wavelet theory. This construction is taken from Offin and Oskolkov (1993). Let δ be a fixed number, $0 < \delta \le 1/3$, and let $\hat{\varphi}(\lambda) = \hat{\varphi}_\delta(\lambda)$, $\lambda \in \mathbb{R}$, be a sufficiently smooth function (for simplicity, real-valued and even) that is equal to 1 for $|\lambda| \le (1-\delta)/2$ and equal to 0 for $|\lambda| > (1+\delta)/2$ and is such that the integral translates of its square constitute a partition of unity:

$$\sum_{k \in \mathbb{Z}} (\hat{\varphi}(\lambda + k))^2 = 1, \qquad \lambda \in \mathbb{R}. \tag{7.6.15}$$

It is known that condition (7.6.15) is equivalent to the following property. The set of functions $\Phi := \{\varphi(\cdot + l)\}_{l \in \mathbb{Z}}$, where

$$\varphi(x) = \int_{\mathbb{R}} \hat{\varphi}(\lambda) e^{2\pi i \lambda x} d\lambda,$$

is an orthonormal system on $\mathbb{R}$:

$$\int_{\mathbb{R}} \varphi(x+k)\varphi(x+l)dx = \delta_{k,l}, \qquad k, l \in \mathbb{Z}. \tag{7.6.16}$$

Following Offin and Oskolkov (1993) define

$$\theta(\lambda) := \left(((\hat{\varphi}(\lambda/2))^2 - (\hat{\varphi}(\lambda))^2\right)^{1/2}$$

and consider, for $n = 0, 1, \dots$, the trigonometric polynomials

$$\Psi_n(x) := 2^{-n/2} \sum_{k \in \mathbb{Z}} \theta(k 2^{-n}) e^{2\pi i k x}.$$

Further, introduce the following dyadic translates of Ψ_n:

$$\Psi_{n,j}(x) := \Psi_n(x - (j + 1/2)2^{-n})$$

and define a sequence of polynomials $\{T_k\}_{k=0}^{\infty}$ for which

$$T_0(x) := 1, \qquad T_k(x) := \Psi_{n,j}(x)$$

if $k = 2^n + j$, $n = 0, 1, \dots$, and $0 \le j < 2^n$. Note that T_k is a trigonometric polynomial such that

$$\hat{T}_k(\mathrm{v}) = 0 \qquad \text{if } |\mathrm{v}| \ge 2^n(1+\delta) \text{ or } |\mathrm{v}| \le 2^{n-1}(1-\delta). \tag{7.6.17}$$

It is proved in Offin and Oskolkov (1993) that the system $\{T_k\}_{k=0}^{\infty}$ is a complete orthonormal basis in all L_p, $1 \le p \le \infty$ (here, L_∞ stands for the space of continuous functions), of 1-periodic functions. Also, it is proved in Offin and Oskolkov (1993) that

$$|\Psi_n(x)| \leq C(\kappa,\delta)2^{n/2}(2^n|\sin \pi x|+1)^{-\kappa}, \tag{7.6.18}$$

with κ determined by the smoothness of $\hat{\varphi}(\lambda)$. In particular, we can always make $\kappa > 1$ assuming that $\hat{\varphi}(\lambda)$ is smooth enough. It is more convenient, however, for us to consider 2π-periodic functions. We define $\mathscr{V} := \{v_k\}_{k=0}^{\infty}$ with $v_k(x) := T_k(x/(2\pi))$ for $x \in [0, 2\pi)$.

In the multivariate case of $\mathbf{x} = (x_1,\dots,x_d)$ we define the system $\mathscr{V}^d$ as the tensor product of the univariate systems $\mathscr{V}$. Namely, $\mathscr{V}^d := \{v_{\mathbf{k}}(\mathbf{x})\}_{\mathbf{k}\in\mathbb{Z}_+^d}$, where

$$v_{\mathbf{k}}(\mathbf{x}) := \prod_{i=1}^{d} v_{k_i}(x_i), \qquad \mathbf{k} = (k_1,\dots,k_d).$$

Before we proceed to an m-term approximation with respect to the system $\mathscr{V}^d$ we will show how this system can be used for proving the correct upper bounds for the orthowidths in the cases $(q,p) = (1,1)$ and $(q,p) = (\infty,\infty)$. For the univariate system $\mathscr{V}$ denote

$$S_n(f,\mathscr{V}) := \sum_{k=1}^{n} \langle f, v_k\rangle v_k.$$

Then, as mentioned above, it is proved in Offin and Oskolkov (1993) that the system $\mathscr{V}$ is an orthonormal basis in all L_p, $1 \leq p \leq \infty$. Therefore,

$$\|S_n(f,\mathscr{V})\|_p \leq C\|f\|_p \tag{7.6.19}$$

for all p, including $p = 1$ and $p = \infty$, which are our main interest at present. This property and the definition of $\hat{\varphi}$, which implies that $\hat{\varphi}(\lambda)$ is equal to 1 for $|\lambda| \leq (1-\delta)/2$ and equal to 0 for $|\lambda| > (1+\delta)/2$, lead to the following approximation result.

Proposition 7.6.11 *For any f from the class $W^a_{p,\alpha}$, $a > 0$, we have*

$$\|f - S_n(f,\mathscr{V})\|_p \leq C_1(a,p)n^{-a}.$$

We now use the Smolyak algorithm with respect to the system $\mathscr{V}^d$ (see §4.5). We set $Y_s := S_{2^s-1}(\cdot,\mathscr{V})$. Then by Proposition 7.6.11 the family of operators $\{Y_s\}_{s=0}^{\infty}$ satisfies property (1) from §4.5 for all $a > 0$ and by (7.6.19) it satisfies property (2) from §4.5 for $b = 0$. Thus Theorem 4.5.1 implies the following result:

Theorem 7.6.12 *For $f \in \mathbf{H}^r_p$, $r > 0$, we have*

$$\|f - S_{Q_n}(f,\mathscr{V}^d)\|_p \ll 2^{-rn}n^{d-1}, \qquad S_{Q_n}(f,\mathscr{V}^d) := \sum_{\mathbf{k}\in Q_n\cap\mathbb{N}_0^d} \langle f, v_{\mathbf{k}}\rangle v_{\mathbf{k}}.$$

It is clear that $S_{Q_n}(\cdot, \mathcal{V}^d)$ is the operator of orthogonal projection onto $\operatorname{span}(v_{\mathbf{k}}, \mathbf{k} \in Q_n \cap \mathbb{N}_0^d)$. Therefore, we have the following corollary of Theorem 7.6.12 for the orthowidths.

Corollary 7.6.13 *Suppose that $r > 0$ and either $p = 1$ or $p = \infty$; then*

$$\varphi_m(\mathbf{W}^r_{p,\alpha}, L_p) \asymp \varphi_m(\mathbf{H}^r_p, L_p) \asymp m^{-r}(\log m)^{(d-1)(r+1)}.$$

Proof The upper bounds follow from Theorem 7.6.12 and the embedding $\mathbf{W}^r_{p,\alpha} \subset \mathbf{H}^r_p C$. The lower bounds follow from Theorem 5.2.2. □

We now return to our main topic of this section. Property (7.6.18) implies the following simple lemma.

Lemma 7.6.14 *We have*

$$\Big\| \sum_{\mathbf{k} \in \rho^+(\mathbf{s})} a_{\mathbf{k}} v_{\mathbf{k}} \Big\|_\infty \le C(d, \kappa, \delta) 2^{\|\mathbf{s}\|_1/2} \max_{\mathbf{k}} |a_{\mathbf{k}}|.$$

We will use the notation

$$f_{\mathbf{k}} := \langle f, v_{\mathbf{k}} \rangle = (2\pi)^{-d} \int_{\mathbb{T}^d} f(\mathbf{x}) v_{\mathbf{k}}(\mathbf{x}) d\mathbf{x}.$$

Write

$$Q_n^+ := \left\{ \mathbf{k} = (k_1, \dots, k_d) \in Q_n : k_i \ge 0, i = 1, 2, \dots \right\}, \qquad \theta_n := \{\mathbf{s} : \|\mathbf{s}\|_1 = n\},$$

$$\mathcal{V}(Q_n) := \left\{ f : f = \sum_{\mathbf{k} \in Q_n^+} c_{\mathbf{k}} v_{\mathbf{k}} \right\}, \qquad \mathcal{V}(Q_n)_A := \left\{ f \in \mathcal{V}(Q_n) : \sum_{\mathbf{k} \in Q_n^+} |f_{\mathbf{k}}| \le 1 \right\}.$$

We now prove three inequalities for $f \in \mathcal{V}(Q_n)$ in the case $d = 2$. Theorem 7.6.15 below is a generalized version of the small ball inequality for the system $\mathcal{V}^2$.

Theorem 7.6.15 *Let $d = 2$. For any $f \in \mathcal{V}(Q_n)$ we have, for $l \le n$,*

$$\sum_{\mathbf{s} \in \theta_l} \Big\| \sum_{\mathbf{k} \in \rho^+(\mathbf{s})} f_{\mathbf{k}} v_{\mathbf{k}} \Big\|_1 \le C(6 + n - l) \|f\|_\infty,$$

where the constant C may depend on the choice of $\hat{\varphi}$.

Theorem 7.6.16 *Let $d = 2$. For any $f \in \mathcal{V}(Q_n)$ we have*

$$\sum_{\mathbf{k} \in Q_n^+} |f_{\mathbf{k}}| \le C 2^{n/2} \|f\|_\infty,$$

where the constant C may depend on the choice of $\hat{\varphi}$.

Theorem 7.6.17 *Let $d = 2$. For any $f \in \mathscr{V}(Q_n)$ we have*

$$\sum_{\mathbf{k}\in Q_n^+} |f_{\mathbf{k}}| \le C|Q_n|^{1/2}\|f\|_1,$$

where the constant C may depend on the choice of $\hat{\varphi}$.

It will be convenient for us to prove these theorems under the assumption that f is a real function. Clearly, this is sufficient. The proofs of Theorems 7.6.15 and 7.6.17 are based on the Riesz products for the hyperbolic cross polynomials (see Temlyakov, 1980b, 1986c, 1995a, 1998a). Relation (7.6.17) implies that for s_1 and s_2 greater than 3 the function $v_{\mathbf{k}}$ with $\mathbf{k} \in \rho^+(\mathbf{s})$ may have nonzero Fourier coefficients $\hat{v}_{\mathbf{k}}(\mathbf{m})$ only for

$$\mathbf{m} \in \rho'(\mathbf{s}) := \{\mathbf{m} = (m_1, m_2) : (1-\delta)2^{s_i-2} < |m_i| < (1+\delta)2^{s_i-1}, i = 1,2\}.$$

In other words,

$$v_{\mathbf{k}} \in \mathscr{T}(\rho'(\mathbf{s})), \qquad \mathbf{k} \in \rho^+(\mathbf{s}).$$

We introduce some more notation. For any two integers $a \ge 1$ and $0 \le b < a$, we shall denote by $\mathrm{AP}(a,b)$ an arithmetical progression of the form $al + b$, $l = 0, 1, \dots$ Set

$$H_n(a,b) := \{\mathbf{s} = (s_1, s_2) : \mathbf{s} \in \mathbb{Z}_+^2,\ \|\mathbf{s}\|_1 = n,\ s_1, s_2 \ge a,\ s_1 \in \mathrm{AP}(a,b)\}.$$

For a subspace Y in $L_2(\mathbb{T}^d)$ we denote by $Y^\perp$ its orthogonal complement. We need the following lemma on the Riesz product, which is a variant of Lemma 2.1 from Temlyakov (1995a).

Lemma 7.6.18 *Take any trigonometric polynomials $t_{\mathbf{s}} \in \mathscr{T}(\rho'(\mathbf{s}))$ and form the function*

$$\Phi(\mathbf{x}) := \prod_{\mathbf{s}\in H_n(a,b)} (1 + t_{\mathbf{s}}).$$

Then, for any $a \ge 6$ and any $0 \le b < a$, this function admits the representation

$$\Phi(\mathbf{x}) = 1 + \sum_{\mathbf{s}\in H_n(a,b)} t_{\mathbf{s}}(\mathbf{x}) + g(\mathbf{x}),$$

with $g \in \mathscr{T}(Q_{n+a-6})^\perp$.

Proof We will prove that, for $\mathbf{k} = (k_1, k_2)$ such that $|k_1 k_2| \le 2^{n+a-6}$, we have $\hat{g}(\mathbf{k}) = 0$. This proof follows Temlyakov (1980b). Let $w(kt)$ denote either $\cos kt$ or $\sin kt$. Then $g(\mathbf{x})$ contains terms of the form

$$h(\mathbf{x}) = c\prod_{i=1}^{m} w(k_1^i x_1) w(k_2^i x_2), \qquad \mathbf{k}^i \in \rho'(\mathbf{s}^i),$$

with all $\mathbf{s}^i$, $i = 1,\dots,m$, $m \geq 2$, distinct. For the sake of simplicity of notation we assume that $s_1^1 > s_1^2 > \cdots > s_1^m$. Then for $h(\mathbf{x})$ the frequencies with respect to x_1 have the form

$$k_1 = k_1^1 \pm k_1^2 \pm \cdots \pm k_1^m.$$

Therefore, for $\delta \leq 1/3$ and $a \geq 6$ we obtain

$$k_1 > (1-\delta)2^{s_1^1-2} - \sum_{i\geq 1}(1+\delta)2^{s_1^1-1-ai} > 2^{s_1^1-3}.$$

In the same way, for frequencies of the function $h(\mathbf{x})$ with respect to x_2 we get $k_2 > 2^{s_2^m-3}$. Consequently,

$$k_1 k_2 > 2^{s_1^1+s_2^m-6}.$$

In order to complete the proof it remains to observe that for all terms $h(\mathbf{x})$ of the function $g(\mathbf{x})$ we have $m \geq 2$, which implies $s_1^1 + s_2^m \geq n + a$. The lemma is proved. □

Proof of Theorem 7.6.15 For a rectangle $R \subset \mathbb{Z}_+^2$ write

$$S_R(f,\mathscr{V}) := \sum_{\mathbf{k}\in R} f_{\mathbf{k}} v_{\mathbf{k}}, \qquad \delta_{\mathbf{s}}(f,\mathscr{V}) := S_{\rho^+(\mathbf{s})}(f,\mathscr{V}).$$

It is proved in Offin and Oskolkov (1993) that $\|S_R\|_{L_\infty \to L_\infty} \leq B$, where B may depend only on the function $\hat{\varphi}$. Define

$$t_{\mathbf{s}} := S_{\rho^+(\mathbf{s})}(\operatorname{sign} \delta_{\mathbf{s}}(f,\mathscr{V}))B^{-1}.$$

Then $t_{\mathbf{s}} \in \mathscr{T}(\rho'(\mathbf{s}))$ and $\|t_{\mathbf{s}}\|_\infty \leq 1$. By Lemma 7.6.18 with n replaced by l and $a = 6+n-l$, where n is from Theorem 7.6.15, we obtain

$$\Phi(\mathbf{x}) = 1 + \sum_{\mathbf{s}\in H_l(6+n-l,b)} t_{\mathbf{s}}(\mathbf{x}) + g(\mathbf{x})$$

with $g \in \mathscr{T}(Q_n)^\perp$. Clearly, $\|\Phi\|_1 = 1$. Therefore, on the one hand

$$|\langle f, \Phi - 1\rangle| \leq 2\|f\|_\infty;$$

on the other hand,

$$\begin{aligned}
\langle f, \Phi - 1\rangle &= \sum_{\mathbf{s}\in H_l(6+n-l,b)} \langle f, t_{\mathbf{s}}\rangle = \sum_{\mathbf{s}\in H_l(6+n-l,b)} \langle \delta_{\mathbf{s}}(f,\mathscr{V}), t_{\mathbf{s}}\rangle \\
&= \sum_{\mathbf{s}\in H_l(6+n-l,b)} \langle \delta_{\mathbf{s}}(f,\mathscr{V}), \operatorname{sign} \delta_{\mathbf{s}}(f,\mathscr{V})\rangle B^{-1} \\
&= B^{-1} \sum_{\mathbf{s}\in H_l(6+n-l,b)} \|\delta_{\mathbf{s}}(f,\mathscr{V})\|_1.
\end{aligned}$$

Thus, for each $0 \le b < 6+n-l$ we have

$$\sum_{\mathbf{s}\in H_l(6+n-l,b)} \|\delta_{\mathbf{s}}(f,\mathscr{V})\|_1 \le 2B\|f\|_\infty.$$

This easily implies Theorem 7.6.15. □

Proof of Theorem 7.6.16 Theorem 7.6.16 is a corollary of Theorem 7.6.15. Indeed, by Lemma 7.6.14 we get

$$\sum_{\mathbf{k}\in\rho^+(\mathbf{s})} |f_{\mathbf{k}}| = \left\langle \delta_{\mathbf{s}}(f,\mathscr{V}), \sum_{\mathbf{k}\in\rho^+(\mathbf{s})} (\operatorname{sign} f_{\mathbf{k}})v_{\mathbf{k}} \right\rangle \le C\|\delta_{\mathbf{s}}(f,\mathscr{V})\|_1 2^{\|\mathbf{s}\|_1/2}.$$

Thus, by Theorem 7.6.15 we get

$$\sum_{\mathbf{k}\in Q_n^+} |f_{\mathbf{k}}| \le C\sum_{l\le n}(6+n-l)2^{l/2}\|f\|_\infty \ll 2^{n/2}\|f\|_\infty,$$

which completes the proof of Theorem 7.6.16. □

Proof of Theorem 7.6.17 We begin with some auxiliary results. The following simple remark is from Temlyakov (1998a).

Remark 7.6.19 For any real numbers y_l such that $|y_l| \le 1$, $l = 1,\dots,N$, we have ($i^2 = -1$)

$$\left|\prod_{l=1}^{N}\left(1+\frac{iy_l}{\sqrt{N}}\right)\right| \le C,$$

where C is an absolute constant.

We now prove two lemmas, which are analogs of Lemmas 2.2 and 2.3 from Temlyakov (1998a). Write

$$E_{Q_n}^{\perp}(f)_p := \inf_{g\in\mathscr{T}(Q_n)^{\perp}} \|f-g\|_p.$$

Lemma 7.6.20 *For any function f of the form*

$$f = \sum_{\mathbf{s}\in H_n(a,b)} t_{\mathbf{s}}$$

with $a \ge 6$, $0 \le b < a$, where $t_{\mathbf{s}}$, $\mathbf{s} \in H_n(a,b)$, is a real trigonometric polynomial in $\mathscr{T}(\rho'(\mathbf{s}))$ such that $\|t_{\mathbf{s}}\|_\infty \le 1$, we have

$$E_{Q_{n+a-6}}^{\perp}(f)_\infty \le C(1+n/a)^{1/2},$$

with C depending only on $\hat{\varphi}$.

Proof Let us form the function

$$\mathrm{RP}(f) := \operatorname{Im} \prod_{\mathbf{s}\in H_n(a,b)} \left(1+it_{\mathbf{s}}(1+n/a)^{-1/2}\right),$$

which is an analog of the Riesz product. Then by Remark 7.6.19 we have

$$\|\mathrm{RP}(f)\|_\infty \le C. \tag{7.6.20}$$

Lemma 7.6.18 provides the representation

$$\mathrm{RP}(f) = (1+n/a)^{-1/2} \sum_{\mathbf{s}\in H_n(a,b)} t_{\mathbf{s}} + g, \qquad g \in \mathscr{T}(Q_{n+a-6}). \tag{7.6.21}$$

Combining (7.6.20) and (7.6.21) we obtain the lemma. □

Remark 7.6.21 It is clear that in Lemma 7.6.20 we can drop the assumption that the $t_{\mathbf{s}}$ are real polynomials.

Lemma 7.6.22 *For any function f of the form*

$$f = \sum_{\mathbf{s}\in\theta_n} t_{\mathbf{s}}, \qquad t_{\mathbf{s}} \in \mathscr{T}(\rho'(\mathbf{s})), \qquad \|t_{\mathbf{s}}\|_\infty \le 1,$$

we have, for any $a \ge 6$,

$$E^\perp_{Q_{n+a-6}}(f)_\infty \le Ca(1+n/a)^{1/2}.$$

Proof Let us introduce some more notation. Write

$$\theta_{n,a} := \{\mathbf{s} \in \theta_n : \text{ either } s_1 < a \text{ or } s_2 < a\}.$$

Then

$$f = \sum_{\mathbf{s}\in\theta_n} t_{\mathbf{s}} = \sum_{\mathbf{s}\in\theta_{n,a}} t_{\mathbf{s}} + \sum_{b=0}^{a-1} \sum_{\mathbf{s}\in H_n(a,b)} t_{\mathbf{s}}$$

and

$$E^\perp_{Q_{n+a-6}}(f)_\infty \le \sum_{\mathbf{s}\in\theta_{n,a}} \|t_{\mathbf{s}}\|_\infty + \sum_{b=0}^{a-1} E^\perp_{Q_{n+a-6}}\Bigg(\sum_{\mathbf{s}\in H_n(a,b)} t_{\mathbf{s}}\Bigg)_\infty.$$

Using the assumption $\|t_{\mathbf{s}}\|_\infty \le 1$, Lemma 7.6.20, and Remark 7.6.21 we obtain the required estimate. □

We now proceed to the proof of Theorem 7.6.17. For $l \in [0,n]$ consider

$$t^1_{\mathbf{s}} := \sum_{\mathbf{k}\in\rho^+(\mathbf{s})} (\operatorname{sign} f_{\mathbf{k}})v_{\mathbf{k}}, \qquad M_l := \max_{\mathbf{s}\in\theta_l} \|t^1_{\mathbf{s}}\|_\infty, \qquad t_{\mathbf{s}} := t^1_{\mathbf{s}}/M_l.$$

By Lemma 7.6.14,

$$M_l \ll 2^{l/2}.$$

Applying Lemma 7.6.22 with $a = 6+n-l$ leads to

$$\sum_{\mathbf{s}\in\theta_l}\sum_{\mathbf{k}\in\rho^+(\mathbf{s})}|f_{\mathbf{k}}| = \left\langle f, \sum_{\mathbf{s}\in\theta_l} t_{\mathbf{s}}^1\right\rangle = M_l\left\langle f, \sum_{\mathbf{s}\in\theta_l} t_{\mathbf{s}}\right\rangle$$

$$\ll 2^{l/2}\left\|E_{Q_n}^{\perp}\left(\sum_{\mathbf{s}\in\theta_l} t_{\mathbf{s}}\right)\right\|_\infty \|f\|_1 \ll 2^{l/2}(6+n-l)\left(1+\frac{n}{6+n-l}\right)^{1/2}.$$

Summing over $l \le n$ completes the proof of Theorem 7.6.17. □

Lemma 7.6.23 *Let $2 \le p < \infty$ and let $\mathscr{V}_n^1 := \{v_{\mathbf{k}} : \mathbf{k} \in Q_n^+\}$. Then*

$$\sigma_m(\mathscr{V}(Q_n)_A, \mathscr{V}_n^1)_p \ll |Q_n|^{1/2-1/p} m^{1/p-1}. \tag{7.6.22}$$

Proof Note that for $f \in \mathscr{V}(Q_n)_A$ we easily obtain that $\|f\|_2 \le 1$ and $\|f\|_\infty \ll 2^{n/2}$, which, in turn, imply for $2 \le p \le \infty$ that

$$\|f\|_p \ll 2^{n(1/2-1/p)}. \tag{7.6.23}$$

Thus, it is sufficient to prove (7.6.22) for large enough m.

First, we will prove the lemma for ΔQ_n instead of Q_n. Then $f \in \mathscr{V}(\Delta Q_n)_A$ has a representation

$$f = \sum_{\mathbf{k}\in\Delta Q_n} f_{\mathbf{k}} v_{\mathbf{k}}, \qquad \sum_{\mathbf{k}\in\Delta Q_n} |f_{\mathbf{k}}| \le 1, \qquad \Delta Q_n^+ := Q_n^+ \setminus Q_{n-1}^+.$$

Using the fact that the system $\mathscr{V}^d$ is orthonormal we obtain by Lemma 7.6.6 with $m_1 := [m/2]$ that

$$\sigma_{m_1}(f, \mathscr{V}^d)_2 \le (m_1+1)^{-1/2}. \tag{7.6.24}$$

For a set Λ denote

$$\mathscr{V}(\Lambda)_q := \left\{ f : f = \sum_{\mathbf{k}\in\Lambda\cap\mathbb{Z}_+^d} f_{\mathbf{k}} v_{\mathbf{k}}, \ \|f\|_q \le 1 \right\}.$$

Next, we estimate the best m_1-term approximation of $g \in \mathscr{V}(\Delta Q_n)_2$ in L_p, $2 < p < \infty$. We apply Lemma 7.6.6 to each set of $g_{\mathbf{k}}$, $\mathbf{k} \in \rho^+(\mathbf{s})$, $\|\mathbf{s}\|_1 = n$, with $m_{\mathbf{s}} := [m_1/m_n]$, $m_n := [|\Delta Q_n| 2^{-n}]$, assuming that $n \ge C$ where the absolute constant C is large enough to guarantee that $m_n \ge 1$. Denote by $G_{\mathbf{s}}$ the set of cardinality $|G_{\mathbf{s}}| = m_{\mathbf{s}}$ of $\mathbf{k}$ from $\rho^+(\mathbf{s})$ with the largest $|g_{\mathbf{k}}|$. Then, by Lemmas 7.6.6 and 7.6.14, we obtain

$$\left\| \sum_{\mathbf{k}\in\rho^+(\mathbf{s})\setminus G_{\mathbf{s}}} g_{\mathbf{k}} v_{\mathbf{k}} \right\|_\infty \ll 2^{n/2}(m_{\mathbf{s}}+1)^{-1/2}\|\delta_{\mathbf{s}}(g,\mathscr{V})\|_2.$$

Applying the following simple inequality for $2 \le p \le \infty$,

$$\|f\|_p \le \|f\|_2^{\alpha}\|f\|_\infty^{1-\alpha}, \qquad \alpha = 2/p,$$

we obtain

$$\left\| \sum_{\mathbf{k}\in\rho(\mathbf{s})\setminus G_{\mathbf{s}}} g_{\mathbf{k}} v_{\mathbf{k}} \right\|_p \ll (|\Delta Q_n|/m_1)^{1/2-1/p} \|\delta_{\mathbf{s}}(g,\mathscr{V})\|_2. \tag{7.6.25}$$

Inequality (7.6.8) easily implies a similar inequality for $\mathscr{V}^d$ for $2 \le p < \infty$:

$$\|f\|_p \ll \left(\sum_{\mathbf{s}} \|\delta_{\mathbf{s}}(f,\mathscr{V})\|_p^2 \right)^{1/2}. \tag{7.6.26}$$

Combining (7.6.24), (7.6.25), and (7.6.26) completes the proof of the lemma in the case of ΔQ_n.

We now derive the general case of Q_n from the above considered case of ΔQ_l considered above. Set

$$\mu := \frac{1}{2}\left(\frac{1}{2} - \frac{1}{p}\right)\left(1 - \frac{1}{p}\right)^{-1}$$

and denote by l_0 the smallest l satisfying

$$m2^{-\mu(n-l)} \ge 1. \tag{7.6.27}$$

Then for $f \in \mathscr{V}(Q_n)_A$ we have, by (7.6.23) and (7.6.27),

$$f_0 := \sum_{\mathbf{k}\in Q_{l_0}^+} f_{\mathbf{k}} v_{\mathbf{k}}, \qquad \|f_0\|_p \ll 2^{l_0(1/2-1/p)} \ll 2^{n(1/2-1/p)} m^{2(1/p-1)}. \tag{7.6.28}$$

For $l > l_0$ define $m_l := [m2^{-\mu(n-l)}] \ge 1$. Then

$$m' := \sum_{l_0 < l \le n} m_l \le C(p)m$$

and

$$\sigma_{m'}(f - f_0, \mathscr{V}_n^1)_p \ll \sum_{l_0 < l \le n} |\Delta Q_l|^{1/2-1/p} m_l^{1/p-1} \ll |Q_n|^{1/2-1/p} m^{1/p-1}. \tag{7.6.29}$$

Combining (7.6.28) and (7.6.29) completes the proof of the lemma. □

Lemma 7.6.23 and Theorem 7.6.17 imply the following.

Lemma 7.6.24 *Let $2 \le p < \infty$ and let $\mathscr{V}_n^1 := \{v_{\mathbf{k}} : \mathbf{k} \in Q_n\}$. Then, for $d = 2$,*

$$\sigma_m(\mathscr{V}(Q_n)_1, \mathscr{V}_n^1)_p \ll (|Q_n|/m)^{1-1/p}.$$

Lemma 7.6.23 and Theorem 7.6.16 imply:

Lemma 7.6.25 *Let $2 \le p < \infty$ and let $\mathscr{V}_n^1 := \{v_{\mathbf{k}} : \mathbf{k} \in Q_n\}$. Then, for $d = 2$,*

$$\sigma_m(\mathscr{V}(Q_n)_\infty, \mathscr{V}_n^1)_p \ll n^{-1/2}(|Q_n|/m)^{1-1/p}.$$

We now apply the second step of the strategy described at the beginning of this chapter. Theorem 7.4.3, Remark 7.4.5, and Lemma 7.6.24 imply the following lemma.

Lemma 7.6.26 *Let $2 \le p < \infty$ and $\beta := 1 - 1/p$. Then, for $d = 2$,*

$$\varepsilon_k(\mathscr{V}(Q_n)_1, L_p) \ll \begin{cases} (|Q_n|/k)^\beta (\log(4|Q_n|/k))^\beta, & k \le 2|Q_n|, \\ 2^{-k/(2|Q_n|)}, & k \ge 2|Q_n|. \end{cases}$$

Theorem 7.4.3, Remark 7.4.5 and Lemma 7.6.25 imply the following lemma.

Lemma 7.6.27 *Let $2 \le p < \infty$ and $\beta := 1 - 1/p$. Then for, $d = 2$,*

$$\varepsilon_k(\mathscr{V}(Q_n)_\infty, L_p) \ll \begin{cases} n^{-1/2}(|Q_n|/k)^\beta (\log(4|Q_n|/k))^\beta, & k \le 2|Q_n|, \\ n^{-1/2} 2^{-k/(2|Q_n|)}, & k \ge 2|Q_n|. \end{cases}$$

The fact $\mathscr{T}(Q_n) \subset \mathscr{V}(Q_{n+6})$, Corollary 7.2.2 and inequality (7.6.13) allow us to derive the following results from Lemmas 7.6.26 and 7.6.27.

Theorem 7.6.28 *Let $1 < p < \infty$ and $\beta := 1 - 1/p$. Then, for $d = 2$,*

$$\varepsilon_k(\mathscr{T}(Q_n)_1, L_p) \ll \begin{cases} (|Q_n|/k)^\beta (\log(4|Q_n|/k))^\beta, & k \le 2|Q_n|, \\ 2^{-k/(2|Q_n|)}, & k \ge 2|Q_n|. \end{cases}$$

Theorem 7.6.29 *Let $2 \le p < \infty$ and $\beta := 1 - 1/p$. Then, for $d = 2$,*

$$\varepsilon_k(\mathscr{T}(Q_n)_\infty, L_p) \ll \begin{cases} n^{-1/2}(|Q_n|/k)^\beta (\log(4|Q_n|/k))^\beta, & k \le 2|Q_n|, \\ n^{-1/2} 2^{-k/(2|Q_n|)}, & k \ge 2|Q_n|. \end{cases}$$

7.6.5 Hyperbolic Crosses. Upper Bounds for $p = \infty$

We now discuss a more difficult and more interesting case, that of $p = \infty$. Denote

$$\|f\|_A := \sum_{\mathbf{k}} |\hat{f}(\mathbf{k})|, \qquad \hat{f}(\mathbf{k}) := (2\pi)^{-d} \int_{\mathbb{T}^d} f(\mathbf{x}) e^{-i(\mathbf{k},\mathbf{x})} d\mathbf{x}.$$

Theorem 7.6.30, which follows below, is from Temlyakov (2015a) (and see Theorem 9.2.8). We will use it in this section. As above, let

$$\Pi(\mathbf{N}, d) := \{(a_1, \dots, a_d) \in \mathbb{Z}^d : |a_j| \le N_j,\ j = 1, \dots, d\},$$

where the N_j are nonnegative integers and $\mathbf{N} := (N_1, \dots, N_d)$. We write

$$\mathscr{T}(\mathbf{N}, d) := \mathscr{T}(\Pi(\mathbf{N}, d)) = \left\{ t : t = \sum_{\mathbf{k} \in \Pi(\mathbf{N}, d)} c_{\mathbf{k}} e^{i(\mathbf{k},\mathbf{x})} \right\}.$$

Then

$$\dim \mathcal{T}(\mathbf{N},d) = \prod_{j=1}^{d}(2N_j+1) =: \vartheta(\mathbf{N}).$$

For a nonnegative integer m, denote $\overline{m} := \max(m,1)$.

Theorem 7.6.30 *There exist constructive greedy-type approximation methods $G_m^p(\cdot)$ which provide m-term polynomials with respect to $\mathcal{T}^d$ having the following properties: for $2 \le p < \infty$,*

$$\|f - G_m^p(f)\|_p \le C_1(d)(\overline{m})^{-1/2}p^{1/2}\|f\|_A;$$

and, for $p = \infty$, $f \in \mathcal{T}(\mathbf{N},d)$,

$$\|f - G_m^\infty(f)\|_\infty \le C_3(d)(\overline{m})^{-1/2}(\ln \vartheta(\mathbf{N}))^{1/2}\|f\|_A.$$

We will use a version of Theorem 7.6.30 which follows from the proof of Theorem 7.6.30 in Temlyakov (2015a) and the simple inequality

$$\|f\|_A \le \|f\|_2(\#\{\mathbf{k} : \hat{f}(\mathbf{k}) \neq 0\})^{1/2}.$$

Theorem 7.6.31 *Let $\Lambda \subset \Pi(\mathbf{N},d)$ with $N_j = 2^n$, $j = 1,\dots,d$. There exist constructive greedy-type approximation methods $G_m^\infty(\cdot)$ which provide m-term polynomials with respect to $\mathcal{T}^d$ having the following properties: for $f \in \mathcal{T}(\Lambda)$ we have $G_m^\infty(f) \in \mathcal{T}(\Lambda)$ and*

$$\|f - G_m^\infty(f)\|_\infty \le C_4(d)(\overline{m})^{-1/2}n^{1/2}|\Lambda|^{1/2}\|f\|_2.$$

We now prove the following lemma. Let $\mathcal{D}_n^2 := \{u_\mathbf{k} : \mathbf{k} \in Q_n\} \cup \{e^{i(\mathbf{k},\mathbf{x})} : \mathbf{k} \in Q_n\}$.

Lemma 7.6.32 *Let $1 < q \le 2$. Then*

$$\sigma_m(\mathcal{T}(Q_n)_q, \mathcal{D}_n^2)_\infty \ll n^{1/2}(|Q_n|/m)^{1/q}.$$

Proof Take $f \in \mathcal{T}(Q_n)$. First applying Lemma 7.6.7 with $p = 2$ and $[m/2]$ and then applying Theorem 7.6.31 with $\Lambda = Q_n$ and $[m/2]$ we obtain

$$\sigma_m(f, \mathcal{D}_n^2)_\infty \ll n^{1/2}(|Q_n|/m)^{1/q}\|f\|_q,$$

which proves the lemma. □

Theorem 7.4.3, Remark 7.4.5, and Lemma 7.6.32 imply the following estimates.

Theorem 7.6.33 *Let $1 < q \le 2$. Then*

$$\varepsilon_k(\mathcal{T}(Q_n)_q, L_\infty) \ll \begin{cases} n^{1/2}(|Q_n|/k)^{1/q}(\log(4|Q_n|/k))^{1/q}, & k \le 2|Q_n|, \\ n^{1/2}2^{-k/(2|Q_n|)}, & k \ge 2|Q_n|. \end{cases}$$

Let us discuss the case $q = 1$, which is not covered by Theorem 7.6.33. For this case we restrict ourselves to $d = 2$. Let $\mathscr{V}_n^2 := \{v_{\mathbf{k}} : \mathbf{k} \in Q_n\} \cup \{e^{i(\mathbf{k},\mathbf{x})} : \mathbf{k} \in Q_{n+6}\}$. Then we have the following analog of Lemma 7.6.32.

Lemma 7.6.34 *We have*

$$\sigma_m(\mathscr{V}(Q_n)_1, \mathscr{V}_n^2)_\infty \ll n^{1/2}|Q_n|/m.$$

Proof Take $f \in \mathscr{V}(Q_n)$. First applying Lemma 7.6.24 with $p = 2$ and $[m/2]$ and then applying Theorem 7.6.31 with $\Lambda = Q_{n+6}$ and $[m/2]$ we obtain

$$\sigma_m(f, \mathscr{V}_n^2)_\infty \ll n^{1/2}(|Q_n|/m)\|f\|_q,$$

which proves the lemma. □

In the same way as above we derive the following result on entropy numbers from Lemma 7.6.34.

Theorem 7.6.35 *We have*

$$\varepsilon_k(\mathscr{T}(Q_n)_1, L_\infty) \ll \begin{cases} n^{1/2}(|Q_n|/k)\log(4|Q_n|/k), & k \le 2|Q_n|, \\ n^{1/2}2^{-k/(2|Q_n|)}, & k \ge 2|Q_n|. \end{cases}$$

7.7 Entropy Numbers for the W-Type Function Classes

7.7.1 Formulation of Results

In this subsection we use results from §7.6 to prove some lower and upper bounds for the entropy numbers of function classes. We begin with the $\mathbf{W}^r_{q,\alpha}$ classes.

Theorem 7.7.1 *For* $1 < q, p < \infty$ *and* $r > (1/q - 1/p)_+$ *we have*

$$\varepsilon_k(\mathbf{W}^r_{q,\alpha}, L_p) \ll k^{-r}(\log k)^{r(d-1)}.$$

Theorem 7.7.2 *For* $r > 0$ *and* $1 \le q \le \infty$ *we have*

$$\varepsilon_k(\mathbf{W}^r_{q,\alpha}, L_1) \gg k^{-r}(\log k)^{r(d-1)}.$$

The above two theorems provide the order of $\varepsilon_k(\mathbf{W}^r_{q,\alpha}, L_p)$ for all $1 < p, q < \infty$ (see the detailed discussion in §7.9 below). Along with the classes $\mathbf{W}^r_q$ it is natural to consider some more general classes. We proceed to the definition of these classes. For $f \in L_1$ define

$$f_l := \sum_{\|\mathbf{s}\|_1 = l} \delta_{\mathbf{s}}(f), \qquad l \in \mathbb{N}_0, \qquad \mathbb{N}_0 := \mathbb{N} \cup \{0\}.$$

Consider the class

$$\mathbf{W}^{a,b}_q := \{f : \|f_l\|_q \le 2^{-al}(\bar{l})^{(d-1)b}\}, \qquad \bar{l} := \max(l,1).$$

Define

$$\|f\|_{\mathbf{W}^{a,b}_q} := \sup_l \|f_l\|_q 2^{al}(\bar{l})^{-(d-1)b}.$$

It is easy to see (for example, from Theorem 4.4.9) that the class $\mathbf{W}^r_q$ is embedded in the class $\mathbf{W}^{r,0}_q$ for $1 < q < \infty$.

The definition of the classes $\mathbf{W}^{a,b}_q$ is based on the dyadic blocks $\delta_s(f)$ of f. As we know it is convenient in the case $1 < q < \infty$ but brings problems in the cases $q = 1$ and $q = \infty$. For this reason, we define a slight modification of the classes $\mathbf{W}^{a,b}_q$. Consider a class $\bar{\mathbf{W}}^{a,b}_q$ which consists of functions f with a representation

$$f = \sum_{n=1}^{\infty} t_n, \qquad t_n \in \mathscr{T}(Q_n), \qquad \|t_n\|_q \le 2^{-an} n^{b(d-1)}.$$

It is easy to see that in the case $1 < q < \infty$ the classes $\bar{\mathbf{W}}^{a,b}_q$ and $\mathbf{W}^{a,b}_q$ are equivalent. The embedding of $\mathbf{W}^{a,b}_q$ into $\bar{\mathbf{W}}^{a,b}_q$ is obvious and the opposite embedding follows from the inequality for $f \in \bar{\mathbf{W}}^{a,b}_q$:

$$\|f_l\|_q = \|(S_{Q_l} - S_{Q_{l-1}})(f)\|_q \ll \sum_{n \ge l-d} \|t_n\|_q \ll 2^{-al}(\bar{l})^{b(d-1)}.$$

In the cases $q = 1$ and $q = \infty$ the classes $\bar{\mathbf{W}}^{a,b}_q$ are wider than $\mathbf{W}^{a,b}_q$.

The classes $\mathbf{W}^{a,b}_q$ and $\bar{\mathbf{W}}^{a,b}_q$ provide control of smoothness at two scales: a controls the power-type smoothness and b controls the logarithmic-scale smoothness. Similar classes with power-type and logarithmic scales of smoothness are studied in the book Triebel (2010).

Theorem 7.7.3 *Let $1 < q < p < \infty$ and $a > \beta := 1/q - 1/p$. Then*

$$\varepsilon_k(\mathbf{W}^{a,b}_q, L_p) \asymp \varepsilon_k(\bar{\mathbf{W}}^{a,b}_q, L_p) \asymp k^{-a}(\log k)^{(d-1)(a+b)}.$$

The following version of Theorem 7.7.2 holds for the $\mathbf{W}^{a,b}$-type classes.

Theorem 7.7.4 *For $a > 0$ and $1 \le q < \infty$ one has*

$$\varepsilon_k(\mathbf{W}^{a,b}_q, L_1) \gg k^{-a}(\log k)^{(a+b)(d-1)}.$$

In the case $q = \infty$ we have a weaker lower bound.

Theorem 7.7.5 *We have for all $d \ge 2$*

$$\varepsilon_k(\mathbf{W}^{a,b}_\infty, L_1) \gg k^{-a}(\log k)^{(d-1)(a+b)-1/2}.$$

In the case $p = q$ the correct order is known for all $1 \le q \le \infty$.

Theorem 7.7.6 *We have for all $d \geq 2$, $1 \leq q \leq \infty$, $a > 0$*

$$\varepsilon_k(\mathbf{W}^{a,b}_q, L_q) \asymp \varepsilon_k(\bar{\mathbf{W}}^{a,b}_q, L_q) \asymp k^{-a}(\log k)^{(d-1)(a+b)}. \tag{7.7.1}$$

The following theorems cover the case $q = 1$. These theorems hold only for $d = 2$.

Theorem 7.7.7 *Let $1 \leq p < \infty$ and $r > \max(1/2, 1-1/p)$. Then, for $d = 2$,*

$$\varepsilon_k(\mathbf{W}^r_{1,\alpha}, L_p) \asymp k^{-r}(\log k)^{r+1/2}.$$

Theorem 7.7.8 *Let $1 \leq p < \infty$ and $a > \max(1/2, 1-1/p)$. Then, for $d = 2$,*

$$\varepsilon_k(\mathbf{W}^{a,b}_1, L_p) \asymp \varepsilon_k(\bar{\mathbf{W}}^{a,b}_1, L_p) \asymp k^{-a}(\log k)^{a+b}.$$

Theorem 7.7.9 *Let $r > 1$. Then, for $d = 2$,*

$$\varepsilon_k(\mathbf{W}^r_{1,0}, L_\infty) \asymp k^{-r}(\log k)^{r+1}.$$

Theorem 7.7.10 *Let $d = 2$ and $a > 1$. Then*

$$\varepsilon_k(\mathbf{W}^{a,b}_1, L_\infty) \asymp \varepsilon_k(\bar{\mathbf{W}}^{a,b}_1, L_\infty) \asymp k^{-a}(\log k)^{a+b+1/2}.$$

We now formulate a theorem for the case $q = \infty$. It provides the correct order in the case $d = 2$.

Theorem 7.7.11 *We have, for $d = 2$, $1 \leq p < \infty$, $a > \max(1/2, 1-1/p)$,*

$$\varepsilon_k(\mathbf{W}^{a,b}_\infty, L_p) \asymp \varepsilon_k(\bar{\mathbf{W}}^{a,b}_\infty, L_p) \asymp k^{-a}(\log k)^{a+b-1/2}. \tag{7.7.2}$$

Finally, we present some results in the case $p = \infty$. The following theorem provides the upper bound for all d.

Theorem 7.7.12 *Let $1 < q \leq 2$ and $a > 1/q$. Then*

$$\varepsilon_k(\bar{\mathbf{W}}^{a,b}_q, L_\infty) \ll k^{-a}(\log k)^{(d-1)(a+b)+1/2}.$$

The next two theorems are for the case $d = 2$.

Theorem 7.7.13 *Let $1 < q \leq \infty$ and $r > \max(1/q, 1/2)$. Then, for $d = 2$,*

$$\varepsilon_k(\mathbf{W}^r_{q,\alpha}, L_\infty) \asymp k^{-r}(\log k)^{r+1/2}.$$

Theorem 7.7.14 *Let $1 < q < \infty$ and $a > \max(1/q, 1/2)$. Then, for $d = 2$,*

$$\varepsilon_k(\mathbf{W}^{a,b}_q, L_\infty) \asymp \varepsilon_k(\bar{\mathbf{W}}^{a,b}_q, L_\infty) \asymp k^{-a}(\log k)^{a+b+1/2}.$$

We illustrate the above results on the behavior of the entropy numbers $\varepsilon_k(\mathbf{W}^r_{q,\alpha}, L_p)$ in Figure 7.1. It is in the form

$$\varepsilon_k(\mathbf{W}^r_{q,\alpha}, L_p) \asymp (k^{-1}(\log k)^{d-1})^r (\log k)^{(d-1)w(q,p)}$$

under certain conditions on r, which we do not specify. For $(q, p) \in [1, \infty]^2$, represented by the point $(1/q, 1/p) \in [0,1]^2$ we give the order of the $\varepsilon_k(\mathbf{W}^r_{q,\alpha}, L_p)$ by indicating the parameter $w(q, p)$ and also give a reference to the theorem which establishes that relation. In those cases when the order of the $\varepsilon_k(\mathbf{W}^r_{q,\alpha}, L_p)$ is not known we refer to the corresponding open problem.

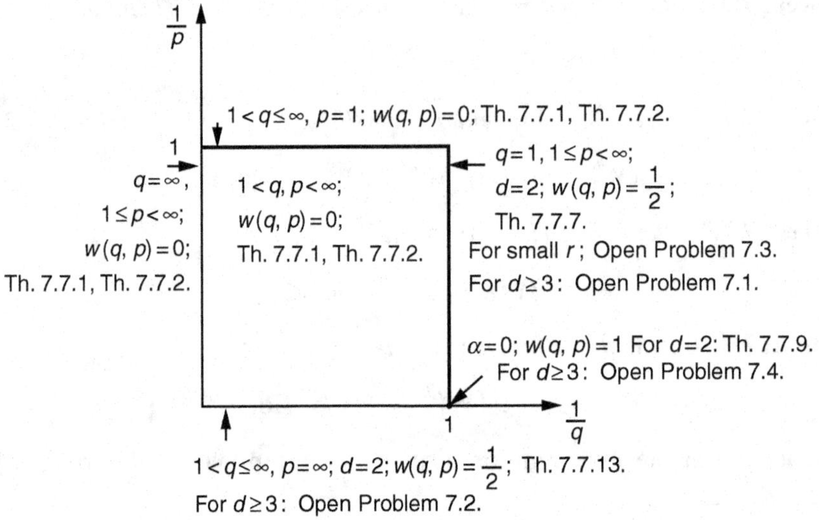

Figure 7.1 The entropy numbers of the $\mathbf{W}$ classes.

7.7.2 The Lower Bounds

In this subsection we prove the lower bounds in the above theorems.

Proof of the lower bounds in Theorems 7.7.2 and 7.7.4 The lower bounds in these theorems do not depend on q (the dependence is in the hidden constant). Therefore, the result is stronger when q is larger. In Theorem 7.7.2 the strongest statement is that for $q = \infty$. We will prove this statement in two steps. First, the required lower bound in Theorem 7.7.2 for all $1 < q < \infty$ follows from Lemma 7.6.3 and the Bernstein inequality in L_q for $\mathscr{T}(Q_n)$. Second, the lower bound for $q = \infty$ follows from the lower bound already proved for $q < \infty$ and from Theorem 7.7.1, with the help of Theorem 7.3.3 with $u = 2$, $p = \infty$. Theorem 7.3.3 is applied to the operator S giving the convolution with the kernel $F_r(\mathbf{x}, \alpha)$.

The required lower bound in Theorem 7.7.4 for all $1 < q < \infty$ follows from Lemma 7.6.3. □

Proof of the lower bounds in Theorem 7.7.3 The corresponding lower bounds follow from Theorem 7.7.4. □

Proof of Theorem 7.7.5 It follows from the definition of $\mathbf{W}^{a,b}_\infty$ that

$$2^{-an}n^{b(d-1)}\mathscr{T}(\Delta Q_n)_\infty \subset \mathbf{W}^{a,b}_\infty.$$

To complete the proof it remains to apply Lemma 7.6.4. □

Proof of the lower bounds in Theorems 7.7.6 and 7.7.8 We will prove the lower bounds for the $\mathbf{W}^{a,b}_q$ class with $1 \le q \le \infty$ and any d. This lower bound is derived from the well-known simple inequality (see Corollary 7.2.2)

$$N_\varepsilon(B_X, X) \ge \varepsilon^{-D} \tag{7.7.3}$$

for any D-dimensional real Banach space X. Consider as a Banach space X the space $\mathscr{T}(\Delta Q_n)$ with L_q-norm. Clearly, it can be seen as a D-dimensional real Banach space with $D = 2|\Delta Q_n|$. It follows from the definition of $\mathbf{W}^{a,b}_q$ that

$$2^{-an}n^{b(d-1)}\mathscr{T}(\Delta Q_n)_q \subset \mathbf{W}^{a,b}_q. \tag{7.7.4}$$

Take $k = 2|\Delta Q_n|$. Then (7.7.3) implies that

$$\varepsilon_k\left(\mathscr{T}(\Delta Q_n)_q, L_q \cap \mathscr{T}(\Delta Q_n)\right) \gg 1. \tag{7.7.5}$$

We now use one more well-known fact from entropy theory: Proposition 7.1.2. This and inequality (7.7.5) imply that

$$\varepsilon_k(\mathscr{T}(\Delta Q_n)_q, L_q) \gg 1. \tag{7.7.6}$$

Taking into account (7.7.4) and the fact $k \asymp 2^n n^{d-1}$ we derive from (7.7.6) the required lower bound for the $\mathbf{W}^{a,b}_q$.

The lower bounds in Theorems 7.7.6 and 7.7.8 are proved. □

Proof of the lower bounds in Theorem 7.7.7 We begin with the lower bound for $p = 2$. We use Theorem 7.2.1. Then for any set F with existing Lebesgue measure $\mathrm{vol}(F)$ we have

$$N_\varepsilon(F, E) \ge \varepsilon^{-d}\frac{\mathrm{vol}(F)}{\mathrm{vol}(B_E)}. \tag{7.7.7}$$

For a fixed natural number n consider the orthogonal projector $S_{\Delta Q_n}$ onto $\mathscr{T}(\Delta Q_n)$. Then, for any m,

$$\varepsilon_m(\mathbf{W}^r_{1,\alpha}, L_2) \ge \varepsilon_m\left(S_{\Delta Q_n}(\mathbf{W}^r_{1,\alpha}), L_2 \cap \mathscr{T}(\Delta Q_n)\right). \tag{7.7.8}$$

Next, it is easy to understand that

$$\begin{aligned} S_{\Delta Q_n}(\mathbf{W}^r_{1,\alpha}) = \{f \in \mathscr{T}(\Delta Q_n) : f = F_r(\cdot,\alpha) * \varphi(\cdot),\\ \varphi \in \mathscr{T}(\Delta Q_n),\ E^{\perp}_{\Delta Q_n}(\varphi)_1 \le 1\}. \end{aligned}$$

We observe that the operator of convolution with $F_0(x,\alpha)$ defined on $\mathscr{T}(\Delta Q_n)$

induces an orthogonal operator in the space $\mathbb{R}^{2|\Delta Q_n|}$ of Fourier coefficients $A(f)$. Therefore

$$\operatorname{vol}(\{A(f) : f \in S_{\Delta Q_n}(\mathbf{W}^r_{1,\alpha})\})^{(2|\Delta Q_n|)^{-1}} \gg 2^{-rn}(\operatorname{vol}((B^{\perp}_{\Delta Q_n}(L_1)))^{(2|\Delta Q_n|)^{-1}}.$$

Applying Theorem 7.5.12 we get

$$\operatorname{vol}(\{A(f) : f \in S_{\Delta Q_n}(\mathbf{W}^r_{1,\alpha})\})^{(2|\Delta Q_n|)^{-1}} \gg 2^{-n(r+1/2)}. \tag{7.7.9}$$

Further,

$$(\operatorname{vol}\{A(f) : f \in \mathcal{T}(\Delta Q_n),\ \|f\|_2 \le 1\})^{(2|\Delta Q_n|)^{-1}} \ll (2^n n)^{-1/2}. \tag{7.7.10}$$

Thus, the relations (7.7.7)–(7.7.10) imply that

$$N_\varepsilon(\mathbf{W}^r_{1,\alpha}, L_2)^{(2|\Delta Q_n|)^{-1}} \gg \varepsilon^{-1} 2^{-rn} n^{1/2}. \tag{7.7.11}$$

Specifying $m = 2|\Delta Q_n|$, we obtain from (7.7.11)

$$\varepsilon_m \gg 2^{-rn} n^{1/2} \asymp m^{-r}(\log m)^{r+1/2}.$$

It is clear that the case of general m follows from the special case $m = 2|\Delta Q_n|$, $n \in \mathbb{N}$, which has been considered above. So, we have established the lower estimate for $p = 2$. It implies the corresponding lower estimate for all $p \ge 2$.

Let us prove the lower estimate for $p = 1$. We use the following interpolation inequality for the entropy numbers (see Remark 7.3.2)

$$\varepsilon_{2m}(\mathbf{W}^r_{1,\alpha}, L_2) \le 2\varepsilon_m(\mathbf{W}^r_{1,\alpha}, L_1)^{(p-2)/(2(p-1))} \varepsilon_m(\mathbf{W}^r_{1,\alpha}, L_p)^{p/(2(p-1))} \tag{7.7.12}$$

with $p > 2$ such that $1 - 1/p < r$. The lower estimate for the left-hand side of (7.7.12) has been proved above. The upper estimate for $\varepsilon_m(\mathbf{W}^r_{1,\alpha}, L_p)$, $r > 1 - 1/p$, will be proved below. Substituting these estimates into (7.7.12) we obtain the required lower estimate for the $\varepsilon_m(\mathbf{W}^r_{1,\alpha}, L_1)$. This completes the proof of the lower estimate in Theorem 7.7.7. □

Proof of the lower bounds in Theorem 7.7.9 As above, let $M_\varepsilon(F, E)$ denote the maximal number of points $x_i \in F$ such that $\|x_i - x_j\|_E > \varepsilon$, $i \ne j$. The following simple inequality is well known (see Theorem 7.1.1):

$$N_\varepsilon(F, E) \le M_\varepsilon(F, E) \le N_{\varepsilon/2}(F, E). \tag{7.7.13}$$

As in the above case we will carry out the proof for an m of a special form, $m = 2|\Delta Q_n|$. Using Theorem 7.5.12 and the relation (7.7.7) we find the following analog of (7.7.11):

$$N_\varepsilon(\mathcal{T}(\Delta Q_n)^{\perp}_1, L_2)^{(2|\Delta Q_n|)^{-1}} \gg \varepsilon^{-1} n^{1/2}, \tag{7.7.14}$$

where

$$\mathcal{T}(\Delta Q_n)^{\perp}_1 = \{f \in \mathcal{T}(\Delta Q_n) : E^{\perp}_{\Delta Q_n}(f)_1 \le 1\}.$$

By (7.7.13) and (7.7.14) we conclude that there are 2^m polynomials $\{t_j\}_{j=1}^{2^m}$ from $\mathscr{T}(\Delta Q_n)$ such that

$$E^{\perp}_{\Delta Q_n}(t_j)_1 \leq 1, \qquad j = 1,\dots,2^m, \tag{7.7.15}$$

$$\|t_i - t_j\|_2^2 \gg n, \quad i \neq j. \tag{7.7.16}$$

Let $t_j^{\perp} \in \mathscr{T}(\Delta Q_n)^{\perp}$, $j = 1,\dots,2^m$, be such that

$$\|t_j - t_j^{\perp}\|_1 \leq 2. \tag{7.7.17}$$

Consider the following collection of functions:

$$\varphi_j := (t_j - t_j^{\perp})/2, \qquad f_j := F_r(\cdot,0) * \varphi_j(\cdot), \qquad j = 1,\dots,2^m.$$

Then

$$f_j \in \mathbf{W}^r_{1,0}, \qquad j = 1,\dots,2^m.$$

We now estimate from below the quantities $\|f_i - f_j\|_\infty$ for $i \neq j$. Consider the inner products

$$a_{ij} := \langle f_i - f_j, \varphi_i - \varphi_j \rangle.$$

On the one hand, by (7.7.17) we have

$$a_{ij} \leq 2\|f_i - f_j\|_\infty. \tag{7.7.18}$$

On the other hand,

$$a_{ij} = \sum_{\mathbf{k}} \hat{F}_r(\mathbf{k},0)|\hat{\varphi}_i(\mathbf{k}) - \hat{\varphi}_j(\mathbf{k})|^2 \gg 2^{-rn}\|t_i - t_j\|_2^2. \tag{7.7.19}$$

Thus by (7.7.16), (7.7.18), and (7.7.19) we get

$$\|f_i - f_j\|_\infty \gg 2^{-rn} n, \qquad i \neq j.$$

Therefore

$$\varepsilon_m(\mathbf{W}^r_{1,0}, L_\infty) \gg 2^{-rn} n \asymp m^{-r}(\log m)^{r+1}.$$

This completes the proof of Theorem 7.7.9. □

Proof of the lower bounds in Theorem 7.7.10 We now prove the lower bound for $\varepsilon_k(\mathbf{W}^{a,b}_1, L_\infty)$. This proof is somewhat similar to the proof of the lower bounds in Theorem 7.7.8. Instead of (7.7.3) we now use the inequality (see Theorem 7.2.1)

$$N_\varepsilon(B_Y, X) \geq \varepsilon^{-D} \frac{\mathrm{vol}(B_Y)}{\mathrm{vol}(B_X)} \tag{7.7.20}$$

with $B_Y := B_{\Delta Q_n}(L_1)$ and $B_X := B_{\Delta Q_n}(L_\infty)$. It follows from the definition of $\mathbf{W}_1^{a,b}$ that

$$2^{-an} n^{b(d-1)} \mathscr{T}(\Delta Q_n)_1 \subset \mathbf{W}_1^{a,b}. \tag{7.7.21}$$

Take $k = 2|\Delta Q_n|$. Then (7.7.20), Theorem 7.5.12, and (7.5.20) imply that

$$\varepsilon_k(\mathscr{T}(\Delta Q_n)_1, L_\infty \cap \mathscr{T}(\Delta Q_n)) \gg n^{1/2}. \tag{7.7.22}$$

Proposition 7.1.2 and inequality (7.7.22) imply

$$\varepsilon_k(\mathscr{T}(\Delta Q_n)_1, L_\infty) \gg n^{1/2}. \tag{7.7.23}$$

Taking into account (7.7.21) and the fact that $k \asymp 2^n n^{d-1}$ we derive from (7.7.23) the required lower bound for the $\mathbf{W}_1^{a,b}$.

The lower bounds in Theorem 7.7.10 are proved. $\square$

Proof of the lower bounds in Theorem 7.7.11 The required lower bounds follow from Theorem 7.7.5. $\square$

Proof of the lower bounds in Theorems 7.7.13 and 7.7.14 In the case $1 < q < \infty$ the proofs repeat the proof of Theorem 7.7.10 with $B_Y := B_{\Delta Q_n}(L_q)$ instead of $B_Y := B_{\Delta Q_n}(L_1)$. To embed $2^{-rn}\mathscr{T}(\Delta Q_n)_q$ into $\mathbf{W}^r_{q,\alpha}$ we use the Bernstein inequality.

Let us prove the lower bound for the $\varepsilon_k(\mathbf{W}^r_{\infty,\alpha}, L_\infty)$. For this purpose we use Theorem 7.3.3 with operator $S = I^r_\alpha$, where

$$I^r_\alpha \varphi := F_r(\cdot, \alpha) * \varphi(\cdot).$$

Then Theorem 7.3.3 with $q = 4$, $u = 2$, and $p = \infty$ gives

$$\varepsilon_{2k}(\mathbf{W}^r_{4,\alpha}, L_\infty) \le 2\varepsilon_k(\mathbf{W}^r_{2,\alpha}, L_\infty)^{1/2} \varepsilon_k(\mathbf{W}^r_{\infty,\alpha}, L_\infty)^{1/2}. \tag{7.7.24}$$

From the part of the lower bounds already proved in Theorem 7.7.13 we obtain

$$\varepsilon_{2k}(\mathbf{W}^r_{4,\alpha}, L_\infty) \gg k^{-r}(\log k)^{r+1/2}. \tag{7.7.25}$$

The upper bound in Theorem 7.7.13, which is proved below, gives for $r > 1/2$

$$\varepsilon_k(\mathbf{W}^r_{2,\alpha}, L_\infty) \ll k^{-r}(\log k)^{r+1/2}. \tag{7.7.26}$$

Substituting the bounds (7.7.25) and (7.7.26) into (7.7.24) we get the required lower bound. $\square$

7.7.3 The Upper Bounds

In this subsection we provide a proof of the upper bounds in the above theorems. We begin with a general scheme. Let X and Y be two Banach spaces. We discuss the problem of estimating the entropy numbers of an approximation class, defined in the space X, in the norm of the space Y. Suppose that a sequence of finite-dimensional subspaces $X_n \subset X$, $n = 1,\dots,$ is given. Define the following class:

$$\bar{\mathbf{W}}^{a,b}_X := \bar{\mathbf{W}}^{a,b}_X\{X_n\}$$
$$:= \left\{ f \in X : f = \sum_{n=1}^{\infty} f_n,\ f_n \in X_n, \|f_n\|_X \le 2^{-an} n^b,\ n = 1,2,\dots \right\}.$$

In particular,

$$\bar{\mathbf{W}}^{a,b}_q = \bar{\mathbf{W}}^{a,b(d-1)}_{L_q}\{\mathscr{T}(Q_n)\}.$$

Write $D_n := \dim X_n$ and assume that for the unit balls $B(X_n) := \{f \in X_n : \|f\|_X \le 1\}$ we have the following upper bounds for the entropy numbers: for real α and nonnegative γ and $\beta \in (0,1]$,

$$\varepsilon_k(B(X_n),Y) \ll n^{\alpha} \begin{cases} (D_n/(k+1))^{\beta}(\log(4D_n/(k+1)))^{\gamma}, & k \le 2D_n, \\ 2^{-k/(2D_n)}, & k \ge 2D_n. \end{cases} \tag{7.7.27}$$

Theorem 7.7.15 *Assume that $D_n \asymp 2^n n^c$, $c \ge 0$, and $a > \beta$ and that subspaces $\{X_n\}$ satisfy (7.7.27). Then*

$$\varepsilon_k(\bar{\mathbf{W}}^{a,b}_X\{X_n\},Y) \ll k^{-a}(\log k)^{ac+b+\alpha}. \tag{7.7.28}$$

Proof For a given k let n be such that $k \asymp D_n \asymp 2^n n^c$. It follows from the definition of the class $\bar{\mathbf{W}}^{a,b}_X$ that

$$\varepsilon_k(\bar{\mathbf{W}}^{a,b}_X,Y) \le \sum_{l=1}^{\infty} 2^{-al} l^b \varepsilon_{k_l}(B(X_l),Y),$$

provided that $\sum_{l=1}^{\infty} k_l \le k$. For $l < n$ we define $k_l := [3a(n-l)D_l/\beta]$. Then $\sum_{l=1}^{n-1} k_l \ll k$ and, by our assumption regarding (7.7.27),

$$\sum_{l=1}^{n-1} 2^{-al} l^b \varepsilon_{k_l}(B(X_l),Y) \ll \sum_{l=1}^{n} 2^{-al} l^{b+\alpha} 2^{-k_l/(2D_l)}$$
$$\ll 2^{-an} n^{b+\alpha} \ll k^{-a}(\log k)^{ac+b+\alpha}.$$

For $l \geq n$ we define $k_l := [D_n 2^{\mu(n-l)}]$, $\mu := (a-\beta)/(2\beta)$. Then $\sum_{l\geq n} k_l \ll k$. Therefore, by (7.7.27) we get

$$\sum_{l\geq n} 2^{-al} l^b \varepsilon_{k_l}(B(X_l),Y) \ll \sum_{l\geq n} 2^{-al} l^{b+\alpha} 2^{\mu(l-n)\beta}(D_l/D_n)^\beta (l-n)^\gamma$$
$$\ll 2^{-an} n^{b+\alpha} \ll k^{-a}(\log k)^{ac+b+\alpha}.$$

Thus we have proved that

$$\varepsilon_{Ck}(\bar{\mathbf{W}}^{a,b}_X, Y) \ll k^{-a}(\log k)^{ac+b+\alpha}. \tag{7.7.29}$$

Taking into account that the right-hand side in (7.7.29) decays polynomially, we conclude that the upper bound in (7.7.28) holds. □

Remark 7.7.16 In the case $Y = X$, Theorem 7.7.15 holds without assumption (7.7.27). It is sufficient to use Corollary 7.2.2.

We now proceed to the proofs of the upper bounds in the theorems listed above. The proofs use Theorem 7.7.15, which is a general result on the $\bar{\mathbf{W}}^{a,b}_X\{X_n\}$. We apply this theorem in the case $X = L_q$, $1 \leq q \leq \infty$, $X_n = \mathscr{T}(Q_n)$. An important ingredient of Theorem 7.7.15 is the assumption that the subspaces $\{X_n\}$ satisfy (7.7.27). The main work of this chapter is devoted to establishing (7.7.27) in the case $X_n = \mathscr{T}(Q_n)$, $X = L_q$, $Y = L_p$ for different parameters $1 \leq q, p \leq \infty$. We now indicate which results are used to obtain the versions of (7.7.27) needed for the proof of the upper bounds in the corresponding theorem.

Proof of the upper bounds in Theorem 7.7.1 The upper bounds in Theorem 7.7.3 for the classes $\mathbf{W}^{r,0}_q$, which we prove below, imply Theorem 7.7.1 by embedding: for $1 < q < \infty$ we have $\mathbf{W}^r_{q,\alpha} \hookrightarrow \mathbf{W}^{r,0}_q$. □

Proof of the upper bounds in Theorem 7.7.3 Theorem 7.7.15 and Lemma 7.6.10 imply the required upper bounds. □

Proof of the upper bounds in Theorem 7.7.6 This follows from Remark 7.7.16. □

Proof of the upper bounds in Theorem 7.7.7 The required upper bounds follow from the embedding $\mathbf{W}^r_{1,\alpha} \hookrightarrow \mathbf{H}^r_1$ and Theorem 7.8.2, which we will prove below. □

Proof of the upper bounds in Theorem 7.7.8 The case $p = 1$ follows from Remark 7.7.16. The case $1 < p < \infty$ follows from Theorem 7.6.28, which gives (7.7.27) for $\alpha = 0$, $\beta = \gamma = 1 - 1/p$. □

Proof of the upper bounds in Theorem 7.7.9 The required upper bounds follow from the embedding $\mathbf{W}^r_{1,\alpha} \hookrightarrow \mathbf{H}^r_1$ and Theorem 7.8.3, which we will prove below. □

Proof of the upper bounds in Theorem 7.7.10 The result follows from Theorem 7.6.35, which gives (7.7.27) for $\alpha = 1/2$, $\beta = \gamma = 1$. □

Proof of the upper bounds in Theorem 7.7.11 This follows from Theorem 7.6.29, which gives (7.7.27) for $\alpha = -1/2$ and $\beta = \gamma = 1 - 1/p$, for $p \in [2, \infty)$. □

Proof of Theorem 7.7.12 The result follows from Theorem 7.6.33, which gives (7.7.27) for $\alpha = 1/2$, $\beta = \gamma = 1/q$. □

Proof of the upper bounds in Theorem 7.7.13 The required upper bounds follow from the embedding $\mathbf{W}^r_{q,\alpha} \hookrightarrow \mathbf{W}^{r,0}_q$, $1 < q < \infty$, and Theorem 7.7.12. □

Proof of the upper bounds in Theorem 7.7.14 The required upper bounds follow from Theorem 7.7.12. □

7.8 Entropy Numbers for the H-Type Function Classes

In this subsection we use the technique for studying $\mathbf{W}$-type function classes developed above to prove the lower and upper bounds for the entropy numbers of the classes $\mathbf{H}^r_q$ in L_p. We prove the following lower bound with the help of the volume-estimates technique.

Theorem 7.8.1 *For $r > 0$ we have*

$$\varepsilon_k(\mathbf{H}^r_\infty, L_1) \gg k^{-r}(\log k)^{(d-1)(r+1/2)}.$$

We prove the corresponding upper bounds for all $1 \le q, p \le \infty$ except $p = \infty$, under certain assumptions on the smoothness r.

Theorem 7.8.2 *Let $1 \le q \le \infty$ and $p < \infty$. We have, for $r > (1/q - 1/p)_+$ in the case where either $2 \le p < \infty$ or $1 \le p < 2 \le q \le \infty$, and for $r > 1/q - 1/2$ in the case $1 \le q, p < 2$*

$$\varepsilon_k(\mathbf{H}^r_q, L_p) \ll k^{-r}(\log k)^{(d-1)(r+1/2)}.$$

In the extreme case $p = \infty$ we will prove the following bound.

Theorem 7.8.3 *For $1 \le q \le 2$, $r > 1/q$, we have*

$$\varepsilon_k(\mathbf{H}^r_q, L_\infty) \ll k^{-r}(\log k)^{(d-1)(r+1/2)+1/2}.$$

The following theorem shows that in the case $d = 2$ Theorem 7.8.3 is sharp (in the sense of order).

Theorem 7.8.4 *In the case $d = 2$, for any $1 \le q \le \infty$, $r > 1/q$ we have*

$$\varepsilon_k(\mathbf{H}^r_q, L_\infty) \asymp k^{-r}(\log k)^{r+1}.$$

We illustrate the above results on the behavior of the $\varepsilon_k(\mathbf{H}^r_q, L_p)$ in Figure 7.2. They are in the form

$$\varepsilon_k(\mathbf{H}^r_q, L_p) \asymp k^{-r}(\log k)^{(d-1)(r+1/2)+z(q,p)}$$

under certain conditions on r, which we do not specify. For $(q,p) \in [1,\infty]^2$, represented by the point $(1/q, 1/p) \in [0,1]^2$, we give the order of $\varepsilon_k(\mathbf{H}^r_q, L_p)$, by indicating the parameter $z(q,p)$, and a reference to the theorem which establishes that relation. In those cases when the order of $\varepsilon_k(\mathbf{H}^r_q, L_p)$ is not known we refer to the corresponding open problem.

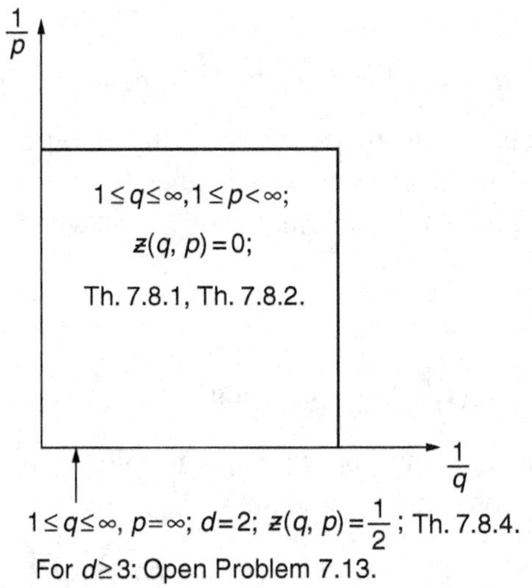

Figure 7.2 The entropy numbers of the **H** classes.

Proof of Theorem 7.8.1 As the proof of Theorem 7.7.5, the present proof of Theorem 7.8.1 is based on a lower bound for the entropy numbers of an appropriate unit ball of the hyperbolic cross polynomials. Define the norm

$$\|f\|_{\mathbf{H}_q} := \sup \|\delta_{\mathbf{s}}(f)\|_q.$$

Consider the following modifications of the hyperbolic layers ΔQ_n: for a number

$n \geq 2d$ define

$$\theta_n^2 := \{\mathbf{s} : \|\mathbf{s}\|_1 = n,\, s_j \geq 2,\, j = 1,\dots,d,\},$$
$$\rho^+(\mathbf{s}) := \{\mathbf{k} : 2^{s_j-1} \leq k_j < 2^{s_j},\, j = 1,\dots,d\},$$
$$\Delta Q_n^2 := \bigcup_{\mathbf{s}\in\theta_n^2} \rho^+(\mathbf{s}).$$

Then

$$\mathrm{vol}(B_{\Delta Q_n^2}(\mathbf{H}_\infty)) = \prod_{\mathbf{s}\in\theta_n^2} \mathrm{vol}(B_{\Pi(2^{\mathbf{s}-1},2^{\mathbf{s}}-\mathbf{1},d)}(L_\infty)).$$

For each $\mathbf{s} \in \theta_n^2$ we have $|\Pi(2^{\mathbf{s}-\mathbf{1}}, 2^{\mathbf{s}} - \mathbf{1}, d)| = 2^{n-d}$. By Theorem 7.5.2 we then obtain

$$\mathrm{vol}(B_{\Delta Q_n^2}(\mathbf{H}_\infty))^{(2|\Delta Q_n^2|)^{-1}} \gg 2^{-n/2}. \tag{7.8.1}$$

The following lemma is an analog of Lemmas 7.6.3 and 7.6.4.

Lemma 7.8.5 *Let $\Lambda := \Delta Q_n^2$ and $D = 2|\Lambda|$. Then*

$$\varepsilon_D(\mathscr{T}(\Lambda)_{\mathbf{H}_\infty}, L_1) \gg n^{(d-1)/2}.$$

Proof This proof is similar to the proof of lower bounds in Lemma 7.6.1. We use the following inequality from Theorem 7.2.1

$$N_\varepsilon(B_Y, X) \geq \varepsilon^{-D} \frac{\mathrm{vol}(B_Y)}{\mathrm{vol}(B_X)} \tag{7.8.2}$$

with $B_Y := B_\Lambda(\mathbf{H}_\infty)$ and $B_X := B_\Lambda(L_1)$. Relations (7.8.1), (7.8.2), and Theorem 7.5.3 imply that

$$\varepsilon_k(\mathscr{T}(\Lambda)_{\mathbf{H}_\infty}, L_1 \cap \mathscr{T}(\Lambda)) \gg n^{(d-1)/2}. \tag{7.8.3}$$

Proposition 7.1.2 and inequality (7.8.3) imply

$$\varepsilon_k(\mathscr{T}(\Lambda)_{\mathbf{H}_\infty}, L_1) \gg n^{(d-1)/2}. \qquad \square$$

We complete the proof of Theorem 7.8.1 by using Lemma 7.8.5 and the fact that for some $c(d) > 0$ we have

$$c(d)2^{-rn}\mathscr{T}(\Lambda)_{\mathbf{H}_\infty} \subset \mathbf{H}_\infty^r.$$

Note that $D \asymp 2^n n^{d-1}$. $\square$

Proof of Theorem 7.8.2 We consider three cases:

(I) $2 \leq q \leq \infty$, $p < \infty$;
(II) $1 < q \leq 2 \leq p < \infty$;
(III) $q = 1$, $2 \leq p < \infty$.

In all three cases we prove the required upper bounds under the assumption $r > \beta_+ := (1/q - 1/p)_+$. Then we explain how other cases covered by Theorem 7.8.2 follow from the above three cases.

We begin with case (I). It is clear that it suffices to prove the required bound for $2 \le q \le p < \infty$, $r > \beta := 1/q - 1/p$. In the case $2 \le q < \infty$ we have an embedding $\mathbf{H}^r_q \hookrightarrow \mathbf{W}^{r,1/2}_q$. Indeed, if $f \in \mathbf{H}^r_q$ then $\|\delta_{\mathbf{s}}(f)\|_q \ll 2^{-r\|\mathbf{s}\|_1}$ and by Corollary A.3.5 we obtain

$$\Big\| \sum_{\|\mathbf{s}\|_1 = n} \delta_{\mathbf{s}}(f) \Big\|_q \ll 2^{-rn} n^{(d-1)/2},$$

which proves the above embedding. It remains to apply Theorems 7.7.3 and 7.7.6 with $a = r$ and $b = 1/2$.

We now proceed to case (II). Arguing in the same way as in Lemma 7.6.5 we obtain the following bound for $f \in \mathscr{T}(\Delta Q_n)_{\mathbf{H}_q}$:

$$\Big\| \sum_{\mathbf{k} \in \rho(\mathbf{s}) \setminus G_{\mathbf{s}}} f_{\mathbf{k}} u_{\mathbf{k}} \Big\|_p \ll (|\Delta Q_n|/m)^{\beta} \|\delta_{\mathbf{s}}(f)\|_q \le (|\Delta Q_n|/m)^{\beta}.$$

This implies the following analog of Lemma 7.6.7.

Lemma 7.8.6 *Let $1 < q \le 2 \le p < \infty$. Let $\mathscr{D}'_n := \{u_{\mathbf{k}} : \mathbf{k} \in Q_n\}$. Then*

$$\sigma_m(\mathscr{T}(Q_n)_{\mathbf{H}_q}, \mathscr{D}'_n)_p \ll (|Q_n|/m)^{\beta} n^{(d-1)/2}, \qquad \beta = 1/q - 1/p.$$

This lemma, in turn, implies the following analog of Lemma 7.6.8.

Lemma 7.8.7 *Let $1 < q \le 2 \le p < \infty$ and $\beta := 1/q - 1/p$. Then*

$$\varepsilon_k(\mathscr{T}(Q_n)_{\mathbf{H}_q}, L_p) \ll n^{(d-1)/2} \begin{cases} (|Q_n|/k)^{\beta} (\log(4|Q_n|/k))^{\beta}, & k \le 2|Q'_n|, \\ 2^{-k/(2|Q_n|)}, & k \ge 2|Q_n|. \end{cases}$$

The proof of Theorem 7.8.2 in the case $1 < q \le 2 \le p < \infty$ is completed in the same way as that in which the upper bounds in Theorem 7.7.3 were proved.

Let us proceed to case (III). The idea of the proof in this case is very similar to that in the case (II). However, here we have some technical difficulties. It was convenient to use the dictionary $\mathscr{D}'_n$ built out of functions $u_{\mathbf{k}}$. It works well for $1 < q < \infty$ but does not work for $q = 1$, because the functions $u_{\mathbf{k}}$ are built out of Dirichlet kernels. This the main new ingredient in studying the case $q = 1$ is the use of a dictionary built out of de la Valleé Poussin kernels instead of Dirichlet kernels. We now point out the necessary changes in the proof. First, in the case $q = 1$ the assumption $f \in \mathbf{H}^r_1$ implies that $\|A_{\mathbf{s}}(f)\|_1 \ll 2^{-r\|\mathbf{s}\|_1}$ instead of $\|\delta_{\mathbf{s}}(f)\|_1 \ll 2^{-r\|\mathbf{s}\|_1}$. As a result we need to work with larger dyadic blocks than $\rho(\mathbf{s})$. The following lemma is useful in this regard.

Lemma 7.8.8 *For $c > 0$ denote*

$$\rho(\mathbf{s},c) := \bigcup_{\mathbf{u}:\|\mathbf{u}-\mathbf{s}\|_\infty \le c} \rho(\mathbf{u}).$$

Then for $t_\mathbf{s} \in \mathscr{T}(\rho(\mathbf{s},c))$ we have, for $2 \le p < \infty$,

$$\left\| \sum_{\mathbf{s}\in\theta_n} t_\mathbf{s} \right\|_p \le C(d,p,c) \left(\sum_{\mathbf{s}\in\theta_n} \|t_\mathbf{s}\|_p^2 \right)^{1/2}.$$

Proof For any $\mathbf{u}$ there is a finite number of $\mathbf{s}$, which depends on d and c in such a way that $\delta_\mathbf{u}(t_\mathbf{s}) \neq 0$. Using this fact we write, by Corollary A.3.5,

$$\begin{aligned}\left\| \sum_{\mathbf{s}\in\theta_n} t_\mathbf{s} \right\|_p^2 &\ll \sum_\mathbf{u} \left\| \delta_\mathbf{u}\left(\sum_{\mathbf{s}\in\theta_n} t_\mathbf{s} \right) \right\|_p^2 \ll \sum_\mathbf{u} \sum_{\mathbf{s}\in\theta_n:\|\mathbf{s}-\mathbf{u}\|_\infty \le c} \|\delta_\mathbf{u}(t_\mathbf{s})\|_p^2 \\ &= \sum_{\mathbf{s}\in\theta_n} \sum_{\mathbf{u}:\|\mathbf{s}-\mathbf{u}\|_\infty \le c} \|\delta_\mathbf{u}(t_\mathbf{s})\|_p^2 \ll \sum_{\mathbf{s}\in\theta_n} \|t_\mathbf{s}\|_p^2. \end{aligned} \tag{7.8.4}$$

In the same way as in §7.6.3 it is convenient to work with "signed" dyadic blocks marked by ε. For $s \in \mathbb{N}$ denote

$$a^+(s) := \{k : 2^{s-2} < k < 2^s\}, \qquad a^-(s) := \{-k : 2^{s-2} < k < 2^s\},$$

and for $s = 0$ denote

$$a^+(0) = a^-(0) = a(0) := \{0\}.$$

In the multivariate case, for $\mathbf{s} = (s_1,\dots,s_d)$ and $\varepsilon = (\varepsilon_1,\dots,\varepsilon_d)$, denote

$$a^\varepsilon(\mathbf{s}) := a^{\varepsilon_1}(s_1) \times \cdots \times a^{\varepsilon_d}(s_d).$$

Let us illustrate the argument in the case $\varepsilon = (+,\dots,+)$. For other ε it proceeds in exactly the same way. For convenience we now drop ε from the notation. We estimate the entropy numbers of the following compact sets:

$$Y_n := \left\{ f = \sum_{\mathbf{s}\in\theta_n} t_\mathbf{s},\, t_\mathbf{s} \in \mathscr{T}(a(\mathbf{s})),\, \|t_\mathbf{s}\|_1 \le 1 \right\}.$$

Write

$$\mathbf{k}(\mathbf{s}) := (k_1(\mathbf{s}),\dots,k_d(\mathbf{s})) := ([2^{s_1-1}],\dots,[2^{s_d-1}]).$$

For a fixed n consider the following dictionary:

$$\mathscr{D}_n^w := \{w_\mathbf{s}(\mathbf{x} - \mathbf{x}(\mathbf{n}))\}_{\mathbf{s}\in\theta_n,\mathbf{n}\in P'(\mathbf{k}(\mathbf{s}))},$$

where

$$w_\mathbf{s}(\mathbf{x}) := e^{i(\mathbf{k}(\mathbf{s}),\mathbf{x})} \mathscr{V}_{\mathbf{k}(\mathbf{s})}(\mathbf{x}).$$

Then, for $t_{\mathbf{s}} \in \mathscr{T}(a(\mathbf{s}))$, we have a representation

$$t_{\mathbf{s}}(\mathbf{x}) = \sum_{\mathbf{n} \in P'(\mathbf{k}(\mathbf{s}))} c(\mathbf{s},\mathbf{n}) w_{\mathbf{s}}(\mathbf{x} - \mathbf{x}(\mathbf{n})).$$

By the Marcinkiewicz theorem we get

$$\sum_{\mathbf{n} \in P'(\mathbf{k}(\mathbf{s}))} |c(\mathbf{s},\mathbf{n})| \ll \|t_{\mathbf{s}}\|_1. \tag{7.8.5}$$

We now argue as in the proof of Lemma 7.6.5. We apply Lemma 7.6.6 to each set of $c(\mathbf{s},\mathbf{n})$, $\mathbf{n} \in P'(\mathbf{k}(\mathbf{s}))$, $\|\mathbf{s}\|_1 = n$, with $m_{\mathbf{s}} := [m/m_n]$. Denote by $G_{\mathbf{s}}$ the set, of cardinality $|G_{\mathbf{s}}| = m_{\mathbf{s}}$, of $\mathbf{n}$ from $P'(\mathbf{k}(\mathbf{s}))$ with the largest $|c(\mathbf{s},\mathbf{n})|$. Then, by the Marcinkiewicz theorem, we obtain

$$\left\| \sum_{\mathbf{n} \in P'(\mathbf{k}(\mathbf{s})) \setminus G_{\mathbf{s}}} c(\mathbf{s},\mathbf{n}) w_{\mathbf{s}}(\mathbf{x} - \mathbf{x}(\mathbf{n})) \right\|_p \asymp 2^{n(1-1/p)} \left(\sum_{\mathbf{n} \in P'(\mathbf{k}(\mathbf{s})) \setminus G_{\mathbf{s}}} |c(\mathbf{s},\mathbf{n})|^p \right)^{1/p}.$$

Applying Lemma 7.6.6 with $q = 1$ we continue the right-hand side as

$$\ll 2^{n(1-1/p)} (m_{\mathbf{s}} + 1)^{-\beta} \sum_{\mathbf{n} \in P'(\mathbf{k}(\mathbf{s}))} |c(\mathbf{s},\mathbf{n})|, \qquad \beta = 1 - 1/p.$$

Using (7.8.5) we obtain

$$\left\| \sum_{\mathbf{n} \in P'(\mathbf{k}(\mathbf{s})) \setminus G_{\mathbf{s}}} c(\mathbf{s},\mathbf{n}) w_{\mathbf{s}}(\mathbf{x} - \mathbf{x}(\mathbf{n})) \right\|_p \ll (|\Delta Q_n|/m)^{\beta} \|t_{\mathbf{s}}\|_1 \le (|\Delta Q_n|/m)^{\beta}.$$

Our compact set Y_n is contained in $\mathscr{T}(\Delta Q_n^a)$, where

$$\Delta Q_n^a := \bigcup_{\mathbf{s} \in \theta_n} a(\mathbf{s}).$$

However, the elements $w_{\mathbf{s}}(\mathbf{x} - \mathbf{x}(\mathbf{n}))$ of the dictionary are not necessarily in $\mathscr{T}(\Delta Q_n^a)$. Therefore we consider a new dictionary $\mathscr{D}_n^a := \{S_{\Delta Q_n^a}(w), w \in \mathscr{D}_n^w\}$. In the case $1 < p < \infty$ it does not affect the approximation properties of elements from Y_n. In the same way as above we obtain the following lemma.

Lemma 7.8.9 *Let $q = 1$, $2 \le p < \infty$ and $\beta := 1 - 1/p$. Then*

$$\varepsilon_k(Y_n, L_p) \ll n^{(d-1)/2} \begin{cases} (|\Delta Q_n^a|/k)^{\beta} (\log(C(d)|\Delta Q_n^a|/k))^{\beta}, & k \le 2|\Delta Q_n^a|, \\ 2^{-k/(2|\Delta Q_n^a|)}, & k \ge 2|\Delta Q_n^a|. \end{cases}$$

Using Lemma 7.8.9, we can complete the proof of Theorem 7.8.2 in the case $q = 1$, $2 \le p < \infty$ in the same way as the upper bounds in Theorem 7.7.3 were proved.

We now show how the remaining cases covered by Theorem 7.8.2 follow from the three cases considered above. The case $1 \le q \le \infty$, $2 \le p < \infty$ is covered by cases (I)–(III). The case $1 \le p < 2$, $2 \le q \le \infty$ follows from the case $2 \le q = p < \infty$

covered above. The remaining case, $1 \le p < 2$, $1 \le q < 2$, follows from the case $1 \le q < p = 2$. This is why we need a stronger restriction $r > 1/q - 1/2$ in this case. The proof of Theorem 7.8.2 is complete. □

Proof of Theorem 7.8.3 In this proof we combine the technique from the above proof of cases (II) and (III) of Theorem 7.8.2 with the technique from §7.6.5.

First, consider the case $1 < q \le 2$. Let $\mathscr{D}_n^2 := \{u_{\mathbf{k}} : \mathbf{k} \in Q_n\} \cup \{e^{i(\mathbf{k},\mathbf{x})} : \mathbf{k} \in Q_n\}$.

Lemma 7.8.10 *Let* $1 < q \le 2$. *Then*

$$\sigma_m(\mathscr{T}(Q_n)_{\mathbf{H}_q}, \mathscr{D}_n^2)_\infty \ll n^{d/2}(|Q_n|/m)^{1/q}.$$

Proof Take $f \in \mathscr{T}(Q_n)$. First applying Lemma 7.8.6 with $p = 2$ and $[m/2]$ and then applying Theorem 7.6.31 with $\Lambda = Q_n$ and $[m/2]$ we obtain

$$\sigma_m(f, \mathscr{D}_n^2)_\infty \ll n^{d/2}(|Q_n|/m)^{1/q}\|f\|_{\mathbf{H}_q},$$

which proves the lemma. □

Theorem 7.4.3, Remark 7.4.5, and Lemma 7.8.10 imply the following estimates.

Lemma 7.8.11 *Let* $1 < q \le 2$. *Then*

$$\varepsilon_k(\mathscr{T}(Q_n)_{\mathbf{H}_q}, L_\infty) \ll n^{d/2}\begin{cases} (|Q_n|/k)^{1/q}(\log(4|Q_n|/k))^{1/q}, & k \le 2|Q_n|, \\ 2^{-k/(2|Q_n|)}, & k \ge 2|Q_n|. \end{cases}$$

Using Lemma 7.8.11 we can complete the proof of Theorem 7.8.3 in the case $1 < q \le 2$ in the same way as Theorem 7.7.12 was proved above.

Second, consider the case $q = 1$. This proof repeats that in the case $1 < q \le 2$ but with the dictionary $\mathscr{D}_n^2 := \{u_{\mathbf{k}} : \mathbf{k} \in Q_n\} \cup \{e^{i(\mathbf{k},\mathbf{x})} : \mathbf{k} \in Q_n\}$ replaced by the dictionary $\mathscr{D}_n^3 := \mathscr{D}_n^a \cup \{e^{i(\mathbf{k},\mathbf{x})} : \mathbf{k} \in \Delta Q_n^a\}$. As a result we obtain the following bounds for the entropy numbers.

Lemma 7.8.12 *Let* $q = 1$. *Then*

$$\varepsilon_k(Y_n, L_\infty) \ll n^{d/2}\begin{cases} (|\Delta Q_n^a|/k)\log(C(d)|\Delta Q_n^a|/k), & k \le 2|\Delta Q_n^a|, \\ 2^{-k/(2|\Delta Q_n^a|)}, & k \ge 2|\Delta Q_n^a|. \end{cases}$$

Using Lemma 7.8.12 completes the proof of Theorem 7.8.3 in the case $q = 1$ in the same way as Theorem 7.7.12 was proved above. Theorem 7.8.3 is therefore proved. □

Proof of Theorem 7.8.4 Under the extra assumption $r > \max(1/2, 1/q)$, the upper bound follows from Theorem 7.8.3; one can prove this upper bound under the assumption $r > 1/q$ using a method based directly on discretization and on Corollary 7.4.6. We do not present the details of this proof here.

The proof of the lower bound repeats the proof of Theorem 7.8.1. Instead of Lemma 7.8.5 we get the following lemma.

Lemma 7.8.13 *Let $\Lambda := \Delta Q_n^2$ and $D = 2|\Lambda|$. Then, in the case $d = 2$, we have*

$$\varepsilon_D(\mathscr{T}(\Lambda)_{\mathbf{H}_\infty}, L_\infty) \gg n.$$

Proof This proof is similar to the proof of Lemma 7.8.5 above. Instead of Theorem 7.5.3 we use Theorem 7.5.12 (it is easy to check that Theorem 7.5.12 holds with ΔQ_n replaced by ΔQ_n^2).

We use the inequality (7.8.2) with $B_Y := B_\Lambda(\mathbf{H}_\infty)$ and $B_X := B_\Lambda(L_\infty)$. Relations (7.8.1), (7.8.2), and Theorem 7.5.12 imply that

$$\varepsilon_k(\mathscr{T}(\Lambda)_{\mathbf{H}_\infty}, L_\infty \cap \mathscr{T}(\Lambda)) \gg n. \tag{7.8.6}$$

Proposition 7.1.2 and inequality (7.8.6) imply

$$\varepsilon_k(\mathscr{T}(\Lambda)_{\mathbf{H}_\infty}, L_\infty) \gg n. \qquad \square$$

Using Lemma 7.8.13 we complete the proof of Theorem 7.8.4 in the same way as in the proof of Theorem 7.8.1. $\square$

7.9 Discussion and Open Problems

Three types of mixed-smoothness classes, $\mathbf{W}^r_{q,\alpha}$, $\mathbf{W}^{a,b}_q$, and $\mathbf{H}^r_q$, were studied in this chapter. We now summarize the above results and formulate some open problems. We begin with the most classical classes, $\mathbf{W}^r_{q,\alpha}$, of functions with bounded mixed derivative.

First, we discuss the results obtained for all d. Theorems 7.7.1 and 7.7.2 provide the correct order of $\varepsilon_k(\mathbf{W}^r_{q,\alpha}, L_p)$ for all $1 < q \le \infty$ and $1 \le p < \infty$ under a minimal restriction on r, i.e., $r > (1/q - 1/p)_+$. The following two cases are not settled for $d > 2$.

Open Problem 7.1 Find the correct order in k of the entropy numbers $\varepsilon_k(\mathbf{W}^r_{1,\alpha}, L_p)$ for $1 \le p \le \infty$, $r > 1 - 1/p$, in the case $d \ge 3$.

Open Problem 7.2 Find the correct order in k of the entropy numbers $\varepsilon_k(\mathbf{W}^r_{q,\alpha}, L_\infty)$ for $1 \le q \le \infty$, $r > 1/q$, in the case $d \ge 3$.

Second, we point out that in the case $d = 2$ the correct order of the $\varepsilon_k(\mathbf{W}^r_{q,\alpha}, L_p)$ is known for all $1 \le q, p \le \infty$, with the exception of some special cases and some cases of small smoothness. Indeed, Theorem 7.7.7 shows that Open Problem 7.1 is settled for all $1 \le p < \infty$ in the case $d = 2$, except for the case of small smoothness. Theorem 7.7.9 partially solves the problem for $p = \infty$; it gives the correct order for a special case, $\alpha = 0$.

Open Problem 7.3 Find the correct order in k of the entropy numbers $\varepsilon_k(\mathbf{W}^r_{1,\alpha}, L_p)$ for $1 < p < 2$, $1 - 1/p < r \le 1/2$, in the case $d = 2$.

Open Problem 7.4 Find the correct order in k of the entropy numbers $\varepsilon_k(\mathbf{W}^r_{1,\alpha}, L_\infty)$, $\alpha \in (0, 2\pi)$, in the case $d = 2$.

Theorem 7.7.13 shows that Open Problem 7.2 is settled for all $1 < q \le 2$, $r > 1/q$, and for $2 < q \le \infty$, $r > 1/2$ in the case $d = 2$. Thus, only the case of small smoothness, $1/q < r \le 1/2$, is left open for $\varepsilon_k(\mathbf{W}^r_{q,\alpha}, L_\infty)$.

Open Problem 7.5 Find the correct order in k of the entropy numbers $\varepsilon_k(\mathbf{W}^r_{q,\alpha}, L_\infty)$ for $2 < q \le \infty$ in the case of small smoothness, $1/q < r \le 1/2$, for $d = 2$.

We now continue with the approximation classes $\mathbf{W}^{a,b}_q$ and $\bar{\mathbf{W}}^{a,b}_q$. These classes were introduced in Temlyakov (2015b) and their entropy numbers were studied in Temlyakov (2016b). As above, first we discuss results obtained for all d. Theorems 7.7.3 and 7.7.4 provide the right order of $\mathbf{W}^{a,b}_q$ and $\bar{\mathbf{W}}^{a,b}_q$ for all $1 < q < \infty$ and $1 \le p < \infty$ under a minimal restriction on a: $a > (1/q - 1/p)_+$. Theorem 7.7.6 gives the corresponding right orders for all $1 \le q = p \le \infty$. The following three cases are not settled for $d > 2$.

Open Problem 7.6 Find the correct order in k of the entropy numbers $\varepsilon_k(\mathbf{W}^{a,b}_1, L_p)$ and $\varepsilon_k(\bar{\mathbf{W}}^{a,b}_1, L_p)$ for $1 < p \le \infty$, in the case $d \ge 3$.

Open Problem 7.7 Find the correct order in k of the entropy numbers $\varepsilon_k(\mathbf{W}^{a,b}_q, L_\infty)$ and $\varepsilon_k(\bar{\mathbf{W}}^{a,b}_q, L_\infty)$ for $1 \le q < \infty$, $r > 1/q$, in the case $d \ge 3$.

Open Problem 7.8 Find the correct order in k of the entropy numbers $\varepsilon_k(\mathbf{W}^{a,b}_\infty, L_p)$ and $\varepsilon_k(\bar{\mathbf{W}}^{a,b}_\infty, L_p)$ for $1 \le p < \infty$, in the case $d \ge 3$.

As the second step we discuss the results in the case $d = 2$. Here, the right orders of the corresponding entropy numbers are known for all $1 \le q, p \le \infty$, with the exception of some cases of small smoothness. Theorems 7.7.8 and 7.7.10 completely cover the situation addressed in Open problem 7.6 for $d = 2$, except for the small-smoothness case. Theorem 7.7.11 covers the case addressed in Open Problem 7.8 for $d = 2$, except the small smoothness situation. Theorem 7.7.14 covers the situation addressed in Open Problem 7.7 for $d = 2$, except for the small-smoothness case .

Open Problem 7.9 Find the correct order in k of the entropy numbers $\varepsilon_k(\mathbf{W}^{a,b}_1, L_p)$ for $1 < p < 2$, $1 - 1/p < a \le 1/2$, in the case $d = 2$.

Open Problem 7.10 Find the correct order in k of the entropy numbers $\varepsilon_k(\mathbf{W}^{a,b}_\infty, L_p)$ for $1 \le p < \infty$, $0 < a \le \max(1/2, 1 - 1/p)$, in the case $d = 2$.

Open Problem 7.11 Find the correct order in k of the entropy numbers $\varepsilon_k(\mathbf{W}^{a,b}_q, L_\infty)$ for $2 < q < \infty$, $1/q < a \le 1/2$, in the case $d = 2$.

Theorems 7.8.1 and 7.8.2 provide the correct order of the entropy numbers $\varepsilon_k(\mathbf{H}^r_q, L_p)$ for all $1 \le q \le \infty$, $1 \le p < \infty$ under certain assumptions on r. Some further results for the case of small smoothness can be found in Belinsky (1998).

Open Problem 7.12 Find the correct order in k of the entropy numbers $\varepsilon_k(\mathbf{H}^r_1, L_p)$ for $1 < p < 2$ in the case of small smoothness, $1 - 1/p < r \le 1/2$.

Theorem 7.8.4 gives the correct order of the $\varepsilon_k(\mathbf{H}^r_q, L_\infty)$ for all $1 \le q \le \infty$ under the assumption $r > 1/q$. Thus, the case $p = \infty$ is settled for $d = 2$.

Open Problem 7.13 Find the correct order in k of the entropy numbers $\varepsilon_k(\mathbf{H}^r_q, L_\infty)$, for $1 \le q \le \infty$, $r > 1/q$, in the case $d \ge 3$.

Let us now make some comments on the relation between the classes $\mathbf{W}^r_{q,\alpha}$, $\mathbf{W}^{a,b}_q$, and $\mathbf{H}^r_q$. As we already pointed out above we have the following embeddings: $\mathbf{W}^r_{q,\alpha} \hookrightarrow \mathbf{W}^{r,0}_q$, $1 < q < \infty$, and $\mathbf{H}^r_q \hookrightarrow \mathbf{W}^{r,1/2}_q$, $2 \le q < \infty$. Thus, in this sense, the classes $\mathbf{W}^{a,b}_q$ are larger than the corresponding classes $\mathbf{W}^r_{q,\alpha}$ and $\mathbf{H}^r_q$. However, the above proofs show that the upper bounds for the entropy numbers of the classes $\mathbf{W}^{a,b}_q$ provide upper bounds that are correct in the sense of order for the entropy numbers of the classes $\mathbf{W}^r_{q,\alpha}$ and $\mathbf{H}^r_q$. This gives an additional motivation for the study of the classes $\mathbf{W}^{a,b}_q$.

In the case of the extreme values $q = 1$ and $q = \infty$, classes $\mathbf{W}^r_{q,\alpha}$ and $\mathbf{W}^{r,0}_q$ are very different. Let us begin with $q = 1$ and make some comments on Theorem 7.7.8. We point out that the correct order of $\varepsilon_k(\mathbf{W}^r_{1,\alpha}, L_p)$ is not known for $d > 2$. We confine ourselves to the case $d = 2$. Theorem 7.7.7 states that for $r > \max(1/2, 1 - 1/p)$,

$$\varepsilon_k(\mathbf{W}^r_{1,\alpha}, L_p) \asymp k^{-r} (\log k)^{r+1/2}, \qquad 1 \le p < \infty. \tag{7.9.1}$$

Theorem 7.7.8 gives, for $r > \max(1/2, 1 - 1/p)$,

$$\varepsilon_k(\mathbf{W}^{r,0}_1, L_p) \asymp k^{-r} (\log k)^r, \qquad 1 \le p < \infty. \tag{7.9.2}$$

This shows that, in the sense of entropy numbers, the class $\mathbf{W}^{r,0}_1$ is smaller than $\mathbf{W}^r_{1,\alpha}$. It is interesting to compare (7.9.1) and (7.9.2) with the known estimates in the case $1 < q, p < \infty$, i.e.,

$$\varepsilon_k(\mathbf{W}^r_{q,\alpha}, L_p) \asymp \varepsilon_k(\mathbf{W}^{r,0}_q, L_p) \asymp k^{-r} (\log k)^r, \qquad 1 \le p < \infty. \tag{7.9.3}$$

Relation (7.9.3) holds for the case $d = 2$ and a corresponding expression is known for the general case (see Theorems 7.7.1 and 7.7.2 above, and also (7.10.2) and its discussion below, and see Section 3.6 of Temlyakov (2011) for the corresponding results and historical comments). Relations (7.9.2) and (7.9.3) show that in the sense of entropy numbers the class $\mathbf{W}^{r,0}_1$ behaves as a limiting case of the classes $\mathbf{W}^r_{q,\alpha}$ when $q \to 1$.

The proofs of the upper bounds in Theorems 7.7.8 and 7.7.10 are based on the new and powerful greedy approximation technique. In particular, Theorem 7.7.12 gives the same upper bound as in (7.10.6) for the class $\bar{\mathbf{W}}_1^{r,0}$, which is wider than any of the classes $\mathbf{W}^r_{q,\alpha}$, $q > 1$.

Let us make some comments in the case $q = \infty$. Theorem 7.7.11 uncovers an interesting new phenomenon. Comparing (7.7.2) with (7.7.1), we see that the entropy numbers of the class $\bar{\mathbf{W}}^{a,b}_\infty$ in L_p-space have different rates of decay in the cases $1 \le p < \infty$ and $p = \infty$. We note that in the proof of the upper bounds in this new phenomenon we used the Riesz products for the hyperbolic crosses. This technique works well in the case $d = 2$ but we do not know how to extend it to the general case, $d > 2$. This difficulty is of the same nature as the corresponding difficulty in generalizing the small ball inequality from $d = 2$ to $d > 2$ (see Temlyakov, 2011, Chapter 3, for further discussion). We have already mentioned above that in studying the entropy numbers of function classes the discretization technique is useful. Classically, the Marcinkiewicz theorem serves as a powerful tool for discretizing the L_p-norm of a trigonometric polynomial. It works well in the multivariate case for trigonometric polynomials with frequencies from a parallelepiped. However, there is no analog of Marcinkiewicz' theorem for hyperbolic cross polynomials (see the discussion at the end of §7.5 and Kashin and Temlyakov, 2003 and Dinh Dung *et al.*, 2016, Section 2.5, for further discussion). Thus, in order to overcome this difficulty, we developed a new technique in §7.6 for estimating the entropy numbers of the unit balls of the hyperbolic cross polynomials. The most interesting results are obtained for dimension $d = 2$. It would be very interesting to extend these results to the case $d > 2$. It is a challenging open problem.

7.10 Some Historical Comments

It is well known that, in the univariate case,

$$\varepsilon_k(W^r_{q,\alpha}, L_p) \asymp k^{-r} \tag{7.10.1}$$

holds for all $1 \le q, p \le \infty$ and $r > (1/q - 1/p)_+$. We note that the condition $r > (1/q - 1/p)_+$ is necessary and sufficient for the compact embedding of $W^r_{q,\alpha}$ into L_p. Thus (7.10.1) provides a complete description of the decay rate of $\varepsilon_k(W^r_{q,\alpha}, L_p)$ in the univariate case. We point out that (7.10.1) shows that the rate of decay of $\varepsilon_k(W^r_{q,\alpha}, L_p)$ depends only on r and does not depend on q and p. In this sense the strongest upper bound (for $r > 1$) is $\varepsilon_k(W^r_{1,\alpha}, L_\infty) \ll k^{-r}$ and the strongest lower bound is $\varepsilon_k(W^r_{\infty,\alpha}, L_1) \gg k^{-r}$.

There are different generalizations of the classes $W^r_{q,\alpha}$ to the case of multivariate functions. In this section we discuss only the known results for the classes $\mathbf{W}^r_{q,\alpha}$ of functions with bounded mixed derivative. For further discussions see Temlyakov (2011), Chapter 3, and Dinh Dung *et al.* (2016).

The problem of estimating $\varepsilon_k(\mathbf{W}^r_{q,\alpha}, L_p)$ has a long history. The first result on the correct order of $\varepsilon_k(\mathbf{W}^r_{2,\alpha}, L_2)$ was obtained by Smolyak (1960). Later (see Temlyakov, 1988a, 1989d and Theorems 7.7.1, 7.7.2 above) it was established that

$$\varepsilon_k(\mathbf{W}^r_{q,\alpha}, L_p) \asymp k^{-r}(\log k)^{r(d-1)} \tag{7.10.2}$$

holds for all $1 < q, p < \infty$, $r > 1$. The case $1 < q = p < \infty$, $r > 0$ was established by Dinh Dung (1985). Belinsky (1998) extended (7.10.2) to the case $r > (1/q - 1/p)_+$ when $1 < q, p < \infty$.

It is known in approximation theory (see Temlyakov, 1993b), and we saw in Chapters 4–6, that investigation of the asymptotic characteristics of the classes $\mathbf{W}^r_{q,\alpha}$ in L_p becomes more difficult when q or p takes the value 1 or ∞ than when $1 < q, p < \infty$. This turns out to be the case for $\varepsilon_k(\mathbf{W}^r_{q,\alpha}, L_p)$ too. It was discovered that in some of these extreme cases (where q or p equals 1 or ∞) relation (7.10.2) holds and in other cases it does not hold. We will describe the picture in some detail. It was proved in Temlyakov (1989d) that (7.10.2) holds for $p = 1$, $1 < q < \infty$, $r > 0$. It was also proved that (7.10.2) holds for $p = 1$, $q = \infty$ (see Belinsky, 1998, for $r > 1/2$ and Kashin and Temlyakov, 1995, for $r > 0$). Summarizing, we state that (7.10.2) holds for $1 < q, p < \infty$ and $p = 1$, $1 < q \le \infty$ for all d (with appropriate restrictions on r). This easily implies that (7.10.2) also holds for $q = \infty$, $1 \le p < \infty$. For all other pairs (q, p), namely, for $p = \infty$, $1 \le q \le \infty$ and $q = 1$, $1 \le p \le \infty$ the decay rate of $\varepsilon_k(\mathbf{W}^r_{q,\alpha}, L_p)$ is not known in the case $d > 2$. It is an outstanding open problem.

In the case $d = 2$ this problem is essentially solved. We now cite the corresponding results. The first result on the correct order of $\varepsilon_k(\mathbf{W}^r_{q,\alpha}, L_p)$ in the case $p = \infty$ was obtained by Kuelbs and Li (1993) for $q = 2$, $r = 1$. It was proved in Temlyakov (1995a) that

$$\varepsilon_k(\mathbf{W}^r_{q,\alpha}, L_\infty) \asymp k^{-r}(\log k)^{r+1/2} \tag{7.10.3}$$

holds for $1 < q < \infty$, $r > 1$. We note that the upper bound in (7.10.3) was proved under the condition $r > 1$ and the lower bound in (7.10.3) was proved under the condition $r > 1/q$. Belinsky (1998) proved the upper bound in (7.10.3) for $1 < q < \infty$ under the condition $r > \max(1/q, 1/2)$. Relation (7.10.3) for $q = \infty$ under the assumption that $r > 1/2$ was proved in Temlyakov (1998a).

The case $q = 1$, $1 \le p \le \infty$ was settled by Kashin and Temlyakov (2003). The authors proved that

$$\varepsilon_k(\mathbf{W}^r_{1,\alpha}, L_p) \asymp k^{-r}(\log k)^{r+1/2} \tag{7.10.4}$$

holds for $1 \le p < \infty$, $r > \max(1/2, 1 - 1/p)$ and that

$$\varepsilon_k(\mathbf{W}^r_{1,0}, L_\infty) \asymp k^{-r}(\log k)^{r+1}, \quad r > 1. \tag{7.10.5}$$

Let us make an observation on the base of the above discussion. In the univariate case the entropy numbers $\varepsilon_k(W^r_{q,\alpha}, L_p)$ have the same order of decay with respect to k for all pairs (q,p), $1 \le q,p \le \infty$. In the case $d=2$ we have three different orders of decay of $\varepsilon_k(\mathbf{W}^r_{q,\alpha}, L_p)$, which depend on the pair (q,p). For instance, in the case $1 < q,p < \infty$ it is $k^{-r}(\log k)^r$, in the case $q=1$, $1<p<\infty$, it is $k^{-r}(\log k)^{r+1/2}$, and in the case $q=1$, $p=\infty$ it is $k^{-r}(\log k)^{r+1}$.

Above we discussed results on the correct order of decay of the entropy numbers. Clearly, each order relation $\asymp$ is a combination of the upper bound $\ll$ and the corresponding lower bound $\gg$. We now briefly discuss the methods that were used above for proving upper and lower bounds. The upper bounds in Theorem 7.7.1 were proved by the standard method of reduction by discretization to estimates of the entropy numbers of finite-dimensional sets. Here the results of Höllig (1980), Maiorov (1978) or Schütt (1984), are applied. It is clear from the above discussion that it is sufficient to prove the lower bound in (7.10.2) in the case $p=1$. The proof of the lower bound in this case (see Theorem 7.7.2) is more difficult and is based on nontrivial estimates of the volumes of the sets of Fourier coefficients of bounded trigonometric polynomials. Theorem 7.5.1 plays a key role in this method.

An analog of the upper bound in (7.10.3) for any d was obtained by Belinsky (1998): for $q>1$ and $r > \max(1/q, 1/2)$ we have

$$\varepsilon_k(\mathbf{W}^r_{q,\alpha}, L_\infty) \ll k^{-r}(\log k)^{(d-1)r+1/2}. \tag{7.10.6}$$

That proof is based on Theorem 7.10.1 below. Let $|\cdot| := \|\cdot\|_2$ denote the ℓ_2^D-norm and let B_2^D be a unit ball in ℓ_2^D. Denote by S^{D-1} the boundary of B_2^D. We define by $d\sigma(x)$ the normalized $(D-1)$-dimensional measure on S^{D-1}. Consider another norm $\|\cdot\|$ on $\mathbb{R}^D$ and denote by X the space $\mathbb{R}^D$ equipped with $\|\cdot\|$.

Theorem 7.10.1 *Let X be $\mathbb{R}^D$ equipped with $\|\cdot\|$ and let*

$$M_X := \int_{S^{D-1}} \|x\| d\sigma(x).$$

Then we have

$$\varepsilon_k(B_2^D, X) \ll M_X \begin{cases} (D/k)^{1/2}, & k \le D, \\ 2^{-k/D}, & k \ge D. \end{cases}$$

Kuelbs and Li (1993) discovered the fact that there is a tight relationship between the small ball problem and the behavior of the entropy $H_\varepsilon(\mathbf{W}^1_{2,\alpha}, L_\infty)$. Using results obtained by Lifshits and Tsirelson (1986), by Bass (1988), and by Talagrand (1994) for the small ball problem, they proved that

$$\varepsilon_k(\mathbf{W}^1_{2,\alpha}, L_\infty) \asymp k^{-1}(\ln k)^{3/2}. \tag{7.10.7}$$

The proof of the most difficult part of (7.10.7) – the lower bound – is based on a special inequality, now known as the Small Ball Inequality, for the Haar polynomials, proved by Talagrand (1994) (see Temlyakov, 1995b, for a simple proof).

We discussed above the known results on the rate of decay of $\varepsilon_k(\mathbf{W}^r_{q,\alpha}, L_p)$. In the case $d = 2$ the picture is almost complete. For $d > 2$ the situation is fundamentally different. The problem of the correct order of decay of $\varepsilon_k(\mathbf{W}^r_{q,\alpha}, L_p)$ is still open for $q = 1$, $1 \le p \le \infty$ and $p = \infty$, $1 \le q \le \infty$. In particular, it is open in the case $q = 2$, $p = \infty$, $r = 1$, which is related to the small ball problem. We now discuss in a little more detail the case $p = \infty$, $1 \le q \le \infty$. We pointed out above that in the case $d = 2$ the proof of lower bounds (the most difficult part) was based on the small ball inequalities for the Haar system for $r = 1$ and for the trigonometric system for all r. The existing conjecture is that

$$\varepsilon_k(\mathbf{W}^r_{q,\alpha}, L_\infty) \asymp k^{-r}(\ln k)^{(d-1)r+1/2}, \qquad 1 < q < \infty, \tag{7.10.8}$$

for large enough r. The upper bound in (7.10.8) follows from (7.10.6). It is known that the corresponding lower bound in (7.10.8) would follow from the d-dimensional version of the small ball inequality for the trigonometric system.

8
Greedy Approximation

8.1 Introduction

Estimation of a solution that satisfies a certain optimality criterion is the goal of many engineering applications. In many contemporary problems one would often like to obtain an approximation to the solution using a sparse linear combination of elements of a given system (a dictionary). This chapter is devoted to the theoretical aspects of sparse approximation. The main motivation for the study of sparse approximation is that many real-world signals can be well approximated by sparse signals. Sparse approximation automatically implies a need for nonlinear approximation, in particular, for greedy approximation. We give a brief description of a sparse approximation problem. In a general setting we are working in a Banach space X with a redundant system of elements forming a dictionary $\mathscr{D}$. There is a solid justification for the importance of the Banach-space setting in numerical analysis in general, and in sparse approximation in particular (see, for instance, the preface of Temlyakov, 2011, or Savu and Temlyakov, 2013). An element (function, signal) $f \in X$ is said to be K-sparse with respect to $\mathscr{D}$ if it has a representation $f = \sum_{i=1}^{K} x_i g_i$, $g_i \in \mathscr{D}$, $i = 1, \ldots, K$. The set of all K-sparse elements is denoted by $\Sigma_K(\mathscr{D})$. For a given element f we introduce the error of the best K-term approximation:

$$\sigma_K(f, \mathscr{D}) := \inf_{a \in \Sigma_K(\mathscr{D})} \|f - a\|.$$

We are interested in the following fundamental problem of sparse approximation.

Problem How can we design a practical algorithm that builds sparse approximations comparable to the best K-term approximations?

Clearly, the most difficult part of this problem is to identify the dictionary elements which are good for the K-term approximation. In the case of an orthonormal basis in a Hilbert space the recipe is evident: pick the elements from the expansion whose coefficient have the largest absolute value. This idea leads to a greedy algorithm, with respect to a basis in a Banach space, that we call the thresholding greedy algorithm (TGA). It turns out that this algorithm works excellently for bases like the univariate Haar basis. However, it does not work well for the trigonometric system, where interference between harmonics is essential. Recently, we discovered that another greedy algorithm – the weak Chebyshev greedy algorithm (WCGA) – works well for both the trigonometric system and the Haar basis. Results for both the TGA and the WCGA algorithms are presented in this chapter. We begin our discussion with the TGA.

Let a Banach space X, with a basis $\Psi = \{\psi_k\}_{k=1}^{\infty}$, be given. We assume that $\|\psi_k\| \geq C > 0$, $k = 1,2,\ldots$, and consider the following theoretical greedy algorithm. For a given element $f \in X$ we consider the expansion

$$f = \sum_{k=1}^{\infty} c_k(f,\Psi)\psi_k. \tag{8.1.1}$$

For an element $f \in X$ we say that a permutation ρ of the positive integers c_k is decreasing if

$$|c_{k_1}(f,\Psi)| \geq |c_{k_2}(f,\Psi)| \geq \cdots, \tag{8.1.2}$$

where $\rho(j) = k_j$, $j = 1,2,\ldots$, and write $\rho \in D(f)$. If the inequalities are strict in (8.1.2), then $D(f)$ consists of only one permutation. We define the mth greedy approximant of f, with respect to the basis Ψ corresponding to a permutation $\rho \in D(f)$, by the formula

$$G_m(f) := G_m(f,\Psi) := G_m(f,\Psi,\rho) := \sum_{j=1}^{m} c_{k_j}(f,\Psi)\psi_{k_j}.$$

The above simple algorithm $G_m(\cdot,\Psi)$ describes the theoretical scheme for the m-term approximation of an element f. We call this algorithm the thresholding greedy algorithm or simply the greedy algorithm (GA). In order to understand the efficiency of this algorithm we need to compare its accuracy with the best-possible accuracy when the approximant is a linear combination of m terms from Ψ. We define the best m-term approximation with respect to Ψ as follows:

$$\sigma_m(f) := \sigma_m(f,\Psi)_X := \inf_{c_k,\Lambda} \left\| f - \sum_{k\in\Lambda} c_k\psi_k \right\|_X,$$

where the infimum is taken over the coefficients c_k and sets of indices Λ with cardinality $|\Lambda| = m$. The best we can achieve with the algorithm G_m is

$$\|f - G_m(f,\Psi,\rho)\|_X = \sigma_m(f,\Psi)_X$$

or the slightly weaker approximation

$$\|f - G_m(f,\Psi,\rho)\|_X \leq G\sigma_m(f,\Psi)_X, \tag{8.1.3}$$

for all elements $f \in X$ and with a constant $G = C(X, \Psi)$ that is independent of f and m. It is clear that, when $X = H$ is a Hilbert space and $\mathscr{B}$ is an orthonormal basis, we have

$$\|f - G_m(f, \mathscr{B}, \rho)\|_H = \sigma_m(f, \mathscr{B})_H.$$

The concept of a greedy basis was introduced in Konyagin and Temlyakov (1999).

Definition 8.1.1 We call a basis Ψ greedy if, for every $f \in X$, there exists a permutation $\rho \in D(f)$ such that

$$\|f - G_m(f, \Psi, \rho)\|_X \le C\sigma_m(f, \Psi)_X,$$

with a constant C that is independent of f and m.

Lebesgue (1909) proved the following inequality: for any 2π-periodic continuous function f we have

$$\|f - S_n(f)\|_\infty \le \left(4 + \frac{4}{\pi^2} \ln n\right) E_n(f)_\infty, \tag{8.1.4}$$

where $S_n(f)$ is the nth partial sum of the Fourier series of f and $E_n(f)_\infty$ is the error of the best approximation of f by the trigonometric polynomials of order n in the uniform norm $\|\cdot\|_\infty$. *Lebesque inequality* (8.1.4) relates the error of a particular method (S_n) of approximation by the trigonometric polynomials of order n to the best-possible error $E_n(f)_\infty$ of approximation by the trigonometric polynomials of order n. By a *Lebesgue-type inequality* we mean an inequality that provides an upper estimate for the error of a particular method of approximation of f by elements of a special form, say $\mathscr{A}$, in relation to the best-possible approximation of f by elements of the form $\mathscr{A}$. In the case of approximation with regard to bases (or minimal systems), Lebesgue-type inequalities are known both in linear and in nonlinear settings (see the surveys Konyagin and Temlyakov, 2002 and Temlyakov, 2003a, 2008).

By Definition 8.1.1 greedy bases are those for which we have ideal (up to a multiplicative constant) Lebesgue inequalities for greedy approximation. In §8.2 we obtain Lebesgue-type inequalities for greedy approximation with respect to the trigonometric system. In §8.3 we study Lebesgue-type inequalities for greedy approximation with respect to the Haar basis and prove that the Haar basis is a greedy basis for L_p, $1 < p < \infty$. In §8.4 we obtain Lebesgue-type inequalities for multivariate systems of special structure. We consider systems $\Psi^d := \Psi \times \cdots \times \Psi$ (d times) which are tensor products of the univariate systems Ψ, concentrating on the case when Ψ is a greedy basis for $L_p([0,1))$, $1 < p < \infty$. In §§8.2–8.4 we obtain Lebesgue-type inequalities for the TGA; §8.7 is devoted to Lebesgue-type inequalities for the WCGA.

In §8.2 we consider the case where $X = L_p(\mathbb{T}^d)$, $1 \le p \le \infty$, $\Psi = \mathcal{T}^d := \{e^{i(\mathbf{k},\mathbf{x})}\}_{\mathbf{k}\in\mathbb{Z}^d}$ is the trigonometric system. We make a remark about an approximation of one special function by trigonometric polynomials that shows the advantage of nonlinear approximation over linear approximation.

Let us denote, for $f \in L_p(\mathbb{T})$,

$$E_n(f,\mathcal{T})_p := \inf_{c_k, |k|\le n} \left\| f(x) - \sum_{|k|\le n} c_k e^{ikx} \right\|_p .$$

Both de la Vallée Poussin (1908) and Bernstein (1912) proved that

$$E_n(|\sin x|, \mathcal{T})_\infty \asymp n^{-1}.$$

Ismagilov (1974) proved that

$$\sigma_n(|\sin x|, \mathcal{T})_\infty \le C_\varepsilon n^{-6/5+\varepsilon}$$

with arbitrary $\varepsilon > 0$. Later Maiorov (1978) proved that

$$\sigma_n(|\sin x|, \mathcal{T})_\infty \asymp n^{-3/2}.$$

These results showed the advantage of nonlinear approximation over linear approximation for typical individual functions. Now, when we know that efficiency of the m-term best approximation is satisfactory the following important problem arises. Construct an algorithm which realizes a satisfactory m-term approximation. It is clear from the definition of $\sigma_m(f,\mathcal{T}^d)_p$ that a satisfactory algorithm will be nonlinear. In §8.2 we concentrate on the efficiency of the thresholding greedy algorithm. We prove there the following theorem (Temlyakov, 1998c).

Theorem 8.1.2 *For each $f \in L_p(\mathbb{T}^d)$ we have*

$$\|f - G_m(f,\mathcal{T}^d)\|_p \le (1 + 3m^{h(p)})\sigma_m(f,\mathcal{T}^d)_p, \qquad 1 \le p \le \infty,$$

where $h(p) := |1/2 - 1/p|$.

In §8.3 we discuss another important class of bases, wavelet-type bases. We discuss in detail the simplest representative of such bases, the Haar basis. Denote by $\mathcal{H} := \{H_k\}_{k=1}^\infty$ the Haar basis on $[0,1)$, normalized in $L_2(0,1)$: $H_1 = 1$ on $[0,1)$ and, for $k = 2^n + l$, $n = 0,1,\dots$, and $l = 1,2,\dots,2^n$,

$$H_k = \begin{cases} 2^{n/2}, & x \in [(2l-2)2^{-n-1}, (2l-1)2^{-n-1}), \\ -2^{n/2}, & x \in [(2l-1)2^{-n-1}, 2l2^{-n-1}), \\ 0, & \text{otherwise}. \end{cases}$$

Write $\mathcal{H}_p := \{H_{k,p}\}_{k=1}^\infty$ for the Haar basis $\mathcal{H}$ renormalized in $L_p(0,1)$. We will use the following definition of the L_p-equivalence of bases. We say that $\Psi = \{\psi_k\}_{k=1}^\infty$ is

L_p-equivalent to $\Phi = \{\phi_k\}_{k=1}^{\infty}$, if for any finite set Λ and any coefficients c_k, $k \in \Lambda$, we have

$$C_1(p,\Psi,\Phi)\left\|\sum_{k\in\Lambda} c_k\phi_k\right\|_p \le \left\|\sum_{k\in\Lambda} c_k\psi_k\right\|_p \le C_2(p,\Psi,\Phi)\left\|\sum_{k\in\Lambda} c_k\phi_k\right\|_p, \tag{8.1.5}$$

with two positive constants $C_1(p,\Psi,\Phi), C_2(p,\Psi,\Phi)$ which may depend on p, Ψ, and Φ. For a discussion of sufficient conditions for Ψ to be L_p-equivalent to $\mathscr{H}$ see, for instance, Temlyakov (2011), Section 1.10. We will prove the following theorem in §8.3 (see Temlyakov, 1998e).

Theorem 8.1.3 *Let $1 < p < \infty$ and let a basis Ψ be L_p-equivalent to the Haar basis $\mathscr{H}_p$. Then, for any $f \in L_p(0,1)$ and any $\rho \in D(f)$, we have*

$$\|f - G_m(f,\Psi,\rho)\|_p \le C(p,\Psi)\sigma_m(f,\Psi)_p, \tag{8.1.6}$$

with a constant $C(p,\Psi)$ that is independent of f, ρ, and m.

Theorem 8.1.3 shows that each basis Ψ which is L_p-equivalent to the univariate Haar basis $\mathscr{H}_p$ is a greedy basis for $L_p(0,1)$, $1 < p < \infty$. We note that in the case of a Hilbert space each orthonormal basis is a greedy basis.

We now recall the definitions of unconditional and democratic bases.

Definition 8.1.4 A basis $\Psi = \{\psi_k\}_{k=1}^{\infty}$ of a Banach space X is said to be unconditional if, for every choice of signs $\theta = \{\theta_k\}_{k=1}^{\infty}$, $\theta_k = 1$ or -1, $k = 1,2,\dots$, the linear operator M_θ defined by $M_\theta(\sum_{k=1}^{\infty} a_k\psi_k) = \sum_{k=1}^{\infty} a_k\theta_k\psi_k$ is a bounded operator from X into X.

Definition 8.1.5 We say that a basis $\Psi = \{\psi_k\}_{k=1}^{\infty}$ is a democratic basis for X if there exists a constant $D := D(X,\Psi)$ such that, for any two finite sets of indices P and Q with the same cardinality $|P| = |Q|$, we have $\|\sum_{k\in P}\psi_k\| \le D\|\sum_{k\in Q}\psi_k\|$.

The following theorem was proved in Konyagin and Temlyakov (1999).

Theorem 8.1.6 *A normalized basis is greedy if and only if it is unconditional and democratic.*

This theorem gives a characterization of greedy bases. Further investigations (Temlyakov, 1998d, Cohen *et al.*, 2000, Dilworth *et al.*, 2003b, Kerkyacharian and Picard, 2006, Gribonval and Nielsen, 2001, Temlyakov, 2002b, and Kamont and Temlyakov, 2004) showed that the concept of greedy bases is very useful in the direct and inverse theorems of nonlinear approximation and also in applications in statistics. It was noticed in Dilworth *et al.* (2003b) that the proof of Theorem 8.1.6 from Konyagin and Temlyakov (1999) works also for a basis that is not assumed to be normalized (they assumed instead that $\inf_n \|\psi_n\| > 0$). In §8.5 we prove a more general version of Theorem 8.1.6. We refer the reader for further detailed discussion of greedy-type bases to the book Temlyakov (2015c).

In §8.6 we consider the general setting of greedy approximation in Banach spaces. We demonstrate in §8.7 that the weak Chebyshev greedy algorithm (WCGA), which we define shortly, enables the construction of sparse approximations comparable with best m-term approximations.

Let X be a real Banach space with norm $\|\cdot\| := \|\cdot\|_X$. We say that a set of elements (functions) $\mathscr{D}$ from X is a dictionary if each $g \in \mathscr{D}$ has norm bounded by one ($\|g\| \leq 1$), and the closure of span $\mathscr{D}$ is X. For a nonzero element $g \in X$ we let F_g denote a norming (peak) functional for g:

$$\|F_g\|_{X^*} = 1, \qquad F_g(g) = \|g\|_X.$$

The existence of such a functional is guaranteed by the Hahn–Banach theorem.

Let $\tau := \{t_k\}_{k=1}^{\infty}$ be a given weakness sequence of nonnegative numbers $t_k \leq 1$, $k = 1, \ldots$ We define the weak Chebyshev greedy algorithm (see Temlyakov, 2001) as a generalization, for Banach spaces, of the weak orthogonal matching pursuit (WOMP). In a Hilbert space the WCGA coincides with the WOMP. The WOMP is very popular in signal processing, in particular, in compressed sensing. In approximation theory the WOMP is called the weak orthogonal greedy algorithm (WOGA). We study the WCGA in detail in §§8.6 and 8.7.

Weak Chebyshev greedy algorithm. Let f_0 be given. Then, for each $m \geq 1$, we have the following inductive definition.

(1) Let $\varphi_m := \varphi_m^{c,\tau} \in \mathscr{D}$ be any element satisfying

$$|F_{f_{m-1}}(\varphi_m)| \geq t_m \sup_{g \in \mathscr{D}} |F_{f_{m-1}}(g)|.$$

(2) Define

$$\Phi_m := \Phi_m^{\tau} := \operatorname{span}\{\varphi_j\}_{j=1}^{m},$$

and define $G_m := G_m^{c,\tau}$ to be the best approximant to f_0 from Φ_m.

(3) Let

$$f_m := f_m^{c,\tau} := f_0 - G_m.$$

In §8.7 we consider only the case $t_k = t \in (0,1]$, $k = 1, 2, \ldots$

The trigonometric system is a classical system that is known to be difficult to treat. In §8.7 we study, among other problems, that of nonlinear sparse approximation with respect to the trigonometric system. Let $\mathscr{RT}$ denote the real trigonometric system $1, \sin 2\pi x, \cos 2\pi x, \ldots$ on $[0,1]$ and let $\mathscr{RT}_p$ to be its version normalized in $L_p(0,1)$. Denote by $\mathscr{RT}_p^d := \mathscr{RT}_p \times \cdots \times \mathscr{RT}_p$ the d-variate trigonometric system. We need to consider a real trigonometric system because the WCGA has been well studied for real Banach spaces. In order to illustrate the performance of the WCGA we discuss in this section the above mentioned problem for a real trigonometric system. We prove in §8.7 the following Lebesgue-type inequality for the WCGA from Temlyakov (2014); see Corollary 8.7.22 below.

Theorem 8.1.7 *Let $\mathscr{D}$ be a real d-variate trigonometric system normalized in L_p, $2 \le p < \infty$. Then, for any $f_0 \in L_p$, the WCGA with weakness parameter t gives*

$$\|f_{C(t,p,d)m\ln(m+1)}\|_p \le C\sigma_m(f_0,\mathscr{D})_p. \tag{8.1.7}$$

Open Problem 7.1 from Temlyakov (2003a), p. 91, asks whether (8.1.7) holds without an extra $\ln(m+1)$ factor. Theorem 8.1.7 is the first result on Lebesgue-type inequalities for the WCGA with respect to the trigonometric system. It provides a progress in solving the above-mentioned open problem, but the problem remains open.

Theorem 8.1.7 shows that the WCGA is very appropriate for the trigonometric system. We show in Proposition 8.7.21 that an analog of (8.1.7) holds for uniformly bounded orthogonal systems. We note that it is known (see Temlyakov, 2011, and Theorem 8.1.3 above) that the TGA is very suitable for bases that are L_p-equivalent to the Haar basis, $1 < p < \infty$.

The proof of Theorem 8.1.7 uses a technique developed in Temlyakov (2014) for proving Lebesgue-type inequalities for redundant dictionaries with special properties. We present these results in §8.7. They are an extension of earlier work of the Livshitz and Temlyakov (2014). In §8.7.4 we test the power of general results on specific dictionaries, namely, on bases. In §8.7.4 we provide a number of examples, including the trigonometric system, where the technique from §8.7.2 can be successfully applied. In particular, results in §8.7 demonstrate that the general technique from §8.7.2 provides almost optimal m-term approximation results for uniformly bounded orthogonal systems (see Proposition 8.7.21). Proposition 8.7.34 shows that an extra assumption, that a uniformly bounded orthogonal system Ψ forms a quasi-greedy basis, allows us to improve inequality (8.1.7) to

$$\|f_{C(t,p,\Psi)m\ln\ln(m+3)}\|_p \le C\sigma_m(f_0,\Psi)_p.$$

In this chapter we discuss in detail two greedy-type algorithms, the TGA and the WCGA defined above. Both these algorithms use a greedy step in building an approximant. The greedy step of the TGA at the mth iteration involves finding the largest $|c_k(f_{m-1})|$, which is equivalent to finding the mth largest $|c_k(f)|$. For instance, in the case of $\Psi = \mathscr{T}^d$ this step means finding the Fourier coefficient of f_{m-1} that is largest in absolute value. The greedy step of the WCGA is more complicated. In this step we look for $\varphi_m \in \mathscr{D}$ satisfying

$$|F_{f_{m-1}}(\varphi_m)| \geq t_m \sup_{g\in\mathscr{D}} |F_{f_{m-1}}(g)|.$$

The greedy step is based on the norming functional $F_{f_{m-1}}$. The existence of such a functional is guaranteed by the Hahn–Banach theorem but it could be a hard task to build such a functional. The norming functional F_f is a linear functional (in other words it is an element of the space X^* dual to X), which can be explicitly written down in some cases. In a Hilbert space, F_f can be identified with $f\|f\|^{-1}$. In a real L_p, $1 < p < \infty$, it can be identified with $f|f|^{p-2}\|f\|_p^{1-p}$. An important advantage of L_p-spaces is the simple and explicit form of the norming functional F_f of a function $f \in L_p(\Omega)$. The F_f acts as follows (for real L_p-spaces):

$$F_f(g) = \int_\Omega \|f\|_p^{1-p}|f(\mathbf{x})|^{p-2}f(\mathbf{x})g(\mathbf{x})d\mu.$$

Thus, for instance, in the case $\mathscr{D} = \mathscr{R}\mathscr{T}^d$ the WCGA should find at the mth iteration the Fourier coefficient of $f_{m-1}(\mathbf{x})|f_{m-1}(\mathbf{x})|^{p-2}$ that is largest in absolute value; This is somewhat similar to the corresponding step of the TGA. Finally, we stress that the TGA works only for bases and that there is no analog of the TGA for redundant dictionaries. The WCGA works for any dictionary.

8.2 The Trigonometric System

In this section we prove Theorem 8.1.2. We formulate it again here for convenience.

Theorem 8.2.1 *For each $f \in L_p(\mathbb{T}^d)$ we have*

$$\|f - G_m(f,\mathscr{T}^d)\|_p \leq (1+3m^{h(p)})\sigma_m(f,\mathscr{T}^d)_p, \qquad 1 \leq p \leq \infty,$$

where $h(p) := |1/2 - 1/p|$.

Proof We treat separately the two cases $1 \leq p \leq 2$ and $2 \leq p \leq \infty$. Before splitting the discussion into these two cases we prove an auxiliary statement for $1 \leq p \leq \infty$. Here we use the notation

$$\hat{f}(\mathbf{k}) := (2\pi)^{-d}\int_{\mathbb{T}^d} f(\mathbf{x})e^{-i(\mathbf{k},\mathbf{x})}d\mathbf{x}.$$

Lemma 8.2.2 *Let $\Lambda \subset \mathbb{Z}^d$ be a finite subset with cardinality $|\Lambda| = m$. Then, for the operator S_Λ defined on $L_1(\mathbb{T}^d)$ by*

$$S_\Lambda(f) := \sum_{\mathbf{k}\in\Lambda} \hat{f}(\mathbf{k})e^{i(\mathbf{k},\mathbf{x})},$$

we have, for all $1 \leq p \leq \infty$,

$$\|S_\Lambda(f)\|_p \leq m^{h(p)}\|f\|_p. \tag{8.2.1}$$

Proof For a given linear operator A denote by $\|A\|_{a\to b}$ the norm of this operator as an operator from $L_a(\mathbb{T}^d)$ to $L_b(\mathbb{T}^d)$. Then it is obvious that

$$\|S_\Lambda\|_{2\to 2} = 1. \tag{8.2.2}$$

Consider

$$\mathscr{D}_\Lambda(\mathbf{x}) := \sum_{\mathbf{k}\in\Lambda} e^{i(\mathbf{k},\mathbf{x})}; \tag{8.2.3}$$

then

$$S_\Lambda(f) = f * \mathscr{D}_\Lambda := (2\pi)^{-d} \int_{\mathbb{T}^d} f(\mathbf{x}-\mathbf{y})\mathscr{D}_\Lambda(\mathbf{y})d\mathbf{y}$$

and, for $p=1$ or $p=\infty$, we have

$$\|S_\Lambda\|_{p\to p} \le \|\mathscr{D}_\Lambda\|_1 \le \|\mathscr{D}_\Lambda\|_2 = m^{1/2}. \tag{8.2.4}$$

The relations (8.2.2) and (8.2.4) and the Riesz–Thorin theorem (see Theorem A.3.2 and also Zygmund, 1959) imply (8.2.1). □

We now return to the proof of Theorem 8.2.1.

Case 1: $2 \le p \le \infty$ Take any function $f \in L_p(\mathbb{T}^d)$. Let t_m be a trigonometric polynomial which realizes the best m-term approximation to f in $L_p(\mathbb{T}^d)$. For the existence of t_m see Theorem 1.7 from Temlyakov (2011), p. 10. Denote by Λ the set of frequencies of t_m, i.e. $\Lambda := \{\mathbf{k} : \hat{t}_m(\mathbf{k}) \neq 0\}$; then $|\Lambda| \le m$. Denote by Λ' the set of frequencies of $G_m(f) := G_m(f, \mathscr{T}^d)$; then $|\Lambda'| = m$. Let us use the representation

$$f - G_m(f) = f - S_{\Lambda'}(f) = f - S_\Lambda(f) + S_\Lambda(f) - S_{\Lambda'}(f).$$

From this representation we derive

$$\|f - G_m(f)\|_p \le \|f - S_\Lambda(f)\|_p + \|S_\Lambda(f) - S_{\Lambda'}(f)\|_p. \tag{8.2.5}$$

We use Lemma 8.2.2 to estimate the first term on the right-hand side of (8.2.5):

$$\|f - S_\Lambda(f)\|_p = \|f - t_m - S_\Lambda(f - t_m)\|_p \le (1 + m^{h(p)})\sigma_m(f, \mathscr{T}^d)_p. \tag{8.2.6}$$

In estimating the second term in (8.2.5) we use the well-known inequality $\|f\|_2 \le \|f\|_p$ for $2 \le p \le \infty$ and the following lemma.

Lemma 8.2.3 *Let $\Lambda \subset \mathbb{Z}^d$ be a finite subset with cardinality $|\Lambda| = n$. Then, for $2 \le p \le \infty$, we have*

$$\|S_\Lambda(f)\|_p \le n^{h(p)} \|S_\Lambda(f)\|_2 \le n^{h(p)} \|f\|_2. \tag{8.2.7}$$

Proof For $p = \infty$ we have

$$\|S_\Lambda(f)\|_\infty \le \sum_{\mathbf{k}\in\Lambda} |\hat{f}(\mathbf{k})| \le n^{1/2}\left(\sum_{\mathbf{k}\in\Lambda} |\hat{f}(\mathbf{k})|^2\right)^{1/2} \le n^{1/2}\|S_\Lambda(f)\|_2. \tag{8.2.8}$$

For $2 < p < \infty$ we use (8.2.8) and the following well-known inequality:

$$\|g\|_p \le \|g\|_2^{2/p}\|g\|_\infty^{1-2/p}.$$

□

We now continue to estimate $\|S_\Lambda(f) - S_{\Lambda'}(f)\|_p$. Using Lemma 8.2.3 we get

$$\begin{aligned}\|S_\Lambda(f) - S_{\Lambda'}(f)\|_p &= \|S_{\Lambda\setminus\Lambda'}(f) - S_{\Lambda'\setminus\Lambda}(f)\|_p \le \|S_{\Lambda\setminus\Lambda'}(f)\|_p + \|S_{\Lambda'\setminus\Lambda}(f)\|_p \\ &\le m^{h(p)}(\|S_{\Lambda\setminus\Lambda'}(f)\|_2 + \|S_{\Lambda'\setminus\Lambda}(f)\|_2). \end{aligned} \tag{8.2.9}$$

The definition of Λ' and the relations $|\Lambda'| = m$, $|\Lambda| \le m$ imply that

$$\|S_{\Lambda\setminus\Lambda'}(f)\|_2 \le \|S_{\Lambda'\setminus\Lambda}(f)\|_2. \tag{8.2.10}$$

Finally, we have

$$\|S_{\Lambda'\setminus\Lambda}(f)\|_2 \le \|f - S_\Lambda(f)\|_2 \le \|f - t_m\|_2 \le \|f - t_m\|_p = \sigma_m(f, \mathcal{T}^d)_p. \tag{8.2.11}$$

Combining the relations (8.2.9)–(8.2.11) we get

$$\|S_\Lambda(f) - S_{\Lambda'}(f)\|_p \le 2m^{h(p)}\sigma_m(f, \mathcal{T}^d)_p. \tag{8.2.12}$$

The relations (8.2.5), (8.2.6) and (8.2.12) result in

$$\|f - G_m(f)\|_p \le (1 + 3m^{h(p)})\sigma_m(f, \mathcal{T}^d)_p.$$

This completes the proof of Theorem 8.2.1 in the case $2 \le p \le \infty$.

Case 2: $1 \le p \le 2$ We keep the notation of case 1. Again we start with the inequality (8.2.5). Next, the inequality (8.2.6) holds for $1 \le p \le 2$ also because it is based on Lemma 8.2.2, which covers the whole range $1 \le p \le \infty$ of the parameter p. Thus, it remains to estimate $\|S_\Lambda(f) - S_{\Lambda'}(f)\|_p$. Using the inequality $\|f\|_p \le \|f\|_2$ we get

$$\begin{aligned}\|S_\Lambda(f) - S_{\Lambda'}(f)\|_p &= \|S_{\Lambda\setminus\Lambda'}(f) - S_{\Lambda'\setminus\Lambda}(f)\|_p \le \|S_{\Lambda\setminus\Lambda'}(f)\|_p + \|S_{\Lambda'\setminus\Lambda}(f)\|_p \\ &\le \|S_{\Lambda\setminus\Lambda'}(f)\|_2 + \|S_{\Lambda'\setminus\Lambda}(f)\|_2. \end{aligned} \tag{8.2.13}$$

In order to estimate $\|S_{\Lambda'\setminus\Lambda}(f)\|_2$ we use the part of the Hausdorff–Young theorem (see Theorem A.3.1) which states that

$$\|(\hat{f}(\mathbf{k}))_{\mathbf{k}\in\mathbb{Z}^d}\|_{\ell_{p'}} \le \|f\|_p, \qquad 1 \le p \le 2, \qquad p' := \frac{p}{p-1}.$$

We have

$$\begin{aligned}
\|S_{\Lambda'\setminus\Lambda}(f)\|_2 &= \|(\hat{f}(\mathbf{k}))_{\mathbf{k}\in\Lambda'\setminus\Lambda}\|_{\ell_2} \le |\Lambda'\setminus\Lambda|^{1/p-1/2}\|(\hat{f}(\mathbf{k}))_{\mathbf{k}\in\Lambda'\setminus\Lambda}\|_{\ell_{p'}} \\
&\le m^{h(p)}\|(\hat{f}(\mathbf{k})-\hat{t}_m(\mathbf{k}))_{\mathbf{k}\in\mathbb{Z}^d}\|_{\ell_{p'}} \le m^{h(p)}\|f-t_m\|_p \\
&= m^{h(p)}\sigma_m(f,\mathcal{T}^d)_p. \qquad (8.2.14)
\end{aligned}$$

Gathering together (8.2.5), (8.2.6), (8.2.10), (8.2.13), and (8.2.14) we get

$$\|f-G_m(f)\|_p \le (1+3m^{h(p)})\sigma_m(f,\mathcal{T}^d)_p,$$

which completes the proof of Theorem 8.2.1. □

Remark 8.2.4 Lemma 8.2.2 implies that, for all $1\le p\le\infty$,

$$\|G_m(f)\|_p \le m^{h(p)}\|f\|_p. \qquad (8.2.15)$$

Remark 8.2.5 There is a positive absolute constant C such that, for each m and $1\le p\le\infty$ there exists a function $f\neq 0$ with the property

$$\|G_m(f)\|_p \ge Cm^{h(p)}\|f\|_p.$$

Remark 8.2.6 The trivial inequality $\sigma_m(f,\mathcal{T}^d)_p \le \|f\|_p$ and Remark 8.2.5 show that the factor $m^{h(p)}$ in Theorem 8.2.1 is sharp in the sense of growth order.

Remark 8.2.7 Using Remark 8.2.5 it is easy to construct for each $p\neq 2$ a function $f\in L_p(\mathbb{T})$ such that the sequence $\{\|G_m(f)\|_p\}_{m=1}^{\infty}$ is not bounded.

Remarks 8.2.5–8.2.7 show that the TGA does not work well for the trigonometric system. Here are two more results in this direction from Temlyakov (2003a).

Theorem 8.2.8 *There exists a continuous function f such that $G_m(f,\mathcal{T})$ does not converge to f in L_p for any $p>2$.*

Theorem 8.2.9 *There exists a function f that belongs to any L_p, $p<2$, such that $G_m(f,\mathcal{T})$ does not converge to f in measure.*

We now make some remarks about possible generalizations of Theorem 8.2.1. Reviewing the proof of Theorem 8.2.1 one verifies that all the arguments used hold true for any orthonormal system $\{\phi_j\}_{j=1}^{\infty}$ of uniformly bounded functions $\|\phi_j\|_\infty \le M$, $j=1,2,\dots$ The only differences are that instead of the Hausdorff–Young theorem we use the Riesz theorem and that the constants in Lemmas 8.2.2 and 8.2.3 depend on M. Let us formulate the corresponding analog of Theorem 8.2.1. Let $\Phi := \{\phi_j\}_{j=1}^{\infty}$ be an orthonormal system in $L_2(\mathbb{T}^d)$ such that $\|\phi_j\|_\infty \le M$, $j=1,2,\dots$

Theorem 8.2.10 *For any orthonormal system $\Phi=\{\phi_j\}_{j=1}^{\infty}$ of uniformly bounded functions $\|\phi_j\|_\infty\le M$ there exists a constant $C(M)$ such that*

$$\|f-G_m(f,\Phi)\|_p \le C(M)m^{h(p)}\sigma_m(f,\Phi)_p, \quad 1\le p\le\infty,$$

where $h(p) := |1/2 - 1/p|$.

8.3 Wavelet Bases

In this section it will be convenient for us to index elements of bases by dyadic intervals $\psi_1 =: \psi_{[0,1]}$ and

$$\psi_{2^n+l} =: \psi_I, \qquad I = [(l-1)2^{-n}, l2^{-n}).$$

We note that there is another natural greedy-type algorithm based on ordering the $\|c_k(f,\Psi)\psi_k\|$ instead of the absolute values of coefficients. In this case we do not need the restriction $\|\psi_k\| \geq C > 0$, $k = 1,2,\dots$ Let $\Lambda_m(f)$ be a set of indices such that

$$\min_{k\in\Lambda_m(f)} \|c_k(f,\Psi)\psi_k\| \geq \max_{k\notin\Lambda_m(f)} \|c_k(f,\Psi)\psi_k\|.$$

We define $G_m^X(f,\Psi)$ by the formula

$$G_m^X(f,\Psi) := S_{\Lambda_m(f)}(f,\Psi), \qquad \text{where } S_E(f) := S_E(f,\Psi) := \sum_{k\in E} c_k(f,\Psi)\psi_k.$$

It is clear that for a normalized basis ($\|\psi_k\| = 1$, $k = 1,2,\dots$) the above greedy algorithm $G_m^X(\cdot,\Psi)$ coincides with the TGA. It is also clear that the above greedy algorithm $G_m^X(\cdot,\Psi)$ can be considered as a greedy algorithm $G_m(\cdot,\Psi')$ with $\Psi' := \{\psi_k/\|\psi_k\|\}_{k=1}^{\infty}$ a normalized version of the Ψ. Thus, we will concentrate on studying the algorithm $G_m(\cdot,\Psi)$. In the above definition of $G_m(\cdot,\Psi)$ we impose an extra condition on the basis Ψ, i.e., $\inf_k \|\psi_k\| > 0$. This restriction allows us to define $G_m(f,\Psi)$ for all $f \in X$. We begin by proving Theorem 8.3.1 below (see Temlyakov, 1998e) and note that Theorem 8.1.3 from §8.1 follows from Theorem 8.3.1 by a simple renormalization argument.

Theorem 8.3.1 *Let* $1 < p < \infty$ *and let a basis* $\Psi := \{\psi_I\}_I$ *be* L_p*-equivalent to* $\mathcal{H}$. *Then, for any* $f \in L_p$, *we have*

$$\|f - G_m^p(f,\Psi)\|_p \leq C(p,\Psi)\sigma_m(f,\Psi)_p$$

with $G_m^p(f,\Psi) := G_m^{L_p}(f,\Psi)$.

Proof Let us take a parameter $0 < t \leq 1$ and consider the following greedy-type algorithm $G^{p,t}$ with regard to the Haar system. For the Haar basis $\mathcal{H}$ we define

$$c_I(f) := \langle f, H_I\rangle = \int_0^1 f(x)H_I(x)dx.$$

Denote by $\Lambda_m(t)$ any set of m dyadic intervals such that

$$\min_{I\in\Lambda_m(t)} \|c_I(f)H_I\|_p \geq t \max_{J\notin\Lambda_m(t)} \|c_J(f)H_J\|_p, \tag{8.3.1}$$

and define

$$G_m^{p,t}(f) := G_m^{p,t}(f, \mathcal{H}) := \sum_{I \in \Lambda_m(t)} c_I(f) H_I. \tag{8.3.2}$$

For a given function $f \in L_p$ we define

$$g(f) := \sum_I c_I(f, \Psi) H_I. \tag{8.3.3}$$

It is clear that $g(f) \in L_p$ and

$$\sigma_m(g(f), \mathcal{H})_p \le C_1(p)^{-1} \sigma_m(f, \Psi)_p; \tag{8.3.4}$$

here and later on we use the brief notation $C_i(p) := C_i(p, \Psi, \mathcal{H})$, $i = 1, 2$, for the constants from (8.1.5). Let

$$G_m^p(f, \Psi) = \sum_{I \in \Lambda_m} c_I(f, \Psi) \psi_I.$$

Next, for any two intervals $I \in \Lambda_m$, $J \notin \Lambda_m$, by the definition of Λ_m we have

$$\|c_I(f, \Psi) \psi_I\|_p \ge \|c_J(f, \Psi) \psi_J\|_p.$$

Using (8.1.5) we then obtain

$$\begin{aligned} \|c_I(g(f)) H_I\|_p &= \|c_I(f, \Psi) H_I\|_p \ge C_2(p)^{-1} \|c_I(f, \Psi) \psi_I\|_p \\ &\ge C_2(p)^{-1} \|c_J(f, \Psi) \psi_J\|_p \ge C_1(p) C_2(p)^{-1} \|c_J(g(f)) H_J\|_p. \end{aligned} \tag{8.3.5}$$

Writing $t = C_1(p) C_2(p)^{-1}$, this inequality implies that for any m we can find a set $\Lambda_m(t)$ such that $\Lambda_m(t) = \Lambda_m$ and, therefore,

$$\|f - G_m^p(f, \Psi)\|_p \le C_2(p) \|g(f) - G_m^{p,t}(g(f))\|_p. \tag{8.3.6}$$

The relations (8.3.4) and (8.3.6) show that Theorem 8.3.1 follows from Theorem 8.3.2 below, which we will prove shortly. □

Theorem 8.3.2 *Let* $1 < p < \infty$ *and* $0 < t \le 1$. *Then, for any* $g \in L_p$, *we have*

$$\|g - G_m^{p,t}(g, \mathcal{H})\|_p \le C(p,t) \sigma_m(g, \mathcal{H})_p.$$

Proof The Littlewood–Paley theorem for the Haar system gives, for $1 < p < \infty$,

$$C_3(p) \left\| \left(\sum_I |c_I(g) H_I|^2 \right)^{1/2} \right\|_p \le \|g\|_p \le C_4(p) \left\| \left(\sum_I |c_I(g) H_I|^2 \right)^{1/2} \right\|_p. \tag{8.3.7}$$

First we formulate two simple corollaries from (8.3.7):

$$\|g\|_p \le C_5(p)\Big(\sum_I \|c_I(g)H_I\|_p^p\Big)^{1/p}, \qquad 1 < p \le 2, \tag{8.3.8}$$

$$\|g\|_p \le C_6(p)\Big(\sum_I \|c_I(g)H_I\|_p^2\Big)^{1/2}, \qquad 2 \le p < \infty. \tag{8.3.9}$$

The inequalities dual to (8.3.8) and (8.3.9) are

$$\|g\|_p \ge C_7(p)\Big(\sum_I \|c_I(g)H_I\|_p^2\Big)^{1/2}, \qquad 1 < p \le 2, \tag{8.3.10}$$

$$\|g\|_p \ge C_8(p)\Big(\sum_I \|c_I(g)H_I\|_p^p\Big)^{1/p}, \qquad 2 \le p < \infty. \tag{8.3.11}$$

We now proceed to the proof of Theorem 8.3.2. Let T_m be an m-term Haar polynomial of best m-term approximation to g in L_p (for the existence see Baishanski, 1983, Dubinin, 1997 and also Theorems 1.8 and 1.9 from Temlyakov, 2011):

$$T_m = \sum_{I\in\Lambda} a_I H_I, \qquad |\Lambda| = m.$$

For any finite set Q of dyadic intervals we denote by S_Q the projector

$$S_Q(f) := \sum_{I\in Q} c_I(f)H_I.$$

From (8.3.7) we get

$$\begin{aligned}\|g - S_\Lambda(g)\|_p &= \|g - T_m - S_\Lambda(g - T_m)\|_p \le \|\mathrm{Id} - S_\Lambda\|_{p\to p}\sigma_m(g,\mathcal{H})_p \\ &\le C_4(p)C_3(p)^{-1}\sigma_m(g,\mathcal{H})_p,\end{aligned} \tag{8.3.12}$$

where Id denotes the identify operator. Further, we have

$$G_m^{p,t}(g) = S_{\Lambda_m(t)}(g)$$

and

$$\|g - G_m^{p,t}(g)\|_p \le \|g - S_\Lambda(g)\|_p + \|S_\Lambda(g) - S_{\Lambda_m(t)}(g)\|_p. \tag{8.3.13}$$

The first term on the right-hand side of (8.3.13) is estimated in (8.3.12). We now estimate the second term. We represent it in the form

$$S_\Lambda(g) - S_{\Lambda_m(t)}(g) = S_{\Lambda\setminus\Lambda_m(t)}(g) - S_{\Lambda_m(t)\setminus\Lambda}(g)$$

and remark that we get a similar result to (8.3.12):

$$\|S_{\Lambda_m(t)\setminus\Lambda}(g)\|_p \le C_9(p)\sigma_m(g,\mathcal{H})_p. \tag{8.3.14}$$

The key point of the proof of Theorem 8.3.2 is the estimate

$$\|S_{\Lambda \setminus \Lambda_m(t)}(g)\|_p \leq C(p,t)\|S_{\Lambda_m(t) \setminus \Lambda}(g)\|_p, \tag{8.3.15}$$

which will be derived from the following two lemmas.

Lemma 8.3.3 *Consider*

$$f = \sum_{I \in Q} c_I H_I, \qquad |Q| = N.$$

Let $1 \leq p < \infty$. Assume that

$$\|c_I H_I\|_p \leq 1, \qquad I \in Q. \tag{8.3.16}$$

Then

$$\|f\|_p \leq C_{10}(p)N^{1/p}.$$

Lemma 8.3.4 *Consider*

$$f = \sum_{I \in Q} c_I H_I, \qquad |Q| = N.$$

Let $1 < p \leq \infty$. Assume that

$$\|c_I H_I\|_p \geq 1, \qquad I \in Q.$$

Then

$$\|f\|_p \geq C_{11}(p)N^{1/p}.$$

Proof First we prove Lemma 8.3.3. We note that in the case $1 < p \leq 2$ it follows from (8.3.8). We will now give a proof of this lemma for all $1 \leq p < \infty$. We have

$$\|c_I H_I\|_p = |c_I||I|^{1/p-1/2}.$$

The assumption (8.3.16) implies that

$$|c_I| \leq |I|^{1/2-1/p}.$$

Next, we have

$$\|f\|_p \leq \left\| \sum_{I \in Q} |c_I H_I| \right\|_p \leq \left\| \sum_{I \in Q} |I|^{-1/p} \chi_I(x) \right\|_p, \tag{8.3.17}$$

where $\chi_I(x)$ is a characteristic function for the interval I:

$$\chi_I(x) = \begin{cases} 1, & x \in I, \\ 0, & x \notin I. \end{cases}$$

In order to proceed further we need another simple lemma. Statements similar to Lemma 8.3.5 below are often used in the theory of wavelets (see for instance Hsiao *et al.*, 1994).

Lemma 8.3.5 *Let $n_1 < n_2 < \cdots < n_s$ be integers and let $E_j \subset [0,1]$ be measurable sets, $j = 1,\ldots,s$. Then for any $0 < q < \infty$ we have*

$$\int_0^1 \left(\sum_{j=1}^{s} 2^{n_j/q}\chi_{E_j}(x)\right)^q dx \le C_{12}(q)\sum_{j=1}^{s} 2^{n_j}|E_j|.$$

Proof Write

$$F(x) := \sum_{j=1}^{s} 2^{n_j/q}\chi_{E_j}(x)$$

and estimate it on the sets

$$E_l^- := E_l \setminus \bigcup_{k=l+1}^{s} E_k, \qquad l = 1,\ldots,s-1, \qquad E_s^- := E_s.$$

We have, for $x \in E_l^-$,

$$F(x) \le \sum_{j=1}^{l} 2^{n_j/q} \le C(q)2^{n_l/q}.$$

Therefore,

$$\int_0^1 F(x)^q dx \le C(q)^q \sum_{l=1}^{s} 2^{n_l}|E_l^-| \le C(q)^q \sum_{l=1}^{s} 2^{n_l}|E_l|,$$

which proves the lemma. □

We return to the proof of Lemma 8.3.3. Denote by $n_1 < n_2 < \cdots < n_s$ all integers such that there is an $I \in Q$ with $|I| = 2^{-n_j}$. Introduce the sets

$$E_j := \bigcup_{I \in Q; |I| = 2^{-n_j}} I.$$

Then the number N of elements in Q can be written in the form

$$N = \sum_{j=1}^{s} |E_j|2^{n_j}. \tag{8.3.18}$$

Using this notation, the right-hand side of (8.3.17) can be rewritten as

$$Y := \left(\int_0^1 \left(\sum_{j=1}^{s} 2^{n_j/p}\chi_{E_j}(x)\right)^p dx\right)^{1/p}.$$

Applying Lemma 8.3.5 with $q = p$ we get

$$\|f\|_p \le Y \le C_{13}(p)\left(\sum_{j=1}^{s} |E_j|2^{n_j}\right)^{1/p} = C_{13}(p)N^{1/p}.$$

At the last step we used (8.3.18). Lemma 8.3.3 is proved. □

We now prove Lemma 8.3.4. We can derive it from Lemma 8.3.3. Define

$$u := \sum_{I\in Q} \bar{c}_I |c_I|^{-1} |I|^{1/p-1/2} H_I,$$

where the overbar means the complex conjugate. Then, for $p' = p/(p-1)$, we have

$$\|\bar{c}_I |c_I|^{-1} |I|^{1/p-1/2} H_I\|_{p'} = 1$$

and, by Lemma 8.3.3,

$$\|u\|_{p'} \le C_{10}(p) N^{1/p'}. \tag{8.3.19}$$

Consider $\langle f, u\rangle$. We have on the one hand

$$\langle f, u\rangle = \sum_{I\in Q} |c_I||I|^{1/p-1/2} = \sum_{I\in Q} \|c_I H_I\|_p \ge N \tag{8.3.20}$$

and on the other hand

$$\langle f, u\rangle \le \|f\|_p \|u\|_{p'}. \tag{8.3.21}$$

Combining (8.3.19)–(8.3.21) we get Lemma 8.3.4. □

We can now complete the proof of Theorem 8.3.2. It remains to prove inequality (8.3.15). Write

$$A := \max_{I\in\Lambda\setminus\Lambda_m(t)} \|c_I(g) H_I\|_p,$$

and

$$B := \min_{I\in\Lambda_m(t)\setminus\Lambda} \|c_I(g) H_I\|_p.$$

Then, by the definition of $\Lambda_m(t)$, we have

$$B \ge tA. \tag{8.3.22}$$

Using Lemma 8.3.3 we get

$$\|S_{\Lambda\setminus\Lambda_m(t)}(g)\|_p \le AC_{10}(p)|\Lambda\setminus\Lambda_m(t)|^{1/p} \le t^{-1} BC_{10}(p)|\Lambda\setminus\Lambda_m(t)|^{1/p}. \tag{8.3.23}$$

From Lemma 8.3.4 we have

$$\|S_{\Lambda_m(t)\setminus\Lambda}(g)\|_p \ge BC_{11}(p)|\Lambda_m(t)\setminus\Lambda|^{1/p}. \tag{8.3.24}$$

Taking into account that $|\Lambda_m(t)\setminus\Lambda| = |\Lambda\setminus\Lambda_m(t)|$ we obtain from (8.3.23) and (8.3.24) inequality (8.3.15).

The proof of Theorem 8.3.2 is complete.

8.4 Some Inequalities for the Tensor Product of Greedy Bases

8.4.1 Introduction

In this subsection we study properties of bases that are important in nonlinear m-term approximation with respect to these bases. We begin with a brief historical survey that provides a motivation for our investigation. Also, this research is motivated by applications in nonparametric statistics. The following important property of the Haar basis (that the Haar basis is democratic) was established in §8.3: for any Λ, $|\Lambda| = m$, one has

$$C_1(p)m^{1/p} \leq \left\| \sum_{k\in\Lambda} H_{k,p} \right\|_p \leq C_2(p)m^{1/p}, \qquad 1 < p < \infty. \tag{8.4.1}$$

Our main interest in this subsection is to study multivariate bases. There are two standard ways to build a multivariate Haar basis. One way is based on multiresolution analysis. In this way we obtain a multivariate Haar basis consisting of functions whose supports are dyadic cubes. In this case the theory of greedy approximation is parallel to the univariate case (see Temlyakov, 1998e, Cohen *et al.*, 2000). In this section we use the tensor product of univariate bases as a way of building a multivariate basis.

We define a multivariate Haar basis $\mathcal{H}_p^d$ as the tensor product of univariate Haar bases: $\mathcal{H}_p^d := \mathcal{H}_p \times \cdots \times \mathcal{H}_p$; $H_{\mathbf{n},p}(x) := H_{n_1,p}(x_1) \cdots H_{n_d,p}(x_d)$, $\mathbf{x} = (x_1, \dots, x_d)$, $\mathbf{n} = (n_1, \dots, n_d)$. The supports of the functions $H_{\mathbf{n},p}$ are arbitrary dyadic parallelepipeds (intervals). It is known (see Temlyakov, 2002a) that the tensor product structure of multivariate wavelet bases makes them universal for the approximation of anisotropic smoothness classes with different anisotropies. It is also known that the study of such bases is more difficult than the study of univariate bases. In many cases we need to develop new technique and in some cases we encounter new phenomena. For instance, it turns out that property (8.4.1) does not hold for a multivariate Haar basis $\mathcal{H}_p^d$ for $p \neq 2$ (see Temlyakov, 2002b, for a detailed discussion). It is known from Temlyakov (1998d), Wojtaszczyk (2000), and Kamont and Temlyakov (2004) that the function

$$\mu(m, \mathcal{H}_p^d) := \sup_{k\leq m} \left(\sup_{\Lambda:|\Lambda|=k} \left\| \sum_{\mathbf{n}\in\Lambda} H_{\mathbf{n},p} \right\|_p \Big/ \inf_{\Lambda:|\Lambda|=k} \left\| \sum_{\mathbf{n}\in\Lambda} H_{\mathbf{n},p} \right\|_p \right)$$

plays a very important role in estimates of the m-term greedy approximation in terms of the best m-term approximation. For instance (see Temlyakov, 1998d),

$$\|f - G_m^{L_p}(f, \mathcal{H}_p^d)\|_p \leq C(p,d)\mu(m, \mathcal{H}_p^d)\sigma_m(f, \mathcal{H}_p^d)_p, \qquad 1 < p < \infty. \tag{8.4.2}$$

Both the greedy approximant $G_m^{L_p}(f, \mathcal{H}_p^d)$ and the best m-term approximation

$\sigma_m(f,\mathscr{H}_p^d)_p$ are defined above. The following theorem gives, in particular, upper estimates for the $\mu(m,\mathscr{H}_p^d)$.

Theorem 8.4.1 *Let $1<p<\infty$. Then, for any Λ, $|\Lambda|=m$, we have, for $2\le p<\infty$,*

$$C^1_{p,d}m^{1/p}\min_{\mathbf{n}\in\Lambda}|c_{\mathbf{n}}|\le\left\|\sum_{\mathbf{n}\in\Lambda}c_{\mathbf{n}}H_{\mathbf{n},p}\right\|_p\le C^2_{p,d}m^{1/p}(\log m)^{h(p,d)}\max_{\mathbf{n}\in\Lambda}|c_{\mathbf{n}}|$$

and, for $1<p\le 2$,

$$C^3_{p,d}m^{1/p}(\log m)^{-h(p,d)}\min_{\mathbf{n}\in\Lambda}|c_{\mathbf{n}}|\le\left\|\sum_{\mathbf{n}\in\Lambda}c_{\mathbf{n}}H_{\mathbf{n},p}\right\|_p\le C^4_{p,d}m^{1/p}\max_{\mathbf{n}\in\Lambda}|c_{\mathbf{n}}|,$$

where $h(p,d):=(d-1)|1/2-1/p|$.

Theorem 8.4.1 for $d=1$, $1<p<\infty$ was proved in Temlyakov (1998e) (see also Lemmas 8.3.3 and 8.3.4 above); in the case $d=2$, $4/3\le p\le 4$ it was proved in Temlyakov (1998d). The general case was proved in Wojtaszczyk (2000). It is known (Temlyakov, 2002b) that the extra log factors in Theorem 8.4.1 are sharp.

In §8.4.2 we will generalize Theorem 8.4.1 to the case of a basis that is a tensor product of greedy bases. We now give the corresponding definitions and introduce some notation. We will do this in a general setting.

Let Ψ be a normalized basis for $L_p([0,1))$. For the space $L_p([0,1)^d)$ we define $\Psi^d:=\Psi\times\cdots\times\Psi$ (d times), $\psi_{\mathbf{n}}(\mathbf{x}):=\psi_{n_1}(x_1)\cdots\psi_{n_d}(x_d)$, $\mathbf{x}=(x_1,\dots,x_d)$, $\mathbf{n}=(n_1,\dots,n_d)$. The following theorem will be proved in §8.4.2 using a proof scheme similar to that from Wojtaszczyk (2000).

Theorem 8.4.2 *Let $1<p<\infty$ and let Ψ be a greedy basis for $L_p([0,1))$. Then, for any Λ, $|\Lambda|=m$, we have, for $2\le p<\infty$,*

$$C^5_{p,d}m^{1/p}\min_{\mathbf{n}\in\Lambda}|c_{\mathbf{n}}|\le\left\|\sum_{\mathbf{n}\in\Lambda}c_{\mathbf{n}}\psi_{\mathbf{n}}\right\|_p\le C^6_{p,d}m^{1/p}(\log m)^{h(p,d)}\max_{\mathbf{n}\in\Lambda}|c_{\mathbf{n}}|$$

and, for $1<p\le 2$,

$$C^7_{p,d}m^{1/p}(\log m)^{-h(p,d)}\min_{\mathbf{n}\in\Lambda}|c_{\mathbf{n}}|\le\left\|\sum_{\mathbf{n}\in\Lambda}c_{\mathbf{n}}\psi_{\mathbf{n}}\right\|_p\le C^8_{p,d}m^{1/p}\max_{\mathbf{n}\in\Lambda}|c_{\mathbf{n}}|,$$

where $h(p,d):=(d-1)|1/2-1/p|$.

The inequality (8.4.2) was extended in Wojtaszczyk (2000) to a normalized unconditional basis Ψ for X instead of $\mathscr{H}_p^d$, for $L_p([0,1)^d)$. Therefore, as a corollary of Theorem 8.4.2 we obtain the following inequality for a greedy basis Ψ, for $L_p([0,1))$,

$$\|f-G_m^{L_p}(f,\Psi^d)\|_p\le C(\Psi,d,p)(\log m)^{h(p,d)}\sigma_m(f,\Psi^d)_p,\quad 1<p<\infty. \quad (8.4.3)$$

In §8.4.3 we will prove a generalization of Theorem 8.4.1 to the case of $H_{\mathbf{n},q}$ instead of $H_{\mathbf{n},p}$. It will be convenient for us to enumerate the Haar system by dyadic intervals. We use the following notation: $h_{[0,1]} := H_{1,\infty}$; $h_{[(l-1)2^{-n},l2^{-n})} := H_{2^n+l,\infty}$, $l = 1,\dots,2^n$, $n = 0,1,\dots$; $h_I(\mathbf{x}) := h_{I_1}(x_1)\cdots h_{I_d}(x_d)$, $I = I_1 \times \cdots \times I_d$.

An interesting generalization of m-term approximation was considered in Cohen *et al.* (2000). Let $\Psi = \{\psi_I\}_I$ be a basis indexed by dyadic intervals. Take a number α and assign to each index set Λ the following measure:

$$\Phi_\alpha(\Lambda) := \sum_{I\in\Lambda} |I|^\alpha.$$

In the case $\alpha = 0$ we get $\Phi_0(\Lambda) = |\Lambda|$. An analog of best m-term approximation is the following:

$$\inf_{\Lambda:\Phi_\alpha(\Lambda)\le m}\ \inf_{c_I,I\in\Lambda}\left\|f - \sum_{I\in\Lambda} c_I\psi_I\right\|_p.$$

A detailed study of this type of approximation (a restricted approximation) can be found in Cohen *et al.* (2000). The following theorem, proved in §8.4.3, provides inequalities useful in the study of restricted approximation in the case of the $\mathcal{H}_p^d$.

Theorem 8.4.3 *Let $1 < p < \infty$. Then, for any $a > 0$ and any Λ, $|\Lambda| = m$, we have, for $2 \le p < \infty$,*

$$\sum_{I\in\Lambda}\left\||I|^{-a}h_I\right\|_p^p \ll \left\|\sum_{I\in\Lambda}|I|^{-a}h_I\right\|_p^p \ll (\log m)^{(1/2-1/p)p(d-1)}\sum_{I\in\Lambda}\left\||I|^{-a}h_I\right\|_p^p \quad (8.4.4)$$

and, for $1 < p \le 2$,

$$(\log m)^{(1/2-1/p)p(d-1)}\sum_{I\in\Lambda}\left\||I|^{-a}h_I\right\|_p^p \ll \left\|\sum_{I\in\Lambda}|I|^{-a}h_I\right\|_p^p \ll \sum_{I\in\Lambda}\left\||I|^{-a}h_I\right\|_p^p. \quad (8.4.5)$$

Here, the symbol $\ll$ means that the corresponding inequality has an extra factor that does not depend on m and Λ. We note that Theorem 8.4.3 in the case $a = 1/p$ coincides with Theorem 8.4.1. Theorem 8.4.3 in the case $d = 1$ was proved in Cohen *et al.* (2000).

8.4.2 Proof of Theorem 8.4.2

This proof goes by induction. We first prove some inequalities in the univariate case. We need some known facts. There is a result in functional analysis (Kadec and Pelczynski, 1962; Lindenstrauss and Tzafriri, 1979) which says that for any unconditional basis $B = (b_k)$ of $L_p([0,1)^d)$, normalized so that $\|b_k\|_p = 1$, there is a subsequence k_j, $j = 1, 2, \ldots$, such that (b_{k_j}) satisfies

$$\left\| \sum_{j=1}^{\infty} \alpha_{k_j} b_{k_j} \right\|_p^p \asymp \sum_{j=1}^{\infty} |\alpha_{k_j}|^p.$$

It follows that, for any democratic and unconditional basis B for $L_p([0,1)^d)$, we have

$$\left\| \sum_{k \in \Lambda} b_k \right\|_p \asymp (|\Lambda|)^{1/p},$$

where the constants of equivalency depend at most on B and p. For an unconditional democratic basis B in L_p, the above results combine to show that

$$C_1 \min_{k \in \Lambda} |a_k| \, (|\Lambda|)^{1/p} \le \left\| \sum_{k \in \Lambda} a_k b_k \right\|_p \le C_2 \max_{k \in \Lambda} |a_k| \, (|\Lambda|)^{1/p} \tag{8.4.6}$$

for any finite set Λ with $C_1, C_2 > 0$ independent of Λ and $\{a_k\}$. This proves Theorem 8.4.2 for $d = 1$, $1 < p < \infty$.

We will often use the following well-known lemma (see Lindenstrauss and Tzafriri, 1979, p. 73).

Lemma 8.4.4 *For any finite collection $\{f_s\}$ of functions in L_p, $1 \le p \le \infty$, we have*

$$\left(\sum_s \|f_s\|_p^{p_l} \right)^{1/p_l} \le \left\| \left(\sum_s |f_s|^2 \right)^{1/2} \right\|_p \le \left(\sum_s \|f_s\|_p^{p_u} \right)^{1/p_u}, \tag{8.4.7}$$

with $p_l := \max(2, p)$ and $p_u := \min(2, p)$.

We note that by Theorem 8.1.6 a greedy basis Ψ is unconditional. It is known that the tensor product of unconditional bases for $L_p([0,1))$, $1 < p < \infty$, is an unconditional basis for $L_p([0,1)^d)$. Therefore, for any $1 < p < \infty$ and any $\{a_{\mathbf{n}}\}$ we have

$$C_1(p,d) \left\| \left(\sum_{\mathbf{n}} |a_{\mathbf{n}} \psi_{\mathbf{n}}|^2 \right)^{1/2} \right\|_p \le \left\| \sum_{\mathbf{n}} a_{\mathbf{n}} \psi_{\mathbf{n}} \right\|_p$$
$$\le C_2(p,d) \left\| \left(\sum_{\mathbf{n}} |a_{\mathbf{n}} \psi_{\mathbf{n}}|^2 \right)^{1/2} \right\|_p, \tag{8.4.8}$$

and also, for any set of disjoint Λ_j, we have

$$C_3(p,d)\left\|\left(\sum_j\left|\sum_{\mathbf{n}\in\Lambda_j} a_{\mathbf{n}}\psi_{\mathbf{n}}\right|^2\right)^{1/2}\right\|_p \le \left\|\sum_j\sum_{\mathbf{n}\in\Lambda_j} a_{\mathbf{n}}\psi_{\mathbf{n}}\right\|_p$$
$$\le C_4(p,d)\left\|\left(\sum_j\left|\sum_{\mathbf{n}\in\Lambda_j} a_{\mathbf{n}}\psi_{\mathbf{n}}\right|^2\right)^{1/2}\right\|_p. \quad (8.4.9)$$

Lemma 8.4.5 *Let $2 \le p < \infty$ and let Ψ be a greedy basis for $L_p([0,1))$. Then, for any finite Λ, $|\Lambda| = m$, and any coefficients $\{a_k\}$ we have*

$$\left(\sum_{k\in\Lambda}|a_k|^p\right)^{1/p} \ll \left\|\sum_{k\in\Lambda}a_k\psi_k\right\|_p \ll (\log m)^{1/2-1/p}\left(\sum_{k\in\Lambda}|a_k|^p\right)^{1/p}.$$

Proof The lower estimate follows from (8.4.8) and Lemma 8.4.4. We now prove the upper estimate. Let

$$|a_{k_1}| \ge |a_{k_2}| \ge \cdots, \qquad k_j \in \Lambda, \qquad j = 1,2,\dots,m.$$

For notational convenience we set $a_{k_j} = 0$ for $j > m$. Writing

$$f_s := \sum_{j=2^s}^{2^{s+1}-1} a_{k_j}\psi_{k_j}, \quad (8.4.10)$$

we get, for n such that $2^n \le m < 2^{n+1}$,

$$f := \sum_{k\in\Lambda}a_k\psi_k = \sum_{s=0}^{n} f_s. \quad (8.4.11)$$

By (8.4.9) and Lemma 8.4.4 we obtain

$$\|f\|_p \ll \left(\sum_{s=0}^{n}\|f_s\|_p^2\right)^{1/2}.$$

Next, by (8.4.6)

$$\|f_s\|_p \ll |a_{k_{2^s}}|2^{s/p}.$$

Thus

$$\|f\|_p \ll \left(\sum_{s=0}^{n}|a_{k_{2^s}}|^2 2^{2s/p}\right)^{1/2}.$$

By Hölder's inequality with parameter $p/2$ we continue the above expression:

$$\le \left(\sum_{s=0}^{n}|a_{k_{2^s}}|^p 2^s\right)^{1/p}\left(\sum_{s=0}^{n}1\right)^{(1-2/p)/2} \ll (\log m)^{1/2-1/p}\left(\sum_{k\in\Lambda}|a_k|^p\right)^{1/p}. \quad \square$$

Lemma 8.4.6 *Let $1 < p \le 2$ and let Ψ be a greedy basis for $L_p([0,1))$. Then, for any finite Λ, $|\Lambda| = m$, and any coefficients $\{a_k\}$ we have*

$$(\log m)^{1/2-1/p}\left(\sum_{k\in\Lambda}|a_k|^p\right)^{1/p} \ll \left\|\sum_{k\in\Lambda}a_k\psi_k\right\|_p \ll \left(\sum_{k\in\Lambda}|a_k|^p\right)^{1/p}.$$

Proof The upper estimate follows from (8.4.8) and Lemma 8.4.4. We now proceed to the lower estimate. Using the notation in (8.4.10) and (8.4.11), by (8.4.9), (8.4.7), and (8.4.6) we obtain

$$\|f\|_p \gg \left(\sum_{s=0}^{n}\|f_s\|_p^2\right)^{1/2} \gg \left(\sum_{s=0}^{n}|a_{k_{2^{s+1}}}|^2 2^{2s/p}\right)^{1/2}.$$

Next, by Hölder's inequality with parameter $2/p$ we get

$$\sum_{s=0}^{n}|a_{k_{2^{s+1}}}|^p 2^s \le \left(\sum_{s=0}^{n}|a_{k_{2^{s+1}}}|^2 2^{2s/p}\right)^{p/2}(n+1)^{1-p/2}.$$

Therefore

$$\|f\|_p \gg \left(\sum_{s=0}^{n}|a_{k_{2^{s+1}}}|^p 2^s\right)^{1/p} n^{1/2-1/p} \gg (\log m)^{1/2-1/p}\left(\sum_{k\in\Lambda}|a_k|^p\right)^{1/p}. \quad \square$$

Proof of Theorem 8.4.2 We obtain the lower estimate for $2 \le p < \infty$ and the upper estimate for $1 < p \le 2$ from (8.4.8) and Lemma 8.4.4. It remains to prove Theorem 8.4.2 in the following cases: for $2 \le p < \infty$, the upper estimate; for $1 < p \le 2$, the lower estimate. We mentioned above that the assumption that Ψ is a greedy basis for $L_p([0,1))$ implies that Ψ^d is an unconditional basis for $L_p([0,1)^d)$. Therefore, it is sufficient to prove Theorem 8.4.2 in the particular case $c_{\mathbf{n}} = 1$, $\mathbf{n} \in \Lambda$. We first prove the upper estimate in the case $2 \le p < \infty$. Let

$$\Lambda_d := \{n_d : \exists \mathbf{k} \in \Lambda \text{ with } k_d = n_d\},$$
$$\Lambda(n_d) := \{(k_1,\dots,k_{d-1}) : (k_1,\dots,k_{d-1},n_d) \in \Lambda\}.$$

Then we have, by Lemma 8.4.5,

$$\left\|\sum_{n_d\in\Lambda_d}\psi_{n_d}(x_d)\left(\sum_{(n_1,\dots,n_{d-1})\in\Lambda(n_d)}\psi_{n_1}(x_1)\cdots\psi_{n_{d-1}}(x_{d-1})\right)\right\|_p^p$$
$$\ll (\log m)^{(1/2-1/p)p}\sum_{n_d\in\Lambda_d}\left\|\sum_{(n_1,\dots,n_{d-1})\in\Lambda(n_d)}\psi_{n_1}(x_1)\cdots\psi_{n_{d-1}}(x_{d-1}))\right\|_p^p.$$

We continue, by the induction assumption, as follows:

$$\ll (\log m)^{(1/2-1/p)p}\Big(\sum_{n_d\in\Lambda_d}|\Lambda(n_d)|(\log m)^{(1/2-1/p)p(d-2)}\Big)$$
$$= m(\log m)^{(1/2-1/p)(d-1)p}.$$

We now proceed to the lower estimate in the case $1 < p \le 2$. By Lemma 8.4.6 we get

$$\Big\|\sum_{n_d\in\Lambda_d}\psi_{n_d}(x_d)\Big(\sum_{(n_1,\dots,n_{d-1})\in\Lambda(n_d)}\psi_{n_1}(x_1)\cdots\psi_{n_{d-1}}(x_{d-1})\Big)\Big\|_p^p$$
$$\gg (\log m)^{(1/2-1/p)p}\sum_{n_d\in\Lambda_d}\Big\|\sum_{(n_1,\dots,n_{d-1})\in\Lambda(n_d)}\psi_{n_1}(x_1)\cdots\psi_{n_{d-1}}(x_{d-1}))\Big\|_p^p.$$

We again continue, by the induction assumption, as follows:

$$\gg (\log m)^{(1/2-1/p)p}\Big(\sum_{n_d\in\Lambda_d}|\Lambda(n_d)|(\log m)^{(1/2-1/p)p(d-2)}\Big)$$
$$= m(\log m)^{(1/2-1/p)(d-1)p}.$$

The proof of Theorem 8.4.2 is complete. □

8.4.3 Proof of Theorem 8.4.3

The lower estimate in the case $2 \le p < \infty$ and the upper estimate in the case $1 < p \le 2$ follow from (8.4.8) and Lemma 8.4.4. We first note that the lower estimate in the case $1 < p \le 2$ follows from the upper estimate in the case $2 \le p < \infty$ by the duality argument. Indeed, let us assume that (8.4.4) has been proved. Let $q \in (1,2]$. Denote $p := q/(q-1) \in [2,\infty)$. We have

$$\sum_{I\in\Lambda}\||I|^{-a}h_I\|_q^q = \sum_{I\in\Lambda}|I|^{-aq+1} = \Big\langle\sum_{I\in\Lambda}|I|^{-a}h_I, \sum_{I\in\Lambda}|I|^{-a(q-1)}h_I\Big\rangle$$
$$\le \Big\|\sum_{I\in\Lambda}|I|^{-a}h_I\Big\|_q\Big\|\sum_{I\in\Lambda}|I|^{-a(q-1)}h_I\Big\|_p.$$

Using (8.4.4) we continue:

$$\ll \Big\|\sum_{I\in\Lambda}|I|^{-a}h_I\Big\|_q(\log m)^{(1/2-1/p)(d-1)}\Big(\sum_{I\in\Lambda}\Big\||I|^{-a(q-1)}h_I\Big\|_p^p\Big)^{1/p}$$
$$= \Big\|\sum_{I\in\Lambda}|I|^{-a}h_I\Big\|_q(\log m)^{(1/2-1/p)(d-1)}\Big(\sum_{I\in\Lambda}|I|^{-aq+1}\Big)^{1/p}.$$

This implies the lower estimate in (8.4.5).

It remains to prove the upper estimate in (8.4.4). We will carry out the proof by induction. First, consider the univariate case. We have

$$\sum_I \| |I|^{-a} h_I \|_p^p = \sum_I |I|^{-ap+1}$$

and, by (8.4.8),

$$\left\| \sum_I |I|^{-a} h_I \right\|_p^p \ll \int_0^1 \left(\sum_I (|I|^{-a} h_I)^2 \right)^{p/2} = \int_0^1 \left(\sum_{j=1}^s 2^{2an_j} \chi_{E_j} \right)^{p/2}$$

for some $n_1 < n_2 < \cdots < n_s$ and $E_j \subset [0,1]$, $j = 1, \dots, s$. By an analog of Lemma 8.3.5 (see also Lemma 2.3 from Temlyakov, 1998e) that follows from its proof we continue:

$$\ll \sum_{j=1}^s 2^{2n_j a(p/2)} |E_j| = \sum_{j=1}^s 2^{n_j ap} |E_j| = \sum_I |I|^{-ap+1}.$$

We proceed to the multivariate case. Let

$$\Lambda_d := \{I_d : \exists J \in \Lambda \text{ with } J_d = I_d\},$$
$$\Lambda(I_d) := \{(J_1, \dots, J_{d-1}) : (J_1, \dots, J_{d-1}, I_d) \in \Lambda\}.$$

Using the fact (Temlyakov, 1998c) that the univariate Haar basis is a greedy basis for $L_p([0,1))$, $1 < p < \infty$, we get, by Lemma 8.4.5,

$$\left\| \sum_{I_d \in \Lambda_d} |I_d|^{-a} h_{I_d}(x_d) \left(\sum_{(J_1,\dots,J_{d-1}) \in \Lambda(I_d)} |J_1|^{-a} h_{J_1}(x_1) \cdots |J_{d-1}|^{-a} h_{J_{d-1}}(x_{d-1}) \right) \right\|_p^p$$
$$\ll (\log m)^{(1/2-1/p)p} \sum_{I_d \in \Lambda_d} \| |I_d|^{-a} h_{I_d}(x_d) \|_p^p$$
$$\times \left\| \sum_{(J_1,\dots,J_{d-1}) \in \Lambda(I_d)} |J_1|^{-a} h_{J_1}(x_1) \cdots |J_{d-1}|^{-a} h_{J_{d-1}}(x_{d-1}) \right\|_p^p.$$

By the induction assumption we continue:

$$\ll (\log m)^{(1/2-1/p)p(d-1)} \sum_{I_d \in \Lambda_d} \| |I_d|^{-a} h_{I_d}(x_d) \|_p^p$$
$$\times \left(\sum_{(J_1,\dots,J_{d-1}) \in \Lambda(I_d)} \| |J_1|^{-a} h_{J_1}(x_1) \|_p^p \cdots \| |J_{d-1}|^{-a} h_{J_{d-1}}(x_{d-1}) \|_p^p \right)$$
$$= (\log m)^{(1/2-1/p)p(d-1)} \sum_{I \in \Lambda} \| |I|^{-a} h_I \|_p^p.$$

8.5 Weight-Greedy Bases

Let Ψ be a basis for X. If $\inf_n \|\psi_n\| > 0$ then $c_n(f) \to 0$ as $n \to \infty$, where

$$f = \sum_{n=1}^{\infty} c_n(f)\psi_n.$$

Next, we rearrange the coefficients $\{c_n(f)\}$ in a decreasing way:

$$|c_{n_1}(f)| \geq |c_{n_2}(f)| \geq \cdots$$

and define the mth greedy approximant as

$$G_m(f,\Psi) := \sum_{k=1}^{m} c_{n_k}(f)\psi_{n_k}. \tag{8.5.1}$$

In the case $\inf_n \|\psi_n\| = 0$ we define $G_m(f,\Psi)$ by (8.5.1) for an f of the form

$$f = \sum_{n\in Y} c_n(f)\psi_n, \qquad |Y| < \infty. \tag{8.5.2}$$

Let a weight sequence $w = \{w_n\}_{n=1}^{\infty}$, $w_n > 0$, be given. For $\Lambda \subset \mathbb{N}$ denote $w(\Lambda) := \sum_{n\in\Lambda} w_n$. For a positive real number $v > 0$ define

$$\sigma_v^w(f,\Psi) := \inf_{\{b_n\},\Lambda:w(\Lambda)\leq v} \left\| f - \sum_{n\in\Lambda} b_n\psi_n \right\|,$$

where Λ are finite.

Definition 8.5.1 We say that a basis Ψ is weight-greedy (w-greedy) if, for any $f \in X$ in the case $\inf_n \|\psi_n\| > 0$ or for any $f \in X$ of the form (8.5.2) in the case $\inf_n \|\psi_n\| = 0$, we have

$$\|f - G_m(f,\Psi)\| \leq C_G \sigma_{w(\Lambda_m)}^w(f,\Psi),$$

where

$$G_m(f,\Psi) = \sum_{n\in\Lambda_m} c_n(f)\psi_n, \qquad |\Lambda_m| = m.$$

Definition 8.5.2 We say a basis Ψ is weight-democratic (w-democratic) if, for any finite $A, B \subset \mathbb{N}$ such that $w(A) \leq w(B)$, we have

$$\left\| \sum_{n\in A} \psi_n \right\| \leq C_D \left\| \sum_{n\in B} \psi_n \right\|.$$

Theorem 8.5.3 *A basis Ψ is a w-greedy basis if and only if it is unconditional and w-democratic.*

Proof We prove the theorem in two parts.

Part I We first prove the implication

$$\text{unconditional} + w\text{-democratic} \quad \Rightarrow \quad w\text{-greedy}.$$

Let f be any element or of the form (8.5.2) if $\inf_n \|\psi_n\| = 0$. Consider

$$G_m(f,\Psi) = \sum_{n\in Q} c_n(f)\psi_n =: S_Q(f).$$

We take any finite set $P \subset \mathbb{N}$ satisfying $w(P) \le w(Q)$. Then, our assumption $w_n > 0$, $n \in \mathbb{N}$ implies that either $P = Q$ or $Q \setminus P$ is nonempty. Write

$$\sigma_P(f,\Psi) := \inf_{\{b_n\}} \left\| f - \sum_{n\in P} b_n \psi_n \right\|.$$

Then, by the unconditionality of Ψ, we have

$$\|f - S_P(f)\| \le K\sigma_P(f,\Psi). \tag{8.5.3}$$

This (with $P = Q$) completes the proof in the case $\sigma^w_{w(Q)}(f,\Psi) = \sigma_Q(f,\Psi)$. Suppose that $\sigma^w_{w(Q)}(f,\Psi) < \sigma_Q(f,\Psi)$. Clearly, we now may consider only those P that satisfy the following two conditions:

$$w(P) \le w(Q) \qquad \text{and} \qquad \sigma_P(f,\Psi) < \sigma_Q(f,\Psi).$$

For a P satisfying the above conditions we have $Q \setminus P \neq \emptyset$. We estimate

$$\|f - S_Q(f)\| \le \|f - S_P(f)\| + \|S_P(f) - S_Q(f)\|. \tag{8.5.4}$$

Also, we have

$$S_P(f) - S_Q(f) = S_{P\setminus Q}(f) - S_{Q\setminus P}(f). \tag{8.5.5}$$

We get a result similar to (8.5.3):

$$\|S_{Q\setminus P}(f)\| \le K\sigma_P(f,\Psi). \tag{8.5.6}$$

It remains to estimate $\|S_{P\setminus Q}(f)\|$. We have, by the unconditionality and w-democracy of Ψ,

$$\begin{aligned}\|S_{P\setminus Q}(f)\| &\le 2K \max_{n\in P\setminus Q} |c_n(f)| \left\| \sum_{n\in P\setminus Q} \psi_n \right\| \\ &\le 2KC_D \min_{n\in Q\setminus P} |c_n(f)| \left\| \sum_{n\in Q\setminus P} \psi_n \right\| \le 4K^2 C_D \|S_{Q\setminus P}(f)\|. \end{aligned} \tag{8.5.7}$$

Combining (8.5.3)–(8.5.7) completes the proof of part I.

Remark 8.5.4 Suppose that Ψ, instead of being w-democratic, satisfies the following inequality:

$$\left\|\sum_{n\in A}\psi_n\right\| \le K(N)\left\|\sum_{n\in B}\psi_n\right\|$$

for all $A,B\subset\mathbb{N}$, $w(A)\le w(B)\le N$. Then part I of the proof gives

$$\|f-G_m(f,\Psi)\|\le CK(w(Q))\sigma^w_{w(Q)}(f,\Psi).$$

Part II We now prove the implication

$$w\text{-greedy} \quad\Rightarrow\quad \text{unconditional} + w\text{-democratic}.$$

IIa. We begin with

$$w\text{-greedy} \quad\Rightarrow\quad \text{unconditional}.$$

We will prove a slightly stronger statement.

Lemma 8.5.5 *Let Ψ be a basis such that, for any f of the form* (8.5.2), *we have*

$$\|f-G_m(f,\Psi)\|\le C\sigma_\Lambda(f,\Psi),$$

where

$$G_m(f,\Psi)=\sum_{n\in\Lambda}c_n(f)\psi_n.$$

Then Ψ is unconditional.

Proof It is clear that it is sufficient to prove that there exists a constant C_0 such that, for any finite Λ and any f of the form (8.5.2), we have

$$\|S_\Lambda(f)\|\le C_0\|f\|.$$

Let f and Λ be given and $\Lambda\subset[1,M]$. Consider

$$f_M := S_{[1,M]}(f).$$

Then $\|f_M\|\le C_B\|f\|$. We take a $b>\max_{1\le n\le M}|c_n(f)|$ and define a new function

$$g := f_M - S_\Lambda(f_M) + b\sum_{n\in\Lambda}\psi_n.$$

Then

$$G_m(g,\Psi)=b\sum_{n\in\Lambda}\psi_n,\qquad m:=|\Lambda|,$$

and

$$\sigma_\Lambda(g,\Psi)\le\|f_M\|.$$

Thus, by the assumption of the lemma we have

$$\|f_M - S_\Lambda(f_M)\| = \|g - G_m(g,\Psi)\| \le C\sigma_\Lambda(g,\Psi) \le C\|f_M\|.$$

Therefore

$$\|S_\Lambda(f)\| = \|S_\Lambda(f_M)\| \le C_0\|f\|. \qquad \square$$

IIb. It remains to prove the implication

$$w\text{-greedy} \quad \Rightarrow \quad w\text{-democratic}.$$

First, let $A, B \subset \mathbb{N}$, $w(A) \le w(B)$, be such that $A \cap B = \emptyset$. Consider

$$f := \sum_{n\in A} \psi_n + (1+\varepsilon)\sum_{n\in B} \psi_n, \qquad \varepsilon > 0.$$

Then

$$G_m(f,\Psi) = (1+\varepsilon)\sum_{n\in B} \psi_n, \qquad m := |B|$$

and

$$\sigma_A(f,\Psi) \le \left\|\sum_{n\in B} \psi_n\right\|(1+\varepsilon).$$

Therefore, by the w-greedy assumption we get

$$\left\|\sum_{n\in A} \psi_n\right\| \le C(1+\varepsilon)\left\|\sum_{n\in B} \psi_n\right\|.$$

Now let A, B be arbitrary and finite, $w(A) \le w(B)$. Then, using the unconditionality of Ψ, which has already been proven in IIa and the above part of IIb, we obtain

$$\left\|\sum_{n\in A} \psi_n\right\| \le \left\|\sum_{n\in A\setminus B} \psi_n\right\| + \left\|\sum_{n\in A\cap B} \psi_n\right\| \le C\left\|\sum_{n\in B\setminus A} \psi_n\right\| + K\left\|\sum_{n\in B} \psi_n\right\|$$
$$\le C_1\left\|\sum_{n\in B} \psi_n\right\|.$$

This completes the proof of Theorem 8.5.3. $\square$

8.6 The Weak Chebyshev Greedy Algorithm

Let X be a Banach space with norm $\|\cdot\|$. We say that a set of elements (functions) $\mathcal{D}$ from X is a dictionary if each $g \in \mathcal{D}$ has norm bounded by one ($\|g\| \le 1$) and the closure of span $\mathcal{D}$ is X. A dictionary is symmetric if

$$g \in \mathcal{D} \qquad \text{implies} \qquad -g \in \mathcal{D}.$$

We denote the closure (in X) of the convex hull of $\mathscr{D}$ by $A_1(\mathscr{D})$. For a dictionary $\mathscr{D}$, denote by $\mathscr{D}^{\pm} := \{\pm g : g \in \mathscr{D}\}$ its symmetrized version. In this section we study the weak Chebyshev greedy algorithms defined in the introduction to this chapter. We begin with a simple property of the WCGA.

Remark 8.6.1 It follows from the definition of the WCGA that the sequence $\{\|f_m\|\}$ is a nonincreasing sequence.

When X is a real Banach space and the modulus of smoothness of X is defined as

$$\rho(u) := \frac{1}{2} \sup_{x,y;\|x\|=\|y\|=1} \left| \|x+uy\| + \|x-uy\| - 2 \right|, \tag{8.6.1}$$

then a uniformly smooth Banach space is one with $\rho(u)/u \to 0$ when $u \to 0$.

We proceed to a theorem on the convergence of the WCGA. In the formulation of this theorem we need a special sequence which is defined for a given modulus of smoothness $\rho(u)$ and a given $\tau = \{t_k\}_{k=1}^{\infty}$.

Definition 8.6.2 Let $\rho(u)$ be an even convex function on $(-\infty,\infty)$ with the properties $\rho(2) \geq 1$ and

$$\lim_{u\to 0} \rho(u)/u = 0.$$

For any $\tau = \{t_k\}_{k=1}^{\infty}$, $0 < t_k \leq 1$, and $0 < \theta \leq 1/2$ we define $\xi_m := \xi_m(\rho,\tau,\theta)$ as a number u satisfying the equation

$$\rho(u) = \theta t_m u. \tag{8.6.2}$$

Remark 8.6.3 Our assumptions on $\rho(u)$ imply that the function

$$s(u) := \rho(u)/u, \qquad u \neq 0, \qquad s(0) = 0,$$

is a continuous increasing function on $[0,\infty)$ with $s(2) \geq 1/2$. Thus (8.6.2) has a unique solution $\xi_m = s^{-1}(\theta t_m)$ such that $0 < \xi_m \leq 2$.

The following theorem from Temlyakov (2001) gives a sufficient condition for convergence of the WCGA.

Theorem 8.6.4 *Let X be a uniformly smooth Banach space with modulus of smoothness $\rho(u)$. Assume that a sequence $\tau := \{t_k\}_{k=1}^{\infty}$ satisfies the condition that for any $\theta > 0$ we have*

$$\sum_{m=1}^{\infty} t_m \xi_m(\rho,\tau,\theta) = \infty.$$

Then, for any $f \in X$, we have

$$\lim_{m\to\infty} \|f_m^{c,\tau}\| = 0.$$

Corollary 8.6.5 *Let a Banach space X have a modulus of smoothness $\rho(u)$ of power type $1 < q \le 2$; that is, $\rho(u) \le \gamma u^q$. Assume that*

$$\sum_{m=1}^{\infty} t_m^p = \infty, \qquad p = \frac{q}{q-1}. \tag{8.6.3}$$

Then the WCGA converges for any $f \in X$.

Proof Denote $\rho^q(u) := \gamma u^q$. Then

$$\frac{\rho(u)}{u} \le \frac{\rho^q(u)}{u},$$

and therefore, for any $\theta > 0$, we have

$$\xi_m(\rho, \tau, \theta) \ge \xi_m(\rho^q, \tau, \theta).$$

For ρ^q we get from the definition of ξ_m that

$$\xi_m(\rho^q, \tau, \theta) = (\theta t_m/\gamma)^{1/(q-1)}.$$

Thus (8.6.3) implies that

$$\sum_{m=1}^{\infty} t_m \xi_m(\rho, \tau, \theta) \ge \sum_{m=1}^{\infty} t_m \xi_m(\rho^q, \tau, \theta) \asymp \sum_{m=1}^{\infty} t_m^p = \infty.$$

It remains to apply Theorem 8.6.4. $\square$

The following theorem from Temlyakov (2001) gives the rate of convergence of the WCGA for f in $A_1(\mathcal{D}^{\pm})$.

Theorem 8.6.6 *Let X be a uniformly smooth Banach space with modulus of smoothness $\rho(u) \le \gamma u^q$, $1 < q \le 2$. Then, for a sequence $\tau := \{t_k\}_{k=1}^{\infty}$, $t_k \le 1$, $k = 1, 2, \dots$, we have for any $f \in A_1(\mathcal{D}^{\pm})$ that*

$$\|f_m^{c,\tau}\| \le C(q, \gamma)\left(1 + \sum_{k=1}^{m} t_k^p\right)^{-1/p}, \qquad p := \frac{q}{q-1},$$

with a constant $C(q, \gamma)$ which may depend only on q and γ.

We will use the following two simple and well-known lemmas in the proof of the above two theorems.

Lemma 8.6.7 *Let X be a uniformly smooth Banach space and L be a finite-dimensional subspace of X. For any $f \in X \setminus L$ let f_L denote the best approximant of f from L. Then we have*

$$F_{f-f_L}(\phi) = 0$$

for any $\phi \in L$.

Proof Let us assume the contrary, that there is a $\phi \in L$ such that $\|\phi\| = 1$ and

$$F_{f-f_L}(\phi) = \beta > 0.$$

For any λ we have, from the definition of $\rho(u)$, that

$$\|f - f_L - \lambda\phi\| + \|f - f_L + \lambda\phi\| \le 2\|f - f_L\|\left(1 + \rho\left(\frac{\lambda}{\|f - f_L\|}\right)\right). \tag{8.6.4}$$

Next,

$$\|f - f_L + \lambda\phi\| \ge F_{f-f_L}(f - f_L + \lambda\phi) = \|f - f_L\| + \lambda\beta. \tag{8.6.5}$$

Combining (8.6.4) and (8.6.5) we get

$$\|f - f_L - \lambda\phi\| \le \|f - f_L\|\left(1 - \frac{\lambda\beta}{\|f - f_L\|} + 2\rho\left(\frac{\lambda}{\|f - f_L\|}\right)\right). \tag{8.6.6}$$

Taking into account that $\rho(u) = o(u)$, we can find a $\lambda' > 0$ such that

$$\left(1 - \frac{\lambda'\beta}{\|f - f_L\|} + 2\rho\left(\frac{\lambda'}{\|f - f_L\|}\right)\right) < 1.$$

Then (8.6.6) gives

$$\|f - f_L - \lambda'\phi\| < \|f - f_L\|,$$

which contradicts the assumption that $f_L \in L$ is the best approximant of f. □

Lemma 8.6.8 *For any bounded linear functional F and any dictionary $\mathcal{D}$, we have*

$$\sup_{g\in\mathcal{D}} F(g) = \sup_{f\in A_1(\mathcal{D})} F(f).$$

Proof The inequality

$$\sup_{g\in\mathcal{D}} F(g) \le \sup_{f\in A_1(\mathcal{D})} F(f)$$

is obvious. We will prove the opposite inequality. Take any $f \in A_1(\mathcal{D})$. Then, for any $\varepsilon > 0$, there exist $g_1^\varepsilon, \dots, g_N^\varepsilon \in \mathcal{D}$ and numbers $a_1^\varepsilon, \dots, a_N^\varepsilon$ such that $a_i^\varepsilon > 0$, $a_1^\varepsilon + \dots + a_N^\varepsilon = 1$ and

$$\left\|f - \sum_{i=1}^N a_i^\varepsilon g_i^\varepsilon\right\| \le \varepsilon.$$

Thus

$$F(f) \le \|F\|\varepsilon + F\left(\sum_{i=1}^N a_i^\varepsilon g_i^\varepsilon\right) \le \varepsilon\|F\| + \sup_{g\in\mathcal{D}} F(g)$$

which proves the lemma. □

We will also need one more lemma from Temlyakov (2001).

Lemma 8.6.9 *Let X be a uniformly smooth Banach space with modulus of smoothness $\rho(u)$. Take a number $\varepsilon \geq 0$ and two elements f, f^ε from X such that*

$$\|f - f^\varepsilon\| \leq \varepsilon, \qquad f^\varepsilon/A(\varepsilon) \in A_1(\mathscr{D}^\pm),$$

with some number $A(\varepsilon) > 0$. Then we have

$$\|f_m^{c,\tau}\| \leq \|f_{m-1}^{c,\tau}\| \inf_{\lambda \geq 0} \left(1 - \lambda t_m A(\varepsilon)^{-1}\left(1 - \frac{\varepsilon}{\|f_{m-1}^{c,\tau}\|}\right) + 2\rho\left(\frac{\lambda}{\|f_{m-1}^{c,\tau}\|}\right)\right),$$

for $m = 1, 2, \ldots$

Proof We have, for any λ,

$$\|f_{m-1} - \lambda\varphi_m\| + \|f_{m-1} + \lambda\varphi_m\| \leq 2\|f_{m-1}\|\left(1 + \rho\left(\frac{\lambda}{\|f_{m-1}\|}\right)\right) \tag{8.6.7}$$

and by (1) from the definition of the WCGA in §8.1 and Lemma 8.6.8 we get

$$\begin{aligned}|F_{f_{m-1}}(\varphi_m)| &\geq t_m \sup_{g\in\mathscr{D}} |F_{f_{m-1}}(g)| = t_m \sup_{g\in\mathscr{D}^\pm} F_{f_{m-1}}(g) = t_m \sup_{\phi\in A_1(\mathscr{D}^\pm)} F_{f_{m-1}}(\phi)\\ &\geq t_m A(\varepsilon)^{-1} F_{f_{m-1}}(f^\varepsilon).\end{aligned}$$

By Lemma 8.6.7 we obtain

$$\begin{aligned}F_{f_{m-1}}(f^\varepsilon) &= F_{f_{m-1}}(f + f^\varepsilon - f) \geq F_{f_{m-1}}(f) - \varepsilon = F_{f_{m-1}}(f_{m-1}) - \varepsilon\\ &= \|f_{m-1}\| - \varepsilon.\end{aligned}$$

We have either $|F_{f_{m-1}}(\varphi_m)| = F_{f_{m-1}}(\varphi_m)$ or $|F_{f_{m-1}}(\varphi_m)| = F_{f_{m-1}}(-\varphi_m)$. We present the argument in the case $|F_{f_{m-1}}(\varphi_m)| = F_{f_{m-1}}(\varphi_m)$. The other case is similar. Thus, as in (8.6.6) we get from (8.6.7)

$$\begin{aligned}\|f_m\| &\leq \inf_{\lambda\geq 0} \|f_{m-1} - \lambda\varphi_m\|\\ &\leq \|f_{m-1}\| \inf_{\lambda\geq 0}\left(1 - \lambda t_m A(\varepsilon)^{-1}\left(1 - \frac{\varepsilon}{\|f_{m-1}\|}\right) + 2\rho\left(\frac{\lambda}{\|f_{m-1}\|}\right)\right),\end{aligned} \tag{8.6.8}$$

which proves the lemma. □

Proof of Theorem 8.6.4 The definition of the WCGA implies that $\{\|f_m\|\}$ is a nonincreasing sequence. Therefore we have

$$\lim_{m\to\infty} \|f_m\| = \alpha.$$

We will prove that $\alpha = 0$ by contradiction. Assume that, on the contrary, $\alpha > 0$. Then, for any m, we have

$$\|f_m\| \geq \alpha.$$

We set $\varepsilon = \alpha/2$ and find f^ε such that

$$\|f - f^\varepsilon\| \le \varepsilon \qquad \text{and} \qquad f^\varepsilon/A(\varepsilon) \in A_1(\mathscr{D}^\pm),$$

for some $A(\varepsilon)$. Then, by Lemma 8.6.9 we get

$$\|f_m\| \le \|f_{m-1}\| \inf_\lambda \left(1 - \lambda t_m A(\varepsilon)^{-1}/2 + 2\rho(\lambda/\alpha)\right).$$

Let us specify $\theta := \alpha/(8A(\varepsilon))$, and take $\lambda = \alpha \xi_m(\rho, \tau, \theta)$. Then we obtain

$$\|f_m\| \le \|f_{m-1}\|(1 - 2\theta t_m \xi_m).$$

The assumption

$$\sum_{m=1}^{\infty} t_m \xi_m = \infty$$

implies that

$$\|f_m\| \to 0 \qquad \text{as} \qquad m \to \infty.$$

We have reached a contradiction, which proves Theorem 8.6.4. □

We proceed to the proof of Theorem 8.6.6.

Proof of Theorem 8.6.6 By Lemma 8.6.9 with $\varepsilon = 0$ and $A(\varepsilon) = 1$ we have, for $f \in A_1(\mathscr{D}^\pm)$, that

$$\|f_m\| \le \|f_{m-1}\| \inf_{\lambda \ge 0} \left(1 - \lambda t_m + 2\gamma \left(\frac{\lambda}{\|f_{m-1}\|}\right)^q\right). \tag{8.6.9}$$

Now deduce λ from the equation

$$\frac{1}{2}\lambda t_m = 2\gamma \left(\frac{\lambda}{\|f_{m-1}\|}\right)^q;$$

this implies that

$$\lambda = \|f_{m-1}\|^{q/(q-1)} (4\gamma)^{-1/(q-1)} t_m^{1/(q-1)}.$$

Let

$$A_q := 2(4\gamma)^{1/(q-1)}.$$

Using the notation $p := q/(q-1)$ we get from (8.6.9)

$$\|f_m\| \le \|f_{m-1}\| \left(1 - \frac{\lambda t_m}{2}\right) = \|f_{m-1}\|(1 - t_m^p \|f_{m-1}\|^p / A_q).$$

Raising both sides of this inequality to the power p and taking into account the inequality $x^r \le x$ for $r \ge 1$, $0 \le x \le 1$, we obtain

$$\|f_m\|^p \le \|f_{m-1}\|^p (1 - t_m^p \|f_{m-1}\|^p / A_q).$$

By an analog of Lemma 2.16 of Temlyakov (2011) (see also Temlyakov, 2000a, Lemma 3.1), using the estimate $\|f\|^p \leq 1 < A_q$ we get

$$\|f_m\|^p \leq A_q \left(1 + \sum_{n=1}^{m} t_n^p\right)^{-1},$$

which implies that

$$\|f_m\| \leq C(q,\gamma) \left(1 + \sum_{n=1}^{m} t_n^p\right)^{-1/p}.$$

Theorem 8.6.6 is now proved. □

As a typical example of a uniformly smooth Banach space we will use the space L_p, $1 < p < \infty$. It is well known (see, for instance, Donahue *et al.*, 1997, Lemma B.1) that, in the case $X = L_p$, $1 \leq p < \infty$, we have

$$\rho(u) \leq u^p/p \qquad \text{if } 1 \leq p \leq 2 \qquad \text{and} \qquad \rho(u) \leq (p-1)u^2/2 \qquad \text{if } 2 \leq p < \infty. \tag{8.6.10}$$

It is also known (see Lindenstrauss and Tzafriri, 1979, p. 63) that, for any X with $\dim X = \infty$, one has

$$\rho(u) \geq (1+u^2)^{1/2} - 1$$

and, for every X with $\dim X \geq 2$,

$$\rho(u) \geq Cu^2, \qquad C > 0.$$

This limits the power-type modulus of smoothness of nontrivial Banach spaces to the case u^q, $1 \leq q \leq 2$.

Remark 8.6.10 It follows from the above proof of Theorem 8.6.6 that $C(q,\gamma) \leq C\gamma^{1/q}$. In particular, in the case $X = L_p$, inequality (8.6.10) implies $C(q,\gamma) \leq Cp^{1/2}$ for $2 \leq p < \infty$.

Proposition 8.6.11 *The condition (8.6.3) in Corollary 8.6.5 is sharp.*

Proof Let $1 < q \leq 2$. Consider $X = \ell_q$. It is known (Lindenstrauss and Tzafriri, 1979, p. 67) that ℓ_q, $1 < q \leq 2$, is a uniformly smooth Banach space with modulus of smoothness $\rho(u)$ of power type q. Denote $p := q/(q-1)$ and take any $\{t_k\}_{k=1}^{\infty}$, $0 < t_k \leq 1$, such that

$$\sum_{k=1}^{\infty} t_k^p < \infty. \tag{8.6.11}$$

Choose $\mathscr{D}$ as a standard basis $\{e_j\}_{j=1}^{\infty}$, $e_j := (0,\dots,0,1,0,\dots)$, for ℓ_q. Consider a realization of the WCGA for

$$f := \left(1, t_1^{1/(q-1)}, t_2^{1/(q-1)}, \dots\right).$$

First, (8.6.11) guarantees that $f \in \ell_q$. Next, it is well known that F_f can be identified as

$$F_f = (1, t_1, t_2, \dots) \Big/ \left(1 + \sum_{k=1}^{\infty} t_k^p\right)^{1/p} \in \ell_p.$$

At the first step of the WCGA we pick $\varphi_1 = e_2$ and get

$$f_1 = \left(1, 0, t_2^{1/(q-1)}, \dots\right).$$

We continue with f replaced by f_1 and so on. After m steps we get

$$f_m = \left(1, 0, \dots, 0, t_{m+1}^{1/(q-1)}, \dots\right).$$

It is clear that for all m we have $\|f_m\|_{\ell_q} \geq 1$. □

8.7 Sparse Approximation With Respect to General Dictionaries

8.7.1 Introduction

We have already made some comments on results included in this section in the introduction to this chapter. We now complement those remarks by a more detailed discussion. This section contains recent results constituting breakthrough remarks in constructive sparse approximation. In all cases discussed here the new technique is based on greedy approximation. In §8.7.2 we concentrate on breakthrough results from Livshitz and Temlyakov (2014) and Temlyakov (2014). In these papers we extended a fundamental result of Zhang (2011) on the Lebesgue-type inequality for restricted isometry property (RIP) dictionaries in a Hilbert space (see Theorem 8.7.9 and Definition 8.7.7 below) in several directions. We found new, more general conditions than the RIP conditions on a dictionary, which still guarantee the Lebesgue-type inequalities in a Hilbert space setting. We generalized these conditions to a Banach space setting and proved the Lebesgue-type inequalities for dictionaries satisfying those conditions. To illustrate the power of the new conditions we applied this new technique to bases instead of redundant dictionaries. In particular, this technique gave very strong results for the trigonometric system; see Theorem 8.1.7 mentioned in the introduction to this chapter.

In a general setting we are working in a Banach space X with a redundant system of elements $\mathcal{D}$ (a dictionary $\mathcal{D}$). An element (function, signal) $f \in X$ is said to be m-sparse with respect to $\mathcal{D}$ if it has a representation $f = \sum_{i=1}^m x_i g_i$, $g_i \in \mathcal{D}$, $i = 1, \dots, m$. The set of all m-sparse elements is denoted by $\Sigma_m(\mathcal{D})$. For a given element f_0 we introduce the error of the best m-term approximation $\sigma_m(f_0, \mathcal{D}) := \inf_{f \in \Sigma_m(\mathcal{D})} \|f_0 - f\|$. As we pointed out earlier, we are interested in the following fundamental problem of sparse approximation.

Problem How do we design a practical algorithm that builds sparse approximations that are comparable to the best m-term approximations?

In a general setting we study an algorithm (i.e., an approximation method)

$$\mathscr{A} = \{A_m(\cdot, \mathscr{D})\}_{m=1}^{\infty}$$

with respect to a given dictionary $\mathscr{D}$. The sequence of mappings $A_m(\cdot, \mathscr{D})$ defined on X satisfies the following condition: for any $f \in X$, $A_m(f, \mathscr{D}) \in \Sigma_m(\mathscr{D})$. In other words, A_m provides an m-term approximant with respect to $\mathscr{D}$. It is clear that for any $f \in X$ and any m we have $\|f - A_m(f, \mathscr{D})\| \geq \sigma_m(f, \mathscr{D})$. We are interested in pairs $(\mathscr{D}, \mathscr{A})$ for which the algorithm $\mathscr{A}$ provides an approximation close to the best m-term approximation. We introduce the corresponding definitions.

Definition 8.7.1 We say that $\mathscr{D}$ is an almost greedy dictionary with respect to $\mathscr{A}$ if there exist two constants C_1 and C_2 such that, for any $f \in X$, we have

$$\|f - A_{C_1 m}(f, \mathscr{D})\| \leq C_2 \sigma_m(f, \mathscr{D}). \tag{8.7.1}$$

If $\mathscr{D}$ is an almost greedy dictionary with respect to $\mathscr{A}$ then $\mathscr{A}$ provides almost ideal sparse approximation. It provides a $C_1 m$-term approximant as good (up to a constant C_2) as the ideal m-term approximant for every $f \in X$. In the case $C_1 = 1$ we call $\mathscr{D}$ a greedy dictionary. We also need a more general definition. Let $\phi(u)$ be a function such that $\phi(u) \geq 1$.

Definition 8.7.2 We say that $\mathscr{D}$ is a ϕ-greedy dictionary with respect to $\mathscr{A}$ if there exists a constant C_3 such that, for any $f \in X$, we have

$$\|f - A_{\phi(m)m}(f, \mathscr{D})\| \leq C_3 \sigma_m(f, \mathscr{D}). \tag{8.7.2}$$

If $\mathscr{D} = \Psi$ is a basis then in the above definitions we replace “dictionary” by “basis”. Inequalities of the form (8.7.1) and (8.7.2) are called Lebesgue-type inequalities.

In the case $\mathscr{A} = \{G_m(\cdot, \Psi)\}_{m=1}^{\infty}$ of the thresholding greedy algorithm (TGA), the theory of greedy and almost greedy bases is well developed (see Temlyakov, 2011, 2015c). We recall that in the case of a normalized basis $\Psi = \{\psi_k\}_{k=1}^{\infty}$ of a Banach space X the TGA at the mth iteration gives an approximant

$$G_m(f, \Psi) := \sum_{j=1}^{m} c_{k_j} \psi_{k_j}, \qquad f = \sum_{k=1}^{\infty} c_k \psi_k, \qquad |c_{k_1}| \geq |c_{k_2}| \geq \cdots.$$

In particular, it is known (see Theorem 8.3.2 above and Temlyakov, 2011, p. 17) that the univariate Haar basis is a greedy basis with respect to the TGA for all L_p, $1 < p < \infty$. Also, it is known that the TGA does not work well for the trigonometric system (see §8.2).

It was demonstrated in Temlyakov (2014) that the weak Chebyshev greedy algorithm (WCGA), which we will define shortly, is a solution to the above problem for a special class of dictionaries. In this section we use a version of the WCGA with $\tau = \{t\}$, $t \in (0,1]$. For the reader's convenience we repeat the (see §8.1) definition here.

Weak Chebyshev greedy algorithm Let f_0 be given. Then, for each $m \geq 1$, we have the following inductive definition.

(1) Let $\varphi_m := \varphi_m^{c,t} \in \mathscr{D}$ be any element satisfying

$$|F_{f_{m-1}}(\varphi_m)| \geq t \sup_{g \in \mathscr{D}} |F_{f_{m-1}}(g)|.$$

(2) Define $\Phi_m := \Phi_m^t := \operatorname{span}\{\varphi_j\}_{j=1}^m$, and define $G_m := G_m^{c,t}$ to be the best approximant to f_0 from Φ_m.
(3) Let $f_m := f_m^{c,t} := f_0 - G_m$.

We note that the properties of a given basis with respect to the TGA and WCGA could be very different. For instance, the class of quasi-greedy bases (with respect to TGA) – that is, the class of bases Ψ for which $G_m(f, \Psi)$ converges for each $f \in X$ – is a rather narrow subset of all bases. It is close, in a certain sense, to the set of unconditional bases. The situation is absolutely different for the WCGA. If X is uniformly smooth then the WCGA converges, for each $f \in X$, with respect to any dictionary in X (see Theorem 8.6.4 above and Temlyakov, 2011, Chapter 6).

Theorem 8.1.7 shows that the WCGA is very well designed for the trigonometric system. We showed in Temlyakov (2014) that an analog of (8.1.7) holds for uniformly bounded orthogonal systems. The proof of Theorem 8.1.7 uses a technique developed in compressed sensing for proving the Lebesgue-type inequalities for redundant dictionaries with special properties. The first results on Lebesgue-type inequalities were proved for incoherent dictionaries (see Temlyakov, 2011, for a detailed discussion). Then a number of results were proved for dictionaries satisfying the restricted isometry property (RIP) assumption. The incoherence assumption on a dictionary is stronger than the RIP assumption. The corresponding Lebesgue-type inequalities for orthogonal matching pursuit (OMP) under the RIP assumption were not known for a while. As a result, new greedy-type algorithms were introduced and the exact recovery of sparse signals and Lebesgue-type inequalities were proved for these algorithms: regularized orthogonal matching pursuit (see Needell and Vershynin, 2009), compressive sampling matching pursuit (CoSaMP) (see Needell and Tropp, 2009), and subspace pursuit (SP) (see Dai and Milenkovic, 2009). The OMP is simpler than CoSaMP and SP; however, at the time of the invention of CoSaMP and SP these algorithms provided exact recovery of sparse signals and Lebesgue-type inequalities for dictionaries satisfying the restricted isometry property (see Needell and Tropp, 2009, and Dai and Milenkovic, 2009). The corresponding results for the OMP were not known at that time. Later, a breakthrough result in this direction was obtained by Zhang (2011) (for further results see Foucart, 2012, and Wang and Shim, 2012). In particular, Zhang proved that if $\mathscr{D}$ satisfies RIP then the OMP recovers exactly all m-sparse signals within Cm iterations. We now mention some papers on the Lebesgue-type inequalities for the OMP with respect to an incoherent dictionary: Donoho *et al.* (2007); Gilbert *et al.* (2003), Livshitz (2012); Temlyakov and Zheltov (2011); Tropp (2004). In Livshitz and Temlyakov (2014) and Temlyakov (2014) Zhang's technique was developed to obtain recovery results and Lebesgue-type inequalities in the Banach space setting.

Theorem 8.1.7 guarantees that the WCGA works very well for each individual function f. It is a constructive method, which provides after $\asymp m \ln m$ iterations an error comparable with $\sigma_m(f, \mathscr{D})$. Here are two important points. First, in order to guarantee a rate of decay of errors $\|f_n\|$ of the WCGA we would like to know how the smoothness assumptions on f_0 affect the rate of decay of $\sigma_m(f_0, \mathscr{D})$. Second, if as we believe one cannot get rid of $\ln m$ in Theorem 8.1.7 then it would be nice to find a constructive method which provides for a certain smoothness class the order of the best m-term approximation after m iterations. Thus, as a complement to Theorem 8.1.7 we would like to obtain results which relate the rate of decay of $\sigma_m(f, \mathscr{T}^d)_p$ to some smoothness-type properties of f. In Chapter 9 we will concentrate on constructive methods of m-term approximation. We measure smoothness in terms of mixed derivatives and mixed differences. We note that the function classes with bounded mixed derivatives are not only interesting and challenging objects in approximation theory but are also important in numerical computations.

We discuss here the problem of sparse approximation. This problem is closely connected with the problem of the recovery of sparse functions (signals). In the sparse recovery problem we assume that an unknown function f is sparse with respect to a given dictionary and that we want to recover it. This problem was a starting point for compressed sensing theory (see Temlyakov, 2011, Chapter 5). In particular, the celebrated contribution of Candes, Tao, and Donoho (see Temlyakov, 2011, Chapter 5) was to show that recovery can be made by the ℓ_1-minimization algorithm. We stress that this ℓ_1-minimization algorithm works for the exact recovery of sparse signals. It does not provide sparse approximation. The greedy-type algorithms discussed in this chapter provide sparse approximation satisfying Lebesgue-type inequalities. It is clear that the Lebesgue-type inequalities (8.7.1) and (8.7.2) guarantee the exact recovery of sparse signals.

8.7.2 Lebesgue-Type Inequalities. General Results

An important advantage of the WCGA is its convergence and rate-of-convergence properties. The WCGA is well defined for all m. Moreover, it is known (see Theorem 8.6.4 and Temlyakov, 2001, 2011) that the WCGA with weakness parameter $t \in (0,1]$ converges for all f_0 in all uniformly smooth Banach spaces with respect to any dictionary. We discuss here the Lebesgue-type inequalities for the WCGA with weakness parameter $t \in (0,1]$. This discussion is based on the papers Livshitz and Temlyakov (2014) and Temlyakov (2014). For notational convenience we consider here a countable dictionary $\mathscr{D} = \{g_i\}_{i=1}^{\infty}$. The following assumptions **A1** and **A2** were used in Livshitz and Temlyakov (2014). For a given f_0 let a sparse element (signal)

$$f := f^{\varepsilon} = \sum_{i \in T} x_i g_i, \qquad g_i \in \mathscr{D},$$

be such that $\|f_0 - f^{\varepsilon}\| \leq \varepsilon$ and $|T| = K$. For $A \subset T$ denote

$$f_A := f_A^{\varepsilon} := \sum_{i \in A} x_i g_i.$$

A1 We say that $f = \sum_{i \in T} x_i g_i$ satisfies a Nikol'skii-type $\ell_1 X$ inequality with parameter r if, for any $A \subset T$,

$$\sum_{i \in A} |x_i| \leq C_1 |A|^r \|f_A\|. \tag{8.7.3}$$

We say that a dictionary $\mathscr{D}$ has the Nikol'skii-type $\ell_1 X$ property with parameters K, r if any K-sparse element satisfies a Nikol'skii-type $\ell_1 X$ inequality with parameter r.

A2 We say that $f = \sum_{i \in T} x_i g_i$ has the incoherence property with parameters D and U if for any $A \subset T$ and any Λ such that $A \cap \Lambda = \emptyset$, $|A| + |\Lambda| \leq D$, we have for any $\{c_i\}$

$$\left\| f_A - \sum_{i \in \Lambda} c_i g_i \right\| \geq U^{-1} \|f_A\|. \tag{8.7.4}$$

We say that a dictionary $\mathscr{D}$ is (K,D)-unconditional with a constant U if, for any $f = \sum_{i \in T} x_i g_i$ with $|T| \leq K$, inequality (8.7.4) holds.

The term *unconditional* in **A2** is justified by the following remark. The above definition of a (K,D)-unconditional dictionary is equivalent to the following definition. Let $\mathscr{D}$ be such that any subsystem of D distinct elements $e_1, \ldots, e_D$ from $\mathscr{D}$ is linearly independent and, for any $A \subset [1,D]$ with $|A| \leq K$ and any coefficients $\{c_i\}$, we have

$$\left\| \sum_{i \in A} c_i e_i \right\| \leq U \left\| \sum_{i=1}^{D} c_i e_i \right\|.$$

It is convenient for us to use the following assumption, **A3**, introduced in Temlyakov (2014), which is a corollary of assumptions **A1** and **A2**.

A3 We say that $f=\sum_{i\in T}x_i g_i$ has the ℓ_1-incoherence property with parameters D, V, and r if, for any $A\subset T$ and any Λ such that $A\cap\Lambda=\emptyset$, $|A|+|\Lambda|\le D$, we have for any $\{c_i\}$

$$\sum_{i\in A}|x_i|\le V|A|^r\left\|f_A-\sum_{i\in\Lambda}c_i g_i\right\|. \tag{8.7.5}$$

A dictionary $\mathscr{D}$ has the ℓ_1-incoherence property with parameters K, D, V, and r if, for any $A\subset B$, $|A|\le K$, $|B|\le D$, we have for any $\{c_i\}_{i\in B}$

$$\sum_{i\in A}|c_i|\le V|A|^r\left\|\sum_{i\in B}c_i g_i\right\|.$$

It is clear that **A1** and **A2** imply **A3** with $V=C_1U$. Also, **A3** implies **A1** with $C_1=V$ and **A2** with $U=VK^r$. Obviously, we can restrict ourselves to $r\le 1$.

We now give a simple remark that widens the collection of dictionaries satisfying the above properties **A1**, **A2**, and **A3**.

Definition 8.7.3 Let $\mathscr{D}^1=\{g_i^1\}$ and $\mathscr{D}^2=\{g_i^2\}$ be countable dictionaries. We say that $\mathscr{D}^2$ D-dominates $\mathscr{D}^1$ (with a constant B) if, for any set Λ, $|\Lambda|\le D$, of indices and any coefficients $\{c_i\}$, we have

$$\left\|\sum_{i\in\Lambda}c_i g_i^1\right\|\le B\left\|\sum_{i\in\Lambda}c_i g_i^2\right\|.$$

In such a case we write $\mathscr{D}^1\prec\mathscr{D}^2$ or, more specifically, $\mathscr{D}^1\le B\mathscr{D}^2$.

In the case $\mathscr{D}^1\le E_1^{-1}\mathscr{D}^2$ and $\mathscr{D}^2\le E_2\mathscr{D}^1$ we say that $\mathscr{D}^1$ and $\mathscr{D}^2$ are D-equivalent (with constants E_1 and E_2) and write $\mathscr{D}^1\approx\mathscr{D}^2$ or, more specifically, $E_1\mathscr{D}^1\le\mathscr{D}^2\le E_2\mathscr{D}^1$.

Proposition 8.7.4 *Assume that $\mathscr{D}^1$ has one of the properties **A1** or **A3**. If $\mathscr{D}^2$ D-dominates $\mathscr{D}^1$ (with a constant B) then $\mathscr{D}^2$ has the same property as $\mathscr{D}^1$: **A1** with $C_1^2=C_1^1B$ or **A3** with $V^2=V^1B$.*

Proof In both the cases **A1** and **A3** the proof is the same. We demonstrate the case **A3**. Let $f=\sum_{i\in T}x_i g_i^2$. Then by the property **A3** of $\mathscr{D}^1$ we have

$$\sum_{i\in A}|x_i|\le V^1|A|^r\left\|\sum_{i\in A}x_i g_i^1-\sum_{i\in\Lambda}c_i g_i^1\right\|\le V^1B|A|^r\left\|\sum_{i\in A}x_i g_i^2-\sum_{i\in\Lambda}c_i g_i^2\right\|. \qquad\square$$

Proposition 8.7.5 *Assume $\mathscr{D}^1$ has the property **A2**. If $\mathscr{D}^1$ and $\mathscr{D}^2$ are D-equivalent (with constants E_1 and E_2) then $\mathscr{D}^2$ has the property **A2** with $U^2=U^1E_2/E_1$.*

Proof Let $f = \sum_{i\in T} x_i g_i^2$. Then by $\mathscr{D}^1 \approx \mathscr{D}^2$ and the **A2** property of $\mathscr{D}^1$ we have

$$\left\|\sum_{i\in A} x_i g_i^2 - \sum_{i\in\Lambda} c_i g_i^2\right\| \geq E_1 \left\|\sum_{i\in A} x_i g_i^1 - \sum_{i\in\Lambda} c_i g_i^1\right\| \geq E_1(U^1)^{-1}\left\|\sum_{i\in A} x_i g_i^1\right\|$$
$$\geq (E_1/E_2)(U^1)^{-1}\left\|\sum_{i\in A} x_i g_i^2\right\|. \qquad \square$$

We now proceed to the main results of Livshitz and Temlyakov (2014) and Temlyakov (2014) on the WCGA with respect to redundant dictionaries. The following theorem, from Temlyakov (2014), was proved in Livshitz and Temlyakov (2014) in the case $q = 2$.

Theorem 8.7.6 *Let X be a Banach space with $\rho(u) \leq \gamma u^q$, $1 < q \leq 2$. Suppose that a K-sparse f^ε satisfies* **A1, A2** *and $\|f_0 - f^\varepsilon\| \leq \varepsilon$. Assume that $rq' \geq 1$. Then the WCGA, with weakness parameter t applied to f_0 gives*

$$\|f_{C(t,\gamma,C_1)U^{q'}\ln(U+1)K^{rq'}}\| \leq C\varepsilon \qquad \text{for } K + C(t,\gamma,C_1)U^{q'}\ln(U+1)K^{rq'} \leq D$$

with C an absolute constant.

It was pointed out in Livshitz and Temlyakov (2014) that Theorem 8.7.6 implies a corollary for Hilbert spaces that gives sufficient conditions, which are somewhat weaker than the known RIP conditions on $\mathscr{D}$, for a Lebesgue-type inequality to hold. We now formulate the corresponding definitions and results. Let $\mathscr{D}$ be a Riesz dictionary with depth D and parameter $\delta \in (0,1)$. This class of dictionaries is a generalization of the class of classical Riesz bases. We give a definition in a general Hilbert space (see Temlyakov, 2011, p. 306).

Definition 8.7.7 A dictionary $\mathscr{D}$ is called a Riesz dictionary with depth D and parameter $\delta \in (0,1)$ if, for any D distinct elements $e_1,\dots,e_D$ of the dictionary and any coefficients $a = (a_1,\dots,a_D)$, we have

$$(1-\delta)\|a\|_2^2 \leq \left\|\sum_{i=1}^{D} a_i e_i\right\|^2 \leq (1+\delta)\|a\|_2^2. \tag{8.7.6}$$

We denote the class of Riesz dictionaries with depth D and parameter $\delta \in (0,1)$ by $R(D,\delta)$.

The term "Riesz dictionary with depth D and parameter $\delta \in (0,1)$" is another name for a dictionary satisfying the restricted isometry property (RIP) with parameters D and δ. The following simple lemma holds.

Lemma 8.7.8 *Let $\mathscr{D} \in R(D,\delta)$ and let $e_j \in \mathscr{D}$, $j = 1,\dots,s$. For $f = \sum_{i=1}^{s} a_i e_i$ and $A \subset \{1,\dots,s\}$ denote*

$$S_A(f) := \sum_{i\in A} a_i e_i.$$

If $s \leq D$ then

$$\|S_A(f)\|^2 \leq (1+\delta)(1-\delta)^{-1}\|f\|^2.$$

Lemma 8.7.8 implies that if $\mathscr{D} \in R(D,\delta)$ then it is (D,D)-unconditional with a constant $U = (1+\delta)^{1/2}(1-\delta)^{-1/2}$.

Theorem 8.7.9 *Let X be a Hilbert space. Suppose that a K-sparse f^{ε} satisfies* **A2** *and $\|f_0 - f^{\varepsilon}\| \leq \varepsilon$. Then the WOMP with weakness parameter t applied to f_0 gives*

$$\|f_{C(t,U)K}\| \leq C\varepsilon \qquad \text{for } K + C(t,U)K \leq D,$$

with C an absolute constant.

Theorem 8.7.9 implies the following corollaries.

Corollary 8.7.10 *Let X be a Hilbert space. Suppose that any K-sparse f satisfies* **A2**. *Then the WOMP with weakness parameter t applied to f_0 gives*

$$\|f_{C(t,U)K}\| \leq C\sigma_K(f_0,\mathscr{D}) \qquad \text{for } K + C(t,U)K \leq D,$$

with C an absolute constant.

Corollary 8.7.11 *Let X be a Hilbert space. Suppose that $\mathscr{D} \in R(D,\delta)$. Then the WOMP with weakness parameter t applied to f_0 gives*

$$\|f_{C(t,\delta)K}\| \leq C\sigma_K(f_0,\mathscr{D}) \qquad \text{for } K + C(t,\delta)K \leq D,$$

with C an absolute constant.

It was emphasized in Livshitz and Temlyakov (2014) that in Theorem 8.7.6 we impose conditions on an individual function f^{ε}. It may happen that the dictionary does not have the Nikol'skii $\ell_1 X$ property and (K,D)-unconditionality but the given f_0 can be approximated by an f^{ε} which does satisfy assumptions **A1** and **A2**. Even in the case of a Hilbert space the above results from Livshitz and Temlyakov (2014) add something new to a study based on the RIP property of a dictionary. First, Theorem 8.7.9 shows that it is sufficient to impose assumption **A2** on f^{ε} in order to obtain exact recovery and the Lebesgue-type inequality results. Second, Corollary 8.7.10 shows that the condition **A2**, which is weaker than the RIP condition, is sufficient for exact recovery and the Lebesgue-type inequality results. Third, Corollary 8.7.11 shows that even if we impose our assumptions in terms of the RIP we do not need to assume that $\delta < \delta_0$. In fact, the result works for all $\delta < 1$, with parameters depending on δ.

Theorem 8.7.6 follows from a combination of Theorems 8.7.12 and 8.7.13. In the case $q = 2$ these theorems were proved in Livshitz and Temlyakov (2014) and, in the general case $q \in (1,2]$, in Temlyakov (2014).

Theorem 8.7.12 *Let X be a Banach space with $\rho(u) \le \gamma u^q$, $1 < q \le 2$. Suppose that for a given f_0 we have $\|f_0 - f^\varepsilon\| \le \varepsilon$ with K-sparse $f := f^\varepsilon$ satisfying* **A3**. *Then, for any $k \ge 0$, we have for $K + m \le D$*

$$\|f_m\| \le \|f_k\| \exp\left(-\frac{c_1(m-k)}{K^{rq'}}\right) + 2\varepsilon, \qquad q' := \frac{q}{q-1},$$

where

$$c_1 := (t^{q'})/(2(16\gamma)^{1/(q-1)}V^{q'}).$$

In all theorems that follow we assume $rq' \ge 1$.

Theorem 8.7.13 *Let X be a Banach space with $\rho(u) \le \gamma u^q$, $1 < q \le 2$. Suppose that a K-sparse f^ε satisfies* **A1**, **A2**, *and $\|f_0 - f^\varepsilon\| \le \varepsilon$. Then the WCGA with weakness parameter t applied to f_0 gives*

$$\|f_{C'U^{q'}\ln(U+1)K^{rq'}}\| \le CU\varepsilon \qquad \text{for } K + C'U^{q'}\ln(U+1)K^{rq'} \le D,$$

with C an absolute constant and $C' = C_2(q)\gamma^{1/(q-1)}C_1^{q'}t^{-q'}$.

We can formulate an immediate corollary of Theorem 8.7.13 for $\varepsilon = 0$.

Corollary 8.7.14 *Let X be a Banach space with $\rho(u) \le \gamma u^q$. Suppose that a K-sparse f satisfies* **A1**, *and* **A2**. *Then the WCGA with weakness parameter t applied to f recovers it exactly after $C'U^{q'}\ln(U+1)K^{rq'}$ iterations under the condition $K + C'U^{q'}\ln(U+1)K^{rq'} \le D$.*

We formulate versions of Theorem 8.7.13 either with assumptions **A1**, and **A2** replaced by a single assumption **A3** or with assumptions **A1** and **A2** replaced by two assumptions, **A2** and **A3**. The corresponding modified proofs go as in the proof of Theorem 8.7.12, to be found in §8.7.3 below.

Theorem 8.7.15 *Let X be a Banach space with $\rho(u) \le \gamma u^q$, $1 < q \le 2$. Suppose that a K-sparse f^ε satisfies* **A3** *and that $\|f_0 - f^\varepsilon\| \le \varepsilon$. Then the WCGA with weakness parameter t applied to f_0 gives*

$$\|f_{C(t,\gamma,q)V^{q'}\ln(VK)K^{rq'}}\| \le CVK^r\varepsilon \qquad \text{for } K + C(t,\gamma,q)V^{q'}\ln(VK)K^{rq'} \le D$$

with C an absolute constant and $C(t,\gamma,q) = C_2(q)\gamma^{1/(q-1)}t^{-q'}$.

Theorem 8.7.16 *Let X be a Banach space with $\rho(u) \le \gamma u^q$, $1 < q \le 2$. Suppose that a K-sparse f^ε satisfies* **A2** *and* **A3** *and that $\|f_0 - f^\varepsilon\| \le \varepsilon$. Then the WCGA with weakness parameter t applied to f_0 gives*

$$\|f_{C(t,\gamma,q)V^{q'}\ln(U+1)K^{rq'}}\| \le CU\varepsilon \qquad \text{for } K + C(t,\gamma,q)V^{q'}\ln(U+1)K^{rq'} \le D$$

with C an absolute constant and $C(t,\gamma,q) = C_2(q)\gamma^{1/(q-1)}t^{-q'}$.

Theorems 8.7.15 and 8.7.12 imply the following analog of Theorem 8.7.6.

Theorem 8.7.17 *Let X be a Banach space with $\rho(u) \le \gamma u^q$, $1 < q \le 2$. Suppose that a K-sparse f^ε satisfies* **A3** *and $\|f_0 - f^\varepsilon\| \le \varepsilon$. Then the WCGA with weakness parameter t applied to f_0 gives*

$$\|f_{C(t,\gamma,q)V^{q'}\ln(VK)K^{rq'}}\| \le C\varepsilon \qquad \text{for } K + C(t,\gamma,q)V^{q'}\ln(VK)K^{rq'} \le D$$

with C an absolute constant and $C(t,\gamma,q) = C_2(q)\gamma^{1/(q-1)}t^{-q'}$.

The following version of Theorems 8.7.6 and 8.7.17 is also useful in applications. It follows from Theorems 8.7.16 and 8.7.12.

Theorem 8.7.18 *Let X be a Banach space with $\rho(u) \le \gamma u^q$, $1 < q \le 2$. Suppose that a K-sparse f^ε satisfies* **A2** *and* **A3** *and that $\|f_0 - f^\varepsilon\| \le \varepsilon$. Then the WCGA with weakness parameter t applied to f_0 gives*

$$\|f_{C(t,\gamma,q)V^{q'}\ln(U+1)K^{rq'}}\| \le C\varepsilon \qquad \text{for } K + C(t,\gamma,q)V^{q'}\ln(U+1)K^{rq'} \le D$$

with C an absolute constant and $C(t,\gamma,q) = C_2(q)\gamma^{1/(q-1)}t^{-q'}$.

8.7.3 Proofs

We begin with a proof of Theorem 8.7.12.

Proof of Theorem 8.7.12 Let

$$f := f^\varepsilon = \sum_{i\in T} x_i g_i, \qquad |T| = K, \qquad g_i \in \mathscr{D}.$$

Denote by T^m the set of indices of $g_j \in \mathscr{D}$, $j \in T$, chosen by the WCGA after m iterations; thus $\Gamma^m := T \setminus T^m$. Denote by $A_1(\mathscr{D}^\pm)$ the closure in X of the convex hull of the symmetrized dictionary $\mathscr{D}^\pm := \{\pm g, g \in \mathscr{D}\}$. We will bound $\|f_m\|$ from above. Assume that $\|f_{m-1}\| \ge \varepsilon$. Let $m > k$. We bound from below

$$S_m := \sup_{\phi\in A_1(\mathscr{D}^\pm)} |F_{f_{m-1}}(\phi)|.$$

Denote $A_m := \Gamma^{m-1}$. Then

$$S_m \ge F_{f_{m-1}}(f_{A_m}/\|f_{A_m}\|_1),$$

where $\|f_A\|_1 := \sum_{i\in A}|x_i|$. Next, by Lemma 8.6.7 (see also Lemma 6.9, p. 342, from Temlyakov, 2011) we obtain

$$F_{f_{m-1}}(f_{A_m}) = F_{f_{m-1}}(f^\varepsilon) \ge \|f_{m-1}\| - \varepsilon.$$

Thus

$$S_m \ge \|f_{A_m}\|_1^{-1}(\|f_{m-1}\| - \varepsilon). \tag{8.7.7}$$

From the definition of the modulus of smoothness we have, for any λ,

$$\|f_{m-1}-\lambda\varphi_m\|+\|f_{m-1}+\lambda\varphi_m\|\le 2\|f_{m-1}\|\left(1+\rho\left(\frac{\lambda}{\|f_{m-1}\|}\right)\right) \tag{8.7.8}$$

and by (1) from the definition of the WCGA and Lemma 8.6.8 (see also Lemma 6.10 from Temlyakov, 2011, p. 343), we get

$$|F_{f_{m-1}}(\varphi_m)|\ge t\sup_{g\in\mathscr{D}}|F_{f_{m-1}}(g)|=t\sup_{\phi\in A_1(\mathscr{D}^{\pm})}|F_{f_{m-1}}(\phi)|=tS_m.$$

Then either $F_{f_{m-1}}(\varphi_m)\ge tS_m$ or $F_{f_{m-1}}(-\varphi_m)\ge tS_m$. Both cases can be treated in the same way. We will demonstrate the case $F_{f_{m-1}}(\varphi_m)\ge tS_m$. We have, for $\lambda\ge 0$,

$$\|f_{m-1}+\lambda\varphi_m\|\ge F_{f_{m-1}}(f_{m-1}+\lambda\varphi_m)\ge\|f_{m-1}\|+\lambda tS_m.$$

From this and from (8.7.8) we obtain

$$\|f_m\|\le\|f_{m-1}-\lambda\varphi_m\|\le\|f_{m-1}\|+\inf_{\lambda\ge 0}(-\lambda tS_m+2\|f_{m-1}\|\rho(\lambda/\|f_{m-1}\|)).$$

We discuss here the case $\rho(u)\le\gamma u^q$. Using (8.7.7) we get

$$\|f_m\|\le\|f_{m-1}\|\left(1-\frac{\lambda t}{\|f_{A_m}\|_1}+2\gamma\frac{\lambda^q}{\|f_{m-1}\|^q}\right)+\frac{\epsilon\lambda t}{\|f_{A_m}\|_1}.$$

Let λ_1 be a solution of

$$\frac{\lambda t}{2\|f_{A_m}\|_1}=2\gamma\frac{\lambda^q}{\|f_{m-1}\|^q},\qquad \lambda_1=\left(\frac{t\|f_{m-1}\|^q}{4\gamma\|f_{A_m}\|_1}\right)^{1/(q-1)}.$$

Our assumption (8.7.5) gives

$$\begin{aligned}\|f_{A_m}\|_1&=\|(f^\epsilon-G_{m-1})_{A_m}\|_1\le VK^r\|f^\epsilon-G_{m-1}\|\\&\le VK^r(\|f_0-G_{m-1}\|+\|f_0-f^\epsilon\|)\le VK^r(\|f_{m-1}\|+\epsilon).\end{aligned} \tag{8.7.9}$$

Now, specify

$$\lambda=\left(\frac{t\|f_{A_m}\|_1^{q-1}}{16\gamma(VK^r)^q}\right)^{1/(q-1)}.$$

Then, using $\|f_{m-1}\|\ge\epsilon$ we get

$$\left(\frac{\lambda}{\lambda_1}\right)^{q-1}=\frac{\|f_{A_m}\|_1^q}{4\|f_{m-1}\|^q(VK^r)^q}\le 1$$

and obtain

$$\|f_m\|\le\|f_{m-1}\|\left(1-\frac{t^{q'}}{2(16\gamma)^{1/(q-1)}(VK^r)^{q'}}\right)+\frac{\epsilon t^{q'}}{(16\gamma)^{1/(q-1)}(VK^r)^{q'}}. \tag{8.7.10}$$

Denote

$$c_1 := \frac{t^{q'}}{2(16\gamma)^{1/(q-1)}V^{q'}}.$$

Then

$$\|f_m\| \le \|f_k\| \exp\left(-\frac{c_1(m-k)}{K^{rq'}}\right) + 2\varepsilon. \qquad \square$$

We proceed to a proof of Theorem 8.7.13.

Proof of Theorem 8.7.13 Modifications of the following proof in the style of the above proof of Theorem 8.7.12 give Theorems 8.7.15 and 8.7.16. We begin with a brief description of the structure of the proof. We are given f_0 and $f := f^\varepsilon$, such that $\|f_0 - f\| \le \varepsilon$ and f is K-sparse, satisfying **A1** and **A2** (see §8.7.2). We apply the WCGA to f_0, and we control how many dictionary elements g_i from the representation of f

$$f := f^\varepsilon := \sum_{i \in T} x_i g_i$$

are picked up by the WCGA after m iterations. As above denote by T^m the set of indices $i \in T$ such that g_i has been picked up by the WCGA at one of the first m iterations. Denote $\Gamma^m := T \setminus T^m$. It is clear that if $\Gamma^m = \emptyset$ then $\|f_m\| \le \varepsilon$ because in this case $f \in \Phi_m$.

Our analysis goes as follows. For a residual f_k we assume that Γ^k is nonempty. Then we prove that after $N(k)$ iterations we arrive at a residual $f_{k'}$, $k' = k + N(k)$, such that either

$$\|f_{k'}\| \le CU\varepsilon \tag{8.7.11}$$

or

$$|\Gamma^{k'}| < |\Gamma^k| - 2^{L-2}, \tag{8.7.12}$$

for some natural number L. An important fact is that for the number $N(k)$ of iterations we have a bound

$$N(k) \le \beta 2^{aL}, \qquad a := rq'. \tag{8.7.13}$$

Next, we prove that if we begin with $k = 0$ and apply the above argument to the sequence of residuals $f_0, f_{k_1}, \dots, f_{k_s}$ then, after not more than $N := 2^{2a+1}\beta K^a$ iterations, we obtain either $\|f_N\| \le CU\varepsilon$ or $\Gamma^N = \emptyset$, which in turn implies that $\|f_N\| \le \varepsilon$.

We now proceed to a detailed argument. The following corollary of (8.7.4) will be often used: for $m \le D - K$ and $A \subset \Gamma^m$ we have

$$\|f_A\| \le U(\|f_m\| + \varepsilon). \tag{8.7.14}$$

This follows from the fact that $f_A - f + G_m$ has the form $\sum_{i \in \Lambda} c_i g_i$, with Λ satisfying $|A| + |\Lambda| \le D$, $A \cap \Lambda = \emptyset$, and from our assumption that $\|f - f_0\| \le \varepsilon$.

The following lemma plays a key role in the proof.

Lemma 8.7.19 *Let f satisfy **A1** and **A2** and let $A \subset \Gamma^k$ be nonempty. Denote $B := \Gamma^k \setminus A$. Then for any $m \in (k, D-K]$ we have either $\|f_{m-1}\| \le \varepsilon$ or*

$$\|f_m\| \le \|f_{m-1}\|(1-u) + 2u(\|f_B\| + \varepsilon), \tag{8.7.15}$$

where

$$u := c_1|A|^{-rq'}, \qquad c_1 := \frac{t^{q'}}{2(16\gamma)^{1/(q-1)}(C_1U)^{q'}},$$

*with C_1 and U from **A1** and **A2**.*

Proof As above in the proof of Theorem 8.7.12 we bound S_m from below. It is clear that $S_m \ge 0$. Denote $A(m) := A \cap \Gamma^{m-1}$. Then

$$S_m \ge F_{f_{m-1}}(f_{A(m)}/\|f_{A(m)}\|_1).$$

Next,

$$F_{f_{m-1}}(f_{A(m)}) = F_{f_{m-1}}(f_{A(m)} + f_B - f_B).$$

We have $f_{A(m)} + f_B = f^\varepsilon - f_\Lambda$ with $F_{f_{m-1}}(f_\Lambda) = 0$. Moreover, it is easy to see that $F_{f_{m-1}}(f^\varepsilon) \ge \|f_{m-1}\| - \varepsilon$. Therefore,

$$F_{f_{m-1}}(f_{A(m)} + f_B - f_B) \ge \|f_{m-1}\| - \varepsilon - \|f_B\|.$$

Thus

$$S_m \ge \|f_{A(m)}\|_1^{-1} \max(0, \|f_{m-1}\| - \varepsilon - \|f_B\|).$$

By (8.7.3) we get

$$\|f_{A(m)}\|_1 \le C_1|A(m)|^r\|f_{A(m)}\| \le C_1|A|^r\|f_{A(m)}\|.$$

Then

$$S_m \ge \frac{\|f_{m-1}\| - \|f_B\| - \varepsilon}{C_1|A|^r\|f_{A(m)}\|}. \tag{8.7.16}$$

From the definition of the modulus of smoothness we have, for any λ,

$$\|f_{m-1} - \lambda\varphi_m\| + \|f_{m-1} + \lambda\varphi_m\| \le 2\|f_{m-1}\|\left(1 + \rho\left(\frac{\lambda}{\|f_{m-1}\|}\right)\right)$$

and by (1) from the definition of the WCGA and Lemma 8.6.8 (see also Lemma 6.10 from Temlyakov, 2011, p. 343), we get

$$|F_{f_{m-1}}(\varphi_m)| \ge t \sup_{g \in \mathscr{D}} |F_{f_{m-1}}(g)| = t \sup_{\phi \in A_1(\mathscr{D}^\pm)} |F_{f_{m-1}}(\phi)|.$$

From this we obtain

$$\|f_m\| \le \|f_{m-1}\| + \inf_{\lambda \ge 0}(-\lambda t S_m + 2\|f_{m-1}\|\rho(\lambda/\|f_{m-1}\|)).$$

We discuss here the case $\rho(u) \le \gamma u^q$. Using (8.7.16) we get, for any $\lambda \ge 0$,

$$\|f_m\| \le \|f_{m-1}\| \left(1 - \frac{\lambda t}{C_1|A|^r\|f_{A(m)}\|} + 2\gamma\frac{\lambda^q}{\|f_{m-1}\|^q}\right) + \frac{\lambda t(\|f_B\|+\varepsilon)}{C_1|A|^r\|f_{A(m)}\|}.$$

Let λ_1 be a solution of

$$\frac{\lambda t}{2C_1|A|^r\|f_{A(m)}\|} = 2\gamma\frac{\lambda^q}{\|f_{m-1}\|^q}, \qquad \lambda_1 = \left(\frac{t\|f_{m-1}\|^q}{4\gamma C_1|A|^r\|f_{A(m)}\|}\right)^{1/(q-1)}.$$

Inequality (8.7.14) gives

$$\|f_{A(m)}\| \le U(\|f_{m-1}\|+\varepsilon).$$

Now specify

$$\lambda = \left(\frac{t\|f_{A(m)}\|^{q-1}}{16\gamma C_1|A|^r U^q}\right)^{1/(q-1)}.$$

Then $\lambda \le \lambda_1$ and we obtain

$$\|f_m\| \le \|f_{m-1}\| \left(1 - \frac{t^{q'}}{2(16\gamma)^{1/(q-1)}(C_1U|A|^r)^{q'}}\right) + \frac{t^{q'}(\|f_B\|+\varepsilon)}{(16\gamma)^{1/(q-1)}(C_1|A|^rU)^{q'}}. \tag{8.7.17}$$

□

For simplicity of notation we consider separately the case $|\Gamma^k| \ge 2$ and the case $|\Gamma^k| = 1$. We begin with the generic case $|\Gamma^k| \ge 2$ and apply Lemma 8.7.19 with different pairs A_j, B_j, which we now construct. Let n be a natural number such that

$$2^{n-1} < |\Gamma^k| \le 2^n.$$

For $j = 1, 2, \dots, n, n+1$ consider the pairs of sets A_j, B_j: $A_{n+1} = \Gamma^k$, $B_{n+1} = \emptyset$; for $j \le n$, $A_j := \Gamma^k \setminus B_j$ with $B_j \subset \Gamma^k$ such that $|B_j| \ge |\Gamma^k| - 2^{j-1}$ and, for any set $J \subset \Gamma^k$ with $|J| \ge |\Gamma^k| - 2^{j-1}$, we have

$$\|f_{B_j}\| \le \|f_J\|.$$

We note that the above definition implies that $|A_j| \le 2^{j-1}$ and that if, for some $Q \subset \Gamma^k$ we have

$$\|f_Q\| < \|f_{B_j}\|, \qquad \text{then } |Q| < |\Gamma^k| - 2^{j-1}. \tag{8.7.18}$$

Set $B_0 := \Gamma^k$. Note that property (8.7.18) is obvious for $j = 0$.

Let $j_0 \in [1, n]$ be an index such that if $j_0 = 1$ then $B_1 \ne \Gamma^k$ and if $j_0 \ge 2$ then

$$B_1 = B_2 = \cdots = B_{j_0-1} = \Gamma^k, \qquad B_{j_0} \ne \Gamma^k.$$

For a given $b > 1$, to be specified later, denote by $L := L(b)$ the index such that $(B_0 := \Gamma^k)$

$$\|f_{B_0}\| < b\|f_{B_{j_0}}\|,$$
$$\|f_{B_{j_0}}\| < b\|f_{B_{j_0+1}}\|,$$
$$\vdots$$
$$\|f_{B_{L-2}}\| < b\|f_{B_{L-1}}\|,$$
$$\|f_{B_{L-1}}\| \geq b\|f_{B_L}\|.$$

Then

$$\|f_{B_j}\| \leq b^{L-1-j}\|f_{B_{L-1}}\|, \qquad j = j_0, \dots, L, \tag{8.7.19}$$

and

$$\|f_{B_0}\| = \cdots = \|f_{B_{j_0-1}}\| \leq b^{L-j_0}\|f_{B_{L-1}}\|. \tag{8.7.20}$$

Clearly, $L \leq n+1$.

Define $m_0 := \cdots m_{j_0-1} := k$ and, inductively,

$$m_j = m_{j-1} + [\beta|A_j|^{rq'}], \qquad j = j_0, \dots, L,$$

where $[x]$ denotes as before the integer part of x. Let us take a parameter β which satisfies the following inequalities, with c_1 from Lemma 8.7.19:

$$\beta \geq 1, \qquad e^{-c_1\beta/2} < 1/2, \qquad 16Ue^{-c_1\beta/2} < 1. \tag{8.7.21}$$

We note that the inequality $\beta \geq 1$ implies that

$$[\beta|A_j|^{rq'}] \geq \beta|A_j|^{rq'}/2.$$

Taking into account that $rq' \geq 1$ and $|A_j| \geq 1$ we obtain

$$m_j \geq m_{j-1} + 1.$$

At iterations from $m_{j-1}+1$ to m_j we apply Lemma 8.7.19 with $A = A_j$ and obtain from (8.7.15) that either $\|f_{m-1}\| \leq \varepsilon$ or

$$\|f_m\| \leq \|f_{m-1}\|(1-u) + 2u(\|f_{B_j}\| + \varepsilon), \qquad u := c_1|A_j|^{-rq'}.$$

Using $1 - u \leq e^{-u}$ and $\sum_{k=0}^{\infty}(1-u)^k = 1/u$ we then derive

$$\|f_{m_j}\| \leq \|f_{m_{j-1}}\|e^{-c_1\beta/2} + 2(\|f_{B_j}\| + \varepsilon). \tag{8.7.22}$$

We continue this process up to $j = L$. Denote $\eta := e^{-c_1\beta/2}$. Then either $\|f_{m_L}\| \le \varepsilon$ or

$$\|f_{m_L}\| \le \|f_k\|\eta^{L-j_0+1} + 2\sum_{j=j_0}^{L}(\|f_{B_j}\| + \varepsilon)\eta^{L-j}.$$

We will bound the $\|f_k\|$. It follows from the definition of f_k that $\|f_k\|$ is the error of the best approximation of f_0 by the subspace Φ_k. Representing f_0 as $f + f_0 - f$ we see that $\|f_k\|$ is not greater than the error of the best approximation of f by the subspace Φ_k plus $\|f_0 - f\|$. This implies that $\|f_k\| \le \|f_{B_0}\| + \varepsilon$. Therefore we continue the above relation as

$$\le (\|f_{B_0}\| + \varepsilon)\eta^{L-j_0+1} + 2\sum_{j=j_0}^{L}(\|f_{B_{L-1}}\|(\eta b)^{L-j}b^{-1} + \varepsilon\eta^{L-j})$$

$$\le b^{-1}\|f_{B_{L-1}}\|\left((\eta b)^{L-j_0+1} + 2\sum_{j=j_0}^{L}(\eta b)^{L-j}\right) + \frac{2\varepsilon}{1-\eta}.$$

Our choice of β guarantees that $\eta < 1/2$. Choose $b = 1/(2\eta)$. Then

$$\|f_{m_L}\| \le \|f_{B_{L-1}}\| 8e^{-c_1\beta/2} + 4\varepsilon. \tag{8.7.23}$$

By (8.7.14) we get

$$\|f_{\Gamma^{m_L}}\| \le U(\|f_{m_L}\| + \varepsilon) \le U\left(\|f_{B_{L-1}}\| 8e^{-c_1\beta/2} + 5\varepsilon\right).$$

If $\|f_{B_{L-1}}\| \le 10U\varepsilon$ then, by (8.7.23),

$$\|f_{m_L}\| \le CU\varepsilon, \qquad C = 44.$$

If $\|f_{B_{L-1}}\| \ge 10U\varepsilon$ then, by our choice of β, we have $16Ue^{-c_1\beta/2} < 1$ and

$$U\left(\|f_{B_{L-1}}\| 8e^{-c_1\beta/2} + 5\varepsilon\right) < \|f_{B_{L-1}}\|.$$

Therefore

$$\|f_{\Gamma^{m_L}}\| < \|f_{B_{L-1}}\|.$$

This implies that

$$|\Gamma^{m_L}| < |\Gamma^k| - 2^{L-2}.$$

In the above proof our assumption $j_0 \le n$ is equivalent to the assumption that $B_n \ne \Gamma^k$. We now consider the case $B_n = \Gamma^k$ and, therefore, $B_j = \Gamma^k$, $j = 0, 1, \dots, n$. This means that $\|f_{\Gamma^k}\| \le \|f_J\|$ for any J with $|J| \ge |\Gamma^k| - 2^{n-1}$. Therefore, if for some $Q \subset \Gamma^k$ we have

$$\|f_Q\| < \|f_{\Gamma^k}\|, \qquad \text{then } |Q| < |\Gamma^k| - 2^{n-1}. \tag{8.7.24}$$

In this case we set $m_0 := k$ and

$$m_1 := k + [\beta|\Gamma^k|^{rq'}].$$

Then, by Lemma 8.7.19 with $A = \Gamma^k$ we obtain, as in (8.7.22),

$$\|f_{m_1}\| \leq \|f_{m_0}\|e^{-c_1\beta/2} + 2\varepsilon \leq \|f_{\Gamma^k}\|e^{-c_1\beta/2} + 3\varepsilon. \tag{8.7.25}$$

By (8.7.14) we get

$$\|f_{\Gamma^{m_1}}\| \leq U(\|f_{m_1}\| + \varepsilon) \leq U(\|f_{\Gamma^k}\|e^{-c_1\beta/2} + 4\varepsilon).$$

If $\|f_{\Gamma^k}\| \leq 8U\varepsilon$ then, by (8.7.25),

$$\|f_{m_1}\| \leq 7U\varepsilon.$$

If $\|f_{\Gamma^k}\| \geq 8U\varepsilon$ then, by our choice of β, we have $2Ue^{-c_1\beta/2} < 1$ and

$$\|f_{\Gamma^{m_1}}\| \leq U(\|f_{\Gamma^k}\|e^{-c_1\beta/2} + 4\varepsilon) < \|f_{\Gamma^k}\|. \tag{8.7.26}$$

This implies that

$$|\Gamma^{m_1}| < |\Gamma^k| - 2^{n-1}.$$

It remains to consider the case $|\Gamma^k| = 1$. By the above argument, where we used Lemma 8.7.19 with $A = \Gamma^k$ we obtain (8.7.26). In the case $|\Gamma^k| = 1$ inequality (8.7.26) implies $\Gamma^{m_1} = \emptyset$, which completes the proof in this case.

We now complete the proof of Theorem 8.7.13. We begin with f_0 and apply the above argument (with $k = 0$). As a result we either get the required inequality or we reduce the cardinality of the support of f from $|T| = K$ to $|\Gamma^{m_{L_1}}| < |T| - 2^{L_1-2}$ (the WCGA picks up at least 2^{L_1-2} dictionary elements g_i from the representation of f), $m_{L_1} \leq \beta 2^{aL_1}$, $a := rq'$. We continue the process and build a sequence m_{L_j} such that $m_{L_j} \leq \beta 2^{aL_j}$ and after m_{L_j} iterations we reduce the support by at least 2^{L_j-2}. We also note that $m_{L_j} \leq \beta 2^{2a}K^a$. We continue this process until the following inequality is satisfied for the first time:

$$m_{L_1} + \cdots + m_{L_s} \geq 2^{2a}\beta K^a. \tag{8.7.27}$$

Then, clearly,

$$m_{L_1} + \cdots + m_{L_s} \leq 2^{2a+1}\beta K^a.$$

Using the inequality

$$(a_1 + \cdots + a_s)^\theta \leq a_1^\theta + \cdots + a_s^\theta, \qquad a_j \geq 0, \qquad \theta \in (0,1],$$

we derive from (8.7.27)

$$\begin{aligned} 2^{L_1-2}+\cdots+2^{L_s-2} &\geq \left(2^{a(L_1-2)}+\cdots+2^{a(L_s-2)}\right)^{1/a} \\ &\geq 2^{-2}\left(2^{aL_1}+\cdots+2^{aL_s}\right)^{1/a} \\ &\geq 2^{-2}\left((\beta)^{-1}(m_{L_1}+\cdots+m_{L_s})\right)^{1/a} \geq K. \end{aligned}$$

Thus, after not more than $N := 2^{2a+1}\beta K^a$ iterations we either get the required inequality or we recover f exactly (the WCGA picks up all the dictionary elements g_i from the representation of f), and thus $\|f_N\| \leq \|f_0 - f\| \leq \varepsilon$. □

Proof of Theorem 8.7.15 We begin with the version of Lemma 8.7.19 that is used in this proof.

Lemma 8.7.20 *Let f satisfy* **A3** *and let $A \subset \Gamma^k$ be nonempty. Denote $B := \Gamma^k \setminus A$. Then, for any $m \in (k, D-K]$, we have either $\|f_{m-1}\| \leq \varepsilon$ or*

$$\|f_m\| \leq \|f_{m-1}\|(1-u) + 2u(\|f_B\| + \varepsilon), \tag{8.7.28}$$

where

$$u := c_2|A|^{-rq'}, \qquad c_2 := \frac{t^{q'}}{2(16\gamma)^{1/(q-1)}V^{q'}},$$

with r and V from **A3**.

Proof The proof combines the proofs of Theorem 8.7.12 and Lemma 8.7.19. As in the proof of Lemma 8.7.19 we denote $A(m) := A \cap \Gamma^{m-1}$ and get

$$S_m \geq \|f_{A(m)}\|_1^{-1} \max(0, \|f_{m-1}\| - \varepsilon - \|f_B\|).$$

From this, in the same way as in the proof of Theorem 8.7.12 we obtain for any $\lambda \geq 0$

$$\|f_m\| \leq \|f_{m-1}\| \left(1 - \frac{\lambda t}{\|f_{A(m)}\|_1} + 2\gamma \frac{\lambda^q}{\|f_{m-1}\|^q}\right) + \frac{\lambda t(\|f_B\| + \varepsilon)}{\|f_{A(m)}\|_1}.$$

Using the definition of $A(m)$ we bound, by **A3**, as follows:

$$\begin{aligned} \|f_{A(m)}\|_1 = \sum_{i \in A(m)} |x_i| &\leq V|A(m)|^r \|f_{A(m)} + f - f_{A(m)} - G_{m-1}\| \\ &\leq V|A|^r \|f - G_{m-1}\| \leq V|A|^r(\|f_{m-1}\| + \varepsilon). \end{aligned}$$

This inequality is a variant of inequality (8.7.9) with K replaced by $|A|$. Arguing as in the proof of Theorem 8.7.12 but with K replaced by $|A|$ we obtain the required inequality, which is the corresponding modification (with K replaced by $|A|$ and ε by $\|f_B\| + \varepsilon$) of (8.7.10). □

The rest of the proof repeats the proof of Theorem 8.7.13 using Lemma 8.7.20 instead of Lemma 8.7.19 and using the fact that **A3** implies **A2** with $U = VK^r \le VK$. □

Proof of Theorem 8.7.16 We can repeat the proof of Theorem 8.7.13 with the use of Lemma 8.7.20 instead of Lemma 8.7.19. □

8.7.4 Examples

In this subsection, following Temlyakov (2014), we discuss applications of the theorems from §§8.7.2 for specific dictionaries $\mathscr{D}$. Mostly, $\mathscr{D}$ will be a basis Ψ for X. Because of that we use m instead of K in the notation of sparse approximation. In some examples we take $X = L_p$, $2 \le p < \infty$. Then it is known that $\rho(u) \le \gamma u^2$ with $\gamma = (p-1)/2$. In other examples we take $X = L_p$, $1 < p \le 2$. Then it is known that $\rho(u) \le \gamma u^p$, with $\gamma = 1/p$.

Proposition 8.7.21 *Let Ψ be a uniformly bounded orthogonal system normalized in $L_p(\Omega)$, $2 \le p < \infty$, where Ω is a bounded domain. Then we have*

$$\|f_{C(t,p,\Omega)m\ln(m+1)}\|_p \le C\sigma_m(f_0,\Psi)_p. \tag{8.7.29}$$

Proof The proof of Proposition 8.7.21 is based on Theorem 8.7.17. Let Ψ be a uniformly bounded orthogonal system normalized in $L_p(\Omega)$, $2 \le p < \infty$, and let Ω be a bounded domain. Then we have

$$C_1(\Omega,p)\|\psi_j\|_2 \le \|\psi_j\|_p \le C_2(\Omega,p)\|\psi_j\|_2, \qquad j = 1,2\dots$$

Next, for $f = \sum_i c_i(f)\psi_i$,

$$\begin{aligned}\sum_{i\in A}|c_i(f)| &= \left\langle f, \sum_{i\in A}(\operatorname{sign} c_i(f))\psi_i\|\psi_i\|_2^{-2}\right\rangle \\ &\le \|f\|_2\left\|\sum_{i\in A}(\operatorname{sign} c_i(f))\psi_i\|\psi_i\|_2^{-2}\right\|_2 \le C_3(\Omega,p)|A|^{1/2}\|f\|_p.\end{aligned}$$

Therefore Ψ satisfies **A3** with $D = \infty$, $V = C_3(\Omega,p)$, and $r = 1/2$. Theorem 8.7.17 gives

$$\|f_{C(t,p,D)m\ln(m+1)}\|_p \le C\sigma_m(f_0,\Psi)_p. \tag{8.7.30}$$

□

Corollary 8.7.22 *Let Ψ be the real d-variate trigonometric system normalized in L_p, $2 \le p < \infty$. Then Proposition 8.7.21 applies and gives, for any $f_0 \in L_p$,*

$$\|f_{C(t,p,d)m\ln(m+1)}\|_p \le C\sigma_m(f_0,\Psi)_p. \tag{8.7.31}$$

We note that (8.7.31) provides some progress towards solving Open Problem 7.1 from Temlyakov (2003a), p. 91.

Proposition 8.7.23 *Let Ψ be a uniformly bounded orthogonal system normalized in $L_p(\Omega)$, $1 < p \le 2$, where Ω is a bounded domain. Then we have*

$$\|f_{C(t,p,\Omega)m^{p'-1}\ln(m+1)}\|_p \le C\sigma_m(f_0,\Psi)_p. \tag{8.7.32}$$

Proof The proof of Proposition 8.7.23 is based on Theorem 8.7.17. Let Ψ be a uniformly bounded orthogonal system normalized in $L_p(\Omega)$, $1 < p \le 2$, and let Ω be a bounded domain. Then we have

$$C_1(\Omega,p)\|\psi_j\|_2 \le \|\psi_j\|_p \le C_2(\Omega,p)\|\psi_j\|_2, \qquad j = 1,2\dots$$

Next, for $f = \sum_i c_i(f)\psi_i$,

$$\sum_{i\in A}|c_i(f)| = \left\langle f, \sum_{i\in A}(\operatorname{sign} c_i(f))\psi_i\|\psi_i\|_2^{-2}\right\rangle \le \|f\|_p \left\|\sum_{i\in A}(\operatorname{sign} c_i(f))\psi_i\|\psi_i\|_2^{-2}\right\|_{p'}$$
$$\le C_4(\Omega,p)|A|^{1-1/p'}\|f\|_p.$$

Therefore Ψ satisfies **A3** with $D = \infty$, $V = C_4(\Omega,p)$, and $r = 1 - 1/p'$. Theorem 8.7.17 gives

$$\|f_{C(t,p,D)m^{p'-1}\ln(m+1)}\|_p \le C\sigma_m(f_0,\Psi)_p. \tag{8.7.33}$$

□

Corollary 8.7.24 *Let Ψ be a real d-variate trigonometric system normalized in L_p, $1 < p \le 2$. Then Proposition 8.7.23 applies and gives, for any $f_0 \in L_p$,*

$$\|f_{C(t,p,d)m^{p'-1}\ln(m+1)}\|_p \le C\sigma_m(f_0,\Psi)_p. \tag{8.7.34}$$

Proposition 8.7.25 *Let Ψ be a multivariate Haar basis $\mathcal{H}_p^d = \mathcal{H}_p \times \cdots \times \mathcal{H}_p$, normalized in L_p, $2 \le p < \infty$. Then*

$$\|f_{C(t,p,d)m^{2/p'}}\|_p \le C\sigma_m(f_0,\mathcal{H}_p^d)_p. \tag{8.7.35}$$

Proof The proof of Proposition 8.7.25 is based on Theorem 8.7.13. Inequality (8.7.35) provides some progress in Open Problem 7.2 from Temlyakov (2003a), p. 91, in the case $2 < p < \infty$. Let Ψ be the multivariate Haar basis $\mathcal{H}_p^d = \mathcal{H}_p \times \cdots \times \mathcal{H}_p$ normalized in L_p, $2 \le p < \infty$. It is an unconditional basis and therefore $U \le C(p,d)$. Next, for any A,

$$\left\|\sum_{i\in A}x_iH_{i,p}\right\|_p \ge C(p,d)\left(\sum_{i\in A}|x_i|^p\right)^{1/p} \ge C(p,d)|A|^{1/p-1}\sum_{i\in A}|x_i|.$$

Therefore, we can take $r = 1/p'$. Theorem 8.7.13 gives

$$\|f_{C(t,p,d)m^{2/p'}}\|_p \le C\sigma_m(f_0, \mathcal{H}_p^d)_p. \tag{8.7.36}$$

□

Proposition 8.7.26 *Let Ψ be the univariate Haar basis $\mathcal{H}_p$ normalized in L_p, $1 < p \le 2$. Then*

$$\|f_{C(t,p)m}\|_p \le C\sigma_m(f_0, \mathcal{H}_p)_p. \tag{8.7.37}$$

Proof The proof of Proposition 8.7.26 is based on Theorem 8.7.18. Inequality (8.7.37) solves the Open Problem 7.2 from Temlyakov (2003a), p. 92, in the case $1 < p \le 2$. Let Ψ be the univariate Haar basis $\mathcal{H}_p = \{H_{I,p}\}_I$, normalized in L_p, $1 < p \le 2$, where $H_{I,p}$ are the Haar functions indexed by dyadic intervals of support of the $H_{I,p}$ (we index function 1 by $[0,1]$ and the first Haar function by $[0,1)$). Then, for any finite set A of dyadic intervals, we have for $f = \sum_I c_I(f) H_{I,p}$

$$\sum_{I \in A} |c_I| = \langle f, f_A^* \rangle, \qquad f_A^* := \sum_{I \in A} (\operatorname{sign} c_I(f)) H_{I,p} \|H_{I,p}\|_2^{-2}.$$

Therefore

$$\sum_{I \in A} |c_I| \le \|f\|_p \|f_A^*\|_{p'}.$$

It is easy to check that

$$\|H_{I,p}\|_{p'} \|H_{I,p}\|_2^{-2} = |I|^{-1/p} |I|^{1/p'} |I|^{-(1-2/p)} = 1.$$

By Lemma 8.3.3 we get

$$\|f_A^*\|_{p'} \le C(p)|A|^{1/p'};$$

thus

$$\sum_{I \in A} |c_I| \le C(p)|A|^{1/p'} \|f\|_p.$$

This means that $\mathcal{H}_p$ satisfies **A3** with $V = C(p)$ and $r = 1/p'$. Also, it is an unconditional basis and therefore satisfies **A2** with $U = C(p)$. It is known that L_p-space with $1 < p \le 2$ has a modulus of smoothness $\rho(u) \le \gamma u^p$. Therefore, Theorem 8.7.18 applies in this case and gives

$$\|f_{C(t,p)m}\|_p \le C\sigma_m(f_0, \mathcal{H}_p)_p. \tag{8.7.38}$$

□

Proposition 8.7.27 *Let X be a Banach space with $\rho(u) \le \gamma u^2$. Assume that Ψ is a normalized Schauder basis for X. Then*

$$\|f_{C(t,X,\Psi)m^2 \ln m}\| \le C\sigma_m(f_0, \Psi). \tag{8.7.39}$$

Proof The proof is based on Theorem 8.7.17. We have, for any $f = \sum_i c_i(f)\psi_i$,

$$\sum_{i\in A} |c_i(f)| \leq C(\Psi)|A|\|f\|.$$

This implies that Ψ satisfies **A3** with $D = \infty$, $V = C(\Psi)$, $r = 1$, and any K. Theorem 8.7.17 gives

$$\|f_{C(t,X,\Psi)m^2 \ln m}\| \leq C\sigma_m(f_0, \Psi). \tag{8.7.40}$$

□

We note that the above bound still works if we replace the assumption that Ψ is a Schauder basis by the assumption that a dictionary $\mathscr{D}$ is $(1,D)$-unconditional with constant U. Then we obtain

$$\|f_{C(t,\gamma,U)K^2 \ln K}\| \leq C\sigma_K(f_0, \Psi), \qquad \text{for } K + C(t,\gamma,U)K^2 \ln K \leq D.$$

Proposition 8.7.28 *Let X be a Banach space with $\rho(u) \leq \gamma u^q$, $1 < q \leq 2$. Assume that Ψ is a normalized Schauder basis for X. Then*

$$\|f_{C(t,X,\Psi)m^{q'} \ln m}\| \leq C\sigma_m(f_0, \Psi). \tag{8.7.41}$$

Proof The proof is again based on Theorem 8.7.17. For any $f = \sum_i c_i(f)\psi_i$ we have

$$\sum_{i\in A} |c_i(f)| \leq C(\Psi)|A|\|f\|.$$

This implies that Ψ satisfies **A3** with $D = \infty$, $V = C(\Psi)$, $r = 1$, and any T. Theorem 8.7.17 gives

$$\|f_{C(t,X,\Psi)m^{q'} \ln m}\| \leq C\sigma_m(f_0, \Psi). \tag{8.7.42}$$

□

We note that, again, the above bound still works if we replace the assumption that Ψ is a Schauder basis by the assumption that a dictionary $\mathscr{D}$ is $(1,D)$-unconditional with constant U. Then we obtain

$$\|f_{C(t,\gamma,q,U)K^{q'} \ln K}\| \leq C\sigma_K(f_0, \mathscr{D}), \qquad \text{for } K + C(t,\gamma,q,U)K^{q'} \ln K \leq D.$$

We now discuss the application of the general results of §8.7.2 to quasi-greedy bases. We begin with a brief introduction to the theory of quasi-greedy bases. The reader can find a detailed presentation of this theory in Temlyakov (2015c). Let X be an infinite-dimensional separable Banach space with norm $\|\cdot\| := \|\cdot\|_X$ and let $\Psi := \{\psi_m\}_{m=1}^{\infty}$ be a normalized basis for X.

Definition 8.7.29 The basis Ψ is called quasi-greedy if there exists some constant C such that

$$\sup_m \|G_m(f,\Psi)\| \le C\|f\|.$$

The concept of a quasi-greedy basis was introduced in Konyagin and Temlyakov (1999). Subsequently, Wojtaszczyk (2000) proved that these are precisely the bases for which the TGA converges, i.e.,

$$\lim_{n\to\infty} G_n(f) = f.$$

The following lemma is from Dilworth *et al.* (2003a) (see also Dilworth *et al.*, 2012, and Garrigós *et al.*, 2013, for further discussion).

Lemma 8.7.30 *Let Ψ be a quasi-greedy basis of X. Then, for any finite set of indices Λ, we have, for all $f \in X$,*

$$\|S_\Lambda(f,\Psi)\| \le C\ln(|\Lambda|+1)\|f\|,$$

where for $f = \sum_{k=1}^{\infty} c_k(f)\psi_k$ we write $S_\Lambda(f,\Psi) := \sum_{k\in\Lambda} c_k(f)\psi_k$.

We now formulate a result, Theorem 8.7.31 below, about quasi-greedy bases in L_p-spaces. The theorem is from Temlyakov *et al.* (2011). We note that in the case $p = 2$ it was proved in Wojtaszczyk (2000). First, we give some notation. For a given element $f \in X$ we consider the expansion

$$f = \sum_{k=1}^{\infty} c_k(f)\psi_k$$

and the decreasing rearrangement of its coefficients

$$|c_{k_1}(f)| \ge |c_{k_2}(f)| \ge \cdots.$$

Write

$$a_n(f) := |c_{k_n}(f)|.$$

Theorem 8.7.31 *Let $\Psi = \{\psi_m\}_{m=1}^{\infty}$ be a quasi-greedy basis of L_p-space, $1 < p < \infty$. Then for each $f \in X$ we have*

$$C_1(p)\sup_n n^{1/p}a_n(f) \le \|f\|_p \le C_2(p)\sum_{n=1}^{\infty} n^{-1/2}a_n(f), \qquad 2 \le p < \infty,$$

$$C_3(p)\sup_n n^{1/2}a_n(f) \le \|f\|_p \le C_4(p)\sum_{n=1}^{\infty} n^{1/p-1}a_n(f), \qquad 1 < p \le 2.$$

We now proceed to apply the general results from §8.7.2 to quasi-greedy bases.

Proposition 8.7.32 *Let Ψ be a normalized quasi-greedy basis for L_p, $2 \le p < \infty$. Then*

$$\|f_{C(t,p)m^{2(1-1/p)}\ln(m+1)}\| \le C\sigma_m(f_0,\Psi). \tag{8.7.43}$$

Proof Again, the proof is based on Theorem 8.7.17. Let Ψ be a normalized quasi-greedy basis for L_p, $2 \le p < \infty$. Theorem 8.7.31 implies that, for any $f = \sum_i c_i(f)\psi_i$,

$$\sum_{i\in A} |c_i(f)| \le \sum_{n=1}^{|A|} a_n(f) \le C_1(p)^{-1} \sum_{n=1}^{|A|} n^{-1/p}\|f\|_p \le C(p)|A|^{1-1/p}\|f\|_p.$$

This means that Ψ satisfies **A3** with $D = \infty$, $V = C(p)$, and $r = 1 - 1/p$. Theorem 8.7.17 gives

$$\|f_{C(t,p)m^{2(1-1/p)}\ln(m+1)}\| \le C\sigma_m(f_0,\Psi). \tag{8.7.44}$$

□

Proposition 8.7.33 *Let Ψ be a normalized quasi-greedy basis for L_p, $1 < p \le 2$. Then*

$$\|f_{C(t,p)m^{p'/2}\ln(m+1)}\| \le C\sigma_m(f_0,\Psi). \tag{8.7.45}$$

Proof Once again proof is based on Theorem 8.7.17. Let Ψ be a normalized quasi-greedy basis for L_p, $1 < p \le 2$. Theorem 8.7.31 implies that, for any $f = \sum_i c_i(f)\psi_i$,

$$\sum_{i\in A} |c_i(f)| \le \sum_{n=1}^{|A|} a_n(f) \le C_3(p)^{-1} \sum_{n=1}^{|A|} n^{-1/2}\|f\|_p \le C(p)|A|^{1/2}\|f\|_p.$$

This means that Ψ satisfies **A3** with $D = \infty$, $V = C(p)$, and $r = 1/2$. Theorem 8.7.17 gives

$$\|f_{C(t,p)m^{p'/2}\ln(m+1)}\| \le C\sigma_m(f_0,\Psi). \tag{8.7.46}$$

□

Proposition 8.7.34 *Let Ψ be a normalized uniformly bounded orthogonal quasi-greedy basis for L_p, $2 \le p < \infty$. Then*

$$\|f_{C(t,p,\Psi)m\ln\ln(m+3)}\|_p \le C\sigma_m(f_0,\Psi)_p. \tag{8.7.47}$$

Proof The proof is based on Theorem 8.7.18. Let Ψ be a normalized uniformly bounded orthogonal quasi-greedy basis for L_p, $2 \le p < \infty$. For the existence of such bases see Nielsen (2007). Then orthogonality implies that Ψ satisfies **A3** with $V = C(\Psi, p)$ and $r = 1/2$. We obtain from Lemma 8.7.30 that Ψ is (K,∞)-unconditional with $U \le C\ln(K+1)$. Theorem 8.7.18 gives

$$\|f_{C(t,p,\Psi)m\ln\ln(m+3)}\|_p \le C\sigma_m(f_0,\Psi)_p. \tag{8.7.48}$$

□

Proposition 8.7.35 *Let Ψ be a normalized uniformly bounded orthogonal quasi-greedy basis for L_p, $1 < p \le 2$. Then*

$$\|f_{C(t,p,\Psi)m^{p'/2}\ln\ln(m+3)}\|_p \le C\sigma_m(f_0,\Psi)_p. \tag{8.7.49}$$

Proof The proof is again based on Theorem 8.7.18. Let Ψ be a normalized uniformly bounded orthogonal quasi-greedy basis for L_p, $1 < p \le 2$. Then orthogonality implies that we can take $r = 1/2$. We obtain from Lemma 8.7.30 that Ψ is (K,∞)-unconditional with $U \le C\ln(K+1)$. Theorem 8.7.18 gives

$$\|f_{C(t,p,\Psi)m^{p'/2}\ln\ln(m+3)}\|_p \le C\sigma_m(f_0,\Psi)_p. \tag{8.7.50}$$

□

8.7.5 Discussion

Proposition 8.7.26 is the first result about almost greedy bases with respect to the WCGA in Banach spaces. It shows that the univariate Haar basis is an almost greedy basis with respect to the WCGA in L_p-space for $1 < p \le 2$. Proposition 8.7.21 shows that uniformly bounded orthogonal bases are ϕ-greedy bases with respect to the WCGA with $\phi(u) = C(t,p,\Omega)\ln(u+1)$ in L_p-space for $2 \le p < \infty$. We do not know whether these bases are almost greedy with respect to the WCGA. They are good candidates for that, however.

It is known (see Theorem 8.3.2 above and also Temlyakov, 2011, p. 17) that the univariate Haar basis is a greedy basis with respect to the TGA for all L_p, $1 < p < \infty$. Proposition 8.7.25 shows only that it is a ϕ-greedy basis with respect to the WCGA with $\phi(u) = C(t,p)u^{1-2/p}$ in L_p-space for $2 \le p < \infty$. It is much weaker than the corresponding results for the $\mathscr{H}_p$, $1 < p \le 2$, and for the trigonometric system, $2 \le p < \infty$ (see Corollary 8.7.22). We do not know whether this result on the Haar basis can be substantially improved. At the level of today's technique we observe that the univariate Haar basis is ideal (i.e., it is a greedy basis) for the TGA in L_p, $1 < p < \infty$, and is an almost ideal (an almost greedy basis) for the WCGA in L_p, $1 < p \le 2$, and that the trigonometric system is very appropriate for the WCGA in L_p, $2 \le p < \infty$.

Corollary 8.7.24 shows that our results for the trigonometric system in L_p, $1 <$

$p < 2$, are not as strong as for $2 \le p < \infty$. We do not know whether this is due to a lack of appropriate technique or whether it reflects the nature of the WCGA in relation to the trigonometric system.

We note that Propositions 8.7.4 and 8.7.5 can be used to formulate the above propositions for more general bases. In such cases we use Propositions 8.7.4 and 8.7.5 with $D = \infty$. In Propositions 8.7.21, 8.7.23, 8.7.32, and 8.7.33, where we used Theorem 8.7.17, we can replace the basis Ψ by a basis Φ which dominates the basis Ψ. In Propositions 8.7.25, 8.7.26, 8.7.34, and 8.7.35, where we used either Theorem 8.7.13 or 8.7.18 we can replace the basis Ψ by a basis Φ, which is equivalent to the basis Ψ.

It is interesting to compare Theorem 8.7.12 with the following known result. The following theorem provides the rate of convergence (see Temlyakov, 2011, p. 347).

Theorem 8.7.36 *Let X be a uniformly smooth Banach space with modulus of smoothness $\rho(u) \le \gamma u^q$, $1 < q \le 2$. Take a number $\varepsilon \ge 0$ and two elements f_0, f^ε from X such that*

$$\|f_0 - f^\varepsilon\| \le \varepsilon, \qquad f^\varepsilon/A(\varepsilon) \in A_1(\mathscr{D}^\pm),$$

for some number $A(\varepsilon) > 0$. Then for the WCGA we have

$$\|f_m^{c,t}\| \le \max\left(2\varepsilon, C(q,\gamma)(A(\varepsilon)+\varepsilon)t(1+m)^{1/q-1}\right).$$

Both Theorem 8.7.36 and Theorem 8.7.12 provide stability of the WCGA with respect to noise. In order to apply them to noisy data we interpret f_0 as a noisy version of a signal and f^ε as a noiseless version of the signal. Then, the assumption $f^\varepsilon/A(\varepsilon) \in A_1(\mathscr{D}^\pm)$ describes our smoothness assumption on the noiseless signal and the assumption $f^\varepsilon \in \Sigma_K(\mathscr{D})$ describes our structural assumption on the noiseless signal. In fact, Theorem 8.7.36 simultaneously takes care of two issues: noisy data and approximation in an interpolation space. It can be applied to the approximation of f_0 under the assumption that f_0 belongs to an interpolation space between X and the space generated by the $A_1(\mathscr{D}^\pm)$-norm (the atomic norm).

8.8 Open Problems

There are many open problems in the theory of greedy approximation. We refer the reader to the papers Temlyakov (2003a, 2007) and to the book Temlyakov (2011) for open theoretical problems in greedy approximation. Here we concentrate on open problems related to multivariate approximation. The discussion in §8.7.5 showed that in the majority of cases we do not know the optimal ϕ such that a basis from a given collection of bases is ϕ-greedy with respect to the WCGA. We now formulate some of these open problems. We assume the version WCGA(t), $t \in (0,1]$, of the weak Chebyshev greedy algorithm with weakness parameter t.

Open Problem 8.1 For a Banach space L_p, $1 < p < \infty$, characterize almost greedy bases with respect to the WCGA.

Open Problem 8.2 Is $\mathscr{R}\mathscr{T}_p^d$ an almost greedy basis, with respect to the WCGA, in $L_p(\mathbb{T}^d)$, $1 < p < \infty$?

Open Problem 8.3 Is $\mathscr{H}_p$ an almost greedy basis, with respect to the WCGA, in L_p, $2 < p < \infty$?

Open Problem 8.4 Is $\mathscr{H}_p^d$, $d \geq 2$, an almost greedy basis, with respect to the WCGA, in L_p, $1 < p < \infty$?

Open Problem 8.5 For each L_p, $1 < p < \infty$, find the best ϕ such that any Schauder basis is ϕ-greedy with respect to the WCGA.

Open Problem 8.6 For each L_p, $1 < p < \infty$, find the best ϕ such that any unconditional basis is ϕ-greedy with respect to the WCGA.

Open Problem 8.7 Is there a greedy-type algorithm $\mathscr{A}$ such that the multivariate Haar system $\mathscr{H}_p^d$ is an almost greedy basis of L_p, $1 < p < \infty$, with respect to $\mathscr{A}$?

Open Problem 8.8 Characterize Banach spaces X such that the WCGA(t) converges for every $\mathscr{D}$ and every $f \in X$.

9

Sparse Approximation

9.1 Introduction

Our main interest in this chapter is to study some approximation problems for classes of functions with mixed smoothness. We use techniques, based on a combination of results from the hyperbolic cross approximation (see Chapter 4) obtained in the 1980s and 1990s, and recent results on greedy approximation to obtain sharp estimates for best m-term approximation with respect to the trigonometric system. We demonstrated in Chapter 8 (see also Temlyakov, 2014) that the weak Chebyshev greedy algorithm (WCGA) is very good for m-term approximation with respect to a special class of dictionaries, in particular, for the trigonometric system. We proved in Chapter 8 the following Lebesgue-type inequality for the WCGA.

Theorem 9.1.1 *Let $\mathscr{D}$ be the real d-variate trigonometric system normalized in L_p, $2 \leq p < \infty$. Then, for any $f \in L_p$, the WCGA with weakness parameter t gives*

$$\|f_{C(t,p,d)m\ln(m+1)}\|_p \leq C\sigma_m(f,\mathscr{D})_p. \tag{9.1.1}$$

The above Lebesgue-type inequality guarantees that the WCGA works very well for each individual function f. As a complement to this inequality we would like to obtain results which relate the rate of decay of $\sigma_m(f,\mathscr{T}^d)_p$ to some smoothness-type properties of f. This is the main goal of this chapter. Smoothness is measured in terms of mixed derivatives and mixed differences. We note that function classes with bounded mixed derivatives are not only interesting and challenging objects for approximation theory but are also important in numerical computations.

The sparse trigonometric approximation of periodic functions began in the paper of Stechkin (1955), who used it in a criterion for the absolute convergence of trigonometric series. Ismagilov (1974) found nontrivial estimates for the m-term approximation of functions with singularities of the type $|x|$ and gave interesting and important applications to the widths of Sobolev classes. He used a deterministic method based on number-theoretical constructions. His method was developed by Maiorov (1978), who used a method based on Gaussian sums. Further strong results were obtained in DeVore and Temlyakov (1995) with the help of a nonconstructive result from finite-dimensional Banach spaces due to Gluskin (1989). Another powerful nonconstructive method, which was based on a probabilistic argument, was used by Makovoz (1984) and by Belinskii (1998a). Different methods were created in Temlyakov (1986b), Kashin and Temlyakov (1994), and Temlyakov (1998b, 2013) for proving lower bounds for function classes. It was discovered in Dilworth *et al.* (2002) and Temlyakov (2005) that greedy algorithms can be used for constructive m-term approximation with respect to the trigonometric system. We demonstrate in this chapter how greedy algorithms can be used to prove optimal or best known upper bounds for the m-term approximation of classes of functions with mixed smoothness. They provide a simple and powerful method of proving upper bounds. However, we encounter difficulties in using this method for small smoothness. For this reason we first study, in §9.2, the case of large smoothness and then in §9.3 the case of small smoothness.

We now give some detailed historical comments with the emphasis on methods of approximation. The problem concerns the trigonometric m-term approximation in the uniform norm. It is convenient for us to deal with both 1-periodic and 2π-periodic functions here. The first result that indicated an advantage of m-term approximation with respect to the real trigonometric system $\mathscr{RT}$ over approximation by trigonometric polynomials of order m was due to Ismagilov (1974):

$$\sigma_m(|\sin 2\pi x|, \mathscr{RT})_\infty \le C_\varepsilon m^{-6/5+\varepsilon}, \qquad \text{for any } \varepsilon > 0. \tag{9.1.2}$$

Maiorov (1978) improved the estimate (9.1.2) to

$$\sigma_m(|\sin 2\pi x|, \mathscr{RT})_\infty \asymp m^{-3/2}. \tag{9.1.3}$$

Both Ismagilov (1974) and Maiorov (1978) used constructive methods to obtain the estimates (9.1.2) and (9.1.3). Maiorov (1978) applied a number-theoretical method based on Gaussian sums. The key point of that technique can be formulated in terms of the best m-term approximation of trigonometric polynomials. Let $\mathscr{RT}(N)$ be the subspace of real trigonometric polynomials of order N. Using Gaussian sums one can prove (constructively) the estimate (see Lemma 2.1.9)

$$\sigma_m(t, \mathscr{RT})_\infty \le C N^{3/2} m^{-1} \|t\|_1, \qquad t \in \mathscr{RT}(N). \tag{9.1.4}$$

Denote

$$\left\| a_0 + \sum_{k=1}^{N} (a_k \cos k2\pi x + b_k \sin k2\pi x) \right\|_A := |a_0| + \sum_{k=1}^{N} (|a_k| + |b_k|).$$

We note that by the simple inequality

$$\|t\|_A \leq CN\|t\|_1, \qquad t \in \mathscr{RT}(N),$$

the estimate (9.1.4) follows from the estimate

$$\sigma_m(t, \mathscr{RT})_\infty \leq C(N^{1/2}/m)\|t\|_A, \qquad t \in \mathscr{RT}(N). \tag{9.1.5}$$

Thus, (9.1.5) is stronger than (9.1.4). The following estimate was proved in DeVore and Temlyakov (1995):

$$\sigma_m(t, \mathscr{RT})_\infty \leq Cm^{-1/2}(\ln(1+N/m))^{1/2}\|t\|_A, \qquad t \in \mathscr{RT}(N). \tag{9.1.6}$$

In a way (9.1.6) is actually much stronger than (9.1.5) and (9.1.4). The proof of the estimate (9.1.6) from DeVore and Temlyakov (1995) is not constructive. It used a nonconstructive theorem of Gluskin (1989). Belinskii (1998a) used a probabilistic method to prove the following inequality for $2 \leq p < \infty$:

$$\sigma_m(t, \mathscr{RT})_\infty \leq C(N/m)^{1/p}(\ln(1+N/m))^{1/p}\|t\|_p, \qquad t \in \mathscr{RT}(N).$$

His proof is nonconstructive as well. Breakthrough results in constructive m-term approximation were obtained by the application of the general theory of greedy approximation in Banach spaces. It was pointed out in Dilworth *et al.* (2002) that the weak Chebyshev greedy algorithm (see Chapter 8) provides in the univariate case a constructive proof of the inequality

$$\sigma_m(f, \mathscr{T}^d)_p \leq C(p)m^{-1/2}\|f\|_A, \qquad p \in [2, \infty).$$

Here

$$\|f\|_A := \sum_{\mathbf{k}} |\hat{f}(\mathbf{k})|, \qquad \hat{f}(\mathbf{k}) := (2\pi)^{-d} \int_{\mathbb{T}^d} f(\mathbf{x}) e^{-i(\mathbf{k},\mathbf{x})} d\mathbf{x}.$$

The following result on constructive approximation in L_∞ is from Temlyakov (2005).

Theorem 9.1.2 *There exists a constructive method $A(N,m)$ such that, for any $t \in \mathscr{RT}(N)$, it provides an m-term trigonometric polynomial $A(N,m)(t)$ with the following approximation property:*

$$\|t - A(N,m)(t)\|_\infty \leq Cm^{-1/2}(\ln(1+N/m))^{1/2}\|t\|_A$$

with C an absolute constant.

An interesting phenomenon specific to the multivariate m-term approximation was discovered in Temlyakov (1986b, c). It was established that

$$\sigma_m(\mathbf{W}^r_q)_p := \sup_{f \in \mathbf{W}^r_q} \sigma_m(f, \mathscr{T}^d)_p$$

decays faster than the Kolmogorov width $d_m(\mathbf{W}^r_q, L_p)$ for $1 < q < p \le 2$. The proof of the upper bounds for the $\sigma_m(\mathbf{W}^r_q)_p$, $1 < q \le p \le 2$, $r > 2(1/q - 1/p)$, in Temlyakov (1986b, c) is constructive. It is based on the fundamental embedding inequality proved in Chapter 3 (see Theorem 3.3.6). For the reader's convenience we formulate it below in slightly different notation, as Theorem 9.1.9. This theorem is often used in the approximation of classes with mixed smoothness. We use it many times in this chapter.

A very interesting and difficult case for m-term approximation is approximation in L_p for, $p > 2$. Makovoz (1984) used the probabilistic Rosenthal inequality for m-term approximation in L_p, $2 < p < \infty$. Later, Belinskii (1987) used the Rosenthal inequality technique to prove the following lemma.

Lemma 9.1.3 *Let $2 < p < \infty$. For any trigonometric polynomial*

$$t(\theta_n, x) := \sum_{j=1}^{n} c_j e^{ik_j x}, \qquad \theta_n := \{k_j\}_{j=1}^n,$$

and any $m \le n$, there exists $t(\theta_m, x)$ with $\theta_m \subset \theta_n$ such that

$$\|t(\theta_n, x) - t(\theta_m, x)\|_p \le C(p)(n/m)^{1/2}\|t(\theta_n, x)\|_2.$$

Lemma 9.1.3 and its multivariate versions have been used in a number of papers on m-term trigonometric approximation in L_p, $2 < p < \infty$ (see, for instance, Romanyuk, 2003, and references therein). The use of Lemma 9.1.3 allows researchers to obtain the right orders of $\sigma_m(\mathbf{W})_p$ for different function classes $\mathbf{W}$ in L_p, $2 < p < \infty$. However, this approach does not provide a constructive method of approximation. Another nonconstructive method for m-term trigonometric approximation, which is more powerful than the probabilistic method discussed above, was suggested in DeVore and Temlyakov (1995); this method is based on a nonconstructive result from finite-dimensional geometry due to Gluskin (1989).

The main results of §9.2.1 are the following two theorems. We use the notation $\beta := \beta(q, p) := 1/q - 1/p$ and $\eta := \eta(q) := 1/q - 1/2$. In the case of the trigonometric system $\mathcal{T}^d$, we drop $\mathcal{T}^d$ from the notation:

$$\sigma_m(\mathbf{W})_p := \sigma_m(\mathbf{W}, \mathcal{T}^d)_p.$$

Theorem 9.1.4 *We have*

$$\sigma_m(\mathbf{W}^r_q)_p \asymp \begin{cases} m^{-r+\beta}(\log m)^{(d-1)(r-2\beta)}, & 1 < q \le p \le 2,\ r > 2\beta, \\ m^{-r+\eta}(\log m)^{(d-1)(r-2\eta)}, & 1 < q \le 2 \le p < \infty,\ r > 1/q, \\ m^{-r}(\log m)^{r(d-1)}, & 2 \le q \le p < \infty,\ r > 1/2. \end{cases}$$

Theorem 9.1.5 *We have*

$$\sigma_m(\mathbf{W}^r_q)_\infty \ll \begin{cases} m^{-r+\eta}(\log m)^{(d-1)(r-2\eta)+1/2}, & 1 < q \le 2,\ r > 1/q, \\ m^{-r}(\log m)^{r(d-1)+1/2}, & 2 \le q < \infty,\ r > 1/2. \end{cases}$$

The case $1 < q \le p \le 2$ in Theorem 9.1.4, which corresponds to the first line, was proved in Temlyakov (1986b) (see also Temlyakov, 1986c, Chapter 4). The proofs from Temlyakov (1986b, c) are constructive. In §9.2.1 we concentrate on the case $p \ge 2$. We will use recently developed techniques on greedy approximation in Banach spaces to prove Theorems 9.1.4 and 9.1.5. It is important that greedy approximation allows us not only to prove the above theorems but also to provide a constructive way for building the corresponding m-term approximants. We give a precise formulation.

Theorem 9.1.6 *For $p \in (1,\infty)$ and $\mu > 0$, constructive methods $A_m(f,p,\mu)$ exist which provide for $f \in \mathbf{W}^r_q$ an m-term approximation such that*

$$\|f - A_m(f,p,\mu)\|_p \ll \begin{cases} m^{-r+\beta}(\log m)^{(d-1)(r-2\beta)}, & 1 < q \le p \le 2,\ r > 2\beta+\mu, \\ m^{-r+\eta}(\log m)^{(d-1)(r-2\eta)}, & 1 < q \le 2 \le p < \infty,\ r > 1/q+\mu, \\ m^{-r}(\log m)^{r(d-1)}, & 2 \le q \le p < \infty,\ r > 1/2+\mu. \end{cases}$$

A similar modification of Theorem 9.1.5 holds for $p = \infty$. We do not have lower bounds corresponding to the upper bounds in Theorem 9.1.5 in the case of approximation in the uniform norm L_∞.

As a direct corollary of Theorems 9.1.1 and 9.1.4 we obtain the following result.

Theorem 9.1.7 *Let $p \in [2,\infty)$. Apply the WCGA with weakness parameter $t \in (0,1]$ to $f \in L_p$ with respect to the real trigonometric system $\mathscr{RT}^d_p$. If $f \in \mathbf{W}^r_q$ then we have*

$$\|f_m\|_p \ll \begin{cases} m^{-r+\eta}(\log m)^{(d-1)(r-2\eta)+r-\eta}, & 1 < q \le 2,\ r > 1/q, \\ m^{-r}(\log m)^{rd}, & 2 \le q < \infty,\ r > 1/2. \end{cases}$$

For the reader's convenience we formulate some known results from hyperbolic cross approximation theory, which will be systematically used in our analysis. Along with Corollary A.3.5 we will use the following corollary of the Littlewood–Paley theorem A.3.3.

Corollary 9.1.8 *Let $1 < p < \infty$. Denote $p_* := \max(p,2)$ and $p^* := \min(p,2)$. Then, for $f \in L_p$, we have*

$$C_3(p,d)\left(\sum_{\mathbf{s}} \|\delta_{\mathbf{s}}(f)\|_p^{p_*}\right)^{1/p_*} \le \|f\|_p \le C_4(p,d)\left(\sum_{\mathbf{s}} \|\delta_{\mathbf{s}}(f)\|_p^{p^*}\right)^{1/p^*}.$$

We now proceed to estimate $\|f\|_u$ in terms of the array $\{\|\delta_{\mathbf{s}}(f)\|_v\}$. Here and below u and v are scalars such that $1 \le u, v \le \infty$. Let an array $\varepsilon = \{\varepsilon_{\mathbf{s}}\}$ be given, where $\varepsilon_{\mathbf{s}} \ge 0$, $\mathbf{s} = (s_1,\dots,s_d)$, and the s_j are nonnegative integers, $j = 1,\dots,d$. We denote by $G(\varepsilon,v)$ and $F(\varepsilon,v)$ the following sets of functions $(1 \le v \le \infty)$:

$$G(\varepsilon,v) := \{f \in L_v : \|\delta_{\mathbf{s}}(f)\|_v \le \varepsilon_{\mathbf{s}} \text{ for all } \mathbf{s}\},$$
$$F(\varepsilon,v) := \{f \in L_v : \|\delta_{\mathbf{s}}(f)\|_v \ge \varepsilon_{\mathbf{s}} \text{ for all } \mathbf{s}\}.$$

The following theorem is from Temlyakov (1985d) (see also Temlyakov, 1986c, p. 29). For the special case $v = 2$ see Temlyakov (1986b and 1986c, p. 86).

Theorem 9.1.9 *The following relations hold:*

$$\sup_{f\in G(\varepsilon,v)} \|f\|_u \asymp \left(\sum_{\mathbf{s}} \varepsilon_{\mathbf{s}}^u 2^{\|\mathbf{s}\|_1(u/v-1)}\right)^{1/u}, \qquad 1 \le v < u < \infty, \tag{9.1.7}$$

$$\inf_{f\in F(\varepsilon,v)} \|f\|_u \asymp \left(\sum_{\mathbf{s}} \varepsilon_{\mathbf{s}}^u 2^{\|\mathbf{s}\|_1(u/v-1)}\right)^{1/u}, \qquad 1 < u < v \le \infty, \tag{9.1.8}$$

with constants that are independent of ε.

We will need a corollary of Theorem 9.1.9 (see Temlyakov, 1986c, Chapter 1, Theorem 2.2), which we formulate as a theorem.

Theorem 9.1.10 *Let* $1 < q \le 2$. *For any* $t \in \mathscr{T}(N)$ *we have*

$$\|t\|_A := \sum_{\mathbf{k}} |\hat{t}(\mathbf{k})| \le C(q,d)N^{1/q}(\log N)^{(d-1)(1-1/q)}\|t\|_q.$$

The following Nikol'skii-type inequalities for the hyperbolic cross trigonometric polynomials are from Temlyakov (1986c), Chapter 1, §2 (see also Theorem 4.3.17 in the present text).

Theorem 9.1.11 *Let* $1 \le q < p < \infty$. *For any* $t \in \mathscr{T}(N)$ *we have*

$$\|t\|_p \le C(q,p,d)N^{\beta}\|t\|_q, \qquad \beta = 1/q - 1/p.$$

9.2 Constructive Sparse Trigonometric Approximation

9.2.1 Sparse Approximation

For a Banach space X we define as above the modulus of smoothness

$$\rho(u) := \sup_{\|x\|=\|y\|=1} \left(\frac{1}{2}(\|x+uy\| + \|x-uy\|) - 1\right).$$

A uniformly smooth Banach space has the property

$$\lim_{u\to 0} \rho(u)/u = 0.$$

It is well known (see for instance Donahue *et al.*, 1997, Lemma B.1) that in the case $X = L_p$, $1 \le p < \infty$, we have

$$\rho(u) \le \begin{cases} u^p/p & \text{if } 1 \le p \le 2, \\ (p-1)u^2/2 & \text{if } 2 \le p < \infty. \end{cases} \tag{9.2.1}$$

Denote by $A_1(\mathcal{D}) := A_1(\mathcal{D}, X)$ the closure in X of the convex hull of $\mathcal{D}$. In this chapter we use the WCGA, studied in Chapter 8, with a special weakness sequence $\tau = \{t_k\}_{k=1}^{\infty}$, for $t_k = t \in (0,1]$, $k = 1,2,\ldots$ In this case Theorem 8.6.6 reads as follows.

Theorem 9.2.1 *Let X be a uniformly smooth Banach space with modulus of smoothness $\rho(u) \le \gamma u^q$, $1 < q \le 2$. Then for $t \in (0,1]$ we have that for any $f \in A_1(\mathcal{D}^{\pm})$,*

$$\|f - G_m^{c,t}(f, \mathcal{D})\| \le C(q,\gamma)(1 + mt^p)^{-1/p}, \qquad p := \frac{q}{q-1},$$

where the constant $C(q,\gamma)$ may depend only on q and γ.

Sometimes we need to control the dependence of $C(q,\gamma)$ on the parameter γ. The following remark is sufficient for our purposes (see Remark 8.6.10).

Remark 9.2.2 It follows from the proof of Theorem 9.2.1 that $C(q,\gamma) \le C(q)\gamma^{1/q}$.

We note that step (2) of the WCGA (see §8.1) makes it difficult to control the coefficients of the approximant – they are obtained through the Chebyshev projection of f onto Φ_m. This motivates us to consider the incremental algorithm with schedule ε, IA(ε), which gives explicit coefficients of the approximant. An advantage of the IA(ε) over other greedy-type algorithms is that it gives precise control of the coefficients of the approximant. For all approximants $G_m^{i,\varepsilon}$ we have the property $\|G_m^{i,\varepsilon}\|_A = 1$. Moreover, we know that all nonzero coefficients of the approximant have the form a/m, where a is a natural number.

We proceed to the incremental greedy algorithm (see Temlyakov, 2005, and 2011, Chapter 6). Let $\varepsilon = \{\varepsilon_n\}_{n=1}^{\infty}$, $\varepsilon_n > 0$, $n = 1,2,\ldots$ We note that the incremental greedy algorithm belongs to the family of relaxed greedy algorithms (see Temlyakov, 2011, Chapter 6).

Incremental algorithm with schedule ε (IA(ε)) Denote $f_0^{i,\varepsilon} := f$ and $G_0^{i,\varepsilon} := 0$. Then for each $m \ge 1$ we have the following inductive definition.

(1) $\varphi_m^{i,\varepsilon} \in \mathcal{D}$ is any element satisfying

$$F_{f_{m-1}^{i,\varepsilon}}(\varphi_m^{i,\varepsilon} - f) \ge -\varepsilon_m.$$

(2) Define

$$G_m^{i,\varepsilon} := (1-1/m)G_{m-1}^{i,\varepsilon} + \varphi_m^{i,\varepsilon}/m.$$

(3) Let

$$f_m^{i,\varepsilon} := f - G_m^{i,\varepsilon}.$$

In order to be able to run the IA(ε) for all iterations, we need the existence of an element $\varphi_m^{i,\varepsilon} \in \mathcal{D}$ at step (1) of the algorithm for all m. It is clear that the following condition guarantees such an existence.

Condition B We say that for a given dictionary $\mathcal{D}$ an element f satisfies condition B if, for all $F \in X^*$, we have

$$F(f) \le \sup_{g\in\mathcal{D}} F(g).$$

It is well known (see, for instance, Temlyakov, 2011, p. 343) that any $f \in A_1(\mathcal{D})$ satisfies condition B. For completeness we give a simple argument here. Take any $f \in A_1(\mathcal{D})$. Then, for any $\delta > 0$, there exist $g_1^\delta, \dots, g_N^\delta \in \mathcal{D}$ and numbers $a_1^\delta, \dots, a_N^\delta$ such that $a_i^\delta > 0$, $a_1^\delta + \cdots + a_N^\delta = 1$, and

$$\left\| f - \sum_{i=1}^{N} a_i^\delta g_i^\delta \right\| \le \delta.$$

Thus

$$F(f) \le \|F\|\delta + F\left(\sum_{i=1}^{N} a_i^\delta g_i^\delta\right) \le \delta\|F\| + \sup_{g\in\mathcal{D}} F(g),$$

which proves Condition B.

We note that Condition B is equivalent to the property $f \in A_1(\mathcal{D})$. Indeed, as we showed above, the property $f \in A_1(\mathcal{D})$ implies Condition B. Let us now show that Condition B implies that $f \in A_1(\mathcal{D})$. Assuming the contrary, $f \notin A_1(\mathcal{D})$, then, by the separation theorem for convex bodies, we can find an $F \in X^*$ such that

$$F(f) > \sup_{\phi\in A_1(\mathcal{D})} F(\phi) \ge \sup_{g\in\mathcal{D}} F(g),$$

which contradicts Condition B.

We will formulate results on the IA(ε) in terms of Condition B because in applications it is easy to check Condition B.

Theorem 9.2.3 *Let X be a uniformly smooth Banach space with modulus of smoothness $\rho(u) \le \gamma u^q$, $1 < q \le 2$. Define*

$$\varepsilon_n := \beta\gamma^{1/q} n^{-1/p}, \qquad p = \frac{q}{q-1}, \qquad n = 1, 2, \dots$$

Then, for every f satisfying Condition B we have

$$\|f_m^{i,\varepsilon}\| \le C(\beta)\gamma^{1/q} m^{-1/p}, \qquad m = 1,2,\dots$$

For the case $f \in A_1(\mathscr{D})$ this theorem is proved in Temlyakov (2005) (see also Temlyakov, 2011, Chapter 6). As we mentioned above, Condition B is equivalent to $f \in A_1(\mathscr{D})$.

We now give some applications of Theorem 9.2.3 in the construction of special polynomials. We begin with a general result.

Theorem 9.2.4 *Let X be a uniformly smooth Banach space with modulus of smoothness $\rho(u) \le \gamma u^q$, $1 < q \le 2$. For any n elements $\varphi_1, \varphi_2, \dots, \varphi_n$, $\|\varphi_j\| \le 1$, $j = 1,\dots,n$, there exist a subset $\Lambda \subset [1,n]$ of cardinality $|\Lambda| \le m < n$ and natural numbers a_j, $j \in \Lambda$, such that*

$$\left\| \frac{1}{n}\sum_{j=1}^{n} \varphi_j - \sum_{j\in\Lambda} \frac{a_j}{m}\varphi_j \right\|_X \le C\gamma^{1/q} m^{1/q-1}, \qquad \sum_{j\in\Lambda} a_j = m.$$

Proof For a given set $\varphi_1, \varphi_2, \dots, \varphi_n$ consider a new Banach space

$$X_n := \operatorname{span}(\varphi_1, \varphi_2, \dots, \varphi_n)$$

with norm $\|\cdot\|_X$. In the space X_n consider the dictionary $\mathscr{D}_n := \{\varphi_j\}_{j=1}^n$. Then the space X_n is a uniformly smooth Banach space with modulus of smoothness $\rho(u) \le \gamma u^q$, $1 < q \le 2$, and $f := n^{-1}\sum_{j=1}^n \varphi_j \in A_1(\mathscr{D}_n)$. Applying the IA($\varepsilon$) to f with respect to $\mathscr{D}_n$ we obtain by Theorem 9.2.3 after m iterations

$$\left\| f - \sum_{k=1}^{m} \frac{1}{m}\varphi_{j_k} \right\|_X \le C\gamma^{1/q} m^{1/q-1},$$

where φ_{j_k} is obtained at the kth iteration of the IA(ε). Clearly, $\sum_{k=1}^m m^{-1}\varphi_{j_k}$ can be written in the form $\sum_{j\in\Lambda}(a_j/m)\varphi_j$ with $|\Lambda| \le m$. □

Corollary 9.2.5 *Let $m \in \mathbb{N}$ and $n = 2m$. For any n trigonometric polynomials $\varphi_j \in \mathscr{RT}(N)$, $\|\varphi_j\|_\infty \le 1$, $j = 1,\dots,n$ with $N \le n^b$, $b \in (0,\infty)$, there exist a set Λ and natural numbers a_j, $j \in \Lambda$, such that $|\Lambda| \le m$, $\sum_{j\in\Lambda} a_j = m$, and*

$$\left\| \frac{1}{n}\sum_{j=1}^{n} \varphi_j - \sum_{j\in\Lambda} \frac{a_j}{m}\varphi_j \right\|_\infty \le C(b)(\ln m)^{1/2} m^{-1/2}. \tag{9.2.2}$$

Proof First, we apply Theorem 9.2.4 with $X = L_p$, $2 \le p < \infty$. It is well known (see for instance Donahue *et al.*, 1997, Lemma B.1) that in the case $X = L_p$, $1 \le p < \infty$, we have

$$\rho(u) \le \begin{cases} u^p/p & \text{if } 1 \le p \le 2, \\ (p-1)u^2/2 & \text{if } 2 \le p < \infty. \end{cases} \tag{9.2.3}$$

Using (9.2.3) we get

$$\left\| \frac{1}{n}\sum_{j=1}^{n}\varphi_j - \sum_{j\in\Lambda(p)} \frac{a_j(p)}{m}\varphi_j \right\|_p \le Cp^{1/2}m^{-1/2}, \qquad \sum_{j\in\Lambda(p)} a_j(p) = m, \tag{9.2.4}$$

with $|\Lambda(p)| \le m$.

Second, by the Nikol'skii inequality (see Chapter 1), for a trigonometric polynomial t of order N one has

$$\|t\|_p \le CN^{1/q-1/p}\|t\|_q, \qquad 1 \le q < p \le \infty.$$

Thus we obtain from (9.2.4)

$$\left\| \frac{1}{n}\sum_{j=1}^{n}\varphi_j - \sum_{j\in\Lambda(p)} \frac{a_j(p)}{m}\varphi_j \right\|_\infty$$
$$\le CN^{1/p}\left\| \frac{1}{n}\sum_{j=1}^{n}\varphi_j - \sum_{j\in\Lambda(p)} \frac{a_j(p)}{m}\varphi_j \right\|_p \le Cp^{1/2}N^{1/p}m^{-1/2}.$$

Choosing $p \asymp \ln N \asymp \ln m$ we obtain (9.2.2). □

We note that Corollary 9.2.5 provides a construction of analogs of the Rudin–Shapiro polynomials (see, for instance, Temlyakov, 2011, p. 155) in a much more general situation than in the case of the Rudin–Shapiro polynomials themselves, albeit with a slightly weaker bound, which contains an extra $(\ln m)^{1/2}$ factor.

We can derive the following result from Theorem 9.2.3.

Theorem 9.2.6 *For any $t \in \mathscr{RT}(\mathbf{N},d)$, the IA($\varepsilon$) applied to $f := t/\|t\|_A$ provides after m iterations an m-term trigonometric polynomial $G_m(t) := G_m^{i,\varepsilon}(f)\|t\|_A$, which belongs to $\mathscr{RT}(\mathbf{N},d)$, with the following approximation property:*

$$\|t - G_m(t)\|_\infty \le C(d)(\bar{m})^{-1/2}(\ln \vartheta(\mathbf{N}))^{1/2}\|t\|_A,$$
$$\bar{m} := \max(1,m), \qquad \|G_m(t)\|_A = \|t\|_A,$$

with a constant $C(d)$ which may depend only on d.

Proof It is clear that it is sufficient to prove Theorem 9.2.6 for $t \in \mathscr{RT}(\mathbf{N},d)$ with $\|t\|_A = 1$. Then $t \in A_1\left((\mathscr{RT}(\mathbf{N},d)\cap\mathscr{RT}^d)^\pm, L_p\right)$ for all $p \in [2,\infty)$. Now, applying Theorem 9.2.3 with $X = L_p$ and $\mathscr{D}^\pm$, where $\mathscr{D} := \{\varphi_1,\varphi_2,\dots,\varphi_n\}$, $n = \vartheta(\mathbf{N})$, for the real trigonometric system

$$\varphi_l := \prod_{j\in E}\cos k_jx_j \prod_{j\in[1,d]\setminus E} \sin k_jx_j,$$

we obtain that

$$\left\| t - \sum_{j\in\Lambda}\frac{a_j}{m}\varphi_j \right\|_p \le C\gamma^{1/2}m^{-1/2}, \qquad \sum_{j\in\Lambda}|a_j| = m, \tag{9.2.5}$$

where $\sum_{j\in\Lambda}(a_j/m)\varphi_j$ is $G_m^{i,\varepsilon}(t)$. By (9.2.1) we find that $\gamma \le p/2$. Next, by the Nikol'skii inequality we get from (9.2.5)

$$\left\|t - \sum_{j\in\Lambda}\frac{a_j}{m}\varphi_j\right\|_\infty \le C(d)n^{1/p}\left\|t - \sum_{j\in\Lambda}\frac{a_j}{m}\varphi_j\right\|_p \le C(d)p^{1/2}n^{1/p}m^{-1/2}.$$

Choosing $p \asymp \ln n$ we obtain the desired bound in the theorem. □

We point out that the above proof of Theorem 9.2.6 gives the following statement.

Theorem 9.2.7 *Let $2 \le p < \infty$. For any $t \in \mathscr{RT}(\mathbf{N},d)$, the* IA($\varepsilon$) *applied to $f := t/\|t\|_A$ provides after m iterations an m-term trigonometric polynomial $G_m(t) := G_m^{i,\varepsilon}(f)\|t\|_A$, which belongs to $\mathscr{RT}(\mathbf{N},d)$, with the following approximation property:*

$$\|t - G_m(t)\|_p \le C(d)(\bar{m})^{-1/2}p^{1/2}\|t\|_A, \qquad \bar{m} := \max(1,m), \qquad \|G_m(t)\|_A = \|t\|_A,$$

with a constant $C(d)$, which may depend only on d.

We note that the implementation of the IA(ε) depends on the dictionary and the ambient space X. For example, for $d = 1$ the IA(ε) from Theorem 9.2.6 acts with respect to the real trigonometric system $1, \cos x, \sin x, \dots, \cos Nx, \sin Nx$ in the space $X = L_p$ with $p \asymp \ln N$.

The above theorems 9.2.6 and 9.2.7 are formulated for m-term approximation with respect to the real trigonometric system because the general Theorem 9.2.3 was proved for real Banach spaces. Clearly, as a corollary of Theorems 9.2.6 and 9.2.7 we obtain corresponding results for the complex trigonometric system $\mathscr{T}^d := \{e^{i(\mathbf{k},\mathbf{x})}\}_{\mathbf{k}\in\mathbb{Z}^d}$. As above, denote $\bar{m} := \max(1,m)$.

Theorem 9.2.8 *There exist constructive greedy-type approximation methods $G_m^p(\cdot)$ which provide m-term polynomials with respect to $\mathscr{T}^d$ with the following properties: for $2 \le p < \infty$,*

$$\|f - G_m^p(f)\|_p \le C_1(d)(\bar{m})^{-1/2}p^{1/2}\|f\|_A, \qquad \|G_m^p(f)\|_A \le C_2(d)\|f\|_A \quad (9.2.6)$$

and, for $p = \infty$, $f \in \mathscr{T}(\mathbf{N},d)$, we have $G_m^\infty(f) \in \mathscr{T}(\mathbf{N},d)$, $\|G_m^\infty(f)\|_A \le C_4(d)\|f\|_A$, and

$$\|f - G_m^\infty(f)\|_\infty \le C_3(d)(\bar{m})^{-1/2}(\ln\vartheta(\mathbf{N}))^{1/2}\|f\|_A. \qquad (9.2.7)$$

In this chapter we will apply Theorem 9.2.8 for the m-term approximation in L_p, $2 < p \le \infty$, of functions with mixed smoothness. We begin with the case $1 < p \le 2$ and then discuss the case $p \in (2,\infty]$.

The following theorem was proved in Temlyakov (1986b) (see also Temlyakov,

1986c, Chapter 4). The proofs given there are constructive. We use again the notation $\beta := \beta(q,p) := 1/q - 1/p$ and $\eta := \eta(q) := 1/q - 1/2$.

Theorem 9.2.9 *Let $1 < q \le p \le 2$, $r > 2\beta$. Then*

$$\sigma_m(\mathbf{W}^r_q)_p \asymp m^{-r+\beta}(\log m)^{(d-1)(r-2\beta)}.$$

Proof We begin with the upper bounds. In the case $\beta = 0$, which means $p = q$, the corresponding upper bounds follow from Theorem 4.4.9. Thus, we consider here the case $\beta > 0$. Take an $n \in \mathbb{N}$ and include in approximation

$$S_{Q_n}(f) := \sum_{\mathbf{s}:\|\mathbf{s}\|_1 \le n} \delta_{\mathbf{s}}(f).$$

Let $\kappa > 1$ be such that $r - (1+\kappa)\beta > 0$. Specify

$$m_l := [2^{\kappa(n-l)} l^{d-1}], \qquad l = n, n+1, \ldots$$

Let N denote l such that $m_N > 0$ and $m_{N+1} = 0$. It is easy to see that

$$2^N \asymp 2^n n^{(d-1)/\kappa}. \tag{9.2.8}$$

For $l \in (n,N]$ include in the approximation the m_l blocks $\delta_{\mathbf{s}}(f)$, $\|\mathbf{s}\|_1 = l$, with largest $\|\delta_{\mathbf{s}}(f)\|_p$. Denote this set of indices $\mathbf{s}$ by G_l. Then, by relation (9.1.8) of Theorem 9.1.9 with parameters $u = q$ and $v = p$ and by the assumption $f \in \mathbf{W}^r_{q,\alpha}$, we obtain

$$\left(\sum_{\mathbf{s}:\|\mathbf{s}\|_1 = l} \|\delta_{\mathbf{s}}(f)\|_p^q 2^{-l\beta q}\right)^{1/q} \ll \|f_l\|_q \le 2^{-rl}. \tag{9.2.9}$$

We now need the following well-known simple lemma (see, for instance, Temlyakov, 1986c, p. 92 and also Lemma 7.6.6).

Lemma 9.2.10 *Let $a_1 \ge a_2 \ge \cdots \ge a_M \ge 0$ and $1 \le q \le p \le \infty$. Then, for all $m \le M$, we have*

$$\left(\sum_{k=m}^{M} a_k^p\right)^{1/p} \le m^{-\beta}\left(\sum_{k=1}^{M} a_k^q\right)^{1/q}.$$

Applying Lemma 9.2.10 to $\{\|\delta_{\mathbf{s}}(f)\|_p\}$ we obtain

$$\left(\sum_{\mathbf{s}:\|\mathbf{s}\|_1 = l, \mathbf{s} \notin G_l} \|\delta_{\mathbf{s}}(f)\|_p^p\right)^{1/p} \ll (m_l+1)^{-\beta} 2^{-(r-\beta)l}. \tag{9.2.10}$$

Next, using Corollary 9.1.8 we derive from (9.2.10)

$$\left\|\sum_{\mathbf{s}:\|\mathbf{s}\|_1 = l, \mathbf{s} \notin G_l} \delta_{\mathbf{s}}(f)\right\|_p \ll \left(\sum_{\mathbf{s}:\|\mathbf{s}\|_1 = l, \mathbf{s} \notin G_l} \|\delta_{\mathbf{s}}(f)\|_p^p\right)^{1/p} \ll (m_l+1)^{-\beta} 2^{-(r-\beta)l}. \tag{9.2.11}$$

Denote

$$f_l' := \sum_{\mathbf{s}:\|\mathbf{s}\|_1=l, \mathbf{s}\notin G_l} \delta_{\mathbf{s}}(f).$$

Then (9.2.11) implies

$$\left\| \sum_{n<l\le N} f_l' \right\|_p \ll \sum_{n<l\le N} (m_l+1)^{-\beta} 2^{-(r-\beta)l} \ll 2^{-(r-\beta)n} n^{-\beta(d-1)}. \qquad (9.2.12)$$

The approximant

$$A_m(f) := S_{Q_n}(f) + \sum_{n<l\le N} \sum_{\mathbf{s}\in G_l} \delta_{\mathbf{s}}(f)$$

has at most m terms, where

$$m \ll |Q_n| + \sum_{n<l\le N} 2^l m_l \ll 2^n n^{d-1}. \qquad (9.2.13)$$

By Theorem 9.1.11 we have

$$\left\| \sum_{l>N} f_l \right\|_p \ll \sum_{l>N} \|f_l\|_p \ll \sum_{l>N} \|f_l\|_q 2^{\beta l} \ll 2^{-(r-\beta)N}. \qquad (9.2.14)$$

Combining (9.2.12) with (9.2.14) and taking into account (9.2.13) we obtain

$$\|f - A_m(f)\|_p \ll m^{-(r-\beta)} (\log m)^{(r-2\beta)(d-1)}.$$

This completes the proof of the upper bounds.

It remains to prove the matching lower bound in the case $1 < p \le 2$. Denote

$$\theta_n := \{\mathbf{s} \in \mathbb{N}^d : \|\mathbf{s}\|_1 = n\}, \qquad \Delta Q_n := \bigcup_{\mathbf{s}\in\theta_n} \rho(\mathbf{s}).$$

Let $K_m := \{\mathbf{k}^j\}_{j=1}^m$ be given. Choose n such that it is the minimal number to satisfy

$$|\Delta Q_n| \ge 4m.$$

Clearly

$$2^n n^{d-1} \asymp m.$$

Denote

$$\theta_n' := \{\mathbf{s} \in \theta_n : |K_m \cap \rho(\mathbf{s})| \le |\rho(\mathbf{s})|/2\}.$$

Note that for $\mathbf{s} \in \mathbb{N}^d$ we have $|\rho(\mathbf{s})| = 2^n$. Then

$$(|\theta_n| - |\theta_n'|)2^n/2 \le m \le |\theta_n| 2^n/4,$$

which implies

$$|\theta_n'| \ge |\theta_n|/2.$$

Define the function

$$f_n(\mathbf{x}) := \sum_{\mathbf{k}\in\Delta Q_n} e^{i(\mathbf{k},\mathbf{x})}.$$

It is easy to check that, for $1 < q < \infty$ (see Lemma 4.2.1 and Remark 4.2.2),

$$\|f_n\|_{\mathbf{W}^r_{q,\alpha}} \ll 2^{n(r+1-1/q)} n^{(d-1)/q}. \tag{9.2.15}$$

By (9.1.8) with $v = q = 2$, $p = u$, $1 < u < 2$, we obtain, for any $t = \sum_{j=1}^m c_j e^{i(\mathbf{k}^j,\mathbf{x})}$,

$$\|f_n - t\|_p \gg \left(\sum_{\mathbf{s}\in\theta'_n} \|\delta_{\mathbf{s}}(f_n - t)\|_2^p 2^{n(p/2-1)}\right)^{1/p} \gg \left(\sum_{\mathbf{s}\in\theta'_n} 2^{pn/2} 2^{n(p/2-1)}\right)^{1/p}$$
$$\gg 2^{n(1-1/p)} n^{(d-1)/p}.$$

This relation and (9.2.15) give the required lower bound for $1 < p < 2$. The above argument gives the lower bound in the case $p = 2$ without the use of Theorem 9.1.9; it is sufficient to use the Parseval identity.

The proof of Theorem 9.2.9 is complete. □

First, we extend Theorem 9.2.9 to the case $1 < q \le p < \infty$.

Theorem 9.2.11 *We have*

$$\sigma_m(\mathbf{W}^r_q)_p \asymp \begin{cases} m^{-r+\beta}(\log m)^{(d-1)(r-2\beta)}, & 1 < q \le p \le 2,\ r > 2\beta, \\ m^{-r+\eta}(\log m)^{(d-1)(r-2\eta)}, & 1 < q \le 2 \le p < \infty,\ r > 1/q, \\ m^{-r}(\log m)^{r(d-1)}, & 2 \le q \le p < \infty,\ r > 1/2. \end{cases}$$

Proof The case $p \le 2$, which corresponds to the first line of the right-hand side, follows from Theorem 9.2.9. We note that, in the case $p > 2$, Theorem 9.2.11 is proved in Temlyakov (1998b). However, the proof there is not constructive; it uses a nonconstructive result from DeVore and Temlyakov (1995). We provide here a constructive proof which is based on greedy algorithms. Also, this proof works under weaker conditions on r, that is: $r > 1/q$ instead of $r > 1/q + \eta$ for $1 < q \le 2$. The following lemma plays the key role in the proof.

Lemma 9.2.12 *Define, for $f \in L_1$,*

$$f_l := \sum_{\|\mathbf{s}\|_1 = l} \delta_{\mathbf{s}}(f), \qquad l \in \mathbb{N}_0, \qquad \mathbb{N}_0 := \mathbb{N} \cup \{0\}.$$

Consider the class

$$\mathbf{W}^{a,b}_A := \{f : \|f_l\|_A \le 2^{-al} l^{(d-1)b}\}.$$

Then, for $2 \le p \le \infty$ and $0 < \mu < a$, there is a constructive method $A_m(\cdot, p, \mu)$ based on greedy algorithms which provides the following bound for $f \in \mathbf{W}^{a,b}_A$:

$$\|f - A_m(f,p,\mu)\|_p \ll m^{-a-1/2}(\log m)^{(d-1)(a+b)}, \qquad 2 \le p < \infty, \tag{9.2.16}$$

$$\|f - A_m(f,\infty,\mu)\|_\infty \ll m^{-a-1/2}(\log m)^{(d-1)(a+b)+1/2}. \tag{9.2.17}$$

Proof We prove the lemma for $m \asymp 2^n n^{d-1}$, $n \in \mathbb{N}$. Let $f \in \mathbf{W}^{a,b}_A$. We will approximate f_l in L_p. By Theorem 9.2.8 we obtain, for $p \in [2,\infty)$,

$$\|f_l - G^p_{m_l}(f_l)\|_p \ll (\bar{m}_l)^{-1/2}\|f_l\|_A \ll (\bar{m}_l)^{-1/2} 2^{-al} l^{(d-1)b}. \tag{9.2.18}$$

We take $\mu \in (0,a)$ and specify

$$m_l := [2^{n-\mu(l-n)} l^{d-1}], \qquad l = n, n+1, \ldots$$

In addition we include in the approximant

$$S_n(f) := \sum_{\|\mathbf{s}\|_1 \le n} \delta_{\mathbf{s}}(f).$$

Define

$$A_m(f,p,\mu) := S_n(f) + \sum_{l>n} G^p_{m_l}(f_l).$$

Thus we have built an m-term approximant of f, with

$$m \ll 2^n n^{d-1} + \sum_{l \ge n} m_l \ll 2^n n^{d-1}.$$

The error of this approximation in L_p is bounded from above by

$$\begin{aligned} \|f - A_m(f,p,\mu)\|_p &\le \sum_{l \ge n} \|f_l - G^p_{m_l}(f_l)\|_p \ll \sum_{l \ge n} (\bar{m}_l)^{-1/2} 2^{-al} l^{(d-1)b} \\ &\ll \sum_{l \ge n} 2^{(n-\mu(l-n))/2} l^{-(d-1)/2} 2^{-al} l^{(d-1)b} \\ &\ll 2^{-n(a+1/2)} n^{(d-1)(b-1/2)}. \end{aligned}$$

This completes the proof of the lemma in the case $2 \le p < \infty$.

Let us discuss the case $p = \infty$. The proof repeats that given for the above case $p < \infty$, with the following change. Instead of using (9.2.6) for estimating an m_l-term approximation of f_l in L_p we use (9.2.7) to estimate an m_l-term approximation of f_l in L_∞. Then the bound (9.2.18) is replaced by

$$\|f_l - G^\infty_{m_l}(f_l)\|_\infty \ll (\bar{m}_l)^{-1/2}(\ln 2^l)^{1/2}\|f_l\|_A \ll (\bar{m}_l)^{-1/2} l^{1/2} 2^{-al} l^{(d-1)b}. \tag{9.2.19}$$

The extra factor $l^{1/2}$ in (9.2.19) gives an extra factor $(\log m)^{1/2}$ in (9.2.17). □

We can now complete the proof of Theorem 9.2.11. First, consider the case $1 < q \le 2 \le p < \infty$. It is well known (see, for instance, Temlyakov, 1986c, p. 34, Theorem 2.1, and also Theorem 4.4.9) that, for $f \in \mathbf{W}^r_q$, one has

$$\|f_l\|_q \ll 2^{-lr}. \tag{9.2.20}$$

Theorem 9.1.10 implies that

$$\|f_l\|_A \ll 2^{-(r-1/q)l} l^{(d-1)(1-1/q)}.$$

Therefore, it is sufficient to use Lemma 9.2.12 with $a = r - 1/q$ and $b = 1 - 1/q$ to obtain the upper bounds.

Second, the upper bounds in the case $2 \le q \le p < \infty$ follow from the above case $1 < q \le 2 \le p < \infty$ with $q = 2$. The lower bounds follow from Theorem 9.2.9 for $p = 2$. The lower bounds in the case $2 \le q \le p < \infty$ follow from known results for the case $1 < p \le q < \infty$ given in Kashin and Temlyakov (1994) (and see Theorem 9.2.23 below). □

Let us discuss the case $p = \infty$. In the same way as that in which Theorem 9.2.11 was derived from (9.2.16) in Lemma 9.2.12, the following upper bounds in the case $p = \infty$ can be derived from (9.2.17) in Lemma 9.2.12.

Theorem 9.2.13 *We have*

$$\sigma_m(\mathbf{W}^r_q)_\infty \ll \begin{cases} m^{-r+\eta}(\log m)^{(d-1)(r-2\eta)+1/2}, & 1 < q \le 2,\ r > 1/q, \\ m^{-r}(\log m)^{r(d-1)+1/2}, & 2 \le q < \infty,\ r > 1/2. \end{cases}$$

The upper bounds are provided by a constructive method $A_m(\cdot, \infty, \mu)$ based on greedy algorithms.

Consider the case $\sigma_m(\mathbf{W}^r_{1,\alpha})_p$, which is not covered by Theorems 9.2.9 and 9.2.11. The function $F_r(\mathbf{x}, \alpha)$ belongs to the closure in L_p of $\mathbf{W}^r_{1,\alpha}$, $r > 1 - 1/p$, and therefore on the one hand,

$$\sigma_m(\mathbf{W}^r_{1,\alpha})_p \ge \sigma_m(F_r(\mathbf{x}, \alpha))_p.$$

On the other hand, it follows from the definition of $\mathbf{W}^r_{1,\alpha}$ that, for any $f \in \mathbf{W}^r_{1,\alpha}$,

$$\sigma_m(f)_p \le \sigma_m(F_r(\mathbf{x}, \alpha))_p.$$

Thus,

$$\sigma_m(\mathbf{W}^r_{1,\alpha})_p = \sigma_m(F_r(\mathbf{x}, \alpha))_p. \tag{9.2.21}$$

We now prove some results on $\sigma_m(F_r)_p$.

Theorem 9.2.14 *We have*

$$\sigma_m(F_r)_p \asymp \begin{cases} m^{-r+1-1/p}(\log m)^{(d-1)(r-1+2/p)}, & 1 < p \le 2,\ r > 1-1/p, \\ m^{-r+1/2}(\log m)^{r(d-1)}, & 2 \le p < \infty,\ r > 1. \end{cases}$$

The upper bounds are provided by a constructive method $A_m(\cdot, p, \mu)$ based on greedy algorithms.

Proof We begin with the case $1 < p \le 2$. The following error bound for approximation by the hyperbolic cross polynomials is known (see, for instance, Temlyakov, 1986c, p. 38, and Theorem 4.4.1)

$$E_{Q_n}(F_r)_p := \inf_{t \in \mathcal{T}(Q_n)} \|F_r - t\|_p \ll 2^{-n(r-1+1/p)} n^{(d-1)/p}. \tag{9.2.22}$$

Taking into account that $|Q_n| \asymp 2^n n^{d-1}$, we obtain from (9.2.22) the required upper bound in the case $1 < p \le 2$. Thus, it remains to prove the corresponding lower bound in the case $1 < p \le 2$. Denote

$$\theta_n := \{\mathbf{s} \in \mathbb{N}^d : \|\mathbf{s}\|_1 = n\}, \qquad \Delta Q_n := \bigcup_{\mathbf{s} \in \theta_n} \rho(\mathbf{s}).$$

Let $K_m := \{\mathbf{k}^j\}_{j=1}^m$ be given. Choose n such that it is the minimal number to satisfy

$$|\Delta Q_n| \ge 4m.$$

Clearly

$$2^n n^{d-1} \asymp m.$$

Denote

$$\theta_n' := \{\mathbf{s} \in \theta_n : |K_m \cap \rho(\mathbf{s})| \le |\rho(\mathbf{s})|/2\}.$$

Note that for $\mathbf{s} \in \mathbb{N}^d$ we have $|\rho(\mathbf{s})| = 2^n$. Then

$$(|\theta_n| - |\theta_n'|)2^n/2 \le m \le |\theta_n| 2^n/4,$$

which implies that

$$|\theta_n'| \ge |\theta_n|/2.$$

By (9.1.8) with $v = q = 2$, $p = u$, $1 < u < 2$, we obtain, for any $t = \sum_{j=1}^m c_j e^{i(\mathbf{k}^j, \mathbf{x})}$,

$$\|F_r - t\|_p \gg \left(\sum_{\mathbf{s} \in \theta_n'} \|\delta_{\mathbf{s}}(F_r - t)\|_2^p 2^{n(p/2-1)} \right)^{1/p}$$
$$\gg \left(\sum_{\mathbf{s} \in \theta_n'} 2^{pn(-r+1/2)} 2^{n(p/2-1)} \right)^{1/p} \gg 2^{-n(r-1+1/p)} n^{(d-1)/p}.$$

This gives the required lower bound for $1 < p < 2$. As before the above argument gives the lower bound in the case $p = 2$ without the use of Theorem 9.1.9; it is sufficient to use the Parseval identity.

We now proceed to the case $2 \le p < \infty$. The analysis here is similar to that in the proof of Theorem 9.2.11. We get, for $F_r^l := \sum_{\|\mathbf{s}\|_1 = l} \delta_{\mathbf{s}}(F_r)$,

$$\|F_r^l\|_A \ll 2^{-lr} 2^l l^{d-1}.$$

The required upper bound follows from Lemma 9.2.12 with $a = r - 1$ and $b = 1$.

The lower bound follows from the case $p = 2$. The theorem in proved. □

In the same way that a modification of the proof of Theorem 9.2.11 gave Theorem 9.2.13, the corresponding modification of the proof of Theorem 9.2.14 gives the following result.

Theorem 9.2.15 *We have*

$$\sigma_m(F_r)_\infty \ll m^{-r+1/2} (\log m)^{r(d-1)+1/2}, \qquad r > 1.$$

The bounds are provided by a constructive method $A_m(\cdot, \infty, \mu)$ based on greedy algorithms.

We now proceed to consider the classes $\mathbf{H}_q^r$ and $\mathbf{B}_{q,\theta}^r$. Define

$$\|f\|_{\mathbf{H}_q^r} := \sup_{\mathbf{s}} \|\delta_{\mathbf{s}}(f)\|_q 2^{r\|\mathbf{s}\|_1}$$

and, for $1 \le \theta < \infty$, define

$$\|f\|_{\mathbf{B}_{q,\theta}^r} := \left(\sum_{\mathbf{s}} \left(\|\delta_{\mathbf{s}}(f)\|_q 2^{r\|\mathbf{s}\|_1} \right)^\theta \right)^{1/\theta}.$$

We will write $\mathbf{B}_{q,\infty}^r := \mathbf{H}_q^r$. With a small abuse of notation, denote the corresponding unit ball by

$$\mathbf{B}_{q,\theta}^r := \{f : \|f\|_{\mathbf{B}_{q,\theta}^r} \le 1\}.$$

It will be convenient for us to use the following slight modification of the classes $\mathbf{B}_{q,\theta}^r$. Define

$$\|f\|_{\mathbf{H}_{q,\theta}^r} := \sup_n \left(\sum_{\mathbf{s}:\|\mathbf{s}\|_1 = n} \left(\|\delta_{\mathbf{s}}(f)\|_q 2^{r\|\mathbf{s}\|_1} \right)^\theta \right)^{1/\theta}$$

and

$$\mathbf{H}_{q,\theta}^r := \{f : \|f\|_{\mathbf{H}_{q,\theta}^r} \le 1\}.$$

The best m-term approximations of the classes $\mathbf{B}^r_{q,\theta}$ were studied in detail by Romanyuk (2003). The following theorem was proved in Temlyakov (1986b) (see also Temlyakov, 1986c, Chapter 4). The proofs in Temlyakov (1986b, c) are constructive.

Theorem 9.2.16 *Let $1 < q \le p \le 2$, $r > \beta$. Then*

$$\sigma_m(\mathbf{H}^r_q)_p \asymp m^{-r+\beta}(\log m)^{(d-1)(r-\beta+1/p)}.$$

Proof The upper bounds follow from Theorem 4.4.10. The lower bounds follow from the proof of Theorem 9.2.9. Indeed, instead of (9.2.15) we get

$$\|f_n\|_{\mathbf{H}^r_q} \ll 2^{n(r+1-1/q)}. \tag{9.2.23}$$

This and the lower bound

$$\|f_n - t\|_p \gg 2^{n(1-1/p)} n^{(d-1)/p}$$

obtained in the proof of Theorem 9.2.9 give the required lower bound. □

The following analog of Theorem 9.2.11 for the classes $\mathbf{H}^r_q$ was proved in Romanyuk (2003). In the case $p > 2$ that proof is not constructive.

Theorem 9.2.17 *We have*

$$\sigma_m(\mathbf{H}^r_q)_p \asymp \begin{cases} m^{-r+\beta}(\log m)^{(d-1)(r-\beta+1/p)}, & 1 < q \le p \le 2,\ r > \beta, \\ m^{-r+\eta}(\log m)^{(d-1)(r-1/q+1)}, & 1 < q \le 2 \le p < \infty,\ r > 1/q, \\ m^{-r}(\log m)^{(d-1)(r+1/2)}, & 2 \le q \le p < \infty,\ r > 1/2. \end{cases}$$

Proposition 9.2.18 *The upper bounds in Theorem 9.2.17 are provided by a constructive method $A_m(\cdot, p, \mu)$ based on greedy algorithms.*

Proof The case $p \le 2$, which corresponds to the first line, follows from Theorem 9.2.16. We now consider $p \ge 2$. From the definition of the classes $\mathbf{H}^r_q$ for $1 < q < \infty$ we find that

$$f \in \mathbf{H}^r_q \quad \Longleftrightarrow \quad \|\delta_{\mathbf{s}}(f)\|_q \le 2^{-r\|\mathbf{s}\|_1}.$$

Next,

$$\|\delta_{\mathbf{s}}(f)\|_A \ll 2^{\|\mathbf{s}\|_1/q}\|\delta_{\mathbf{s}}(f)\|_q.$$

Therefore, for $f \in \mathbf{H}^r_q$ we obtain

$$\|f_l\|_A \ll 2^{-(r-1/q)l} l^{d-1}.$$

Applying Lemma 9.2.12 with $a = r - 1/q$, $b = 1$ in the case $1 < q \le 2 \le p < \infty$, we obtain the required upper bounds. The upper bounds in the case $2 \le q \le p < \infty$

follow from the above case with $q=2$. The lower bounds in the case $1<q\le 2\le p<\infty$ follow from Theorem 9.2.16.

The lower bounds in the case $2\le q\le p<\infty$ follow from known results given in Kashin and Temlyakov (1994) (see Theorem 9.2.24 below). □

Let us now consider the case $p=\infty$:

Theorem 9.2.19 *We have*

$$\sigma_m(\mathbf{H}^r_q)_\infty \ll \begin{cases} m^{-r+\eta}(\log m)^{(d-1)(r-1/q+1)+1/2}, & 1<q\le 2,\ r>1/q, \\ m^{-r}(\log m)^{(r+1/2)(d-1)+1/2}, & 2\le q<\infty,\ r>1/2. \end{cases}$$

The upper bounds are provided by a constructive method $A_m(\cdot,\infty,\mu)$ based on greedy algorithms.

For a nonconstructive proof of the bounds in Theorem 9.2.19 when $2\le q<\infty$ see Belinskii (1989).

We now proceed to the classes $\mathbf{B}^r_{q,\theta}$. There is the following extension of Theorem 9.2.17 (see Romanyuk, 2003).

Theorem 9.2.20 *We have*

$$\sigma_m(\mathbf{B}^r_{q,\theta})_p$$
$$\asymp \begin{cases} m^{-r+\beta}(\log m)^{(d-1)(r-\beta+1/p-1/\theta)_+}, & 1<q\le p\le 2,\ r>\beta, \\ m^{-r+\eta}(\log m)^{(d-1)(r-1/q+1-1/\theta)}, & 1<q\le 2\le p<\infty,\ r>1/q, \\ m^{-r}(\log m)^{(d-1)(r+1/2-1/\theta)}, & 2\le q\le p<\infty,\ r>1/2. \end{cases}$$

Proposition 9.2.21 *The upper bounds in Theorem 9.2.20 are provided by a constructive method based on greedy algorithms.*

Proof In the same constructive way that Theorems 9.2.9 and 9.2.16 were proved in Temlyakov (1986c) one can prove the proposition as it relates to the first line of Theorem 9.2.20 (see Romanyuk, 2003, Theorem 3.1).

We now consider $p\ge 2$. In this case we will prove the corresponding error bounds for both the classes $\mathbf{B}^r_{q,\theta}$ and $\mathbf{H}^r_{q,\theta}$. We get from the definition of the classes $\mathbf{H}^r_{q,\theta}$, for $1<q<\infty$;

$$f\in\mathbf{H}^r_{q,\theta} \quad \Longleftrightarrow \quad \left(\sum_{\|\mathbf{s}\|_1=l}\|\delta_{\mathbf{s}}(f)\|_q^\theta\right)^{1/\theta}\le 2^{-rl}.$$

Next,

$$\|f_l\|_A \leq \sum_{\|\mathbf{s}\|_1=l} \|\delta_{\mathbf{s}}(f)\|_A \ll 2^{l/q} \sum_{\|\mathbf{s}\|_1=l} \|\delta_{\mathbf{s}}(f)\|_q$$

$$\ll 2^{l/q} l^{(d-1)(1-1/\theta)} \left(\sum_{\|\mathbf{s}\|_1=l} \|\delta_{\mathbf{s}}(f)\|_q^\theta \right)^{1/\theta} \ll 2^{-l(r-1/q)} l^{(d-1)(1-1/\theta)}.$$

Therefore, for $f \in \mathbf{H}^r_{q,\theta}$ we obtain

$$\|f_l\|_A \ll 2^{-(r-1/q)l} l^{(d-1)(1-1/\theta)}.$$

Applying Lemma 9.2.12 with $a = r - 1/q$ and $b = 1 - 1/\theta$ in the case $1 < q \leq 2 \leq p < \infty$, we obtain the required upper bounds. The upper bounds in the case $2 \leq q \leq p < \infty$ follow from the above case with $q = 2$. The lower bounds in the case $1 < q \leq 2 \leq p < \infty$ follow from the case $1 < q \leq 2$, $p = 2$.

It was proved in Kashin and Temlyakov (1994) that (as can easily be derived from Theorem 7.4.3 and Lemma 7.8.5)

$$\sigma_m(\mathbf{H}^r_\infty \cap \mathscr{T}(\Delta Q_n))_p \gg m^{-r}(\log m)^{(d-1)(r+1/2)} \tag{9.2.24}$$

for some n such that $m \asymp 2^n n^{d-1}$. It is easy to see that, for any $f \in \mathbf{H}^r_\infty \cap \mathscr{T}(\Delta Q_n)$, we have

$$\|f\|_{\mathbf{B}^r_{q,\theta}} \ll n^{(d-1)/\theta}. \tag{9.2.25}$$

Relations (9.2.24) and (9.2.25) imply the lower bound when $2 \leq q \leq p < \infty$. The proposition is proved. □

We now consider the case $p = \infty$:

Theorem 9.2.22 *We have*

$$\sigma_m(\mathbf{H}^r_{q,\theta})_\infty \ll \begin{cases} m^{-r+\eta}(\log m)^{(d-1)(r-1/q+1-1/\theta)+1/2}, & 1 < q \leq 2,\ r > 1/q, \\ m^{-r}(\log m)^{(r+1/2-1/\theta)(d-1)+1/2}, & 2 \leq q < \infty,\ r > 1/2. \end{cases}$$

The upper bounds in this theorem are provided by a constructive method $A_m(\cdot, \infty, \mu)$ based on greedy algorithms.

We now formulate some known results in the case $1 < p \leq q \leq \infty$.

Theorem 9.2.23 *Let $1 < p \leq q < \infty$, $r > 0$. Then*

$$\sigma_m(\mathbf{W}^r_q)_p \asymp m^{-r}(\log m)^{(d-1)r}.$$

The upper bound in Theorem 9.2.23 follows from the error bounds for approximation by the hyperbolic cross polynomials (see Temlyakov, 1986c, Chapter 2, §2, and Theorem 4.4.9):

$$E_{Q_n}(\mathbf{W}^r_q)_q \ll 2^{-rn}, \qquad 1 < q < \infty.$$

The lower bound in Theorem 9.2.23 was proved in Kashin and Temlyakov (1994) (it can be easily derived from Theorem 7.4.3 and Lemma 7.6.3).

The following result for $\mathbf{H}^r_q$ classes is known.

Theorem 9.2.24 *Let $p \le q$, $2 \le q \le \infty$, $1 < p < \infty$, $r > 0$. Then*

$$\sigma_m(\mathbf{H}^r_q)_p \asymp m^{-r}(\log m)^{(d-1)(r+1/2)}.$$

The lower bound for all $p > 1$,

$$\sigma_m(\mathbf{H}^r_\infty)_p \gg m^{-r}(\log m)^{(d-1)(r+1/2)},$$

was obtained in Kashin and Temlyakov (1994) (it can be easily derived from Theorem 7.4.3 and Lemma 7.8.5). The corresponding upper bounds follow from approximation by the hyperbolic cross polynomials (see Temlyakov, 1986c, Chapter 2, Theorem 2.2, and Theorem 4.4.10 above):

$$E_{Q_n}(\mathbf{H}^r_q)_q := \sup_{f \in \mathbf{H}^r_q} E_{Q_n}(f)_q \asymp n^{(d-1)/2} 2^{-rn}, \qquad 2 \le q < \infty.$$

The following result for **B** classes was proved in Romanyuk (2003).

Theorem 9.2.25 *Let $1 < p \le q < \infty$, $2 \le q < \infty$, $1 < p < \infty$, $r > 0$. Then*

$$\sigma_m(\mathbf{B}^r_{q,\theta})_p \asymp m^{-r}(\log m)^{(d-1)(r+1/2-1/\theta)_+}.$$

9.2.2 Discussion

As we stressed in the title of this section and in the introduction, we are interested in constructive methods of m-term approximation with respect to the trigonometric system. Theorem 9.1.1 basically solves this problem for approximation in L_p, $2 \le p < \infty$. We do not have a similar result for approximation in L_p, $1 < p < 2$. The corresponding Lebesgue-type inequality from Temlyakov (2014) (see Corollary 8.7.24) gives, for $1 < p < 2$,

$$\|f_{C(t,p,d)m^{p'-1}\log(m+1)}\|_p \le C\sigma_m(f)_p,$$

which is much weaker than Theorem 9.1.1. It would be interesting to obtain satisfactory Lebesgue-type inequalities in the case $1 < p < 2$ for either the WCGA or some other constructive methods.

The main results of this section were on the m-term approximation in the case $2 \le p \le \infty$. For $p \in [2,\infty)$ the situation is satisfactory: we have a universal algorithm (WCGA) which provides almost optimal (up to an extra $(\log m)^{C(r,d)}$ factor) m-term approximation for all the classes $\mathbf{W}^r_q$, $\mathbf{H}^r_q$, and $\mathbf{B}^r_{q,\theta}$. Also, there are constructive methods, based on greedy algorithms, which provide the optimal rate for the above classes. However, the upper bounds in, say, Theorem 9.1.6 hold for a smoothness r

larger than that required for the embedding of $\mathbf{W}^r_q$ into L_p. It is of great interest to find constructive methods which provide the correct orders of decay of $\sigma_m(\mathbf{W}^r_q)_p$, $\sigma_m(\mathbf{H}^r_q)_p$, and $\sigma_m(\mathbf{B}^r_{q,\theta})_p$ for small values of the smoothness. We will address this issue in §9.3.

The case $p = \infty$ (approximation in the uniform norm) is very interesting and difficult. The space $C(\mathbb{T}^d)$ (in our notation $L_\infty(\mathbb{T}^d)$) is not a smooth Banach space. Therefore, in the case of approximation in L_∞ the existing greedy approximation theory does not apply directly. In particular, there is no analog of Theorem 9.1.1 in the case $p = \infty$. However, for function classes with mixed smoothness there is a way around this problem. As is demonstrated in the proof of Theorem 9.2.6 we can use greedy algorithms in L_p for large p to obtain bounds on m-term approximation in L_∞. The price we pay for this trick is an extra $(\log m)^{1/2}$ factor in the error bound. This extra factor results from the factor $p^{1/2}$ in the error bounds of approximation by greedy algorithms in L_p, $2 \le p < \infty$ (see Remark 9.2.2 and Theorem 9.2.3). An extra $(\log m)^{1/2}$ appears, as a result of different techniques, in other upper bounds of the asymptotic characteristics of classes of functions with mixed smoothness, when we go from $p < \infty$ to $p = \infty$ (see, for instance, Temlyakov, 2011, §3.6). Unfortunately, we do not have corresponding lower bounds for our upper bounds for m-term approximation in L_∞. A very special case in Theorem 9.3.20 below could be interpreted as a hint that we cannot get rid of that extra $(\log m)^{1/2}$ for approximation in L_∞.

We have discussed isotropic classes of functions with mixed smoothness. In isotropic classes, the smoothness assumptions are the same for each variable. In the hyperbolic cross approximation theory, anisotropic classes of functions with mixed smoothness are of interest and importance. We will give the corresponding definitions. Let $\mathbf{r} = (r_1, \dots, r_d)$ be such that $0 < r_1 = r_2 = \cdots = r_\nu < r_{\nu+1} \le r_{\nu+2} \le \cdots \le r_d$ with $1 \le \nu \le d$. For $\mathbf{x} = (x_1, \dots, x_d)$, denote

$$F_{\mathbf{r}}(\mathbf{x}) := \prod_{j=1}^{d} F_{r_j}(x_j)$$

and

$$\mathbf{W}^{\mathbf{r}}_p := \{f : f = \varphi * F_{\mathbf{r}},\ \|\varphi\|_p \le 1\}.$$

We now proceed to consider the classes $\mathbf{H}^{\mathbf{r}}_q$ and $\mathbf{B}^{\mathbf{r}}_{q,\theta}$. Define

$$\|f\|_{\mathbf{H}^{\mathbf{r}}_q} := \sup_{\mathbf{s}} \|\delta_{\mathbf{s}}(f)\|_q 2^{(\mathbf{r},\mathbf{s})}$$

and, for $1 \le \theta < \infty$, define

$$\|f\|_{\mathbf{B}^{\mathbf{r}}_{q,\theta}} := \left(\sum_{\mathbf{s}} \left(\|\delta_{\mathbf{s}}(f)\|_q 2^{(\mathbf{r},\mathbf{s})}\right)^{\theta}\right)^{1/\theta}.$$

We will write $\mathbf{B}^{\mathbf{r}}_{q,\infty} := \mathbf{H}^{\mathbf{r}}_q$. Denote the corresponding unit ball by

$$\mathbf{B}^{\mathbf{r}}_{q,\theta} := \{f : \|f\|_{\mathbf{B}^{\mathbf{r}}_{q,\theta}} \le 1\}.$$

It is known that in many problems involving the estimation of asymptotic characteristics, the anisotropic classes of functions of d variables with mixed smoothness behave in the same way as the isotropic classes of functions of ν variables (see, for instance, Temlyakov, 1986c). It is clear that this statement holds for the lower bounds. To prove it for the upper bounds one needs to develop, in some cases, a special technique. The techniques developed in this section work for anisotropic classes as well. For instance, the main lemma 9.2.12 is replaced by the following lemma.

Lemma 9.2.26 *Denote $r := r_1$. Define, for $f \in L_1$,*

$$f_{l,\mathbf{r}} := \sum_{\mathbf{s}: rl \le (\mathbf{r},\mathbf{s}) < r(l+1)} \delta_{\mathbf{s}}(f), \qquad l \in \mathbb{N}_0.$$

Consider the class

$$\mathbf{W}^{\mathbf{r},a,b}_A := \{f : \|f_{l,\mathbf{r}}\|_A \le 2^{-al} l^{(\nu-1)b}\}.$$

Then for $2 \le p < \infty$ and $a > 0$ there is a constructive method based on greedy algorithms which provides the bound

$$\sigma_m(\mathbf{W}^{\mathbf{r},a,b}_A)_p \ll m^{-a-1/2}(\log m)^{(\nu-1)(a+b)}. \tag{9.2.26}$$

For $p = \infty$ we have

$$\sigma_m(\mathbf{W}^{\mathbf{r},a,b}_A)_\infty \ll m^{-a-1/2}(\log m)^{(\nu-1)(a+b)+1/2}. \tag{9.2.27}$$

Proposition 9.2.27 *The results of §9.2 hold for the anisotropic classes of functions with mixed smoothness with $r = r_1$ and d replaced by ν.*

9.3 Constructive Sparse Trigonometric Approximation for Small Smoothness

9.3.1 Introduction

The main goal of this subsection is to extend the results on the m-term trigonometric approximation in L_p of the classes $\mathbf{W}^r_q$ of functions with mixed derivatives of order r that are bounded in L_q to the case of small smoothness r. The theory of sparse approximation with respect to the trigonometric system has a long and interesting history (see §9.1).

The main results of this section are in §9.3.3, where we consider m-term approximation in L_p with $p \in (2,\infty)$. Here is a typical result from §9.3.3.

Theorem 9.3.1 *Let $1 < q \le 2 < p < \infty$, $\beta := 1/q - 1/p$, and $\beta p' < r < 1/q$, where $p' := p/(p-1)$. Then we have*

$$\sigma_m(\mathbf{W}^r_q)_p \asymp m^{-(r-\beta)p/2}(\log m)^{(d-1)(r(p-1)-\beta p)}.$$

The upper bounds are achieved by a constructive greedy-type algorithm.

Theorem 9.3.1 complements the known result from Temlyakov (2015a) (see Theorem 9.2.11) for large smoothness. Let $1 < q \le 2 < p < \infty$ and $r > 1/q$. Then we have

$$\sigma_m(\mathbf{W}^r_q)_p \asymp m^{-r+\eta}(\log m)^{(d-1)(r-2\eta)}, \qquad \eta := 1/q - 1/2.$$

The upper bounds are achieved by a constructive greedy-type algorithm.

In §9.3.3 we also consider the case $r = 1/q$ and more general smoothness classes $\mathbf{W}^{a,b}_q$, to be defined shortly. We now introduce some more notation. Let $\mathbf{s} = (s_1, \dots, s_d)$ be a vector whose coordinates are nonnegative integers. For $f \in L_1(\mathbb{T}^d)$, denote as above

$$\delta_{\mathbf{s}}(f) := \delta_{\mathbf{s}}(f, \mathbf{x}) := \sum_{\mathbf{k} \in \rho(\mathbf{s})} \hat{f}(\mathbf{k}) e^{i(\mathbf{k},\mathbf{x})}.$$

Let G be a finite set of points in $\mathbb{Z}^d$; we denote

$$\mathscr{T}(G) := \left\{ t : t(\mathbf{x}) = \sum_{\mathbf{k} \in G} c_{\mathbf{k}} e^{i(\mathbf{k},\mathbf{x})} \right\}, \qquad S_G(f) := \sum_{\mathbf{k} \in G} \hat{f}(\mathbf{k}) e^{i(\mathbf{k},\mathbf{x})}.$$

Along with the classes $\mathbf{W}^r_q$ it is natural to consider some more general classes, which were defined in §7.7. We recall the definition of these classes. Define, for $f \in L_1$,

$$f_l := \sum_{\|\mathbf{s}\|_1 = l} \delta_{\mathbf{s}}(f), \qquad l \in \mathbb{N}_0, \qquad \mathbb{N}_0 := \mathbb{N} \cup \{0\}.$$

Consider the class

$$\mathbf{W}^{a,b}_q := \{ f : \|f_l\|_q \le 2^{-al} (\bar{l})^{(d-1)b} \}, \qquad \bar{l} := \max(l, 1).$$

Define

$$\|f\|_{\mathbf{W}^{a,b}_q} := \sup_l \|f_l\|_q 2^{al} (\bar{l})^{-(d-1)b}.$$

It is well known that the class $\mathbf{W}^r_q$ is embedded in the class $\mathbf{W}^{r,0}_q$ for $1 < q < \infty$. The classes $\mathbf{W}^{a,b}_q$ provide control of smoothness at two scales: a controls the power-type smoothness and b controls the logarithmic-scale smoothness. Similar classes with power and logarithmic scales of smoothness were studied in the book by Triebel (2010).

In §9.3.2 we discuss the case $1 < q \le p \le 2$. We use the technique developed in

Temlyakov (1986b, c). The main results of §9.3.2 are the following two theorems. We use the notation $\beta := \beta(q,p) := 1/q - 1/p$ and $\eta := \eta(q) := 1/q - 1/2$.

Theorem 9.3.2 *Let $1 < q \le p \le 2$. We have*

$$\sigma_m(\mathbf{W}^{a,b}_q)_p \asymp \begin{cases} m^{-a+\beta}(\log m)^{(d-1)(a+b-2\beta)}, & a > 2\beta, \\ m^{-a+\beta}(\log m)^{(d-1)b}, & \beta < a < 2\beta, \\ m^{-\beta}(\log m)^{(d-1)b}(\log\log m)^{1/q}, & a = 2\beta. \end{cases}$$

Theorem 9.3.3 *Let $1 < q \le p \le 2$, $r > \beta$. We have*

$$\sigma_m(\mathbf{W}^r_q)_p \asymp m^{-r+\beta}(\log m)^{(d-1)(r-2\beta)_+}.$$

For the case $r > 2\beta$ Theorem 9.3.3 was proved in Temlyakov (1986b, c) (see Theorem 9.2.9 above) and, as is pointed out in Romanyuk (2003), in the case $\beta < r \le 2\beta$ the order of $\sigma_m(\mathbf{W}^r_q)_p$ was obtained in Belinskii (1988). For completeness, we present a detailed proof of Theorem 9.3.3 in §9.3.2 (moreover, the author could not find the paper Belinskii, 1988).

In §9.3.4 we consider the case $q = 1$. It is known that the analysis of the approximation properties of classes $\mathbf{W}^r_q$ in L_p, in the case of extreme values when at least one of q and p takes the value 1 or ∞, is a difficult problem. In §§9.3.2 and 9.3.3 we only study the case $1 < q, p < \infty$. The results in §9.3.4 are not as complete as those in §§9.3.2 and 9.3.3. We prove some upper bounds in §9.3.4. These upper bounds are nontrivial and they are based on deep results from hyperbolic cross approximation theory. We also show that the results presented in §9.3.4 are optimal up to a factor $(\log m)^{\varepsilon}$, with arbitrarily small $\varepsilon > 0$.

9.3.2 The Case $1 < q \le p \le 2$

Proof of Theorem 9.3.2 In the case $1 < q = p \le 2$ the upper bounds follow from approximation by partial sums $S_{Q_n}(\cdot)$. The corresponding lower bounds follow from the proof of the lower bounds of Theorem 2.1 from Temlyakov (1986c), Chapter 4 (see Theorem 9.2.9 above). We now assume that $\beta > 0$. The case $a > 2\beta$ in Theorem 9.3.2, which corresponds to the first line of the result, was proved for the classes $\mathbf{W}^r_q$ in Temlyakov (1986b) (see also Temlyakov, 1986c, Chapter 4, and Theorem 9.2.9 above). In that proof the assumption $f \in \mathbf{W}^r_q$ was used to claim that $\|f_l\|_q \ll 2^{-rl}$, which implies that $\|f\|_{\mathbf{W}^{r,0}_q} < \infty$. Thus, that proof gives the required upper bound for the class $\mathbf{W}^{r,0}_q$. That same proof gives the corresponding upper bound for the class $\mathbf{W}^{a,b}_q$, $a > 2\beta$, for all b. The proofs from Temlyakov (1986b, c) are constructive.

Consider now the case $\beta < a < 2\beta$. The proof of the upper bounds in this case uses ideas from Temlyakov (1986b, c). Take an $n \in \mathbb{N}$ and include in the approximation

$$S_{Q_n}(f) := \sum_{\mathbf{s}:\|\mathbf{s}\|_1 \le n} \delta_{\mathbf{s}}(f).$$

Choose N such that

$$2^N \asymp 2^n n^{d-1}$$

and for $l \in (n, N]$ include in the approximation m_l blocks $\delta_{\mathbf{s}}(f)$, $\|\mathbf{s}\|_1 = l$, with the largest $\|\delta_{\mathbf{s}}(f)\|_p$. Denote this set of indices $\mathbf{s}$ by G_l. Then, by relation (9.1.8) of Theorem 9.1.9 with parameters $u = q$ and $v = p$ and by the assumption $f \in \mathbf{W}^{a,b}_q$, we obtain

$$\left(\sum_{\mathbf{s}:\|\mathbf{s}\|_1 = l} \|\delta_{\mathbf{s}}(f)\|_p^q 2^{-l\beta q} \right)^{1/q} \ll \|f_l\|_q \le 2^{-al} l^{(d-1)b}. \tag{9.3.1}$$

Applying Lemma 9.2.10 to $\{\|\delta_{\mathbf{s}}(f)\|_p\}$ we obtain

$$\left(\sum_{\mathbf{s}:\|\mathbf{s}\|_1 = l, \mathbf{s} \notin G_l} \|\delta_{\mathbf{s}}(f)\|_p^p \right)^{1/p} \ll (m_l + 1)^{-\beta} 2^{-(a-\beta)l} l^{(d-1)b}. \tag{9.3.2}$$

Next, using Corollary 9.1.8 we derive from (9.3.2)

$$\left\| \sum_{\mathbf{s}:\|\mathbf{s}\|_1 = l, \mathbf{s} \notin G_l} \delta_{\mathbf{s}}(f) \right\|_p \ll \left(\sum_{\mathbf{s}:\|\mathbf{s}\|_1 = l, \mathbf{s} \notin G_l} \|\delta_{\mathbf{s}}(f)\|_p^p \right)^{1/p} \ll (m_l + 1)^{-\beta} 2^{-(a-\beta)l} l^{(d-1)b}. \tag{9.3.3}$$

Denote

$$f_l' := \sum_{\mathbf{s}:\|\mathbf{s}\|_1 = l, \mathbf{s} \notin G_l} \delta_{\mathbf{s}}(f).$$

Let $\kappa > 0$ be such that $a - \beta < \kappa\beta < \beta$. Specify

$$m_l := [2^{\kappa(N-l)}].$$

Then (9.3.3) implies that

$$\left\| \sum_{n < l \le N} f_l' \right\|_p \ll \sum_{n < l \le N} (m_l + 1)^{-\beta} 2^{-(a-\beta)l} l^{(d-1)b} \ll 2^{-(a-\beta)N} N^{(d-1)b}. \tag{9.3.4}$$

The approximant

$$A_m(f) := S_{Q_n}(f) + \sum_{n < l \le N} \sum_{\mathbf{s} \in G_l} \delta_{\mathbf{s}}(f)$$

has at most m terms, where

$$m \ll |Q_n| + \sum_{n < l \le N} 2^l m_l \ll 2^N. \tag{9.3.5}$$

By Theorem 9.1.11 we have

$$\left\|\sum_{l>N} f_l\right\|_p \ll \sum_{l>N} \|f_l\|_p \ll \sum_{l>N} \|f_l\|_q 2^{\beta l} \ll 2^{-(a-\beta)N} N^{(d-1)b}. \tag{9.3.6}$$

Combining (9.3.4) with (9.3.6) and taking into account (9.3.5) we obtain

$$\|f - A_m(f)\|_p \ll m^{-(a-\beta)} (\log m)^{b(d-1)}.$$

This completes the proof of the upper bounds in the case $\beta < a < 2\beta$.

The proof of the lower bounds in the case $\beta < a < 2\beta$ is straightforward. For a given m let $N \in \mathbb{N}$ be the smallest number satisfying $2^N \geq 2m$. Take $\mathbf{s} \in \mathbb{N}^d$ such that $\|\mathbf{s}\|_1 = N$ and consider

$$f := f_{\mathbf{s}} := 2^{-(a+1-1/q)N} N^{(d-1)b} \sum_{\mathbf{k} \in \rho(\mathbf{s})} e^{i(\mathbf{k},\mathbf{x})}.$$

Then

$$\|f\|_q \ll 2^{-aN} N^{(d-1)b}$$

and, therefore, $\|f\|_{\mathbf{W}^{a,b}_q} \ll 1$.

Let $K_m := \{\mathbf{k}^j\}_{j=1}^m$ be given. Then, for any g of the form

$$g = \sum_{\mathbf{k} \in K_m} c_{\mathbf{k}} e^{i(\mathbf{k},\mathbf{x})},$$

we have by Corollary A.3.4 and Theorem 9.1.11

$$\begin{aligned}\|f - g\|_p &\gg \|f_{\mathbf{s}} - S_{\rho(\mathbf{s})}(g)\|_p \gg 2^{(1/2-1/p)N} \|f_{\mathbf{s}} - S_{\rho(\mathbf{s})}(g)\|_2 \\ &\gg 2^{-(a-\beta)N} N^{(d-1)b} \gg m^{-a+\beta} (\log m)^{(d-1)b}.\end{aligned}$$

We now proceed to the case $a = 2\beta$. We begin with the upper bounds. The proof is as in the case $\beta < a < 2\beta$. As above, we choose N such that $2^N \asymp 2^n n^{d-1}$. Then $N - n \asymp \log n$. For $l \in (n, N]$ set $m_l = 2^{N-l}$. Then, as above,

$$m \ll |Q_n| + \sum_{n<l\leq N} 2^l m_l \ll 2^N (N-n). \tag{9.3.7}$$

In the same way that (9.3.6) was established we get

$$\left\|\sum_{l>N} f_l\right\|_p \ll \sum_{l>N} \|f_l\|_p \ll \sum_{l>N} \|f_l\|_q 2^{\beta l} \ll 2^{-\beta N} N^{(d-1)b}. \tag{9.3.8}$$

By (9.3.3) we have

$$\|f_l'\|_p \ll (m_l+1)^{-\beta} 2^{-\beta l} l^{(d-1)b} \tag{9.3.9}$$

and, using (9.3.9), we obtain

$$\left\| \sum_{n<l\le N} f_l' \right\|_p \ll \left(\sum_{n<l\le N} \|f_l'\|_p^p \right)^{1/p} \ll \left(\sum_{n<l\le N} \left((m_l+1)^{-\beta} 2^{-\beta l} l^{(d-1)b} \right)^p \right)^{1/p}$$
$$\ll 2^{-\beta N} N^{(d-1)b} (N-n)^{1/p}. \tag{9.3.10}$$

Relations (9.3.7), (9.3.8), and (9.3.10) imply the required upper bound.

We now prove the lower bounds in the case $a = 2\beta$. Let N be as above. For $l \in (n,N]$ choose an arbitrary set B_l of $\mathbf{s}$ such that $\|\mathbf{s}\|_1 = l$ and $|B_l| = m_l := 2^{N-l}$. Consider f such that $f_l = 0$ for $l \notin (n,N]$ and, for $l \in (n,N]$,

$$f_l := 2^{-(2\beta+1-1/q)l} l^{(d-1)b} m_l^{-1/q} \sum_{\mathbf{s}\in B_l} \sum_{\mathbf{k}\in\rho(\mathbf{s})} e^{i(\mathbf{k},\mathbf{x})}.$$

Then, by Corollary 9.1.8,

$$\|f_l\|_q \ll 2^{-2\beta l} l^{(d-1)b}$$

and therefore $\|f\|_{\mathbf{W}^{2\beta,b}_q} \ll 1$. We now prove the lower bound for the $\sigma_m(f)_p$ with $m < 2^N(N-n)/8$. Let $K_m := \{\mathbf{k}^j\}_{j=1}^m$ be given. Write

$$L := \left\{ l \in (n,N] : \left| K_m \cap \bigcup_{\mathbf{s}\in B_l} \rho(\mathbf{s}) \right| \le 2^N/4 \right\}.$$

Then

$$(N-n-|L|)2^N/4 \le m \le 2^N(N-n)/8,$$

which implies that

$$|L| \ge (N-n)/2.$$

Take $l \in L$. Denote

$$K_m^l := K_m \cap \bigcup_{\mathbf{s}\in B_l} \rho(\mathbf{s})$$

and

$$B_l' := \{\mathbf{s} \in B_l : |K_m^l \cap \rho(\mathbf{s})| \le |\rho(\mathbf{s})|/2\}.$$

As above we derive that

$$|B_l'| \ge |B_l|/2.$$

Let g be any polynomial of the form

$$g = \sum_{\mathbf{k}\in K_m} c_{\mathbf{k}} e^{i(\mathbf{k},\mathbf{x})}.$$

By relation (9.1.8) with parameters $u = p$ and $v = 2$ we get

$$\|f-g\|_p \gg \left(\sum_{n<l\le N}\sum_{\mathbf{s}\in B_l}\left(\|\delta_{\mathbf{s}}(f-g)\|_2 2^{l(1/2-1/p)}\right)^p\right)^{1/p}$$
$$\gg \left(\sum_{l\in L}\sum_{\mathbf{s}\in B_l'}\left(\|\delta_{\mathbf{s}}(f-g)\|_2 2^{l(1/2-1/p)}\right)^p\right)^{1/p}$$
$$\gg \left(\sum_{l\in L}\sum_{\mathbf{s}\in B_l'}\left(2^{-(2\beta+1-1/q)l}l^{(d-1)b}2^{-(N-l)/q}2^{l/2}2^{l(1/2-1/p)}\right)^p\right)^{1/p}$$
$$\gg 2^{-\beta N}N^{(d-1)b}|L|^{1/p} \gg 2^{-\beta N}N^{(d-1)b}(N-n)^{1/p}.$$

Taking into account that $2^N \asymp 2^n n^{d-1}$ and $m \le 2^N(N-n)/8$ completes the proof for the lower bounds. Theorem 9.3.2 is proved. $\square$

Proof of Theorem 9.3.3 In the case $r \ne 2\beta$ the upper bounds in Theorem 9.3.3 follow from Theorem 9.3.2 by the embedding of $\mathbf{W}^r_q$ into $\mathbf{W}^{r,0}_q$. It turns out that in the case $r = 2\beta$ this approach does not give a sharp upper bound. We now prove the corresponding upper bound in the case $r = 2\beta$. We begin with an analog of Lemma 9.2.10.

Lemma 9.3.4 *Let $\{w_j\}_{j=1}^M$ be a set of positive weights. Let $a_1 \ge a_2 \ge \cdots \ge a_M \ge 0$ and $1 \le q \le p \le \infty$. Then, for all $m \le M$, we have*

$$\left(\sum_{k=m}^M a_k^p w_k\right)^{1/p} \le \left(\sum_{k=1}^m w_k\right)^{-\beta}\left(\sum_{k=1}^M a_k^q w_k\right)^{1/q}.$$

Proof The monotonicity of $\{a_k\}_{k=1}^M$ implies that

$$a_m^q \sum_{k=1}^m w_k \le \sum_{k=1}^M a_k^q w_k$$

and

$$a_m \le \left(\sum_{k=1}^m w_k\right)^{-1/q}\left(\sum_{k=1}^M a_k^q w_k\right)^{1/q}.$$

Therefore

$$\left(\sum_{k=m}^M a_k^p w_k\right)^{1/p} \le a_m^{(p-q)/p}\left(\sum_{k=m}^M a_k^q w_k\right)^{1/p} \le \left(\sum_{k=1}^m w_k\right)^{-\beta}\left(\sum_{k=1}^M a_k^q w_k\right)^{1/q}. \quad \square$$

The lemma is proved.

We now prove the upper bound in the case $r = 2\beta$. Let $n \in \mathbb{N}$ and, as above, let N be such that $2^N \asymp 2^n n^{d-1}$. For $f \in \mathbf{W}^r_q$ we include in the approximation $S_{Q_n}(f)$

and we approximate

$$g := g(f) := \sum_{\mathbf{s}\in\Delta(n,N)} \delta_{\mathbf{s}}(f), \qquad \Delta(n,N) := \{\mathbf{s} : n < \|\mathbf{s}\|_1 \le N\}.$$

Using relation (9.1.8) with parameters $u = q$, $v = p$ and taking into account that $r = 2\beta$, we obtain

$$\left(\sum_{\mathbf{s}\in\Delta(n,N)} \left(2^{\beta\|\mathbf{s}\|_1}\|\delta_{\mathbf{s}}(f)\|_p\right)^q\right)^{1/q} \ll \|f\|_{\mathbf{W}^r_q}. \tag{9.3.11}$$

We need to apply Lemma 9.3.4. Consider $\{v_{\mathbf{s}}\}_{\mathbf{s}\in\Delta(n,N)}$, $v_{\mathbf{s}} := \|\delta_{\mathbf{s}}(f)\|_p 2^{-\|\mathbf{s}\|_1/p}$ with weights $w_{\mathbf{s}} := 2^{\|\mathbf{s}\|_1}$. Then (9.3.11) gives

$$\left(\sum_{\mathbf{s}\in\Delta(n,N)} v_{\mathbf{s}}^q w_{\mathbf{s}}\right)^{1/q} \ll \|f\|_{\mathbf{W}^r_q}.$$

Choose the k largest $v_{\mathbf{s}}$ and denote the corresponding set of indices $\mathbf{s}$ by $G(k)$. By Lemma 9.3.4 we obtain from the above estimate

$$\left(\sum_{\mathbf{s}\in\Delta(n,N)\setminus G(k)} v_{\mathbf{s}}^p w_{\mathbf{s}}\right)^{1/p} \ll \left(\sum_{\mathbf{s}\in G(k)} w_{\mathbf{s}}\right)^{-\beta} \|f\|_{\mathbf{W}^r_q}. \tag{9.3.12}$$

By Corollary 9.1.8 we find that

$$\left\|\sum_{\mathbf{s}\in\Delta(n,N)\setminus G(k)} \delta_{\mathbf{s}}(f)\right\|_p \ll \left(\sum_{\mathbf{s}\in\Delta(n,N)\setminus G(k)} \|\delta_{\mathbf{s}}(f)\|_p^p\right)^{1/p}. \tag{9.3.13}$$

Combining (9.3.13) with (9.3.12) we obtain

$$\begin{aligned}\left(\sum_{\mathbf{s}\in\Delta(n,N)\setminus G(k)} \|\delta_{\mathbf{s}}(f)\|_p^p\right)^{1/p} &= \left(\sum_{\mathbf{s}\in\Delta(n,N)\setminus G(k)} v_{\mathbf{s}}^p w_{\mathbf{s}}\right)^{1/p} \\ &\ll \left(\sum_{\mathbf{s}\in G(k)} w_{\mathbf{s}}\right)^{-\beta} \|f\|_{\mathbf{W}^r_q}.\end{aligned}$$

Choose k such that

$$2^N \le \sum_{\mathbf{s}\in G(k)} w_{\mathbf{s}} < 2^{N+1}.$$

In this way we construct an m-term approximation of f with $m \ll 2^N$ and error

$$\left\|f - S_{Q_n}(f) - \sum_{\mathbf{s}\in G(k)} \delta_{\mathbf{s}}(f)\right\|_p \ll 2^{-\beta N} \ll m^{-\beta}.$$

The upper bounds in Theorem 9.3.3 are proved.

The lower bounds in the case $r > 2\beta$ are proved in Theorem 9.2.9 (see also Temlyakov, 1986c). The lower bounds in the case $\beta < r \le 2\beta$ follow from the univariate case. Theorem 9.3.3 is proved. □

9.3.3 The Case $1 < q \le 2 < p < \infty$

The main goal of this subsection is to prove Theorem 9.3.1. We reformulate it here for convenience.

Theorem 9.3.5 *Let $1 < q \le 2 < p < \infty$, $\beta := 1/q - 1/p$, and $\beta p' < r < 1/q$, where $p' := p/(p-1)$. Then we have*

$$\sigma_m(\mathbf{W}^r_q)_p \asymp m^{-(r-\beta)p/2}(\log m)^{(d-1)(r(p-1)-\beta p)}.$$

The upper bounds are achieved by a constructive greedy-type algorithm.

Proof We will prove the upper bounds for a wider class, $\mathbf{W}^{r,0}_q$. Let $f \in \mathbf{W}^{r,0}_q$. Let $n \in \mathbb{N}$. We will build an m-term approximation with $m = 2|Q_n| \asymp 2^n n^{d-1}$. We include in the approximation $S_{Q_n}(f)$. We split the remainder function into two functions:

$$f - S_{Q_n}(f) = g_A + g_0.$$

Now we use Theorem 9.2.8 to approximate g_A and we approximate g_0 by 0. We now describe the construction of g_A. First, we choose $N \in (n, Cn]$, $C = C(p,d)$, which will be specified later on, and include in g_A

$$g^1_A := \sum_{n<l\le N} f_l.$$

Then by Theorem 9.1.10 we have

$$\begin{aligned}\|g^1_A\|_A = \sum_{n<l\le N} \|f_l\|_A &\ll \sum_{n<l\le N} \|f_l\|_q 2^{l/q} l^{(d-1)(1-1/q)} \\ &\ll \sum_{n<l\le N} 2^{l(1/q-r)} l^{(d-1)(1-1/q)} \\ &\ll 2^{N(1/q-r)} N^{(d-1)(1-1/q)}. \end{aligned} \tag{9.3.14}$$

Next, for $l > N$ define

$$u_l := [n^{d-1} 2^{\kappa(N-l)}]$$

with κ satisfying

$$\frac{1/q - r}{1 - 1/q} < \kappa < \frac{r-\beta}{\beta}.$$

Such a κ exists because our assumption $r > \beta p'$ is equivalent to the inequality

$$\frac{1/q - r}{1 - 1/q} < \frac{r-\beta}{\beta}.$$

Denote by $G(l)$ the set of indices $\mathbf{s}$, $\|\mathbf{s}\|_1 = l$, of cardinality $|G(l)| = u_l$, with the largest value of $\|\delta_{\mathbf{s}}(f_l)\|_2$. Second, we include in g_A

$$g_A^2 := \sum_{l>N} \sum_{\mathbf{s}\in G(l)} \delta_{\mathbf{s}}(f_l).$$

It is clear that there is only a finite number of nonzero terms in the above sum. We have

$$\begin{aligned}\|g_A^2\|_A = \sum_{l>N} \Big\| \sum_{\mathbf{s}\in G(l)} \delta_{\mathbf{s}}(f_l) \Big\|_A &\ll \sum_{l>N} \|f_l\|_q 2^{l/q} |G(l)|^{1-1/q} \\ &\ll \sum_{l>N} 2^{l(1/q-r)} n^{(d-1)(1-1/q)} 2^{\kappa(N-l)(1-1/q)} \\ &\ll n^{(d-1)(1-1/q)} 2^{\kappa N(1-1/q)} \sum_{l>N} 2^{l(1/q-r-\kappa(1-1/q))}.\end{aligned}$$

By our choice of κ we have $1/q - r - \kappa(1-1/q) < 0$ and, therefore, we continue as follows:

$$\ll N^{(d-1)(1-1/q)} 2^{N(1/q-r)}. \tag{9.3.15}$$

By Theorem 9.2.8 we obtain from (9.3.14) and (9.3.15)

$$\|g_A - G^p_{m/2}(g_A)\|_p \ll m^{-1/2} N^{(d-1)(1-1/q)} 2^{N(1/q-r)}. \tag{9.3.16}$$

We now bound the $\|g_0\|_p$. Denote

$$f_l^o := f_l - \sum_{\mathbf{s}\in G(l)} \delta_{\mathbf{s}}(f_l).$$

By relation (9.1.8) with parameters $u = q$ and $v = 2$ we get

$$\left(\sum_{\|\mathbf{s}\|_1 = l} \left(\|\delta_{\mathbf{s}}(f_l)\|_2 2^{\|\mathbf{s}\|_1(1/2-1/q)} \right)^q \right)^{1/q} \ll 2^{-rl}$$

and

$$\left(\sum_{\|\mathbf{s}\|_1 = l} \|\delta_{\mathbf{s}}(f_l)\|_2^q \right)^{1/q} \ll 2^{l(-r+1/q-1/2)}.$$

By Lemma 9.2.10 we obtain

$$\left(\sum_{\mathbf{s}\notin G(l)} \|\delta_{\mathbf{s}}(f_l)\|_2^p \right)^{1/p} \ll (u_l+1)^{-\beta} 2^{l(-r+1/q-1/2)}.$$

By relation (9.1.7) with parameters $u = p$ and $v = 2$ we then obtain

$$\|f_l^o\|_p \ll \left(\sum_{\mathbf{s}\notin G(l)} \left(\|\delta_{\mathbf{s}}(f_l)\|_2 2^{\|\mathbf{s}\|_1(1/2-1/p)} \right)^p \right)^{1/p} \ll 2^{-(r-\beta)l} (u_l+1)^{-\beta}.$$

Thus

$$\|g_O\|_p \le \sum_{l>N} \|f_l^o\|_p \ll \sum_{l>N} 2^{-(r-\beta)l} n^{-\beta(d-1)} 2^{-\beta\kappa(N-l)}$$
$$\ll n^{-\beta(d-1)} 2^{-\beta\kappa N} \sum_{l>N} 2^{-(r-\beta-\beta\kappa)l}. \qquad (9.3.17)$$

By our choice of κ we have $r-\beta-\beta\kappa>0$. Therefore, (9.3.17) gives

$$\|g_O\|_p \ll 2^{-(r-\beta)N} n^{-\beta(d-1)}. \qquad (9.3.18)$$

We now obtain N from the condition

$$2^{(1/q-r)N} n^{(d-1)(1-1/q)} m^{-1/2} \asymp 2^{-(r-\beta)N} n^{-\beta(d-1)}.$$

This is equivalent to

$$2^N \asymp 2^{np/2} n^{(d-1)(1-p/2)}$$

or, in terms of m,

$$2^N \asymp m^{p/2} (\log m)^{(d-1)(1-p)}. \qquad (9.3.19)$$

As a result (9.3.16) and (9.3.18) give us the following upper bound for the approximation error:

$$\|f - S_{Q_n}(f) - G^p_{m/2}(g_A)\|_p \ll 2^{-(r-\beta)N} n^{-\beta(d-1)}$$
$$\ll m^{-(r-\beta)p/2} (\log m)^{(d-1)(r(p-1)-\beta p)}.$$

This completes the proof of the upper bounds.

We proceed to the lower bounds. For a given m choose N as in (9.3.19). Consider the function

$$g(\mathbf{x}) := \sum_{\|\mathbf{s}\|_1=N} \sum_{\mathbf{k}\in\rho(\mathbf{s})} e^{i(\mathbf{k},\mathbf{x})} = \sum_{\mathbf{k}\in\Delta Q_N} e^{i(\mathbf{k},\mathbf{x})}, \qquad \Delta Q_N := Q_N \setminus Q_{N-1}.$$

It is known, and it is easy to derive from Theorem 9.1.9, that

$$\|g\|_q \asymp 2^{N(1-1/q)} N^{(d-1)/q}, \qquad 1<q<\infty. \qquad (9.3.20)$$

We now estimate the $\sigma_m(g)_p$ from below. Take any set K_m of m frequencies $\mathbf{k}$. Consider an additional function

$$h(\mathbf{x}) := \sum_{\mathbf{k}\in\Delta Q_N\setminus K_m} e^{i(\mathbf{k},\mathbf{x})}.$$

For any polynomial t with frequencies from K_m we have

$$\langle g-t, h\rangle \le \|g-t\|_p \|h\|_{p'} \qquad (9.3.21)$$

and

$$\langle g-t,h\rangle = \langle g,h\rangle = \sum_{\mathbf{k}\in\Delta Q_N\setminus K_m} 1 = |\Delta Q_N \setminus K_m|. \tag{9.3.22}$$

From our choice (9.3.19) of N it is clear that, asymptotically,

$$|\Delta Q_N \setminus K_m| \geq |\Delta Q_N| - m \gg 2^N N^{d-1}.$$

Next, we have

$$\begin{aligned}\|h\|_{p'} &\leq \|g\|_{p'} + \|g-h\|_{p'} \leq \|g\|_{p'} + \|g-h\|_2 \\ &\ll 2^{N/p} N^{(d-1)/p'} + m^{1/2} \ll m^{1/2}.\end{aligned}$$

Thus, (9.3.21) and (9.3.22) yield

$$\sigma_m(g)_p \gg 2^N N^{d-1} m^{-1/2}.$$

We have, from (9.3.20),

$$\|g^{(r)}\|_q \ll 2^{N(r+1-1/q)} N^{(d-1)/q}.$$

Therefore

$$\begin{aligned}\sigma_m(\mathbf{W}^r_q)_p &\gg 2^{N(1/q-r)} N^{(d-1)(1-1/q)} m^{-1/2} \\ &\asymp m^{-(r-\beta)p/2} (\log m)^{(d-1)(r(p-1)-\beta p)}.\end{aligned}$$

This proves the lower bounds of the theorem. □

The above proof of Theorem 9.3.5 gives the correct order of $\sigma_m(\mathbf{W}^{a,b}_q)_p$ for $\beta p' < a < 1/q$ and all b. We formulate this as a theorem.

Theorem 9.3.6 *Let $1 < q \leq 2 < p < \infty$ and $\beta p' < a < 1/q$. Then we have*

$$\sigma_m(\mathbf{W}^{a,b}_q)_p \asymp m^{-(a-\beta)p/2} (\log m)^{(d-1)(b+a(p-1)-\beta p)}.$$

The upper bounds are achieved by a constructive greedy-type algorithm.

For $f \in \mathbf{W}^{r,0}_q$ one has, for $r > \beta$,

$$\left\| \sum_{l>N} f_l \right\|_p \leq \sum_{l>N} \|f_l\|_p \ll \sum_{l>N} \|f_l\|_q 2^{\beta l} \ll 2^{-(r-\beta)N}. \tag{9.3.23}$$

In the proof of Theorem 9.3.5 we constructed g^2_A and g_0. This resulted in a better error estimate for the m-term approximation of the tail $\sum_{l>N} f_l$ than the simple bound (9.3.23). We found the error $\ll 2^{-(r-\beta)N} n^{-\beta(d-1)}$. We obtain the same improvement of the error if, in addition to the assumption $f \in \mathbf{W}^{r,0}_q$, we assume that $f \in \mathbf{W}^{r-\beta,-\beta}_p$. For $f \in \mathbf{W}^{r-\beta,-\beta}_p$ we have

$$\left\| \sum_{l>N} f_l \right\|_p \ll 2^{-(r-\beta)N} N^{-\beta(d-1)}.$$

We now formulate a theorem which follows from the proof of Theorem 9.3.5.

Theorem 9.3.7 *Let $1 < q \le 2 < p < \infty$ and $\beta < a < 1/q$. Then we have*

$$\sigma_m \left(\mathbf{W}^{a,b}_q \cap \mathbf{W}^{a-\beta,b-\beta}_p \right)_p \asymp m^{-(a-\beta)p/2} (\log m)^{(d-1)(b+a(p-1)-\beta p)}.$$

The upper bounds are achieved by a constructive greedy-type algorithm.

We note that the class $\mathbf{H}^r_q$ (see the definition in §4.2.2), $1 < q \le 2$, is embedded into the class $\mathbf{W}^{r,b}_q \cap \mathbf{W}^{r-\beta,b-\beta}_p$ with $b = 1/q$. This follows from Corollary 9.1.8 and Theorem 9.1.9. The following theorem holds.

Theorem 9.3.8 *Let $1 < q \le 2 < p < \infty$ and $\beta < a < 1/q$. Then we have*

$$\sigma_m(\mathbf{H}^r_q)_p \asymp m^{-(r-\beta)p/2} (\log m)^{(d-1)(1/q+r(p-1)-\beta p)}.$$

The upper bounds are achieved by a constructive greedy-type algorithm.

The order of $\sigma_m(\mathbf{H}^r_q)_p$ is known (see Romanyuk, 2003). However, the corresponding upper bounds in that paper are proved by a nonconstructive method of approximation.

We now proceed to the case $a = 1/q$.

Theorem 9.3.9 *Let $1 < q \le 2 < p < \infty$ and $a = 1/q$. Then we have*

$$\sigma_m(\mathbf{W}^{1/q,b}_q)_p \asymp m^{-1/2} (\log m)^{(d-1)(b+1-1/q)+1}.$$

The upper bounds are achieved by a constructive greedy-type algorithm.

Proof The proof goes along the lines of the proof of Theorem 9.3.5. We use the same notation as above. We begin with the upper bounds. In the case $a = 1/q$ the bound (9.3.14) reads

$$\|g^1_A\|_A \ll (N-n) N^{(d-1)(b+1-1/q)}. \tag{9.3.24}$$

We repeat the argument from the proof of Theorem 9.3.5 for g^2_A and g_0 with $\kappa \in (0, (a-\beta)/\beta)$ and conclude that

$$\|g^2_A\|_A \ll N^{(d-1)(b+1-1/q)}, \tag{9.3.25}$$

$$\|g_0\|_p \ll 2^{-(a-\beta)N} n^{(b-\beta)(d-1)}. \tag{9.3.26}$$

Choosing N as in (9.3.19) and applying Theorem 9.2.8 we obtain from (9.3.24)–(9.3.26)

$$\|f - S_{Q_n}(f) - G^p_{m/2}(g_A)\|_p \ll (N-n)N^{(d-1)(b+1-1/q)}m^{-1/2}$$
$$\ll m^{-1/2}(\log m)^{(d-1)(b+1-1/q)+1}.$$

This completes the proof of the upper bounds.

We proceed to the lower bounds. For a given m choose N as in (9.3.19). Consider the function

$$g(\mathbf{x}) := \sum_{n<l\le N} 2^{-l/q} l^{b(d-1)} \left(2^{l(1-1/q)} l^{(d-1)/q}\right)^{-1} \sum_{\mathbf{k}\in\Delta Q_l} e^{i(\mathbf{k},\mathbf{x})}, \quad \Delta Q_l := Q_l \setminus Q_{l-1}.$$

Then $\|g\|_{\mathbf{W}^{1/q,b}_q} \ll 1$.

We now estimate the $\sigma_m(g)_p$ from below. Take any set K_m of m frequencies $\mathbf{k}$. Consider an additional function

$$h(\mathbf{x}) := \sum_{\mathbf{k}\in\Delta(n,N)\setminus K_m} e^{i(\mathbf{k},\mathbf{x})}.$$

For any polynomial t with frequencies from K_m we have

$$|\langle g-t,h\rangle| \le \|g-t\|_p \|h\|_{p'} \tag{9.3.27}$$

and

$$\langle g-t,h\rangle = \langle g,h\rangle = \sum_{\mathbf{k}\in\Delta(n,N)\setminus K_m} \hat{g}(\mathbf{k}). \tag{9.3.28}$$

From our choice (9.3.19) of N it is clear that, asymptotically,

$$\sum_{\mathbf{k}\in\Delta(n,N)\setminus K_m} \hat{g}(\mathbf{k}) \gg (N-n)N^{(d-1)(b+1-1/q)}.$$

Next, we have

$$\|h\|_{p'} \le \|g\|_{p'} + \|g-h\|_{p'} \le \|g\|_{p'} + \|g-h\|_2$$
$$\ll 2^{N/p} N^{(d-1)/p'} + m^{1/2} \ll m^{1/2}.$$

Thus, (9.3.27) and (9.3.28) yield

$$\sigma_m(g)_p \gg (N-n)N^{(d-1)(b+1-1/q)}m^{-1/2}.$$

Therefore

$$\sigma_m(\mathbf{W}^{1/q,b}_q)_p \gg (N-n)N^{(d-1)(b+1-1/q)}m^{-1/2}$$
$$\asymp m^{-1/2}(\log m)^{(d-1)(b+1-1/q)+1}.$$

This proves the lower bounds. □

The above proof of Theorem 9.3.9 can be adjusted to prove the following results for the $\mathbf{W}_q^{1/q}$ classes.

Theorem 9.3.10 *Let $1 < q \le 2 < p < \infty$. Then we have*

$$\sigma_m(\mathbf{W}_q^{1/q})_p \asymp m^{-1/2}(\log m)^{d(1-1/q)}.$$

The upper bounds are achieved by a constructive greedy-type algorithm.

Proof We need the following analog of Theorem 9.1.10.

Lemma 9.3.11 *For $t \in \mathscr{T}(\Delta(n,N))$ we have, for $1 < q \le 2$,*

$$\|t\|_A \ll N^{(d-1)(1-1/q)}(N-n)^{1-1/q}\|t\|_{\mathbf{W}_q^{1/q}}.$$

Proof Let $r = 1/q$ and

$$t = \varphi * F_r, \qquad \|\varphi\|_q = \|t\|_{\mathbf{W}_q^r}.$$

Then

$$\|t\|_A \le \|\varphi\|_q \left\| \sum_{\mathbf{k}\in\Delta(n,N)} \varepsilon_{\mathbf{k}} \hat{F}_r(\mathbf{k}) e^{i(\mathbf{k},\mathbf{x})} \right\|_{q'}, \qquad |\varepsilon_{\mathbf{k}}| = 1. \tag{9.3.29}$$

Using relation (9.1.7) with parameters $u = q'$ and $v = 2$ we obtain

$$\left\| \sum_{\mathbf{k}\in\Delta(n,N)} \varepsilon_{\mathbf{k}} \hat{F}_r(\mathbf{k}) e^{i(\mathbf{k},\mathbf{x})} \right\|_{q'}$$
$$\ll \left(\sum_{n<l\le N} \left(2^{-lr}2^{l/2}2^{l(1/2-1/q')}\right)^{q'} l^{d-1} \right)^{1/q'} \ll (N-n)^{1/q'} N^{(d-1)/q'}. \tag{9.3.30}$$

Combining (9.3.29) and (9.3.30) completes the proof of the lemma. □

We return to the proof of Theorem 9.3.10. The proof goes along the lines of the proof of Theorem 9.3.9. We use the same notation as above and begin with the upper bounds. In our case Lemma 9.3.11 implies the following analog of the bound (9.3.24):

$$\|g_A^1\|_A \ll (N-n)^{1-1/q} N^{(d-1)(1-1/q)}. \tag{9.3.31}$$

Also, we have the bounds (9.3.25) and (9.3.26) with $a = 1/q$, $b = 0$. Choosing N as in (9.3.19) and applying Theorem 9.2.8 we obtain, from (9.3.31), (9.3.25), and (9.3.26),

$$\begin{aligned}\|f - S_{Q_n}(f) - G^p_{m/2}(g_A)\|_p &\ll (N-n)^{1-1/q} N^{(d-1)(1-1/q)} m^{-1/2} \\ &\ll m^{-1/2}(\log m)^{(d-1)(1-1/q)+1-1/q} \\ &= m^{-1/2}(\log m)^{d(1-1/q)}.\end{aligned}$$

This completes the proof of upper bounds.

The lower bounds follow from the same example (with $b = 0$) that was used in the proof of Theorem 9.3.9. In this case instead of $\|g\|_{\mathbf{W}_q^{1/q,0}} \ll 1$ we have

$$\|g\|_{\mathbf{W}_q^{1/q}} \ll (N-n)^{1/q}$$

which brings the bound

$$\begin{aligned}\sigma_m(\mathbf{W}_q^{1/q})_p &\gg (N-n)^{1-1/q} N^{(d-1)(1-1/q)} m^{-1/2} \\ &\asymp m^{-1/2}(\log m)^{d(1-1/q)}.\end{aligned}$$

This proves the lower bounds. Theorem 9.3.10 is proved. □

9.3.4 The Case $q = 1$

We begin with the case $2 \le p < \infty$.

Theorem 9.3.12 *For any $\varepsilon > 0$ we have, for $2 \le p < \infty$,*

$$\sigma_m(\mathbf{W}_1^{a,b})_p \ll \begin{cases} m^{-a+1/2}(\log m)^{(d-1)(a-1+b)+\varepsilon}, & a > 1, \\ m^{-(a-\beta)p/2}(\log m)^{(d-1)b+\varepsilon}, & \beta < a < 1,\ \beta = 1-1/p, \\ m^{-1/2}(\log m)^{(d-1)b+1+\varepsilon}, & a = 1, \end{cases}$$

with multiplicative constants that are allowed to depend on ε, d, and p.

The upper bounds are achieved by a constructive greedy-type algorithm.

Proof For large smoothness, $a > 1$, the next lemma, from Temlyakov (2015a) (see Lemma 9.2.12 above), plays the key role in the proof.

Lemma 9.3.13 *Define, for $f \in L_1$,*

$$f_l := \sum_{\|\mathbf{s}\|_1 = l} \delta_{\mathbf{s}}(f), \qquad l \in \mathbb{N}_0, \qquad \mathbb{N}_0 := \mathbb{N} \cup \{0\}.$$

Consider the class

$$\mathbf{W}_A^{a,b} := \{f : \|f_l\|_A \le 2^{-al}(\bar{l})^{(d-1)b}\}.$$

Then, for $2 \le p < \infty$ and $0 < \mu < a$, there is a constructive method $A_m(\cdot, p, \mu)$ based on greedy algorithms which provides an m-term approximation with respect to $\mathcal{T}^d$ with the following bound for $f \in \mathbf{W}_A^{a,b}$:

$$\|f - A_m(f,p,\mu)\|_p \ll m^{-a-1/2}(\log m)^{(d-1)(a+b)}, \qquad 2 \le p < \infty. \tag{9.3.32}$$

We also need the following version of Theorem 9.1.10 in the case $q = 1$ (see Temlyakov, 1986c, Chapter 1, Section 2).

Lemma 9.3.14 *For any $\varepsilon > 0$ there is a $C(\varepsilon, d)$ such that, for each $t \in \mathscr{T}(N)$, we have*

$$\|t\|_A \le C(\varepsilon, d) N (\log N)^{\varepsilon} \|t\|_1.$$

Let $f \in \mathbf{W}_1^{a,b}$, $a > 1$. By Lemma 9.3.14 we get

$$\|f_l\|_A \ll 2^{-l(a-1)} l^{b(d-1)+\varepsilon}$$

with a multiplicative constant that is allowed to depend on ε and d. Setting $b(\varepsilon) := b + \varepsilon/(d-1)$, we obtain

$$\|f\|_{\mathbf{W}_A^{a-1,b(\varepsilon)}} \ll 1.$$

Lemma 9.3.13 gives a constructive proof of

$$\sigma_m(f)_p \ll m^{-a+1/2} (\log m)^{(d-1)(a-1+b)+\varepsilon}.$$

This proves the first inequality in Theorem 9.3.12.

Consider now the case $\beta < a < 1$. The argument in this case is close to the proof of Theorem 9.3.5. We use the same notation but now define

$$g_A := \sum_{n<l\le N} f_l, \qquad g_0 := \sum_{l>N} f_l.$$

By Lemma 9.3.14 we get

$$\|g_A\|_A \ll \sum_{n<l\le N} 2^{l(1-a)} l^{b(d-1)+\varepsilon} \ll 2^{N(1-a)} N^{b(d-1)+\varepsilon}. \tag{9.3.33}$$

By Theorem 9.2.8 we obtain

$$\sigma_m(g_A)_p \ll m^{-1/2} 2^{N(1-a)} N^{b(d-1)+\varepsilon}. \tag{9.3.34}$$

Using Theorem 9.1.11 leads to

$$\|g_0\|_p \le \sum_{l>N} \|f_l\|_p \le \sum_{l>N} \|f_l\|_1 2^{\beta l} \ll 2^{-(a-\beta)N} N^{b(d-1)}. \tag{9.3.35}$$

Now choose N such that

$$m^{-1/2} 2^{N(1-a)} \asymp 2^{-(a-\beta)N};$$

that is,

$$2^N \asymp m^{p/2}. \tag{9.3.36}$$

This, combined with (9.3.34) and (9.3.35), gives the error bound

$$\sigma_m(f)_p \ll m^{-(a-\beta)p/2} (\log m)^{b(d-1)+\varepsilon}.$$

This proves the required bound in the second case.

In the case $a = 1$ we get, as in (9.3.33),

$$\|g_A\|_A \ll N^{b(d-1)+1+\varepsilon}.$$

Determing N from (9.3.36), we obtain

$$\sigma_m(f)_p \ll m^{-1/2}(\log m)^{b(d-1)+1+\varepsilon}.$$

This completes the proof of Theorem 9.3.12. □

We note that in the case $a \le 1$ the corresponding lower bounds with $\varepsilon = 0$ follow from the univariate case (see Belinskii, 1987). We now prove the lower bounds for $a > 1$. It is sufficient to prove them for $p = 2$. Let m be given and n be such that $2^n n^{d-1} \asymp m$ and $m \le c(d)|\Delta Q_n|$, with small enough $c(d) > 0$. Let

$$\mathcal{K}_{N-1}(x) := \sum_{|k|\le N} (1 - |k|/N) e^{ikx} = \frac{(\sin(Nx/2))^2}{N(\sin(x/2))^2}$$

be the univariate Fejér kernel. The Fejér kernel $\mathcal{K}_{N-1}$ is an even nonnegative trigonometric polynomial in $\mathcal{T}(N-1)$. In the multivariate case define

$$\mathcal{K}_{\mathbf{N}-\mathbf{1}}(\mathbf{x}) := \prod_{j=1}^{d} \mathcal{K}_{N_j-1}(x_j), \qquad \mathbf{N} = (N_1, \ldots, N_d).$$

Then the $\mathcal{K}_{\mathbf{N}-\mathbf{1}}$ are nonnegative trigonometric polynomials from $\mathcal{T}(\mathbf{N}-\mathbf{1}, d)$ which have the following property:

$$\|\mathcal{K}_{\mathbf{N}-\mathbf{1}}\|_1 = 1. \tag{9.3.37}$$

Consider the function

$$g(\mathbf{x}) := \sum_{\|\mathbf{s}\|_1 = n} \mathcal{K}_{2^{\mathbf{s}-2}-\mathbf{1}}(\mathbf{x}) e^{i(2^{\mathbf{s}}-2^{\mathbf{s}-2},\mathbf{x})}.$$

Then, by (9.3.37),

$$\|g\|_1 \ll n^{d-1}. \tag{9.3.38}$$

Take any set K_m of m frequencies. It is clear that for small enough $c(d)$ we have

$$\sigma_m(g)_2^2 \ge \sum_{\mathbf{k}\in \Delta Q_n \setminus K_m} |\hat{g}(\mathbf{k})|^2 \gg |\Delta Q_n|. \tag{9.3.39}$$

Relations (9.3.38) and (9.3.39) imply that

$$\sigma_m(\mathbf{W}_1^{a,b})_2 \gg 2^{n(1/2-a)} n^{-(d-1)/2+b(d-1)} \asymp m^{-a+1/2}(\log m)^{(d-1)(b+a-1)}.$$

Consider now the case $q = 1$, $1 < p \le 2$. We need a version of the relation (9.1.8) from Theorem 9.1.9 adjusted to our case.

Lemma 9.3.15 *Let $1 < p < \infty$. For any $\varepsilon > 0$ there exists a constant $C(\varepsilon, d, p)$ such that, for each $t \in \mathscr{T}(Q_n)$, we have*

$$\sum_{\|\mathbf{s}\|_1 \le n} \|\delta_{\mathbf{s}}(t)\|_p \le C(\varepsilon, d, p) n^{\varepsilon} 2^{\beta n} \|t\|_1, \qquad \beta = 1 - 1/p.$$

Proof Choose $u \in (1, p)$ such that $u' := u/(u-1) > d/\varepsilon$. By the Hölder inequality and Theorem 9.1.9 we get

$$\begin{aligned}\sum_{\|\mathbf{s}\|_1 \le n} \|\delta_{\mathbf{s}}(t)\|_p &\le \left(\sum_{\|\mathbf{s}\|_1 \le n} 1\right)^{1/u'} \left(\sum_{\|\mathbf{s}\|_1 \le n} \|\delta_{\mathbf{s}}(t)\|_p^u\right)^{1/u} \\ &\le n^{\varepsilon} 2^{n(1/u - 1/p)} \left(\sum_{\|\mathbf{s}\|_1 \le n} \|\delta_{\mathbf{s}}(t)\|_p^u 2^{\|\mathbf{s}\|_1 (1/p - 1/u) u}\right)^{1/u} \\ &\ll n^{\varepsilon} 2^{n(1/u - 1/p)} \|t\|_u.\end{aligned}$$

Continuing the right-hand side by Theorem 9.1.11, we obtain

$$\ll n^{\varepsilon} 2^{n(1/u - 1/p)} 2^{n(1 - 1/u)} \|t\|_1 = n^{\varepsilon} 2^{n\beta} \|t\|_1. \qquad \square$$

Theorem 9.3.16 *Let $1 < p \le 2$. For any $\varepsilon > 0$ we have*

$$\sigma_m(\mathbf{W}_1^{a,b})_p \ll \begin{cases} m^{-a+\beta} (\log m)^{(d-1)(a+b-2\beta)+\varepsilon}, & a > 2\beta, \\ m^{-a+\beta} (\log m)^{(d-1)b+\varepsilon}, & \beta < a \le 2\beta. \end{cases}$$

Proof We begin with the case of large smoothness. Let $f \in \mathbf{W}_1^{a,b}$, $a > 2\beta$. Then, by Lemma 9.3.15,

$$\sum_{\|\mathbf{s}\|_1 = l} \|\delta_{\mathbf{s}}(f_l)\|_p \ll 2^{-(a-\beta)l} l^{b(d-1)+\varepsilon}. \tag{9.3.40}$$

As in the proof of Theorem 9.3.2, denote by G_l the set of indices $\mathbf{s}$ having the m_l largest $\|\delta_{\mathbf{s}}(f_l)\|_p$. Then, by Corollary 9.1.8 and Lemma 9.2.10, we obtain for $f_l' := \sum_{\mathbf{s} \notin G_l} \delta_{\mathbf{s}}(f_l)$,

$$\begin{aligned}\|f_l'\|_p &\ll \left(\sum_{\mathbf{s} \notin G_l} \|\delta_{\mathbf{s}}(f_l)\|_p^p\right)^{1/p} \\ &\ll (m_l + 1)^{-\beta} \sum_{\|\mathbf{s}\|_1 = l} \|\delta_{\mathbf{s}}(f_l)\|_p \\ &\le (m_l + 1)^{-\beta} 2^{-(a-\beta)l} l^{b(d-1)+\varepsilon}.\end{aligned} \tag{9.3.41}$$

For $l > n$, let

$$m_l := [2^{-\kappa(l-n)} n^{d-1}],$$

where $\kappa > 1$ is such that $a > (1+\kappa)\beta$. We define the m-term approximant

$$A_m(f) := S_{Q_n}(f) + \sum_{l>n} \sum_{\mathbf{s}\in G_l} \delta_{\mathbf{s}}(f_l).$$

Then

$$m \le |Q_n| + \sum_{l>n} 2^l m_l \ll 2^n n^{d-1}. \tag{9.3.42}$$

For the approximation error, we obtain from (9.3.41)

$$\|f - A_m(f)\|_p \le \sum_{l>n} \|f_l'\|_p \ll 2^{-(a-\beta)n} n^{(b-\beta)(d-1)+\varepsilon}. \tag{9.3.43}$$

Relations (9.3.42) and (9.3.43) imply the required upper bound.

In the case $\beta < a \le 2\beta$ the proof repeats the corresponding argument from the proof of Theorem 9.3.2. Instead of (9.3.1) we use (9.3.40). Also, in the case $a = 2\beta$ the factor $\log\log m$ is included in $(\log m)^{\varepsilon}$.

The lower bounds with $\varepsilon = 0$ in the case of small smoothness, $a \le 2\beta$, follow from the univariate case. We now consider the case of large smoothness, $a > 2\beta$. In the case $p = 2$ it is established in the proof of Theorem 9.3.12. We use the same example to prove the lower bounds for $p < 2$. Instead of (9.3.39), by Theorem 9.1.9 we obtain

$$\sigma_m(g)_p^p \gg \sum_{\|\mathbf{s}\|_1=n} \left(\left(\sum_{\mathbf{k}\in\rho(\mathbf{s})\setminus K_m} |\hat{g}(\mathbf{k})|^2 \right)^{1/2} 2^{n(1/2-1/p)} \right)^p \gg n^{d-1} 2^{n(p-1)}. \tag{9.3.44}$$

Relations (9.3.38) and (9.3.44) imply

$$\sigma_m(\mathbf{W}_1^{a,b})_p \gg 2^{n(1-1/p-a)} n^{(d-1)(b+1/p-1)} \asymp m^{-a+\beta} (\log m)^{(d-1)(a+b-1)}.$$

The theorem is proved. □

We recall the definition of the class $\bar{\mathbf{W}}_q^{a,b}$ (see §7.7), which consists of functions f with a representation

$$f = \sum_{n=1}^{\infty} t_n, \qquad t_n \in \mathscr{T}(Q_n), \qquad \|t_n\|_q \le 2^{-an} n^{b(d-1)}.$$

It is easy to see that, in the case $1 < q < \infty$, the classes $\bar{\mathbf{W}}_q^{a,b}$ and $\mathbf{W}_q^{a,b}$ are equivalent. In the case $q = 1$ the classes $\bar{\mathbf{W}}_1^{a,b}$ are wider than $\mathbf{W}_1^{a,b}$. However, the results of this section hold for these classes as well.

Remark 9.3.17 Theorems 9.3.12 and 9.3.16 hold for the class $\bar{\mathbf{W}}_1^{a,b}$ as well as $\mathbf{W}_1^{a,b}$.

9.3.5 The Entropy Technique

In this subsection we use known results on entropy numbers to prove a lower bound in the case of functions of two variables. We note that this is of interest for small smoothness, $r < 1/2$. We will use the following well-known simple theorem (see Theorem 7.1.1).

Theorem 9.3.18 *For any compact set A we have*

$$M_{2\varepsilon}(A) \le N_\varepsilon(A) \le M_\varepsilon(A). \tag{9.3.45}$$

The following theorem (see Theorem 7.4.3) is from Temlyakov (2013). We reformulate it here for the reader's convenience.

Theorem 9.3.19 *Let a compact $F \subset X$ be such that there exists a normalized system $\mathscr{D}$, $|\mathscr{D}| = N$, and a number $r > 0$ such that*

$$\sigma_k(F,\mathscr{D})_X \le k^{-r}, \qquad k \le N.$$

Then, for $k \le N$,

$$\varepsilon_k(F,X) \le C(r)\left(\frac{\log(2N/k)}{k}\right)^r. \tag{9.3.46}$$

We use the above theorem to prove the following lower bound for best m-term approximations.

Theorem 9.3.20 *In the case $d = 2$ the following lower bound holds for any $q < \infty$, $r > 1/q$:*

$$\sigma_m(\mathbf{W}^r_q)_\infty \gg m^{-r}(\log m)^{1/2}.$$

Proof We will use a special inequality from Temlyakov (1995a) (see (4.2.20)) called the small ball inequality. For an even number n define

$$Y_n := \{\mathbf{s} = (2n_1, 2n_2),\ n_1 + n_2 = n/2\}.$$

Then, for any coefficients $\{c_{\mathbf{k}}\}$,

$$\left\| \sum_{\mathbf{s}\in Y_n} \sum_{\mathbf{k}\in\rho(\mathbf{s})} c_{\mathbf{k}} e^{i(\mathbf{k},\mathbf{x})} \right\|_\infty \ge C \sum_{\mathbf{s}\in Y_n} \left\| \sum_{\mathbf{k}\in\rho(\mathbf{s})} c_{\mathbf{k}} e^{i(\mathbf{k},\mathbf{x})} \right\|_1, \tag{9.3.47}$$

where C is a positive number. Inequality (9.3.47) plays a key role in the proof of lower bounds for entropy numbers.

Take any even $n \in \mathbb{N}$, depending on m, which will be chosen later. Consider the following compact:

$$F(Y_n)_\infty := \left\{ t = \sum_{\mathbf{s}\in Y_n} t_{\mathbf{s}} : t_{\mathbf{s}} \in \mathscr{T}(\rho(\mathbf{s})),\ \|t_{\mathbf{s}}\|_\infty \le 1 \right\}.$$

The known results on the volumes of sets of Fourier coefficients of trigonometric polynomials imply the following lemma (see Temlyakov, 1989d, 1995a).

Lemma 9.3.21 *There exist $2^{n2^{n-1}}$ functions $f_j \in F(Y_n)_\infty$, $j = 1, \dots, 2^{n2^{n-1}}$, such that, for $i \neq j$,*

$$\|f_i - f_j\|_2 \gg n^{1/2}.$$

We now show that for the f_j from Lemma 9.3.21 we have

$$\|f_i - f_j\|_\infty \gg n.$$

Indeed, for any $f \in F(Y_n)_\infty$ we have

$$\|f\|_2^2 = \sum_{\mathbf{s} \in Y_n} \|t_\mathbf{s}\|_2^2 \le \sum_{\mathbf{s} \in Y_n} \|t_\mathbf{s}\|_1 \|t_\mathbf{s}\|_\infty \le \sum_{\mathbf{s} \in Y_n} \|t_\mathbf{s}\|_1.$$

It remains to apply the small ball inequality (9.3.47). Therefore, for $k = n2^{n-1}$ we obtain, using Theorem 9.3.18,

$$\varepsilon_k(F(Y_n)_\infty)_\infty \gg \log k.$$

We now use Theorem 9.3.19. We specify $F := F(Y_n)_\infty$, $\mathscr{D} := \{e^{i(\mathbf{k},\mathbf{x})} : \|\mathbf{k}\|_\infty \le 2^{n+1}\}$, $X := L_\infty$. It is clear that, for $l \ge \dim \mathscr{T}(Y_n) \asymp n2^n \asymp k$, we have

$$\sigma_l(F, \mathscr{D}) = 0.$$

Also, for any $f \in F$ we have

$$\|f\|_\infty \le n/2 \ll \log k.$$

Denote

$$B := \max_l l^r \sigma_l(F, \mathscr{D})_\infty.$$

By Theorem 9.3.19 we obtain

$$\log k \ll B n^r k^{-r} \qquad \text{and} \qquad B \gg n^{-r} k^r \log k.$$

This implies that there is an $l \asymp k$ such that

$$\sigma_l(F, \mathscr{D})_\infty \gg n^{-r} \log k \asymp (\log k)^{1-r}.$$

Next, it is clear that, for any m,

$$\sigma_m(F, \mathscr{D})_\infty \ll \sigma_m(F, \mathscr{T})_\infty.$$

Further, by the Littlewood–Paley theorem there is a $c_1(q) > 0$ such that

$$c_1(q) n^{-1/2} 2^{-rn} F \subset \mathbf{W}^r_q, \qquad q < \infty.$$

This completes the proof of the theorem. □

9.3.6 Discussion

The effect of *small smoothness* in the behavior of the asymptotic characteristics of smoothness classes was discovered by Kashin (1981). He proved that the rate of decay of the Kolmogorov widths $d_n(W^r_1, L_p)$ of the univariate classes W^r_1 depends on r differently in the range $1 - 1/p < r < 1$ (small smoothness) and in the range $r > 1$. For further results see Kulanin (1985). Belinskii (1987) studied the univariate m-term trigonometric approximation and observed the small-smoothness effect in that setting. Romanyuk (2003) conducted a detailed study of the m-term trigonometric approximation of classes of multivariate functions with small mixed smoothness. The Besov classes $\mathbf{B}^r_{q,\theta}$ were studied in Romanyuk (2003). Define

$$\|f\|_{\mathbf{H}^r_q} := \sup_{\mathbf{s}} \|\delta_{\mathbf{s}}(f)\|_q 2^{r\|\mathbf{s}\|_1}$$

and, for $1 \le \theta < \infty$, define

$$\|f\|_{\mathbf{B}^r_{q,\theta}} := \left(\sum_{\mathbf{s}} \left(\|\delta_{\mathbf{s}}(f)\|_q 2^{r\|\mathbf{s}\|_1}\right)^\theta\right)^{1/\theta}.$$

We write $\mathbf{B}^r_{q,\infty} := \mathbf{H}^r_q$. With a small abuse of notation, denote the corresponding unit ball by

$$\mathbf{B}^r_{q,\theta} := \{f : \|f\|_{\mathbf{B}^r_{q,\theta}} \le 1\}.$$

Lemma 9.1.3 was used in Romanyuk (2003) in the case of approximation in L_p, $2 < p < \infty$. This makes the corresponding results in that paper nonconstructive. We note that the bound for the m-term approximation error in Lemma 9.1.3 follows from Theorem 9.2.8, and the extra property $\Theta_m \subset \Theta_n$ in Lemma 9.1.3 follows from the proof of Theorem 9.2.8 in Temlyakov (2015a). Thus, Theorem 9.2.8 makes Lemma 9.1.3 constructive and, therefore, the nonconstructive results from Romanyuk (2003), which are based on Lemma 9.1.3, are made constructive in this way. Also, the use of Theorem 9.2.8 is technically easier than the use of Lemma 9.1.3. For instance, in the proof of the upper bounds in Theorem 9.3.5 we estimate $\|g_A\|_A$ in a rather simple way, because of the additivity property of the norm $\|\cdot\|_A$, and then apply Theorem 9.2.8 to g_A. Typically, in Romanyuk (2003), Lemma 9.1.3 is applied to individual dyadic blocks $\delta_{\mathbf{s}}(f)$, with $m_{\mathbf{s}}$ depending on the norm of the $\delta_{\mathbf{s}}(f)$. It would be interesting to see how much the technique based on Theorem 9.2.8 could simplify the study of $\sigma_m(\mathbf{B}^r_{q,\theta})_p$.

Let us make some comparison of our results on the $\mathbf{W}^r_q$ classes with the known results on the $\mathbf{B}^r_{q,\theta}$ classes. It follows from Corollary 9.1.8 that, for $1 < q \le 2$, we have

$$\|f\|_{\mathbf{B}^r_{q,2}} \ll \|f\|_{\mathbf{W}^r_q} \ll \|f\|_{\mathbf{B}^r_{q,q}}.$$

For example, in the case $\beta p' < r < 1/q$ Theorem 9.3.5 gives

$$\sigma_m(\mathbf{W}^r_q)_p \asymp m^{-(r-\beta)p/2}(\log m)^{(d-1)(r(p-1)-\beta p)}. \tag{9.3.48}$$

The corresponding results from Romanyuk (2003) give

$$\sigma_m(\mathbf{B}^r_{q,\theta})_p \asymp m^{-(r-\beta)p/2}(\log m)^{(d-1)((r-1/q)(p-1)+1-1/\theta)}. \tag{9.3.49}$$

In the case $\theta = q$ the right-hand sides of (9.3.48) and (9.3.49) coincide. This means that our results (for the upper bounds) for a wider class $\mathbf{W}^r_q$ imply the corresponding results for the smaller class $\mathbf{B}^r_{q,q}$. Relations (9.3.48) and (9.3.49) show that $\sigma_m(\mathbf{W}^r_q)_p$ and $\sigma_m(\mathbf{B}^r_{q,2})_p$ have different orders.

As we pointed out in the introduction to this chapter, the main novelty of our approach is in providing constructive algorithms for optimal m-term trigonometric approximation on classes with small mixed smoothness. This is achieved by using Theorem 9.2.8. The use of Theorem 9.2.8 is simpler than the use of Lemma 9.1.3, traditionally used in this area of research. In addition to the traditional use of Theorem 9.1.9, which goes back to Temlyakov (1986b, c), we use other deep results from the hyperbolic cross approximation theory: Theorems 9.1.10 and 9.1.11 and Lemma 9.3.14. We also prove a new result, Lemma 9.3.15. These results allow us to treat the case $q = 1$ (see §9.3.4).

A number of interesting unresolved problems on m-term trigonometric approximation are discussed in Temlyakov (2015a), Section 6. The paper Temlyakov (2015b) makes progress in some of them. For instance, Theorems 9.3.1 and 9.3.10 in the present text cover the case $\beta p' < r \le 1/q$ for the constructive m-term approximation of $\mathbf{W}^r_q$ classes. The case $\beta < r \le \beta p'$ is still open. There is no progress on small-smoothness classes in the case $2 \le q < p < \infty$. In the case $q = 1$, the results presented in §9.3.4 are optimal up to a factor $(\log m)^\varepsilon$ with arbitrarily small $\varepsilon > 0$. It would be interesting to find the correct orders of $\sigma_m(\mathbf{W}^{a,b}_1)_p$ and the correct orders for the constructive m-term approximation of these classes.

The reader can find detailed discussions of greedy algorithms in Banach spaces in Temlyakov (2011) and of their applications for the m-term trigonometric approximation in Dilworth *et al.* (2002) and Temlyakov (2005, 2014, 2015a).

9.4 Open Problems

It is well known that the extreme cases, when one of the parameters p or q takes the value 1 or ∞, are difficult in the hyperbolic cross approximation theory. Often, the study of these cases requires special techniques. Many problems which involve the extreme values of parameters are still open. Also, the case of small smoothness is still open in many settings.

Open Problem 9.1 Find a constructive method which provides the order of $\sigma_m(\mathbf{W}^r_q)_p$, $2 \le q \le p < \infty$, $\beta < r \le 1/2$.

Open Problem 9.2 Find the order of $\sigma_m(\mathbf{W}^r_q)_\infty$, $1 \le q \le \infty$, $r > 1/p$.

Open Problem 9.3 Find the order of $\sigma_m(\mathbf{W}^r_q)_1$, $1 \le q \le \infty$, $r > 0$.

Open Problem 9.4 Find the order of $\sigma_m(\mathbf{W}^r_\infty)_p$, $1 \le p \le \infty$, $r > 0$.

We have formulated the above problems for the **W** classes. Those problems are open for the **H** and **B** classes as well. In addition the following problem is open for the **H** and **B** classes.

Open Problem 9.5 Find the order of $\sigma_m(\mathbf{H}^r_q)_p$ and $\sigma_m(\mathbf{B}^r_{q,\theta})_p$ for $1 \le p < q \le 2$, $r > 0$.

9.5 Concluding Remarks

The main part of this book is devoted to the study of classes of functions with mixed smoothness – the **W**-type and the **H**-type classes. For convenience we call these classes *classes with mixed smoothness.* As the reader can see, the study of classes with mixed smoothness is much more difficult than, say, the study of the anisotropic Sobolev and Nikol'skii classes. Certainly, a natural question is: why study classes with mixed smoothness? We present here some arguments which partially answer the above question. We give two kinds of arguments – *a priori* and *a posteriori*. We begin with the *a priori* arguments.

Korobov studied the numerical integration of functions from the classes

$$\mathbf{E}^r_d(C) := \left\{ f \in L_1(\mathbb{T}^d) : |\hat{f}(\mathbf{k})| \le C \prod_{j=1}^{d} (\max\{1, |k_j|\}^{-r} \right\}.$$

One of Korobov's motivations for studying the classes $\mathbf{E}^r_d(C)$ was related to the numerical solution of integral equations. Let $K(x,y)$ be the kernel of the integral operator J_K. Then the kernel of $(J_K)^d$ is given by

$$K^d(x,y) = \int_{\mathbb{T}^{d-1}} K(x,x_1)K(x_1,x_2)\cdots K(x_{d-1},y)dx_1\cdots dx_{d-1}.$$

The smoothness properties of $K(x,y)$ are naturally transformed into mixed smoothness properties of $K(x,x_1)K(x_1,x_2)\cdots K(x_{d-1},y)$. In the simplest case, where $f_j(t)$ satisfies $\|f_j'\|_\infty \le 1$, $j = 1,2,\dots,d$, we obtain

$$\|(f_1(x_1)\cdots f_d(x_d))^{(1,1,\dots,1)}\|_\infty \le 1.$$

This is an *a priori* argument about the importance of classes with mixed smoothness. There are other *a priori* arguments in support of the importance of classes with mixed smoothness. For instance, the work of Yserentant (2010) on the regularity of the eigenfunctions of the electronic Schrödinger operator and the recent paper Triebel (2015) on the regularity of the solutions of the Navier–Stokes equations, show that mixed regularity plays a fundamental role in mathematical physics. This makes approximation techniques developed for classes of functions with bounded mixed derivatives a proper choice for the numerical treatment of those problems.

We now give some *a posteriori* arguments. These arguments can be formulated in the following general way. Methods developed for the approximation of classes with mixed smoothness are very good in different senses. We briefly discuss this important point, beginning with numerical integration. It was immediately understood that a trivial generalization of the univariate quadrature formulas with equidistant nodes to cubature formulas with rectangular grids does not work for classes with mixed smoothness. As a result, different fundamental methods of numerical integration were constructed: the Korobov cubature formulas (in particular, the Fibonacci cubature formulas), the Smolyak cubature formulas, and the Frolov cubature formulas. These nontrivial constructions are very useful in practical numerical integration, especially when the dimension of the model is moderate (≤ 40). In approximating the multivariate functions the following fundamental methods have been designed: the hyperbolic cross approximation (in particular, the approximation of f by $S_{Q_n}(f)$) and the Smolyak-type recovering algorithms.

Here is one more important *a posteriori* argument in favor of the detailed study of classes with mixed smoothness. It turns out that the study of mixed-smoothness classes is directly related to deep problems in other areas of mathematics. The numerical integration of these classes is closely related to discrepancy theory and to nonlinear approximation with respect to special redundant dictionaries. Estimates of the entropy numbers of classes with mixed smoothness are closely related to (and in some cases equivalent to) the small ball problem from probability theory. Moreover, the study of classes with mixed smoothness requires new techniques based on deep results and ideas from other areas of mathematics. We list some of these areas: geometry, with delicate estimates of the volumes of special convex bodies; functional analysis, with the heavy use of duality arguments; number theory, including both simple methods based on congruence relations and deep methods based on the theory of irreducible polynomials; harmonic analysis, including both classical methods such as the Littlewood–Paley theory and Marcinkiewicz multipliers and also recently invented new approaches based on Riesz products for hyperbolic crosses and the small ball inequality; greedy approximation, including approximation in Banach spaces with respect to redundant dictionaries.

Finally, we illustrate the following general observation, which gives an *a posteriori* argument. Methods of approximation which are optimal in the sense of order for the mixed-smoothness classes are universal for the collection of anisotropic smoothness classes. This gives an *a posteriori* justification for the thorough study of classes of functions with mixed smoothness. The phenomenon of saturation is well known in approximation theory (DeVore and Lorentz, 1993, Chapter 11). The classical example of a saturation method is the Fejér operator for the approximation of univariate periodic functions. In the case of a sequence of the Fejér operators K_n, saturation means that the approximation order resulting from the use of the operators K_n does not improve over the rate $1/n$ even if we increase the smoothness of the functions under approximation. Methods (algorithms) that do not have the saturation property are called unsaturated. The reader can find a detailed discussion of unsaturated algorithms in approximation theory and in numerical analysis in the survey by Babenko (1985). We point out that the concept of smoothness becomes more complicated in the multivariate case than it is in the univariate case: in the multivariate case a function may have different smoothness properties in different coordinate directions. In other words, a function may belong to different anisotropic smoothness classes (see the Sobolev and Nikol'skii classes in Chapter 3). It is known (see Chapter 3 and Temlyakov, 1993b) that the approximation characteristics of anisotropic smoothness classes depend on the average smoothness and the optimal approximation methods depend on the anisotropy of the classes. This motivated a study in Temlyakov (1988c) of the existence of an approximation method that can be used for all anisotropic smoothness classes. This is a problem of the existence of a universal method of approximation. We note that the universality concept in learning theory is very important and it is close to the concepts of adaptation and distribution-free estimation in nonparametric statistics (Györfy *et al.*, 2002, Binev *et al.*, 2005, Temlyakov, 2006).

For illustration we present here a discussion of only one of the known results on universal cubature formulas.

Let a vector $\mathbf{r} = (r_1,\dots,r_d)$, $r_j > 0$, and a number m be given. Denote $g(\mathbf{r}) := (\sum_{j=1}^d r_j^{-1})^{-1}$. We define numbers $N_j := \max\left([m^{\rho_j}],1\right)$, $\rho_j := g(\mathbf{r})/r_j$, $j = 1,\dots,d$ and the cubature formula

$$q_m(f,\mathbf{r}) := q_{\mathbf{N}}(f), \qquad \mathbf{N} := (N_1,\dots,N_d),$$

$$q_{\mathbf{N}}(f) := \left(\prod_{j=1}^d N_j\right)^{-1} \sum_{j_d=1}^{N_d} \cdots \sum_{j_1=1}^{N_1} f(2\pi j_1/N_1,\dots,2\pi j_d/N_d).$$

It is known (see Chapter 3 of Bakhvalov, 1959, and Temlyakov, 1993b) that, for $g(\mathbf{r}) > 1/p$,

$$\kappa_m(H_p^{\mathbf{r}}) \asymp q_m(H_p^{\mathbf{r}},\mathbf{r}) \asymp m^{-g(\mathbf{r})}, \qquad 1 \le p \le \infty,$$

where

$$q_m(\mathbf{W},\mathbf{r}) := \sup_{f\in\mathbf{W}} \left| q_m(f,\mathbf{r}) - (2\pi)^{-d} \int_{\mathbb{T}^d} f(\mathbf{x})d\mathbf{x} \right|.$$

We note that the cubature formula $q_m(\cdot,\mathbf{r})$ depends essentially on the anisotropic class defined by the vector $\mathbf{r}$. It is known (see Chapter 6) that the Fibonacci cubature formulas are optimal (in the sense of order) among all cubature formulas:

$$\text{for } g(\mathbf{r}) > 1/p, \qquad \kappa_{b_n}(H_p^{\mathbf{r}}) \asymp \Phi_n(H_p^{\mathbf{r}}) \asymp b_n^{-g(\mathbf{r})}.$$

Thus, the Fibonacci cubature formulas are universal for the collection $\{H_p^{\mathbf{r}} : 1 \leq p \leq \infty, g(\mathbf{r}) > 1/p\}$ in the following sense. The quantity $\Phi_n(\cdot)$ does not depend on the vector $\mathbf{r}$ or the parameter p, and it provides an optimal (in the sense of order) error bound for each class $H_p^{\mathbf{r}}$ from the collection.

We note that function classes with mixed smoothness are not only an interesting and challenging object for approximation theory but are important in numerical computation. Bungartz and Griebel and their research groups (see Bungartz and Griebel, 2004, 1999, and Griebel, 2006) have used approximation methods designed for these classes in elliptic variational problems.

Appendix

Classical Inequalities

This appendix contains some well-known results in analysis and one result in the geometry of convex bodies. These results are systematically used in the book. A number are proved, for the sake of completeness of the treatment. At the end of this appendix we give some notation.

A.1 The Spaces $L_{\mathbf{p}}$ and Some Inequalities

A.1.1 A Property of the Modulus of Continuity

Let $f(\mathbf{x})$, $\mathbf{x} = (x_1,\dots,x_d)$, be a measurable almost everywhere finite function which is 2π-periodic in each variable. In the case $d = 1$ we write $f \in L_p$ for $1 \le p < \infty$ if

$$\|f\|_p := \left((2\pi)^{-1}\int_{-\pi}^{\pi}|f(x)|^p dx\right)^{1/p} < \infty,$$

where the integral is considered as a Lebesgue integral. In the case $d > 1$, $\mathbf{p} = (p_1,\dots,p_d)$, $1 \le p_j < \infty$, $j = 1,\dots,d$, $f \in L_{\mathbf{p}}$ we have

$$\|f\|_{\mathbf{p}} := \left((2\pi)^{-1}\int_{-\pi}^{\pi}\left(\cdots(2\pi)^{-1}\int_{-\pi}^{\pi}\left((2\pi)^{-1}\int_{-\pi}^{\pi}|f(\mathbf{x})|^{p_1}dx_1\right)^{p_2/p_1}\right.\right.$$
$$\left.\left.\times dx_2\cdots\right)^{p_d/p_{d-1}} dx_d\right)^{1/p_d} < \infty.$$

In the case $\mathbf{p} = \infty$ it will be convenient for us to assume that the space L_∞ is the space of continuous functions and that

$$\|f\|_\infty := \sup_{\mathbf{x}} \|f(\mathbf{x})\|.$$

For $f \in L_{\mathbf{p}}$ we define the modulus of continuity in $L_{\mathbf{p}}$,

$$\omega(f,\delta)_{\mathbf{p}} := \sup_{|\mathbf{y}|\le\delta} \|f(\cdot+\mathbf{y}) - f(\cdot)\|_{\mathbf{p}}.$$

Theorem A.1.1 *Let $\mathbf{1} \leq \mathbf{p} < \infty$ (a vector inequality means that the corresponding inequality holds for each coordinate) or $\mathbf{p} = \infty$. Then $\omega(f,\delta) \to 0$ for $\delta \to 0$.*

Proof In the case $\mathbf{p} = \infty$ the theorem follows from the uniform continuity of a function which is continuous on a compact. Let $\mathbf{1} \leq \mathbf{p} < \infty$. We first prove an auxiliary statement.

Lemma A.1.2 *Let N be a natural number and*

$$f^N(\mathbf{x}) := \begin{cases} f(\mathbf{x}), & \text{if } |f(\mathbf{x})| > N, \\ 0 & \text{otherwise.} \end{cases}$$

Then, for $f \in L_{\mathbf{p}}$, $1 \leq \mathbf{p} < \infty$,

$$\lim_{N \to \infty} \|f^N\|_{\mathbf{p}} = 0.$$

Proof For $d = 1$ the lemma follows from the definition of the Lebesgue integral. In the general case we can prove the lemma by induction. Let the lemma be valid for $d-1$ and let $\mathbf{x}^d := (x_1,\dots,x_{d-1})$ and $\mathbf{p}^d := (p_1,\dots,p_{d-1})$; from the inclusion $f \in L_{\mathbf{p}}$ it follows that $\left\|f(\cdot,x_d)\right\|_{\mathbf{p}^d} = \varphi(x_d) \in L_{p_d}$. Consequently, for almost all x_d, $f(\mathbf{x}^d,x_d)$ belongs to $L_{\mathbf{p}^d}$ and, by the induction hypothesis,

$$\lim_{N \to \infty} \left\|f^N(\cdot,x_d)\right\|_{\mathbf{p}^d} = 0.$$

Further, $\left\|f^N(\cdot,x_d)\right\|_{\mathbf{p}^d} \leq \varphi(x_d) \in L_{p_d}$. Thus, applying the Lebesgue theorem about the limit passing under the integral sign, we obtain the lemma for dimension d. □

Corollary A.1.3 *Let $f \in L_{\mathbf{p}}$, $\mathbf{1} \leq \mathbf{p} < \infty$. Then*

$$\lim_{t \to 0} \sup_{|E| \leq t} \left\|f(\mathbf{x})\chi_E(\mathbf{x})\right\|_{\mathbf{p}} = 0,$$

where χ_E is the characteristic function of a measurable set E and $|E|$ denotes the measure of the set E.

We now conclude the proof of Theorem A.1.1. We use the Lusin theorem, which gives that for any $\varepsilon > 0$ and for a measurable almost everywhere finite $f(\mathbf{x})$ there is a continuous $g(\mathbf{x})$ such that

$$\text{measure}\{\mathbf{x} : f(\mathbf{x}) \neq g(\mathbf{x})\} < \varepsilon.$$

The theorem follows from Corollary A.1.3 and the Lusin theorem. □

A.1.2 Some Notation

We present here some well-known inequalities. First, let us introduce some notation.

For $1 \le p \le \infty$ we denote by p' the dual exponent, that is, the number (or ∞) such that $1/p + 1/p' = 1$. For a vector $\mathbf{1} \le \mathbf{p} \le \infty$ we denote $\mathbf{p}' := (p_1', \dots, p_d')$ and $\mathbf{1}/\mathbf{p} := (1/p_1, \dots, 1/p_d)$.

For the sake of brevity we shall write $\int f d\mu$ instead of $(2\pi)^{-d} \int_{\mathbb{T}^d} f(\mathbf{x}) d\mathbf{x}$, where $\mathbb{T}^d$ equals $[-\pi, \pi]^d$ or $[0, 2\pi]^d$ and μ means the normalized Lebesgue measure on $\mathbb{T}^d$. When $\mathbf{p} = \mathbf{1}p = (p, \dots, p)$ we shall write the scalar p instead of the vector $\mathbf{p}$.

A.1.3 The Hölder Inequality

Let $1 \le p \le \infty$, $f_1 \in L_p$, $f_2 \in L_{p'}$. Then $f_1 f_2 \in L_1$ and

$$\int |f_1 f_2| d\mu \le \|f_1\|_p \|f_2\|_{p'}. \tag{A.1.1}$$

Proof The inequality (A.1.1) is evident for $p = 1$ and $p = \infty$. Let $1 < p < \infty$. We will consider the function $y = x^{p-1}$ defined on $[0, a]$ and the inverse function $x = y^{1/(p-1)}$ defined on $[0, b]$. Then, calculating the areas of the figures $[0, a] \times [0, b]$,

$$G_1 := \{(x, y) : 0 \le x \le a,\ 0 \le y \le x^{p-1}\},$$
$$G_2 := \{(x, y) : 0 \le y \le b,\ 0 \le x \le y^{1/(p-1)}\},$$

we get

$$ab \le |G_1| + |G_2| = a^p/p + b^{p'}/p'. \tag{A.1.2}$$

Substituting $a = |f_1|/\|f_1\|_p$ and $b = |f_2|/\|f_2\|_{p'}$ into the relation (A.1.2) and integrating, we get the relation (A.1.1). □

A.1.4 The Hölder Inequality for a Vector

As a consequence of the relation (A.1.1) we obtain the Hölder inequality for a vector $\mathbf{1} \le \mathbf{p} \le \infty$:

$$\int |f_1 f_2| d\mu \le \|f_1\|_{\mathbf{p}} \|f_2\|_{\mathbf{p}'}.$$

A.1.5 The Hölder Inequality for Several Functions

Let $1 \le p_i \le \infty$, $i = 1, \dots, m$, $1/p_1 + \cdots + 1/p_m = 1$, $f_i \in L_{p_i}$, $i = 1, \dots, m$. Then $f_1 \cdots f_m \in L_1$ and

$$\int |f_1 \cdots f_m| d\mu \le \|f_1\|_{p_1} \cdots \|f_m\|_{p_m}. \tag{A.1.3}$$

Proof The proof will be carried out by induction. For $m = 2$ it is the Hölder inequality. Suppose that relation (A.1.3) has been proved for $m-1$. We can assume without loss of generality that $p_m > 1$. Applying the Hölder inequality for $g_1 := f_1 \cdots f_{m-1}$ and $g_2 := f_m$ with exponents p'_m and p_m we get

$$\int |f_1 \cdots f_m| d\mu \le \|f_1 \cdots f_{m-1}\|_{p'_m} \|f_m\|_{p_m}.$$

Denote $q_i := p_i/p'_m$, $i = 1, \dots, m-1$. Then $1/q_1 + \cdots + 1/q_{m-1} = 1$. Using the induction hypothesis, we get

$$\|f_1 \cdots f_{m-1}\|_{p'_m} \le \left(\prod_{i=1}^{m-1} \left\| |f_i|^{p'_m} \right\|_{q_i} \right)^{1/p'_m} = \prod_{i=1}^{m-1} \|f_i\|_{p_i},$$

which implies (A.1.3). □

A.1.6 The Monotonicity of L_p-Norms

Let $1 \le q \le p \le \infty$; then

$$\|f\|_q \le \|f\|_p \tag{A.1.4}$$

and, for $\mathbf{1} \le \mathbf{q} \le \mathbf{p} \le \infty$,

$$\|f\|_{\mathbf{q}} \le \|f\|_{\mathbf{p}}. \tag{A.1.5}$$

Proof Clearly, it suffices to prove (A.1.4). We set $a := p/q$ and apply the Hölder inequality with exponents a and a' to the functions $f_1 := |f|^q$ and $f_2 := 1$. Then

$$\|f\|_q \le \left\| |f|^q \right\|_a^{1/q} = \|f\|_p.$$ □

A.1.7 Interpolation Inequality

Let $1 \le a < p < b \le \infty$, $\theta := (1/p - 1/b)(1/a - 1/b)^{-1}$; then

$$\|f\|_p \le \|f\|_a^{\theta} |f\|_b^{1-\theta}. \tag{A.1.6}$$

Proof In the case $b = \infty$ we have

$$\|f\|_p = \left(\int |f|^{p-a} |f|^a d\mu \right)^{1/p} \le \|f\|_a^{a/p} \|f\|_\infty^{(1-a/p)}.$$

Assume therefore that $b < \infty$. We set $1/q := p\theta/a$; then $1/q' = p(1-\theta)/b$ and $p = a/q + b/q'$. Applying the Hölder inequality with exponents q and q' to the functions $f_1 = |f|^{a/q}$ and $f_2 = |f|^{b/q'}$ we get

$$\|f\|_p^p \le \left(\int |f|^a d\mu\right)^{1/q} \left(\int |f|^b d\mu\right)^{1/q'} = \|f\|_a^{p\theta} \|f\|_b^{p(1-\theta)},$$

which implies the inequality (A.1.6). □

A.1.8 The Hölder Inequality for Sums

From the inequality (A.1.1) we easily obtain the Hölder inequality for sums:

$$\sum_{i=1}^{N} |a_i b_i| \le \left(\sum_{i=1}^{N} |a_i|^p\right)^{1/p} \left(\sum_{i=1}^{N} |b_i|^{p'}\right)^{1/p'}, \qquad 1 \le p \le \infty.$$

We remark that in this inequality one can take $N = \infty$.

A.1.9 The Minkowski Inequality

Let $1 \le p \le \infty$, $f \in L_p$, $i = 1, \dots, m$. Then

$$\left\| \sum_{i=1}^{m} f_i \right\|_p \le \sum_{i=1}^{m} \|f_i\|_p. \tag{A.1.7}$$

Proof Clearly, it suffices to prove (A.1.7) for $m = 2$. For $p = 1$ and $p = \infty$ the inequality (A.1.7) is evident. Assume then that $1 < p < \infty$. Using the Hölder inequality for sums it is easy to verify that $S = f_1 + f_2 \in L_p$. Further,

$$\int |S|^p d\mu \le \int |S|^{p-1} |f_1| d\mu + \int |S|^{p-1} |f_2| d\mu.$$

Applying the Hölder inequality with exponents p' and p we get

$$\|S\|_p^p \le \|S\|_p^{p/p'} (\|f_1\|_p + \|f_2\|_p),$$

which implies (A.1.7).

In the case of a vector $\mathbf{1} \le \mathbf{p} \le \infty$, the inequality

$$\left\| \sum_{i=1}^{m} f_i \right\|_{\mathbf{p}} \le \sum_{i=1}^{m} \|f_i\|_{\mathbf{p}}, \tag{A.1.8}$$

follows from (A.1.7). □

Generalized Minkowski Inequality It is possible to deduce the generalized Minkowski inequality from the Minkowski inequality. Let $\mathbf{1} \le \mathbf{p} \le \infty$; then

$$\left\| \int \varphi(\cdot,\mathbf{y})d\mu(\mathbf{y}) \right\|_{\mathbf{p}} \le \int \|\varphi(\cdot,\mathbf{y})\|_{\mathbf{p}} d\mu(\mathbf{y}). \tag{A.1.9}$$

A.1.10 A Vector Norms Inequality

Let $1 \le q \le p \le \infty$. Then

$$\left(\int \left(\int |f(\mathbf{x},\mathbf{y})|^q d\mu(\mathbf{y}) \right)^{p/q} d\mu(\mathbf{x}) \right)^{1/p} \tag{A.1.10}$$

$$\le \left(\int \left(\int |f(\mathbf{x},\mathbf{y})|^p d\mu(\mathbf{x}) \right)^{q/p} d\mu(\mathbf{y}) \right)^{1/q}. \tag{A.1.11}$$

Proof Setting $\varphi := |f|^q$ and $\mathbf{p} = (p/q,\dots,p/q)$, inequality (A.1.11) follows from (A.1.9). □

A.1.11 The Young Inequality

Let p, q, and a be real numbers satisfying the conditions

$$1 \le p \le q \le \infty, \qquad 1 - 1/p + 1/q = 1/a. \tag{A.1.12}$$

Let $f \in L_p$ and $K \in L_a$ be 2π-periodic functions of a single variable. Let us consider the convolution of these functions:

$$J(x) := (2\pi)^{-1} \int_{-\pi}^{\pi} K(x-y)f(y)dy =: K * f.$$

Then

$$\|J\|_q \le \|K\|_a \|f\|_p. \tag{A.1.13}$$

Proof In the case $q = \infty$ the relation (A.1.13) follows from the Hölder inequality. Let $q < \infty$. We first consider the case $1 < p < q$, $a < q$. Let us represent the function $|Kf|$ in the form

$$|Kf| = \left(|K|^a |f|^p\right)^{1/q} |K|^{1-a/q} |f|^{1-p/q}. \tag{A.1.14}$$

We apply the Hölder inequality for three functions with exponents $p_1 := q$, $p_2 := (1/a - 1/q)^{-1}$, $p_3 := (1/p - 1/q)^{-1}$. We find

$$|J(x)| \le \left(\int |K(x-y)|^a |f(y)|^p d\mu(y) \right)^{1/q} \|K\|_a^{1-a/q} \|f\|_p^{1-p/q}. \tag{A.1.15}$$

Raising both sides of (A.1.15) to the power q and integrating we obtain the inequality (A.1.13).

It remains to consider the case where either $a = q$ or $p = q$. Let $p = q$; then $a = 1$. We have

$$J(x) = (2\pi)^{-1} \int_{-\pi}^{\pi} K(u) f(x-u) du.$$

Applying the generalized Minkowski inequality we get

$$\|J\|_p \le \|f\|_p \int |K(u)| d\mu(u) = \|f\|_p \|K\|_1.$$

Now let $a = q$ and $p = 1$. Clearly, in this case the required inequality is obtained in the same way as above. $\square$

The Young inequality for vectors **p**, **q**, **a** Let $\mathbf{1} \le \mathbf{p} \le \mathbf{q} \le \infty$, $\mathbf{1} - 1/\mathbf{p} + 1/\mathbf{q} = \mathbf{1}/\mathbf{a}$, and

$$J(\mathbf{x}) := \int K(\mathbf{x}-\mathbf{y}) f(\mathbf{y}) d\mu(\mathbf{y}) =: K * f.$$

Then

$$\|J\|_{\mathbf{q}} \le \|K\|_{\mathbf{a}} \|f\|_{\mathbf{p}}. \tag{A.1.16}$$

Proof The inequality (A.1.16) can be obtained by sequential application of the inequality (A.1.13) with the help of the following analog of the generalized Minkowski inequality ($\mathbf{x} \in \mathbb{T}^d$, $\mathbf{y} \in \mathbb{T}^d$):

$$\left\| \int \varphi(\cdot, \mathbf{y}) d\mu(\mathbf{y}) \right\|_{\mathbf{q}} \le \left\| \int \cdots \left\| \int \left\| \int \varphi(\cdot, \mathbf{y}) d\mu(y_1) \right\|_{q_1} d\mu(y_2) \right\|_{q_2} \cdots d\mu(y_d) \right\|_{q_d}.$$

$\square$

A.1.12 The Abel Inequality

For nonnegative and nonincreasing $v_1, \dots, v_n$ we have

$$\left| \sum_{i=1}^{n} u_i v_i \right| \le v_1 \max_k \left| \sum_{i=1}^{k} u_i \right|. \tag{A.1.17}$$

This inequality is obtained easily from the following formula:

$$\sum_{i=1}^{n} u_i v_i = \sum_{\nu=1}^{n-1} (v_\nu - v_{\nu+1}) \sum_{i=1}^{\nu} u_i + v_n \sum_{i=1}^{n} u_i, \tag{A.1.18}$$

which is called the Abel transformation.

It is well known that a space of continuous functions that are 2π-periodic in each variable and equipped with the uniform norm $\|\cdot\|_\infty$ is a Banach space. It will be convenient for us to denote it by L_∞.

Let $1 \le p < \infty$; then $\|\cdot\|_p$ is the norm if we do not distinguish measurable equivalent functions which may not coincide on a set of measure zero. This follows from the Minkowski inequality. The space L_p, $1 \le p < \infty$, is a Banach space. Indeed, let $\{f_n\}_{n=1}^\infty$ be a Cauchy sequence in L_p. We can find a subsequence $\{n_k\}_{n=1}^\infty$ such that $\|f_{n_{k+1}} - f_{n_k}\|_p \le 2^{-k}$. Then, by the Levi theorem, we find that the series

$$f_{n_1} + \sum_{k=1}^{\infty} (f_{n_{k+1}} - f_{n_k}) = f$$

converges to f almost everywhere, that is, $\{f_{n_k}\}$ converges to f almost everywhere. Furthermore, applying the Fatou theorem to the sequences $\{|f_{n_k}|^p\}_{k=1}^\infty$ and $\{|f_{n_k} - f_{n_m}|^p\}_{k=m+1}^\infty$ we find that $f \in L_p$ and $\|f - f_{n_m}\|_p \to 0$ for $m \to \infty$. It then follows easily that the Cauchy sequence $\{f_n\}_{n=1}^\infty$ converges to f in the space L_p.

As mentioned above, functions in L_p are defined up to equivalence. We will assume that we are dealing with a continuous function f if it is equivalent to a continuous function.

Along with the spaces L_p we use the spaces ℓ_p, $1 \le p \le \infty$, of sequences $\mathbf{z} = \{z_k\}_{k=1}^\infty$ equipped with the norm

$$\|\mathbf{z}\|_p := \|\mathbf{z}\|_{\ell_p} := \left(\sum_{k=1}^{\infty} |z_k|^p\right)^{1/p}, \qquad 1 \le p < \infty,$$

$$\|\mathbf{z}\|_\infty := \|\mathbf{z}\|_{\ell_\infty} := \sup_k |z_k|.$$

The spaces ℓ_p are Banach spaces. For sequences $\mathbf{z}$ and $\mathbf{w}$ (or for vectors) we write

$$(\mathbf{z}, \mathbf{w}) := \sum_{j=1}^{\infty} z_j \overline{w}_j.$$

A.2 Duality in L_p-Spaces

A.2.1 Dual Norms

Let $f \in L_p$, $g \in L_{p'}$. We denote

$$\langle f, g \rangle := (2\pi)^{-d} \int_{\mathbb{T}^d} f(\mathbf{x})\overline{g(\mathbf{x})} d\mathbf{x} = \int f \overline{g} d\mu,$$

where $\overline{z}$ is the complex conjugate to the number z.

Theorem A.2.1 *Let $1 \le p \le \infty$ and $f \in L_p$; then*

$$\|f\|_p = \sup_{g \in L_{p'}, \|g\|_{p'} \le 1} |\langle f, g\rangle|.$$

Proof The estimate

$$|\langle f, g\rangle| \le \|f\|_p$$

for g, such that $\|g\|_{p'} \le 1$, follows from the Hölder inequality.

For $1 < p < \infty$ we let, with $\|f\|_p > 0$,

$$\overline{g} := |f|^{p-1}(\operatorname{sign} f) \,/\, \|f\|_p^{p-1},$$

where

$$\operatorname{sign} z := \begin{cases} \overline{z}/|z|, & z \ne 0, \\ 0, & z = 0. \end{cases}$$

Then

$$\|g\|_{p'} = 1, \qquad \langle f, g\rangle = \|f\|_p,$$

which implies the theorem in this case.

Let $p = \infty$. As mentioned, L_∞ is the space of continuous functions. Consequently, there is a point $\mathbf{x}^0 \in \mathbb{T}^d$ such that

$$\|f\|_\infty = \left|f(\mathbf{x}^0)\right|.$$

We assume $\varphi_\varepsilon(\mathbf{x})$, $0 < \varepsilon \le 1$, to be 2π-periodic in each variable and such that

$$\varphi_\varepsilon(\mathbf{x}) = \begin{cases} (2\pi/\varepsilon)^d & \text{for } |\mathbf{x}_j - \mathbf{x}_j^0| \le \varepsilon/2,\ j = 1, \dots, d, \\ 0 & \text{otherwise}. \end{cases}$$

Then $\|\varphi_\varepsilon\|_1 = 1$ and, for $\overline{g}_\varepsilon := \varphi_\varepsilon \operatorname{sign} f(\mathbf{x}^0)$, $\|g_\varepsilon\|_1 \le 1$, we have

$$\left|f(\mathbf{x}^0)\right| = \lim_{\varepsilon \to 0} \langle f, g_\varepsilon\rangle,$$

which proves the theorem for this case.

Now let $p = 1$. We set $\overline{g} := \operatorname{sign} f$. By the Lusin theorem for an arbitrary $\varepsilon > 0$ we can find a continuous g_ε such that $|g_\varepsilon| \le 1$ and $|E_\varepsilon| \le \varepsilon$, where

$$E_\varepsilon := \left\{\mathbf{x} : g_\varepsilon(\mathbf{x}) \ne g(\mathbf{x})\right\}.$$

Then

$$\|f\|_1 = \langle f, g\rangle = \int_{\mathbb{T}^d \setminus E_\varepsilon} f g_\varepsilon \, d\mu + \int_{E_\varepsilon} f g \, d\mu,$$

which implies that

$$\int f g_\varepsilon \, d\mu \geq \|f\|_1 - \left| \int_{E_\varepsilon} f g \, d\mu \right| - \left| \int_{E_\varepsilon} f g_\varepsilon \, d\mu \right|. \tag{A.2.1}$$

The case $p = 1$ now follows from (A.2.1) by virtue of Corollary A.1.3. The theorem is proved. □

Remark A.2.2 A statement analogous to Theorem A.2.1 is valid for the spaces ℓ_p:

$$\|\mathbf{z}\|_{\ell_p} = \sup_{\|\mathbf{w}\|_{\ell_{p'}} \leq 1} |(\mathbf{z}, \mathbf{w})|, \qquad 1 \leq p \leq \infty.$$

A.2.2 The Nikol'skii Duality Theorem

Let F be a complex linear normed space and F^* be the dual (conjugate) space to F; that is, elements of F^* are bounded linear functionals φ defined on F with the norm

$$\|\varphi\| := \sup_{f \in F; \|f\| \leq 1} \big|\varphi(f)\big|.$$

Let $\Phi := \{\varphi_k\}_{k=1}^n$ be a set of functionals from F^*. Denote

$$F_\Phi := \big\{f \in F : \varphi_k(f) = 0, \ k = 1, \dots, n\big\}.$$

Theorem A.2.3 (The Nikol'skii duality theorem) *Let $\Phi = \{\varphi_k\}_{k=1}^n$ be a fixed system of functionals from F^*. Then, for any $\varphi \in F^*$,*

$$\inf_{c_1, \dots, c_n} \left\| \varphi - \sum_{k=1}^n c_k \varphi_k \right\| = \sup_{f \in F_\Phi; \|f\| \leq 1} \big|\varphi(f)\big|. \tag{A.2.2}$$

Proof Let us denote the left-hand side of (A.2.2) by a and the right-hand side of (A.2.2) by b. From the relation

$$\big|\varphi(f)\big| = \left| \left(\varphi - \sum_{k=1}^n c_k \varphi_k \right)(f) \right| \leq \left\| \varphi - \sum_{k=1}^n c_k \varphi_k \right\|,$$

which is valid for any $f \in F_\Phi$, $\|f\| \leq 1$, it follows that $b \leq a$. We will prove the inverse inequality. Clearly, we can assume that the system of functionals $\varphi_1, \dots, \varphi_n$ is linearly independent.

Lemma A.2.4 *Let $\varphi_1, \dots, \varphi_n \in F^*$ be linearly independent. There exists a set of elements $f_1, \dots, f_n \in F$ which is biorthogonal to $\varphi_1, \dots, \varphi_n$, that is, $\varphi_i(f_j) = 0$ for $1 \leq i \neq j \leq n$ and $\varphi_i(f_i) = 1$, $i = 1, \dots, n$.*

Proof The proof will be carried out by induction. The case $n = 1$ is evident. Let us assume that a biorthogonal system can be constructed if the number of functionals is less than n. Clearly, it suffices to prove the existence of $f_1 \in F$ such that

$$\varphi_1(f_1) = 1, \qquad \varphi_k(f_1) = 0, \qquad k = 2, \dots, n.$$

Let $\Phi_1 := \{\varphi_k\}_{k=2}^n$ and $\{g_k\}_{k=2}^n$ be a system biorthogonal to Φ_1. It is sufficient to prove the existence of $f_1 \in F_{\Phi_1}$ such that $\varphi_1(f_1) \neq 0$ Let us assume the contrary, that is, for any $f \in F_{\Phi_1}$ we have $\varphi_1(f) = 0$. We will show that this contradicts the linear independence of the functionals $\varphi_1, \dots, \varphi_n$. Let $f \in F$; then

$$f - \sum_{k=2}^{n} \varphi_k(f) g_k \in F_{\Phi_1}$$

and

$$\varphi_1\left(f - \sum_{k=2}^{n} \varphi_k(f) g_k\right) = 0,$$

which implies that

$$\varphi_1(f) = \sum_{k=2}^{n} \varphi_1(g_k)\varphi_k(f).$$

Consequently,

$$\varphi_1 = \sum_{k=2}^{n} \varphi_1(g_k)\varphi_k,$$

which contradicts the linear independence of $\varphi_1, \dots .\varphi_n$. The lemma is proved. □

We continue the proof of the theorem. Let $\varphi \in F^*$. Along with φ we consider a contraction φ_Φ of φ to the subspace F_Φ, that is, a linear bounded functional φ_Φ, defined on F_Φ, such that $\varphi_\Phi(f) = \varphi(f)$ for all $f \in F_\Phi$. Any functional

$$\psi = \varphi - \sum_{k=1}^{n} c_k \varphi_k \tag{A.2.3}$$

is a continuation of φ_Φ to F. We will now prove that each continuation of a functional φ_Φ from F_Φ to F has the form (A.2.3). We use Lemma A.2.4. Let the system $f_1, \dots, f_n$ be biorthogonal to Φ; then, for any $f \in F$,

$$f - \sum_{k=1}^{n} \varphi_k(f) f_k \in F_\Phi.$$

Consequently, for any continuation ψ of the functional φ_Φ we have

$$\psi\left(f - \sum_{k=1}^{n} \varphi_k(f) f_k\right) = \varphi\left(f - \sum_{k=1}^{n} \varphi_k(f) f_k\right),$$

which implies that

$$\psi(f) = \varphi(f) + \sum_{k=1}^{n} \big(\psi_k(f_k) - \varphi(f_k)\big)\varphi_k(f).$$

Thus, the representation (A.2.3) is valid for ψ.

Let ψ be a continuation of the functional φ_Φ such that $\|\psi\| = \|\varphi_\Phi\|$. The existence of such a continuation follows from the Hahn–Banach theorem. Then

$$\|\psi\| = \left\| \varphi - \sum_{k=1}^{n} c_k \varphi_k \right\| = \|\varphi_\Phi\| = \sup_{f \in F_\Phi; \|f\| \le 1} |\varphi(f)|,$$

that is, $a \le b$; this concludes the proof of the theorem. □

Theorem A.2.5 *Let $\varphi, \varphi_1, \ldots .\varphi_n \in L_p$, $1 \le p \le \infty$; then*

$$\inf_{c_1,\ldots,c_n} \left\| \varphi - \sum_{k=1}^{n} c_k \varphi_k \right\|_p = \sup_{\|g\|_{p'} \le 1; \langle \varphi_k, g\rangle = 0,\, k=1,\ldots,n} |\langle \varphi, g\rangle|.$$

Proof This theorem follows from Theorems A.2.1 and A.2.3. Indeed, let us consider a function $\varphi \in L_p$ as a functional φ acting on $L_{p'}$ by the formula $\varphi(f) = \langle f, \varphi\rangle$. Then, by Theorem A.2.1, we have $\|\varphi\| = \|\varphi(\cdot)\|_p$. It remains to apply Theorem A.2.3. □

A.3 Fourier Series of Functions in L_p

A.3.1 The Hausdorff–Young Theorem

For a function $f \in L_1$ we define the Fourier coefficients

$$\hat{f}(\mathbf{k}) := (2\pi)^{-d} \int_{\mathbb{T}^d} f(\mathbf{x}) e^{-i(\mathbf{k},\mathbf{x})} d\mathbf{x} = \langle f, e^{i(\mathbf{k},\mathbf{x})}\rangle. \tag{A.3.1}$$

We have the well-known Parseval identity, which states that, for any $f \in L_2$,

$$\|f\|_2 = \left(\sum_{\mathbf{k}} |\hat{f}(\mathbf{k})|^2 \right)^{1/2}, \tag{A.3.2}$$

and the Riesz–Fischer theorem: if $\sum_{\mathbf{k}} |c_{\mathbf{k}}|^2 < \infty$ then

$$f(\mathbf{x}) = \sum_{\mathbf{k}} c_{\mathbf{k}} e^{i(\mathbf{k},\mathbf{x})} \in L_2 \qquad \text{and} \qquad \hat{f}(\mathbf{k}) = c_{\mathbf{k}}.$$

In the space L_p, $1 < p < \infty$, the following statement holds.

Theorem A.3.1 (The Hausdorff–Young theorem) *Let $1 < p \le 2$; then, for any $f \in L_p$,*

$$\left(\sum_{\mathbf{k}} |\hat{f}(\mathbf{k})|^{p'} \right)^{1/p'} \le \|f\|_p. \tag{A.3.3}$$

If a sequence $\{c_{\mathbf{k}}\}$ *is such that* $\sum_{\mathbf{k}} |c_{\mathbf{k}}|^p < \infty$ *then there exists a function* $f \in L_{p'}$ *for which* $\hat{f}(\mathbf{k}) = c_{\mathbf{k}}$ *and*

$$\|f\|_{p'} \leq \left(\sum_{\mathbf{k}} |\hat{f}(\mathbf{k})|^p \right)^{1/p}. \tag{A.3.4}$$

We will derive this theorem from the following interpolation theorem, which is a special case of the general Riesz–Thorin theorem.

Denote the norm of an operator T acting from a Banach space E to a Banach space F by

$$\|T\|_{E \to F} = \sup_{\|f\|_E \leq 1} \|Tf\|_F.$$

Theorem A.3.2 (The Riesz–Thorin theorem) *Let* E_q *be either* L_q *or* ℓ_q *and* F_p *be either* L_p *or* ℓ_p, *and for* $1 \leq q_i, p_i \leq \infty$, *let*

$$\|T\|_{E_{q_i} \to F_{p_i}} \leq M_i, \qquad i = 1, 2.$$

Then, for all $0 < \theta < 1$,

$$\|T\|_{E_q \to F_p} \leq M_1^\theta M_2^{1-\theta},$$

where

$$1/q = \theta/q_1 + (1-\theta)/q_2, \qquad 1/p = \theta/p_1 + (1-\theta)/p_2.$$

Proof of Theorem A.3.1 We first prove the relation (A.3.3). Let us consider an operator T which maps a function $f \in L_1$ to the sequence $\{\hat{f}(\mathbf{k})\}$ of its Fourier coefficients. Then by (A.3.2) for $f \in L_2$ we have

$$\|Tf\|_{\ell_2} = \|f\|_2 \tag{A.3.5}$$

and obviously, for $f \in L_1$,

$$\|Tf\|_{\ell_\infty} \leq \|f\|_1. \tag{A.3.6}$$

Let $1/p = \theta/1 + (1-\theta)/2$; then $1/p' = \theta/\infty + (1-\theta)/2$ and relation (A.3.3) follows from (A.3.5), (A.3.6), and Theorem A.3.2 with

$$E_q := L_q, \qquad q_1 := 1, \qquad q_2 := 2;$$
$$F_p := l_p, \qquad p_1 := \infty, \qquad p_2 := 2.$$

We now prove relation (A.3.4). Clearly, by the completeness of the space $L_{p'}$ it is sufficient to prove (A.3.4) in the case when a finite number of the $c_{\mathbf{k}}$ are nonzero. Let this be the case and let

$$f = \sum_{\mathbf{k}} c_{\mathbf{k}} e^{i(\mathbf{k},\mathbf{x})}.$$

By Theorem A.2.1 we have

$$\|f\|_{p'} = \sup_{\|g\|_p \le 1} |\langle f, g\rangle| = \sup_{\|g\|_p \le 1} \left| \sum_{\mathbf{k}} c_{\mathbf{k}} \overline{\hat{g}(\mathbf{k})} \right|. \tag{A.3.7}$$

Applying the Hölder inequality and relation (A.3.3) we see that (A.3.7) gives

$$\|f\|_{p'} \le \sup_{\|g\|_p \le 1} \left(\sum_{\mathbf{k}} |c_{\mathbf{k}}|^p \right)^{1/p} \left(\sum_{\mathbf{k}} |\hat{g}(\mathbf{k})|^{p'} \right)^{1/p'} \le \left(\sum_{\mathbf{k}} |c_{\mathbf{k}}|^p \right)^{1/p}.$$

Relation (A.3.4) is proved. □

A.3.2 A Few Fundamental Theorems

Let [y] be the integral part of the real number y, that is, the largest integer $[y]$ such that $[y] \le y$. For a vector $\mathbf{s} = (s_1, \dots, s_d)$ with nonnegative integer coordinates we define the set $\rho(\mathbf{s})$ of vectors $\mathbf{k}$ with integer coordinates:

$$\rho(\mathbf{s}) := \left\{ \mathbf{k} = (k_1, \dots, k_d) : [2^{s_j - 1}] \le |k_j| < 2^{s_j},\ j = 1, \dots, d \right\}.$$

For $f \in L_1$ denote

$$\delta_{\mathbf{s}}(f, \mathbf{x}) := \sum_{\mathbf{k} \in \rho(\mathbf{s})} \hat{f}(\mathbf{k}) e^{i(\mathbf{k}, \mathbf{x})}.$$

Theorem A.3.3 (The Littlewood–Paley theorem) *Let $1 < p < \infty$. There exist positive numbers $C_1(d,p)$ and $C_2(d,p)$, which depend on d and p, such that, for each function $f \in L_p$,*

$$C_1(d,p)\|f\|_p \le \left\| \left(\sum_{\mathbf{s}} |\delta_{\mathbf{s}}(f, \mathbf{x})|^2 \right)^{1/2} \right\|_p \le C_2(d,p)\|f\|_p.$$

Corollary A.3.4 *Let G be a finite set of vectors* **s** *and let the operator S_G map a function $f \in L_p$, $p > 1$, to the function*

$$S_G(f) := \sum_{\mathbf{s} \in G} \delta_{\mathbf{s}}(f).$$

Then

$$\|S_G\|_{L_p \to L_p} \le C(d,p), \qquad 1 < p < \infty.$$

For the sake of brevity we shall write $\|T\|_{L_q \to L_p} = \|T\|_{q \to p}$.

Corollary A.3.5 *Let $p^* := \min(p, 2)$; then, for $f \in L_p$, we have*

$$\|f\|_p \le C(d,p) \left(\sum_{\mathbf{s}} \|\delta_{\mathbf{s}}(f, \mathbf{x})\|_p^{p^*} \right)^{1/p^*}, \qquad 1 < p < \infty.$$

Proof Let $2 \le p < \infty$; then by Theorem A.3.3,

$$\|f\|_p \le C(d,p)\left\|\left(\sum_{\mathbf{s}}|\delta_{\mathbf{s}}(f,\mathbf{x})|^2\right)^{1/2}\right\|_p = C(d,p)\left\|\sum_{\mathbf{s}}|\delta_{\mathbf{s}}(f,\mathbf{x})|^2\right\|_{p/2}^{1/2}$$

$$\le C(d,p)\left(\sum_{\mathbf{s}}\left\||\delta_{\mathbf{s}}(f,\mathbf{x})|^2\right\|_{p/2}\right)^{1/2} = C(d,p)\left(\sum_{\mathbf{s}}\|\delta_{\mathbf{s}}(f,\mathbf{x})\|_p^2\right)^{1/2}.$$

Let $1 < p \le 2$. Using the inequality $|a+b|^k \le |a|^k + |b|^k$, which holds for $0 \le k \le 1$, from Theorem A.3.3 we find, by the Fatou theorem,

$$\|f\|_p^p \le C(d,p)(2\pi)^{-d}\int_{\mathbb{T}^d}\left(\sum_{\mathbf{s}}|\delta_{\mathbf{s}}(f,\mathbf{x})|^2\right)^{p/2} d\mathbf{x} \le C(d,p)\sum_{\mathbf{s}}\|\delta_{\mathbf{s}}(f,\mathbf{x})\|_p^p. \quad \square$$

Theorem A.3.6 (The Marcinkiewicz multiplier theorem) *Suppose that $\lambda_0, \lambda_1, \dots$ are Marcinkiewicz multipliers, that is, they satisfy the conditions*

$$|\lambda_n| \le M, \qquad n = 0, \pm 1, \dots, \qquad \sum_{l=\pm 2^\nu}^{\pm(2^{\nu+1}-1)} |\lambda_l - \lambda_{l+1}| \le M, \qquad \nu = 0, 1, \dots,$$

where M is a number. Then the operator Λ which maps a function f to the function

$$\sum_k \lambda_k \hat{f}(k) e^{ikx},$$

is bounded as an operator from L_p to L_p for $1 < p < \infty$.

Theorem A.3.7 (The Hardy–Littlewood inequality) *Let $1 < q < p < \infty$,*

$$\mu := 1 - 1/q + 1/p, \qquad \|f\|_{L_p(R)} := \left(\int_{-\infty}^{\infty}|f|^q dx\right)^{1/q} < \infty,$$

and

$$J(x) := \int_{-\infty}^{\infty} f(y)|x-y|^{-\mu} dy.$$

Then the inequality

$$\|J\|_{L_p(R)} \le C(q,p)\|f\|_{L_q(R)}$$

holds.

Corollary A.3.8 *Let $1 < q < p < \infty$, $\beta := 1/q - 1/p$. Then the operator A_β, which maps a function $f \in L_q$ into the function*

$$A_\beta(f) := \sum_{\mathbf{k}} \hat{f}(\mathbf{k})\left(\prod_{j=1}^d \max(1, |k_j|)\right)^{-\beta} e^{i(\mathbf{k},\mathbf{x})}$$

is bounded as an operator from L_q to L_p.

We will formulate a result in the theory of convex sets which we use in proving the existence of special trigonometric polynomials.

Theorem A.3.9 *Let $B \subset R^n$ be a convex centrally symmetric set, $\mathbf{u}$ be some unit vector from R^n, and $B_\alpha = \{\mathbf{x} \in B : (\mathbf{x}, \mathbf{u}) = \alpha\}$ be sections of B by hyperplanes of the dimension $n-1$ orthogonal to $\mathbf{u}$. Then the $(n-1)$-dimensional volume of B_α is nonincreasing in $[0, +\infty)$ as a function of α.*

This theorem is a corollary of the Brunn theorem (see Hadwiger, 1957).

The following theorem is a well-known result in Hilbert space operator theory. It is an easy corollary of the Schmidt representation theorem.

Theorem A.3.10 *Let A be a compact operator in a Hilbert space H. Denote by B_H the unit ball of H and by $A(B_H) := \{Ax, x \in B_H\}$ the image of B_H under a linear mapping A. Then*

$$d_n(A(B_H), H) = s_{n+1}(A),$$

where $s_j(A) := (\lambda_j(AA^*))^{1/2}$, $j = 1, 2, \ldots$, are the singular numbers of the operator A.

A.4 Some Notation

We list here some of the most often used notation.

Numbers We use the following standard notations for sets of numbers:

$$\mathbb{N} := \{1, 2, \ldots\} \text{ is the set of natural numbers}, \qquad \mathbb{N}_0 := \mathbb{N} \cup \{0\};$$

$$\mathbb{N}^d := \mathbb{N} \times \cdots \times \mathbb{N}, \qquad \mathbb{N}_0^d := \mathbb{N}_0 \times \cdots \times \mathbb{N}_0, \qquad d \text{ times};$$

$$\mathbb{Z} := \{\pm n : n \in \mathbb{N}_0\}, \qquad \mathbb{Z}_+ := \mathbb{N}_0, \qquad \mathbb{Z}^d := \mathbb{Z} \times \cdots \times \mathbb{Z};$$

$$\mathbb{R} \text{ is the set of real numbers}, \qquad \mathbb{R}^d := \mathbb{R} \times \cdots \times \mathbb{R};$$

$$\mathbb{T} := [0, 2\pi) \text{ or } \mathbb{T} := [-\pi, \pi), \qquad \mathbb{T}^d := \mathbb{T} \times \cdots \times \mathbb{T}.$$

We use two logarithms: ln for natural logarithms to the base e and log for logarithms to the base 2. Clearly, in all order-type inequalities which contain a logarithms factor $(\ln m)^a$ we could equally well use $\log m$. However, in some proof it is convenient to use ln and in others log.

Polynomials Here is some notation for the trigonometric polynomials. For a finite set $G \in \mathbb{Z}^d$,

$$\mathscr{T}(G) := \left\{ f(\mathbf{x}) : f(\mathbf{x}) = \sum_{\mathbf{k}\in G} c_{\mathbf{k}} e^{i(\mathbf{k},\mathbf{x})} \right\}, \qquad \mathscr{D}_G(\mathbf{x}) := \sum_{\mathbf{k}\in G} e^{i(\mathbf{k},\mathbf{x})},$$

$$S_G(f)(\mathbf{x}) := \sum_{\mathbf{k}\in G} \hat{f}(\mathbf{k}) e^{i(\mathbf{k},\mathbf{x})} = (f * \mathscr{D}_G)(\mathbf{x}).$$

Here are typical examples of sets G used in the book: the first is

$$\Pi(\mathbf{N},d) := \left\{ \mathbf{a} \in \mathbb{R}^d : |a_j| \le N_j,\ j = 1,\dots,d \right\},$$

where the N_j are nonnegative integers. We use the abbreviated notation

$$\mathscr{T}(\mathbf{N},d) := \mathscr{T}(\Pi(\mathbf{N},d)).$$

The set of real trigonometric polynomials with harmonics from $\Pi(\mathbf{N},d)$ is denoted by $\mathscr{RT}(\mathbf{N},d)$.

For the second example, the following notation is often used in the hyperbolic cross approximation:

$$\Gamma(N) := \left\{ \mathbf{k} : \prod_{j=1}^{d} \max(|k_j|,1) \le N \right\}.$$

We use the shortened notation $\mathscr{T}(N) := \mathscr{T}(\Gamma(N))$, $S_N(f) := S_{\Gamma(N)}(f)$. For $\mathbf{s} \in \mathbb{N}_0^d$ denote

$$\rho(\mathbf{s}) := \{\mathbf{k} : [2^{s_j-1}] \le |k_j| < 2^{s_j},\ j = 1,2,\dots,d\}, \qquad Q_n := \cup_{\|\mathbf{s}\|_1 \le n} \rho(\mathbf{s}).$$

For $S_{\rho(\mathbf{s})}(f)$ we use a special notation:

$$\delta_{\mathbf{s}}(f) := S_{\rho(\mathbf{s})}(f) = \sum_{\mathbf{k}\in\rho(\mathbf{s})} \hat{f}(\mathbf{k}) e^{i(\mathbf{k},\mathbf{x})}.$$

Sometimes we use the notation $\theta_n := \{\mathbf{s} \in \mathbb{N}_0^d : \|\mathbf{s}\|_1 = n\}$. We often use modifications of $\rho(\mathbf{s})$, $\Gamma(N)$, Q_n, θ_n which are defined and used locally.

Function classes We use the following notation for function classes. Let us begin with the univariate case. For $r > 0$ and $\alpha \in \mathbb{R}$ the functions of a single variable

$$F_r(x,\alpha) := 1 + 2\sum_{k=1}^{\infty} k^{-r} \cos(kx - \alpha\pi/2)$$

are called the *Bernoulli kernels*. Sometimes it is convenient to use their modifications $F_r^0(x,\alpha) := F_r(x,\alpha) - 1$. We define the following operator in the univariate space L_1:

$$(I_\alpha^r \phi)(x) := (2\pi)^{-1} \int_0^{2\pi} F_r(x-y,\alpha)\phi(y)dy.$$

Denote by $W_{q,\alpha}^r B$, $r > 0$, $\alpha \in \mathbb{R}$, $1 \le q \le \infty$, the class of functions $f(x)$ representable in the form

$$f = I_\alpha^r \phi, \qquad \|\phi\|_q \le B. \tag{A.4.1}$$

For functions representable in the form (A.4.1) with some q and B we define

$$D_\alpha^r f = \phi.$$

In the case $B = 1$ we drop B from the notation.

The Sobolev class $W_{\mathbf{q},\alpha}^{\mathbf{r}} B$, $\mathbf{r} = (r_1,\dots,r_d)$, $r_j > 0$, $\mathbf{q} = (q_1,\dots,q_d)$, and $\alpha = (\alpha_1,\dots,\alpha_d)$ consists of functions $f(\mathbf{x})$ which have the following integral representation for each $1 \le j \le d$:

$$f(\mathbf{x}) = (2\pi)^{-1} \int_0^{2\pi} \varphi_j(x_1,\dots,x_{j-1},y,x_{j+1},\dots,x_d) F_{r_j}(x_j - y, \alpha_j)dy,$$

$$\|\varphi_j\|_{\mathbf{q}} \le B. \tag{A.4.2}$$

Sometimes we denote

$$\varphi_j(\mathbf{x}) := f_j^{(r_j)}(\mathbf{x},\alpha_j).$$

The Nikol'skii class $H_{\mathbf{q}}^{\mathbf{r}} B$, $\mathbf{r} = (r_1,\dots,r_d)$, and $\mathbf{q} = (q_1,\dots,q_d)$, is the set of functions $f \in L_{\mathbf{q}}$ such that, for each $l_j := [r_j] + 1$, $j = 1,\dots,d$, the following relations hold:

$$\|f\|_{\mathbf{q}} \le B, \qquad \|\Delta_h^{l_j,j} f\|_{\mathbf{q}} \le B|h|^{r_j}, \qquad j = 1,\dots,d,$$

where $\Delta_h^{l,j}$ is the lth difference with step h in the variable x_j. In the case $B = 1$ we do not include B in the notation for the Sobolev and Nikol'skii classes. It is usual to call these classes isotropic in the case $\mathbf{r} = r\mathbf{1}$ and anisotropic in the general case.

In the study of anisotropic function classes in Chapter 3 we used the following notation. Let

$$\mathbf{r} := (r_1,\dots,r_d), \qquad r_j > 0, \qquad j = 1,\dots,d,$$

$$g(\mathbf{r}) := \left(\sum_{j=1}^{d} r_j^{-1}\right)^{-1},$$

$$\mathbf{v} := \mathbf{v}(\mathbf{r}) := g(\mathbf{r})/\mathbf{r} = \big(g(\mathbf{r})/r_1,\dots,g(\mathbf{r})/r_d\big),$$

$$2^{\mathbf{v}n} := (2^{v_1 n},\dots,2^{v_d n}),$$

$$[2^{\mathbf{v}n}] := \big([2^{v_1 n}],\dots,[2^{v_d n}]\big),$$

$$\mathscr{T}^{\mathbf{r}}(n) := \mathscr{T}\big([2^{\mathbf{v}n}],d\big),$$

$$E_n^{\mathbf{r}}(f)_{\mathbf{p}} := E_{\mathscr{T}^{\mathbf{r}}(n)}(f)_{\mathbf{p}} := \inf_{t\in\mathscr{T}^{\mathbf{r}}(n)} \|f-t\|_{\mathbf{p}}.$$

Also, let:

$$V(f,\mathbf{r},n) := V(\mathbf{r},n)(f) := f * \mathscr{V}_{[2^{\mathbf{v}n}]},$$

$$A(f,\mathbf{r},0) := V(f,\mathbf{r},0),$$

$$A(f,\mathbf{r},n) := V(f,\mathbf{r},n) - V(f,\mathbf{r},n-1), \qquad n = 1,2,\dots$$

The multivariate analogs of the Bernoulli kernels are

$$F_r(\mathbf{x},\alpha) := \prod_{j=1}^{d} F_r(x_j,\alpha_j), \qquad F_r^0(\mathbf{x},\alpha_j) := \prod_{j=1}^{d} F_r^0(x_j,\alpha_j).$$

In the multivariate case let $I^r_{(\alpha_1,\dots,\alpha_d)} := \prod_{j=1}^d I^r_{\alpha_j}$. Denote

$$\mathbf{W}^r_{q,\alpha} := \{f \in L_q(\mathbb{T}^d) : f = F_r(\cdot,\alpha) * \phi(\cdot),\ \|\phi\|_q \le 1\}.$$

The quantity $\mathbf{W}^r_{q,\alpha}$ is the unit ball of the space $\mathbf{W}^r_{q,\alpha}$ of functions with bounded mixed derivative. We write, for $f \in \mathbf{W}^r_{q,\alpha}$,

$$\|f\|_{\mathbf{W}^r_{q,\alpha}} := \|\phi\|_q, \qquad \phi(\mathbf{x}) := f^{(r)}(\mathbf{x},\alpha),$$

the (r,α)-derivative of f. For trigonometric polynomials $f \in \mathscr{T}([-N,N]^d)$ we have

$$f^{(r)}(\mathbf{x},\alpha) = D^r_\alpha(f) := f(\mathbf{x}) * \prod_{j=1}^{d}\left(1 + 2\sum_{k_j=-N}^{N} \cos(k_j x_j + \alpha_j \pi/2)\right).$$

The classes $\mathbf{H}^r_q$ were defined in §4.4.2. Let $\mathbf{t} = (t_1,\dots,t_d)$ and $\Delta^l_{\mathbf{t}} f(\mathbf{x})$ be the mixed lth difference with step t_j in the variable x_j, that is,

$$\Delta^l_{\mathbf{t}} f(\mathbf{x}) := \Delta^l_{t_d} \cdots \Delta^l_{t_1} f(x_1,\dots,x_d).$$

Let e be a subset of natural numbers in $[1,d]$. We denote

$$\Delta_{\mathbf{t}}^l(e) = \prod_{j\in e} \Delta_{t_j}^l, \qquad \Delta_{\mathbf{t}}^l(\varnothing) = I.$$

We define the class $\mathbf{H}_{q,l}^r B$, $l > r$, as the set of $f \in L_q$ such that, for any e,

$$\left\|\Delta_{\mathbf{t}}^l(e) f(\mathbf{x})\right\|_q \leq B \prod_{j\in e} |t_j|^r. \tag{A.4.3}$$

In the case $B = 1$, as before we drop the B.

The following classes were defined in §7.7:

$$\mathbf{W}_q^{a,b} := \{f : \|f_l\|_q \leq 2^{-al} (\bar{l})^{(d-1)b}\}, \qquad \bar{l} := \max(l,1),$$

where

$$f_l := \sum_{\|\mathbf{s}\|_1 = l} \delta_{\mathbf{s}}(f) = \sum_{\mathbf{k}\in Q_l \setminus Q_{l-1}} \hat{f}(\mathbf{k}) e^{i(\mathbf{k},\mathbf{x})}.$$

Define

$$\|f\|_{\mathbf{W}_q^{a,b}} := \sup_l \|f_l\|_q 2^{al} (\bar{l})^{-(d-1)b}.$$

The class $\bar{\mathbf{W}}_q^{a,b}$ consists of functions f with representation

$$f = \sum_{n=1}^{\infty} t_n, \qquad t_n \in \mathscr{T}(Q_n), \qquad \|t_n\|_q \leq 2^{-an} n^{b(d-1)}.$$

References

Akhiezer, N.I. (1965). *Lectures in Approximation Theory*. Nauka, 1965; English translation of 1st edition published by Ungar, 1956.

Akhiezer, N.I. and M.G. Krein (1937). On the best approximation of differentiable periodic functions by trigonometric sums. *Dokl. Akad. Nauk SSSR*, **15** 107–112.

Andrianov, A.V. and V.N. Temlyakov (1997). On two methods of generalization of properties of univariate function systems to their tensor product. *Trudy MIAN*, **219** 32–43; English translation in *Proc. Steklov Inst. Math.*, **219** 25–35.

Babadzhanov, S.B. and V.M. Tikhomirov (1967). On widths of a certain class in the L_p-spaces $(p \geq 1)$. *Izv. Akad. Nauk UzSSR Ser. Fiz.– Mat. Nauk*, **11** 24–30.

Babenko, K.I. (1985). Some problems in approximation theory and numerical analysis. *Russian Math. Surveys*, **40** 1–30.

Babenko, K.I. (1960a). On the approximation of periodic functions of several variables by trigonometric polynomials. *Dokl. Akad. Nauk SSSR*, **132** 247–250; English translation in *Soviet Math. Dokl.*, **1** (1960).

Babenko, K.I. (1960b). On the approximation of a certain class of periodic functions of several variables by trigonometric polynomials. *Dokl. Akad. Nauk SSSR*, **132** 982–985; English translation in *Soviet Math. Dokl.*, **1** (1960).

Baishanski, B.M. (1983). Approximation by polynomials of given length. *Illinois J. Math.*, **27** 449–458.

Bakhvalov, N.S. (1959). On the approximate computation of multiple integrals. *Vestnik Moskov. Univ. Ser. Mat. Mekh. Astr. Fiz. Khim.*, **4** 3–18.

Bakhvalov, N.S. (1963a). Embedding theorems for classes of functions with several bounded derivatives. *Vestnik Moskov. Univ. Ser. Mat. Mekh.*, **3** 7–16.

Bakhvalov, N.S. (1963b). Optimal convergence bounds for quadrature processes and integration methods of Monte Carlo type for classes of functions. *Zh. Vychisl. Mat. i Mat. Fiz. Suppl.*, **4** 5–63.

Bakhvalov, N.S. (1972). Lower estimates of asymptotic characteristics of classes of functions with dominating mixed derivative. *Matem. Zametki*, **12** 655–664; English translation in *Math. Notes*, **12** (1972).

Bary, N.K. (1961). *Trigonometric Series*. Nauka. English translation, Pergamon Press, 1964.

Bass, R.F. (1988). Probability estimates for multiparameter Brownian processes. *Ann. Probab.*, **16** 251–264.

Beck, J. and W. Chen (1987). *Irregularities of Distribution*. Cambridge University Press.

Belinskii, E.S. (1987). Approximation by a "floating" system of exponentials on classes of smooth periodic functions. *Matem. Sb.*, **132** 20–27; English translation in *Math. USSR Sb.*, **60** (1988).

Belinskii, E.S. (1988). Approximation by a "floating" system of exponentials on classes of periodic functions with bounded mixed derivative. In *Research on the Theory of Functions of Many Real Variables*. Yaroslavl' State University, 16–33 (in Russian).

Belinskii, E.S. (1989). Approximation of functions of several variables by trigonometric polynomials with given number of harmonics, and estimates of ε-entropy. *Anal. Math.*, **15** 67–74.

Belinskii, E.S. (1998a). Decomposition theorems and approximation by a "floating" system of exponentials. *Trans. Amer. Math. Soc.*, **350** 43–53.

Belinskii, E.S. (1998b). Estimates of entropy numbers and Gaussian measures for classes of functions with bounded mixed derivative. *J. Approx. Theory*, **93** 114–127.

Bernstein, S.N. (1912). Sur la valeur asymptotique de la meilleure approximation de $|x|$. *Comptes Rendus*, **154** 184–186.

Bernstein, S.N. (1914). Sur la meilleure approximation de $|x|$ par des polynomes des degrés donnés. *Acta Math.*, **37** 1–57.

Bernstein, S.N. (1952). *Collected Works, Vols. I and II.* Akad. Nauk SSSR.

Bilyk, D. and M. Lacey (2008). On the small ball inequality in three dimensions. *Duke Math J.*, **143** 81–115.

Bilyk, D., M. Lacey, and A. Vagharshakyan (2008). On the small ball inequality in all dimensions. *J. Funct. Analysis*, **254** 2470–2502.

Binev, P., A. Cohen, W. Dahmen, R. DeVore, and V.N. Temlyakov (2005). Universal algorithms for learning theory. Part I: Piecewise constant functions. *J. Machine Learning Theory*, **6** 1297–1321.

Bourgain, J. and V.D. Milman (1987). New volume ratio properties for convex symmetric bodies in $\mathbb{R}^n$. *Invent. Math.*, **88** 319–340.

Bugrov, Ya. S. (1964). Approximation of a class of functions with a dominating mixed derivative. *Mat. Sb.*, **64** 410–418.

Bungartz, H.-J. and M. Griebel (1999). A note on the complexity of solving Poisson's equation for spaces of bounded mixed derivatives. *J. Complexity*, **15** 167–199.

Bungartz, H.-J. and M. Griebel (2004). Sparse grids. *Acta Numerica*, **13** 147–269.

Bykovskii, V.A. (1985). On the correct order of the error of optimal cubature formulas in spaces with dominating derivative, and on quadratic deviations of grids. Preprint, Computing Center, Far-Eastern Scientific Center, Acad. Sci. USSR, Vladivostok.

Bykovskii, V.A. (1995). Estimates of deviations of optimal lattices in the L_p-norm and the theory of quadrature formulas. Preprint, Applied Mathematics Institute, Far-Eastern Scientific Center, Acad. Sci. Russia, Khabarovsk.

Cassels, J.W.S. (1971). *An Introduction to the Geometry of Numbers*. Springer-Verlag.

Chazelle, B. (2000). *The Discrepancy Method.* Cambridge University Press.

Chebyshev, P.L. (1854). Théorie des mecanismes connus sous le nom de parallélogrammes. *Mem. Présentés à l'Acad. Imp. Sci. St.-Pétersbourg par Divers Savants*, **7** 539–568.

Chen, W.W.L. (1980). On irregularities of distribution. *Mathematika*, **27** 153–170.

Cohen, A., R.A. DeVore, and R. Hochmuth (2000). Restricted nonlinear approximation. *Constructive Approx.*, **16** 85–113.

Dai, W. and O. Milenkovic (2009). Subspace pursuit for compressive sensing signal reconstruction, *IEEE Trans. Inf. Theory*, **55** 2230–2249.

Davenport, H. (1956). Note on irregularities of distribution. *Mathematika*, **3** 131–135.

Davis, G., S. Mallat, and M. Avellaneda (1997). Adaptive greedy approximations. *Constructive Approx.*, **13** 57–98.

de la Vallée Poussin, Ch. (1908). Sur la convergence des formules d'interpolation entre ordonées equidistantes. *Bull. Acad. Belgique* **4** 403–410.

de la Vallée Poussin, Ch. (1919). *Lecons sur l'Approximation des Fonctions d'une Variable Réelle*. Gauthier-Villars, Paris, 1919; 2nd edition published by Chelsea Publishing Co., 1970.

DeVore, R.A. (1998). Nonlinear approximation. *Acta Numerica* **7** 51–150.

DeVore, R.A and G.G. Lorentz (1993). *Constructive Approximation*. Springer-Verlag.

DeVore, R.A. and V.N. Temlyakov (1995). Nonlinear approximation by trigonometric sums. *J. Fourier Anal. Applic.*, **2** 29–48.

Dilworth, S.J., D. Kutzarova, and V.N. Temlyakov (2002). Convergence of some greedy algorithms in Banach spaces. *J. Fourier Anal. Applic.* **8** 489–505.

Dilworth, S.J., N.J. Kalton, and Denka Kutzarova (2003a). On the existence of almost greedy bases in Banach spaces. *Studia Math.*, **158** 67–101.

Dilworth, S.J., N.J. Kalton, Denka Kutzarova, and V.N. Temlyakov (2003b). The thresholding greedy algorithm, greedy bases, and duality. *Constructive Approx.*, **19** 575–597.

Dilworth, S.J., M. Soto-Bajo, and V.N. Temlyakov (2012). Quasi-greedy bases and Lebesgue-type inequalities. *Stud. Math.*, **211** 41–69.

Dinh Zung [Dinh Dung] (1984). Approximation of classes of functions on the torus defined by a mixed modulus of continuity. In *Constructive Theory of Functions (Proc. Internat. Conf., Varna, 1984)*. Bulgarian Academy of Science, 43–48.

Dinh Dung (1985). Approximation of multivariate functions by means of harmonic analysis. Dissertation, Moscow, MGU.

Dinh Zung [Dinh Dung] (1986). Approximation by trigonometric polynomials of functions of several variables on the torus. *Mat. Sb.*, **131** 251–271; English translation in *Mat. Sb.*, **59**.

Dinh Dung (1991). On optimal recovery of multivariate periodic functions, In *Proc. ICM-90 Satellite Conf. on Harmonic Analysis*, S. Igary (ed). Springer-Verlag, 96–105.

Dinh Dung and T. Ullrich (2014). Lower bounds for the integration error for multivariate functions with mixed smoothness and optimal Fibonacci cubature for functions on the square. *Math. Nachr.*, **288** 743–762.

Ding Dung, V.N. Temlyakov, and T. Ullrich (2016). Hyperbolic cross approximation. arXiv:1601.03978v1 [math.NA], accessed 15 Jan 2016.

Donahue, M., L. Gurvits, C. Darken, and E. Sontag (1997). Rate of convex approximation in non-Hilbert spaces. *Constructive Approx.*, **13** 187–220.

Donoho, D., M. Elad, and V.N. Temlyakov (2007). On the Lebesgue type inequalities for greedy approximation. *J. Approximation Theory*, **147** 185–195.

Dubinin, V.V. (1992). Cubature formulas for classes of functions with bounded mixed difference. *Mat. Sb.*, **183**; English translation in *Mat. Sb.*, **76** 283–292.

Dubinin, V.V. (1997). Greedy algorithms and applications. Ph.D. thesis, University of South Carolina, 1997.

Dzyadyk, V.K. (1977). *Introduction to the Theory of Uniform Approximation of Functions by Polynomials*. Nauka.

Favard, J. (1937). Sur les meilleurs procédés d'approximation de certaines classes de fonctions par des polynomes trigonometriques. *Bull. Sci. Math.*, **61** 209–224; 243–256.

Foucart, S. (2012). Sparse recovery algorithms: sufficient conditions in terms of restricted isometry constants. In *Proc. Conf. on Approximation Theory XIII: San Antonio, 2010*, 65–77.

Franke, J. (1986). On the spaces $F^s_{p,q}$ of Triebel–Lizorkin type: pointwise multipliers and spaces on domains. *Math. Nachr.* **125** 29–68.

Fredholm, I. (1903). Sur une classe d'equations fonctionnelles. *Acta Math.*, **27** 365–390.

Frolov, K.K. (1976). Upper bounds on the error of quadrature formulas on classes of functions. *Dokl. Akad. Nauk SSSR*, **231** 818–821; English translation in *Soviet Math. Dokl.*, **17**.

Frolov, K.K. (1979). Quadrature formulas on classes of functions. PhD dissertation, Vychisl. Tsentr Academy Nauk SSSR.

Frolov, K.K. (1980). An upper estimate of the discrepancy in the L_p-metric, $2 \le p < \infty$. *Dokl. Akad. Nauk SSSR*, **252** 805–807; English translation in *Soviet Math. Dokl.*, **21**.

Galeev, E.M. (1978). Approximation of classes of functions with several bounded derivatives by Fourier sums. *Matem. Zametki*, **23** 197–212; English translation in *Math. Notes*, **23**.

Galeev, E.M. (1982). Order estimates of derivatives of the multidimensional periodic Dirichlet α-kernel in a mixed norm. *Mat. Sb.*, **117**(159) 32–43; English translation in *Mat. Sb.*, **45**.

Galeev, E.M. (1984). Kolmogorov widths of certain classes of periodic functions of several variables. In *Constructive Theory of Functions (Proc. Internat. Conf., Varna, 1984)*. Publ. House Bulgarian Acad. Sci. 27–32.

Galeev, E.M. (1985). Kolmogorov widths of the classes $W^{\bar{\alpha}}_p$ and $H^{\bar{\alpha}}_p$ of periodic functions of several variables in the space L_q. *Izv. Akad. Nauk SSSR*, **49** 916–934; English translation in *Math. Izv. Acad. Sci. USSR* **27** (1986).

Galeev, E.M. (1988). Orders of orthogonal projection widths of classes of periodic functions of one and several variables. *Matem. Zametki*, **43** 197–211; English translation in *Math. Notes*, **43**.

Galeev, E.M. (1990). Kolmogorov widths of classes of periodic functions of one and several variables. *Izv. Akad. Nauk SSSR*, **54** 418–430; English translation in *Math. Izv. Acad. of Sciences USSR*, **36** (1991).

Garnaev, A.Yu. and E.D. Gluskin (1984). On widths of the Euclidean ball. *Dokl. Akad. Nauk SSSR*, **277** 1048–1052; English translation in *Soviet Math. Dokl.*, **30**.

Gao, F., C-K. Ing, and Y. Yang (2013). Metric entropy and sparse linear approximation of ℓ_q-hulls for $0 < q \le 1$. *J. Approx. Theory*, **166** 42–55.

Garrigós, G., E. Hernández, and T. Oikhberg (2013). Lebesgue type inequalities for quasi-greedy bases. *Constr. Approx.*, **38** 447–479.

Gilbert, A.C., S. Muthukrishnan and M.J. Strauss (2003). Approximation of functions over redundant dictionaries using coherence. In *The 14th Annual ACM–SIAM Symp. on Discrete Algorithms*. SIAM, 243–252.

Gluskin, E.D. (1974). On a problem concerning widths. *Dokl. Akad. Nauk SSSR*, **219** (1974), 527–530; English translation in *Soviet Math. Dokl.* **15** (1974).

Gluskin, E.D. (1983). Norms of random matrices and widths of finite-dimensional sets. *Mat. Sb.*, **120** (162) 180–189; English translation in *Mat. Sb.*, **48** (1984).

Gluskin, E.D. (1989). Extremal properties of orthogonal parallelpipeds and their application to the geometry of Banach spaces. *Mat. Sb.*, **64** 85–96.

Gribonval, R. and M. Nielsen (2001). Some remarks on non-linear approximation with Schauder bases. *East J. Approx.*, **7** 267–285.

Griebel, M. (2006). Sparse grids and related approximation schemes for higher dimensional problems. In *Proc. Conf. on Foundations of Computational Mathematics, Santander 2005*, pp. 106–161. London Mathematical Society Lecture Notes Series, vol. 331, Cambridge University Press.

Györfy, L., M. Kohler, A. Krzyzak, and H. Walk (2002). *A Distribution-Free Theory of Non-Parametric Regression*. Springer-Verlag.

Hadwiger, H. (1957). *Vorlesungen über Inhalt, Oberflüsche und Isoperimetrie*. Springer-Verlag.

Halász, G. (1981). On Roth's method in the theory of irregularities of point distributions. *In Proc. Conf. on Recent Progress in Analytic Number Theory Vol. 2 (Durham, 1979)*, Academic Press, 79–94.

Halton, J.H. and S.K. Zaremba (1969). The extreme and L_2 discrepancies of some plane sets. *Monats. Math.*, **73** 316–328.

Hardy, G.H. and J.E. Littlewood (1928). Some properties of fractional integrals. I. *Math. Zeit.*, **27** 565– 606.

Hardy, G.H. and J.E. Littlewood (1966). In *Collected Papers of G. Hardy, vol. 1*, Clarendon Press, 113–114.

Heinrich, S., E. Novak, G. Wasilkowski and H. Wozniakowski (2001). The inverse of the star-discrepancy depends linearly on the dimension. *Acta Arithmetica*, **96** 279–302.

Hlawka, E. (1962). Zur angenaherten Berechnung mehrfacher Integrale. *Monats. Math.*, **B66** 140–151.

Höllig, K. (1980). Diameters of classes of smooth functions. In *Quantitative Approximation*, Academic Press, 163–176.

Hsiao, C.C., B. Jawerth, B.J. Lucier, and X. Yu (1994). Near optimal compression of orthogonal wavelet expansions. In *Wavelets: Mathematics and Applications*, CRC, 425–446.

Ismagilov, R.S. (1974). Widths of sets in normed linear spaces and the approximation of functions by trigonometric polynomials. *Uspekhi Mat. Nauk*, **29** 161–178; English translation in *Russian Math. Surveys*, **29** (1974).

Jackson, D. (1911). Über die Genauigkeit der Annaherung stegiger Function durch ganze rationale Functionen gegebenen Grader und trigonometrishe Summen gegebenen Ordmund. Dissertation, Göttingen.

Jackson, D. (1933). Certain problems of closest approximation. *Bull. Amer. Math. Soc.*, **39** 889–906.

Jawerth, B. (1977). Some observations on Besov and Lizorkin–Triebel spaces. *Math. Scand.* **40** 94–104.

Kadec, M.I. and A. Pelczynski (1962). Bases, lacunary sequences, and complemented subspaces in the spaces L_p. *Studia Math.*, **21** 161–176.

Kamont, A. and V.N. Temlyakov (2004). Greedy approximation and the multivariate Haar system. *Studia Math.*, **161**(3) 199–223.

Kashin, B.S. (1977). Widths of certain finite-dimensional sets and classes of smooth functions. *Izv. AN SSSR*, **41** 334–351; English translation in *Math. Izv.*, **11** (1977).

Kashin, B.S. (1980). On certain properties of the space of trigonometric polynomials with the uniform norm. *Trudy Mat. Inst. Steklov*, **145** 111–116; English translation in *Proc. Steklov Inst. Math.*, **145** (1981).

Kashin, B.S. (1981). Widths of Sobolev classes of small-order smoothness. *Vestnik Moskov. Univ., Ser. Mat. Mekh.*, **5** 50–54; English translation in *Moscow Univ. Math. Bull.*, **5** 62–66.

Kashin, B.S. and V.N. Temlyakov (1994). On best m-terms approximations and the entropy of sets in the space L^1. *Mat. Zametki*, **56** 57–86; English translation in *Math. Notes*, **56** 1137–1157.

Kashin, B.S. and V.N. Temlyakov (1995). Estimate of approximate characteristics for classes of functions with bounded mixed derivative. *Math. Notes*, **58** 1340–1342.

Kashin, B.S. and V.N. Temlyakov (2003). The volume estimates and their applications. *East J. Approx.*, **9** 469–485.

Kashin, B.S. and V.N. Temlyakov (2008). On a norm and approximate characteristics of classes of multivariate functions. *J. Math. Sci.*, **155** 57–80.

Keng, Hua Loo and Wang Yuan (1981). *Applications of Number Theory to Numerical Analysis*. Springer-Verlag.

Kerkyacharian, G. and D. Picard (2006). Nonlinear approximation and Muckenhoupt weights. *Constructive Approx.*, **24** 123–156.

Kolmogorov, A.N. (1936). Uber die beste Annäherung von Funktionen einer Funktionklasse. *Ann. Math.*, **37** 107–111.

Kolmogorov, A.N. (1985). *Selected Papers, Mathematics and Mechanics*. Nauka, Moscow.

Konyagin, S.V. and V.N. Temlyakov (1999). A remark on greedy approximation in Banach spaces. *East. J. Approx.*, **5** 365-379.

Konyagin, S.V. and V.N. Temlyakov (2002). Greedy approximation with regard to bases and general minimal systems. *Serdica Math. J.*, **28** 305–328.

Konyushkov, A.A. (1958). Best approximations by trigonometric polynomials and Fourier coefficients. *Mat. Sb.*, **44** 53–84.

Korobov, N.M. (1959). On the approximate computation of multiple integrals. *Dokl. Akad. Nauk SSSR*, **124** 1207–1210.

Korobov, N.M. (1963). *Number-Theoretic Methods in Numerical Analysis*. Fizmatgis.

Kuelbs, J. and W.V. Li (1993). Metric entropy and the small ball problem for Gaussian measures. *J. Functional Analysis*, **116** 133–157.

Kuipers, L. and H. Niederreiter (1974). *Uniform Distribution of Sequences*. Wiley.

Kulanin, E.D. (1985) On widths of functional classes of small smoothness. *Dokl. Bulgarian Acad. Sci.*, **41** 1601–1602.

Kushpel', A.K. (1989). Estimates of the widths of classes of analytic functions. *Ukrainian Math. Jour.*, **41** 567–570; English translation in *Ukr. Math. Jour.*, **41** (1989).

Kushpel', A.K. (1990). Estimation of the widths of classes of smooth functions in the space L_q. *Ukrainian Math. Jour.*, **42** 279–280; English translation in *Ukr. Math. Jour.*, **42** (1990).

Lebesgue, H. (1909). Sur les intégrales singuliéres. *Ann. Fac. Sci. Univ. Toulouse (3)*, **1** 25–117.

Lebesgue, H. (1910). Sur la represantation trigonometrique approchée des fonctions satisfaisants une condition de Lipschitz. *Bull. Soc. Math. France*, **38** 184–210.

Lifshits, M.A. and B.S. Tsirelson (1986). Small deviations of Gaussian fields. *Teor. Probab. Appl.*, **31** 557–558.

Lindenstrauss, J. and L. Tzafriri (1979). *Classical Banach Spaces I,II*. Springer-Verlag.

Livshitz, E.D. (2012). On the optimality of the orthogonal greedy algorithms for μ-coherent dictionaries. *J. Approx. Theory*, **164**(5) 668–681.

Livshitz, E.D. and V.N. Temlyakov (2014). Sparse approximation and recovery by greedy algorithms, *IEEE Transactions on Information Theory*, **60** 3989–4000.

Maiorov, V.E. (1975). Discretization of the diameter problem, *Uspekhi Matem. Nauk*, **30** 179–180.

Maiorov, V.E. (1978). On various widths of the class H_p^r in the space L_q. *Izv. Akad. Nauk SSSR Ser. Mat.*, **42** 773–788; English translation in *Math. USSR-Izv.*, **13** (1979).

Maiorov, V.E. (1986). Trigonometric diameters of the Sobolev classes W_p^r in the space L_q. *Math. Notes* **40** 590–597.

Makovoz, Yu.I. (1972). On a method of estimation from below of diameters of sets in Banach spaces. *Mat. Sb.*, **87**, 136–142; English translation in *Mat. Sb.*, **16** (1972).

Makovoz, Y. (1984). On trigonometric n-widths and their generalizations. *J. Approx. Theory*, **41** 361–366.

Matoušek, J. (1999). *Geometric Discrepancy*. Springer-Verlag.

Mityagin, B.S. (1962). Approximation of functions in the spaces L^p and C on the torus, *Mat. Sb.*, **58** 397–414.

Needell, D. and J.A. Tropp (2009). CoSaMP: iterative signal recovery from incomplete and inaccurate samples. *Appl. Comput. Harmonic Anal.*, **26** 301–321.

Needell, D. and R. Vershynin (2009). Uniform uncertainty principle and signal recovery via orthogonal matching pursuit. *Found. Comp. Math.*, **9** 317–334.

Niederreiter, H., R.F. Tichy and G. Turnwald (1990). An inequality for differences of distribution functions. *Arch. Math.*, **54** 166–172.

Nielsen, M. (2007). An example of an almost greedy uniformly bounded orthonormal basis for $L_p(0,1)$. *J. Approx. Theory*, **149** 188–192.

Nikol'skaya, N.S. (1974). Approximation of differentiable functions of several variables by Fourier sums in the L_p-metric., *Sibirsk. Mat. Zh.*, **15** 395–412; English translation in *Siberian Math. J.*, **15** (1974).

Nikol'skaya, N.S. (1975). Approximation of periodic functions in the class $S^r_p\star$ by Fourier sums. *Sibirsk. Mat. Zh.*, **16** 761–780; English translation in *Siberian Math. J.*, **16** (1975).

Nikol'skii, S.M. (1951). Inequalities for entire functions of exponential type and their use in the theory of differentiable functions of several variables. *Trudy MIAN*, **38** 244–278; English translation in *Amer. Math. Soc. Transl. (2)*, **80** (1969).

Nikol'skii, S.M. (1963). Functions with dominating mixed derivative satisfying a multiple Hölder condition. *Sibirsk. Mat. Zh.*, **4** 1342–1364; English translation in *Amer. Math. Soc. Transl. (2)*, **102** (1973).

Nikol'skii, S.M. (1969). *Approximation of Functions of Several Variables and Imbedding Theorems*. Nauka; English translation published by Springer, 1975.

Nikol'skii, S.M. (1979). *Quadrature Formulas*. Nauka.

Novak, E. (1988). *Deterministic and Stochastic Error Bounds in Numerical Analysis*, Lecture Notes in Mathematics, **1349**, Springer-Verlag.

Offin, D. and K. Oskolkov (1993). A note on orthonormal polynomial bases and wavelets. *Constructive Approx.*, **9** 319–325.

Quade, E. (1937). Trigonometric approximation in the mean. *Duke Math J.*, **3** 529–543.

Romanyuk, A.S. (2003). Best M-term trigonometric approximations of Besov classes of periodic functions of several variables. *Izvestia RAN, Ser. Mat.*, **67** 61–100; English translation in *Izvestiya Math.*, **67** 265.

Roth, K.F. (1954). On irregularities of distribution. *Mathematika*, **1** 73–79.

Roth, K.F. (1976). On irregularities of distribution. II. *Comm. Pure Appl. Math.*, **29** 749–754.

Roth, K.F. (1979). On irregularities of distribution. III. *Acta Arith.*, **35** 373–384.

Roth, K.F. (1980). On irregularities of distribution. IV. *Acta Arith.*, **37** 67–75.

Rudin, W. (1952). L_2-approximation by partial sums of orthogonal developments. *Duke Math. J.*, **19** 1–4.

Rudin, W. (1959). Some theorems on Fourier coefficients. *Proc. Amer. Math. Soc.*, **10** 855–859.

Savu, D. and V.N. Temlyakov (2013). Lebesgue-type inequalities for greedy approximation in Banach spaces. *IEEE Trans. Inform. Theory*, **58** 1098–1106.

Schmidt, W.M. (1972). Irregularities of distribution, VII. *Acta Arith.*, **21** 45–50.

Schmidt, W.M. (1977a). Irregularities of distribution, X. In *Number Theory and Algebra*. Academic Press, 311–329.

Schmidt, W.M. (1977b). *Lectures on Irregularities of Distribution*. Tata Institute of Fundamental Research.

Schütt, C. (1984). Entropy numbers of diagonal operators between symmetric Banach spaces. *J. Approx. Theory*, **40** 121–128.

Shapiro, H.S. (1951). Extremal problems for polynomials and power series. M.S. thesis, MIT. Massachusetts Institute of Technology.

Skriganov, M.M. (1994). Constructions of uniform distributions in terms of geometry of numbers. *Algebra Anal.*, **6** 200–230.

Smolyak, S.A. (1960). The ε-entropy of the classes $E_s^{\alpha k}(B)$ and $W_s^{\alpha}(B)$ in the metric L_2. *Dokl. Akad. Nauk SSSR*, **131** 30–33.

Smolyak, S.A. (1963). Quadrature and interpolation formulas for tensor products of certain classes of functions. *Dokl. Akad. Nauk SSSR*, **148** 1042–1045; English translation in *Soviet Math. Dokl.*, **4** (1963).

Sobolev, S.L. (1994). *Introduction to the Theory of Cubature Formulas*. Nauka.

Stechkin, S.B. (1951). On the degree of best approximation of continuous functions. *Izvestia AN SSSR, Ser. Mat.*, **15** 219–242.

Stechkin, S.B. (1954). On the best approximation of given classes of functions by arbitrary polynomials. *Uspekhi Mat. Nauk*, **9** 133–134 (in Russian).

Stechkin, S.B. (1955). On absolute convergence of orthogonal series. *Dokl. AN SSSR*, **102** 37–40 (in Russian).

Stepanets, A.I. (1987). *Classification and Approximation of Periodic Functions*. Naukova Dumka.

Talagrand, M. (1994). The small ball problem for the Brownian sheet. *Ann. Probab.*, **22** 1331–1354.

Telyakovskii, S.A. (1963). On estimates of the derivatives of trigonometric polynomials in several variables. *Siberian Math. Zh.*, **4** 1404–1411.

Telyakovskii, S.A. (1964). Some estimates for trigonometric series with quasi-convex coefficients, *Mat. Sb.*, **63** 426–444; English translation in *Amer. Math. Soc. Transl.* **86**.

Telyakovskii, S.A. (1988). Research in the theory of approximation of functions at the Mathematical Institute of the Academy of Sciences, *Trudy MIAN*, **182** 128–179; English translation in *Proc. Steklov Inst. Math.* **1** (1990).

Temlyakov, V.N. (1979). Approximation of periodic functions of several variables with bounded mixed derivative. *Dokl. Akad. Nauk SSSR*, **248** 527–531; English translation in *Soviet Math. Dokl.*, **20** (1979).

Temlyakov, V.N. (1980a). On the approximation of periodic functions of several variables with bounded mixed difference. *Dokl. Akad. Nauk SSSR*, **253** 544–548; English translation in *Soviet Math. Dokl.*, **22** (1980).

Temlyakov, V.N. (1980b). Approximation of periodic functions of several variables with bounded mixed difference. *Mat. Sb.*, **133** 65–85; English translation in Math. USSR Sbornik **41** (1982).

Temlyakov, V.N. (1980c). Approximation of periodic functions of several variables with bounded mixed derivative. *Trudy MIAN*, **156** 233–260; English translation in *Proc. Steklov Inst. Math.*, **2** (1983).

Temlyakov, V.N. (1982a). Widths of some classes of functions of several variables. *Dokl. Akad. Nauk SSSR*, **267** 314–317; English translation in *Soviet Math. Dokl.*, **26**.

Temlyakov, V.N. (1982b). Approximation of functions with a bounded mixed difference by trigonometric polynomials and the widths of some classes of functions. *Izv. Akad. Nauk SSSR*, **46** 171–186; English translation in *Math. Izv. Acad. Sci. USSR*, **20** (1983).

Temlyakov, V.N. (1985a). On the approximate reconstruction of periodic functions of several variables. *Dokl. Akad. Nauk SSSR*, **280** 1310–1313; English translation in *Soviet Math. Dokl.*, **31**.

Temlyakov, V.N. (1985b). Approximate recovery of periodic functions of several variables. *Mat. Sb.*, **128** 256–268; English translation in *Mat. Sb.*, **56** (1987).

Temlyakov, V.N. (1985c). Quadrature formulas and recovery on the values at the knots of number-theoretical nets for classes of functions of small smoothness. *Uspekhi Matem. Nauk*, **40** 203–204.

Temlyakov, V.N. (1985d). Approximation of periodic functions of several variables by trigonometric polynomials, and widths of some classes of functions. *Izv. Akad. Nauk SSSR*, **49** 986–1030; English translation in *Math. USSR Izv.*, **27** (1986).

Temlyakov, V.N. (1985e). On linear bounded methods of approximation of functions. *Dokl. Sem. Inst. Prikl. Mat. Vekua*, **1** 144–147.

Temlyakov, V.N. (1986a). On reconstruction of multivariate periodic functions based on their values at the knots of number-theoretical nets. *Anal. Math.*, **12** 287–305.

Temlyakov, V.N. (1986b). Approximation of periodic functions of several variables by bilinear forms. *Izvestiya AN SSSR*, **50** 137–155; English translation in *Math. USSR Izvestiya*, **28** 133–150.

Temlyakov, V.N. (1986c). Approximation of functions with bounded mixed derivative. *Trudy MIAN*, **178** 1–112; English translation in *Proc. Steklov Inst. Math.*, **1** (1989).

Temlyakov, V.N. (1987). Estimates of the best bilinear approximations of functions of two variables and some of their applications, *Mat. Sb.*, **134** 93–107; English translation in *Math. USSR – Sb* **62** (1989), 95–109.

Temlyakov, V.N. (1988a). On estimates of ε-entropy and widths of classes of functions with bounded mixed derivative or difference. *Dokl. Akad. Nauk SSSR*, **301** 288–291; English translation in *Soviet Math. Dokl.*, **38**, 84–87.

Temlyakov, V.N. (1988b). Estimates of best bilinear approximations of periodic functions. *Trudy Mat. Inst. Steklov*, **181** 250–267; English translation in *Proc. Steklov Inst. Math.*, **4** (1989), 275–293.

Temlyakov, V.N. (1988c). Approximation by elements of a finite-dimensional subspace of functions from various Sobolev or Nikol'skii spaces. *Matem. Zametki*, **43** 770–786; English translation in *Math. Notes*, **43**.

Temlyakov, V.N. (1989a). Error estimates of quadrature formulas for classes of functions with bounded mixed derivative. *Matem. Zametki*, **46** 128–134; English translation in *Math. Notes*, **46**.

Temlyakov, V.N. (1989b). Approximation of functions of several variables by trigonometric polynomials with harmonics from hyperbolic crosses. *Ukrainian Math. J.*, **41** 518–524; English translation in *Ukr. Math. J.*, **41**.

Temlyakov, V.N. (1989c). Bilinear approximation and applications. *Trudy Mat. Inst. Steklov*, **187** 191–215; English translation in *Proc. Steklov Inst. Math.*, **3** (1990), 221–248.

Temlyakov, V.N. (1989d). Estimates of the asymptotic characteristics of classes of functions with bounded mixed derivative or difference. *Trudy Matem. Inst. Steklov*, **189** 138–168; English translation in *Proc. Steklov Inst. Math.*, **4** 161–197.

Temlyakov, V.N. (1990a). On a problem of estimating widths of classes of infinity differentiable functions. *Matem. Zametki* **47** 155–157.

Temlyakov, V.N. (1990b) On a way of obtaining lower estimates for the errors of quadrature formulas. *Matem. Sbornik*, **181** (1990), 1403–1413; English translation in *Math. USSR Sbornik*, **71** (1992).

Temlyakov, V.N. (1991a). On universal cubature formulas. *Dokl. Akad. Nauk SSSR*, **316** 34–47; English translation in *Soviet Math. Dokl.*, **43** (1991), 39–42.

Temlyakov, V.N. (1991b). Error estimates for Fibonacci quadrature formulas for classes of functions with bounded mixed derivative. *Trudy MIAN*, **200** 327–335; English translation in *Proc. Steklov Inst. Math.*, **2** (1993).

Temlyakov, V.N. (1992a). Bilinear approximation and related questions. *Trudy Mat. Inst. Steklov*, **194** 229–248; English translation in *Proc. Steklov Inst. Math.*, **4** (1993), 245–265.

Temlyakov, V.N. (1992b). Estimates of best bilinear approximations of functions and approximation numbers of integral operators. *Mat. Zametki*, **51** 125–134; English translation in *Math. Notes*, **51**, 510–517.

Temlyakov, V.N. (1993a). On approximate recovery of functions with bounded mixed derivative. *J. Complexity*, **9** 41–59.

Temlyakov, V.N. (1993b) *Approximation of Periodic Functions*. Nova Science Publishers.

Temlyakov, V.N. (1994). On error estimates for cubature formulas. *Trudy Matem. Inst. Steklova*, **207** 326–338; English translation in *Proc. Steklov Inst. Math.*, **6**, (1995).

Temlyakov, V.N. (1995a). An inequality for trigonometric polynomials and its application for estimating the entropy numbers. *J. Complexity*, **11** 293–307.

Temlyakov, V.N. (1995b). Some inequalities for multivariate Haar polynomials. *East J. Approx.*, **1** 61–72.

Temlyakov, V.N. (1996). An inequality for trigonometric polynomials and its application for estimating the Kolmogorov widths. *East J. Approx.*, **2** 253–262.

Temlyakov, V.N. (1998a). On two problems in the multivariate approximation. *East J. Approx.*, **4** 505–514.

Temlyakov, V.N. (1998b). Nonlinear Kolmogorov's widths. *Matem. Zametki*, **63**, 891–902.

Temlyakov, V.N. (1998c). Greedy algorithm and m-term trigonometric approximation. *Constr. Approx.*, **14**, 569–587.

Temlyakov, V.N. (1998d). Nonlinear m-term approximation with regard to the multivariate Haar system. *East J. Approx.*, **4** 87–106.

Temlyakov, V.N. (1998e). The best m-term approximation and greedy algorithms, *Advances in Comp. Math.*, **8** 249–265.

Temlyakov, V.N. (2000a). Weak greedy algorithms, *Advances in Comput. Math.*, **12** 213–227.

Temlyakov, V.N. (2000b). Greedy algorithms with regards to multivariate systems with special structure. *Constr. Approx.*, **16** 399–425.

Temlyakov, V.N. (2001). Greedy algorithms in Banach spaces. *Adv. Comput. Math.*, **14** 277–292.

Temlyakov, V.N. (2002a). Universal bases and greedy algorithms for anisotropic function classes. *Constr. Approx.*, **18** 529–550.

Temlyakov, V.N. (2002b). Nonlinear approximation with regard to bases. In *Approximation Theory X*. Vanderbilt University Press 373–402.

Temlyakov, V.N. (2003a). Nonlinear method of approximation. *Found. Compt. Math.*, **3** 33–107.

Temlyakov, V.N. (2003b). Cubature formulas and related questions, *J. Complexity*, **19** 352–391.

Temlyakov, V.N. (2005). Greedy-type approximation in Banach spaces and applications. *Constr. Approx.*, **21** 257–292.

Temlyakov, V.N. (2006). On universal estimators in learning theory. *Trudy MIAN im. VA Steklova*, **255** 256–272; English translation in *Proc. Steklov Inst. Math.*, **255** (2006), 244–259.

Temlyakov, V.N. (2007). Greedy approximation in Banach spaces. In *Banach Spaces and their Applications in Analysis*. de Gruyter, 193–208.

Temlyakov, V.N. (2008). Greedy approximation. *Acta Numerica*, **17** 235–409.

Temlyakov, V.N. (2011). *Greedy Approximation*. Cambridge University Press.

Temlyakov, V.N. (2013). An inequality for the entropy numbers and its application. *J. Approx. Theory*, **173** 110–121.

Temlyakov, V.N. (2014). Sparse approximation and recovery by greedy algorithms in Banach spaces. *Forum of Mathematics, Sigma*, **2** e12, 26 pp.

Temlyakov, V.N. (2015a). Constructive sparse trigonometric approximation and other problems for functions with mixed smoothness. *Matem. Sb.*, **206**, 131–160. ArXiv: 1412.8647v1 [math.NA] 24 December 2014, 1–37.

Temlyakov, V.N. (2015b). Constructive sparse trigonometric approximation for functions with small mixed smoothness. ArXiv: 1503.00282v1 [math.NA] 1 March 2015, 1–30.

Temlyakov, V.N. (2015c). Sparse approximation with bases. In *Proc. Conf. on Advanced Courses in Mathematics CRM Barcelona*, Birkhäuser–Springer.

Temlyakov, V.N. (2016a). Incremental greedy algorithm and its applications in numerical integration. In *Proc. Conf. on Monte Carlo and Quasi-Monte Carlo Methods, Leuven, April 2014*. Springer Proceedings in Mathematics and Statistics, **163** 557–570.

Temlyakov, V.N. (2016b) On the entropy numbers of the mixed smoothness function classes. ArXiv:1602.08712v1 [math.NA] 28 February 2016.

Temlyakov, V.N and P. Zheltov (2011). On performance of greedy algorithms. *J. Approx. Theory*, **163** 1134–1145.

Temlyakov, V.N., Mingrui Yang and Peixin Ye (2011). Greedy approximation with regard to non-greedy bases. *Adv. Comput. Math.*, **34** 319–337.

Tikhomirov, V.M. (1960a). On n-dimensional diameters of certain functional classes. *Dokl. Akad. Nauk SSSR*, **130** 734–737; English translation in *Soviet Math. Dokl.*, **1**.

Tikhomirov, V.M. (1960b). Widths of sets in function spaces and the theory of best approximation. *Uspekhi Matem. Nauk*, **15** 81–120; English translation in *Russian Math. Surveys*, **15**.

Tikhomirov, V.M. (1976). *Some Topics in Approximation Theory*. Moscow State University.

Timan, A.F. (1960). *Theory of Approximation of Functions of a Real Variable*. Phys.–Math. Lit., Moscow, 1960; English translation published by MacMillan, 1963.

Timan, M.F. (1974). On embeddings of the function classes $L_p^{(k)}$. *Izv. Vyssh. Uchebn. Zaved. Mat.*, **10** 61–74; English translation in *Soviet Math. Iz. VUZ.*, **18**.

Triebel, H. (2010). *Bases in Function Spaces, Sampling, Discrepancy, Numerical Integration*. European Mathematical Society.

Triebel, H. (2015). Global solutions of Navier–Stokes equations for large initial data belonging to spaces with dominating mixed smoothness. *J. Complexity*, **31** 147–161.

Trigub, R.M. (1971). Summability and absolute convergence of the Fourier series in total. In *Metric Questions of Theory of Approximation and Mapping*. Naukova Dumka 173–266.

Trigub, R.M. and E.S. Belinsky (2004). *Fourier Analysis and Approximation of Functions*. Kluwer Academic Publishers.

Tropp, J.A. (2004). Greed is good: algorithmic results for sparse approximation. *IEEE Trans. Inform. Theory*, **50** 2231–2242.

Ul'yanov, P.L. (1970). Embedding theorems and relations between best approximations (moduli of continuity) in different metrics, *Mat. Sb.*, **81** (123) (1970), 104–131; English translation in *Mat. Sb.*, **10** (1970).

Uninskii, A.P. (1966). Inequalities in the mixed norm for the trigonometric polynomials and entire functions of finite degree. In *Mater. Vsesoyuzn. Simp. Teor. Vlozhen., Baku.*

van Aardenne-Ehrenfest, T. (1945). Proof of the impossibility of a just distribution of an infinite sequence of points over an interval. *Proc. Kon. Ned. Akad. v. Wetensch*, **48** 266–271.

van der Corput, J.G. (1935a). Verteilungsfunktionen. I. *Proc. Kon. Ned. Akad. v. Wetensch.*, **38** 813–821.

van der Corput, J.G. (1935b). Verteilungsfunktionen. II, *Proc. Kon. Ned. Akad. v. Wetensch.*, **38** 1058–1066.

Vilenkin, I.V. (1967). Plane nets of integration. *Zhur. Vychisl. Mat. i Mat. Fis.*, **7**, 189–196; English translation in *USSR Comp. Math. and Math. Phys.*, **7**, 258–267.

Wang, J. and B. Shim (2012). Improved recovery bounds of orthogonal matching pursuit using restricted isometry property. ArXiv:1211.4293v1 [cs.IT] 19 Nov 2012.

Wojtaszczyk, P. (2000). Greedy algorithm for general biorthogonal systems. *J. Approx. Theory* **107** 293–314.

Yserentant, H. (2010). *Regularity and Approximability of Electronic Wave Functions.* Lecture Notes in Mathematics, Springer.

Zhang, T. (2011). Sparse recovery with orthogonal matching pursuit under RIP, *IEEE Transactions on Information Theory*, **57** 6215–6221.

Zygmund, A. (1959). *Trigonometric Series.* Cambridge University Press.

Index

Printed in the United States
by Baker & Taylor Publisher Services